Lecture Notes in Computer Science 8464

Commenced Publication in 1973
Founding and Former Series Editors:
Gerhard Goos, Juris Hartmanis, and Jan van Leeuwen

Franck van Breugel   Elham Kashefi
Catuscia Palamidessi   Jan Rutten (Eds.)

Horizons
of the Mind

A Tribute to Prakash Panangaden

Essays Dedicated to Prakash Panangaden
on the Occasion of His 60th Birthday

 Springer

Volume Editors

Franck van Breugel
York University
Department of Electrical Engineering and Computer Science
4700 Keele Street, Toronto, ON, M3J 1P3, Canada
E-mail: franck@cse.yorku.ca

Elham Kashefi
University of Edinburgh
School of Informatics, Informatics Forum
10 Crichton Street, Edinburgh, EH8 9LE, UK
E-mail: ekashefi@staffmail.ed.ac.uk

Catuscia Palamidessi
Inria Saclay
Campus de l'École Polytechnique, Bâtiment Alan Turing
1, rue Honoré d'Estienne d'Orves, 91120 Palaiseau, France
E-mail: catuscia@lix.polytechnique.fr

Jan Rutten
CWI
P.O. Box 94079, 1090 GB Amsterdam, The Netherlands
E-mail: janr@cwi.nl

ISSN 0302-9743   e-ISSN 1611-3349
ISBN 978-3-319-06879-4   e-ISBN 978-3-319-06880-0
DOI 10.1007/978-3-319-06880-0
Springer Cham Heidelberg New York Dordrecht London

Library of Congress Control Number: 2014938036

LNCS Sublibrary: SL 1 – Theoretical Computer Science and General Issues

Typesetting: Camera-ready by author, data conversion by Scientific Publishing Services, Chennai, India

Printed on acid-free paper

Springer is part of Springer Science+Business Media (www.springer.com)

Prakash Panangaden

Preface

During the last three decades, Prakash Panangaden has worked on a large variety of topics including probabilistic and concurrent computation, logics and duality, and quantum information and computation. Despite the enormous breadth of his research, Prakash has made significant and deep contributions. For example, he introduced a logic and a real-valued interpretation of the logic to capture equivalence of probabilistic processes quantitatively. This quantitative notion turned out to be closely related to a metric introduced by Nobel laureate Leonid Kantorovich. The fact that Prakash studied physics, both at the undergraduate and graduate level, makes his accomplishments in computer science even more impressive and provides yet another illustration of his ability. What better way to celebrate his 60th birthday than this Festschrift with a large variety of papers attributed to him and a conference, PrakashFest, featuring a diverse range of speakers, held in Oxford in May 2014.

We would like to thank the authors for their contributions to this Festschrift and also for their help with the reviewing process. We are also thankful to the other reviewers. We are grateful to Laurie Hendren for providing us with a variety of input for this Festschrift, ranging from photos to the acknowledgments section of a dissertation. We would like to thank Bart Jacobs and Alexandra Silva for bringing the precise date of Prakash's 60th birthday to our attention. We would also like to thank Alfred Hofmann and Anna Kramer at Springer for their support and help in producing this Festschrift. Samson Abramsky, Karen Barnes, Bob Coecke, Destiny Chen, Michael Mislove, and Jamie Vicary deserve our special thanks for organizing PrakashFest, the conference in honor of Prakash on the occasion of his 60th birthday. We would also like to acknowledge the support for students to attend PrakashFest from the EPSRC Network on Structures in Computing and Physics.

March 2014

Franck van Breugel
Elham Kashefi
Catuscia Palamidessi
Jan Rutten

In 1990, Prakash Panangaden visited the "Centrum Wiskunde en Informatica" (CWI), the Dutch Research Institute for Mathematics and Computer Science, in Amsterdam. At the time, I was a graduate student at CWI. During his visit, Prakash gave several excellent talks. I still distinctly remember his talk on the expressiveness of dataflow primitives. When I applied in 1994 for a fellowship of the Netherlands Organization for Scientific Research to spend one year abroad, visiting McGill University to work with Prakash topped my list. I was his postdoc for two years. Prakash has had a significant impact on me as a researcher, a teacher, and a person. He strengthened my love for mathematics. His own is reflected, among many other things, by his large collection of volumes of the *Graduate Texts in Mathematics* series. Prakash was also instrumental in preparing me for job interviews. In 1995, my daughter Lisa was born in Montreal. Prakash was the first to visit her in the Royal Victoria Hospital. For all these and many other things I am to him for ever thankful.

Franck van Breugel   Toronto, March 2014

I joined Oxford's quantum group in 2003 after my PhD in Imperial College's Computing Department with a physicist as my supervisor. I was told that there was a big-shot professor who would be coming for his sabbatical to Oxford, who is the king of labelled Markov processes and programming language and formal methods and logic and algebra etc. I thought, "Yeah, whatever, I'm a combinatorics person searching to design new quantum algorithms and protocols." In my ignorance, I was totally oblivious to the tipping point in my life about to happen. Prakash arrived, and with his endless charm, generosity, and knowledge enlightened me to the brave new world. In our cosy Wednesday meetings (only now do I realize what privilege I had) we set up the foundation of Measurement Calculus that I am still stuck with; he taught me how science is not separated and he made me fall so deeply in love with the beauty of mathematics that I could not resist falling for his first French logician visitor. Whenever I wonder why I'm doing what I'm doing, I think of Prakash and of the beautiful goddess of knowledge to get inspiration for a life worth living.

Elham Kashefi   Edinburgh, March 2014

I met Prakash for the first time at ICALP 1990 in Warwick, England. He gave one of the best talks and I remember being fascinated by the beauty of its exposition. During a break, a large group of people went outside to enjoy the bright sun of July. People were chatting pleasantly when a student, no doubt excited by the friendly atmosphere in such a gathering of fine minds, asked the name of one of the big shots there. Offended for not being recognized, the latter answered emphatically "I am [...], head of [...] in theoretical computer science!". Silence fell at once. Everybody felt embarrassed, and the poor student was mortified. But Prakash exclaimed, "And don't we all love theoretical computer science!". This was just the right moral statement. I thought, " Wow! this man is not only brilliant, he has a heart too, and he is so witty! As long as there are people like him around, our profession will be a paradise". In all these years, I have met

Prakash at a myriad of events, we became friends, I have had uncountable chats with him, both about science and about life. We have collaborated on various projects, and I am immensely grateful to him for opening my mind to new horizons. But what I am most grateful for is that whenever something happens that makes me feel bad about my profession, I can think of Prakash and what he would say to refocus my attention on the nobility of science and its people.

Catuscia Palamidessi   Paris, March 2014

Prakash and this editor met in a discotheque, in Tampere, Finland. We were all a bit drunk, not so much from alcohol (that as well) but more so from the fact that there seemed to be sun always, everywhere. Science being the solid foundation of our friendship, there must have been some conference nearby. In fact, in the background, ICALP 1988 was taking place.

From the outset it was clear that this was someone special. Able to enjoy life to the fullest while remaining perfectly able to discuss serious science, at any time and anywhere (this included the discotheque). Respectable and enjoyable, and both highly so.

Later there was, all in all, more science than dancing in our various encounters throughout the years to follow (but I do remember various other clubs, including one in Montreal). On the very day of the writing of this little note, I am in the middle of something that builds on joint work with Prakash, and that started with a question he posed. Which illustrates the continuing impact the encounter in Tampere has had.

Many thanks for all of this, Prakash. May you have a long and happy life, and may we remain in touch for many more years to come!

Jan Rutten   Amsterdam, March 2014

Prakash Panangaden's Doctoral Students

Former Students

- Pablo Castro: *On Planning, Prediction and Knowledge Transfer in Fully and Partially Observable Markov Decision Processes*, McGill University, October 2011.

  "Throughout these years, my love of mathematics has greatly increased, in great part due to the many things I learned from Prakash. The clarity with which he explains complex subjects is something I have always admired and aspire to emulate. His insistence on clarity, syntactical and grammatical correctness, both in math and plain English, have definitely improved the way I express myself, and I am very grateful for that."

- Yannick Delbecque: *Quantum Games as Quantum Types*, McGill University, August 2008.

  "I would like to extend my gratitude to my supervisor Prakash Panangaden for his shared wisdom and insight, as well as for his continued support throughout the most difficult stages."

- Norm Ferns: *State-Similarity Metrics for Continuous Markov Decision Processes*, McGill University, October 2007.

  "Firstly, I would like to thank the bestest supervisors in the whole wide world, Prakash Panangaden and Doina Precup. Smarter people I do not know. You have been nothing short of kind and supportive, more patient and understanding with me than warranted, and just all-around swell. I really do appreciate everything. Thank you."

- Ellie D'Hondt: *Distributed Quantum Computation—A Measurement-Based Approach*, Vrije Universiteit Brussels, July 2005.

  "Above all, I am very grateful to my second advisor, Prakash Panangaden. He took me on as a student at a time when he did not know me at all. I was very lucky to meet him and our collaboration has proved to be a fruitful one – and also a lot of fun! I sincerely hope we can continue working together in the future."

- Josée Desharnais: *Labelled Markov Processes*, McGill University, November 1999.

  "First I would like to warmly thank my thesis supervisor, Prakash Panangaden, for his precious enthusiasm, for his invaluable support and patience, for having so strongly believed in me and for his continuous help during the preparation and writing of this thesis. I also want to thank him for having introduced me to the subject of this thesis, for having given me the opportunity to travel and meet people by sending me to many conferences in many beautiful countries."

– Clark Verbrugge: *A Parallel Solution Strategy for Irregular, Dynamic Problems*, McGill University, August 1996.

> "I would like to thank my advisor, Prakash Panangaden, for his support (financial and otherwise), encouragement and invaluable advice; I have learned a great deal from him, and would certainly have never finished this without his guidance."

– Marija Čubrić: *The Behaviour of Dataflow Networks with Finite Buffers*, Concordia University, 1994.

> "First of all, I would like to thank my supervisor, Professor Prakash Panangaden, for his great help—professional and financial. I have learned very much from his interesting lectures, challenging discussions and overall academic performance. I especially thank him for his patience, encouragement and friendship."

– Carol Critchlow: *The Inherent Cost of Achieving Causal Consistency*, Cornell University, August 1991.

> "First, I would like to propose a toast to Prakash Panangaden, whose encouragement and support—academic, moral, and vinous—have been simply invaluable to me throughout my time at Cornell. If not for him, I *really might* be a substitute teacher of remedial algebra for high school students in Winnipeg. So here's to you, Prakash, and *merci bien*, eh?"

– Radhakrishnan Jagadeesan: *Investigations into Abstraction and Concurrency*, Cornell University, August 1991.

> "I would like to thank Prakash Panangaden for being a great thesis advisor. His advice, on matters academic and otherwise, have been influential throughout my stay in Cornell. This thesis would not have been started or finished without his ideas, encouragement and support."

– Charles Elkan: *Flexible Concurrency Control by Reasoning about Database Queries and Updates*, Cornell University, August 1990.

> "The faculty of the Department of Computer Science at Cornell have created an environment that is wonderfully scholarly, friendly, and enterprising. My committee members, official and unofficial, have been inspiring examples–Prakash Panangaden, Dexter Kozen, David McAllester, and Alberto Segre. Thank you for your confidence and your patience."

– Kimberly Taylor: *Knowledge and Inhibition in Asynchronous Distributed Systems*, Cornell University, August 1990.

> "I first and foremost wish to thank Prakash Panangaden for being an unsurpassable advisor, mentor, and friend throughout the past five years. In particular, for listening to and even encouraging the

most nebulous of ideas, for having faith in me long before I had faith in myself, and for allowing me the freedom to develop as a whole person in addition to as a researcher."

- Vasant Shanbhogue: *The Expressiveness of Indeterminate Dataflow Primitives*, Cornell University, May 1990.

  "I chiefly wish to thank my advisor, Prakash Panangaden, for initially suggesting my thesis problem, encouraging me in my work, and suffering me during my moody moments. He has always been very enthusiastic about and receptive to new ideas, and at the same time, he has encouraged the sifting of ideas to separate the gold from the chaff."

- James Russell: *Full Abstraction and Fixed-Point Principles for Indeterminate Computation*, Cornell University, May 1990.

  "First, I would like to thank Prakash Panangaden for being a great advisor. His advice on subjects, academic, professional, and beyond has been invaluable to me throughout my time at Cornell."

- Anne Neirynck: *Static Analysis of Aliases and Side Effects in Higher-Order Languages*, Cornell University, January 1988.

  "Prakash Panangaden has supervised the work described in this thesis, and I am deeply endebted to him for his time, attention, advice and enthusiasm. It has been a pleasure to work with him."

- Michael Schwartzbach: *A Category Theoretic Analysis of Predicative Type Theory*, Cornell University, January 1987.

  "First of all, I want to thank Prakash Panangaden for being an excellent friend and advisor from the first day we both came to Cornell. The success of this project is largely due to his optimism and inspiration."

Current Students

- Gheorghe Comanici: *Markov Decision Processes, Bisimulation and Approximation*, McGill University.

  "I would like to express my sincere gratitude to my current co-supervisor Prakash Panangaden, first and foremost for his great feedback and contribution to my on-going research, as well as his amazing support towards my educational, professional and personal growth. Prakash was there as a teacher when I chose this career path, he was there as a friend when I needed support in staying on track, he is and will be my supervisor in leading me towards my graduation, and I know I am super-lucky for that."

— Anusar Farooqi: *Quantum Information Theory in General Relativity*, McGill University.

"During the first semester of my PhD I went to a Math-Physics-CS social. I was just socializing when I saw a swarm of kids hanging on to, what appeared to be, a professor's every word. I made my way through the swarm. At the centre, Prakash (my friend had informed me that it was her advisor) was holding court. Students were asking him about black holes. Now I have been interested in such things since I was a kid, so I made my way to the front and engaged with Prakash. We talked about quantum entanglement and black holes; specifically whether entanglement still holds between two particles if one of them falls inside a black hole. I still remember his answer: "No one really knows." Later I went back to Prakash and asked him if he would be interested in exploring that question with me as my PhD advisor (along with Professor Niky Kamran in the math department who is an expert in black hole geometries). Prakash was sufficiently intrigued to give it a shot. I'm pretty sure that this is not something that happens everyday. It takes a special kind of professor to ignite a four year research project over a beer. Running into Prakash that fateful evening was one of the best things that ever happened to me."

— Andie Sigler: *Mathematical Structure in Musical Analysis*, McGill University.

"Prakash, jouster of monstrous abstractions, nixer of mathematical misapplications, and mind-expander of luminous categorical domains: nice of Doina to let us borrow her pet mathematician!"

Group photo from the 2014 Bellairs Workshop

Group photo from the 2013 Bellairs Workshop

Prakash Panangaden in the president's chair
at his induction to the Royal Society of Canada

Table of Contents

Layout Randomization and Nondeterminism

Martín Abadi[1], Jérémy Planul[2], and Gordon D. Plotkin[1,3]

[1] Microsoft Research
[2] Department of Computer Science, Stanford University
[3] LFCS, School of Informatics, University of Edinburgh

Abstract. In security, layout randomization is a popular, effective attack mitigation technique. Recent work has aimed to explain it rigorously, focusing on deterministic systems. In this paper, we study layout randomization in the presence of nondeterministic choice. We develop a semantic approach based on denotational models and the induced notions of contextual public observation, characterized by simulation relations. This approach abstracts from language details, and helps manage the delicate interaction between nondeterminism and probability. In particular, memory access probabilities are not independent, but rather depend on a fixed probability distribution over memory layouts; we therefore model probability using random variables rather than any notion of probabilistic powerdomain.

1 Introduction

Randomization has important applications in security, ranging from probabilistic cryptographic schemes [1] to the introduction of artificial diversity in low-level software protection [2]. Developing rigorous models and analyses of the systems that employ randomization can be challenging, not only because of the intrinsic difficulty of reasoning about probabilities but also because these systems typically exhibit many other interesting features. Some of these features, such as assumed bounds on the capabilities and the computational complexity of attackers, stem directly from security considerations. Others, such as nondeterminism, need not be specifically related to security, but arise because of the generality of the ambient computational models, which may for example include nondeterministic scheduling for concurrent programs and for network protocols.

The form of randomization that we explore in this paper is layout randomization in software systems (e.g., [3–5]). Layout randomization refers to a body of widely used techniques that place data and code randomly in memory. In practice, these techniques effectively thwart many attacks that assume knowledge of the location of data and code. Recent research by the authors and others aims to develop rigorous models and proofs for layout randomization [6–9]. The research to date has focused on deterministic, sequential programs. Here, we consider layout randomization for programs that may make nondeterministic choices.

We phrase our study in terms of a high-level language in which variables are abstract (symbolic) locations, and a low-level language in which they are mapped

F. van Breugel et al. (Eds.): Panangaden Festschrift, LNCS 8464, pp. 1–39, 2014.

to random natural-number addresses in memory. Both languages include a standard construct for nondeterministic choice. We give models for the languages. For each language, we also define a contextual implementation relation. Intuitively, a context may represent an attacker, so contextual implementation relations may serve, in particular, for expressing standard security properties. We characterize contextual implementation relations in terms of semantic simulation relations (so-called logical relations). Throughout, the low-level relations are probabilistic. Via the simulation relations, we obtain a semantic correspondence between the high-level and low-level worlds. Basically, simulation relations in one world induce simulation relations in the other, and therefore contextual implementation in one world implies contextual implementation in the other.

Thus, our approach emphasizes semantic constructions. In comparison with prior syntactic work, arguments via models arguably lead to more satisfying security arguments, independent of superficial details of particular languages (as layout randomization is largely language-agnostic in practice). They also help reconcile probabilities and nondeterminism, which have a rich but thorny interaction.

Some of the difficulties of this interaction have been noticed in the past. For instance, in their development of a framework for the analysis of security protocols [10, Section 2.7], Lincoln et al. observed:

> our intention is to design a language of communicating processes so that an adversary expressed by a set of processes is restricted to probabilistic polynomial time. However, if we interpret parallel composition in the standard nondeterministic fashion, then a pair of processes may nondeterministically "guess" any secret information.

They concluded:

> Therefore, although nondeterminism is a useful modeling assumption in studying correctness of concurrent programs, it does not seem helpful for analyzing cryptographic protocols.

Thus, they adopted a form of probabilistic scheduling, and excluded nondeterminism. In further work, Mitchell et al. [11] refined the framework, in particular defining protocol executions by reference to any polynomial-time probabilistic scheduler that operates uniformly over certain kinds of choices. The uniformity prevents collusion between the scheduler and an attacker. Similarly, Canetti et al. [12] resolved nondeterminism by task schedulers, which do not depend on dynamic information generated during probabilistic executions; they thus generated sets of trace distributions, one for each task schedule.

From a semantic perspective, a nondeterministic program denotes a function that produces a set of possible outcomes; equally, a probabilistic program represents a function that produces a distribution over outcomes. Rigorous versions of these statements can be cast in terms of powerdomains and probabilistic powerdomains [13]. In principle, a nondeterministic and probabilistic program may represent either a function producing a set of distributions over outcomes or else one producing a distribution over sets of outcomes. However it seems that only the former option, where nondeterministic choice is resolved before probabilistic choice, leads to a satisfactory theory if, for example, one wishes to retain all the usual laws for both forms of nondeterminism [14–16].

To illustrate these options, imagine a two-player game in which Player I chooses a bit b_I at random, Player II chooses a bit b_{II} nondeterministically, and Player I wins if and only if $b_I = b_{II}$. The system composed of the two players may be seen as producing a set of distributions or a distribution on sets of outcomes.

- With the former view, we can say that, in each possible distribution, Player I wins with probability $1/2$.
- On the other hand, with the latter view, we can say only that, with probability 1, Player I may win and may lose.

The former view is preferable in a variety of security applications, in which we may wish to say that no matter what an attacker does, or how nondeterministic choices are resolved, some expected property holds with high probability.

However, in our work, it does not suffice to resolve nondeterministic choice before probabilistic choice, as we explain in detail below, fundamentally because the probabilistic choices that we treat need not be independent. Instead, we construct a more sophisticated model that employs random variables, here maps from memory layouts to outcomes. The memory layouts form the sample space of the random variables, and, as usual, one works relative to a given distribution over the sample space.

Beyond the study of layout randomization, it seems plausible that an approach analogous to ours could be helpful elsewhere in security analysis. Our models may also be of interest on general grounds, as a contribution to a long line of research on programming-language semantics for languages with nondeterministic and probabilistic choice. Specifically, the models support a treatment of dependent probabilistic choice combined with nondeterminism, which as far as we know has not been addressed in the literature. Finally, the treatment of contextual implementation relations and simulation relations belongs in a long line of research on refinement.

This paper is a full version of a conference paper [17] of the same title. The main differences are that proofs are presented in full (except in some routine or evident cases) and that an inconsistency between the operational and denotational semantics of the low-level language has been corrected by an alteration to its operational semantics.

Contents

In Section 2 we review some preliminary material on cpos.

In Section 3, we consider a high-level language, with abstract locations, standard imperative constructs, and nondeterminism, and describe its denotational and operational semantics. We define a contextual implementation relation with respect to contexts that represent attackers, which we call public contexts; for this purpose, we distinguish public locations, which attackers can access directly, from private locations. We also define a simulation relation, and prove that it coincides with the contextual implementation relation. The main appeal of the simulation relation, as usual, is that it does not require reasoning about all possible contexts.

In Section 4, we similarly develop a lower-level language in which programs may use natural-number memory addresses (rather than abstract locations). Again, we define a denotational semantics, an operational semantics, a contextual implementation relation, and a simulation relation. These definitions are considerably more delicate than those of the high-level language, in particular because they refer to layouts, which map abstract locations to concrete natural-number addresses, and which may be chosen randomly (so we often make probabilistic statements).

In Section 5, we relate the high-level and the low-level languages. We define a simple compilation function that maps from the former to the latter. We then establish that if two high-level commands are in the contextual implementation relation, then their low-level counterparts are also in the contextual implementation relation. The proof leverages simulation relations. In semantics parlance, this result is a full-abstraction theorem; the use of public contexts that represent attackers, however, is motivated by security considerations, and enable us to interpret this theorem as providing a formal security guarantee for the compilation function, modulo a suitable random choice of memory layouts.

Finally, in Section 6 we conclude by discussing some related and further work.

2   Preliminaries on Cpos

We take a cpo to be a partial order P closed under increasing ω-sups, and consider sets to be cpos with the discrete ordering. We write $P_\perp$ for the lift of P, viz. P extended by the addition of a least element, $\perp$. Products $P \times Q$ and function spaces $P \to Q$ (which we may also write as Q^P) are defined as usual, with the function space consisting of all continuous functions (those monotonic functions preserving the ω-lubs).

We use the lower, or Hoare, powerdomain $\mathcal{H}(P)$, which consists of the nonempty, downwards, and ω-sup-closed subsets of P, ordered by inclusion. The lower powerdomain is the simplest of the three powerdomains, and models "may" or "angelic" nondeterminism; the others (upper and convex) may also be worth investigating.

For any nonempty subset X of P, we write $X \downarrow$ for the downwards closure $\{y \mid \exists x \in X. y \leq x\}$ of X. We also write X^* for the downwards and ω-sup closure of X (which is typically the same as $X \downarrow$ in the instances that arise below).

Both $\mathcal{H}(-)$ and $\mathcal{H}(-_\perp)$ are monads (those for lower nondeterminism, and lower nondeterminism and nontermination, respectively). The unit of the former is $x \mapsto \{x\} \downarrow$ and continuous maps $f : P \to \mathcal{H}(Q)$ have extensions $f^\dagger : \mathcal{H}(P) \to \mathcal{H}(Q)$ given by:

$$f^\dagger(X) = (\bigcup_{x \in X} f(x))^*$$

The unit of the latter is $x \mapsto \{x\} \downarrow$ and continuous maps $f : P \to \mathcal{H}(Q_\perp)$ have extensions $f^\dagger : \mathcal{H}(P_\perp) \to \mathcal{H}(Q_\perp)$ given by:

$$f^\dagger(X) = \{\perp\} \cup (\bigcup_{x \in X \setminus \{\perp\}} f(x))^*$$

3   The High-Level Language

In this section, we define our high-level language. In this language, locations are symbolic names, and we use an abstract store to link those locations to their contents, which are natural numbers.

For simplicity, the language lacks data structures and higher-order features. Therefore, locations cannot contain arrays or functions (cf. [9]), except perhaps through encodings. So the language does not provide a direct model of overflows and code-injection attacks, for instance.

There are many other respects in which our languages and their semantics are not maximally expressive, realistic, and complex. They are however convenient for our study of nondeterminism and of the semantic approach to layout randomization.

3.1   Syntax and Informal Semantics

The syntax of the high-level language includes categories for natural-number expressions, boolean expressions, and commands:

$$
\begin{aligned}
e & ::= k \mid !l_{\mathrm{loc}} \mid e + e \mid e * e \\
b & ::= e \leq e \mid \neg b \mid \mathtt{tt} \mid \mathtt{ff} \mid b \vee b \mid b \wedge b \\
c & ::= l_{\mathrm{loc}} := e \mid \mathtt{if}\, b\, \mathtt{then}\, c\, \mathtt{else}\, c \mid \mathtt{skip} \mid c; c \mid c + c \mid \mathtt{while}\, b\, \mathtt{do}\, c
\end{aligned}
$$

where k ranges over numerals, and l over a given finite set of store locations Loc. Natural-number expressions are numerals, dereferencing of memory locations, sums, or products. Boolean expressions are inequalities on natural-number expressions, negations, booleans, disjunctions, or conjunctions. Commands are assignments at a location, conditionals, skip, sequences, nondeterministic choices, or while loops. Command contexts $C[\,]$ are commands with holes; we write $C[c]$ for the command obtained by filling all the holes in $C[\,]$ with c. We further use trivial extensions of this language, in particular with additional boolean and arithmetic expressions.

We assume that the set of store locations Loc is the union of two disjoint sets of locations PubLoc (public locations) and PriLoc (private locations). Let c be a command or a command context. We say that c is public if it does not contain any occurrence of $l_{\text{loc}} := v$ or $!l_{\text{loc}}$ for $l \in \text{PriLoc}$. As in previous work [7], we model attackers by such public commands and command contexts; thus, attackers have direct access to public locations but not, by default, to private locations.

The distinction between public and private locations is directly analogous to that between external and internal state components in automata and other specification formalisms (e.g., [18]). It also resembles distinctions in information-flow systems, which often categorize variables into levels (e.g., [19]), and typically aim to prevent flows of information from "high" to "low" levels. We do not impose any such information-flow constraint: we permit arbitrary patterns of use of public and private locations. Nevertheless, we sometimes use h for a private location and l for a public location, and also associate the symbols H and L with private and public locations, respectively.

3.2   Denotational Semantics

A store s is a function from the finite set Loc of store locations to natural numbers. When Loc consists solely of h and l, for example, we write $(h \mapsto m, l \mapsto n)$ for the store that maps h to m and l to n. A public (private) store is a function from PubLoc (PriLoc) to natural numbers. We write S for the set of stores, S_L for the set of public stores, and S_H for the set of private stores. The following natural functions restrict the store to its public and private locations:

$$S_L \xleftarrow{\ L\ } S \xrightarrow{\ H\ } S_H$$

We write s_L for $L(s)$ and $s =_L s'$ when $s_L = s'_L$, and similarly for H.

The denotational semantics

$$[\![e]\!] : \text{Store} \to \mathbb{N} \qquad [\![b]\!] : \text{Store} \to \mathbb{B}$$

of expressions are defined as usual with, in particular, $[\![!l_{\text{loc}}]\!](s) = s(l)$. The denotational semantics

$$[\![c]\!] : S \to \mathcal{H}(S_\perp)$$

of commands is given in Figure 1, where the semantics of the while loop is the standard least-fixed point one.

$$
\begin{aligned}
&[\![l_{\text{loc}} := e]\!](s) &&= \eta(s[l \mapsto [\![e]\!](s)]) &\qquad& [\![\texttt{skip}]\!](s) &&= \eta(s) \\[4pt]
&[\![\texttt{if } b \texttt{ then } c \texttt{ else } c']\!](s) &&= \begin{cases} [\![c]\!](s) & ([\![b]\!](s) = \texttt{tt}) \\ [\![c']\!](s) & ([\![b]\!](s) = \texttt{ff}) \end{cases} &\qquad& [\![c;c']\!](s) &&= [\![c']\!]^\dagger([\![c]\!](s)) \\
& && && [\![c + c']\!](s) &&= [\![c]\!](s) \cup [\![c']\!](s) \\[4pt]
&[\![\texttt{while } b \texttt{ do } c]\!] &&= \mu\theta : S \to \mathcal{H}(S_\perp).\lambda s : S. \begin{cases} \eta(s) & ([\![b]\!](s) = \texttt{ff}) \\ \theta^\dagger([\![c]\!](s)) & ([\![b]\!](s) = \texttt{tt}) \end{cases}
\end{aligned}
$$

Fig. 1. High-level denotational semantics

Example 1. Consider the two commands:

$$c_0 = (h := 1; l := \neg!l) + (h := 0) \qquad c_1 = (h := 1; l := 1) + (h := 0; l := 0)$$

According to the semantics, $[\![c_0]\!]$ maps any store mapping l to 1 to the set $\{(h \mapsto 1, l \mapsto 0), (h \mapsto 0, l \mapsto 1)\} \downarrow$, and any store where l is 0 to the set $\{(h \mapsto 1, l \mapsto 1), (h \mapsto 0, l \mapsto 0)\} \downarrow$, while $[\![c_1]\!]$ maps any store to the set $\{(h \mapsto 1, l \mapsto 1), (h \mapsto 0, l \mapsto 0)\} \downarrow$. In sum, we may write:

$$[\![c_0]\!](h \mapsto \_, l \mapsto 1) = \{(h \mapsto 1, l \mapsto 0), (h \mapsto 0, l \mapsto 1)\} \downarrow$$
$$[\![c_0]\!](h \mapsto \_, l \mapsto 0) = \{(h \mapsto 1, l \mapsto 1), (h \mapsto 0, l \mapsto 0)\} \downarrow$$
$$[\![c_1]\!](h \mapsto \_, l \mapsto \_) = \{(h \mapsto 1, l \mapsto 1), (h \mapsto 0, l \mapsto 0)\} \downarrow$$

Since the two commands act differently on stores, they do not have the same semantics. However, when one observes only public locations, the apparent behavior of both commands is the same: they non-deterministically write 0 or 1 to l. This similarity will be made rigorous in Example 2. □

3.3 Operational Semantics

The high-level language has a straightforward big-step operational semantics. In this semantics, a high-level state is a pair $\langle c, s \rangle$ of a command and a store or, marking termination, just a store s. The transition relation is a binary relation $\langle c, s \rangle \Rightarrow s$ between such states. Figure 2 gives the rules for $\Rightarrow$. (Note that we treat expressions denotationally; as we wish to focus on commands, this treatment avoids some extra complexity.)

$$\langle l_{\mathrm{loc}} := e, s \rangle \Rightarrow s[l \mapsto [\![e]\!]_s] \qquad \frac{[\![b]\!]_s = \mathsf{tt} \quad \langle c, s \rangle \Rightarrow s'}{\langle \mathsf{if}\, b\, \mathsf{then}\, c\, \mathsf{else}\, c', s \rangle \Rightarrow s'}$$

$$\frac{[\![b]\!]_s = \mathsf{ff} \quad \langle c', s \rangle \Rightarrow s'}{\langle \mathsf{if}\, b\, \mathsf{then}\, c\, \mathsf{else}\, c', s \rangle \Rightarrow s'} \qquad \langle \mathsf{skip}, s \rangle \Rightarrow s \qquad \frac{\langle c, s \rangle \Rightarrow s' \quad \langle c', s' \rangle \Rightarrow s''}{\langle c; c', s \rangle \Rightarrow s''}$$

$$\frac{\langle c, s \rangle \Rightarrow s'}{\langle c + c', s \rangle \Rightarrow s'} \qquad \frac{\langle c', s \rangle \Rightarrow s'}{\langle c + c', s \rangle \Rightarrow s'}$$

$$\frac{[\![b]\!]_s = \mathsf{ff}}{\langle \mathsf{while}\, b\, \mathsf{do}\, c, s \rangle \Rightarrow s} \qquad \frac{[\![b]\!]_s = \mathsf{tt} \quad \langle c, s \rangle \Rightarrow s' \quad \langle \mathsf{while}\, b\, \mathsf{do}\, c, s' \rangle \Rightarrow s''}{\langle \mathsf{while}\, b\, \mathsf{do}\, c, s \rangle \Rightarrow s''}$$

Fig. 2. High-level operational semantics

The following proposition links the operational and denotational semantics of the high-level language.

Proposition 1 (High-level operational/denotational consistency). *For any high-level command c and store s, we have:*

$$[\![c]\!](s) = \{s' | \langle c, s \rangle \Rightarrow s'\} \cup \{\bot\}$$

Proof. In one direction, using rule-induction, one shows that if $\langle c, s \rangle \Rightarrow s'$ then $s' \in [\![c]\!](s)$. In the other direction one shows, by structural induction on loop-free commands, that if $s' \in [\![c]\!](s)$ then $\langle c, s \rangle \Rightarrow s'$. One then establishes the result for all commands, including while loops, by considering their iterates, where loops are unwound a finite number of times. We omit details.

3.4   Implementation Relations and Equivalences

We next define the contextual pre-order that arises from the notion of public observation. We then give an equivalent simulation relation, with which it is easier to work as it does not refer to contexts.

Contextual Pre-order. We introduce a contextual pre-order $\sqsubseteq_L$ on commands. Intuitively, $c \sqsubseteq_L c'$ may be interpreted as saying that c "refines" (or "implements") c', in the sense that the publicly observable outcomes that c can produce are a subset of those that c' permits, in every public context and from every initial store. Thus, let $f = [\![C[c]]\!]$ and $f' = [\![C[c']]\!]$ for an arbitrary public context C, and let s_0 be a store; then for every store s in $f(s_0)$ there is a store s' in $f'(s_0)$ that coincides with s on public locations. Note that we both restrict attention to public contexts and compare s and s' only on public locations.

We define $\sqsubseteq_L$ and some auxiliary relations as follows:

- For $X \in \mathcal{H}(S_\bot)$, we set:

$$X_L = \{s_L \mid s \in X \setminus \{\bot\}\} \cup \{\bot\}$$

- For $f, f' : S \to \mathcal{H}(S_\bot)$, we write that $f \leq_L f'$ when, for every store s_0, we have $f(s_0)_L \leq f'(s_0)_L$.
- Let c and c' be two commands. We write that $c \sqsubseteq_L c'$ when, for every public command context C, we have $[\![C[c]]\!] \leq_L [\![C[c']]\!]$.

Straightforwardly, this contextual pre-order relation yields a notion of contextual equivalence with respect to public contexts.

Simulation. We next give the simulation relation $\preceq$. As in much previous work, one might expect the simulation relation between two commands c and c' to be a relation on stores that respects the observable parts of these stores, and such that if s_0 is related to s_1 and c can go from s_0 to s_0' then there exists s_1' such that s_0' is related to s_1' and c' can go from s_1 to s_1'. In our setting, respecting the observable parts of stores means that related stores give the same values to public locations (much like refinement mappings preserve externally visible state components [18], and low-bisimulations require equivalence on low-security variables [19]).

Although this idea could lead to a sound proof technique for the contextual pre-order, it does not suffice for completeness. Indeed, forward simulations, of the kind just described, are typically incomplete on their own for nondeterministic systems. They can be complemented with techniques such as backward simulation, or generalized (e.g., [18, 20, 21]).

Here we develop one such generalization. Specifically, we use relations on sets of stores. We build them from relations over $\mathcal{H}(S_{H\perp})$ as a way of ensuring the condition that public locations have the same values, mentioned above. We also require other standard closure conditions. Our relations are similar to the ND measures of Klarlund and Schneider [20]. Their work takes place in an automata-theoretic setting; automata consist of states (which, intuitively, are private) and of transitions between those states, labeled by events (which, intuitively, are public). ND measures are mappings from states to sets of finite sets of states, so can be seen as relations between states and finite sets of states. The finiteness requirement, which we do not need, allows a fine-grained treatment of infinite execution paths via König's Lemma.

First, we extend relations R over $\mathcal{H}(S_{H\perp})$ to relations R^+ over $\mathcal{H}(S_\perp)$, as follows. For any $X \in \mathcal{H}(S_\perp)$ and $s \in S_L$, we define $X_s \in \mathcal{H}(S_{H\perp})$ by:

$$X_s = \{s'_H \mid s' \in X, s'_L = s\} \cup \{\perp\}$$

and then we define R^+ by:

$$X R^+ Y \equiv_{\text{def}} \forall s \in S_L.\,(X_s \neq \{\perp\} \Rightarrow Y_s \neq \{\perp\}) \wedge X_s R Y_s$$

If R is reflexive (respectively, is closed under increasing ω-sups; is right-closed under $\leq$; is closed under binary unions) the same holds for R^+. Also, if $X R^+ Y$ then $X_L \leq Y_L$.

For any $f, f' : S_\perp \to \mathcal{H}(S_\perp)$ and relation R over $\mathcal{H}(S_{H\perp})$ we write that $f \preceq_R f'$ when:

$$\forall X, Y \in \mathcal{H}(S_\perp).\, X R^+ Y \Rightarrow f^\dagger(X) R^+ f'^\dagger(Y)$$

If $f \preceq_R f'$ holds then we have $f \leq_L f'$ (as follows from the fact that $X_L \leq Y_L$ holds if $X R^+ Y$ does).

Finally, we write that $f \preceq f'$ if $f \preceq_R f'$ for some reflexive R closed under increasing ω-sups, right-closed under $\leq$, and closed under binary unions.

Contextual Pre-order vs. Simulation. The contextual pre-order coincides with the simulation relation, as we now show. We break the proof into two parts.

Lemma 1. *Let c and c' be two commands of the high-level language such that $[\![c]\!] \preceq [\![c']\!]$. Then $c \sqsubseteq_L c'$ holds.*

Proof. Let c_0 and c_1 be two commands such that $[\![c_0]\!] \preceq_R [\![c_1]\!]$, with R a reflexive relation over $\mathcal{H}(S_{H\perp})$ closed under increasing ω-sups, right-closed under $\leq$, and closed under binary unions, and let C be a public command context. We prove that $[\![C[c_0]]\!] \preceq_R [\![C[c_1]]\!]$ by induction on the size of C, considering the possible forms of C:

1. $l_{\mathtt{loc}} := e$: Suppose that $X R^+ Y$. As e is public, for every s, $[\![e]\!](s)$ only depends on s_L. As l is also public, $[\![l_{\mathtt{loc}} := e]\!](s) = \{s'\} \cup \{\perp\}$, for every s, where $s' =_H s$ and s'_L depends only on s_L. Therefore, for every $s_0 \in S_L$, let $S_0 \subseteq S$ be the (possibly empty) set

$$\{s \in S \mid \exists s' \in S, [\![l_{\mathtt{loc}} := e]\!](s) = \{s'\} \cup \{\perp\} \text{ and } s'_L = s_0\}$$

We then have that $[\![l_{\mathtt{loc}} := e]\!](X)_{s_0} = (\bigcup_{s\in S_0} X_{s_L}) \cup \{\perp\}$ and also that $[\![l_{\mathtt{loc}}:=e]\!](Y)_{s_0} = (\bigcup_{s\in S_0} Y_{s_L}) \cup \{\perp\}$.
To see that $[\![l_{\mathtt{loc}}:=e]\!](X) R^+ [\![l_{\mathtt{loc}}:=e]\!](Y)$, choose $s_0 \in S_L$, and define S_0 as above. Then, if $[\![l_{\mathtt{loc}}:=e]\!](X) \neq \{\perp\}$ there is an $s \in S_0$ such that $X_{s_L} \neq \{\perp\}$. But then, as $X R^+ Y$, we have that $Y_{s_L} \neq \{\perp\}$, and so $[\![l_{\mathtt{loc}}:=e]\!](Y) \neq \{\perp\}$. Finally we have to check that $[\![l_{\mathtt{loc}}:=e]\!](X)_{s_0} R [\![l_{\mathtt{loc}}:=e]\!](Y)_{s_0}$. That follows from the above two formulas, as $X R^+ Y$ and R is closed under countable unions and reflexive.

2. $\mathtt{if}\ b\ \mathtt{then}\ C_{\mathtt{tt}}\ \mathtt{else}\ C_{\mathtt{ff}}$: Suppose that $X R^+ Y$. Define $X_{\mathtt{tt}} \subseteq X$ to be the set $\{s \in X \mid [\![b]\!](s) = \mathtt{tt}\} \cup \{\perp\}$ and define $Y_{\mathtt{tt}}$, $X_{\mathtt{ff}}$, and $Y_{\mathtt{ff}}$ similarly. As b is public, for every s, $[\![b]\!](s)$ only depends on s_L, and so, for any $s_0 \in S_L$ we have:

$$(X_{\mathtt{tt}})_{s_0} = \begin{cases} X_{s_0} \cup \{\perp\} & (\exists s \in X_{\mathtt{tt}}.\, s_0 = s_L) \\ \{\perp\} & (\text{otherwise}) \end{cases}$$

and similar equations hold for $Y_{\mathtt{tt}}$, $X_{\mathtt{ff}}$, and $Y_{\mathtt{ff}}$. We then check that $X_{\mathtt{tt}} R^+ Y_{\mathtt{tt}}$ and $X_{\mathtt{ff}} R^+ Y_{\mathtt{ff}}$ much as in the previous case.
We have

$$[\![\mathtt{if}\ b\ \mathtt{then}\ C_{\mathtt{tt}}[c_0]\ \mathtt{else}\ C_{\mathtt{ff}}[c_0]]\!]^\dagger(X) = [\![C_{\mathtt{tt}}[c_0]]\!]^\dagger(X_{\mathtt{tt}}) \cup [\![C_{\mathtt{ff}}[c_0]]\!]^\dagger(X_{\mathtt{ff}})$$

and similarly for Y. By induction, $[\![C_{\mathtt{tt}}[c_0]]\!]^\dagger(X_{\mathtt{tt}}) R^+ [\![C_{\mathtt{tt}}[c_1]]\!]^\dagger(Y_{\mathtt{tt}})$ and similarly for ff. As R^+ is closed under binary unions, we conclude.

3. skip: The conclusion is immediate as $[\![\mathtt{skip}]\!]^\dagger$ is the identity.

4. $C'; C''$: Here we have:

$$[\![C'[c_0]; C''[c_0]]\!]^\dagger = ([\![C''[c_0]]\!]^\dagger [\![C'[c_0]]\!])^\dagger = [\![C''[c_0]]\!]^\dagger [\![C'[c_0]]\!]^\dagger$$

and the same holds for c_1, and so the conclusion follows using the induction hypothesis.

5. $C' + C''$: Here, as R^+ is closed under binary unions, the conclusion follows using the induction hypothesis.

6. `while b do C_w`: Define iterates $C^{(n)}$ by setting:

$$C^{(0)} = \Omega \qquad C^{(n+1)} = \text{if } b \text{ then skip else } C_w; C^{(n)}$$

where Ω is some command denoting $\bot$. By induction on n, we have $C^{(n)}[c_0] \preceq_R C^{(n)}[c_1]$: the case $n = 0$ follows as we have $\{\bot\}R^+\{\bot\}$, and the induction step follows using the same reasoning as in the second, third, and fourth cases of the proof.

But then, as we have

$$C[c_0] = \bigvee_{n \geq 0} C^{(n)}[c_0]$$

and the same holds for c_1, the conclusion follows using the fact that R^+ is closed under increasing ω-sups.

7. `[ ]`: We have $C[c_0] = c_0$ and $C[c_1] = c_1$, and the conclusion follows using the hypothesis.

This concludes the proof since it follows from $[\![C[c_0]]\!] \preceq_R [\![C[c_1]]\!]$ that $[\![C[c_0]]\!] \leq_L [\![C[c_1]]\!]$.

We need a lemma in order to prove the converse of Lemma 1.

Lemma 2. *Let R_i $(i \geq 0)$ be relations on $\mathcal{H}(S_{H\bot})$ such that if $X R_i Y$ holds then $X \neq \{\bot\}$ implies $Y \neq \{\bot\}$. Let R be the closure of the union of the R_i under increasing ω-sups, binary union, and right-closure under $\leq$. Then R^+ is the closure of the union of the relations R_i^+ under increasing ω-sups, binary union, and right-closure under $\leq$.*

Proof. As $-^+$ is evidently monotone, R^+ contains the R_i^+. Next, we know that if a relation S on $\mathcal{H}(S_{H\bot})$ is closed under any one of increasing ω-sups, binary unions, or right-closure under $\leq$, then so is S^+. So R^+ is closed under increasing ω-sups and binary unions, and right-closed under $\leq$. It is therefore included in the closure of the union of the R_i^+ under increasing ω-sups, binary unions, and right-closure under $\leq$.

For the converse, suppose that $U R^+ W$ to show that U and W are related in the closure of the union of the R_i^+ under increasing ω-sups, binary unions, and right-closure under $\leq$. For any given s in S_L, by definition of $-^+$, $U_s R W_s$, and so, by the definition of R, there is a set $J^{(s)} \subseteq \mathbb{N}$, and relations $X_j^{(s)} R_{i_j^{(s)}} Y_j^{(s)}$ such that $U_s = \bigcup_{j \in J^{(s)}} X_j^{(s)}$ and $W_s \supseteq \bigcup_{j \in J^{(s)}} Y_j^{(s)}$. We may assume without loss of generality that the $J^{(s)}$ are disjoint.

Let

$$J = \bigcup_{s \in S_L} J^{(s)}$$
$$X_j = \{s' \mid s'_H \in X_j^{(s)}, s'_L = s\} \cup \{\bot\} \qquad (j \in J^{(s)})$$
$$Y_j = \{s' \mid s'_H \in Y_j^{(s)}, s'_L = s\} \cup \{\bot\} \qquad (j \in J^{(s)})$$
$$i_j = i_j^{(s)} \qquad (j \in J^{(s)})$$

We verify that $(X_j)_s$ is equal to $X_j^{(s)}$, if j is in $J^{(s)}$, and equal to $\{\bot\}$, otherwise, and similarly for the $(Y_j)_s$. Consequently, $U = \bigcup_j X_j$, $W \supseteq \bigcup_j Y_j$, and for all s in S_L, $(X_j)_s R_{i_j}(Y_j)_s$. Since, by hypothesis, if $(X_j)_s R_{i_j}(Y_j)_s$ holds then $(X_j)_s \neq \{\bot\}$ implies $(Y_j)_s \neq \{\bot\}$, we note that $X_j R_{i_j}^+ Y_j$. We conclude that U and W are related as required.

We also need some notation. Assume a fixed enumeration $x_1 \ldots x_n$ of PubLoc. Then, given high-level commands c_i $(i = 1, \ldots, n)$ we write $[c_x \mid x \in \text{PubLoc}]$ for the high-level command $c_{x_1}; \ldots; c_{x_n}$. As usual, we abbreviate $\text{if } b \text{ then } c \text{ else skip}$ to $\text{if } b \text{ then } c$. We can now show:

Lemma 3. *Let c and c' be two commands of the high-level language such that $c \sqsubseteq_L c'$. Then $[\![c]\!] \preceq [\![c']\!]$ holds.*

Proof. Let c_0 and c_1 be two commands such that $c_0 \sqsubseteq_L c_1$. We define relations R_i $(i \geq 0)$ on $\mathcal{H}(S_{H\bot})$ as follows:

- for every $X \in \mathcal{H}(S_{H\bot})$, we have $X R_0 X$;
- for every $X, Y \in \mathcal{H}(S_\bot)$, such that $X R_i^+ Y$, and for every $s \in S_L$ we have $[\![c_0]\!]^\dagger(X)_s R_{i+1} [\![c_1]\!]^\dagger(Y)_s$.

We first prove by induction on i that, if $X R_i^+ Y$, then, for every $s \in X$ such that $s \neq \bot$, there exist a public command context C and $s_0 \in S$ such that $s \in [\![C[c_0]]\!](s_0)$ and $[\![C[c_1]]\!](s_0)_{s_L} \subseteq Y_{s_L}$.

- Suppose that $X R_0^+ Y$. For every s, we let C be skip and s_0 be s. We have $s \in [\![\text{skip}]\!](s)$ and $[\![\text{skip}]\!](s)_{s_L} = \{s, \bot\}_{s_L} \subseteq X_{s_L} = Y_{s_L}$.
- Suppose that $X R_{i+1}^+ Y$. By definition of $X R_{i+1}^+ Y$, and in particular $X_{s_L} R_{i+1} Y_{s_L}$, there exist $X' R_i^+ Y'$ and $s' \in S_L$ such that $[\![c_0]\!]^\dagger(X')_{s'} = X_{s_L}$ and $[\![c_1]\!]^\dagger(Y')_{s'} = Y_{s_L}$. As s_H in X_{s_L}, by definition of $-^\dagger$, there exist $s'' \in X'$ and $s''' \in S$ such that $s''' \in [\![c_0]\!](s'')$, $s'''_L = s'$ and $s'''_H = s_H$ (note that $s'' \neq \bot$).

  By induction on $X' R_i Y'$ and s'', there exists a public command context C and an $s_0 \in S$ such that both $s'' \in [\![C[c_0]]\!](s_0)$ and $[\![C[c_1]]\!](s_0)_{s''_L} \subseteq Y'_{s''_L}$ hold. We consider the public command context

$$C' =_{\text{def}} C; [\text{if } !x_{\text{loc}} \neq s''_L(x) \text{ then } \Omega \mid x \in \text{PubLoc}];$$
$$[\ ]; [x := s_L(x) \mid x \in \text{PubLoc}]$$

We have s'' in $[\![C[c_0]]\!](s_0)$, so s''' is in

$$[\![C[c_0]; [\text{if } !x_{\text{loc}} \neq s''_L(x) \text{ then } \Omega \mid x \in \text{PubLoc}]; c_0]\!](s_0)$$

so s is in $[\![C'[c_0]]\!](s_0)$.

Also, $[\![C[c_1]]\!](s_0)_{s''_L} \subseteq Y'_{s''_L}$, hence

$$[\![C[c_1]; [\text{if } !x_{\text{loc}} \neq s''_L(x) \text{ then } \Omega \mid x \in \text{PubLoc}]]\!](s_0) \subseteq Y'$$

hence

$$[\![C[c_1]; [\text{if } !x_{\text{loc}} \neq s''_L(x) \text{ then } \Omega \mid x \in \text{PubLoc}]; c_1]\!](s_0)_{s'} \subseteq Y_{s_L}$$

and hence (we rewrite the low variables with their corresponding values in s_L)

$$[\![C'[c_1]]\!](s_0)_{s_L} \subseteq Y_{s_L}$$

We now prove that

$$\forall X, Y \in \mathcal{H}(S_{H\perp}). \, XR_iY \Rightarrow (X \neq \{\perp\} \Rightarrow Y \neq \{\perp\})$$

For $i = 0$, this follows from the definition of R_0. Otherwise, $i = j + 1$, and by definition of (R_i), there exist X', Y', and $s \in S_L$ such that $X'R_j^+Y'$, $X = [\![c_0]\!]^\dagger(X')_s$, and $Y = [\![c_1]\!]^\dagger(Y')_s$. If $X \neq \{\perp\}$, by definition of $-^\dagger$, there exists $s' \in X'$ such that $[\![c_0]\!](s')_s \neq \{\perp\}$ (note that $s' \neq \perp$). As shown above, since $X'R_j^+Y'$, there exist a public command context C and $s_0 \in S$ such that $s' \in [\![C[c_0]]\!](s_0)$ and $[\![C[c_1]]\!](s_0)_{s'_L} \subseteq Y'_{s'_L}$.

We let $C' = C[\;]; [\text{if } !x_{\text{loc}} \neq s'_L(x) \text{ then } \Omega | x \in \text{PubLoc}]; [\;]$. We have $[\![c_0]\!](s')_s \subseteq [\![C'[c_0]]\!](s_0)_s \neq \{\perp\}$. Also, since C' is a public command context, we have $[\![C'[c_0]]\!](s_0) \leq_L [\![C'[c_1]]\!](s_0)$. Hence $[\![C'[c_1]]\!](s_0)_s \neq \{\perp\}$, and we conclude since $[\![C'[c_1]]\!](s_0)_s \subseteq [\![c_1]\!]^\dagger(Y')_s$.

By definition of R_{i+1}, we have

$$X \, R_i^+ \, Y \Rightarrow \forall s \in S_L, [\![c_0]\!]^\dagger(X)_s \, R_{i+1} \, [\![c_1]\!]^\dagger(Y)_s$$

From the result above, we deduce

$$\forall X, Y \in \mathcal{H}(S_\perp). \, XR_i^+Y \Rightarrow [\![c_0]\!]^\dagger(X)R_{i+1}^+[\![c_1]\!]^\dagger(Y) \qquad (*)$$

We now let R be the closure of the union of the R_i increasing ω-sups, right-closure under $\leq$ and closure under binary unions. Note that R is reflexive as it contains R_0. By Lemma 2, we then have that R^+ is the closure of the union of the R_i^+ under increasing ω-sups, right-closure under $\leq$, and closure under binary unions. Since every $[\![c]\!]^\dagger$ is monotone and distributes over these unions, and given the property $(*)$ above, we conclude that $[\![c_0]\!] \preceq_R [\![c_1]\!]$.

Lemmas 1 and 3 give us the desired equivalence:

Theorem 1. *Let c and c' be two commands of the high-level language. Then $c \sqsubseteq_L c'$ holds if and only if $[\![c]\!] \preceq [\![c']\!]$ does.*

Example 2. We can verify that c_0 and c_1, introduced in Example 1, are equivalent (with R the full relation). For instance, take S_0 and S_1 to be $\{(h \mapsto 0, l \mapsto 1)\}\downarrow$ and $\{(h \mapsto 1, l \mapsto 1)\}\downarrow$. We have $S_0R^+S_1$, and:

$$[\![c_0]\!]^\dagger(S_0) = \{(h \mapsto 1, l \mapsto 0), (h \mapsto 0, l \mapsto 1)\}\downarrow$$
$$[\![c_1]\!]^\dagger(S_1) = \{(h \mapsto 1, l \mapsto 1), (h \mapsto 0, l \mapsto 0)\}\downarrow$$

We can then check that:

$$[\![c_0]\!]^\dagger(S_0)R^+[\![c_1]\!]^\dagger(S_1)$$

$\square$

Example 3. In this example, we study the two commands

$$c_2 = \text{if } h = 0 \text{ then } l := 1 \text{ else } (h := 0) + (h := !h - 1)$$
$$c_3 = \text{if } h = 0 \text{ then } l := 1 \text{ else } (h := 0) + \text{skip}$$

which seem to share the same behavior on public variables, but that are inherently different because of their behavior on private variables. According to the semantics, we have:

$$\begin{aligned}
[\![c_2]\!](h \mapsto 0, l \mapsto \_) &= \{(h \mapsto 0, l \mapsto 1)\}\!\downarrow \\
[\![c_2]\!](h \mapsto j + 1, l \mapsto k) &= \{(h \mapsto j, l \mapsto k), (h \mapsto 0, l \mapsto k)\}\!\downarrow \\
[\![c_3]\!](h \mapsto 0, l \mapsto \_) &= \{(h \mapsto 0, l \mapsto 1)\}\!\downarrow \\
[\![c_3]\!](h \mapsto j + 1, l \mapsto k) &= \{(h \mapsto j + 1, l \mapsto k), (h \mapsto 0, l \mapsto k)\}\!\downarrow
\end{aligned}$$

We can verify that $c_2 \preceq_R c_3$, with R defined as the smallest relation that satisfies our conditions (reflexivity, etc.) and such that

$$\{(h \mapsto k)\}R\{(h \mapsto k')\} \qquad \text{for all } k \leq k'$$

For example, let S_0 and S_1 be $\{(h \mapsto 5, l \mapsto 0)\}\!\downarrow$ and $\{(h \mapsto 7, l \mapsto 0)\}\!\downarrow$. Then we have $S_0 R^+ S_1$, and:

$$\begin{aligned}
[\![c_2]\!]^\dagger(S_0) &= \{(h \mapsto 4, l \mapsto 0), (h \mapsto 0, l \mapsto 0)\}\!\downarrow \\
[\![c_3]\!]^\dagger(S_1) &= \{(h \mapsto 7, l \mapsto 0), (h \mapsto 0, l \mapsto 0)\}\!\downarrow
\end{aligned}$$

We can then check that:

$$[\![c_2]\!]^\dagger(S_0)R^+[\![c_3]\!]^\dagger(S_1)$$

However there is no suitable relation R such that $c_3 \preceq_R c_2$. If there were such a relation R, it would be reflexive, so $\{(h \mapsto 1)\} \ R \ \{(h \mapsto 1)\}$. Suppose that $S_0 = \{(h \mapsto 1, l \mapsto 0)\}\!\downarrow$ and that $S_1 = \{(h \mapsto 1, l \mapsto 0)\}\!\downarrow$. We have $S_0 R^+ S_1$, and:

$$\begin{aligned}
[\![c_3]\!]^\dagger(S_0) &= \{(h \mapsto 1, l \mapsto 0), (h \mapsto 0, l \mapsto 0)\}\!\downarrow \\
[\![c_2]\!]^\dagger(S_1) &= \{(h \mapsto 0, l \mapsto 0)\}\!\downarrow
\end{aligned}$$

We need

$$\{(h \mapsto 1, l \mapsto 0), (h \mapsto 0, l \mapsto 0)\}\!\downarrow R^+ \{(h \mapsto 0, l \mapsto 0)\}\!\downarrow$$

hence $\{(h \mapsto 1)\}R\{(h \mapsto 0)\}$. Now take $S_2 = \{(h \mapsto 1, l \mapsto 0)\}\!\downarrow$ and $S_3 = \{(h \mapsto 0, l \mapsto 0)\}\!\downarrow$. We have $S_2 R^+ S_3$, and:

$$\begin{aligned}
[\![c_3]\!]^\dagger(S_2) &= \{(h \mapsto 1, l \mapsto 0), (h \mapsto 0, l \mapsto 0)\}\!\downarrow \\
[\![c_2]\!]^\dagger(S_3) &= \{(h \mapsto 0, l \mapsto 1)\}\!\downarrow
\end{aligned}$$

Since the values of l do not match, we cannot have $[\![c_3]\!]^\dagger(S_2)R^+[\![c_2]\!]^\dagger(S_3)$, hence $c_3 \npreceq_R c_2$.

As predicted by Theorem 1, we also have $c_3 \nleq_L c_2$. Indeed, for $C = \_ ; \_$ and $s_0 = (h \mapsto 1, l \mapsto 0)$, we have $[\![C[c_3]]\!](s_0) \nleq_L [\![C[c_2]]\!](s_0)$. $\square$

4   The Low-Level Language

In this section, we define our low-level language. In this language, we use concrete natural-number addresses for memory. We still use abstract location names, but those are interpreted as natural numbers (according to a memory layout), and can appear in arithmetic expressions.

4.1   Syntax and Informal Semantics

The syntax of the low-level language includes categories for natural-number expressions, boolean expressions, and commands:

$$e ::= k \mid l_{\mathsf{nat}} \mid !e \mid e + e \mid e * e$$
$$b ::= e \leq e \mid \neg b \mid \mathsf{tt} \mid \mathsf{ff} \mid b \vee b \mid b \wedge b$$
$$c ::= e := e \mid \mathsf{if}\, b\, \mathsf{then}\, c\, \mathsf{else}\, c \mid \mathsf{skip} \mid c; c \mid c + c \mid \mathsf{while}\, b\, \mathsf{do}\, c$$

where k ranges over numerals, and l over the finite set of store locations. Boolean expressions are as in the high-level language. Natural-number expressions and commands are also as in the high-level language, except for the inclusion of memory locations among the natural-number expressions, and for the dereferencing construct $!e$ and assignment construct $e := e'$ where e is an arbitrary natural-number expression (not necessarily a location).

Importantly, memory addresses are natural numbers, and a memory is a partial function from those addresses to contents. We assume that accessing an address at which the memory is undefined constitutes an error that stops execution immediately. In this respect, our language relies on the "fatal-error model" of Abadi and Plotkin [7]. With more work, it may be viable to treat also the alternative "recoverable-error model", which permits attacks to continue after such accesses, and therefore requires a bound on the number of such accesses.

4.2   Denotational Semantics

Low-Level Memories, Layouts, and Errors. We assume given a natural number $r > |Loc|$ that specifies the size of the memory. A memory m is a partial function from $\{1, \ldots, r\}$ to the natural numbers; we write Mem for the set of memories. A memory layout w is an injection from Loc to $\{1, \ldots, r\}$; we write $\mathrm{ran}(w)$ for its range. We consider only memory layouts that extend a given public memory layout w_p (an injection from PubLoc to $\{1, \ldots, r\}$), fixed in the remaining of the paper. We let W be the set of those layouts.

The security of layout randomization depends on the randomization itself. We let d be a probability distribution on memory layouts (that extend w_p). When φ is a predicate on memory layouts, we write $P_d(\varphi(w))$ for the probability that $\varphi(w)$ holds with w sampled according to d.

Given a distribution d on layouts, we write δ_d for the minimum probability for a memory address to have no antecedent private location (much as in [7]):

$$\delta_d = \min_{i \in \{1, \ldots, r\} \setminus \mathrm{ran}(w_p)} P_d(i \notin \mathrm{ran}(w))$$

We assume that $\delta_d > 0$. This assumption is reasonable, as $1 - \delta_d$ is the maximum probability for an adversary to guess a private location. For common distributions (e.g., the uniform distribution), δ_d approaches 1 as r grows, indicating that adversaries fail most of the time. We assume d fixed below, and may omit it, writing δ for δ_d.

The denotational semantics of the low-level language uses the "error + nontermination" monad $P_{\xi\perp} =_{\text{def}} (P + \{\xi\})_\perp$, which first adds an "error" element ξ to P and then a least element. As the monad is strong, functions $f : P_1 \times \ldots \times P_n \to Q_{\xi\perp}$ extend to functions $\overline{f}$ on $(P_1)_{\xi\perp} \times \ldots \times (P_n)_{\xi\perp}$, where $\overline{f}(x_1, \ldots, x_n)$ is ξ or $\perp$ if some x_j, but no previous x_i, is; we often write f for $\overline{f}$.

For any memory layout w and store s, we let $w \cdot s$ be the memory defined on $\text{ran}(w)$ by:

$$w \cdot s(i) = s(l) \text{ for } w(l) = i$$

(so that $w \cdot s(w(l)) = s(l)$). The notation $w \cdot s$ extends to $s \in S_{\xi\perp}$, as above, so that $w \cdot \xi = \xi$ and $w \cdot \perp = \perp$. A store projection is a function $\zeta : \text{Mem}^W_{\xi\perp}$ of the form $w \mapsto w \cdot s$, for some $s \in S_{\xi\perp}$; we use the notation $- \cdot s$ to write such store projection functions.

What Should the Denotational Semantics Be? A straightforward semantics might have type:

$$W \times \text{Mem} \to \mathcal{H}(\text{Mem}_{\xi\perp})$$

so that the meaning of a command would be a function from layouts and memories to sets of memories (modulo the use of the "error + nontermination" monad). Using a simple example we now argue that this is unsatisfactory, and arrive at a more satisfactory alternative.

Suppose that there is a unique private location l, no public locations, and that the memory has four addresses, $\{1, 2, 3, 4\}$. We write s_i for the store $(l \mapsto i)$. The 4 possible layouts are $w_i = (l \mapsto i)$, for $i = 1, \ldots, 4$. Assume that d is uniform. Consider the following command:

$$c_4 = (1{:=}1) + (2{:=}1) + (3{:=}1) + (4{:=}1)$$

which nondeterministically guesses an address and attempts to write 1 into it. Intuitively, this command should fail to overwrite l most of the time. However, in a straightforward semantics of the above type we would have:

$$[\![c_4]\!](w_j, w_j \cdot s_0) = \{\xi, w_j \cdot s_1\} \downarrow$$

and we cannot state any quantitative property of the command, only that it sometimes fails and that it sometimes terminates.

One can rewrite the type of this semantics as:

$$\text{Mem} \to \mathcal{H}(\text{Mem}_{\xi\perp})^W$$

and view that as a type of functions that yield an $\mathcal{H}(\text{Mem}_{\xi\perp})$-valued random variable with sample space W (the set of memory layouts) and distribution d.

Thus, in this semantics, the nondeterministic choice is made after the probabilistic one —the wrong way around, as indicated in the Introduction.

It is therefore natural to reverse matters and look for a semantics of type:

$$\text{Mem} \to \mathcal{H}(\text{Mem}^W_{\xi\perp})$$

now yielding a set of $\text{Mem}_{\xi\perp}$-valued random variables—so, making the nondeterministic choice first. Desirable as this may be, there seems to be no good notion of composition of such functions.

Fortunately, this last problem can be overcome by changing the argument type to also be that of $\text{Mem}_{\xi\perp}$-valued random variables:

$$\text{Mem}^W_{\xi\perp} \to \mathcal{H}(\text{Mem}^W_{\xi\perp})$$

It turns out that with this semantics we have:

$$[\![c_4]\!](\zeta_i) = \{\zeta^1_\xi, \zeta^2_\xi, \zeta^3_\xi, \zeta^4_\xi\} \downarrow$$

where $\zeta_i(w) = w \cdot s_i$ and $\zeta^i_\xi(w) = w_i \cdot s_1$ if $w = w_i$ and $= \xi$ otherwise. We can then say that, for every nondeterministic choice, the probability of an error (or nontermination, as we are using the lower powerdomain) is 0.75.

In a further variant of the semantics, one might replace $\text{Mem}_{\xi\perp}$-valued random variables by the corresponding probability distributions on $\text{Mem}_{\xi\perp}$, via the natural map $\text{Ind}_d : \text{Mem}^W_{\xi\perp} \longrightarrow \mathcal{V}(\text{Mem}_{\xi\perp})$ induced by the distribution d on W (where $\mathcal{V}$ is the probabilistic powerdomain monad, see [13]). Such a semantics could have the form:

$$\text{Mem} \to \mathcal{H}_\mathcal{V}(\text{Mem}_{\xi\perp})$$

mapping memories to probability distributions on memories, where $\mathcal{H}_\mathcal{V}$ is a powerdomain for mixed nondeterministic and probabilistic choice as discussed above. However, such an approach would imply (incorrectly) that a new layout is chosen independently for each memory operation, rather than once and for all. In our small example with the single private location l and four addresses, it would not capture that $(1 := 1); (2 := 1)$ will always fail. It would treat the two assignments in $(1 := 1); (2 := 1)$ as two separate guesses that may both succeed. Similarly, it would treat the two assignments in $(1 := 1); (1 := 2)$ as two separate guesses where the second guess may fail to overwrite l even if the first one succeeds. With a layout chosen once and for all, on the other hand, the behavior of the second assignment is completely determined after the first assignment.

Denotational Semantics. The denotational semantics

$$[\![e]\!] : \text{Mem} \times W \to \mathbb{N}_{\xi\perp} \qquad [\![b]\!] : \text{Mem} \times W \to \mathbb{B}_{\xi\perp}$$

of expressions are defined in a standard way. In particular, $[\![l_{\text{nat}}]\!]^w_m = w(l)$, and also $[\![!e]\!]^w_m = m([\![e]\!]^w_m)$, if $[\![e]\!]^w_m \in \text{dom}(m)$, and $= \xi$, otherwise, using an obvious notation for functional application. Note that these semantics never have value $\perp$.

As discussed above, the denotational semantics of commands has type:

$$[\![c]\!]: \mathrm{Mem}^W_{\xi\perp} \to \mathcal{H}(\mathrm{Mem}^W_{\xi\perp})$$

(and we remark that, as W is finite, all increasing chains in $\mathrm{Mem}^W_{\xi\perp}$ are eventually constant, and so for any nonempty subset X of $\mathrm{Mem}^W_{\xi\perp}$ we have $X^* = X \downarrow$). The denotational semantics is defined in Figure 3; it makes use of two auxiliary definitions. We first define:

$$\mathrm{Ass}: \mathrm{Mem}_{\xi\perp} \times \mathbb{N}_{\xi\perp} \times \mathbb{N}_{\xi\perp} \to \mathrm{Mem}_{\xi\perp}$$

by setting $\mathrm{Ass}(m,x,y) = m[x \mapsto y]$ if $x \in \mathrm{dom}(m)$ and $= \xi$, otherwise, for $m \in \mathrm{Mem}$, $x,y \in \mathbb{N}$, and then using the function extension associated to the "error + nontermination" monad. Second, we define

$$\mathrm{Cond}(p, \theta, \theta'): \mathrm{Mem}^W_{\xi\perp} \to \mathcal{H}(\mathrm{Mem}^W_{\xi\perp})$$

for any $p: \mathrm{Mem} \times W \to \mathbb{B}_{\xi\perp}$ and $\theta, \theta': \mathrm{Mem}^W_{\xi\perp} \to \mathcal{H}(\mathrm{Mem}^W_{\xi\perp})$, by:

$$\mathrm{Cond}(p, \theta, \theta')(\zeta) = \{\zeta' \mid \zeta'|_{W_{\zeta,\mathrm{tt}}} \in \theta(\zeta)|_{W_{\zeta,\mathrm{tt}}}, \zeta'|_{W_{\zeta,\mathrm{ff}}} \in \theta'(\zeta)|_{W_{\zeta,\mathrm{ff}}},$$
$$\zeta'(W_{\zeta,\xi}) \subseteq \{\xi\}, \text{ and } \zeta'(W_{\zeta,\perp}) \subseteq \{\perp\}\} \downarrow$$

where $W_{\zeta,t} =_{\mathrm{def}} \{w \mid p(\zeta(w), w) = t\}$, for $t \in \mathbb{B}_{\xi\perp}$, and we apply restriction elementwise to sets of functions.

$$[\![c + c']\!](\zeta) = [\![c]\!](\zeta) \cup [\![c']\!](\zeta) \qquad [\![c; c']\!] = [\![c']\!]^\dagger \circ [\![c]\!] \qquad [\![\mathbf{skip}]\!] = \eta$$

$$[\![e := e']\!](\zeta) \qquad = \eta(\lambda w: W.\, \mathrm{Ass}(\zeta(w), [\![e]\!]^w_{\zeta(w)}, [\![e']\!]^w_{\zeta(w)}))$$

$$[\![\mathbf{if}\, b\, \mathbf{then}\, c\, \mathbf{else}\, c']\!] = \mathrm{Cond}([\![b]\!], [\![c]\!], [\![c']\!])$$

$$[\![\mathbf{while}\, b\, \mathbf{do}\, c]\!] \qquad = \mu\theta: \mathrm{Mem}^W_{\xi\perp} \to \mathcal{H}(\mathrm{Mem}^W_{\xi\perp}).\mathrm{Cond}([\![b]\!], \theta^\dagger \circ [\![c]\!], \eta)$$

Fig. 3. Low-level denotational semantics

Example 4. In this example, we demonstrate our low-level denotational semantics. Consider the command:

$$c_5 = l'_{\mathrm{nat}} := l_{\mathrm{nat}}; (!l'_{\mathrm{nat}}) := 1; l'_{\mathrm{nat}} := 0$$

This command stores the address of location l at location l', then reads the contents of location l' (the address of l) and writes 1 at this address, and finally resets the memory at location l' to 0. Because of this manipulation of memory locations, this command is not the direct translation of a high-level command.
Letting:

$$s_{i,j} = (l \mapsto i, l' \mapsto j) \quad \zeta_{i,j} = -\cdot s_{i,j} \quad \zeta'_i = -\cdot(l \mapsto i, l' \mapsto w(l))$$

we have:

$$[\![l'_{\text{nat}} := l_{\text{nat}}]\!](\zeta_{i,j}) = \{\zeta'_i\}\!\downarrow$$

Note that $\zeta_{i,j}$ is a store projection, but ζ'_i is not. We also have:

$$[\![(!l'_{\text{nat}}) := 1]\!](\zeta'_i) = \{\zeta'_1\}\!\downarrow \qquad [\![l'_{\text{nat}} := 0]\!](\zeta'_1) = \{\zeta_{1,0}\}\!\downarrow$$

In sum, we have:

$$[\![c_5]\!](\zeta_{i,j}) = \{\zeta_{1,0}\}\!\downarrow$$

□

Looking at the type of the semantics

$$[\![c]\!] : \text{Mem}^{W}_{\xi\perp} \to \mathcal{H}(\text{Mem}^{W}_{\xi\perp})$$

one may be concerned that there is no apparent relation between the layouts used in the input to $[\![c]\!]$ and those in its output. However, we note that the semantics could be made parametric. For every $W' \subseteq W$, replace W by W' in the definition of $[\![c]\!]$ to obtain:

$$[\![c]\!]_{W'} : \text{Mem}^{W'}_{\xi\perp} \to \mathcal{H}(\text{Mem}^{W'}_{\xi\perp})$$

There is then a naturality property, that the following diagram commutes for all $W'' \subseteq W' \subseteq W$:

$$
\begin{array}{ccc}
\text{Mem}^{W'}_{\xi\perp} & \xrightarrow{\ [\![c]\!]_{W'}\ } & \mathcal{H}(\text{Mem}^{W'}_{\xi\perp}) \\
\downarrow{\scriptstyle\text{Mem}^{\iota}_{\xi\perp}} & & \downarrow{\scriptstyle\mathcal{H}(\text{Mem}^{\iota}_{\xi\perp})} \\
\text{Mem}^{W''}_{\xi\perp} & \xrightarrow[\ [\![c]\!]_{W''}\ ]{} & \mathcal{H}(\text{Mem}^{W''}_{\xi\perp})
\end{array}
$$

where $\iota : W'' \subseteq W'$ is the inclusion map. Taking $W' = W$ and W'' a singleton yields the expected relation between input and output: the value of a random variable in the output at a layout depends only on the value of the input random variable at that layout. The naturality property suggests re-working the low-level denotational semantics in the category of presheaves over sets of layouts, and this may prove illuminating (see [22] for relevant background).

4.3   Operational Semantics

As a counterpart to the denotational semantics, we give a big-step deterministic operational semantics using oracles to make choices.

The set of oracles Π is ranged over by π and is given by the following grammar:

$$\pi \ ::= \ \varepsilon \mid L\pi \mid R\pi \mid \pi; \pi \mid \text{if}(\pi, \pi)$$

A low-level state σ is:

 – a pair $\langle c, m \rangle$ of a command c and a memory m,
 – a memory m, or
 – the error element ξ.

States of either of the last two forms are called terminal, and written τ. Transitions relate states and terminal states. They are given relative to a layout, and use an oracle to resolve nondeterminism. So we write:

$$w \models \sigma \overset{\pi}{\Rightarrow} \tau$$

Figure 4 gives the rules for this relation.

The rules for conditionals use different oracles for the true and false branches in order to avoid any correlation between the choices made in the two branches.[1] The rules for two commands in sequence also use different oracles, again avoiding correlations, now between the choices made in executing the first command and the choices made in executing the second. The oracles used in the rules for while loops ensure that the operational semantics of a loop and its unrolling are the same. We continue this discussion after Theorem 2 below.

Example 5. Consider the command c_4 introduced in Section 4.2, with added parentheses for disambiguation:

$$c_4 \;=\; (1{:=}1) \;+\; ((2{:=}1) \;+\; ((3{:=}1) \;+\; ((4{:=}1))))$$

We have:

$$
\begin{array}{ll}
w_1 \models \langle c_4, w_1 \cdot s_k \rangle \overset{L}{\Rightarrow} w_1 \cdot s_1 & w_j \models \langle c_4, w_j \cdot s_k \rangle \overset{L}{\Rightarrow} \xi \; (j \neq 1) \\[4pt]
w_2 \models \langle c_4, w_2 \cdot s_k \rangle \overset{RL}{\Rightarrow} w_2 \cdot s_1 & w_j \models \langle c_4, w_j \cdot s_k \rangle \overset{RL}{\Rightarrow} \xi \; (j \neq 2) \\[4pt]
w_3 \models \langle c_4, w_3 \cdot s_k \rangle \overset{RRL}{\Rightarrow} w_3 \cdot s_1 & w_j \models \langle c_4, w_j \cdot s_k \rangle \overset{RRL}{\Rightarrow} \xi \; (j \neq 3) \\[4pt]
w_4 \models \langle c_4, w_4 \cdot s_k \rangle \overset{RRR}{\Rightarrow} w_4 \cdot s_1 & w_j \models \langle c_4, w_j \cdot s_k \rangle \overset{RRR}{\Rightarrow} \xi \; (j \neq 4)
\end{array}
$$

□

The transition relation is deterministic: if $w \models \sigma \overset{\pi}{\Rightarrow} \tau$ and $w \models \sigma \overset{\pi}{\Rightarrow} \tau'$ then $\tau = \tau'$. We can therefore define an evaluation function

$$\mathrm{Eval} : \mathrm{Com} \times W \times \mathrm{Mem} \times \Pi \to \mathrm{Mem}_{\xi\perp}$$

by:

$$\mathrm{Eval}(c, w, m, \pi) = \begin{cases} \tau \;\; (\text{if } w \models \langle c, m \rangle \overset{\pi}{\Rightarrow} \tau) \\ \perp \;\; (\text{otherwise}) \end{cases}$$

---

[1] The rules for, e.g., conditionals differ from those given in [17] which use the same oracle for both branches. The rules in [17] are erroneous in that the resulting operational semantics is not consistent with the denotational semantics in the sense of Theorem 2.

$$\frac{[\![e]\!]_m^w \in \mathrm{dom}(m) \text{ and } [\![e']\!]_m^w \neq \xi}{w \models \langle e := e', m\rangle \xrightarrow{\varepsilon} m[[\![e]\!]_m^w \mapsto [\![e']\!]_m^w]} \qquad \frac{[\![e]\!]_m^w \notin \mathrm{dom}(m) \text{ or } [\![e']\!]_m^w = \xi}{w \models \langle e := e', m\rangle \xrightarrow{\varepsilon} \xi}$$

$$\frac{[\![b]\!]_m^w = \mathrm{tt} \quad \langle c, m\rangle \xrightarrow{\pi} \tau}{w \models \langle \text{if } b \text{ then } c \text{ else } c', m\rangle \xrightarrow{if(\pi,\pi')} \tau} \qquad \frac{[\![b]\!]_m^w = \mathrm{ff} \quad \langle c', m\rangle \xrightarrow{\pi'} \tau}{w \models \langle \text{if } b \text{ then } c \text{ else } c', m\rangle \xrightarrow{if(\pi,\pi')} \tau}$$

$$\frac{[\![b]\!]_m^w = \xi}{w \models \langle \text{if } b \text{ then } c \text{ else } c', m\rangle \xrightarrow{if(\pi,\pi')} \xi} \qquad w \models \langle \text{skip}, m\rangle \xrightarrow{\varepsilon} m$$

$$\frac{w \models \langle c, m\rangle \xrightarrow{\pi} m' \quad \langle c', m'\rangle \xrightarrow{\pi'} \tau}{w \models \langle c; c', m\rangle \xrightarrow{\pi;\pi'} \tau} \qquad \frac{w \models \langle c, m\rangle \xrightarrow{\pi} \xi}{w \models \langle c; c', m\rangle \xrightarrow{\pi;\pi'} \xi} \qquad \frac{w \models \langle c, m\rangle \xrightarrow{\pi} \tau}{w \models \langle c + c', m\rangle \xrightarrow{L\pi} \tau}$$

$$\frac{w \models \langle c', m\rangle \xrightarrow{\pi} \tau}{w \models \langle c + c', m\rangle \xrightarrow{R\pi} \tau} \qquad \frac{[\![b]\!]_m^w = \mathrm{ff}}{w \models \langle \text{while } b \text{ do } c, m\rangle \xrightarrow{if(\pi;\pi',\varepsilon)} m}$$

$$\frac{[\![b]\!]_m^w = \mathrm{tt} \quad \langle c, m\rangle \xrightarrow{\pi} m' \quad \langle \text{while } b \text{ do } c, m'\rangle \xrightarrow{\pi'} \tau}{w \models \langle \text{while } b \text{ do } c, m\rangle \xrightarrow{if(\pi;\pi',\varepsilon)} \tau}$$

$$\frac{[\![b]\!]_m^w = \mathrm{tt} \quad \langle c, m\rangle \xrightarrow{\pi} \xi}{w \models \langle \text{while } b \text{ do } c, m\rangle \xrightarrow{if(\pi;\pi',\varepsilon)} \xi} \qquad \frac{[\![b]\!]_m^w = \xi}{w \models \langle \text{while } b \text{ do } c, m\rangle \xrightarrow{if(\pi;\pi',\varepsilon)} \xi}$$

Fig. 4. Low-level operational semantics

In order to establish the consistency of the operational and denotational semantics we make use of an intermediate denotational semantics

$$[\![c]\!]_i : W \to (\mathrm{Mem} \times \Pi \to \mathrm{Mem}_{\xi\perp})$$

defined by setting for all commands, other than loops:

$$
\begin{aligned}
[\![c + c']\!]_i(w)(m, \pi) &= \begin{cases} [\![c]\!]_i(w)(m, \pi') & (\pi = L\pi') \\ [\![c']\!]_i(w)(m, \pi') & (\pi = R\pi') \end{cases} \\
[\![c; c']\!]_i(w)(m, \pi; \pi') &= \overline{[\![c']\!]_i(w)}([\![c]\!]_i(w)(m, \pi), \pi') \\
[\![\text{skip}]\!]_i(w)(m, \varepsilon) &= m \\
[\![e := e']\!]_i(w)(m, \varepsilon) &= \mathrm{Ass}(w, [\![e]\!]_m^w, [\![e']\!]_m^w) \\
[\![\text{if } b \text{ then } c \text{ else } c']\!]_i(w)(m, if(\pi; \pi')) &= \mathrm{C}([\![b]\!]_m^w, [\![c]\!]_i(w)(m, \pi), [\![c']\!]_i(w)(m, \pi'))
\end{aligned}
$$

and taking $[\![c]\!]_i(w)(m, \pi)$ to be $\perp$ for all other combinations of loop-free commands and oracles, and where we use the error plus nontermination extension of the evident conditional function $C : \mathbb{B} \times P \times P \to P$ for the semantics of conditionals.

For while loops, $[\![\texttt{while } b \texttt{ do } c]\!]_i$ is defined to be

$$\mu\theta.\lambda w.\lambda m, \pi. \begin{cases} C([\![b]\!]_m^w, \overline{\theta(w)}([\![c]\!]_i(w)(m, \pi'), \pi''), m) & (\pi = \textit{if}(\pi'; \pi'', \varepsilon)) \\ \bot & (\text{otherwise}) \end{cases}$$

The following lemma asserts the consistency of the operational semantics and this intermediate denotational semantics. Its proof, which uses rule-induction, structural induction, and consideration of iterates like that of Proposition 1, is omitted.

Lemma 4. *For any low-level command c, layout w, memory m, and oracle π, we have:*

$$[\![c]\!]_i(w)(m, \pi) = \text{Eval}(c, w, m, \pi)$$

Lemma 5. *For any low-level command c and $\zeta \in \text{Mem}_{\xi\bot}^W$, we have:*

$$[\![c]\!](\zeta) = \{\lambda w : W. \overline{[\![c]\!]_i(w)}(\zeta(w), \pi) \mid \pi \in \Pi\} \downarrow$$

Proof. We first extend the language by adding a command Ω, and let it denote the relevant least element in both denotational semantics. With that we establish the result for commands not containing any while loops, proceeding by structural induction:

1. skip, Ω: These two cases are immediate from the definitions of $[\![c]\!]$ and $[\![c]\!]_i$.
2. $e := e'$: We calculate:

$$\{\lambda w : W. \overline{[\![e := e']\!]_i(w)}(\zeta(w), \pi) \mid \pi \in \Pi\} \downarrow$$
$$= \{\lambda w : W. \overline{[\![e := e']\!]_i(w)}(\zeta(w), \varepsilon)\} \downarrow$$
$$= \{\lambda w : W. \text{Ass}(w, [\![e]\!]_{\zeta(w)}^w, [\![e']\!]_{\zeta(w)}^w)\} \downarrow$$
$$= [\![e := e']\!](\zeta)$$

3. $c; c'$: We calculate:

$$[\![c; c']\!](\zeta) = \bigcup\{[\![c']\!](\zeta') \mid \zeta' \in [\![c]\!](\zeta)\} \downarrow$$
$$= \bigcup\{[\![c']\!](\lambda w : W. \overline{[\![c]\!]_i(w)}(\zeta(w), \pi)) \mid \pi \in \Pi\} \downarrow$$
$$= \bigcup\{\{\lambda w : W. \overline{[\![c']\!]_i(w)}(\overline{[\![c]\!]_i(w)}(\zeta(w), \pi), \pi') \mid \pi' \in \Pi \downarrow\} \mid \pi \in \Pi\} \downarrow$$
$$= \{\lambda w : W. \overline{[\![c']\!]_i(w)}(\overline{[\![c]\!]_i(w)}(\zeta(w), \pi), \pi') \mid \pi, \pi' \in \Pi\} \downarrow$$
$$= \{\lambda w : W. \overline{[\![c; c']\!]_i(w)}(\zeta(w), (\pi; \pi')) \mid \pi, \pi' \in \Pi\} \downarrow$$
$$= \{\lambda w : W. \overline{[\![c; c']\!]_i(w)}(\zeta(w), \pi) \mid \pi \in \Pi\} \downarrow$$

4. $c_L + c_R$: We calculate:

$$\{\lambda w\!:\!W.\,\overline{[\![c_L + c_R]\!]_i(w)}(\zeta(w), \pi) \mid \pi \in \Pi\} \downarrow \;=$$
$$\{\lambda w\!:\!W.\,\overline{[\![c_L + c_R]\!]_i(w)}(\zeta(w), L\pi) \mid \pi \in \Pi\} \downarrow$$
$$\cup \{\lambda w\!:\!W.\,\overline{[\![c_L + c_R]\!]_i(w)}(\zeta(w), R\pi) \mid \pi \in \Pi\} \downarrow$$
$$=$$
$$\{\lambda w\!:\!W.\,\overline{[\![c_L]\!]_i(w)}(\zeta(w), \pi) \mid \pi \in \Pi\} \downarrow$$
$$\cup \{\lambda w\!:\!W.\,\overline{[\![c_R]\!]_i(w)}(\zeta(w), \pi) \mid \pi \in \Pi\} \downarrow$$
$$= [\![c_L]\!](\zeta) \cup [\![c_R]\!](\zeta)$$
$$= [\![c_L + c_R]\!](\zeta)$$

5. $\texttt{if } b \texttt{ then } c_{\texttt{tt}} \texttt{ else } c_{\texttt{ff}}$: We have to show that

$$[\![\texttt{if } b \texttt{ then } c_{\texttt{tt}} \texttt{ else } c_{\texttt{ff}}]\!](\zeta) \;=$$
$$\{\lambda w\!:\!W.\,\overline{[\![\texttt{if } b \texttt{ then } c_{\texttt{tt}} \texttt{ else } c_{\texttt{ff}}]\!]_i(w)}(\zeta(w), \pi) \mid \pi \in \Pi\} \downarrow$$

Set $W_{\zeta,t} =_{\text{def}} \{w \mid [\![b]\!]^w_{\zeta(w)} = t\}$, for $t \in \mathbb{B}_{\xi\perp}$. Then note first that ζ' is in the left-hand side if, and only if, there are $\zeta'' \geq \zeta'$, $\zeta_{\texttt{tt}} \in [\![c_{\texttt{tt}}]\!](\zeta)$, and $\zeta_{\texttt{ff}} \in [\![c_{\texttt{ff}}]\!](\zeta)$ s.t.: $\zeta''|_{W_{\zeta,\texttt{tt}}} = \zeta_{\texttt{tt}}|_{W_{\zeta,\texttt{tt}}}, \zeta''|_{W_{\zeta,\texttt{ff}}} = \zeta_{\texttt{ff}}|_{W_{\zeta,\texttt{ff}}}, \zeta''(W_{\zeta,\varepsilon}) \subseteq \{\xi\}$, and $\zeta''(W_{\zeta,\perp}) \subseteq \{\perp\}$.

Using the induction hypothesis, we see that $\zeta_{\texttt{tt}} \in [\![c_{\texttt{tt}}]\!](\zeta)$ if, and only if, for some $\pi_{\texttt{tt}}$, $\zeta_{\texttt{tt}} \leq \lambda w\!:\!W.\,\overline{[\![c_{\texttt{tt}}]\!]_i(w)}(\zeta(w), \pi_{\texttt{tt}})$, and similarly for $\zeta_{\texttt{ff}}$.

We then see that the condition for ζ' to be in the left-hand side is equivalent to the existence of $\zeta'' \geq \zeta'$, $\pi_{\texttt{tt}}$ and $\pi_{\texttt{ff}}$ such that

$$\zeta'' \leq \lambda w\!:\!W.\,\text{C}([\![b]\!]^w_{\zeta(w)}, \overline{[\![c_{\texttt{tt}}]\!]_i(w)}(\zeta(w), \pi_{\texttt{tt}}), \overline{[\![c_{\texttt{ff}}]\!]_i(w)}(\zeta(w), \pi_{\texttt{ff}}))$$

which is equivalent to the condition that ζ' is in the right-hand side.

We can now establish the desired result for general commands c (including Ω). Define iterates $c^{(n)}$ by setting $c^{(0)} = \Omega$ and defining $c^{(n+1)}$ homomorphically, except for while loops, where we put:

$$(\texttt{while } b \texttt{ do } c_w)^{(n+1)} = \texttt{if } b \texttt{ then } c_w^{(n+1)}; (\texttt{while } b \texttt{ do } c_w)^{(n)} \texttt{ else skip}$$

Note that the iterates are all in the sub-language not including loops.

We have that $[\![c^{(n)}]\!]$ is an increasing sequence with lub $[\![c]\!]$, and the same holds for $[\![-]\!]_i$. As the desired result holds for commands not containing while loops, but possibly containing Ω, we can then calculate:

$$[\![c]\!](\zeta) = \bigvee_{n \geq 0} [\![c^{(n)}]\!](\zeta)$$
$$= \bigvee_{n \geq 0} \{\lambda w\!:\!W.\,\overline{[\![c^{(n)}]\!]_i(w)}(\zeta(w), \pi) \mid \pi \in \Pi\} \downarrow$$
$$= \{\lambda w\!:\!W.\,\overline{[\![c^{(n)}]\!]_i(w)}(\zeta(w), \pi) \mid \pi \in \Pi, n \geq 0\} \downarrow$$
$$= \{\lambda w\!:\!W.\,\bigvee_{n \geq 0} \overline{[\![c^{(n)}]\!]_i(w)}(\zeta(w), \pi) \mid \pi \in \Pi\} \downarrow$$
$$= \{\lambda w\!:\!W.\,\overline{\bigvee_{n \geq 0} [\![c^{(n)}]\!]_i(w)}(\zeta(w), \pi) \mid \pi \in \Pi\} \downarrow$$
$$= \{\lambda w\!:\!W.\,\overline{[\![c]\!]_i(w)}(\zeta(w), \pi) \mid \pi \in \Pi\} \downarrow$$

which concludes the proof.

Lemmas 4 and 5 immediately yield the consistency of the operational and denotational semantics:

Theorem 2 (Low-level operational/denotational consistency). *For any low-level command c and $\zeta \in \mathrm{Mem}^W_{\xi\perp}$, we have:*

$$\llbracket c \rrbracket(\zeta) = \{\lambda w \colon W. \overline{\mathrm{Eval}}(c, w, \zeta(w), \pi) \mid \pi \in \Pi\} \downarrow$$

The evaluation function yields operational correlates of the other possible denotational semantics discussed in Section 4.2, similarly, using image or induced distribution functionals. For example, for the first of those semantics, by currying Eval and composing, one obtains:

$$\mathrm{Com} \times W \times \mathrm{Mem} \xrightarrow{\mathrm{curry(Eval)}} \mathrm{Mem}^{\Pi}_{\xi\perp} \xrightarrow{\mathrm{Im}_{\mathrm{Mem}_{\xi\perp}}} \mathcal{P}(\mathrm{Mem}_{\xi\perp})$$

Using such operational correlates, one can verify operational versions of the assertions made in Section 4.2 about the inadequacies of those semantics.

The operational semantics has the peculiarity that the oracles used are independent of the layout but not of the command structure. Allowing the oracle to depend on the layout would amount to making nondeterministic choices after probabilistic ones. The use of the syntactic dependence in the case of conditionals can be seen by considering the example:

$$\texttt{if } b \texttt{ then } (l_{\mathrm{nat}} := 0) + (l_{\mathrm{nat}} := 1) \texttt{ else } (l_{\mathrm{nat}} := 0) + (l_{\mathrm{nat}} := 1)$$

If the same oracle was used for both branches in the operational semantics, then l_{nat} would either always (i.e., for all layouts) be set to 0 or else would always be set to 1; however, for a suitable choice of condition b, the denotational semantics allows the possibility of setting l_{nat} to 0 in some layouts and to 1 in others.

In the case of sequence commands, consider the example

$$(\texttt{if } b \texttt{ then skip else skip}); ((l_{\mathrm{nat}} := 0) + (l_{\mathrm{nat}} := 1))$$

If the oracle chosen for the second command depended on which branch of the conditional was taken, then it could be possible that l_{nat} was sometimes set to 0 and sometimes to 1, whereas to be in accord with the denotational semantics it should either always be set to 0 or always set to 1.

In this connection it is worth noting that the equation

$$(\texttt{if } b \texttt{ then } c_{\mathrm{tt}} \texttt{ else } c_{\mathrm{ff}}); c = \texttt{if } b \texttt{ then } c_{\mathrm{tt}}; c \texttt{ else } c_{\mathrm{ff}}; c$$

which one might naturally expect to hold in the denotational semantics in fact does not. A counterexample can be obtained by taking c_{tt} and c_{ff} to be skip and c to be $(l_{\mathrm{nat}} := 0) + (l_{\mathrm{nat}} := 1)$. The left-hand side is then the command just considered to illustrate the use of oracles for sequence commands, and the right-hand side may sometimes set l_{nat} to 0 and sometimes to 1.

Such subtleties, and more generally the difficulty of both operational and denotational semantics, suggest that these semantics may be attractive subjects for further work. Fortunately, neither the operational semantics nor its relation with the denotational semantics are needed for our main results (which are also those of [17]).

4.4   Implementation Relations and Equivalences

Much as in the high-level language, we define a contextual implementation relation and a simulation relation for the low-level language. The low-level definitions refer to layouts, and in some cases include conditions on induced probabilities.

Contextual Pre-order. Again, the contextual pre-order $c \sqsubseteq_L c'$ may be interpreted as saying that c "refines" (or "implements") c', in the sense that the publicly observable outcomes that c can produce are a subset of those that c' permits, in every public context. In comparison with the definition for the high-level language, however, c and c' are not applied to an arbitrary initial store but rather to a function from layouts to memories (extended with "error + nontermination"), and they produce sets of such functions. We restrict attention to argument functions induced by stores, in the sense that they are store projections of the form $- \cdot s$. Thus, let $f = \llbracket C[c] \rrbracket$ and $f' = \llbracket C[c'] \rrbracket$ for an arbitrary public context C, and let s be a store; then (roughly) for every ζ in $f(- \cdot s)$ there exists ζ' in $f'(- \cdot s)$ such that, for any w, $\zeta(w)$ and $\zeta'(w)$ coincide on public locations.

The treatment of error and nontermination introduces a further complication. Specifically, we allow that ζ produces an error or diverges with sufficient probability ($\geq \delta$), and that ζ' produces an error with sufficient probability ($\geq \delta$), as an alternative to coinciding on public locations.

Therefore, we define $\sqsubseteq_L$ and some auxiliary notation and relations:

- Set PubMem $=_{\mathrm{def}} \mathbb{N}^{\mathrm{ran}(w_p)}$. Then, for any memory m, let m_L in PubMem be the restriction of m to $\mathrm{ran}(w_p)$, extending the notation to $\mathrm{Mem}_{\xi\perp}$ as usual.
- For any $\zeta \in \mathrm{Mem}_{\xi\perp}^W$, we define $\zeta_L \in \mathrm{PubMem}_{\xi\perp}^W$ by $\zeta_L(w) = \zeta(w)_L$.
- For $X, Y \in \mathcal{H}(\mathrm{Mem}_{\xi\perp}^W)$, we write that $X \leq_L Y$ when, for every $\zeta \in X$, there exists $\zeta' \in Y$ such that:
  - $\zeta_L \leq \zeta'_L$, or
  - $P(\zeta(w) \in \{\xi, \perp\}) \geq \delta$ and $P(\zeta'(w) = \xi) \geq \delta$.
- For $f, f' \in \mathrm{Mem}_{\xi\perp}^W \to \mathcal{H}(\mathrm{Mem}_{\xi\perp}^W)$, we write $f \leq_L f'$ when, for all $s \in S$, we have:

$$f(- \cdot s) \leq_L f'(- \cdot s)$$

- Finally, we write $c \sqsubseteq_L c'$ when, for every public command context C, $\llbracket C[c] \rrbracket \leq_L \llbracket C[c'] \rrbracket$.

Simulation. As in the high-level language, we introduce a simulation relation $\preceq$. This relation works only on commands whose outcomes on inputs that are store projections are themselves store projections; nevertheless, simulation remains a useful tool for proofs.

We first define some auxiliary notations:

- We define $\max(X)$, for any $X \in \mathcal{H}(\mathrm{Mem}_{\xi\perp}^W)$, as the set of maximal elements of X. (As W is finite, every element of X is less than a maximal element of X, and $X = \max(X) \downarrow$.)

- We write $S(\zeta)$ for the element of $S_{\xi\perp}$ uniquely determined by a store projection ζ.
- For any cpo P and $\zeta \in P_{\xi\perp}^W$, we define $\zeta_{/\xi}$ by:

$$\zeta_{/\xi} = \begin{cases} w \mapsto \xi \ (\text{if } P(\zeta(w) = \xi) \geq \delta) \\ \zeta \qquad (\text{otherwise}) \end{cases}$$

- For every $X \in \mathcal{H}(\text{Mem}_{\xi\perp}^W)$, we say that X is a store projection set when $\{\zeta_{/\xi} \mid \zeta \in \max(X)\}$ is a set of store projections. Then, we let

$$\chi(X) = S(\{\zeta_{/\xi} \mid \zeta \in \max(X)\}) \cup \{\perp\}$$

Note that $s \in \chi(X)$ if, and only if, $-\cdot s \in X$, and that $\xi \in \chi(X)$ if, and only if, there exists $\zeta \in X$ such that $P(\zeta(w) = \xi) \geq \delta$.

The $\leq_L$ relation restricted to store projection sets has a pleasant characterization. This characterization requires some definitions. First, $-_L$ extends from S to $S_{\xi\perp}$, so that $\perp_L = \perp$ and $\xi_L = \xi$; with that, for any X in $\mathcal{H}(S_{\xi\perp})$, we define X_L in $\mathcal{H}(S_{L\xi\perp})$ to be $\{s_L \mid s \in X\}$.

Fact 3. *Let X and Y be store projection sets. Then:*

$$X \leq_L Y \ \Leftrightarrow \ \chi(X)_L \leq \chi(Y)_L$$

Proof. Let X and Y be store projection sets. Assume first that $X \leq_L Y$, and take a non-bottom element of $\chi(X)_L$. There are two cases. In the first the element is s_L for some $s \in \text{Store}$ such that $\zeta =_{\text{def}} -\cdot s \in X$. As $X \leq_L Y$ we have $\zeta_L \leq \zeta'_L$ for some $\zeta' \in Y$. But then $\zeta' = -\cdot s'$ for some $s' \in \text{Store}$ with $s'_L = s_L$ and so $s_L \in \chi(Y)_L$. In the second case the element is ξ and there is a $\zeta \in X$ such that $P(\zeta(w) = \xi) \geq \delta$. As $X \leq_L Y$ it follows that there is a $\zeta' \in Y$ such that $P(\zeta'(w) = \xi) \geq \delta$, and so $\xi \in \chi(Y)_L$.

For the converse, assume that $\chi(X)_L \leq \chi(Y)_L$. The case $X = \{\perp\}$ is trivial. Otherwise take $\zeta \in X$. Choose $\zeta' \in \max(X)$ such that $\zeta \leq \zeta'$. As X is a store projection set $\neq \{\perp\}$, there are two cases. In the first case ζ' has the form $-\cdot s$. As $\chi(X)_L \leq \chi(Y)_L$ there is a $\zeta'' \in Y$ of the form $-\cdot s'$ where $s'_L = s_L$. We therefore have $\zeta_L \leq \zeta'_L$. In the second case $P(\zeta'(w) = \xi) \geq \delta$ and so $\xi \in \chi(X)$ (and $P(\zeta(w) \in \{\xi, \perp\}) \geq \delta$). As $\chi(X)_L \leq \chi(Y)_L$ it follows that $\xi \in \chi(Y)$, and so there is a $\zeta'' \in Y$ such that $P(\zeta''(w) = \xi) \geq \delta$, which concludes the proof.

Much as in the high-level language, we extend relations R over $\mathcal{H}(S_{H\xi\perp})$ to relations $R^\times$ over $\mathcal{H}(\text{Mem}_{\xi\perp}^W)$. First we extend $-_s$ to $\mathcal{H}(S_{\xi\perp})$ as follows: for $X \in \mathcal{H}(S_{\xi\perp})$ and $s \in S_L$, we let $X_s \in \mathcal{H}(S_{H\xi\perp})$ be $(X \setminus \{\xi\})_s \cup \{\xi \mid \xi \in X\}$. Then, given a relation R over $\mathcal{H}(S_{H\xi\perp})$, we first extend it to a relation R^+ over $\mathcal{H}(S_{\xi\perp})$ by setting

$$X R^+ Y \equiv_{\text{def}} (\xi \in X \Rightarrow \xi \in Y) \ \wedge$$
$$\forall s \in S_L. ((X_s \setminus \{\xi\}) \neq \{\perp\} \Rightarrow (Y_s \setminus \{\xi\}) \neq \{\perp\}) \wedge X_s R Y_s$$

for $X, Y \in \mathcal{H}(S_{\xi\perp})$ and then define $R^\times$ by setting:

$$XR^\times Y \equiv_{\text{def}} X \text{ and } Y \text{ are store projection sets} \wedge \chi(X)R^+\chi(Y)$$

for $X, Y \in \mathcal{H}(\text{Mem}_{\xi\perp}^W)$. (Note that if $R \subseteq \mathcal{H}(S_{H\perp})$, then the high- and low-level definitions of R^+ coincide.)

If R is closed under increasing ω-sups (respectively, is right-closed under $\leq$, is closed under binary unions) the same holds for R^+, and then for $R^\times$ (with $\leq$ restricted to store projection sets). If R is reflexive, then R^+ is and $R^\times$ is reflexive on store projection sets. We also have, much as before, that, for $X, Y \in \mathcal{H}(S_{\xi\perp})$, if XR^+Y then $X_L \leq Y_L$. It then follows from Fact 3 that, for $X, Y \in \mathcal{H}(\text{Mem}_{\xi\perp}^W)$, if $XR^\times Y$ then $X \leq_L Y$.

For any $f, f' : \text{Mem}_{\xi\perp}^W \to \mathcal{H}(\text{Mem}_{\xi\perp}^W)$ and relation R over $\mathcal{H}(S_{H\perp})$ we write that $f \preceq_R f'$ when:

$$\forall X, Y \in \mathcal{H}(\text{Mem}_{\xi\perp}^W). \, XR^\times Y \Rightarrow f^\dagger(X)R^\times f'^\dagger(Y)$$

Finally, we write that $f \preceq f'$ if $f \preceq_R f'$ for some reflexive R closed under increasing ω-sups, right-closed under $\leq$, and closed under binary unions.

Contextual Pre-order vs. Simulation. The contextual pre-order coincides with the simulation relation, but only for commands whose semantics sends store projections to store projection sets. Formally, we say that a given function $f : \text{Mem}_{\xi\perp}^W \to \mathcal{H}(\text{Mem}_{\xi\perp}^W)$ preserves store projections if, for every $s \in S$, $f(-s)$ is a store projection set. The coincidence remains quite useful despite this restriction, which in particular is not an impediment to our overall goal of relating the low-level language to the high-level language.

As in Section 3 we divide the proof of the coincidence into two halves. First, however, we need some preliminary lemmas.

Lemma 6. *For any cpo P, $\zeta \in \text{Mem}_{\xi\perp}^W$, layout w, expression e, boolean expression b, command c, and $x \in \{\perp, \xi\}$ we have:*

$$\zeta(w) = x \Rightarrow \llbracket e \rrbracket_{\zeta(w)}^w = x$$
$$\zeta(w) = x \Rightarrow \llbracket b \rrbracket_{\zeta(w)}^w = x$$
$$\zeta(w) = x \Rightarrow \zeta'(w) = x \quad (\zeta' \in \max(\llbracket c \rrbracket(\zeta)))$$

Proof. For expressions and boolean expressions, the proof is by definition. The proof for commands is then straightforward by structural induction.

Lemma 7. *Let $f : \text{Mem}_{\xi\perp}^W \to \mathcal{H}(\text{Mem}_{\xi\perp}^W)$ be a function that preserves store projections. Let $X \in \mathcal{H}(\text{Mem}_{\xi\perp}^W)$ be a store projection set. Then $f^\dagger(X)$ is a store projection set.*

Proof. We know that $f^\dagger(X) = \{f(\zeta) | \zeta \in \max(X)\} \downarrow$. We need to prove that, for all $\zeta \in \max(X)$ and $\zeta' \in \max(f(\zeta))$, we have that $\zeta'_{/\xi}$ is a store projection.

Since X is a store projection set, we know that $\zeta_{/\xi}$ is a store projection, so we have three possibilities:

- If ζ is a store projection, we conclude since f preserves store projection sets.
- If $\zeta =\perp$, or $\zeta_{/\xi} = \xi$, we conclude using Lemma 6.

Lemma 8. *1. Suppose that e is a public natural-number expression, and that $\zeta \in \mathrm{Mem}^W_{\xi\perp}$ is such that $\zeta_{/\xi}$ is a store projection. Then either:*

- $(w \mapsto \llbracket e \rrbracket^w_{\zeta(w)})_{/\xi} = w \mapsto \xi$,
- $(w \mapsto \llbracket e \rrbracket^w_{\zeta(w)})_{/\xi} =\perp$ and $\zeta_{/\xi} =\perp$, or
- *there exists $n \in \mathbb{N}$ such that $(w \mapsto \llbracket e \rrbracket^w_{\zeta(w)})_{/\xi} = w \mapsto n$.*

Further, $(w \mapsto \llbracket e \rrbracket^w_{\zeta(w)})_{/\xi}$ only depends on $S(\zeta_{/\xi})_L$.

2. Suppose that b is a public boolean expression, and that $\zeta \in \mathrm{Mem}^W_{\xi\perp}$ is such that $\zeta_{/\xi}$ is a store projection. Then either:

- $(w \mapsto \llbracket b \rrbracket^w_{\zeta(w)})_{/\xi} = w \mapsto \xi$,
- $(w \mapsto \llbracket b \rrbracket^w_{\zeta(w)})_{/\xi} =\perp$ and $\zeta_{/\xi} =\perp$, or
- *there exists $t \in \mathbb{B}$ such that $(w \mapsto \llbracket b \rrbracket^w_{\zeta(w)})_{/\xi} = t$.*

Further, $(w \mapsto \llbracket b \rrbracket^w_{\zeta(w)})_{/\xi}$ only depends on $S(\zeta_{/\xi})_L$.

Proof. For the first part, letting e be a public expression, and letting $\zeta \in W \to \mathrm{Mem}_{\xi\perp}$ be such that $\zeta_{/\xi}$ is a store projection, if $S(\zeta_{/\xi}) = \xi$ or $S(\zeta_{/\xi}) =\perp$, we conclude, using Lemma 6. Otherwise $S(\zeta_{/\xi}) \in S$ and the proof is by structural induction on e. Note that as $S(\zeta_{/\xi}) \in S$, $\llbracket e' \rrbracket^w_{\zeta(w)} \neq \perp$ for any e' and w, and so the second case cannot arise when applying the induction hypothesis.

1. k: The conclusion is immediate.
2. l_{nat}: Since l_{nat} is public, and w_p is fixed, $\llbracket e \rrbracket^w_{\zeta(w)} = w_p(l)$ holds for every layout w, and we conclude.
3. $!e$: By the induction hypothesis on e, $(w \mapsto \llbracket e \rrbracket^w_{\zeta(w)})_{/\xi}$ only depends on $S(\zeta_{/\xi})_L$ and either
   - $P(\llbracket e \rrbracket^w_{\zeta(w)} = \xi) \geq \delta$, in which case $P(\llbracket !e \rrbracket^w_{\zeta(w)} = \xi) \geq \delta$ and we conclude

   or
   - there exists an $n \in \mathbb{N}$ such that, for every layout w, $\llbracket e \rrbracket^w_{\zeta(w)} = n$. If $n = w_p(l)$ for some public l, then $\llbracket !e \rrbracket^w_{\zeta(w)} = \zeta(w)(l) = S(\zeta_{/\xi})(l)$, and we conclude. Otherwise, if $n \notin \mathrm{ran}(w_p)$, then we have that $P(\llbracket !e \rrbracket^w_{\zeta(w)} = \xi) = P(n \notin \mathrm{ran}(w)) \geq \delta$, and we conclude.
4. $e + e$ or $e * e$: We conclude by the induction hypothesis.

The proof of the second part is similar; the corresponding induction makes use of the first part in the case when b has the form $e \leq e'$.

We can now prove the first half of the coincidence.

Lemma 9. *Let c and c' be two commands of the low-level language such that $\llbracket c \rrbracket \preceq \llbracket c' \rrbracket$. Then $c \sqsubseteq_L c'$ holds.*

Proof. Let c_0 and c_1 be two commands such that $[\![c_0]\!] \preceq_R [\![c_1]\!]$, with R a reflexive relation closed under increasing ω-sups, right-closed under $\leq$, and closed under binary unions, and let C be a public command context.

We prove by induction on the size of C that $[\![C[c_0]]\!] \preceq_R [\![C[c_1]]\!]$, considering the possible forms of C:

1. $e := e'$: Suppose that $X R^\times Y$. We first do a case study on the semantics of e, e', and $e := e'$. As e and e' are public, for every $\zeta \in \mathrm{Mem}^W_{\xi\perp}$ such that $\zeta_{/\xi}$ is a store projection, we have, by Lemma 8

   - $(w \mapsto [\![e]\!]^w_{\zeta(w)})_{/\xi} = \xi$,
   - $(w \mapsto [\![e]\!]^w_{\zeta(w)})_{/\xi} = \perp$ and $\zeta = \perp$, or
   - there exists $n \in \mathbb{N}$ such that $(w \mapsto [\![e]\!]^w_{\zeta(w)})_{/\xi} = w \mapsto n$.

   and similarly for e'.

   In the bottom case, we have $\zeta = \perp$. In any of the error cases, we have $P([\![e := e']\!]^w_{\zeta(w)} = \xi) \geq \delta$, hence $\chi([\![e := e']\!](\zeta)) = \{w \mapsto \xi\}\!\downarrow$.

   Otherwise, let n and n' be such that $(w \mapsto [\![e]\!]^w_{\zeta(w)})_{/\xi} = w \mapsto n$ and $(w \mapsto [\![e']\!]^w_{\zeta(w)})_{/\xi} = w \mapsto n'$. By Lemma 6, and definition of $()_{/\xi}$ and store projections, there exists $s \in S$ such that $\zeta = - \cdot s$

   - If $n = w_p(l)$ for some public l, then we have:

   $$[\![e := e']\!](\zeta) = \{- \cdot s[l \mapsto n']\}\!\downarrow$$

   We then say that ζ is normal, and write $s^+(\zeta)$ for $(s[l \mapsto n'])_L$.
   - Otherwise, $n \notin \mathrm{ran}(w_p)$, and we have:

   $$P([\![e := e']\!](\zeta)(w) = \xi) = P(n \notin \mathrm{ran}(w)) \geq \delta$$

   By Lemma 8, this analysis only depends on $S(\zeta_{/\xi})_L$.

   We now prove that $[\![e := e']\!](X) R^\times [\![e := e']\!](Y)$. From the case analysis above, we deduce that $[\![e := e']\!](X)$ and $[\![e := e']\!](Y)$ are store projection sets. Also, $\xi \in \chi([\![e := e']\!](X))$ if and only if either there exists $\zeta \in X$ such that $P(\zeta(w) = \xi) \geq \delta$ or else there exists $- \cdot s \in X$ such that $\xi \in \chi([\![e:= e']\!](- \cdot s'))$ if $s'_L = s_L$ and similarly for Y. Since $X R^\times Y$, this proves that $\xi \in \chi([\![e:=e']\!](X)) \Rightarrow \xi \in \chi([\![e:=e']\!](Y))$.

   Further, as can be seen from our case analysis above, for every $s' \in S_L$,

   $$\chi([\![e:=e']\!](X))_{s'} = \bigcup_{\zeta \in \max(X)} \{\chi(X)_{S(\zeta_{/\xi})_L} \mid \zeta \text{ is normal}, s' = s^+(\zeta)\}$$
   $$\cup \{\xi \mid \exists \zeta \in \max(X). \zeta \text{ is not normal}, \zeta \neq \perp\}$$
   $$\cup \{\perp\}$$

   and similarly for Y. By hypothesis, $X R^\times Y$, and so we have both $X \leq_L Y$ (by Fact 3) and $\chi(X) R^+ \chi(Y)$. From the latter we have, for all $s \in S_L$, that $(\chi(X) \setminus \{\xi\})_s \neq \{\perp\} \Rightarrow (\chi(Y) \setminus \{\xi\})_s \neq \{\perp\}$ and also that $\chi(X)_s R \chi(Y)_s$. As R is reflexive, closed under non-empty countable unions, and right-closed under $\leq$, we conclude.

2. $\mathtt{if}\, b\, \mathtt{then}\, C_{\mathtt{tt}}\, \mathtt{else}\, C_{\mathtt{ff}}$: The case where $X = \{\bot\}$ is straightforward, using Lemma 6. Otherwise suppose that $X R^{\times} Y$. As $X \neq \{\bot\}$ we have $Y \neq \{\bot\}$. As b is public, and by Lemma 8, for every $\zeta \neq \bot$ such that $\zeta_{/\xi}$ is a store projection, $(w \mapsto [\![b]\!]^w_{\zeta(w)})/\xi$ only depends on $S(\zeta_{/\xi})_L$, and can only be a boolean $t \in \mathbb{B}$ or ξ, independent of w. Define the store projection set $X_{\mathtt{tt}}$ to be

$$\bigcup_{s \in S} \{\zeta \in X | \zeta = (-\cdot s) \wedge \forall w. [\![b]\!]^w_{ws} = \mathtt{tt}\} \downarrow \ \cup \{\bot\}$$

and define the store projection set X_{ξ} to be

$$\{\zeta \in X \mid \zeta_{/\xi} = w \mapsto \xi\} \downarrow \ \cup \{\bot\}$$

and define the store projection sets $X_{\mathtt{ff}}$, $Y_{\mathtt{tt}}$, $Y_{\mathtt{ff}}$, and Y_{ξ} similarly. We have that $X = X_{\mathtt{tt}} \cup X_{\mathtt{ff}} \cup X_{\xi}$ and that at least one of $X_{\mathtt{tt}}$, $X_{\mathtt{ff}}$, or X_{ξ} is not $\{\bot\}$, and similarly for Y. We also have that

$$[\![\mathtt{if}\, b\, \mathtt{then}\, C_{\mathtt{tt}}[c_0]\, \mathtt{else}\, C_{\mathtt{ff}}[c_0]]\!]^{\dagger}(X) = \\ [\![C_{\mathtt{tt}}[c_0]]\!]^{\dagger}(X_{\mathtt{tt}}) \cup [\![C_{\mathtt{ff}}[c_0]]\!]^{\dagger}(X_{\mathtt{ff}}) \cup X_{\xi} \qquad (*)$$

and similarly for Y.

Similarly to the previous point, we have:

$$\chi(X_{\mathtt{tt}})_{s'} = \bigcup_{s \in S} \{\chi(X)_{s'} \mid \forall w. [\![b]\!]^w_{ws} = \mathtt{tt} \text{ and } s_L = s'\}\downarrow$$

and can check that $X_{\mathtt{tt}} R^{\times} Y_{\mathtt{tt}}$, $X_{\mathtt{ff}} R^{\times} Y_{\mathtt{ff}}$, and $X_{\xi} R^{\times} Y_{\xi}$, making use of the facts that $X \leq_L Y$, and that R is reflexive, closed under non-empty countable unions, and right-closed under $\leq$. We can then conclude using $(*)$ and the fact that R is closed under binary unions.

3. $\mathtt{skip}$, $C'; C''$, $C' + C''$, $\mathtt{while}\, b\, \mathtt{do}\, C_w$, or $[\ ]$: In all these cases the proof is analogous to that of the corresponding parts of the proof of Lemma 1.

This concludes the proof as $[\![C[c_0]]\!] \preceq [\![C[c_1]]\!]$ implies $[\![C[c_0]]\!] \leq_L [\![C[c_1]]\!]$.

We need some further lemmas before proving the second half of the coincidence.

Lemma 10. *Let c be a low-level command such that $[\![c]\!]$ preserves store projection. Let $C[\ ]$ be a public command context. Then $[\![C[c]]\!]$ preserves store projections.*

Proof. The proof is an induction on public command contexts, and is similar to, but simpler than, the proof of Lemma 9.

Let $-^{\times}$ be the map from relations on $\mathcal{H}(S_{\xi\bot})$ to relations on store projection sets left anonymous in the main text. That is, for R a relation on $\mathcal{H}(S_{\xi\bot})$:

$$X R^{\times} Y \equiv_{\mathrm{def}} \chi(X) R \chi(Y)$$

We define $\varpi : S_{\xi\perp} \to \mathcal{H}(\mathrm{Mem}^W_{\xi\perp})$ by:

$$\varpi(\perp) = \{w \mapsto \perp\}\!\downarrow$$
$$\varpi(s) = \{-\cdot s\}\!\downarrow$$
$$\varpi(\xi) = \{\zeta \mid P(\zeta(w) = \xi) \geq \delta\}\!\downarrow$$

Note that, for every $X \in S_{\xi\perp}$, we have $\chi(\varpi(X)) = X$.

Lemma 11. *1. Let R_i ($i \geq 0$) be relations on $\mathcal{H}(S_{H\xi\perp})$ such that if $X R_i Y$ holds, then $\xi \in X$ implies that $\xi \in Y$ and $(X \setminus \{\xi\}) \neq \{\perp\}$ implies that $(Y \setminus \{\xi\}) \neq \{\perp\})$. Let R be the closure of the union of the R_i under increasing ω-sups, binary union, and right-closure under $\leq$. Then R^+ is the closure of the union of the relations R_i^+ under increasing ω-sups, binary unions, and right-closure under $\leq$.*

2. Let R_i ($i \geq 0$) be relations on $\mathcal{H}(S_{\xi\perp})$ and let R be their closure under increasing ω-sups, binary unions, and right-closure under $\leq$. Then R^χ is the closure of the union of the R_i^χ under increasing ω-sups, binary unions, and right-closure under $\leq$ (restricted to store projection sets).

Proof. 1. The proof is almost exactly the same as for Lemma 2. As $-^+$ is evidently monotone, R^+ contains the R_i^+. Next, as we know, if a relation S on $\mathcal{H}(S_{H\xi\perp})$ is closed under any one of increasing ω-sups, binary unions, or right-closure under $\leq$, then so is S^+. So we also have that R^+ is closed under increasing ω-sups and binary unions, and is right-closed under $\leq$. It is therefore included in the closure of the union of the R_i^+ under increasing ω-sups, binary unions, and right-closure under $\leq$.

For the converse, suppose that $U R^+ W$ to show that U and W are related in the closure of the union of the R_i^+ under increasing ω-sups, binary unions, and right-closure under $\leq$. For any given s in S_L, by definition of $-^+$, $U_s R W_s$, and so, by the definition of R, there is a set $J^{(s)} \subseteq \mathbb{N}$, and relations $X_j^{(s)} R_{i_j^{(s)}} Y_j^{(s)}$ such that $U_s = \bigcup_{j \in J^{(s)}} X_j^{(s)}$ and $W_s \supseteq \bigcup_{j \in J^{(s)}} Y_j^{(s)}$. We may assume without loss of generality that the $J^{(s)}$ are disjoint. Let

$$J = \bigcup_{s \in S_L} J^{(s)}$$
$$X_j = \{s' \mid s'_H \in X_j^{(s)}, s'_L = s\} \cup \{\xi \mid \xi \in X_j^{(s)}\} \cup \{\perp\} \qquad (j \in J^{(s)})$$
$$Y_j = \{s' \mid s'_H \in Y_j^{(s)}, s'_L = s\} \cup \{\xi \mid \xi \in Y_j^{(s)}\} \cup \{\perp\} \qquad (j \in J^{(s)})$$
$$i_j = i_j^{(s)} \qquad (j \in J^{(s)})$$

We verify that $(X_j)_s = X_j^{(s)}$, if $j \in J^{(s)}$, and $= \{\perp\}$, otherwise, and similarly for the $(Y_j)_s$. Consequently, $U = \bigcup_j X_j$, $W \supseteq \bigcup_j Y_j$, and for all s in S_L, $(X_j)_s R_{i_j} (Y_j)_s$. Since, by hypothesis, if $X R_i Y$ holds then $\xi \in X$ implies $\xi \in Y$ and $(X \setminus \{\xi\}) \neq \{\perp\}$ implies $(Y \setminus \{\xi\}) \neq \{\perp\}$, we note that $X_j R_{i_j}^+ Y_j$. We conclude that U and W are related as required.

2. As $-^\chi$ is evidently monotone, R^χ contains the R_i^χ. Next one can check that if a relation S on $\mathcal{H}(S_{\xi\perp})$ is closed under either one of increasing ω-sups

or binary unions, then so is S^χ, and that if it is right-closed under $\leq$ then S^χ is right-closed under $\leq$ restricted to store projection sets; to do this one uses the fact that χ is monotone and preserves increasing ω-sups and binary unions. So we also have that R^χ is closed under increasing ω-sups and binary unions, and right-closed under $\leq$ restricted to store projection sets.

For the converse, suppose that $U R^\chi W$ to show that U and W are related in the closure of the union of the R_i^χ under increasing ω-sups, binary unions, and right-closure under $\leq$, restricted to store projection sets.

We have $\chi(U) R \chi(W)$. So, by the definition of R, for some nonempty $J \subseteq \mathbb{N}$, there are relations $X_j R_{i_j} Y_j$ such that both $\chi(U) = \bigcup_j X_j$ and $\chi(W) \supseteq \bigcup_j Y_j$ hold. One can show that, for $j \in J$, $\chi(U \cap \varpi(X_j)) = X_j$ and $\chi(W \cap \varpi(Y_j)) = Y_j$. So we have $(U \cap \varpi(X_j)) R_{i_j}^\chi (W \cap \varpi(Y_j))$, for $j \in J$. So, calculating that:

$$\bigcup_i (U \cap \varpi(X_j)) = U \cap \varpi(\bigcup_i X_j) = U \cap \varpi(\chi(U)) = U$$

and that:

$$\bigcup_i (W \cap \varpi(Y_j)) = W \cap \varpi(\bigcup_i Y_j) \subseteq W \cap \varpi(\chi(W)) = W$$

we see that U and W are related as required.

We can now establish the second half of the coincidence. We will use notations $[c_x \mid x \in \mathrm{PubLoc}]$ and $\mathbf{if}\, b\, \mathbf{then}\, c$ for low-level commands analogous to those used for high-level commands in the proof of Lemma 3.

Lemma 12. *Let c and c' be two commands of the low-level language such that $c \sqsubseteq_L c'$, and $[\![c]\!]$ and $[\![c']\!]$ preserve store projections. Then $[\![c]\!] \preceq [\![c']\!]$ holds.*

Proof. Let c_0 and c_1 be two commands such that $c_0 \sqsubseteq_L c_1$ and that $[\![c_0]\!]$ and $[\![c_1]\!]$ preserve store projections. (The latter property, in combination with Lemmas 10 and 7, allows us to assume that we are always dealing with store projection sets.) We define relations R_i on S_H (for $i \geq 0$) as follows:

- for every $X \in \mathcal{H}(S_{H\perp})$, we have $X R_0 X$;
- for every $X, Y \in \mathcal{H}(\mathrm{Mem}_{\xi\perp}^W)$ such that $X R_i^\chi Y$, we have $(\chi([\![c_0]\!]^\dagger(X)))_s R_{i+1} (\chi([\![c_1]\!]^\dagger(Y)))_s$, forall $s \in S_L$.

We first prove by induction on i that, if $X R_i^\chi Y$, then, for every $s \in S$ such that $\zeta = -\cdot s \in X$, there exists a public command context C and an $s_0 \in S$ such that $\zeta \in ([\![C[c_0]\!]](-\cdot s_0))$ and $\chi([\![C[c_1]\!]](-\cdot s_0))_{s_L} \subseteq \chi(Y)_{s_L}$.

- Suppose $X R_0^\chi Y$. Take $C = \mathbf{skip}$ and $s_0 = s$ (hence $-\cdot s_0 = \zeta$). Then $\zeta \in [\![\mathbf{skip}]\!](\zeta)$ and $\chi([\![\mathbf{skip}]\!](\zeta)) = \{s, \perp\} \subseteq \chi(X) = \chi(Y)$, and we conclude by monotonicity of $-_{s_L}$.
- Suppose that $X R_{i+1}^\chi Y$. By definition of $X R_{i+1}^\chi Y$, and in particular $\chi(X)_{s_L} R_{i+1} \chi(Y)_{s_L}$, there exist $X' R_i^\chi Y'$ and $s' \in S_L$ such that, $\chi([\![c_0]\!]^\dagger(X'))_{s'} = \chi(X)_{s_L}$ and $\chi([\![c_1]\!]^\dagger(Y'))_{s'} = \chi(Y)_{s_L}$.

We have that $\chi(\llbracket c_0 \rrbracket^{\dagger}(X')) = S(\{\zeta''_{/\xi} \mid \zeta'' \in \max(\llbracket c_0 \rrbracket^{\dagger}(X'))\}) \downarrow$ and also $s_H \in \chi(X)_{s_L}$. Hence, using the definition of $-^{\dagger}$, there exist a $\zeta' \in \max(X')$ and a $\zeta'' \in \max(\llbracket c_0 \rrbracket(\zeta'))$ such that $S(\zeta'')_L = s'$ and $S(\zeta'')_H = s_H$. By Lemma 6, as $\zeta'' \neq \perp$ we have $\zeta' \neq \perp$, and as $P(\zeta''(w) = \xi) = 0$, we have $P(\zeta'(w) = \xi) = 0$.

Applying the induction hypothesis to $X' R_i^{\times} Y'$ and $S(\zeta')$, there is a public command context C and an $s_0 \in S$ s.t. $\zeta' \in \llbracket C[c_0] \rrbracket(-\cdot s_0)$ and $\chi(\llbracket C[c_1] \rrbracket(-\cdot s_0))_{S(\zeta')_L} \subseteq \chi(Y')_{S(\zeta')_L}$.

We consider the public command context

$$C' =_{\text{def}} C;$$
$$[\text{if } !x_{\text{nat}} \neq S(\zeta')_L(x) \text{ then } \Omega \mid x \in \text{PubLoc}];$$
$$[\,]; [x := S(\zeta)_L(x) \mid x \in \text{PubLoc}]$$

We have $\zeta' \in \llbracket C[c_0] \rrbracket(-\cdot s_0)$, so ζ'' is in

$$\llbracket C[c_0]; [\text{if } !x_{\text{nat}} \neq S(\zeta')_L(x) \text{ then } \Omega \mid x \in \text{PubLoc}]; c_0 \rrbracket(-\cdot s_0)$$

so ζ is in $\llbracket C'[c_0] \rrbracket(-\cdot s_0)$.

Also, $\chi(\llbracket C[c_1] \rrbracket(-\cdot s_0))_{S(\zeta')_L} \subseteq \chi(Y')_{S(\zeta')_L}$, hence

$$\max(\llbracket C[c_1]; [\text{if } !x_{\text{nat}} \neq S(\zeta')_L(x) \text{ then } \Omega \mid x \in \text{PubLoc}] \rrbracket(-\cdot s_0))_{/\xi}$$
$$\subseteq \max(Y')_{/\xi}$$

hence,

$$\chi(\llbracket C[c_1]; [\text{if } !x_{\text{nat}} \neq S(\zeta')_L(x) \text{ then } \Omega \mid x \in \text{PubLoc}]; c_1 \rrbracket(-\cdot s_0))_{s'}$$
$$\subseteq \chi(\llbracket c_1 \rrbracket^{\dagger}(Y'))_{s'} = \chi(Y)_{s_L}$$

and hence (we replace the low variables by the corresponding values in s_L)

$$\chi(\llbracket C'[c_1] \rrbracket(-\cdot s_0))_{s_L} \subseteq \chi(Y)_{s_L}$$

We now prove that:

$$X R_i^{\times} Y \implies \xi \in \chi(\llbracket c_0 \rrbracket^{\dagger}(X)) \implies \xi \in \chi(\llbracket c_1 \rrbracket^{\dagger}(Y))$$

and that

$$X R_i^{\times} Y \implies \forall s \in S_L. ((\chi(\llbracket c_0 \rrbracket^{\dagger}(X)))_s \setminus \{\xi\}) \neq \{\perp\}$$
$$\implies ((\chi(\llbracket c_1 \rrbracket^{\dagger}(Y)))_s \setminus \{\xi\}) \neq \{\perp\}$$

The two proofs are similar, first, for ξ. If $\xi \in \chi(\llbracket c_0 \rrbracket^{\dagger}(X))$, by definition of $-^{\dagger}$, there exists $\zeta \in \max(X)$ such that $\xi \in \chi(\llbracket c_0 \rrbracket^{\dagger}(\zeta))$. In the case that $P(\zeta(w) = \xi) \geq \delta$, then the proof is by Lemma 6; otherwise $\zeta = -\cdot s$, for some $s \in S$ (it cannot be $\perp$). We know from the above that, since $X R_i^{\times} Y$, there exists a public command context C and an $s_0 \in S$ such that $\zeta \in \llbracket C[c_0] \rrbracket(-\cdot s_0)$ and $\chi(\llbracket C[c_1] \rrbracket(-\cdot s_0))_{s_L} \subseteq \chi(Y)_{s_L}$. We let

$$C' =_{\text{def}} C; [\text{if } !x_{\text{nat}} \neq S(\zeta)_L(x) \text{ then } \Omega \mid x \in \text{PubLoc}]$$

As $\chi(\llbracket C[c_1]\rrbracket(-\cdot s_0))_{s_L} \subseteq \chi(Y)_{s_L}$, $\chi(\llbracket C'[c_1]\rrbracket(-\cdot s_0)) \subseteq \chi(Y)$, and so $\chi(\llbracket C'[c_1]; c_1\rrbracket(-\cdot s_0) \subseteq \chi(\llbracket c_1\rrbracket^\dagger(Y))$.

We have $\xi \in \chi(\llbracket C'[c_0]; c_0\rrbracket(-\cdot s_0))$. Also, as $C'; [\,]$ is a public command context, $\llbracket C'[c_0]; c_0\rrbracket(-\cdot s_0) \leq_L \llbracket C'[c_1]; c_1\rrbracket(-\cdot s_0)$. Hence we have $\xi \in \chi(\llbracket C'[c_1]; c_1\rrbracket(-\cdot s_0))$. This concludes the proof, as we have $\chi(\llbracket C'[c_1]; c_1\rrbracket(-\cdot s_0)) \subseteq \chi(\llbracket c_1\rrbracket^\dagger(Y))$.

Now, for $\perp$. Suppose $\chi(\llbracket c_0\rrbracket^\dagger(X))_s \setminus \{\xi\} \neq \{\perp\}$ for some $s \in S_L$. Then there exists $s' \in S$ such that $s'_L = s_L$ and $-\cdot s' \in \llbracket c_0\rrbracket^\dagger(X)$. By definition of $-^\dagger$, there exists $\zeta \in \max(X)$ such that $-\cdot s' \in \llbracket c_0\rrbracket^\dagger(\zeta)$. If ζ is not a store projection, then there is a contradiction by Lemma 6, otherwise, $\zeta = -\cdot s''$, for some $s'' \in S$.

We know from the above that, since $X R_i^\times Y$, there exists a public command context C and an $s_0 \in S$ such that both $\zeta \in \llbracket C[c_0]\rrbracket(-\cdot s_0)$ and $\chi(\llbracket C[c_1]\rrbracket(-\cdot s_0))_{s''_L} \subseteq \chi(Y)_{s''_L}$ hold. We let

$$C' =_{\text{def}} C; [\text{if } !x_{\text{nat}} \neq S(\zeta)_L(x) \text{ then } \Omega \mid x \in \text{PubLoc}]$$

As $\chi(\llbracket C[c_1]\rrbracket(-\cdot s_0))_{s''_L} \subseteq \chi(Y)_{s''_L}$, $\chi(\llbracket C'[c_1]\rrbracket(-\cdot s_0)) \subseteq \chi(Y)$ holds, and so too, therefore, does $\chi(\llbracket C'[c_1]; c_1\rrbracket(-\cdot s_0) \subseteq \chi(\llbracket c_1\rrbracket^\dagger(Y))$.

We have $-\cdot s' \in \chi(\llbracket C'[c_0]; c_0\rrbracket(-\cdot s_0))$. Also, since $C'; [\,]$ is a public command context, we have

$$\llbracket C'[c_0]; c_0\rrbracket(-\cdot s_0) \leq_L \llbracket C'[c_1]; c_1\rrbracket(-\cdot s_0)$$

Hence we have $-\cdot s''' \in \chi(\llbracket C'[c_1]; c_1\rrbracket(-\cdot s_0))$ for some s''' with $s'''_L = s'_L = s$. As $\chi(\llbracket C'[c_1]; c_1\rrbracket(-\cdot s_0)) \subseteq \chi(\llbracket c_1\rrbracket^\dagger(Y))$, this concludes the proof.

By the definition of R_{i+1}, we have

$$X R_i^\times Y \Rightarrow \forall s \in S_L. (\chi(\llbracket c_0\rrbracket^\dagger(X)))_s R_{i+1}(\chi(\llbracket c_1\rrbracket^\dagger(Y)))_s$$

From both facts and the definition of R_{i+1} we then deduce that

$$\forall X, Y \in \mathcal{H}(\text{Mem}_{\xi\perp}^W). X R_i^\times Y \Rightarrow \llbracket c_0\rrbracket^\dagger(X) R_{i+1}^\times \llbracket c_1\rrbracket^\dagger(Y)$$

We now define R as the closure under increasing ω-sups, right-closure under $\leq$ and closure under binary unions of the union of the R_i. We then conclude, using Lemma 11, that $\llbracket c_0\rrbracket \preceq_R \llbracket c_1\rrbracket$.

Combining Lemmas 9 and 12, we obtain the desired coincidence:

Theorem 4. *Let c and c' be two commands of the low-level language such that $\llbracket c\rrbracket$ and $\llbracket c'\rrbracket$ preserve store projections. Then $c \sqsubseteq_L c'$ holds if and only if $\llbracket c\rrbracket \preceq \llbracket c'\rrbracket$ does.*

Example 6. Suppose that there is only one private location, and consider the two commands:

$$c_4 = (1:=1) + (2:=1) + (3:=1) + (4:=1) \qquad c_6 = (1:=1); (2:=1)$$

As seen above, we have that $\llbracket c_4\rrbracket(\zeta_i) = \{\zeta_\xi^1, \zeta_\xi^2, \zeta_\xi^3, \zeta_\xi^4\}\downarrow$. We also have that $\llbracket c_6\rrbracket(\zeta_i) = \{w \mapsto \xi\}\downarrow$. Since $P(\zeta_\xi^i(w) = \xi) \geq \delta$, we can verify that c_4 and c_6 are equivalent. (Thus, a nondeterministic guess is no better than failure.)   □

5   High and Low

In this section we investigate the relation between the high-level language and the low-level language. Specifically, we define a simple translation from the high-level language to the low-level language, then we study its properties.

We define the compilation of high-level commands c (expressions e, boolean expressions b) to low-level commands $c^\downarrow$ (expressions $e^\downarrow$ and boolean expressions $b^\downarrow$) by setting:

$$(!l_{\text{loc}})^\downarrow = !l_{\text{nat}}$$
$$(l_{\text{loc}} := e)^\downarrow = l_{\text{nat}} := e^\downarrow$$

and proceeding homomorphically in all other cases, for example setting:

$$(e + e')^\downarrow = e^\downarrow + e'^\downarrow$$

Crucially, this compilation function, which is otherwise trivial, transforms high-level memory access to low-level memory access.

We begin with two lemmas about compilation.

Lemma 13.   *1. Let e be a high-level natural-number expression. Then, for every $s \in S$, and $w \in W$,*

$$[\![e^\downarrow]\!]^w_{w \cdot s} = [\![e]\!](s)$$

2. Let b be a high-level boolean expression. Then, for every $s \in S$, and $w \in W$,

$$[\![b^\downarrow]\!]^w_{w \cdot s} = [\![b]\!](s)$$

3. Let c be a high-level command. Then, for every $s \in S$,

$$[\![c^\downarrow]\!](- \cdot s) = \{- \cdot s' \mid s' \in [\![c]\!](s)\}\!\downarrow$$

Proof. The first two parts are straightforward structural inductions on natural number and boolean expressions. For the third we proceed by structural induction on commands:

1. $l_{\text{loc}} := e$: The result is immediate by part 1 and the definition of the semantics.
2. $\text{if } b \text{ then } c_{\text{tt}} \text{ else } c_{\text{ff}}$: By part 2, we have $[\![b]\!](s) = [\![b^\downarrow]\!]^w_{w \cdot s} = t$ with $t \in \mathbb{B}$, hence
   - $[\![c]\!](s) = [\![c_t]\!](s)$, and
   - $[\![c^\downarrow]\!](- \cdot s) = [\![c_t^\downarrow]\!](- \cdot s)$
   The result then follows by applying the induction hypothesis to c_t.
3. skip: Here $\eta(- \cdot s) = \{- \cdot s\}\!\downarrow$ and $\eta(s) = \{s\}\!\downarrow$.
4. $c'; c''$: The result follows from the definition of $-^\dagger$ and applying the induction hypothesis to c' and c''.
5. $c' + c''$: The result follows by applying the induction hypothesis to c and c'.

6. $\texttt{while}\,b\,\texttt{do}\,c_w$: Define iterates $c^{(n)}$ of $\texttt{while}\,b\,\texttt{do}\,c_w$ by setting $c^{(0)} = \Omega$ and $c^{(n+1)} = \texttt{if}\,b\,\texttt{then}\,\texttt{skip}\,\texttt{else}\,c_w; c^{(n)}$, where Ω is some command denoting $\bot$, as does its compilation. Note that the $(c^{(n)})^{\downarrow}$ are the corresponding iterates of $(\texttt{while}\,b\,\texttt{do}\,c_w^{\downarrow})$.

By induction on n, we have

$$[\![(c^{(n)})^{\downarrow}]\!](- \cdot s) = \{- \cdot s' \mid s' \in [\![c^{(n)}]\!](s)\}{\downarrow}$$

as the case $n = 0$ follows from the fact that Ω and its compilation denote (the relevant) $\bot$, and the induction step follows using the same reasoning as in the second, third, and fourth cases.

The conclusion follows, as we have

$$[\![c]\!] = \bigvee_{n \geq 0} [\![c^{(n)}]\!] \quad \text{and} \quad [\![c^{\downarrow}]\!] = \bigvee_{n \geq 0} [\![(c^{(n)})^{\downarrow}]\!]$$

the latter by the above remark on the iterates of $(\texttt{while}\,b\,\texttt{do}\,c_w)^{\downarrow}$.

This concludes the proof.

Lemma 14. *Let c be a high-level command. Then $[\![c^{\downarrow}]\!]$ preserves store projections, and for every store projection set X we have:*

$$\chi([\![c^{\downarrow}]\!]^{\dagger}(X)) = [\![c]\!]^{\dagger}(\chi(X) \setminus \{\xi\}) \cup \{\xi \mid \xi \in \chi(X)\}$$

Proof. This lemma is a straightforward consequence of Lemmas 6 and 13.

Theorem 5 relates the simulation relations of the two languages. It states that a high-level command c simulates another high-level command c, with respect to all public contexts of the high-level language, if and only if the compilation of c simulates the compilation of c', with respect to all public contexts of the low-level language.

Theorem 5. *Let c and c' be two high-level commands. Then $[\![c]\!] \preceq [\![c']\!]$ holds if and only if $[\![c^{\downarrow}]\!] \preceq [\![c'^{\downarrow}]\!]$ does.*

Proof. In one direction, let c and c' be commands such that $[\![c]\!] \preceq_{R_0} [\![c']\!]$, with R_0 a reflexive relation closed under increasing ω-sups, right-closed under $\leq$, and closed under binary unions. Let R be the closure of R_0 in $\mathcal{H}(S_{H\xi\bot})$ by reflexivity, increasing ω-sups, binary union, and right-closure under $\leq$. That is, $X R Y$ holds if both $(X \setminus \{\xi\})\, R_0\, (Y \setminus \{\xi\})$ and $\xi \in X \Rightarrow \xi \in Y$ do. Note that $X R^+ Y$ if $(X \setminus \{\xi\})\, R_0^+\, (Y \setminus \{\xi\})$ and $\xi \in X \Rightarrow \xi \in Y$. Let X and Y in $\mathcal{H}(\text{Mem}_{\xi\bot}^W)$ be such that $X R^{\times} Y$. We have to show that $[\![c^{\downarrow}]\!]^{\dagger}(X) R^{\times} [\![c'^{\downarrow}]\!]^{\dagger}(Y)$. By Lemma 14, $[\![c^{\downarrow}]\!]^{\dagger}(X)$ and $[\![c'^{\downarrow}]\!]^{\dagger}(Y)$ are store projection sets, and so this is equivalent to showing that

$$\chi([\![c^{\downarrow}]\!]^{\dagger}(X))\, R^+ \chi([\![c'^{\downarrow}]\!]^{\dagger}(Y))$$

Using Lemma 14 again, we see that this latter statement is equivalent to:

$$([\![c]\!]^{\dagger}(\chi(X) \setminus \{\xi\}) \cup \{\xi \mid \xi \in \chi(X)\})\, R^+ ([\![c']\!]^{\dagger}(\chi(Y) \setminus \{\xi\}) \cup \{\xi \mid \xi \in \chi(Y)\})$$

which in turn is equivalent, by a previous remark, to

$$[\![c]\!]^\dagger(\chi(X) \setminus \{\xi\}) \, R_0^+ \, [\![c']\!]^\dagger(\chi(Y) \setminus \{\xi\}) \ \wedge \ (\xi \in \chi(X) \Rightarrow \xi \in \chi(Y))$$

As $X R^\times Y$, we have that $\chi(X) \, R^+ \, \chi(Y)$. It follows first that we have that $(\chi(X) \setminus \{\xi\}) \, R_0^+ \, (\chi(Y) \setminus \{\xi\})$, and then $[\![c]\!]^\dagger(\chi(X) \setminus \{\xi\}) \, R_0^+ \, [\![c']\!]^\dagger(\chi(Y) \setminus \{\xi\})$ (as $[\![c]\!] \preceq_{R_0} [\![c']\!]$); and second we have that $\xi \in \chi(X) \Rightarrow \xi \in \chi(Y)$. The conclusion follows.

In the other direction, let c and c' be two commands such that $[\![c^\downarrow]\!] \preceq_R [\![c'^\downarrow]\!]$, with R_0 a reflexive relation closed under increasing ω-sups, right-closed under $\leq$, and closed under binary unions. We let R be the restriction of R_0 to $\mathcal{H}(S_{H\perp})$. That is, $X R Y$ if $X R_0 Y$. Note that $X R^+ Y$ if $X R_0^+ Y$. Let X and Y in $\mathcal{H}(S_\perp)$ be such that $X R^+ Y$. Hence $X R_0^+ Y$. We have $\varpi(X) R_0^\times \varpi(Y)$, hence, by the definition of $\preceq_{R_0}$, we have $[\![c^\downarrow]\!]^\dagger(\varpi(X)) R_0^\times [\![c'^\downarrow]\!]^\dagger(\varpi(Y))$.

By Lemmas 6 and 13, $[\![c^\downarrow]\!]^\dagger(\varpi(X)) = \varpi([\![c]\!]^\dagger(X))$, and similarly for Y. Thus, by the definition of $R_0^\times$, $[\![c]\!]^\dagger(X) R_0^+ [\![c']\!]^\dagger(Y)$, hence $[\![c]\!]^\dagger(X) R^+ [\![c']\!]^\dagger(Y)$, and we conclude.

Our main theorem, Theorem 6, follows from Theorem 5, the two previous theorems, and Lemma 14. Theorem 6 is analogous to Theorem 5, but refers to the contextual pre-orders: a high-level command c implements another high-level command c', with respect to all public contexts of the high-level language, if and only if the compilation of c implements the compilation of c', with respect to all public contexts of the low-level language.

Theorem 6 (Main theorem). *Let c and c' be two high-level commands. Then $c \sqsubseteq_L c'$ holds if and only if $c^\downarrow \sqsubseteq_L c'^\downarrow$ does.*

Theorem 6 follows from Theorem 5, the two previous theorems, and the lemma. The low-level statement is defined in terms of the probability δ that depends on the distribution on memory layouts. When δ is close to 1, the statement indicates that, from the point of view of a public context (that is, an attacker), the compilation of c behaves like an implementation of the compilation of c'. This implementation relation holds despite the fact that the public context may access memory via natural-number addresses, and thereby (with some probability) read or write private data of the commands. The public context may behave adaptively, with memory access patterns chosen dynamically, for instance attempting to exploit correlations in the distribution of memory layouts. The public context may also give "unexpected" values to memory addresses, as in practical attacks; the theorem implies that such behavior is no worse at the low level than at the high level.

For example, for the commands c_0 and c_1 of Example 1, the theorem enables us to compare how their respective compilations behave, in an arbitrary public low-level context. Assuming that δ is close to 1, the theorem basically implies that a low-level attacker that may access memory via natural-number addresses cannot distinguish those compilations. Fundamentally, this property holds simply because the attacker can read or write the location h considered in the example only with low probability.

6   Conclusion

A few recent papers investigate the formal properties of layout randomization, like ours [6–9]. They do not consider nondeterministic choice, and tend to reason operationally. However, the work of Jagadeesan et al. includes some semantic elements that partly encouraged our research; specifically, that work employs trace equivalence as a proof technique for contextual equivalence.

In this paper we develop a semantic approach to the study of layout randomization. Our work concerns nondeterministic languages, for which this approach has proved valuable in reconciling probabilistic choice with nondeterministic choice. However, the approach is potentially more general. In particular, the study of concurrency with nondeterministic scheduling would be an attractive next step. Also, extending our work to higher-order computation presents an interesting challenge.

References

1. Goldwasser, S., Micali, S.: Probabilistic encryption. JCSS 28, 270–299 (1984)
2. Forrest, S., Somayaji, A., Ackley, D.H.: Building diverse computer systems. In: 6th Workshop on Hot Topics in Operating Systems, pp. 67–72 (1997)
3. Druschel, P., Peterson, L.L.: High-performance cross-domain data transfer. Technical Report TR 92-11, Department of Computer Science, The University of Arizona (1992)
4. PaX Project: The PaX project (2004), http://pax.grsecurity.net/
5. Erlingsson, Ú.: Low-level software security: Attacks and defenses. In: Aldini, A., Gorrieri, R. (eds.) FOSAD 2007. LNCS, vol. 4677, pp. 92–134. Springer, Heidelberg (2007)
6. Pucella, R., Schneider, F.B.: Independence from obfuscation: A semantic framework for diversity. Journal of Computer Security 18(5), 701–749 (2010)
7. Abadi, M., Plotkin, G.D.: On protection by layout randomization. ACM Transactions on Information and System Security 15(2), 1–8 (2012)
8. Jagadeesan, R., Pitcher, C., Rathke, J., Riely, J.: Local memory via layout randomization. In: 24th IEEE Computer Security Foundations Symposium, pp. 161–174 (2011)
9. Abadi, M., Planul, J.: On layout randomization for arrays and functions. In: Basin, D., Mitchell, J.C. (eds.) POST 2013. LNCS, vol. 7796, pp. 167–185. Springer, Heidelberg (2013)
10. Lincoln, P., Mitchell, J., Mitchell, M., Scedrov, A.: A probabilistic poly-time framework for protocol analysis. In: Proceedings of the Fifth ACM Conference on Computer and Communications Security, pp. 112–121 (1998)
11. Mitchell, J.C., Ramanathan, A., Scedrov, A., Teague, V.: A probabilistic polynomial-time process calculus for the analysis of cryptographic protocols. TCS 353(1-3), 118–164 (2006)
12. Canetti, R., Cheung, L., Kaynar, D.K., Liskov, M., Lynch, N.A., Pereira, O., Segala, R.: Analyzing security protocols using time-bounded task-pioas. Discrete Event Dynamic Systems 18(1), 111–159 (2008)
13. Gierz, G., Hofmann, K.H., Keimel, K., Lawson, J.D., Mislove, M.W., Scott, D.S.: Continuous lattices and domains. Encyclopaedia of mathematics and its applications, vol. 93. CUP (2003)

14. Mislove, M.W.: On combining probability and nondeterminism. ENTCS 162, 261–265 (2006)
15. Tix, R., Keimel, K., Plotkin, G.D.: Semantic domains for combining probability and non-determinism. ENTCS 222, 3–99 (2009)
16. Goubault-Larrecq, J.: Prevision domains and convex powercones. In: Amadio, R.M. (ed.) FOSSACS 2008. LNCS, vol. 4962, pp. 318–333. Springer, Heidelberg (2008)
17. Abadi, M., Planul, J., Plotkin, G.D.: Layout randomization and nondeterminism. ENTCS 298, 29–50 (2013)
18. Abadi, M., Lamport, L.: The existence of refinement mappings. TCS 82(2), 253–284 (1991)
19. Sabelfeld, A., Sands, D.: Probabilistic noninterference for multi-threaded programs. In: 13th IEEE Computer Security Foundations Workshop, pp. 200–214 (2000)
20. Klarlund, N., Schneider, F.B.: Proving nondeterministically specified safety properties using progress measures. Information and Computation 107(1), 151–170 (1993)
21. de Roever, W.P., Engelhardt, K.: Data Refinement: Model-oriented Proof Theories and their Comparison. Cambridge Tracts in Theo. Comp. Sci., vol. 46. CUP (1998)
22. Jackson, M.: A sheaf theoretic approach to measure theory. PhD thesis, U. Pitt. (2006)

Probabilistic Model Checking
of Labelled Markov Processes
via Finite Approximate Bisimulations

Alessandro Abate[1], Marta Kwiatkowska[1], Gethin Norman[2], and David Parker[3]

[1] Department of Computer Science, University of Oxford, UK
[2] School of Computing Science, University of Glasgow, UK
[3] School of Computer Science, University of Birmingham, UK

Abstract. This paper concerns labelled Markov processes (LMPs), probabilistic models over uncountable state spaces originally introduced by Prakash Panangaden and colleagues. Motivated by the practical application of the LMP framework, we study its formal semantics and the relationship to similar models formulated in control theory. We consider notions of (exact and approximate) probabilistic bisimulation over LMPs and, drawing on methods from both formal verification and control theory, propose a simple technique to compute an approximate probabilistic bisimulation of a given LMP, where the resulting abstraction is characterised as a finite-state labelled Markov chain (LMC). This construction enables the application of automated quantitative verification and policy synthesis techniques over the obtained abstract model, which can be used to perform approximate analysis of the concrete LMP. We illustrate this process through a case study of a multi-room heating system that employs the probabilistic model checker PRISM.

1 Introduction

Labelled Markov processes (LMPs) are a celebrated class of models encompassing concurrency, interaction and probability over uncountable state spaces, originally introduced and studied by Prakash Panangaden and colleagues [14,37]. LMPs evolve sequentially (i.e., in discrete time) over an uncountably infinite state space, according to choices from a finite set of available actions (called labels). They also allow for the possible rejection of a selected action, resulting in termination. LMPs can be viewed as a generalisation of labelled transition systems, allowing state spaces that might be uncountable and that include discrete state spaces as a special case. LMPs also extend related discrete-state probabilistic models from the literature, e.g. [34], and are related to uncountable-state Markov decision processes [38].

The formal semantics of LMPs has been actively studied in the past (see the Related Work section below). One of the earliest contributions is the notion of exact probabilistic bisimulation in [14], obtained as a generalisation of its discrete-state counterpart [34] and used to characterise the LMP model semantics. Exact bisimulation is in general considered a very conservative requirement,

F. van Breugel et al. (Eds.): Panangaden Festschrift, LNCS 8464, pp. 40–58, 2014.

and approximate notions have been consequently developed [15,19], which are based on relaxing the notion of process equivalence or on distance (pseudo-)metrics. These metrics encode exact probabilistic bisimulation, in that the distance between a pair of states is zero if and only if the states are bisimilar. While the exact notion of probabilistic bisimulation can be characterised via a Boolean logic, these approximate notions of probabilistic bisimilarity can be encompassed by real-valued logics, e.g. [30]. In view of their underlying uncountable state spaces, the analysis of LMPs is not tractable, and approximate bisimulation notions can serve as a means to derive abstractions of the original LMPs. Such abstractions, if efficiently computable and finite, can provide a formal basis for approximate verification of LMPs.

Separately from the above work rooted in semantics and logic, models that are closely related to LMPs have also been defined and studied in decision and control [7,28,36]. Of particular interest is the result that quantitative finite abstractions of uncountable-space stochastic processes [2,3] are related to the original, uncountable-state models by notions of approximate probabilistic bisimulations [41]. These notions are characterised via distances between probability measures. Alternatively these formal relations between abstract and concrete models can be established via metrics over trajectories, which are obtained using Lyapunov-like functionals as proposed in [1,31,39], or by randomisation techniques as done in [4]. There is an evident connection between approximation notions and metrics proposed for LMPs and for related models in decision and control, and it is at this intersection that the present contribution unfolds.

In this paper, we build upon existing work on LMPs, with the aim of developing automated verification, as well as optimal policy synthesis, for these models against specifications given in quantitative temporal logic. Drawing on results from the decision and control literature, we give an explicit interpretation of the formal semantics of LMPs. We consider notions of (exact and approximate) probabilistic bisimulation over LMPs, and propose a simple technique to compute an approximate probabilistic bisimulation of a given LMP, where the resulting abstraction is characterised as a finite-state labelled Markov chain (LMC). This enables the direct application of automated quantitative verification techniques over the obtained abstract model by means of the probabilistic model checker PRISM [32], which supports a number of (finite-state) probabilistic models [32,26], including LMCs. We implement an algorithm for computing abstractions of LMPs represented as LMCs and, thanks to the established notion of approximate probabilistic bisimulation, the analysis of the abstraction corresponds to an approximate analysis of the concrete LMP. We illustrate the techniques on a case study of a multi-room heating system, performing both quantitative verification and policy synthesis against a step-bounded variant of the probabilistic temporal logic PCTL [27]. We thus extend the capability of PRISM to also provide analysis methods for (uncountable-state) LMPs, which was not possible previously.

Related Work. Approximation techniques for LMPs can be based on metrics [15,19,42] and coalgebras [43,44]. Approximate notions of probabilistic bisimilarity are formally characterised and computed for finite-state labelled Markov processes in [17]. Metrics are also discussed and employed in [16] and applied to *weak* notions of bisimilarity for finite-state processes, and in [23,24,25] for (finite and infinite) Markov decision processes – in particular, [25] looks at models with uncountable state spaces. The work in [23] is extended by on-the-fly techniques in [12] over finite-state Markov decision processes. LMP approximations are also investigated in [13] and, building on the basis of [17,23], looked at from a different perspective (that of Markov processes as transformers of functions) in [11]. Along the same lines, [33] considers a novel logical characterisation of notions of bisimulations for Markov decision processes. The relationship between exact bisimulation and (CSL) logic is explored in [18] over a continuous-time version of LMPs. Abstractions that are related to Panangaden's finite-state approximants are studied over PCTL properties in [29]. In control theory, the goal of [2,3] is to enable the verification of step-bounded PCTL-like properties [21], as well as time-bounded [41] or unbounded [40] linear-temporal specifications. It is then shown that these approximations are related to the original, uncountable-state models by notions of approximate probabilistic bisimulations [41]. Regarding algorithms for computing abstractions, [9] employs Monte-Carlo techniques for the (approximate) computation of the concepts in [15,19] which relates to the randomised techniques in [4].

Organisation of the Paper. The paper is structured as follows. Section 2 introduces LMPs and discusses two distinct perspectives to their semantic definition. Section 3 discusses notions of exact and approximate probabilistic bisimulations from the literature, with an emphasis on their computability aspects. Section 4 proposes an abstraction procedure that approximates an LMP with an LMC and formally relates the two models. Section 5 describes PRISM model checking of the LMC as a way to study properties of the original LMP. Finally, Section 6 illustrates the technique over a case study.

2   Labelled Markov Processes: Model and Semantics

We consider probabilistic processes defined over uncountable spaces [36], which we assume to be homeomorphic to a Borel subset of a Polish space, namely a metrizable space that is complete (i.e., where every Cauchy sequence converges) and separable (i.e., which contains a countable dense subset). Such a space is endowed with a Borel σ-algebra, which consists of sets that are Borel measurable. The reference metric can be reduced to the Euclidean one.

The uncountable state space is denoted by $\mathcal{S}$, and the associated σ-algebra by $\mathcal{B}(\mathcal{S})$. We also introduce a space of labels (or actions) $\mathcal{U}$, which is assumed to be finite (that is, elements taken from a finite alphabet). For later reference, we extend state and action/label spaces with the additional elements e and $\bar{u}$, respectively, letting $\mathcal{S}^e = \mathcal{S} \cup \{e\}$ and $\mathcal{U}^e = \mathcal{U} \cup \{\bar{u}\}$. We assume a finite set of atomic propositions AP, a function $L : \mathcal{S} \to 2^{\text{AP}}$ which labels states with the propositions that hold in that state, and a reward structure $\mathbf{r} : \mathcal{S} \times \mathcal{U} \to \mathbb{R}_{\geq 0}$, which assigns rewards to state-label pairs over the process.

Processes will evolve in discrete time over the finite interval $[0, N]$ over a sample space $\Omega_{N+1} = \mathcal{S}^{N+1}$, equipped with the canonical product σ-algebra $\mathcal{B}(\Omega_{N+1})$. The selection of labels at each time step depends on a policy (or strategy), which can base its choice on the previous evolution of the process. Formally a policy is a function $\sigma : \{\mathcal{S}^i \mid 1 \leqslant i \leqslant N\} \to \text{dist}(\mathcal{U})$, where $\text{dist}(\mathcal{U})$ is the set of probability distributions over $\mathcal{U}$ and $\sigma(s_0, \ldots, s_k) = \mu$ represents the fact that the policy selects the label u_k in state s_k at time instant k with probability $\mu(u_k)$, given that the states at the previous time instances were $s_0, \ldots, s_{k-1}$.

Under a fixed policy the process is fully probabilistic and we can then reason about the likelihood of events. However, due to the uncountable state space this is not possible for all policies. Following [10], we restrict our attention to so called measurable policies, for which we can define a probability measure, denoted P_s^σ, over the sample space Ω_{N+1} when the initial state of the process equals s.

The following definition is taken from [14,15,19] (these contributions mostly deal with analytic spaces that represent a generalisation of the Borel measurable space we focus on).

Definition 1 (Labelled Markov Process). *A labelled Markov process (LMP) $\mathscr{S}$ is a structure:*

$$(\mathcal{S}, s_0, \mathcal{B}(\mathcal{S}), \{\tau_u \mid u \in \mathcal{U}\}),$$

where $\mathcal{S}$ is the state space, $s_0 \in \mathcal{S}$ is the initial state, $\mathcal{B}(\mathcal{S})$ is the Borel σ-field on $\mathcal{S}$, $\mathcal{U}$ is the set of labels, and for each $u \in \mathcal{U}$:

$$\tau_u : \mathcal{S} \times \mathcal{B}(\mathcal{S}) \longrightarrow [0, 1]$$

is a sub-probability transition function, namely, a set-valued function $\tau_u(s, \cdot)$ that is a sub-probability measure on $\mathcal{B}(\mathcal{S})$ for all $s \in \mathcal{S}$, and such that the function $\tau_u(\cdot, S)$ is measurable for all $S \in \mathcal{B}(\mathcal{S})$. $\qquad\square$

In this work, we will often assume that the initial state s_0 can be any element of $\mathcal{S}$ and thus omit it from the definition. Furthermore, we will implicitly assume that the state space is a standard Borel space, so the LMP $\mathscr{S}$ will often be referred to simply as the pair $(\mathcal{S}, \{\tau_u \mid u \in \mathcal{U}\})$.

It is of interest to explicitly elucidate the underlying semantics of the model that is syntactically characterised in Definition 1. The semantics hinges on how the sub-probability measures are dealt with in the model: we consider two different options, the first drawn from the literature on testing [34], and the second originating from models of decision processes [38]. Recall that we consider finite traces over the discrete domain $[0, N]$ (this is because of the derivation of abstractions that we consider below – an extension of the semantics to infinite traces follows directly). The model is initialised at time $k{=}0$ at state s_0, which is deterministically given or obtained by sampling a given probability distribution π_0, namely $s_0 \sim \pi_0$. At any (discrete) time $0 \leqslant k \leqslant N{-}1$, given a state $s_k \in \mathcal{S}$ and selecting a discrete action $u_k \in \mathcal{U}$, this action is accepted with a probability

$\int_{\mathcal{S}} \tau_{u_k}(s_k, dx)$, whereas it is rejected with probability $1 - \int_{\mathcal{S}} \tau_{u_k}(s_k, dx)$. If the action u_k is rejected, then the model can exhibit two possible behaviours:

1. (Testing process) the dynamics stops, that is, the value s_{k+1} is undefined and the process returns the finite trace

$$((s_0, u_0), (s_1, u_1), \dots, (s_k, u_k));$$

2. (Decision process) a default action $u \in \mathcal{U}^e$ is selected and the process continues its evolution, returning the sample $s_{k+1} \sim \tau_u(s_k, \cdot)$. The default action can, for instance, coincide with the label selected (and accepted) at time $k-1$, i.e. $u = u_{k-1} \in \mathcal{U}$, or with the additional label, i.e. $u = \bar{u}$. At time instant $N-1$, the process completes its course and further returns the trace

$$((s_0, u_0), (s_1, u_1), \dots, (s_k, u_k), (s_{k+1}, u) \dots, (s_{N-1}, u), s_N).$$

Note that the above models can also be endowed with a set of output or observable variables, which are defined over an "observation space" $\mathcal{O}$ via an observation map $h : \mathcal{S} \times \mathcal{U} \to \mathcal{O}$. In the case of "full observation," the map h can simply correspond to the identity and the observation space coincides with the domain of the map. The testing literature often employs a map $h : \mathcal{U} \to \mathcal{O}$, whereas in the decision literature it is customary to consider a map $h : \mathcal{S} \to \mathcal{O}$. That is, the emphasis in the testing literature is on observing actions/labels, whereas in the decision and control field the focus is on observing variables (and thus on the corresponding underlying dynamics of the model).

We elucidate the discussion above by two examples.

Example 1 (Testing process). Consider a fictitious reactive system that takes the shape of a slot or a vending machine, outputting a chosen label and endowed with an internal state with one n-bit memory register retaining a random number. For simplicity, we select a time horizon $N < 2^{n-1}$, so that the internal state never under- or overflows. The state of the machine is $s_k \in \{-2^{n-1}, \dots, 0, \dots, 2^{n-1}\}$, where the index k is a discrete counter initialised at zero. At its k-th use, an operator pushes one of $\mathcal{U} = \{0, 1, 2, \dots, M\}$ buttons u_k, to which the machine responds with probability $\frac{1}{1+e^{-s_k}}$ and, in such a case, resets its state to $s_{k+1} = s_k + u_k \xi_k$, where ξ_k is a fair Bernoulli random variable taking values in $\{-1, 1\}$. On the other hand, if the label/action is not accepted, then the machine gets stuck at state s_k.

Clearly, the dynamics of the process hinges on the external inputs provided by the user (the times of which are not a concern; what matters is the discrete counter k for the input actions). The process generates traces as long as the input actions are accepted. We may be interested in assessing if a given periodic input policy applied within the finite time horizon (for instance, the periodic sequence $(M-1, M, M-1, M, \dots)$) is accepted with probability greater than a given threshold over the model initialised at a given state, where this probability depends on the underlying model. Alternatively, we may be interested in generating a policy that is optimal with respect to a specification over the space of possible labels (for example, we might want the one that minimises the occurrence of choices within the set $\{0, 1, 2\} \subseteq \mathcal{U}$). $\qquad\square$

Example 2 (Decision process). Let us consider a variant of the temperature control model presented in [3,5] and which will be further elaborated upon in Section 6. The temperature of a room is controlled at the discrete time instants $t_k = 0, \delta, 2\delta, \ldots, N\delta$, where $\delta \in \mathbb{R}^+$ represents a given fixed sampling time. The temperature is affected by the heat inflow generated by a heater that is controlled by a thermostat, which at each time instant t_k can either be switched off or set to a level between 1 and M. This choice between heater settings is represented by the labels $\mathcal{U} = \{u_0, u_1, \ldots, u_M\}$ of the LMP, where $u_0 = 0$ and $0 < u_1 < \cdots < u_M$. The (real-valued) temperature s_{k+1} at time t_{k+1} depends on that at time t_k as follows:

$$s_{k+1} = s_k + h(s_k - s_a) + h_{u_k}\zeta(s_k, u_k) + \xi_k,$$

where

$$\zeta(s_k, u_k) = \begin{cases} u_k & \text{w.p. } 1 - s_k \cdot \frac{u_k}{u_M} \cdot \alpha \\ \bar{u} & \text{else,} \end{cases}$$

$\xi_k \sim \mathcal{N}[0, 1]$, s_a represents the constant ambient temperature outside the room, the coefficient h denotes the heat loss, h_{u_k} is the heat inflow when the heater setting corresponds to the label u_k, and α is a normalisation constant.

The quantity $\zeta(s_k, u_k)$ characterises an accepted action (u_k) with a probability that decreases both as the temperature increases (we suppose increasing the temperature has a negative affect through heat-related noise on the correct operation of the thermostat) and as the heater level increases (increasing the heater level puts more stress on the heater, which is then more likely to fail), and conversely provides a default value if the action is not accepted. The default action could feature the heater in the OFF mode (u_0), or the heater stuck to the last viable control value (u_{k-1}). In other words, once an action/label is rejected, the dynamics progresses by adhering to the default action.

We stress that, unlike in the previous example, here the dynamics proceeds regardless of whether the action is accepted or not, since the model variable (s_k) describes a physical quantity with its own dynamics that simply cannot be physically stopped by whatever input choice. Given this model, we may be interested in assessing whether the selected policy satisfies a desired property with a specified probability (similar to the testing case), or in synthesising a policy that maximises the probability of a given specification – say, to maintain the room temperature within a certain comfort interval. Policy synthesis problems appear to be richer for models of decision processes, since the dynamics play a role in a more explicit manner. □

The second semantics (related to a decision process) leads to a reinterpretation of the LMP as a special case of an (infinite-space) MDP [38]. Next, we aim at leveraging this interpretation for *both* semantics: in other words, the two semantics of the LMP can be interpreted in a consistent manner by extending the dynamics and by properly "completing" the sub-stochastic kernels, by means of new absorbing states and additional labels, as described next. This connection has been qualitatively discussed in [37] and is now expounded in detail and newly applied at a semantical level over the different models. Let us start with the testing process. Given a state $s_k \in \mathcal{S}$ and an action $u_k \in \mathcal{U}$, we introduce the binary random variable taking values in the set $\{u_k, \bar{u}\}$ with probability $\{\int_{\mathcal{S}} \tau_{u_k}(s_k, dx), 1 - \int_{\mathcal{S}} \tau_{u_k}(s_k, dx)\}$, respectively. Consider the extended spaces $\mathcal{S}^e$ and $\mathcal{U}^e$. Here e is an absorbing state, namely e is such that $\int_{\mathcal{S}} \tau_u(e, dx) = 0$ for all $u \in \mathcal{U}^e$ (any action selected at state e is rejected), and such that $\tau_{\bar{u}}(s, dx) = \delta_e(dx)$ for all $s \in \mathcal{S}^e$, where $\delta_e(\cdot)$ denotes the Dirac delta function over $\mathcal{S}^e$. We obtain:

$$s_{k+1} \sim \begin{cases} \int_{\mathcal{S}} \tau_{u_k}(s_k, dx)\tau_{u_k}(s_k, \cdot) & \text{if the action is accepted} \\ (1 - \int_{\mathcal{S}} \tau_{u_k}(s_k, dx))\tau_{\bar{u}}(s_k, \cdot) & \text{if the action is rejected.} \end{cases} \tag{1}$$

The labelling map $h : \mathcal{U} \to \mathcal{O}$ is inherited and extended to $\mathcal{U}^e$, so that $h(\bar{u}) = \emptyset$.

Let us now focus on the decision process. Similarly to the testing case, at time k and state s_k, label/action u_k is chosen and this value accepted with a certain probability, else a default value u is given. In the negative instance, the actual value of u depends on the context (see the discussion in the example above) and can correspond to an action within $\mathcal{U}$ (say, the last accepted action) or to the additional action $\bar{u}$ outside this finite set but in $\mathcal{U}^e$. Then, as in the testing case, s_{k+1} is selected according to the probability laws in (1). However, we impose the following condition: once an action is rejected and label u is selected, the very same action is retained deterministically for the remaining part of the time horizon, namely $u_j = u$ for all $k \leqslant j \leqslant N-1$. Essentially, it is as if, for any time instant $k \leqslant j \leqslant N-1$, the action space collapsed into the singleton set $\{u\}$. Notice that, in the decision case, the state space $\mathcal{S}^e$ does not need to be extended; however, the kernel $\tau_{\bar{u}}, \bar{u} \in \mathcal{U}^e \setminus \mathcal{U}$, should be defined and indeed have a non-trivial dynamical meaning if the action $\bar{u}$ is used. Finally, the labelling map $h : \mathcal{S} \to \mathcal{O}$ is inherited from above.

Let us emphasise that, according to the completion procedure described above, LMPs (in general endowed with sub-probability measures) can be considered as special cases of MDPs, which allows connecting with the rich literature on the subject [7,28].

3   Exact and Approximate Probabilistic Bisimulations

We now recall the notions of exact and approximate probabilistic bisimulation for LMPs [14,17]. We also extend these definitions to incorporate the labelling and reward functions introduced in Section 2. We emphasise that both concepts are to be regarded as *strong* notions – we do not consider hidden actions or internal nondeterminism in this work, and thus refrain from dealing with *weak* notions of bisimulation.

Definition 2 ((Exact) Probabilistic Bisimulation). *Consider an LMP $\mathscr{S} = (\mathcal{S}, \{\tau_u \mid u \in \mathcal{U}\})$. An equivalence relation R on $\mathcal{S}$ is a probabilistic bisimulation if, whenever $s_1 R s_2$ for $s_1, s_2 \in \mathcal{S}$, then $L(s_1) = L(s_2)$, $\mathbf{r}(s_1, u) = \mathbf{r}(s_2, u)$ for all $u \in \mathcal{U}$ and, for any $u \in \mathcal{U}$ and set $\tilde{S} \in \mathcal{S}/R$ (which is Borel measurable), it holds that*

$$\tau_u(s_1, \tilde{S}) = \tau_u(s_2, \tilde{S}).$$

A pair of states $s_1, s_2 \in S$ are said to be probabilistically bisimilar if there exists a probabilistic bisimulation R such that $s_1 R s_2$.   □

Observe that the autonomous case of general Markov chains with sub-probability measures, which is characterised by a trivial labels set with a single element, can be obtained as a special case of the above definition.

Let R be a relation on a set A. A set $\tilde{A} \subseteq A$ is said to be R-closed if $R(\tilde{A}) = \{t \mid sRt \wedge s \in \tilde{A}\} \subseteq \tilde{A}$. This notion will be employed shortly – for the moment, note that Definition 2 can be equivalently given by considering the condition on the transition kernel to hold over R-closed measurable sets $\tilde{S} \subseteq S$.

The exact bisimulation relation given above directly extends the corresponding notions for finite Markov chains and Markov decision processes (that is, models characterised by discrete state spaces). However, although intuitive, it can be quite conservative when applied over uncountable state spaces, and procedures to compute such relations over these models are in general deemed to be undecidable. Furthermore, the concept does not appear to accommodate computational robustness [20,45], arguably limiting its applicability to real-world models in engineering and science. These considerations lead to a notion of approximate probabilistic bisimulation with level ε, or simply ε-probabilistic bisimulation [17], as described next.

Definition 3 (Approximate Probabilistic Bisimulation). *Consider an LMP $\mathscr{S} = (S, \{\tau_u \mid u \in \mathcal{U}\})$. A relation R_ε on S is an ε-probabilistic bisimulation relation if, whenever $s_1 R s_2$ for $s_1, s_2 \in S$, then $L(s_1) = L(s_2)$, $\mathbf{r}(s_1, u) = \mathbf{r}(s_2, u)$ for all $u \in \mathcal{U}$ and, for any $u \in \mathcal{U}$ and R_ε-closed set $\tilde{S} \subseteq S$, it holds that*

$$\left| \tau_u(s_1, \tilde{S}) - \tau_u(s_2, \tilde{S}) \right| \leqslant \varepsilon. \tag{2}$$

In this case we say that the two states are ε-probabilistically bisimilar.   □

Unlike the equivalence relation R in the exact case, in general, the relation R_ε does not satisfy the transitive property (the triangle inequality does not hold: each element of a pair of states may be close to a common third element, but map to very different transition measures among each other), and as such is not an equivalence relation [17]. Hence, it induces a cover of the state space S, but not necessarily a partition.

The above notions can be used to relate or compare two separate LMPs, say $\mathscr{S}_1$ and $\mathscr{S}_2$, with the same action space $\mathcal{U}$ by considering an LMP $\mathscr{S}^+$ characterised as follows [37]. The state space S^+ is given by the direct sum of the state spaces of the two processes (i.e. the disjoint union of S_1 and of S_2), where the associated σ-algebra is given by $\mathcal{B}(S_1) \cup \mathcal{B}(S_2)$. The labelling and reward structure combine those for the separate processes, using the fact that the state space is the directed sum of these processes. The transition kernel $\tau_u^+ : S^+ \times \mathcal{B}(S^+) \to [0, 1]$ is such that, for any $u \in \mathcal{U}$, $1 \leqslant i \leqslant 2$, $s_i \in \mathscr{S}_i$, $S^+ \subseteq \mathcal{B}(S_1) \cup \mathcal{B}(S_2)$ we have $\tau_u^+(s_i, S^+) = \tau_u^i(s_i, S^+ \cap S_i)$. The initial states of the composed model are characterised by considering those of the two generating processes with equal likelihood. In the instance of the exact notion, we have the following definition.

Definition 4. *Consider two LMPs $\mathscr{S}_i = (\mathcal{S}_i, \{\tau_u^i \mid u \in \mathcal{U}\})$ where $i = 1, 2$, endowed with the same action space $\mathcal{U}$, and their direct sum $\mathscr{S}^+$. An equivalence relation R on $\mathcal{S}^+$ is a probabilistic bisimulation relation between $\mathscr{S}_1$ and $\mathscr{S}_2$ if, whenever $s_1 R s_2$ for $s_1 \in \mathcal{S}_1$, $s_2 \in \mathcal{S}_2$, then $L(s_1) = L(s_2)$, $\mathbf{r}(s_1, u) = \mathbf{r}(s_2, u)$ for all $u \in \mathcal{U}$ and, for any given $u \in \mathcal{U}$ and R-closed set $\tilde{S}^+ \in \mathcal{B}(\mathcal{S}_1) \cup \mathcal{B}(\mathcal{S}_2)$, it holds that*

$$\tau_u^+(s_1, \tilde{S}^+) = \tau_u^1(s_1, \tilde{S}^+ \cap \mathcal{S}_1) = \tau_u^2(s_2, \tilde{S}^+ \cap \mathcal{S}_2) = \tau_u^+(s_2, \tilde{S}^+).$$

A pair of states $(s_1, s_2) \in \mathcal{S}_1 \times \mathcal{S}_2$ is said to be probabilistically bisimilar if there exists a relation R such that $s_1 R s_2$. Two LMPs $\mathscr{S}_i$ are probabilistically bisimilar if their initial states are.   □

The inequality in (2) can be considered as a correspondence between states in the pair (s_1, s_2) that could result from the existence of a (pseudo-)metric over probability distributions on the state space. This approach has been taken up by a number of articles in the literature, which have introduced metrics as a means to relate two models. Such metrics have been defined based on logical characterizations [15,19,33], categorical notions [43,44], games [17], normed distances over process trajectories [1,31], as well as distances between probability measures [42].

3.1   Computability of Approximate Probabilistic Bisimulations

While for processes over discrete and finite state spaces there exist algorithmic procedures to compute exact [6,34] and approximate [17] probabilistic bisimulations, the computational aspects related to these notions for processes over uncountable state spaces appear to be much harder to deal with. We are of course interested in characterising computationally finite relations, which will be the goal pursued in the next section. Presently, only a few approaches exist to approximate uncountable-space processes with finite-state ones: LMPs [9,11,13,19], (infinite-state) MDPs [33], general Markov chains [29] and stochastic hybrid systems (SHSs) [2,3].

Alternative approaches to check the existence of an approximate probabilistic bisimulations between two models, which hinge on the computation of a function relating the trajectories of the two processes [1,31,39], are limited to models that are both defined over an uncountable space, and do not appear to allow for the constructive synthesis of approximate models from concrete ones. Computation of abstract models, quantitatively related to corresponding concrete ones, is investigated in [4], which leverages randomised approaches and, as such, can enforce similarity requirements only up to a confidence level. Recent work on symbolic abstractions [46] refer to approximation notions over (higher-order) moments on the distance between the trajectory of the abstract model and the solution of the concrete one.

In conclusion, analysing the precision, quality, and scalability properties of constructive approximation techniques for uncountable-state stochastic processes is a major goal with relevant applications that we deem worthwhile pursuing.

4   From LMPs to Finite Labelled Markov Chains

In this section, we introduce an abstraction procedure to approximate a given LMP as a finite-state labelled Markov chain (LMC). The abstraction procedure is based on a discretisation of the state space of the LMP (recall that the space of labels (actions) is finite and as such requires no discretisation) and is inspired by the early work in [3] over (fully-probabilistic) SHS models. It is now extended to account for the presence of actions and to accommodate the LMP framework (with sub-probability measures). The relationship between abstract and concrete models as an approximate probabilistic bisimulation is drawn from results in [41].

Let us start from an LMP $\mathcal{S} = (\mathcal{S}, s_0, \mathcal{B}(\mathcal{S}), \{\tau_u \mid u \in \mathcal{U}\})$, represented via its extended dynamics independently from its actual semantics[1]. We consider a finite partition of the space $\mathcal{S} = \cup_{i=1}^{Q} S_i$ such that $S_i \cap S_j = \emptyset$ for all $1 \leqslant i \neq j \leqslant Q$. In addition, we assume that states in the same element of the partition have the same labelling and reward values, that is, for any $1 \leqslant i \leqslant Q$ we have $L(s)=L(s')$ and $\mathbf{r}(s,u)=\mathbf{r}(s',u)$ for all $s, s' \in S_i$ and $u \in \mathcal{U}$. Let us associate to this partition a finite σ-algebra corresponding to $\sigma(S_1, \ldots, S_Q)$. The finiteness of the introduced σ-algebra, in particular, implies that, for any $1 \leqslant i \leqslant Q$, states $s_1, s_2 \in S_i$, measurable set $S \in \sigma(S_1, \ldots, S_Q)$ and label $u \in \mathcal{U}$, we have:

$$\tau_u(s_1, S) = \tau_u(s_2, S).$$

This follows from the finite structure of the σ-algebra and and the definition of measurability. Let us now select for each $1 \leqslant i \leqslant Q$ a single fixed state $s_i \in S_i$. Using these states, for any label $u \in \mathcal{U}$ we then approximate the kernel τ_u by the matrix $p_u \in [0,1]^Q \times [0,1]^Q$, where for any $1 \leqslant i, j \leqslant Q$:

$$p_u(i,j) = \tau_u(s_i, S_j).$$

Observe that, for any $u \in \mathcal{U}$ and $1 \leqslant i \leqslant Q$, we have $\sum_{j=1}^{Q} p_u(i,j) \leqslant 1$. The structure resulting from this procedure is called a finite labelled Markov chain (LMC). Note that, in general, LMCs do not correspond to (finite-state) MDPs: this correspondence holds only if we have abstracted an LMP that has been "completed" with the procedure described in Section 2, and which as such can be reinterpreted as an uncountable-state MDP.

Let us comment on the procedure above. We have started from the LMP $\mathcal{S} = (\mathcal{S}, \{\tau_u \mid u \in \mathcal{U}\})$, endowed with an uncountable state space $\mathcal{S}$ with the corresponding (uncountable) Borel σ-algebra $\mathcal{B}(\mathcal{S})$. We have partitioned the space $\mathcal{S}$ into a finite quotient made up of uncountable sets S_i, and associated to this finite quotient a finite σ-algebra. We call this intermediate model $\mathcal{S}^f$: the obtained model is still defined over an uncountable state space, but its probabilistic structure is simpler, being characterised by a finite σ-algebra $\sigma(S_1, \ldots, S_Q)$ and piecewise constant kernels – call them $\tau_u^f(s, \cdot)$ – for completeness $\mathcal{S}^f = (\mathcal{S}, s_0, \sigma(S_1, \ldots, S_Q), \{\tau_u^f \mid u \in \mathcal{U}\})$. This latter feature has allowed

---

[1] With slight abuse of notation but for simplicity sake, we avoid referring to extended state and/or action spaces as more proper for "completed" LMP models.

us, in particular, to select an arbitrary state s_i for each of the partition sets S_i, which has led to a finite state space $\mathcal{S}^d = \{s_1, \ldots, s_Q\}$. For each label $u \in \mathcal{U}$ we have introduced the (sub-)probability transition matrix p_u. The new model $\mathscr{S}^d = (\mathcal{S}^d, s_0^d, \sigma(S_1, \ldots, S_Q), \{p_u \mid u \in \mathcal{U}\})$ is an LMC. Here s_0^d is the discrete state in $\mathcal{S}^d$ that corresponds to the quotient set, including the concrete initial condition s_0 of $\mathscr{S}$.

Let us emphasise that, whilst the structure of the state spaces of $\mathscr{S}^f$ and of $\mathscr{S}^d$ are not directly comparable, their probabilistic structure is equivalent and finite – that is, the probability associated to any set in $\mathcal{S}^f$ (the quotient of $\mathcal{S}$) for $\mathscr{S}^f$ is matched by that defined over the finite set of states in $\mathcal{S}^d$ for $\mathscr{S}^d$. The model $\mathscr{S}^f$ allows us to formally relate the concrete, uncountable state-space model $\mathscr{S}$ with the discrete and finite abstraction $\mathscr{S}^d$.

The formal relationship between the concrete and the abstract models can be derived under the following assumption on the regularity of the kernels of $\mathscr{S}$.

Assumption 1. *Consider the LMP $\mathscr{S} = (\mathcal{S}, \{\tau_u \mid u \in \mathcal{U}\})$. For any label $u \in \mathcal{U}$ and states $s', s'', t \in \mathcal{S}$, there exists a positive and finite constant $k(u)$, such that*

$$|T_u(s', t) - T_u(s'', t)| \leqslant k(u)\|s' - s''\|,$$

where T_u is the density associated to the kernel τ_u, which is assumed to admit an integral form so that $\int T_u(s, t)dt = \int \tau_u(s, dt)$ for all $s \in \mathcal{S}$ and $u \in \mathcal{U}$. □

Consider the concrete LMP $\mathscr{S}$, and recall the finite partition for its state space, $\mathcal{S} = \cup_{i=1}^Q S_i$. Let R be the relation over $\mathcal{S}$ such that $s' R s''$ if and only if the states are in the same element of the partition, i.e. there exists $1 \leqslant i \leqslant Q$ such that $s', s'' \in S_i$. The relation R is trivially symmetric and reflexive. Furthermore, if $s' R s''$, then for any $S \in \{S_1, \ldots, S_Q\}$:

$$|\tau_u(s', S) - \tau_u(s'', S)| = \left|\int_S \tau_u(s', dx) - \int_S \tau_u(s'', dx)\right|$$

$$= \left|\int_S T_u(s', x)\, dx - \int_S T_u(s'', x)\, dx\right|$$

$$\leqslant \int_S |T_u(s', x) - T_u(s'', x)|\, dx$$

$$\leqslant \int_S k(u)\|s' - s''\|\, dx$$

$$\leqslant \mathcal{L}(S)k(u)\delta(S), \qquad (3)$$

where $\delta(S) = \sup_{s', s'' \in S} \|s' - s''\|$ denotes the diameter of the partition set S and $\mathcal{L}(S)$ denotes the Lebesgue measure of the set S. By virtue of the inequality established in (3) and Definition 3, we obtain the following result.

Theorem 1. *Consider the LMP $\mathscr{S} = (\mathcal{S}, s_0, \sigma(S_1, \ldots, S_Q), \{\tau_u \mid u \in \mathcal{U}\})$. The introduced relation R is an (approximate) ε-probabilistic bisimulation over $\mathcal{S}$ where*

$$\varepsilon = \max_{u \in \mathcal{U}} \max_{1 \leqslant i \leqslant Q} \mathcal{L}(S_i)k(u)\delta(S_i). \qquad \square$$

From this point on, we assume that we are interested in the dynamics of the LMP over a bounded set $\mathcal{S}$, which allows us to conclude that ε is finite (since its volume $\mathcal{L}(S)$ and its diameter $\delta(S)$ are). More specifically, the approximation level ε can be tuned by reducing the quantity $\delta(\cdot)$, the diameter of the partitions of $\mathcal{S}$. Likewise, better bounds based on local Lipschitz continuity (rather than global, as per Assumption 1) can improve the error, as further elaborated in [21].

We now introduce the model $\mathscr{S}^f$, with its corresponding finite σ-algebra and piecewise constant kernel functions τ_u^f. Working with the same relation R as above, using (3) we have that if $s\,R\,s^f$ and $S \in \{S_1, \ldots, S_Q\}$, then

$$\left|\tau_u(s,S) - \tau_u^f(s^f,S)\right| = |\tau_u(s,S) - \tau_u(s_i,S)| \leqslant \mathcal{L}(S)k(u)\delta(S).$$

This leads us to conclude, via Definition 4, that the LMPs $\mathscr{S}$ and $\mathscr{S}^f$ are ε-probabilistically bisimilar, where ε is taken from Theorem 1. Notice that Definition 4 can be used to relate LMPs with different structures, since it does not require the LMPs to have the same state or probability spaces – the only requirement is that the processes share the same action space. Having argued that the probabilistic structure of $\mathscr{S}^f$ and of $\mathscr{S}^d$ are the same, we proceed now by comparing the LMP $\mathscr{S}$ with the LMC $\mathscr{S}^d$. Consider their direct sum $\mathscr{S}^+$ and relation R where, for $s \in \mathcal{S}$ and $s_i \in \mathcal{S}^d$, we have $s\,R\,s_i$ if and only if $s \in S_i$. Now, any R-closed set S^+ is such that $S^+ \cap \mathcal{S}_d = s_j$ and $S^+ \cap \mathcal{S} = S_j$ for any $1 \leqslant j \leqslant Q$. It therefore follows that

$$\left|\tau_u^+(s,S^+) - \tau_u^+(s_i,S^+)\right| = |\tau_u(s,S_j) - p_u(i,j)|$$
$$= |\tau_u(s,S_j) - \tau_u(s_i,S_j)|$$
$$\leqslant \mathcal{L}(S_j)k(u)\delta(S_i),$$

which leads to the following result.

Theorem 2. *Models $\mathscr{S}$ and $\mathscr{S}^d$ are ε-probabilistically bisimilar.* ◻

Remark 1. The above theorem establishes a formal relationship between $\mathscr{S}$ and $\mathscr{S}^d$ by way of comparing $\mathscr{S}$ with $\mathscr{S}^f$ over the same state space. Unlike most of the mentioned approaches in the LMPs approximations literature, the result comes with a simple procedure to compute the finite abstraction $\mathscr{S}^d$, with a quantitative relationship between the finite abstraction and the original LMP model [3,21]. Thanks to the dependence of the error on the (max) diameter among the partition sets, the approximation level ε can be tuned by selecting a more refined partition of the state space $\mathcal{S}$. Of course this entails obtaining a partition set with larger cardinality by employing smaller partitions. ◻

5   Model Checking Labelled Markov Chains with PRISM

The previous section has described a procedure to approximate an infinite-state LMP by a finite-state LMC. In this section, we show how probabilistic model checking over this finite abstract model can be used to verify properties of the original, concrete uncountable-space LMP.

Probabilistic model checking is a powerful and efficient technique for formally analysing a large variety of quantitative properties of probabilistic models. The properties are specified as formulae in a (probabilistic) temporal logic. In this paper, we use a time-bounded fragment of the logic PCTL [8,27] for discrete-time models, augmented with an operator to reason about costs and rewards [26], although the relationship established in the previous section between LMPs and LMCs in fact preserves a more general class of linear-time properties over a bounded horizon [41].

We will explain our use of probabilistic model checking in the context of (finite-state) LMCs, and subsequently explain the relationship with LMPs. We use logical properties defined according to Φ in the following syntax:

$$\Phi ::= \ \mathsf{P}_{\sim p}\,[\phi\ \mathsf{U}^{\leqslant K}\ \phi]\ \mid\ \mathsf{R}^{\mathbf{r}}_{\sim x}\,[\mathsf{C}^{\leqslant K}\,]$$

$$\phi ::= \ \mathbf{true}\ \mid\ a\ \mid\ \phi\wedge\phi\ \mid\ \neg\phi,$$

where $\sim\ \in\ \{<,\leqslant,>,\geqslant\}$ is a binary comparison operator, $p\in[0,1]$ is a probability bound, $x\in\mathbb{R}_{\geqslant 0}$ is a reward bound, $K\in\mathbb{N}$ is a time bound, $\mathbf{r}$ is a reward structure and a is an atomic proposition. A property $\mathsf{P}_{\sim p}\,[\phi\ \mathsf{U}^{\leqslant K}\ \psi]$ asserts that the probability of ψ becoming true within K time steps, and ϕ remaining true up until that point, satisfies $\sim p$. In standard fashion, we can also reason about (bounded) probabilistic reachability and invariance:

$$\mathsf{P}_{\sim p}\,[\lozenge^{\leqslant K}\phi]\ \overset{\mathrm{def}}{=}\ \mathsf{P}_{\sim p}\,[\mathbf{true}\ \mathsf{U}^{\leqslant K}\phi]$$

$$\mathsf{P}_{\geqslant p}\,[\square^{\leqslant K}\phi]\ \overset{\mathrm{def}}{=}\ \mathsf{P}_{\leqslant 1-p}\,[\lozenge^{\leqslant K}\neg\phi]\,.$$

A property $\mathsf{R}^{\mathbf{r}}_{\sim x}\,[\mathsf{C}^{\leqslant K}]$ asserts that the expected amount of reward (from reward structure $\mathbf{r}$) accumulated over K steps satisfies $\sim x$. State formulae ϕ can identify states according to the atomic propositions that label them, and can be combined by Boolean operations on these propositions.

We define satisfaction of a logical formulae Φ with respect to a state s^d and policy σ^d of an LMC $\mathscr{S}^d$. We write $\mathscr{S}^d, s^d, \sigma^d \models \Phi$ to denote that, starting from state s^d of $\mathscr{S}^d$, and under the control of σ^d, Φ is satisfied. We can then treat the analysis of a formula Φ against a model $\mathscr{S}^d$ in two distinct ways. We can *verify* that a formula Φ is satisfied under all policies of $\mathscr{S}^d$, or we can *synthesise* a single policy that satisfies Φ. In fact, in practice, whichever kind of analysis is required, the most practical solution is to compute the minimum or maximum value, over all policies, for the required property. For example, for an until property $\phi\ \mathsf{U}^{\leqslant K}\ \psi$, we might compute the maximum probability of satisfaction when the initial state is s^d:

$$\mathsf{P}_{\max=?}\,[\phi\ \mathsf{U}^{\leqslant K}\ \psi]\ \overset{\mathrm{def}}{=}\ \max_{\sigma^d}\mathsf{P}^{\sigma^d}_{s^d}\left(\phi\ \mathsf{U}^{\leqslant K}\ \psi\right),$$

where $\mathsf{P}^{\sigma^d}_{s^d}\left(\phi\ \mathsf{U}^{\leqslant K}\ \psi\right)$ is the probability under the policy σ^d when the initial state is s^d of ψ becoming true within K time steps, and ϕ remaining true up until that point.

Computing minimum or maximum probabilities or rewards (and thus checking a property Φ against an LMC) can be performed using existing probabilistic model checking algorithms for Markov decision processes [8,26]. These methods are implemented in the PRISM model checker, which we use for the case study in the next section. When computing optimal values, a corresponding policy (strategy) that achieves them can also be synthesised.

Finally, we discuss how probabilistic model checking of an LMC obtained from an LMP allows us to analyse the original, concrete LMP.

Theorem 3. *Consider a concrete LMP $\mathscr{S}$ and an abstract LMC $\mathscr{S}^d$ which are ε-probabilistic bisimilar. For two ε-probabilistically bisimilar states $s \in \mathcal{S}, s^d \in \mathcal{S}^d$ and until property $\phi \, \mathsf{U}^{\leqslant K} \, \psi$ we have:*

- *for any (measurable) policy σ of $\mathscr{S}$ there exists a policy σ^d of $\mathscr{S}^d$ such that*

$$\left| \mathsf{P}_s^\sigma(\phi \, \mathsf{U}^{\leqslant K} \, \psi) - \mathsf{P}_{s^d}^{\sigma^d}(\phi \, \mathsf{U}^{\leqslant K} \, \psi) \right| \leqslant \varepsilon K,$$

- *for any policy σ^d of $\mathscr{S}^d$ there exists a (measurable) policy σ of $\mathscr{S}$ such that*

$$\left| \mathsf{P}_{s^d}^{\sigma^d}(\phi \, \mathsf{U}^{\leqslant K} \, \psi) - \mathsf{P}_s^\sigma(\phi \, \mathsf{U}^{\leqslant K} \, \psi) \right| \leqslant \varepsilon K.$$

Furthermore, the above bounds apply to the case of optimal policy synthesis, for instance (in the case of maximisation) considering policies σ, σ^d within the same class for $\mathscr{S}$ and $\mathscr{S}^d$, respectively, it holds that

$$\left| \max_{\sigma^d} \mathsf{P}_{s^d}^{\sigma^d}(\phi \, \mathsf{U}^{\leqslant K} \, \psi) - \max_\sigma \mathsf{P}_s^\sigma(\phi \, \mathsf{U}^{\leqslant K} \, \psi) \right| \leqslant \varepsilon K. \qquad \square$$

The above theorem also generalises to expected reward properties and general linear-time properties over a finite horizon, such as bounded linear-time temporal logic (BLTL) or properties expressed as deterministic finite automata.

6   Case Study

This section presents a case study of the multi-room heating benchmark introduced in [22], based on a model proposed by [35] and already discussed in Section 2. The objective is to evaluate the usefulness of probabilistic model checking for the (approximate) verification (and optimisation) of an LMP. The model is an extension of that presented in [5], in that the control set is richer than the binary one considered in the reference, and is also related to that in [3].

We study a model for the control of the temperature evolution of two adjacent rooms. Each room is equipped with a heater and there is a single control which can switch the heaters between $M=10$ different levels of heat flow, with 0 corresponding to the heaters being OFF and 10 to the heaters being ON at full power. The uncountable state space is $\mathbb{R}^2$, modelling the temperature evolution in the two rooms.

As in Section 2, the average temperature of a room evolves according to a stochastic difference equation during the finite time horizon $[0, N]$. As there are now two rooms, following [22] we also include the heat transfer between the rooms in the equations. Letting $\mathbf{s}_k \in \mathbb{R}^2$ denote the temperatures in the rooms at time instant t_k, we have that the equation for room $i \in 1, 2$ (assuming j is the other room) is given by:

$$\mathbf{s}_{k+1}(i) = \mathbf{s}_k(i) + b_i(\mathbf{s}_k(i) - x_a) + a(\mathbf{s}_k(j) - \mathbf{s}_k(i)) + h_{u_k}\zeta(\mathbf{s}_k(i), u_k) + \boldsymbol{\xi}_k(i), \quad (4)$$

where x_a represents the ambient temperature (assumed to be constant) and a the heat transfer rate between the rooms. The quantity b_i is a non-negative constant representing the average heat transfer rate from room i to the ambient and h_{u_k} denotes the heat rate supplied to room i by the corresponding heater at time k. The quantity $\zeta(\mathbf{s}_k(\cdot), u_k)$ characterises an accepted action (u_k) with a probability that, as in Section 2, decreases both as the temperature increases and as the heater level increases. The disturbances $\langle \boldsymbol{\xi}_k(i) \rangle_{0 \leqslant k \leqslant N-1}$ affecting the temperature evolution in room i are assumed to be a sequence of independent identically distributed Gaussian random variables with zero mean and variance ν^2. We also assume that the disturbances affecting the temperature in the two rooms are independent.

The continuous transition kernel τ_u describing the evolution of the uncountable state $\mathbf{s} = (s(1), s(2))$ can easily be derived from (4). Let $\mathcal{N}(\cdot; \mu, V)$ denote the Gaussian measure over $(\mathbb{R}^2, \mathcal{B}(\mathbb{R}^2))$, with mean $\mu \in \mathbb{R}^2$ and covariance matrix $V \in \mathbb{R}^{2 \times 2}$. Then, $\tau_u : \mathcal{B}(\mathbb{R}^2) \times \mathcal{S} \to [0, 1]$ can be expressed as:

$$\tau_u(\cdot \, | s) = \mathcal{N}(\cdot; s + Bs + C, \nu^2 I), \quad (5)$$

where $B \in \mathbb{R}^{2 \times 2}$ with $[B]_{ii} = b_i - a$ and $[B]_{ij} = a$, and $C \in \mathbb{R}^2$ with $[C]_i = -b_i a + u$. With reference to the semantic classification in Section 2, we are dealing here with a decision process.

Let us select a time horizon of $N = 180$ time steps. We are interested in the optimal probability and the corresponding policy that the model dynamics stay within a given "safe" temperature interval in both rooms, say $\mathcal{I} = [17.5, 22.5] \subset \mathcal{S}$ degrees Celsius, and also the optimal expected time and associated policy that the dynamics stay within the interval. We assume that the process is initialised with the temperature in each room being at the mid-point of this interval (if it is initialised outside it, then the associated probability is trivially equal to zero).

We proceed by abstracting the model as a labelled Markov chain [3] as follows. The set $\mathcal{I}$ is partitioned uniformly into $B = 5$ bins or sub-intervals. The labels of the model correspond to choosing the heat-flow level of the heaters for the next time instant. Regarding the atomic propositions and labelling function over the abstract LMC (and concrete LMP), we assign the atomic proposition *safe* to those states where the temperature is within the interval. In addition, to allow the analysis of the time spent within the temperature interval, we use the structure reward $\mathbf{r}$ which assigns the reward 1 to states-label pairs (both of the LMC and LMP) for which the temperature in the state is within the interval and 0 otherwise.

We use PRISM to obtain the minimum and maximum probability of remaining within the *safe* temperature interval over the time horizon, and the minimum and maximum expected time spent in the safe interval up to the horizon. The properties used are $P_{max=?}[\Box^{\leqslant K} safe]$ and $R^{\mathbf{r}}_{max=?}[C^{\leqslant K}]$, as well as the

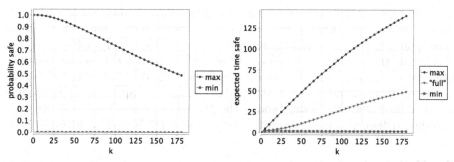

(a) Probability of remaining in safe region   (b) Expected time spent in safe region

Fig. 1. PCTL model checking for the case study

corresponding properties for minimum, rather than maximum, values (see the previous section for details of the notation). The results are presented in Fig. 1(a) and Fig. 1(b), respectively.

The graph plots demonstrate that the minimum probability quickly reaches zero, and that the minimum expected time stabilises as the time horizon increases. Examining with PRISM the policies that obtain these minimum values, we see that the policies coincide and correspond to never switching the heaters on (i.e. setting the heating level to be 0 at each step up until the time horizon). Although at first this may seem the obvious policy for minimising these values, an alternative policy could be to keep the heaters on full at each step (i.e. setting the heating level to 10), as it may be quicker to heat the rooms to above the temperature interval, as opposed to letting the rooms cool to below the interval.

Using PRISM, we find that this alternative approach is actually far less effective in the expected time case, and for small time horizons when considering the probabilistic invariance property. This is due to the fact that it takes much longer to heat a room to above the temperature interval than it does to reach the lower bound by keeping the heaters off. For example for a time bound of 10 minutes, the probability of remaining within the interval equals $1.68e{-}15$ when the heaters are kept off, while if the heaters are on full the probability of remaining within the interval is 0.01562. The fact that there is a chance that the heaters fail at each time step only increases the difference between these policies with regards to remaining within the temperature interval, as it is clearly detrimental to keep the heaters on full, but has no influence when the heaters are kept off. This can be seen in the expected time graph (see Fig. 1(b)), where the expected time of remaining within the temperature interval for the "full" policy keeps increasing while the minimum policy levels off.

In the case of the maximum values for the properties in Fig. 1, we see that for small time horizons there is a very high chance that we can remain within the temperature interval, but as the horizon increases the chance of remaining within the interval drops off. Consider the maximum expected time spent within the interval; this keeps increasing as the horizon increases, but at a lesser rate.

The reason for this behaviour is due to the fact that the heaters can fail and, once a heater fails, there is nothing we can do to stop the temperature in the corresponding room decreasing. Regarding the corresponding policies, we see that, while the heaters are working, the optimal approach is to initially set the heaters to be on full and then lower the heater level as one approaches the upper bound of the temperature interval. In addition, if the temperature in the rooms starts to drop, then the policy repeats the process by setting the heaters to high and then reducing as the temperature nears the upper bound of the interval.

7   Conclusions

This paper has put forward a computable technique to derive finite abstractions of labelled Markov processes (LMPs) in the form of labelled Markov Chains (LMCs), a probabilistic model related to Markov decision processes. The abstract LMC models are shown to correspond to the concrete LMPs via the notion of approximate probabilistic bisimulation. The technique enables the use of PRISM for probabilistic model checking and optimal policy synthesis over the abstract LMCs, extending its current capability to uncountable-state space models. The usefulness of the approach is demonstrated by means of a case study.

Acknowledgments. The authors are supported in part by the ERC Advanced Grant VERIWARE, the EU FP7 project HIERATIC, the EU FP7 project MoVeS, the EU FP7 Marie Curie grant MANTRAS, the EU FP7 project AMBI, and by the NWO VENI grant 016.103.020.

References

1. Abate, A.: A contractivity approach for probabilistic bisimulations of diffusion processes. In: Proc. 48th IEEE Conf. Decision and Control, Shanghai, China, pp. 2230–2235 (December 2009)
2. Abate, A., D'Innocenzo, A., Di Benedetto, M.D.: Approximate abstractions of stochastic hybrid systems. IEEE Transactions on Automatic Control 56(11), 2688–2694 (2011), doi:10.1109/TAC.2011.2160595
3. Abate, A., Katoen, J.-P., Lygeros, J., Prandini, M.: Approximate model checking of stochastic hybrid systems. European Journal of Control 16, 624–641 (2010)
4. Abate, A., Prandini, M.: Approximate abstractions of stochastic systems: A randomized method. In: 2011 50th IEEE Conference on Decision and Control and European Control Conference (CDC-ECC), pp. 4861–4866 (2011)
5. Abate, A., Prandini, M., Lygeros, J., Sastry, S.: Probabilistic reachability and safety for controlled discrete time stochastic hybrid systems. Automatica 44(11), 2724–2734 (2008)
6. Baier, C., Katoen, J.-P.: Principles of model checking. The MIT Press (2008)
7. Bertsekas, D.P., Shreve, S.E.: Stochastic optimal control: The discrete time case, vol. 139. Academic Press (1978)
8. Bianco, A., de Alfaro, L.: Model checking of probabilistic and nondeterministic systems. In: Thiagarajan, P.S. (ed.) FSTTCS 1995. LNCS, vol. 1026, pp. 499–513. Springer, Heidelberg (1995)

9. Bouchard-Cote, A., Ferns, N., Panangaden, P., Precup, D.: An approximation algorithm for labelled Markov processes: towards realistic approximation. In: Proc. 2nd Int. Conf. Quantitative Evaluation of Systems, QEST 2005 (2005)

10. Cattani, S., Segala, R., Kwiatkowska, M., Norman, G.: Stochastic transition systems for continuous state spaces and non-determinism. In: Sassone, V. (ed.) FOSSACS 2005. LNCS, vol. 3441, pp. 125–139. Springer, Heidelberg (2005)

11. Chaput, P., Danos, V., Panangaden, P., Plotkin, G.: Approximating markov processes by averaging. In: Albers, S., Marchetti-Spaccamela, A., Matias, Y., Nikoletseas, S., Thomas, W. (eds.) ICALP 2009, Part II. LNCS, vol. 5556, pp. 127–138. Springer, Heidelberg (2009)

12. Comanici, G., Panangaden, P., Precup, D.: On-the-fly algorithms for bisimulation metrics. In: Proc. 9th Int. Conf. Quantitative Evaluation of Systems (QEST 2012), pp. 681–692 (2012)

13. Danos, V., Desharnais, J., Panangaden, P.: Labelled markov processes: Stronger and faster approximations. Electr. Notes Theor. Comput. Sci. 87, 157–203 (2004)

14. Desharnais, J., Edalat, A., Panangaden, P.: Bisimulation for labelled Markov processes. Information and Computation 179(2), 163–193 (2002)

15. Desharnais, J., Gupta, V., Jagadeesan, R., Panangaden, P.: Metrics for labelled Markov processes. Theoretical Computer Science 318(3), 323–354 (2004)

16. Desharnais, J., Jagadeesan, R., Gupta, V., Panangaden, P.: The metric analogue of weak bisimulation for probabilistic processes. In: Proc. 17th Annual IEEE Symp. Logic in Computer Science (LICS 2002), pp. 413–422 (2002)

17. Desharnais, J., Laviolette, F., Tracol, M.: Approximate analysis of probabilistic processes: logic, simulation and games. In: Proc. 5th Int. Conf. Quantitative Evaluation of SysTems (QEST 2008), pp. 264–273 (September 2008)

18. Desharnais, J., Panangaden, P.: Continuous stochastic logic characterizes bisimulation of continuous-time Markov processes. The Journal of Logic and Algebraic Programming 56(1-2), 99–115 (2003)

19. Desharnais, J., Panangaden, P., Jagadeesan, R., Gupta, V.: Approximating labeled Markov processes. In: Proc. 15th Annual IEEE Symp. Logic in Computer Science (LICS 2000), pp. 95–105 (2000)

20. D'Innocenzo, A., Abate, A., Katoen, J.-P.: Robust PCTL model checking. In: Proc. 15th ACM Int. Conf. Hybrid Systems: computation and control, Beijing, PRC, pp. 275–285 (April 2012)

21. Esmaeil Zadeh Soudjani, S., Abate, A.: Adaptive and sequential gridding procedures for the abstraction and verification of stochastic processes. SIAM Journal on Applied Dynamical Systems 12(2), 921–956 (2013)

22. Fehnker, A., Ivančić, F.: Benchmarks for hybrid systems verification. In: Alur, R., Pappas, G.J. (eds.) HSCC 2004. LNCS, vol. 2993, pp. 326–341. Springer, Heidelberg (2004)

23. Ferns, N., Panangaden, P., Precup, D.: Metrics for finite Markov decision processes. In: Proc. 20th Conf. Uncertainty in Artificial Intelligence (UAI 2004), Banff, Canada, pp. 162–169 (2004)

24. Ferns, N., Panangaden, P., Precup, D.: Metrics for Markov decision processes with infinite state spaces. In: Proc. 21st Conf. Uncertainty in Artificial Intelligence, UAI 2005 (2005)

25. Ferns, N., Panangaden, P., Precup, D.: Bisimulation metrics for continuous Markov decision processes. SIAM Journal of Computing 60(4), 1662–1724 (2011)

26. Forejt, V., Kwiatkowska, M., Norman, G., Parker, D.: Automated verification techniques for probabilistic systems. In: Bernardo, M., Issarny, V. (eds.) SFM 2011. LNCS, vol. 6659, pp. 53–113. Springer, Heidelberg (2011)

27. Hansson, H., Jonsson, B.: A logic for reasoning about time and reliability. Formal Aspects of Computing 6(5), 512–535 (1994)
28. Hernández-Lerma, O., Lasserre, J.B.: Discrete-time Markov control processes, Applications of Mathematics, vol. 30. Springer, New York (1996)
29. Huth, M.: On finite-state approximants for probabilistic computation tree logic. Theoretical Computer Science 346(1), 113–134 (2005)
30. Huth, M., Kwiatkowska, M.Z.: Quantitative analysis and model checking. In: 12th Annual IEEE Symp. Logic in Computer Science, pp. 111–122. IEEE Computer Society (1997)
31. Julius, A.A., Pappas, G.J.: Approximations of stochastic hybrid systems. IEEE Transactions on Automatic Control 54(6), 1193–1203 (2009)
32. Kwiatkowska, M., Norman, G., Parker, D.: PRISM 4.0: Verification of probabilistic real-time systems. In: Gopalakrishnan, G., Qadeer, S. (eds.) CAV 2011. LNCS, vol. 6806, pp. 585–591. Springer, Heidelberg (2011)
33. Larsen, K.G., Mardare, R., Panangaden, P.: Taking it to the limit: Approximate reasoning for markov processes. In: Rovan, B., Sassone, V., Widmayer, P. (eds.) MFCS 2012. LNCS, vol. 7464, pp. 681–692. Springer, Heidelberg (2012)
34. Larsen, K.G., Skou, A.: Bisimulation through probabilistic testing. Information and Computation 94, 1–28 (1991)
35. Malhame, R., Chong, C.-Y.: Electric load model synthesis by diffusion approximation of a high-order hybrid-state stochastic system. IEEE Transactions on Automatic Control 30(9), 854–860 (1985)
36. Meyn, S.P., Tweedie, R.L.: Markov chains and stochastic stability. Communications and Control Engineering Series. Springer, London (1993)
37. Panangaden, P.: Labelled Markov Processes. Imperial College Press (2009)
38. Puterman, M.: Markov decision processes: discrete stochastic dynamic programming. John Wiley & Sons, Inc. (1994)
39. Tkachev, I., Abate, A.: On infinite-horizon probabilistic properties and stochastic bisimulation functions. In: 2011 50th IEEE Conf. Decision and Control and European Control Conference (CDC-ECC), pp. 526–531 (2011)
40. Tkachev, I., Abate, A.: Characterization and computation of infinite horizon specifications over Markov processes. Theoretical Computer Science 515, 1–18 (2013)
41. Tkachev, I., Abate, A.: Formula-free finite abstractions for linear temporal verification of stochastic hybrid systems. In: Proc. 16th ACM Int. Conf. Hybrid Systems: Computation and Control, pp. 283–292 (2013)
42. Tkachev, I., Abate, A.: On approximation metrics for linear temporal model-checking of stochastic systems. In: Proc. 17th ACM Int. Conf. Hybrid Systems: Computation and Control (2014)
43. van Breugel, F., Worrell, J.: An algorithm for quantitative verification of probabilistic transition systems. In: Larsen, K.G., Nielsen, M. (eds.) CONCUR 2001. LNCS, vol. 2154, pp. 336–350. Springer, Heidelberg (2001)
44. van Breugel, F., Worrell, J.: Towards quantitative verification of probabilistic transition systems. In: Orejas, F., Spirakis, P.G., van Leeuwen, J. (eds.) ICALP 2001. LNCS, vol. 2076, pp. 421–432. Springer, Heidelberg (2001)
45. Wimmer, R., Becker, B.: Correctness issues of symbolic bisimulation computation for Markov chains. In: Proc. 15th Int. GI/ITG Conf. Measurement, Modelling, and Evaluation of Computing Systems and Dependability and Fault Tolerance (MMB-DFT), pp. 287–301. Springer, Heidelberg (2010)
46. Zamani, M., Esfahani, P.M., Majumdar, R., Abate, A., Lygeros, J.: Symbolic control of stochastic systems via approximately bisimilar finite abstractions. arXiv: 1302.3868 (2013)

An Operational Interpretation of Negative Probabilities and No-Signalling Models

Samson Abramsky and Adam Brandenburger

Department of Computer Science, University of Oxford
Stern School of Business and Center for Data Science, New York University
samson.abramsky@cs.ox.ac.uk,
adam.brandenburger@stern.nyu.edu

Abstract. Negative probabilities have long been discussed in connection with the foundations of quantum mechanics. We have recently shown that, if signed measures are allowed on the hidden variables, the class of probability models which can be captured by local hidden-variable models are exactly the no-signalling models. However, the question remains of how negative probabilities are to be interpreted. In this paper, we present an operational interpretation of negative probabilities as arising from **standard** probabilities on **signed** events. This leads, by virtue of our previous result, to a systematic scheme for simulating arbitrary no-signalling models.

1 Introduction

Negative probabilities have been discussed in relation to quantum mechanics by many authors, including Wigner, Dirac and Feynman [12,7,8]. For example, Feynman writes:

> The only difference between a probabilistic classical world and the equations of the quantum world is that somehow or other it appears as if the probabilities would have to go negative ...

The separation of quantum from classical physical behaviour in results such as Bell's theorem [3] is expressed in terms of **local realistic models**, in which ontic (or hidden) variables control the behaviour of the system in a classical fashion, satisfying the constraints of **locality** and **realism**. The content of Bell's theorem is exactly that no such model can give rise to the behaviours predicted by quantum mechanics.

However, if we allow negative probabilities on the ontic (or hidden) variables of the model, the situation changes radically.

As a warm-up example, we shall consider the following scenario due to Piponi[1], which, while artificial, is appealingly simple, and does convey some helpful intuitions.

---

[1] See the blog post at
http://blog.sigfpe.com/2008/04/negative-probabilities.html

F. van Breugel et al. (Eds.): Panangaden Festschrift, LNCS 8464, pp. 59–75, 2014.
© Springer International Publishing Switzerland 2014

We shall consider a system comprising two bit registers, A and B. We can perform the following tests or observations on these registers on each run of the system:

- We can read A or B, **but not both.**
- Alternatively, we can observe the value of $A \oplus B$, the **exclusive or** of A and B.

What we find by empirical observation of the system is that, in every run:

1. When we read A, we always get the value 1.
2. When we read B, we always get the value 1.
3. When we observe $A \oplus B$, we always get the value 1.

From 1 and 2, we infer that $A = 1$ and $B = 1$, but this contradicts our observation that $A \oplus B = 1$.

We can try to explain this apparently contradictory behaviour as follows:

- On each run of the system, the registers are set to one of four possible combinations of values:

AB	00	01	10	11

- This joint value is sampled from a probability distribution:

AB	00	01	10	11
	p_1	p_2	p_3	p_4

It is easily seen that no such distribution, where $p_i \geq 0$, $i = 1, \ldots, 4$ and $\sum_i p_i = 1$, can yield the observed behaviour of the system. However, consider the following **signed distribution:**

$$p(00) = -1/2$$
$$p(01) = 1/2$$
$$p(10) = 1/2$$
$$p(11) = 1/2$$

Note that the probability of reading the value 1 for A is

$$p(A = 1) = p(10) + p(11) = 1,$$

and similarly

$$p(B = 1) = p(01) + p(11) = 1.$$

Also,

$$p(A \oplus B = 1) = p(10) + p(01) = 1.$$

Finally, the measure is normalised:

$$p(00) + p(01) + p(10) + p(11) = 1.$$

Also, note that the negative value $p(00) = -1/2$ can never be observed, since we cannot read the values of both A and B in any given run. In fact, the only events which are accessible by direct observation are the following:

$$A = 1 \qquad \{10, 11\}$$
$$A = 0 \qquad \{00, 01\}$$
$$B = 1 \qquad \{01, 11\}$$
$$B = 0 \qquad \{00, 10\}$$
$$A \oplus B = 1 \quad \{10, 01\}$$
$$A \oplus B = 0 \quad \{00, 11\}$$

All of these events get well-defined, non-negative probabilities under our signed distribution, and moreover the complementary pairs of events corresponding to the different possible outcomes for each run, given that a particular choice of quantity to test has been made, yield well-defined probability distributions:

$$p(A = 1) + p(A = 0) \;=\; 1, \quad p(B = 1) + p(B = 0) \;=\; 1, \quad p(A \oplus B = 1) + p(A \oplus B = 0) \;=\; 1.$$

Of course, unless we can give some coherent account of the negative value appearing in the signed measure, it could be said that we have simply explained one mystery in terms of another.

Before addressing this point, we shall turn to a more substantial example, which plays a central rôle in much current work in quantum information and foundations.

We shall now consider a scenario where Alice and Bob each have a choice of two 1-bit registers (or "measurement settings"); say a or a' for Alice, and b or b' for Bob. As before, we shall assume that on each run of the system, Alice can read the value of a **or** a', but not both; and similarly, Bob can only read one of b or b'. (In more physical terms, only one of these quantities can be measured in any given run.) We shall assume that Alice and Bob are spacelike separated, and hence they can perform their measurements or observe their variables independently of each other. By observing multiple runs of this scenario, we obtain probabilities $p(uv|xy)$ of obtaining the joint outcomes $x = u$ and $y = v$ when Alice selects the variable $x \in \{a, a'\}$, and Bob selects $y \in \{b, b'\}$.

Consider the following tabulation of values for these probabilities:

	00	01	10	11
ab	1/2	0	0	1/2
ab'	1/2	0	0	1/2
$a'b$	1/2	0	0	1/2
$a'b'$	0	1/2	1/2	0

The entry in row xy and column uv gives the probability $p(uv|xy)$.

This is the PR-box of Popescu and Rohrlich [10]. As is well-known, it maximises the value of the CHSH expression:

$$\mathbf{E}(ab) + \mathbf{E}(ab') + \mathbf{E}(a'b) - \mathbf{E}(a'b') = 4,$$

where

$$\mathbf{E}(xy) = \sum_{u,v}(-1)^{u+v}p(uv|xy).$$

Thus it exceeds the Tsirelson bound [11] of $2\sqrt{2}$ for the maximum degree of correlation that can be achieved by any bipartite quantum system of this form. It follows that **no quantum system can give rise to this probabilistic model**. At the same time, it still satisfies No-Signalling, and hence is consistent with the relativistic constraints imposed by the spacelike separation of Alice and Bob.

We now consider the analogous form of "explanation" for this model which can be given in the same style as for our previous example. The ontic or hidden variables will assign a definite value to each of the four possible measurements which can be made in this scenario: a, a' b, b'. We shall use the notation $uu'vv'$ for the assignment

$$a \mapsto u, \quad a' \mapsto u', \quad b \mapsto v, \quad b' \mapsto v'.$$

Following Mermin [9], we shall call such assignments **instruction sets**. In this case, there are $2^4 = 16$ instruction sets. We assume that on each run of the system, such an instruction set is sampled according to a probability distribution on this 16-element set. The values observed by Alice and Bob, given their choice of measurement settings, are those prescribed by the instruction set.

Now Bell's theorem tells us that no standard probability distribution on the instruction sets can give rise to the behaviour of the PR Box; while from the Tsirelson bound, we know that the correlations achieved by the PR box exceed those which can be realised by a quantum system.

However, consider the following signed measure on instruction sets [1, Example 5.2]:

$$
\begin{array}{ll}
p(0000) = \;\; 1/2 & p(1000) = -1/2 \\
p(0001) = \quad 0 & p(1001) = \;\; 1/2 \\
p(0010) = -1/2 & p(1010) = \;\; 1/2 \\
p(0011) = \quad 0 & p(1011) = \quad 0 \\
p(0100) = \quad 0 & p(1100) = \quad 0 \\
p(0101) = \quad 0 & p(1101) = \quad 0 \\
p(0110) = \;\; 1/2 & p(1110) = \quad 0 \\
p(0111) = \quad 0 & p(1111) = \quad 0
\end{array}
$$

We can check that this distribution reproduces exactly the probabilities of the PR box. For example:

$$p(00|ab) = p(0000) + p(0001) + p(0100) + p(0101) = 1/2,$$

and similarly $p(uv|xy)$ can be obtained from this signed measure on the instruction sets for all u, v, x, y. Since the "observable events" are exactly those corresponding to these probabilities, the negative values occurring in the signed measure can never be observed. The probabilities of the outcomes for a given choice of measurement settings, corresponding to the rows of the PR box table, form well-defined standard probability distributions.

This is not an isolated result. In [1, Theorem 5.9] it is shown that for a large class of probability models, including Bell scenarios with any numbers of agents, measurement settings and outcomes, and also contextuality scenarios including arbitrary Kochen-Specker configurations, the model can be realised by a signed measure on instruction sets if and only if it is No-Signalling.

But this brings us back to the question: what **are** these negative probabilities? The main purpose of the present paper is to give an operational interpretation of negative probabilities, in a broadly frequentist setting, by means of a simulation in terms of standard probabilities. We shall postpone discussion of the conceptual status of this interpretation to the final section of the paper, after the ideas have been put in place.

The further structure of this paper is as follows. In Section 2, we shall lay out the simple ideas involved in our interpretation of negative probabilities at the level of abstract probability distributions. We shall only consider the case of discrete measures, which will be sufficient for our purposes. In Section 3, we shall review observational scenarios and empirical models in the general setting studied in [1], and the result relating signed measures and No-Signalling models. In Section 4, we shall develop an operational interpretation of hidden-variable models, including those involving negative probabilities. By virtue of the general result on hidden-variable models with signed measures, this yields a uniform scheme for simulating all No-Signalling boxes using only classical resources. Finally, Section 5 concludes with a discussion of the results, and some further directions.

2 Probabilities and Signed Measures

Given a set X, we write $\mathcal{M}(X)$ for the set of (finite-support) **signed probability measures** on X, i.e. the set of maps $m : X \to \mathbb{R}$ of finite support, and such that

$$\sum_{x \in X} m(x) = 1.$$

We extend measures to subsets $S \subseteq X$ by (finite) additivity:

$$m(S) := \sum_{x \in S} m(x).$$

We write $\mathsf{Prob}(X)$ for the subset of $\mathcal{M}(X)$ of measures valued in the non-negative reals; these are just the probability distributions on X with finite support.

These constructions also act on maps. Given a function $f : X \to Y$, we can define

$$\mathcal{M}(f) : \mathcal{M}(X) \longrightarrow \mathcal{M}(Y) :: m \mapsto [y \mapsto \sum_{f(x)=y} m(x)].$$

Thus $\mathcal{M}(f)$ pushes measures on X forwards along f to measures on Y.

Since $\mathcal{M}(f)$ will always map probability distributions to probability distributions, we can define $\mathsf{Prob}(f) := \mathcal{M}(f)|_{\mathsf{Prob}(X)}$. It can easily be checked that these assignments are functorial:

$$\mathcal{M}(g \circ f) = \mathcal{M}(g) \circ \mathcal{M}(f), \quad \mathcal{M}(\mathsf{id}_X) = \mathsf{id}_{\mathcal{M}(X)},$$

and similarly for Prob.

Now we come to the basic idea of our approach, which is to interpret signed measures by "pushing the minus signs inwards". That is, we take basic events to carry an additional bit of information, a sign or "probability charge". Moreover, occurrences of the same underlying event of opposite sign cancel. In this fashion, negative probabilities arise from standard probabilities on signed events.[2]

More formally, given a set X, we take the signed version of X to be the disjoint union of two copies of X, which we can write as

$$X^{\pm} := \{(x, \varsigma) \mid x \in X, \varsigma \in \{+, -\}\}.$$

Also, given a map $f : X \to Y$, we can define a map $f^{\pm} : X^{\pm} \to Y^{\pm}$ by

$$f^{\pm} : (x, \varsigma) \mapsto (f(x), \varsigma), \quad \varsigma \in \{+, -\}.$$

We shall use the notation x^+, x^- rather than $(x, +), (x, -)$. Given $S \subseteq X$, we shall write

$$S^+ := \{x^+ \mid x \in S\}, \quad S^- := \{x^- \mid x \in S\} \subseteq X^{\pm}.$$

The representation of a signed measure on a sample space X by a probability measure on $X^{\pm}$ is formalised by the map

$$\theta_X : \mathcal{M}(X) \longrightarrow \mathsf{Prob}(X^{\pm})$$

given by

$$\theta_X(m)(x^+) = \begin{cases} m(x)/K, & m(x) > 0 \\ 0, & \text{otherwise} \end{cases}$$

$$\theta_X(m)(x^-) = \begin{cases} |m(x)|/K, & m(x) < 0 \\ 0, & \text{otherwise} \end{cases}$$

where $K = \sum_{x \in X} |m(x)|$ is a normalisation constant.

---

[2] This intuitive way of looking at signed measures has, *grosso modo*, appeared in the literature, e.g. in a paper by Burgin [5]. However, the details in [5] are very different to our approach. In particular, the notion of signed relative frequencies in [5], which is defined as a difference of ratios rather than a ratio of differences, is not suitable for our purposes.

Note that the probability measures in the image of θ_X have some special properties. In particular, if we define $W(d)$, for $d \in \mathsf{Prob}(X^\pm)$, by

$$W(d) := d(X^+) - d(X^-),$$

then if $d = \theta_X(m)$, we have

$$W(d) = \sum_{x \in X} m(x)/K = 1/K > 0.$$

We write $\mathsf{Prob}^\pm(X) := \{d \in \mathsf{Prob}(X^\pm) \mid W(d) > 0\}$. Thus θ_X cuts down to a map

$$\theta_X : \mathcal{M}(X) \longrightarrow \mathsf{Prob}^\pm(X).$$

Note also that for any map $f : X \to Y$, and $d \in \mathsf{Prob}(X^\pm)$:

$$W(\mathsf{Prob}(f^\pm)(d)) = W(d).$$

Hence we can extend $\mathsf{Prob}^\pm$ to a functor by $\mathsf{Prob}^\pm(f) := \mathsf{Prob}(f^\pm)$.

We can recover a signed measure on X from a probability distribution in $\mathsf{Prob}^\pm(X)$ by an inverse process to θ_X. Formally, this is given by a map

$$\eta_X : \mathsf{Prob}^\pm(X) \longrightarrow \mathcal{M}(X) :: d \mapsto [x \mapsto (d(x^+) - d(x^-))/W(d)].$$

This map incorporates the idea that positive and negative occurrences of a given event cancel.

The following simple observation will be used in showing the correctness of our simulation scheme in Section 4.2.

Proposition 1. *The following diagram commutes, for all sets X and Y and functions $f : X \to Y$:*

$$
\begin{array}{ccc}
\mathcal{M}(X) & \xrightarrow{\;\mathcal{M}(f)\;} & \mathcal{M}(Y) \\
\downarrow{\scriptstyle \theta_X} & & \uparrow{\scriptstyle \eta_Y} \\
\mathsf{Prob}^\pm(X) & \xrightarrow[\mathsf{Prob}^\pm(f)]{} & \mathsf{Prob}^\pm(Y)
\end{array}
$$

Proof.   We write $d := \theta_X(m)$, and calculate pointwise on $y \in Y$:

$$
\begin{aligned}
\eta_Y \circ \mathsf{Prob}^\pm(f) \circ d(y) &= \frac{d((f^\pm)^{-1}(y^+)) - d((f^\pm)^{-1}(y^-))}{d(X^+) - d(X^-)} \\
&= \sum_{f(x)=y,\, m(x)>0} m(x) - \sum_{f(x)=y,\, m(x)<0} |m(x)| \\
&= \sum_{f(x)=y} m(x) \\
&= \mathcal{M}(f)(m)(y).
\end{aligned}
$$

□

Intuitively, this says that pushing a signed measure m forwards along f can be performed by simulating the measure by a probability distribution $d = \theta_X(m)$ on signed events, pushing d forwards along f, and then interpreting the resulting probability distribution back as a signed measure via the map η_Y.

3   Observational Scenarios and Empirical Models

An **observational scenario** is a structure $(X, \mathcal{U}, O)$, where:

- X is a set of measurements.
- $\mathcal{U}$ is a family of non-empty subsets of X with $\bigcup \mathcal{U} = X$, representing the compatible sets of measurements — those which can be performed together.
- O is a set of measurement outcomes.

For example, in the scenario for the PR box described in the Introduction, we have:

- $X = \{a, a', b, b'\}$.
- The measurement contexts are the choice of measurement settings by Alice and Bob:
$$\{a, b\}, \{a', b\}, \{a, b'\}, \{a', b'\}.$$
- The outcomes are $O = \{0, 1\}$.

An empirical model for such a scenario is a family of probability distributions $\{d_U\}_{U \in \mathcal{U}}$, with $d_U \in \mathsf{Prob}(O^U)$. Here O^U is the set of all functions $s : U \to O$. Such functions represent basic events in the measurement context U, where the measurements in U are performed, and the outcome $s(x)$ is observed for each $x \in U$.

In the case of the PR box, the distributions correspond to the rows of the table, indexed by the measurement contexts xy.

We define an operation of restriction on signed measures, which is a general form of marginalization: if $U \subseteq V$ and $m \in \mathcal{M}(O^V)$, then $m|_U \in \mathcal{M}(O^U)$ is defined by:
$$m|_U(s) = \sum_{t|_U = s} m(t).$$

Note that $m|_U = \mathcal{M}(\rho_U^V)(m)$, where

$$\rho_U^V : O^V \longrightarrow O^U :: s \mapsto s|_U.$$

Also, if $d \in \mathsf{Prob}(O^V)$, then $d|_U \in \mathsf{Prob}(O^U)$.

A model $\{d_U\}_{U \in \mathcal{U}}$ is **compatible** if for all $U, V \in \mathcal{U}$:

$$d_U|_{U \cap V} = d_V|_{U \cap V}.$$

As shown in [1], compatibility can be seen as a general form of **no-signalling**.

A **global section** for an empirical model $\{d_U\}_{U \in \mathcal{U}}$ is a distribution $d \in$ Prob(O^X) such that, for all $U \in \mathcal{U}$:

$$d|_U = d_U.$$

As shown in [1], a global section can be seen as a canonical form of local hidden variable model.

A **signed global section** is a signed measure $m \in \mathcal{M}(O^X)$ such that $m|_U = d_U$ for all $U \in \mathcal{U}$. We have the following result from [1].

Theorem 1. *An empirical model $\{d_U\}_{U \in \mathcal{U}}$ is no-signalling if and only if it has a signed global section.*

4 Operational Interpretation of Hidden-Variable Models

We begin with standard local hidden-variable models. We shall follow the expository scheme due to Mermin phrased in terms of **instruction sets** [9], which is encapsulated in the diagram in Figure 1.

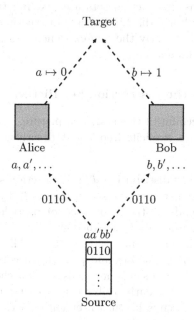

Fig. 1. The Mermin instruction set picture

Here the agents or experimenters Alice and Bob each receive a stream of particles; for each particle, they choose a measurement setting, and observe an outcome. The stream of observed joint outcomes is collected at a target, and provides the statistical data on which the empirical model is based.

In this way of visualising hidden variables, the pairs of particles are generated by some source. Reflecting the idea of local realism, each particle comes with an "instruction set" which specifies an outcome for every possible measurement which can be performed. In the pictured scenario, where we are considering measurement settings a, a' for Alice and b, b' for Bob, such an instruction set will specify an outcome, 0 or 1, for each of these four measurements, and hence is depicted as a string of four bits, which correspond to outcomes for a, a', b, b'. When such a particle arrives e.g. at Alice and she performs the measurement a, the instruction set 0110 dictates that she will observe the outcome 0; while if Bob performs the measurement b on a particle with the same instruction set he will observe the outcome 1. To account for the fact that the same measurements may yield different outcomes, we depict the source as obtaining the stream of particles with inscribed instruction sets by sampling some probability distribution on the space of instruction sets.

The content of Bell's theorem and related no-go results is that there is **no probability distribution on instruction sets** which can account for the probabilistic behaviour which is predicted by quantum mechanics, and highly confirmed by experiment.

In mathematical terms, instruction sets are just functions in O^X; and the non-existence of a probability distribution on instruction sets which recovers the observed behaviour is exactly the non-existence of a global section for the corresponding empirical model.

4.1   Formalisation of the Instruction Set Picture

We now give a formal account of this standard picture.

Firstly, we give a more explicit, frequentist description of the operational reading we have in mind.

1. We fix some probability distribution d on instruction sets.
2. The **source** produces a stream of instruction sets t_n by sampling repeatedly according to d. It sends a stream of pairs of particles inscribed with the corresponding instruction sets t_n to Alice and Bob.
3. **Alice** and **Bob** act independently of the source. Alice receives a stream of particles from the source, chooses and performs a corresponding stream of local measurements a_n, and sends her measurement choices a_n and the outcomes u_n to the target. Similarly, Bob receives a stream of particles, chooses measurements b_n, and sends his measurement choices b_n and the outcomes v_n to the target. Note that the joint outcome $u_n v_n$ of a measurement context $a_n b_n$ specified by an instruction set t_n is given by $u_n = t_n(a_n)$, $v_n = t_n(b_n)$.
4. The **target** receives the streams of measurement choices and outcomes from Alice and Bob. It combines these into a stream τ of joint outcomes, with $\tau_n = u_n v_n$. For each choice of measurement context xy and joint outcome uv, and for each n, it can compute the relative frequency $r_n(uv|xy)$ of uv in the initial segment of τ_{xy} of length n. Here τ_{xy} is the restriction of τ to the subsequence of those τ_i such that $a_i = x$ and $b_i = y$.

5. The limiting value of these relative frequencies $r_n(uv|xy)$ is taken to be the probability $p(uv|xy)$.

We comment on a number of points raised by this description.

- Firstly, note that the rôle of the various agents in this protocol is clearly delineated by the information flows it makes explicit.
  - Alice and Bob cannot predict the outcome, since they must accept unknown particles from the source.
  - Although the instruction sets generated by the source determine the outcomes **given the choice of measurements**, the source cannot predict which measurements Alice and Bob will select.

  These informational independence notions are reflected in the standard assumption of independence between the distribution governing the instruction sets, and the choices made by Alice and Bob. This is usually referred to as Free Choice of Measurements, or λ-independence [6]. Without this assumption, the protocol trivialises, and arbitrary behaviour can be generated [2].
- The target is usually not made explicit. However, since Alice and Bob are assumed to be spacelike separated, in order for it to be possible to obtain empirical data on the correlations between their outcomes, it is necessary to assume that their future light-cones intersect.

We now formalise this informal frequentist account in terms of standard probability theory.

Firstly, we recall standard notation for indicator functions. If $U \subseteq X$, we write $\mathbb{1}_U : X \to \mathbb{R}$ for the function

$$\mathbb{1}_U : x \mapsto \begin{cases} 1, x \in U \\ 0, x \notin U. \end{cases}$$

If we fix a distribution $d \in \mathsf{Prob}(X)$, we can regard $\mathbb{1}_U$ as a random variable with respect to d. Note that, writing $\mathbf{E}(R)$ for the expectation of a random variable R:

$$\mathbf{E}(\mathbb{1}_U) = d(U).$$

We write $X := \{a, a', b, b'\}$, $O := \{0, 1\}$, and $\mathcal{I} := O^X$ for the set of instruction sets. We write $X_A := \{a, a'\}$ for the set of Alice's measurement choices, and similarly $X_B := \{b, b'\}$.

We are given a probability distribution $d \in \mathsf{Prob}(\mathcal{I})$. We also assume a probability distribution d_{AB} on measurement choices xy, which is assumed to be independent of d, reflecting our assumption that Alice and Bob's measurement choices are independent of the source. We shall assume that $d_{AB}(xy) > 0$ for all measurement contexts xy, *i.e.* that all measurements have some chance of being performed. If this were not the case, we could simply exclude measurements which could never be performed from the scenario.

Thus we have a probability distribution μ on $\mathcal{I} \times X_A \times X_B$:

$$\mu(t, x, y) = d(t)d_{AB}(xy).$$

The stream of data at the target comprises items of the form $uvxy \in O \times O \times X_A \times X_B$. This is determined by the instruction set generated at the source and the measurement choices made by Alice and Bob, as specified by the function

$$f : \mathcal{I} \times X_A \times X_B \longrightarrow O \times O \times X_A \times X_B \ :: \ (t, x, y) \mapsto (t(x), t(y), x, y).$$

We can use the functorial action of Prob to push μ forward along f to yield a probability distribution $\nu := \mathsf{Prob}(f)(\mu)$ on $O \times O \times X_A \times X_B$. This can be defined explicitly as follows. Given u, v, x, y, we define $U(uvxy) \subseteq \mathcal{I}$:

$$U(uvxy) \ := \ \{t \in \mathcal{I} \mid t(x) = u \ \wedge \ t(y) = v\}.$$

Now $\nu(uv, xy) = d(U(uvxy))d_{AB}(xy)$. Note that $d(U(uvxy)) = \mathsf{Prob}(\rho^{\mathcal{I}}_{xy})(d)(uv)$.

The conditional probability for the target to observe outcomes uv given measurement choices xy is:

$$\nu(uv|xy) \ = \ \frac{d(U(uvxy))d_{AB}(xy)}{d_{AB}(xy)} \ = \ d(U(uvxy)).$$

Thus the stochastic process at the target for observing outcomes uv given measurement settings xy is modelled by the i.i.d. sequence of random variables $\mathbf{X}_n$, where for all n, $\mathbf{X}_n = \mathbb{1}_{U(uvxy)}$. Note that the conditioning on the choice of measurement settings is represented implicitly by the selection of the infinite subsequence corresponding to the stream τ_{xy} in the informal discussion above. Of course, an i.i.d. sequence is invariant under the selection of arbitrary infinite subsequences.

The relative frequencies observed at the target are represented by the averages of these random variables:

$$r_n(uv|xy) \ = \ \frac{1}{n} \sum_{i=1}^{n} \mathbf{X}_i.$$

Using the Strong Law of Large Numbers [4], we can calculate that

$$p(uv|xy) \ = \ d|_{xy}(uv) \ = \ \mathsf{Prob}(\rho^{\mathcal{I}}_{xy})(d)(uv) \ = \ d(U(uvxy)) \ = \ \mathbf{E}(\mathbb{1}_{U(uvxy)}) \ =_{\text{a.e.}} \ \lim_{n \to \infty} \frac{1}{n} \sum_{i=1}^{n} \mathbf{X}_i.$$

This provides a precise statement of the agreement of the operational protocol with the abstract formulation.

4.2   Signed Probabilities and No-Signalling Models

We now return to Theorem 1 and use our account of negative probabilities to give it an operational interpretation, which we formulate as a refinement of the Mermin picture.

We use the signed version of the Mermin instruction set scenario depicted in Figure 2.

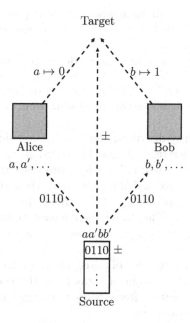

Fig. 2. Signed instruction sets

We have the same picture as before, subject to the following refinements:

- The particles come with an additional bit of information in addition to the instruction set: a **sign**.
- The source repeatedly samples these signed instruction sets according to a (standard) probability distribution, and sends the particles with their instruction sets to Alice and Bob, who choose their measurements and observe outcomes as before.
- The joint outcomes are collected at the target, which also receives the information concerning the signs of the particles.
- The target uses the signs to compute **signed relative frequencies** on the stream of joint outcomes, and hence to recover a signed measure on the joint outcome. In certain cases, this signed measure may in fact be a *bona fide* probability measure.

  The signed relative frequencies incorporate the idea of cancelling positive and negative occurrences of events. The signed relative frequency of an event is the difference between its number of positive and negative occurrences, normalised by the total weight of the ensemble, which is the difference between the total numbers of positive and negative occurrences in the ensemble.

We shall now set out the simulation scheme in more precise terms. The analysis is very similar to the unsigned case, the key difference being the use of signed relative frequencies.

We are given a no-signalling empirical model with probabilities $p(uv|xy)$.

1. By Theorem 1, we can find a signed measure m on the instruction sets which yields the observed probabilities by marginalization: $m|_{xy}(uv) = p(uv|xy)$ for all u, v, x, y.
2. We form the probability distribution $d = \theta_{\mathcal{I}}(m)$ on signed instruction sets.
3. The **source** produces a stream of signed instruction sets s_n^{ς} by repeatedly sampling from d. It sends a stream of particles inscribed with the corresponding instruction sets s_n to Alice and Bob, and sends the signs to the target.
4. **Alice** and **Bob** act independently of the scheme, in exactly the same manner as in the unsigned case. They send their measurement choices a_n, b_n, and the corresponding outcomes u_n, v_n, where $u_n = s_n(a_n)$ and $v_n = s_n(b_n)$, to the target.
5. The **target** receives the streams of outcomes and measurement choices from Alice and Bob, and the stream of signs from the source. It uses these to compute the signed relative frequencies $r_n^{\pm}(uv|xy)$ for joint outcomes uv given measurement choices xy.
6. The limiting value of these signed relative frequencies $r_n^{\pm}(uv|xy)$ is taken to be the probability $p(uv|xy)$.

The stream of data received at the target comprises items of the form

$$(uv)^{\varsigma}xy \in (O \times O)^{\pm} \times X_A \times X_B, \qquad \varsigma \in \{+, -\}.$$

This is determined by the signed instruction set generated at the source, and the measurement choices made by Alice and Bob, as specified by the function

$$g : \mathcal{I}^{\pm} \times X_A \times X_B \longrightarrow (O \times O)^{\pm} \times X_A \times X_B :: (s^{\varsigma}, x, y) \mapsto ((s(x), s(y))^{\varsigma}, x, y).$$

We can use this function to push forward the measure $\mu^{\pm}$ on $\mathcal{I}^{\pm} \times X_A \times X_B$, defined by

$$\mu^{\pm}(s^{\varsigma}, x, y) := d(s^{\varsigma})d_{AB}(xy)$$

to $\nu^{\pm} := \mathsf{Prob}^{\pm}(g)(\mu^{\pm})$ on $(O \times O)^{\pm} \times X_A \times X_B$. This measure is defined explicitly by:

$$\nu^{\pm}((uv)^{\varsigma}xy) = d(U^{\varsigma}(uvxy))d_{AB}(xy).$$

As in the unsigned case, because of the product form of this measure, which corresponds to the independence of the measurement choices from the source distribution, conditioning on measurement choices xy leads to the probability $d(U^{\varsigma}(uvxy))$ for signed outcomes $(uv)^{\varsigma}$.

We define the following i.i.d. sequences of random variables:

$$\mathbf{P}_n^+ := \mathbb{1}_{U^+(uvxy)}, \quad \mathbf{P}_n^- := \mathbb{1}_{U^-(uvxy)}, \quad \mathbf{Q}_n^+ := \mathbb{1}_{\mathcal{I}^+}, \quad \mathbf{Q}_n^- := \mathbb{1}_{\mathcal{I}^-}.$$

The process of forming signed relative frequencies $r_n^\pm(uv|xy)$ at the target is modelled by the sequence of random variables $\mathbf{S}_n$, where:

$$\mathbf{S}_n := \frac{\sum_{i=1}^{n} \mathbf{P}_n^+ - \sum_{i=1}^{n} \mathbf{P}_n^-}{\sum_{i=1}^{n} \mathbf{Q}_n^+ - \sum_{i=1}^{n} \mathbf{Q}_n^-}.$$

The correctness of our simulation is now expressed by the following result.

Theorem 2. *For all* u, v, x, y:

$$\lim_{n\to\infty} \mathbf{S}_n =_{a.e.} p(uv|xy).$$

Proof.   By the Strong Law of Large Numbers,

$$\lim_{n\to\infty} \frac{1}{n} \sum_{i=1}^{n} \mathbf{P}_i^+ =_{a.e.} \mathbf{E}(\mathbb{1}_{U^+(uvxy)}) = d(U^+(uvxy)).$$

Unpacking this more carefully (see e.g. [4, Theorem 6.1, p.85]), the random variables $\mathbf{P}_n$ act on the probability space $\mathsf{Seq} = (\mathcal{I}^\pm)^\omega$, the product of countably many copies of $\mathcal{I}^\pm$, with product measure d^ω. The action is given by:

$$\mathbf{P}_n = \mathbb{1}_{U^+(uvxy)} \circ \pi_n.$$

The Strong Law asserts that, for some set Z_1 of measure zero in Seq, for all $\sigma \in \mathsf{Seq} \setminus Z_1$:

$$\lim_{n\to\infty} \frac{1}{n} \sum_{i=1}^{n} \mathbf{P}_i^+(\sigma) = d(U^+(uvxy)).$$

Similarly, outside sets Z_2, Z_3, Z_4 of measure 0, we have

$$\lim_{n\to\infty} \frac{1}{n} \sum_{i=1}^{n} \mathbf{P}_i^- = d(U^-(uvxy))$$

$$\lim_{n\to\infty} \frac{1}{n} \sum_{i=1}^{n} \mathbf{Q}_i^+ = d(\mathcal{I}^+)$$

$$\lim_{n\to\infty} \frac{1}{n} \sum_{i=1}^{n} \mathbf{Q}_i^- = d(\mathcal{I}^-).$$

Since $d = \theta_{\mathcal{I}}(m)$,

$$d(\mathcal{I}^+) - d(\mathcal{I}^-) = W(d) > 0,$$

and hence for all $\sigma \in \mathsf{Seq} \setminus (Z_3 \cup Z_4)$, for all but finitely many n:

$$\frac{1}{n} \sum_{i=1}^{n} \mathbf{Q}_i^+(\sigma) - \frac{1}{n} \sum_{i=1}^{n} \mathbf{Q}_i^-(\sigma) > 0.$$

Now $Z := Z_1 \cup Z_2 \cup Z_3 \cup Z_4$ has measure 0, and for all $\sigma \in \mathsf{Seq} \setminus Z$, by standard pointwise properties of limits:

$$
\begin{aligned}
\lim_{n \to \infty} \mathbf{S}_n(\sigma) &= \lim_{n \to \infty} \frac{\frac{1}{n}(\sum_{i=1}^{n} \mathbf{P}_i^+(\sigma) - \sum_{i=1}^{n} \mathbf{P}_i^-(\sigma))}{\frac{1}{n}(\sum_{i=1}^{n} \mathbf{Q}_i^+(\sigma) - \sum_{i=1}^{n} \mathbf{Q}_i^-(\sigma))} \\
&= \frac{d(U^+(uvxy)) - d(U^-(uvxy))}{d(\mathcal{I}^+) - d(\mathcal{I}^-)} \\
&= \eta(\mathsf{Prob}^{\pm}(\rho_{xy}^{\mathcal{I}})(\theta_X(m)))(uv) \\
&= \mathcal{M}(\rho_{xy}^{\mathcal{I}})(m)(uv) \qquad \qquad \text{by Proposition 1} \\
&= m|_{xy}(uv) \\
&= p(uv|xy).
\end{aligned}
$$

$\square$

5   Discussion

A first point to make is that the scheme we described in the previous section was formulated for systems of type $(2, 2, 2)$; that is, with two agents, two measurements per agent, and two outcomes per measurement. This was to avoid notational complications. It is clear that the same scheme would apply to Bell-type systems with any numbers of agents, measurement settings and outcomes. It is less clear how to proceed with other kinds of contextuality scenarios, although the result in Theorem 1 certainly applies to such scenarios.

The interpretation we have given of negative probabilities is operational in nature. It can be implemented in a physical scheme as summarized in the signed instruction set diagram in Figure 2. However, one should think of this scheme as a **simulation**, rather than a direct description of a fundamental physical process. The fact that it applies to arbitrary no-signalling systems, including superquantum devices such as PR boxes, which are generally believed to go beyond what is physically realizable, compels caution in this respect.

At the same time, the nature of the simulation, which respects relativistic constraints and uses only classical probabilistic devices, provides interesting food for thought. After all, this is a concrete way of thinking about entanglement, and even superquantum correlations, in terms of familiar-seeming devices: one can e.g. think of the source as generating its stream of signed particles by drawing coloured billiard balls from an urn. The subsequent passages of the particles are entirely classical. The only non-standard element of the process is the cancellation of positive and negative events effected by forming the signed relative frequencies. Can one find some structural features within this mode of description of non-local correlations which can help to delineate the quantum/superquantum boundary?

Among the features which it may be interesting to study from this point of view are the rates of convergence of the stochastic processes described in the previous section. If cancellation of positive events by negative ones can occur with unbounded delays, there may be some form of **retrocausality** hidden in the computation of the signed relative frequencies. Do quantum processes admit bounds on cancellation which ensure that causal anomalies do not arise? It may also be interesting to compare the entropies of the simulating (unsigned) and simulated (signed) processes. Computational efficiency may also provide a useful perspective.

While we certainly do not claim to have solved any mysteries, we hope to have provided a novel way of thinking about some of the mysterious features of the quantum — and even the superquantum — world.

References

1. Abramsky, S., Brandenburger, A.: The sheaf-theoretic structure of non-locality and contextuality. New Journal of Physics 13(2011), 113036 (2011)
2. Abramsky, S., Brandenburger, A., Savochkin, A.: No-Signalling is Equivalent to Free Choice of Measurements. In: Proceedings of QPL 2013 (2013)
3. Bell, J.S.: On the Einstein-Podolsky-Rosen paradox. Physics 1(3), 195–200 (1964)
4. Billingsley, P.: Probability and Measure, 3rd edn. Wiley (2008)
5. Burgin, M.: Interpretations of negative probabilities. arXiv preprint 1008.1287 (2010)
6. Dickson, W.M.: Quantum chance and non-locality. Cambridge University Press (1999)
7. Dirac, P.A.M.: The physical interpretation of quantum mechanics. Proceedings of the Royal Society of London. Series A, Mathematical and Physical Sciences 180(980), 1–40 (1942)
8. Feynman, R.P.: Negative probability. In: Hiley, B.J., Peat, F.D. (eds.) Quantum Implications: Essays in Honour of David Bohm, pp. 235–248. Routledge and Kegan Paul (1987)
9. Mermin, N.D.: Quantum mysteries revisited. Am. J. Phys 58(8), 731–734 (1990)
10. Popescu, S., Rohrlich, D.: Quantum nonlocality as an axiom. Foundations of Physics 24(3), 379–385 (1994)
11. Tsirelson, B.S.: Quantum Generalizations of Bell's Inequality. Letters in Mathematical Physics 4, 93–100 (1980)
12. Wigner, E.: On the quantum correction for thermodynamic equilibrium. Physical Review 40(5), 749 (1932)

Bisimulation on Markov Processes over Arbitrary Measurable Spaces

Giorgio Bacci, Giovanni Bacci, Kim Guldstrand Larsen, and Radu Mardare

Department of Computer Science, Aalborg University, Denmark
{grbacci,giovbacci,kgl,mardare}@cs.aau.dk

Abstract. We introduce a notion of bisimulation on labelled Markov Processes over generic measurable spaces in terms of arbitrary binary relations. Our notion of bisimulation is proven to coincide with the coalgebraic definition of Aczel and Mendler in terms of the Giry functor, which associates with a measurable space its collection of (sub)probability measures. This coalgebraic formulation allows one to relate the concepts of bisimulation and event bisimulation of Danos et al. (i.e., cocongruence) by means of a formal adjunction between the category of bisimulations and a (full sub)category of cocongruences, which gives new insights about the real categorical nature of their results. As a corollary, we obtain sufficient conditions under which state and event bisimilarity coincide.

1 Introduction

The first notion of bisimulation for Markov processes, that are, probabilistic systems with a generic measurable space of states, has been defined categorically by Blute, Desharnais, Edalat, and Panangaden in [1] as a span of zig-zag morphisms, i.e., measurable surjective maps preserving the transition structure of the process. It turned out to be very difficult to prove that the induced notion of bisimilarity is an equivalence and this was only solved under the technical assumption that the state space is analytic. Under this hypothesis in [5] it was proposed a neat logical characterization of bisimilarity, using a much simpler logic than the one previously used for the discrete case.

In [3], Danos, Desharnais, Laviolette, and Panangaden introduced a notion alternative to that of bisimulation, the so called *event bisimulation* (i.e., cocongruence). From a categorical perspective the novelty was that they worked with cospans rather than spans. Remarkably, they were able to give a logical characterization of event bisimilarity without assuming the state space of the Markov processes to be analytic. In addition, they proved that, for analytic spaces, event and state bisimilarity coincide.

It has been always an open question whether the analyticity assumption on the state space can be dropped. In this paper we make a step forward in this direction, providing a notion of state bisimulation on Markov processes over arbitrary measurable spaces. This definition is based on a characterization due to Desharnais et al. [6,7], which mimics the definition probabilistic bisimulation of Larsen and Skou [11] by adding a few measure-theoretic conditions to deal

F. van Breugel et al. (Eds.): Panangaden Festschrift, LNCS 8464, pp. 76–95, 2014.

with the fact that some subsets may not be measurable. Their characterization was given assuming that the bisimulation relation is an equivalence, instead, our definition is expressed in terms of arbitrary binary relations. This mild generalization to binary relation allows us to prove that our definition coincides with the coalgebraic notion of bisimulation of Aczel and Mendler in terms of the Giry functor [8], which associates with a measurable space its collection of subprobability measures. A similar result was proven by de Vink and Rutten [4] who studied Markov processes on ultrametric spaces. However, in [4] the characterization was established assuming that bisimulation relations have a Borel decomposition (which is not a mild assumption). Our proof does not need the existence of Borel decompositions and can be used to refine the result in [4].

Our characterization of probabilistic bisimulation is weaker than the original proposal in [1,5] (which requires the relation to be an equivalence) and weaker than the definition in [15] (which only requires the relation to be reflexive). However, we show that our definition is a generalization of both, and we prove that, when we restrict our definition on the case of single Markov processes, all the results continue to hold, in particular that the class of bisimulations is closed under union and that bisimilarity it is an equivalence.

Another contribution of this paper is a formal coalgebraic analysis of the relationships between the notions of bisimulation and cocongruence on Markov processes. This is done by lifting a standard adjunction that occurs between spans and cospans in categories with pushouts and pullbacks. The lifting to the categories of bisimulations and cocongruences is very simple when the behaviour functor weakly-preserves pullbacks. Although Viglizzo [16] proved that the Giry functor does not enjoy this property, we managed to show that the lifting is possible when we restrict to a suitable (full-)subcategory of congruences. This restriction cannot be avoided, since Terraf [13,14] showed that state and event bisimilarity do not coincide on Markov processes over arbitrary measurable spaces. As a corollary, we establish sufficient conditions under which bisimulation and cocongruence coincide, and as an aside result, this adjunction explains at a more abstract categorical level all the results in [3] that relate state and event bisimulations. To the best of our knowledge, this result is new and, together with the counterexample given by Terraf, completes the comparison between these two notions of equivalence between Markov processes over arbitrary measurable spaces.

2   Preliminaries

Binary Relations. For a binary relation $\mathcal{R} \subseteq X \times Y$ we use $\pi_X \colon \mathcal{R} \to X$ and $\pi_Y \colon \mathcal{R} \to Y$ to denote the canonical projections of $\mathcal{R}$ on X and Y, respectively. Given $\mathcal{R} \subseteq X \times Y$ and $\mathcal{S} \subseteq Y \times Z$, we denote by $\mathcal{R}^{-1} = \{(y, x) \mid (x, y) \in \mathcal{R}\}$ the inverse of $\mathcal{R}$, and by $\mathcal{R}; \mathcal{S} = \{(x, z) \mid (x, y) \in \mathcal{R} \text{ and } (y, z) \in \mathcal{S} \text{ for some } y \in Y\}$ the composition of $\mathcal{R}$ and $\mathcal{S}$. We say that $\mathcal{R} \subseteq X \times Y$ is *z-closed* if, for all $x, x' \in X$ and $y, y' \in Y$, $(x, y), (x', y), (x', y') \in \mathcal{R}$ implies $(x, y') \in \mathcal{R}$; and we denote by $\mathcal{R}^*$, the *z-closure* of $\mathcal{R}$, i.e., the least z-closed relation that contains $\mathcal{R}$.

Note that any equivalence relation is z-closed, indeed one can informally see the z-closure property as a generalization of transitive closure on binary relations.

Measure Theory. A *field* over a set X is a nonempty family of subsets of X closed under complement and union. A σ-algebra over a set X is a field Σ_X such that is closed under countable union. The pair (X, Σ_X) is a *measurable space* and the elements of Σ_X are called *measurable sets*. A *generator* $\mathcal{F}$ for Σ_X is a family of subsets of X such that the smallest σ-algebra containing $\mathcal{F}$, denoted by $\sigma(\mathcal{F})$, is Σ_X.

Let (X, Σ_X), (Y, Σ_Y) be measurable spaces. A function $f \colon X \to Y$ is called *measurable* if $f^{-1}(E) = \{x \mid f(x) \in E\} \in \Sigma_X$, for all $E \in \Sigma_Y$ (notably, if Σ_Y is generated by $\mathcal{F}$, f is measurable iff $f^{-1}(F) \in \Sigma_X$, for all $F \in \mathcal{F}$). The family $\{E \subseteq Y \mid f^{-1}(E) \in \Sigma_X\}$, called the *final σ-algebra w.r.t. f*, is the largest σ-algebra over Y that renders f measurable. Dually, the family $\{f^{-1}(E) \mid E \in \Sigma_Y\}$ is called *initial σ-algebra w.r.t f*, and it is the smallest σ-algebra over X that makes f measurable. Initial and final σ-algebras generalize to families of maps in the obvious way.

A *measure* on (X, Σ_X) is a σ-additive function $\mu \colon \Sigma_X \to [0, \infty]$, that is, $\mu(\bigcup_{i \in I} E_i) = \sum_{i \in I} \mu(E_i)$ for all countable collections $\{E_i\}_{i \in I}$ of pairwise disjoint measurable sets. A measure $\mu \colon \Sigma_X \to [0, \infty]$ is of *(sub)-probability* if $\mu(X) = 1$ (resp. ≤ 1), is *finite* if $\mu(X) < \infty$, and is *σ-finite* if there exists a countable cover $\{E_i\}_{i \in I} \subseteq \Sigma_X$ of X, i.e., $\bigcup_{i \in I} E_i = X$, of measurable sets such that $\mu(E_i) < \infty$, for all $i \in I$. A *pre-measure* on a field $\mathcal{F}$ is a finitely additive function $\mu_0 \colon \mathcal{F} \to [0, \infty]$ with the additional property that whenever $F_0, F_1, F_2, \ldots$ is a countable disjoint collection sets in $\mathcal{F}$ such that $\bigcup_{n \in \mathbb{N}} F_n \in \mathcal{F}$, then $\mu_0(\bigcup_{n \in \mathbb{N}} F_n) = \sum_{n \in \mathbb{N}} \mu_0(F_n)$.

Coalgebras, Bisimulations, and Cocongruences. Let $F \colon \mathbf{C} \to \mathbf{C}$ be an endofunctor on a category $\mathbf{C}$. An *F-coalgebra* is a pair (X, α) consisting of an object X, called *carrier*, and an arrow $\alpha \colon X \to FX$ in $\mathbf{C}$, called *coalgebra structure*. An *homomorphism* between two F-coalgebras (X, α) and (Y, β) is an arrow $f \colon X \to Y$ in $\mathbf{C}$ such that $\alpha \circ f = Ff \circ \beta$. F-coalgebras and homomorphisms between them form a category, denoted by *F-**coalg***.

An *F-bisimulation* (R, f, g) between two F-coalgebras (X, α) and (Y, β) is a span $X \xleftarrow{f} R \xrightarrow{g} Y$ with jointly monic legs (a *monic span*) such that there exists a coalgebra structure $\gamma \colon R \to FR$ making f and g homomorphisms of F-coalgebras. Two F-coalgebras are *bisimilar* if there is a bisimulation between them. A notion alternative to bisimilarity that has been proven very useful in reasoning about probabilistic systems [3], is *cocongruence*. An *F-cocongruence* (K, f, g) between F-coalgebras (X, α) and (Y, β) is an cospan $X \xrightarrow{f} K \xleftarrow{g} Y$ with jointly epic legs (an *epic cospan*) such that there exists a (unique) coalgebra structure $\kappa \colon K \to FK$ on K such that f and g are F-homomorphisms.

3   Labelled Markov Processes and Bisimulation

In this section we recall the notions of labelled Markov kernels and processes, using a notation similar to [2], then we propose a general definition of (state) bisimulation between pairs of Markov kernels.

Let (X, Σ) be a measurable space. We denote by $\Delta(X, \Sigma)$ the set of all sub-probability measures over (X, Σ). For each $E \in \Sigma$ there is a canonical evaluation function $ev_E \colon \Delta(X, \Sigma) \to [0, 1]$, defined by $ev_E(\mu) = \mu(E)$, for all $\mu \in \Delta(X, \Sigma)$, and called *evaluation at E*. By means of these evaluation maps, $\Delta(X, \Sigma)$ can be organized into a measurable space $(\Delta(X, \Sigma), \Sigma_{\Delta(X,\Sigma)})$, where $\Sigma_{\Delta(X,\Sigma)}$ the initial σ-algebra with respect to $\{ev_E \mid E \in \Sigma\}$, i.e., the smallest σ-algebra making ev_E measurable w.r.t. the Borel σ-algebra on $[0, 1]$, for all $E \in \Sigma$. Equivalently, $\Sigma_{\Delta(X,\Sigma)}$ can be also generated by the sets $L_q(E) = \{\mu \in \Delta(X, \Sigma) \mid \mu(E) \geq q\}$, for $q \in [0, 1] \cap \mathbb{Q}$ and $E \in \Sigma$ (see [9]).

Definition 1. *Let (X, Σ) be a measurable space and L a set of action labels. An L-labelled Markov kernel is a tuple $\mathcal{M} = (X, \Sigma, \{\theta_a\}_{a \in L})$ where, for all $a \in L$*

$$\theta_a \colon X \to \Delta(X, \Sigma)$$

is a measurable function, called Markov a-transition function. An L-labelled Markov kernel $\mathcal{M}$ with a distinguished initial state $x \in X$, is said Markov process, and it is denoted by $(\mathcal{M}, x)$.

The labels in L constitute all possible interactions of the processes with the environment: for an action $a \in L$, a current state $x \in X$, and a measurable set $E \in \Sigma_X$, the value $\theta_a(x)(E)$ represents the probability of taking an a-transition from x to arbitrary elements in E.

Before presenting the notions of bisimulation and bisimilarity between labelled Markov kernels, we introduce some preliminary notation.

Definition 2 ($\mathcal{R}$-closed pair). *Let $\mathcal{R} \subseteq X \times Y$ be a relation on the sets X and Y, and $E \subseteq X$, $F \subseteq Y$. A pair (E, F) is $\mathcal{R}$-closed if $\mathcal{R} \cap (E \times Y) = \mathcal{R} \cap (X \times F)$.*

A pair (E, F) is $\mathcal{R}$-closed iff $\pi_X^{-1}(E) = \pi_Y^{-1}(F)$, where π_X, π_Y are the canonical projections on X and Y, respectively. The following lemmas will be useful later in the paper, and are direct consequences of the definition.

Lemma 3. *Let $\mathcal{R}' \subseteq \mathcal{R} \in X \times Y$. If (E, F) is $\mathcal{R}$-closed, then (E, F) is $\mathcal{R}'$-closed.*

Lemma 4. *Let $\mathcal{R} \subseteq X \times X$ be an equivalence relation. If (E, F) is $\mathcal{R}$-closed then $E = F$, moreover E is an union of $\mathcal{R}$-equivalence classes.*

Definition 5 (Bisimulation and Bisimilarity). *Let $\mathcal{M} = (X, \Sigma_X, \{\alpha_a\}_{a \in L})$ and $\mathcal{N} = (Y, \Sigma_Y, \{\beta_a\}_{a \in L})$ be two L-labelled Markov kernels. A binary relation $\mathcal{R} \subseteq X \times Y$ is a (state) bisimulation between $\mathcal{M}$ and $\mathcal{N}$ if, for all $(x, y) \in \mathcal{R}$, $a \in L$, and any pair $E \in \Sigma_X$ and $F \in \Sigma_Y$ such that (E, F) is $\mathcal{R}$-closed*

$$\alpha_a(x)(E) = \beta_a(y)(F).$$

Two L-labelled Markov processes $(\mathcal{M}, x)$ and $(\mathcal{N}, y)$ are (state) *bisimilar, written $x \sim_{(\mathcal{M},\mathcal{N})} y$, if the initial states x and y are related by some bisimulation $\mathcal{R}$ between $\mathcal{M}$ and $\mathcal{N}$.*

Originally, the definition of bisimulation between labelled Markov processes was given in terms of spans of zig-zag morphisms [1]. Later, Danos et al. [3] gave a more direct relational definition, called *state bisimulation*, characterizing the original zig-zag definition in the case of equivalence relations. Their definition differs from Definition 5 only on how $\mathcal{R}$-closed subsets are characterized and in that they require bisimulations to be equivalence relations[1]. Later, in [15], van Breugel et al. proposed a weaker definition, where bisimulation relations are only required to be reflexive. Definition 5 subsumes the definitions of state bisimulation given in [3] and [15] (this is a direct consequence of Lemma 4).

Proposition 6. *Let $\mathcal{F}$ be a family of bisimulation relations between the L-labelled Markov kernels $\mathcal{M} = (X, \Sigma_X, \{\alpha_a\}_{a \in L})$ and $\mathcal{N} = (Y, \Sigma_Y, \{\beta_a\}_{a \in L})$. Then $\bigcup \mathcal{F}$ is a bisimulation.*

Proof. Let $(x, y) \in \bigcup \mathcal{F}$, $a \in L$, and $E \in \Sigma_X$ and $F \in \Sigma_Y$ such that (E, F) is $\bigcup \mathcal{F}$-closed. By $(x, y) \in \bigcup \mathcal{F}$, there exists a bisimulation $\mathcal{R} \subseteq X \times Y$ such that $(x, y) \in \mathcal{R}$. Obviously $\mathcal{R} \subseteq \bigcup \mathcal{F}$, thus, by Lemma 3, (E, F) is $\mathcal{R}$-closed. Since $(x, y) \in \mathcal{R}$ and $\mathcal{R}$ is a bisimulation, we have $\alpha_a(x)(E) = \beta_a(y)(F)$.   □

Corollary 7. $\sim_{(\mathcal{M},\mathcal{N})}$ *is the largest bisimulation between $\mathcal{M}$ and $\mathcal{N}$.*

Proof. By definition $\sim_{(\mathcal{M},\mathcal{N})} = \bigcup \{\mathcal{R} \mid \mathcal{R} \text{ bisimulation between } \mathcal{M} \text{ and } \mathcal{N}\}$, thus, by Lemma 6 it is a bisimulation and in particular it is the largest one.   □

The following results proves that, if bisimilarity is restricted to single labelled Markov kernels, then it is an equivalence.

Theorem 8 (Equivalence). *Let $\mathcal{M} = (X, \Sigma, \{\theta_a\}_{a \in L})$ be an L-labelled Markov kernel. Then, the bisimilarity relation $\sim_{\mathcal{M}} \subseteq X \times X$ on $\mathcal{M}$ is an equivalence.*

Proof. Symmetry: if $\mathcal{R}$ is a bisimulation, so is $\mathcal{R}^{-1}$. Reflexivity: we prove that the identity relation Id_X is a bisimulation, i.e., for all $x \in X$, $a \in A$, and $E, F \in \Sigma$ such that (E, F) is Δ_X-closed, $\theta_a(x)(E) = \theta_a(x)(F)$. This holds trivially by Lemma 4, since Id_X is an equivalence. Transitivity: it suffices to show that, given $\mathcal{R}_1$ and $\mathcal{R}_2$ bisimulations on $\mathcal{M}$, there exists a bisimulation $\mathcal{R}$ on $\mathcal{M}$ that *contains* $\mathcal{R}_1; \mathcal{R}_2$. Let $\mathcal{R}$ be the (unique) smallest equivalence relation containing $\mathcal{R}_1 \cup \mathcal{R}_2$. $\mathcal{R}$ can be defined as $\mathcal{R} = Id_X \cup \bigcup_{n \in \mathbb{N}} \mathcal{S}_n$, where

$$\mathcal{S}_0 \triangleq \mathcal{R}_1 \cup \mathcal{R}_2 \cup \mathcal{R}_1^{-1} \cup \mathcal{R}_2^{-1} \qquad \qquad \mathcal{S}_{n+1} \triangleq \mathcal{S}_n; \mathcal{S}_n.$$

---

[1] Actually, in [3], the definition of state bisimulation is given without mentioning that the relation must be an equivalence, but without that requirement many subsequent results do not hold (e.g. Lemmas 4.1, 4.6, 4.8, Proposition 4.7, and Corollary 4.9). However, looking at the proofs it seems that they were imposing this condition.

It is easy to see that $\mathcal{R}_1; \mathcal{R}_2 \subseteq \mathcal{R}$. We are left to show that $\mathcal{R}$ is a bisimulation. By Lemma 4, it suffices to prove that for all $a \in L$, and measurable set $E \in \Sigma$ such that (E, E) is $\mathcal{R}$-closed, the following implication holds

$$(x, y) \in \mathcal{R} \implies \theta_a(x)(E) = \theta_a(y)(E). \tag{1}$$

If $(x, y) \in \mathcal{R}$, then $(x, y) \in Id_X$ or $(x, y) \in \mathcal{S}_n$ for some $n \geq 0$. If $(x, y) \in Id_X$ then $x = y$, hence (1) trivially holds. Now we show, by induction on $n \geq 0$, that

$$(x, y) \in \mathcal{S}_n \implies \theta_a(x)(E) = \theta_a(y)(E). \tag{2}$$

Base case ($n = 0$): let $j \in \{1, 2\}$. If $(x, y) \in \mathcal{R}_j$, (2) holds since, by hypothesis that $\mathcal{R}_j$ is a bisimulation and by Lemma 3, (E, E) is $\mathcal{R}_j$-closed. If $(x, y) \in \mathcal{R}_j^{-1}$ we have $(y, x) \in \mathcal{R}_j$, hence (2) holds again.

Inductive case ($n + 1$): if $(x, y) \in \mathcal{S}_{n+1}$, then there exists some $z \in X$ such that $(x, z) \in \mathcal{S}_n$ and $(z, y) \in \mathcal{S}_n$. Then, applying the inductive hypothesis twice, we have $\theta_a(x)(E) = \theta_a(z)(E)$ and $\theta_a(z)(E) = \theta_a(y)(E)$, from which (2) follows. $\square$

Remark 9. Theorem 8 has been already stated by van Breugel et al. [15] considering a more restrictive definition of bisimulation than the one given in Definition 1. Although the result is not new, we put the proof here to show that it can be carried out in a much simpler way.

Moreover, notice that, in the proof of Theorem 8 transitivity is verified using a strategy that avoids to prove that bisimulation relations are closed under composition. Indeed, this would have required that (semi-)pullbacks of relations in **Meas** are weakly preserved by the Giry functor. Recently, in [13,14] Terraf showed that this is not the case. The proof is based on the existence of a non-Lebesgue-measurable set V in the open unit interval $(0, 1)$, which is used to define two measures on the σ-algebra extended with V such that they differ in this set. In the light of this, the simplicity of the proof of Theorem 8 is even more remarkable. ∎

By Corollary 7 we have the following direct characterization for bisimilarity.

Proposition 10. *Let* $\mathcal{M} = (X, \Sigma, \{\theta_a\}_{a \in L})$ *be an L-labelled Markov kernel, then, for $x, y \in X$, $x \sim_{\mathcal{M}} y$ iff for all $a \in L$ and $E \in \Sigma$ such that (E, E) is $\sim_{\mathcal{M}}$-closed, $\theta_a(x)(E) = \theta_a(y)(E)$.*

Interestingly, Theorem 8 can be alternatively proven as a corollary of this result; indeed the above characterization implies that $\sim_{\mathcal{M}}$ is an equivalence relation.

Another important property, which will be used later in the paper, is that the class of bisimulations is closed by z-closure.

Lemma 11 ([4]). *If R is a state bisimulation between $(X, \Sigma_X, \{\alpha_a\}_{a \in L})$ and $(Y, \Sigma_Y, \{\beta_a\}_{a \in L})$, then so is R^*, the z-closure of R.*

4   Characterization of the Coalgebraic Bisimulation

In this section, we prove that the notion of state bisimulation (Definition 5) coincides with the abstract coalgebraic definition of Aczel and Mendler.

In order to model Markov processes as coalgebras the most natural choice for a category is **Meas**, the category of measurable spaces and measurable maps. This category is complete and cocomplete: limits and colimits are obtained as in **Set** and endowed, respectively, with initial and final σ-algebra w.r.t. their cone and cocone maps. Hereafter, for the sake of readability, we adopt a notation that makes no distinction between measurable spaces and objects in **Meas**: by the boldface symbol **X** we denote the measurable space (X, Σ_X) (the subscript is used accordingly).

Let L be a set of action labels. L-labelled Markov kernels are standardly modeled as Δ^L-coalgebras, where $(\cdot)^L \colon \mathbf{Meas} \to \mathbf{Meas}$ is the exponential functor and $\Delta \colon \mathbf{Meas} \to \mathbf{Meas}$ is the Giry functor acting on objects **X** and arrows $f \colon \mathbf{X} \to \mathbf{Y}$ as follows, for $\mu \in \Delta(X, \Sigma_X)$

$$\Delta\mathbf{X} = (\Delta(X, \Sigma_X), \Sigma_{\Delta(X,\Sigma_X)}) \qquad \Delta(f)(\mu) = \mu \circ f^{-1}.$$

It is folklore that L-labelled Markov kernels and Δ^L-coalgebras coincide. For $\mathbf{X} = (X, \Sigma_X)$ a measurable space, the correspondence is given by

$$(X, \Sigma_X, \{\theta_a\}_{a \in L}) \mapsto (\mathbf{X}, \alpha \colon \mathbf{X} \to \Delta^L \mathbf{X}), \qquad \text{where } \alpha(x)(a) = \theta_a(x), \quad (3)$$
$$(\mathbf{X}, \alpha \colon \mathbf{X} \to \Delta^L \mathbf{X}) \mapsto (X, \Sigma_X, \{ev_a \circ \alpha\}_{a \in L}). \qquad (4)$$

Where, in (3) α is measurable since $ev_E \circ ev_a \circ \alpha$ is measurable, for all $a \in L$ and $E \in \Sigma$, and in (4), $ev_a \circ \alpha$ is measurable since is the composite of measurable functions. Hereafter, we will make no distinction between Δ^L-coalgebras and L-labelled Markov kernels, and the correspondence above will be used without reference.

Next we relate the notion of Δ^L-bisimulation to the notion of state bisimulation. Recall that in categories with binary products monic spans (R, f, g) are in one-to-one correspondence with monic arrows $R \to X \times Y$. Thus, without loss of generality, we restrict our attention only to relations $R \subseteq X \times Y$ with measurable canonical projections $\pi_X \colon \mathbf{R} \to \mathbf{X}$ and $\pi_Y \colon \mathbf{R} \to \mathbf{Y}$.

Proposition 12. *Let $(\mathbf{X}, \alpha)$ and $(\mathbf{Y}, \beta)$ be Δ^L-coalgebras and $(\mathbf{R}, \pi_X, \pi_Y)$ be a Δ^L-bisimulation between them. Then, $R \subseteq X \times Y$ is a state bisimulation between $(X, \Sigma_X, \{ev_a \circ \alpha\}_{a \in L})$ and $(Y, \Sigma_Y, \{ev_a \circ \beta\}_{a \in L})$.*

Proof. We have to show that for all $(x, y) \in R$, $a \in L$ and $E \in \Sigma_X$, $F \in \Sigma_Y$ such that (E, F) is R-closed, $\alpha(x)(a)(E) = \beta(y)(a)(F)$. Since $(\mathbf{R}, \pi_X, \pi_Y)$ is a Δ^L-bisimulation, there exists a coalgebraic structure $\gamma \colon \mathbf{R} \to \Delta^L \mathbf{R}$ on $\mathbf{R}$ such that π_X and π_Y are Δ^L-homomorphisms.

$$
\begin{aligned}
\alpha(x)(a)(E) &= (\alpha \circ \pi_X)(x,y)(a)(E) && \text{(by def. } \pi_X) \\
&= (\Delta^L \pi_X \circ \gamma)(x,y)(a)(E) && \text{(by } \Delta^L\text{-homomorphism)} \\
&= \Delta\pi_X(\gamma(x,y)(a))(E) && \text{(by def. } Id^L) \\
&= \gamma(x,y)(a) \circ \pi_X^{-1}(E) && \text{(by def. } \Delta) \\
&= \gamma(x,y)(a) \circ \pi_Y^{-1}(F) && \text{(by } (E,F) \text{ } R\text{-closed)} \\
&= \beta(y)(a)(F) && \text{(by reversing the previous steps)}
\end{aligned}
$$

$\square$

The converse of Proposition 12 is more intricate, and we need some preliminary work involving techniques from measure theory in order to formally define a suitable mediating measurable Δ^L-coalgebra structure.

Proposition 13. *Let (X, Σ_X) and (Y, Σ_Y) be measurable spaces, $r_1 \colon R \to X$, $r_2 \colon R \to Y$ be surjective maps, and Σ_R be the initial σ-algebra on R w.r.t. r_1 and r_2, i.e., $\Sigma_R = \sigma(\{r_1^{-1}(E) \mid E \in \Sigma_X\} \cup \{r_2^{-1}(F) \mid F \in \Sigma_Y\})$.*
Then, for any pair of measures μ on (X, Σ_X) and ν on (Y, Σ_Y), such that, for all $E \in \Sigma_X$ and $F \in \Sigma_Y$,

$$
\text{if } r_1^{-1}(E) = r_2^{-1}(F), \text{ then } \mu(E) = \nu(F),
$$

there exists a measure $\mu \wedge \nu$ on (R, Σ_R) such that, for all $E \in \Sigma_X$ and $F \in \Sigma_Y$

$$
(\mu \wedge \nu)(r_1^{-1}(E)) = \mu(E) \qquad \text{and} \qquad (\mu \wedge \nu)(r_2^{-1}(F)) = \mu(F).
$$

Moreover, if μ and ν are σ-finite, $\mu \wedge \nu$ is unique.

The existence and uniqueness of $\mu \wedge \nu$ is guaranteed by the Hahn-Kolmogorov extension theorem. Note that, the conditions imposed on μ and ν are necessary for $\mu \wedge \nu$ to be well-defined (see the appendix for a detailed proof).

Thank to Proposition 13 we can prove the following result, which concludes the correspondence between state bisimulation and Δ^L-bisimulation.

Proposition 14. *Let $R \subseteq X \times Y$ be a state bisimulation between the L-labelled Markov kernels $(X, \Sigma_X, \{\alpha_a\}_{a \in L})$ and $(Y, \Sigma_Y, \{\beta_a\}_{a \in L})$. Then $(\mathbf{R}, \pi_X, \pi_Y)$ is a Δ^L-bisimulation between $(\mathbf{X}, \alpha)$ and $(\mathbf{Y}, \beta)$, where Σ_R is initial w.r.t. π_X and π_Y, and for all $a \in L$, $x \in X$, and $y \in Y$, $\alpha(x)(a) = \alpha_a(x)$ and $\beta(y)(a) = \beta_a(y)$.*

Proof. We have to provide a measurable coalgebra structure $\gamma \colon \mathbf{R} \to \Delta^L \mathbf{R}$ making $\pi_X \colon \mathbf{R} \to \mathbf{X}$ and $\pi_Y \colon \mathbf{R} \to \mathbf{Y}$ Δ^L-homomorphisms.

First notice that, without loss of generality we can assume π_X and π_Y to be surjective. Indeed, if it is not so, we can factorize π_X and π_Y as $\pi_X = m_X \circ e_X$,

$\pi_Y = m_Y \circ e_Y$ such that e_X, e_Y are surjective (epic) and m_X, m_Y are injective (monic), to obtain the following commuting diagrams

$$
\begin{array}{ccccccccc}
\mathbf{X} & \xleftarrow{m_X} & \mathbf{X}' & \xleftarrow{e_X} & \mathbf{R} & \xrightarrow{e_Y} & \mathbf{Y}' & \xrightarrow{m_Y} & \mathbf{Y} \\
{\scriptstyle\alpha}\downarrow & & {\scriptstyle\alpha'}\downarrow & & {\scriptstyle\beta'}\downarrow & & & & \downarrow{\scriptstyle\beta} \\
\Delta^L\mathbf{X} & \xleftarrow{\Delta m_X} & \Delta^L\mathbf{X}' & \xleftarrow{\Delta e_X} & \Delta^L\mathbf{R} & \xrightarrow{\Delta e_Y} & \Delta^L\mathbf{Y}' & \xrightarrow{\Delta m_Y} & \Delta^L\mathbf{Y}
\end{array}
$$

where $\mathbf{X}' = (X', \Sigma_{X'})$, X' is the image $\pi_X(R)$, $\Sigma_{X'}$ is the initial σ-algebra w.r.t. m_X (i.e., $\{m_X^{-1}(E) \mid E \in \Sigma_X\}$), and $\alpha'(x')(a)(m_X^{-1}(E)) = \alpha(m_X(i))(a)(E)$, for all $x' \in X'$ and $E \in \Sigma_X$; (similarly for $\mathbf{Y}$ and β'). Therefore, to find the coalgebra structure γ for making e_X and e_Y Δ^L-homomorphisms, it solves the problem for π_X and π_Y as well.

Recall that (E, F) is R-closed iff $\pi_X^{-1}(E) = \pi_Y^{-1}(F)$. By hypothesis, R is a state bisimulation, so that, for all $(x, y) \in R$, $E \in \Sigma_X$, $F \in \Sigma_Y$, and $a \in L$,

$$\text{if } \pi_X^{-1}(E) = \pi_Y^{-1}(F), \qquad \text{then} \qquad \alpha(a)(x)(E) = \beta(a)(y)(F).$$

For all $(x, y) \in R$ and $a \in L$, by Proposition 13, we define $\gamma((x, y))(a)$ as the unique (sub)probability measure on (R, Σ_R), s.t., for all $E \in \Sigma_X$ and $F \in \Sigma_Y$,

$$\gamma((x, y))(a)(\pi_X^{-1}(E)) = \alpha(x)(a)(E), \quad \text{and} \quad \gamma((x, y))(a)(\pi_Y^{-1}(F)) = \beta(y)(a)(F).$$

By definition of γ, both π_X and π_Y are Δ^L-homomorphisms, indeed,

$$
\begin{aligned}
(\Delta^L \pi_X \circ \gamma)((x, y))(a)(E) &= \gamma((x, y))(a)(\pi_X^{-1}(E)) && \text{(by def. } \Delta^L) \\
&= \alpha(x)(a)(E) && \text{(by def. } \gamma) \\
&= (\alpha \circ \pi_X)((x, y))(a)(E). && \text{(by def. } \pi_X)
\end{aligned}
$$

for all $(x, y) \in R$, $a \in A$, and $E \in \Sigma_X$. The proof for π_Y is similar.

To prove that γ is measurable, by [9, Lemma 4.5] it suffices to show that for any finite union of the form $S = \bigcup_{i=0}^{k}(\pi_X^{-1}(E_i) \cap \pi_Y^{-1}(F_i))$, where $E_i \in \Sigma_X$ and $F_i \in \Sigma_Y$, for $0 \leq i \leq k$, $(ev_a \circ \gamma)^{-1}(L_q(S)) \in \Sigma_R$. We may assume, without loss of generality, that S is given as a disjoint union (otherwise we may represent it taking a disjoint refinement), and by finite additivity, it suffices to consider only the case $S = \pi_X^{-1}(E') \cap \pi_Y^{-1}(F')$, for some $E' \in \Sigma_X$ and $F' \in \Sigma_Y$. According to the definition of γ (see Proposition 13), we have to consider three cases:

− if $\exists E \in \Sigma_X$ such that $S = \pi_X^{-1}(E)$, then

$$
\begin{aligned}
(ev_a \circ \gamma)^{-1}(L_q(S)) &= \{(x, y) \in R \mid \gamma((x, y))(a) \in L_q(S)\} && \text{(by inverse image)} \\
&= \{(x, y) \in R \mid \gamma((x, y))(a)(S) \geq q\} && \text{(by def. } L_q(\cdot)) \\
&= \{(x, y) \in R \mid \alpha(x)(a)(E) \geq q\} && \text{(by def. } \gamma) \\
&= (\pi_X \circ ev_a \circ \alpha)^{-1}(L_q(E)) && \text{(by inverse image)}
\end{aligned}
$$

which is a measurable set in Σ_R, since π_X, ev_a, α are measurable;

– if $\exists F \in \Sigma_Y$, $S = \pi_Y^{-1}(F)$, and $\forall E \in \Sigma_X$, $S \neq \pi_X^{-1}(E)$ one can proceed similarly, replacing in the above derivation E, π_X, and $\alpha(x)$ by F, π_Y and $\beta(y)$, respectively.

– if $\forall E \in \Sigma_X, F \in \Sigma_Y$, $S \neq \pi_X^{-1}(E)$ and $S \neq \pi_Y^{-1}(F)$, then

$$
\begin{aligned}
(ev_a \circ \gamma)^{-1}(L_q(S)) &= \{(x,y) \in R \mid \gamma((x,y))(a) \in L_q(S)\} && \text{(by inverse image)}\\
&= \{(x,y) \in R \mid \gamma((x,y))(a)(S) \geq q\} && \text{(by def. } L_q(\cdot))\\
&= \{(x,y) \in R \mid 0 \geq q\} && \text{(by def. } \gamma)\\
&= R && \text{(by } q \in [0,1] \cap \mathbb{Q})
\end{aligned}
$$

$\square$

Remark 15 (Ultrametric spaces). A similar result was proven by de Vink and Rutten [4] in the setting of ultrametric spaces and non-expansive maps. A characterization for the coalgebraic definition of bisimulation in the continuous case was established under the assumption that the bisimulation relation has a Borel decomposition. Proposition 13 does not need such an extra assumption and it holds for Borel measures as well, so that the proof-strategy of Proposition 14 can be used to drop the assumption in [4, Theorem 5.8]. ■

Theorem 16. *State bisimulation and Δ^L-bisimulation coincide.*

Proof. Direct consequence of Propositions 12, and 14. $\square$

5   Relating Bisimulations and Cocongruences

In this section, we compare the notions of Δ^L-bisimulation and Δ^L-cocongruence. We do this establishing an adjunction between the category of Δ^L-bisimulations and a (suitable) subcategory of Δ^L-cocongruences. As a result of this adjunction, we obtain also sufficient conditions under which the notions of Δ^L-bisimulation and Δ^L-cocongruence coincide.

The correspondence between Δ^L-bisimulation and Δ^L-cocongruence is based on a standard adjunction between span and cospans in categories with pushouts and pullbacks, which we briefly recall. The category $\mathbf{MSp_C}(X, Y)$ has as objects monic spans (R, f, g) between X, Y in $\mathbf{C}$, and arrows $f \colon (R, r_1, r_2) \to (S, s_1, s_2)$ which are morphisms $f \colon R \to S$ in $\mathbf{C}$ such that $s_i \circ f = r_i$, for all $i \in \{1, 2\}$. The category $\mathbf{ECoSp_C}(X, Y)$, of epic cospans between X, Y in $\mathbf{C}$, is defined analogously. In the following, we will omit the subscript $\mathbf{C}$ when the category of reference is understood.

If the category $\mathbf{C}$ has pullbacks and pushouts, we can define two functors: $Pb_{(X,Y)} \colon \mathbf{ECoSp}(X, Y) \to \mathbf{MSp}(X, Y)$ mapping each epic cospan to its pullback, and $Po_{(X,Y)} \colon \mathbf{MSp}(X, Y) \to \mathbf{ECoSp}(X, Y)$ mapping each monic span to its pushout. As for morphisms, let $f \colon (R, r_1, r_2) \to (S, s_1, s_2)$ be a morphism in $\mathbf{MSp}(X,Y)$, then $Po_{(X,Y)}(f)$ is defined as the unique arrow, given by the universal property of pushout, making the following diagram commute

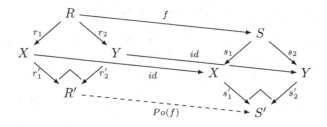

where $Po_{(X,Y)}(R,r_1,r_2) = (R', r_1', r_2')$ and $Po_{(X,Y)}(S, s_1, s_2) = (S', s_1', s_2')$. The action on arrows for $Pb_{(X,Y)}$ is defined similarly, using the universal property of pullbacks. When the domain (X,Y) of the spans and cospans is clear, the subscript in $Po_{(X,Y)}$ and $Pb_{(X,Y)}$ will be omitted. The following is standard:

Lemma 17. *Let $\mathbf{C}$ be a category with pushouts and pullbacks, then*

(i) $Po \dashv Pb$;
(ii) $PoPbPo \cong Po$ *and* $PbPoPb \cong Pb$.

The unit $\eta\colon Id \Rightarrow PbPo$ and counit $\epsilon\colon PoPb \Rightarrow Id$ of the adjunction $Po \dashv Pb$, are given component-wise as follows, for (R, r_1, r_2) in $\mathbf{MSp}(X,Y)$ and (K, k_1, k_2) in $\mathbf{ECoSp}(X,Y)$.

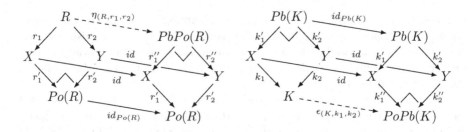

The adjunction $Po \dashv Pb$ induces a monad $(PbPo, \eta, Pb\epsilon Po)$ in $\mathbf{MSp}(X,Y)$ and a comonad $(PoPb, \epsilon, Po\eta Pb)$ in $\mathbf{ECoSp}(X,Y)$, which, by Lemma 17, are idempotent.

Since $\mathbf{Meas}$ has both pushouts and pullbacks, the construction above can be instantiated in this category. Note that, $\mathbf{Meas}$ has binary products and co-products, hence we can identify the categories $\mathbf{MSp}(\mathbf{X}, \mathbf{Y})$ and $\mathbf{ECoSp}(\mathbf{X}, \mathbf{Y})$, respectively, as the categories of relations $R \subseteq X \times Y$ (with measurable canonical projections) and quotients $(X + Y)/E$ (with measurable canonical injections), where E is an equivalence relation on $(X + Y)$.

Moreover, in $\mathbf{Meas}$ it holds that $PoPb \cong Id$ and the composite functor $PbPo$ has the following explicit description. Let $R \subseteq X \times Y$ be an object in $\mathbf{MSp}(\mathbf{X}, \mathbf{Y})$, then $PbPo(R) = R^*$ where $R^* \subseteq X \times Y$ is the *z-closure of R*, with

σ-algebra Σ_{R^*} given as the initial σ-algebra w.r.t. the canonical projections; arrows $f \colon R \to S$ between objects $R, S \subseteq X \times Y$ in $\mathbf{MSp}(\mathbf{X}, \mathbf{Y})$ are mapped to $f^* \colon R^* \to S^*$, the z-closure extension of f, defined in the obvious way.

The monad $PbPo$ in $\mathbf{MSp}(\mathbf{X}, \mathbf{Y})$ has unit $\eta \colon Id \Rightarrow PbPo$ given by the natural inclusion and multiplication $Pb\epsilon Po \colon PbPoPbPo \Rightarrow PbPo$ defined as the "identity" natural transformation. The comonad $PoPb$ in $\mathbf{ECoSp}(\mathbf{X}, \mathbf{Y})$ has counit $\epsilon \colon PoPb \Rightarrow Id$ and comultiplication $Po\eta Pb \colon PoPb \Rightarrow PoPbPoPb$ given as the "identity" natural transformations.

5.1 Adjunction between Bisimulations and Cocongruences

The adjunction $Po \dashv Pb$ over monic span and epic cospans in $\mathbf{Meas}$ can be partially lifted to an adjunction $\overline{Po} \dashv \overline{Pb}$ between the categories of Δ^L-bisimulations and Δ^L-cocongruences. The term "partially" is used since the adjunction can be established only restricting the category of Δ^L-cocongruences to the image given by the lifting $\overline{Po}$. Moreover, we show that the subcategories given by the images of $\overline{Pb}\,\overline{Po}$ and $\overline{Po}\,\overline{Pb}$ are equivalent. This provides sufficient conditions under which the notions of bisimulation and cocongruence coincide.

We denote by $\mathbf{Bisim}((\mathbf{X}, \alpha), (\mathbf{Y}, \beta))$ the category with Δ^L-bisimulations $((\mathbf{R}, \gamma_R), f, g)$ between Δ^L-coalgebras $(\mathbf{X}, \alpha)$ and $(\mathbf{Y}, \beta)$ as objects and arrows $f \colon ((\mathbf{R}, \gamma_R), r_1, r_2) \to ((\mathbf{S}, \gamma_S), s_1, s_2)$ which are morphisms $f \colon \mathbf{R} \to \mathbf{S}$ in $\mathbf{Meas}$ such that $s_i \circ f = r_i$, for all $i \in \{1, 2\}$, and $\gamma_S \circ f = \Delta^L f \circ \gamma_R$, i.e, f is a morphism both in Δ^L-$\mathbf{coalg}$ and $\mathbf{MSp}(\mathbf{X}, \mathbf{Y})$. The category $\mathbf{Cocong}((\mathbf{X}, \alpha), (\mathbf{Y}, \beta))$ of Δ^L-cocongruences between Δ^L-coalgebras $(\mathbf{X}, \alpha)$ and $(\mathbf{Y}, \beta)$ is defined similarly.

The functor $Po \colon \mathbf{MSp}(\mathbf{X}, \mathbf{Y}) \to \mathbf{Cospan}(\mathbf{X}, \mathbf{Y})$ is lifted to the categories of Δ^L-bisimulations and Δ^L-cocongruences by the functor

$$\overline{Po} \colon \mathbf{Bisim}((\mathbf{X}, \alpha), (\mathbf{Y}, \beta)) \to \mathbf{Cocong}((\mathbf{X}, \alpha), (\mathbf{Y}, \beta)),$$

acting on Δ^L-bisimulations as $\overline{Po}(((\mathbf{R}, \gamma_R), r_1, r_2)) = ((Po(\mathbf{R}), \kappa), k_1, k_2)$, where $(Po(\mathbf{R}), k_1, k_2)$ is the pushout of $(\mathbf{R}, r_1, r_2)$ and $\kappa \colon Po(\mathbf{R}) \to \Delta^L Po(\mathbf{R})$ is the unique measurable map given by the universal property of pushouts, making the diagram below commute; and on arrows $f \colon ((\mathbf{R}, \gamma_R), r_1, r_2) \to ((\mathbf{S}, \gamma_S), s_1, s_2)$ as $\overline{Po}(f)$, defined as the unique arrow, given by the universal property of pushouts, making the diagram on the right commute:

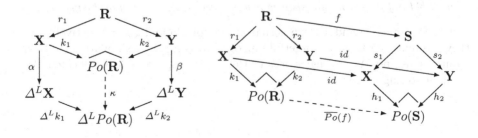

The arrow $\overline{Po}(f)$ is obviously a morphism between the cospans $(Po(\mathbf{R}), k_1, k_2)$ and $(Po(\mathbf{S}), h_1, h_2)$, and can be proved to be also a Δ^L-homomorphism exploiting the universal property of pushouts. Functoriality follows similarly.

Remark 18. The above construction is standard and applies in any category with pushouts independently of the choice of the behaviour functor.

If the behavior functor preserves weak pullbacks, cocongruences give rise to bisimulations via pullbacks (see [10, Prop. 1.2.2]). However, although Δ^L does not preserves weak pullbacks [16], if we restrict our attention only to the full subcategory of Δ^L-cocongruences that are $\overline{Po}$-images of some Δ^L-bisimulation, namely, $\overline{Po}(\mathbf{Bisim}((\mathbf{X}, \alpha), (\mathbf{Y}, \beta)))$, the functor Pb can be lifted as follows.

For $((\mathbf{R}, \gamma_R), r_1, r_2)$ and f objects and arrows in $\mathbf{Bisim}((\mathbf{X}, \alpha)(\mathbf{Y}, \beta))$, respectively, the lifting $\overline{Pb}$ of Pb is defined by

$$\overline{Pb} \colon \overline{Po}\big(\mathbf{Bisim}(((\mathbf{X}, \alpha)(\mathbf{Y}, \beta)))\big) \to \mathbf{Bisim}(((\mathbf{X}, \alpha)(\mathbf{Y}, \beta)))$$

$$\overline{Pb}(\overline{Po}((\mathbf{R}, \gamma_R), r_1, r_2)) = ((\mathbf{R}^*, \gamma_R^*), r_1^*, r_2^*), \qquad \overline{Pb}(\overline{Po}(f)) = f^*,$$

where $PbPo(\mathbf{R}, r_1, r_2) = (\mathbf{R}^*, r_1^*, r_2^*)$, $PbPo(f) = f^*$, and $\gamma_R^* \colon \mathbf{R}^* \to \Delta^L \mathbf{R}^*$ is the unique (sub)probability measure on $\mathbf{R}^*$ (given as in Proposition 13) such that, for all $r \in R^*$, $a \in L$, $E \in \Sigma_X$ and $F \in \Sigma_Y$

$$\gamma_R^*(r)(a)((r_1^*)^{-1}(E)) = \alpha(r_1^*(r))(a)(E),$$

and

$$\gamma_R^*(r)(a)((r_2^*)^{-1}(F)) = \beta(r_2^*(r))(a)(F),$$

Note that the well definition of the measure $\gamma_R^*(r)(a)$ is guaranteed by Lemma 11 and Proposition 14.

Remark 19. In the definition of γ_R^* above, we applied Proposition 13 without ensuring that r_1^* and r_2^* are surjective maps. This is not an issue and it can be solved easily as in the proof of Proposition 14. ∎

The next lemma ensures that the functor $\overline{Pb}$ is well defined.

Lemma 20. *Let $((\mathbf{R}, \gamma_R), r_1, r_2)$ and f be, respectively, an object and an arrow in the category $\mathbf{Bisim}((\mathbf{X}, \alpha)(\mathbf{Y}, \beta))$. Then, the following hold:*

(i) $\overline{Pb}(\overline{Po}((\mathbf{R}, \gamma_R), r_1, r_2))$ is a Δ^L-bisimulation between $(\mathbf{X}, \alpha)$ and $(\mathbf{Y}, \beta)$;
(ii) $\overline{Pb}(\overline{Po}(f))$ is a monic span-morphism and a Δ^L-homomorphism.

The functors $\overline{Po}$ and $\overline{Pb}$ are actual liftings of Po and Pb, respectively, i.e., they commute w.r.t. to the forgetful functors mapping Δ^L-bisimulations to their monic spans and Δ^L-cocongruences to their epic cospans. This is reflected also in the following result:

Theorem 21. *Let* $(\mathbf{X}, \alpha)$ *and* $(\mathbf{Y}, \beta)$ *be* Δ^L-*coalgebras, then*

(i) $\overline{Po} \dashv \overline{Pb}$;

(ii) $\overline{Pb}\,\overline{Po}(\mathbf{Bisim}((\mathbf{X}, \alpha), (\mathbf{Y}, \beta))) \cong \overline{Po}\,\overline{Pb}\,\overline{Po}(\mathbf{Bisim}((\mathbf{X}, \alpha), (\mathbf{Y}, \beta)))$.

Proof. (i) By the universal properties of pushouts, for any pair of bisimulations $((\mathbf{R}, \gamma_R), r_1, r_2)$ and $((\mathbf{S}, \gamma_S), s_1, s_2)$ in $\mathbf{Bisim}((\mathbf{X}, \alpha), (\mathbf{Y}, \beta))$ it holds that,

$$Hom(\overline{Po}((\mathbf{R}, \gamma_R), r_1, r_2), \overline{Po}((\mathbf{S}, \gamma_S), s_1, s_2) \cong$$
$$Hom(((\mathbf{R}, \gamma_R), r_1, r_2), \overline{Pb}\,\overline{Po}((\mathbf{S}, \gamma_S), s_1, s_2),$$

i.e., $\overline{Po}$ is left adjoint to $\overline{Pb}$. (ii) The equivalence is given by the functors $\overline{Po}$ and $\overline{Pb}$ and follows by Proposition 17(ii). □

Theorem 21 provides sufficient conditions under which the notions of bisimulation and cocongruence coincide: (ii) states that the Δ^L-bisimulation formed applying the (closure) operator $\overline{Pb}\,\overline{Po}$ to a Δ^L-bisimulation $((\mathbf{R}, \gamma_R), r_1, r_2)$ is equivalent to the Δ^L-cocongruences obtained applying the operator $\overline{Po}\,\overline{Pb}\,\overline{Po}$.

Related Work. In [3], Danos et al. proposed a notion alternative to bisimulations, the so called *event bisimulation*, being aware that it coincides with cocongruence. Here we recall its definition and try to make a comparison between the results in [3] in connection to Theorem 21.

Definition 22 (Event bisimulation). *Let* $\mathcal{M} = (X, \Sigma_X, \{\theta_a\}_{a \in L})$ *be an* L-*labelled Markov kernel. A sub-σ-algebra* $\Lambda \subseteq \Sigma_X$ *is an* event bisimulation *on* $\mathcal{M}$ *if, for all* $a \in L$, $q \in \mathbb{Q} \cap [0, 1]$, *and* $E \in \Lambda$, $\theta_a^{-1}(L_q(E)) \in \Lambda$.

Any σ-algebra Σ on X induces a notion of separability in the form of an relation $\Re(\Sigma) \subseteq X \times X$ defined by $\Re(\Sigma) = \{(x, y) \mid \forall E \in \Sigma. [x \in E \text{ iff } y \in E]\}$. Moreover, considering only equivalence relations $\mathcal{R} \subseteq X \times X$, they denoted by $\Sigma(\mathcal{R}) = \{E \in \Sigma \mid (E, E) \text{ is } \mathcal{R}\text{-closed}\}$ the set of measurable $\mathcal{R}$-closed sets, which is readily seen to be a σ-algebra on X. The "operator" $\Re(\cdot)$ maps σ-algebras to equivalence relations and, conversely, $\Sigma(\cdot)$ maps equivalence relations to σ-algebras. Moreover, as the next lemma states, under certain circumstances, they can also be thought of as maps between event bisimulations and state bisimulations.

Lemma 23 ([3, Lemma 4.1]). *Let* $\mathcal{M} = (X, \Sigma_X, \{\theta_a\}_{a \in L})$ *be an* L-*labelled Markov kernel. Then,* $\mathcal{R}$ *is a state bisimulation iff* $\Sigma(\mathcal{R})$ *is an event bisimulation.*

Proof. Assume $\mathcal{R}$ is a state bisimulation on $\mathcal{M}$. It is easy to show that $\Sigma(\mathcal{R})$ is a sub-σ-algebra of Σ. It remains to prove that for all $a \in L$ and $q \in \mathbb{Q} \cap [0, 1]$, and $E \in \Sigma(\mathcal{R})$, $\theta_a^{-1}(L_q(E)) \in \Sigma(\mathcal{R})$. By denoting the canonical projections of $\mathcal{R}$, by π_1 and π_2, we have

$$\begin{aligned}
\pi_1^{-1}(\theta_a^{-1}(L_q(E))) &= \{(x, y) \in \mathcal{R} \mid \theta_a(x)(E) \geq q\} && \text{(by pre-image)} \\
&= \{(x, y) \in \mathcal{R} \mid \theta_a(y)(E) \geq q\} && \text{(by } \mathcal{R} \text{ state bisim.)} \\
&= \pi_2^{-1}(\theta_a^{-1}(L_q(E))). && \text{(by pre-image)}
\end{aligned}$$

This proves that $(\theta_a^{-1}(L_q(E)), \theta_a^{-1}(L_q(E)))$ is $\mathcal{R}$-closed, so that $\Sigma(\mathcal{R})$ is an event bisimulation. Conversely, assume that $\Sigma(\mathcal{R})$ is an event bisimulation on $\mathcal{M}$. Then, for every $(x, y) \in \mathcal{R}$, $a \in L$, and $E \in \Sigma(\mathcal{R})$, we have that

$$\text{for all } q \in \mathbb{Q} \cap [0, 1], \quad [\theta_a(x)(E) \geq q \text{ iff } \theta_a(y)(E) \geq q].$$

Since $\mathbb{Q} \cap [0, 1]$ is dense in $[0, 1]$, the above implies $\theta_a(x)(E) = \theta_a(y)(E)$. So $\mathcal{R}$ is a state bisimulation. □

Note that, given that Λ is an event bisimulation, it is not always the case that its induced separability relation $\Re(\Lambda)$ is a state bisimulation. This, somehow, seems in accordance with our restriction to a well-behaved subcategory of cocongruences in the definition of the functor $\overline{Pb}$. Indeed, when one restricts the attention only to state bisimulations $\mathcal{R}$ that are assumed to be equivalence relations, the results in [3] are related to our adjunction as follows:

$$Po(\mathcal{R}) = X/\Re(\Sigma(\mathcal{R})), \qquad\qquad PbPo(\mathcal{R}) = R^* = R.$$

In particular, many of the lemmas and propositions in [3, §4] are consequences of Theorem 21.

6 Conclusions and Future Work

We have proposed a genuinely new characterization of bisimulation in plain mathematical terms, which is proven to be in one-to-one correspondence with the coalgebraic definition of Aczel and Mendler.

Then, the notions of bisimulation and cocongruence (equivalently, event bisimulation) are formally compared establishing an adjunction between the category of coalgebraic bisimulations and a suitable subcategory of cocongruences. By means of this adjunction we provided sufficient conditions under which the notions of bisimulation and cocongruence coincide.

A comparison between bisimulations and cocongruences by means of an adjunction between their categories is interesting not just for Markov processes but, more in general, for any F-coalgebra. Usually, the final bisimulation (i.e., bisimilarity) between two coalgebras is said to be "well behaved" if it coincides with the pullback of the final cocongruence between the same pair of coalgebras. When a final coalgebra exists, the final cocongruence is given by the pair of final homomorphisms and its pullback is called behavioral equivalence (or final semantics). A sufficient condition to ensure that bisimilarity coincides with behavioral equivalence is to require that the behaviour functor (weakly) preserves pullbacks or semi-pullbacks (for instance see [12, Theorem 9.3]). When the behavior functor does not (weakly) preserves pullbacks or semi-pullbacks, one may use cocongruences instead of bisimulations. Another way (the one we have explored in Section 5) is to use adjunctions, which allows one to focus on the well behaved bisimulations by considering some suitable full-subcategories. In this light, Theorem 21 generalizes the results in [3], which are consequences of the existence of the adjunction $\overline{Po} \dashv \overline{Pb}$.

More generally, it would be interesting to study to what extent the weak-pullback preservation assumption on the functor could be removed. For example, in [17] Worrell proved that the category of F-coalgebras over **Set** is complete, provided that the functor F weakly-preserves pullbacks and is bounded. It would be nice to extend this result to general categories, replacing the former assumption with the existence of an adjunction between F-bisimulations and F-cocongruences.

References

1. Blute, R., Desharnais, J., Edalat, A., Panangaden, P.: Bisimulation for labelled Markov processes. In: LICS, pp. 149–158. IEEE Computer Society (1997)
2. Cardelli, L., Mardare, R.: The measurable space of stochastic processes. In: Proc. QEST, pp. 171–180. IEEE Computer Society (2010)
3. Danos, V., Desharnais, J., Laviolette, F., Panangaden, P.: Bisimulation and cocongruence for probabilistic systems. Inf. Comput. 204(4), 503–523 (2006)
4. de Vink, E.P., Rutten, J.J.M.M.: Bisimulation for probabilistic transition systems: A coalgebraic approach. Theor. Comput. Sci. 221(1-2), 271–293 (1999)
5. Desharnais, J., Edalat, A., Panangaden, P.: Bisimulation for labelled Markov processes. Inf. Comput. 179(2), 163–193 (2002)
6. Desharnais, J., Gupta, V., Jagadeesan, R., Panangaden, P.: Approximating labeled markov processes. In: Proc. LICS, pp. 95–106 (2000)
7. Desharnais, J., Gupta, V., Jagadeesan, R., Panangaden, P.: Approximating labelled markov processes. Inf. Comput. 184(1), 160–200 (2003)
8. Giry, M.: A categorical approach to probability theory. In: Banaschewski (ed.) Categorical Aspects of Topology and Analysis. Lecture Notes in Mathematics, vol. 915, pp. 68–85. Springer, Heidelberg (1982) 10.1007/BFb0092872
9. Heifetz, A., Samet, D.: Topology-Free Typology of Beliefs. Journal of Economic Theory 82(2), 324–341 (1998)
10. Kurz, A.: Logics for Coalgebras and Applications to Computer Science. PhD thesis, Ludwig-Maximilians-Universität München (2000)
11. Larsen, K.G., Skou, A.: Bisimulation through probabilistic testing. Information and Computation 94(1), 1–28 (1991)
12. Rutten, J.J.M.M.: Universal coalgebra: a theory of systems. TCS 249(1), 3–80 (2000)
13. Terraf, P.S.: Unprovability of the logical characterization of bisimulation. CoRR, abs/1005.5142 (2010)
14. Terraf, P.S.: Bisimilarity is not borel. CoRR, abs/1211.0967 (2012)
15. van Breugel, F., Mislove, M.W., Ouaknine, J., Worrell, J.: Domain theory, testing and simulation for labelled Markov processes. Theoretical Computer Science 333(1-2), 171–197 (2005)
16. Viglizzo, I.D.: Coalgebras on Measurable Spaces. PhD thesis, Department of Mathematics, Indiana University (2005)
17. Worrell, J.: Toposes of Coalgebras and Hidden Algebras. Electronic Notes in Theoretical Computer Science 11, 212–230 (1998)

A   Technical Proofs

In this appendix we provide the proofs of the technical results used in the paper.

Proof (of Lemma 3). We have to prove that, given $\mathcal{R} \cap (E \times Y) = \mathcal{R} \cap (X \times F)$ and $\mathcal{R}' \subseteq \mathcal{R}$, we have $\mathcal{R}' \cap (E \times Y) = \mathcal{R}' \cap (X \times F)$.
($\subseteq$) Let $(x, y) \in \mathcal{R}'$ and $x \in E$. By $\mathcal{R}' \subseteq \mathcal{R}$, $(x, y) \in \mathcal{R}$. By (E, F) $\mathcal{R}$-closed, we have $y \in F$.
($\supseteq$) Let $(x, y) \in \mathcal{R}'$ and $y \in F$. By $\mathcal{R}' \subseteq \mathcal{R}$, $(x, y) \in \mathcal{R}$. By (E, F) $\mathcal{R}$-closed, we have $x \in E$.   □

Proof (of Lemma 4). We prove only the inclusion $E \subseteq F$, the reverse is similar. Assume $x \in E$. By reflexivity of $\mathcal{R}$, $(x, x) \in \mathcal{R}$. Since (E, F) is $\mathcal{R}$-closed, we have $x \in F$. To prove that E is an union of $\mathcal{R}$-equivalence classes, it suffices to show that if $x \in E$ and $(x, y) \in \mathcal{R}$, then $y \in E$. This easily follows since $E = F$.   □

Proof (of Proposition 10). ($\Rightarrow$) Immediate from Corollary 7. ($\Leftarrow$) Let $x, y \in X$ be such that

$$\forall a \in L, \forall E \in \Sigma \text{ such that } (E, E) \sim\text{-closed}, \quad \theta_a(x)(E) = \theta_a(y)(E). \quad (5)$$

We prove $x \sim y$ showing a bisimulation $\mathcal{R}$ such that $(x, y) \in \mathcal{R}$. Let $\mathcal{R}$ be the smallest equivalence containing $\{(x, y)\}$ and $\sim$, hence $\mathcal{R} = Id_X \cup \bigcup_{n \in \mathbb{N}} \mathcal{S}_n$, where

$$\mathcal{S}_0 \triangleq \{(x, y), (y, x)\} \cup \sim \qquad\qquad \mathcal{S}_{n+1} \triangleq \mathcal{S}_n ; \mathcal{S}_n.$$

By Lemma 4, it suffices to prove that, for all $a \in L$ and $E' \in \Sigma$ such that (E', E') is $\mathcal{R}$-closed, the following holds:

$$(x', y') \in \mathcal{R} \implies \theta_a(x')(E') = \theta_a(y')(E'). \quad (6)$$

If $(x', y') \in \mathcal{R}$, then $(x', y') \in Id_X$ or $(x', y') \in \mathcal{S}_n$ for some $n \geq 0$. If $(x', y') \in Id_X$ (6) holds trivially. Now we show, by induction on $n \geq 0$, that

$$(x', y') \in \mathcal{S}_n \implies \theta_a(x')(E') = \theta_a(y')(E'). \quad (7)$$

Base case ($n = 0$): let $(x', y') \in \sim$. By $\sim \subseteq \mathcal{R}$ and Lemma 3, (E', E') is $\sim$-closed. Thus, by Corollary 7, (7) holds. If $(x', y') \in \{(x, y), (y, x)\}$, then property (5) holds. Again, by Lemma 3, (E', E') is $\sim$-closed, thus (7) holds.
Inductive case ($n + 1$): if $(x', y') \in \mathcal{S}_{n+1}$, then there exists some $z \in X$ such that $(x', z) \in \mathcal{S}_n$ and $(z, y') \in \mathcal{S}_n$. Then, applying the inductive hypothesis twice, we have $\theta_a(x')(E) = \theta_a(z)(E) = \theta_a(y')(E)$.   □

Proof (of Lemma 11). Note that $R^* = \bigcup_{n \in \mathbb{N}} R_n$, for R_n and $n \in \mathbb{N}$ defined as

$$R_0 = R, \qquad\qquad R_{n+1} = R; R_n^{-1}; R.$$

Therefore, that R^* is a bisimulation between the two Markov kernels, directly follows if we can prove that, whenever $(x, y) \in R_n$, then $\alpha_a(E) = \beta_a(F)$, for all $a \in L$, and any pair (E, F) of R_n-closed measurable sets $E \in \Sigma_X$ and $F \in \Sigma_Y$. We proceed by induction on $n \geq 0$.

Base case $(n = 0)$: Clearly, $R_0 = R$ and R is a state bisimulation.

Inductive step $(n > 0)$: Suppose $(x, y) \in R_{n+1}$, then there exist $x' \in X$ and $y' \in Y$ such that $(x, y') \in R$, $(x', y') \in R_n$, $(x', y) \in R$. By Lemma 4, any pair (E, F) which is R_{n+1}-closed is also R_n-closed and R-closed. So that, by hypothesis on R and the inductive hypothesis on R_n, for every $a \in L$, $E \in \Sigma_Y$ and $F \in \Sigma_Y$, we have

$$\alpha_a(x)(E) = \beta_a(y')(F), \quad \alpha_a(x')(E) = \beta_a(y')(F), \quad \alpha_a(x')(E) = \beta_a(y)(F).$$

Therefore, $\alpha_a(x)(E) = \beta_a(y)(F)$.   □

Proof (of Proposition 13). We define $\mu \wedge \nu$ has the Hahn-Kolmogorov extension of a pre-measure defined on a suitable field $\mathcal{F}$ such that $\sigma(\mathcal{F}) = \Sigma_R$. Let $\mathcal{F}$ be the collection of all *finite* unions $\bigcup_{i=0}^{k} G_i$, where $k \in \mathbb{N}$, and for all $i = 0..k$, $G_i = r_1^{-1}(E_i) \cap r_2^{-1}(F_i)$, for some $E_i \in \Sigma_X$ and $F_i \in \Sigma_Y$. Clearly, $\sigma(\mathcal{F}) = \Sigma_R$. To prove that $\mathcal{F}$ is a field we need to show that it is closed under finite intersection and complement (it is already closed under finite union). This is immediate by De Morgan laws and the following equalities:

$$\left(r_1^{-1}(E_i) \cap r_2^{-1}(F_i) \right) \cap \left(r_1^{-1}(E_j) \cap r_2^{-1}(F_j) \right) = r_1^{-1}(E_i \cap E_j) \cap r_2^{-1}(F_i \cap F_j),$$

$$R \setminus \left(r_1^{-1}(E_i) \cap r_2^{-1}(F_i) \right) = r_1^{-1}(X \setminus E_i) \cup r_2^{-1}(Y \setminus F_i).$$

Now we define $\mu \wedge \nu \colon \mathcal{F} \to [0, \infty]$. Note that any element in $S \in \mathcal{F}$ can always be decomposed into a finite union $S = \bigcup_{i=0}^{k} G_i$ of *pair-wise disjoint* sets of the form $G_i = r_1^{-1}(E_i) \cap r_2^{-1}(F_i)$, where $E_i \in \Sigma_X$ and $F_i \in \Sigma_Y$ (hereafter, called G-sets). Then for any G-set we define

$$(\mu \wedge \nu)(G_i) = \begin{cases} \mu(E) & \text{if } \exists E \in \Sigma_X . \, G_i = r_1^{-1}(E) \text{ and } \forall F \in \Sigma_Y . \, G_i \neq r_2^{-1}(F) \\ \nu(E) & \text{if } \forall E \in \Sigma_X . \, G_i \neq r_1^{-1}(E) \text{ and } \exists F \in \Sigma_Y . \, G_i = r_2^{-1}(F) \\ \mu(E) & \text{if } \exists E \in \Sigma_X , F \in \Sigma_Y . \, G_i = r_1^{-1}(E) = r_2^{-1}(F) \\ 0 & \text{otherwise} \end{cases}$$

$$(8)$$

and we define

$$(\mu \wedge \nu)(S) = \sum_{i=0}^{k} (\mu \wedge \nu)(G_i). \tag{9}$$

Note that, in the definition of $\mu \wedge \nu$ on G-sets, the surjectivity of r_1 and r_2 guarantees that if $G_i = r_1^{-1}(E)$ or $G_i = r_2^{-1}(F)$, then E and F are unique. So that, (8) is well-defined. Moreover, the definition on S does not depend on how S is decomposed into a disjoint union. To see this, note that any two representations $\bigcup_{i=0}^{k} G_i$ and $\bigcup_{i=0}^{k'} G_i'$ for S can be decomposed into a common refinement $\bigcup_{i=0}^{k''} G_i''$, so that, by the well definition of $\mu \wedge \nu$ on G-sets they must agree on it. Therefore, $\mu \wedge \nu$ is well-defined on all $\mathcal{F}$, and by construction is finitely additive.

It remains to show that, if $S \in \mathcal{F}$ is the countable disjoint union of sets $S_0, S_1, S_2, \ldots \in \mathcal{F}$, then

$$(\mu \wedge \nu)(S) = \sum_{n \in \mathbb{N}} (\mu \wedge \nu)(S_n).$$

Splitting up S into disjoint G-sets, and restricting S_n to each of these G-sets sets in turn, for all $n \in \mathbb{N}$, by finite additivity of $\mu \wedge \nu$, we may assume without loss of generality that $S = r_1^{-1}(E) \cap r_2^{-1}(F)$, for some $E \in \Sigma_X$ and $F \in \Sigma_Y$. In the same way, by breaking up each S_n into a G-set and using finite additivity of $\mu \wedge \nu$ again, we may assume without loss of generality that each S_n takes the form $S_n = r_1^{-1}(E_n) \cap r_2^{-1}(F_n)$, for some $E_n \in \Sigma_X$ and $F_n \in \Sigma_Y$. By definition of $\mu \wedge \nu$ and σ-additivity of μ and ν, (9) is rewritten as follows

$$(\mu \wedge \nu)(S) = \sum_{n \in \mathbb{N}} (\mu \wedge \nu)(S_n) = (\mu \wedge \nu)(r_1^{-1}(E) \cap r_2^{-1}(F)).$$

This proves that $\mu \wedge \nu \colon \mathcal{F} \to [0, \infty]$ is a pre-measure. By Hahn-Kolmogorov theorem, $\mu \wedge \nu$ can be extended to Σ_R, and since, for all $E \in \Sigma_X$ and $F \in \Sigma_Y$,

$$r_1^{-1}(E) = r_1^{-1}(E) \cap R = r_1^{-1}(E) \cap r_2^{-1}(Y) \in \mathcal{F},$$
$$r_2^{-1}(F) = R \cap r_2^{-1}(F) = r_1^{-1}(X) \cap r_2^{-1}(F) \in \mathcal{F},$$

together with the hypothesis made on μ and ν, i.e.,

$$\text{if } r_1^{-1}(E) = r_2^{-1}(F), \text{ then } \mu(E) = \nu(F),$$

by (8) we have

$$(\mu \wedge \nu)(r_1^{-1}(E)) = \mu(E) \qquad \text{and} \qquad (\mu \wedge \nu)(r_2^{-1}(F)) = \nu(F),$$

therefore, the required conditions are satisfied. If both μ and ν are σ-finite, so is the pre-measure $\mu \wedge \nu \colon \mathcal{F} \to [0, \infty]$, hence its extension on Σ_R is unique. $\quad\square$

Proof (of Lemma 20)

(i) Immediate from Lemma 11 and the correspondence between state bisimulation and Δ^L-bisimulation (Theorem 16).

(ii) Let $f \colon ((\mathbf{R}, \gamma_R), r_1, r_2) \to ((\mathbf{S}, \gamma_S), s_1, s_2)$ in $\mathbf{Bisim}((\mathbf{X}, \alpha), (\mathbf{Y}, \beta))$. By definition $\overline{Pb(Po)}(f) = PbPo(f) = f^*$, hence it is a monic span-morphism from $(\mathbf{R}^*, r_1^*, r_2^*)$ to $(\mathbf{S}^*, s_1^*, s_2^*)$. To prove that f^* is a Δ^L-homomorphism between the coalgebras $(\mathbf{R}^*, \gamma_R^*)$ and $(\mathbf{S}^*, \gamma_S^*)$, we have to show $\gamma_S^* \circ f^* = \Delta^L f^* \circ \gamma_R^*$. By the unicity of the definition of the (sub)probability measures $\gamma_S^*(f^*(r))(a)$ and $\gamma_R^*(r)(a)$, to prove the equality it suffices to show that for arbitrary $r \in R^*$, $a \in L$, $E \in \Sigma_X$, and $F \in \Sigma_Y$,

$$(\gamma_S^* \circ f^*)(r)(a)((s_1^*)^{-1}(E)) = (\Delta^L f^* \circ \gamma_R^*)(r)(a)((s_1^*)^{-1}(E)),$$
$$\text{and}$$
$$(\gamma_S^* \circ f^*)(r)(a)((s_2^*)^{-1}(F)) = (\Delta^L f^* \circ \gamma_R^*)(r)(a)((s_2^*)^{-1}(F)).$$

We prove only the first equality, the other follows similarly.

$$(\gamma_S^* \circ f^*)(r)(a)((s_1^*)^{-1}(E)) =$$

$$
\begin{aligned}
&= \gamma_S^*(f^*(r))(a)((s_1^*)^{-1}(E)) && \text{(composition)} \\
&= \alpha(s_1^* \circ f^*(r))(a)(E) && \text{(by def. } \gamma_S^*) \\
&= \alpha(r_1(r))(a)(E) && \text{(by } f \text{ span-morphism)} \\
&= \gamma_R^*(r)(a)((r_1^*)^{-1}(E)) && \text{(by def. } \gamma_R^*) \\
&= \gamma_R^*(r)(a)((s_1^* \circ f^*)^{-1}(E)) && \text{(by } f^* \text{ span-morphism)} \\
&= \gamma_R^*(r)(a)((f^*)^{-1} \circ (s_1^*)^{-1}(E)) && \text{(by comp. inverses)} \\
&= (\Delta^L f^* \circ \gamma_R^*)(a)((s_1^*)^{-1}(E)) && \text{(by def. } \Delta^L)
\end{aligned}
$$

$$\square$$

Probabilistic Model Checking
for Energy-Utility Analysis*

Christel Baier, Clemens Dubslaff, Joachim Klein,
Sascha Klüppelholz, and Sascha Wunderlich

Institute for Theoretical Computer Science
Technische Universität Dresden, Germany
{baier,dubslaff,klein,klueppel,wunder}@tcs.inf.tu-dresden.de

Abstract. In the context of a multi-disciplinary project, where we contribute with formal methods for reasoning about energy-awareness and
other quantitative aspects of low-level resource management protocols,
we made a series of interesting observations on the strengths and limitations of probabilistic model checking. To our surprise, the operating-
system experts identified several relevant quantitative measures that are
not supported by state-of-the-art probabilistic model checkers. Most notably are conditional probabilities and quantiles. Both are standard in
mathematics and statistics, but research on them in the context of probabilistic model checking is rare. Another deficit of standard probabilistic
model-checking techniques was the lack of methods for establishing properties imposing constraints on the energy-utility ratio.

In this article, we will present formalizations of the above mentioned
quantitative measures, illustrate their significance by means of examples
and sketch computation methods that we developed in our recent work.

1 Introduction

Markovian models can be seen as automata annotated with probabilistic distributions and cost or reward functions to model stochastic phenomena and
resource constraints. Thanks to the Markovian property stating that the future
system behavior only depends on the current state but not on the past, they are
best suited for algorithmic quantitative analysis. This also explains their long
tradition in Computer Science and the increasing interest and relevance to the
research field of probabilistic model checking (PMC). Since its first release more
than 10 years ago, the prominent probabilistic model checker PRISM [32] for
Markovian models and temporal logical specifications, as well as other probabilistic checkers like MRMC [30], PROBDIVINE [10] or the CADP tool set [26],
have been continuously extended by new features and successfully applied in

---

* This work was partly funded by the DFG through the CRC 912 HAEC, the cluster
  of excellence cfAED, the project QuaOS, the Graduiertenkolleg 1763 (QuantLA),
  and the DFG/NWO-project ROCKS and partially by Deutsche Telekom Stiftung,
  the ESF young researcher groups IMData 100098198 and SREX 100111037, and the
  EU-FP-7 grant 295261 (MEALS).

F. van Breugel et al. (Eds.): Panangaden Festschrift, LNCS 8464, pp. 96–123, 2014.
© Springer International Publishing Switzerland 2014

various areas. Examples for areas where PMC is nowadays well established are randomized distributed systems, multimedia and security protocols and systems biology.

In this article, we report on our experiences with the application of PMC in a less well established area, namely the analysis of low-level resource management algorithms. This work has been carried out in the context of multi-disciplinary research projects on advancing electronics and highly-adaptive energy computing where PMC serves as an offline approach to evaluate the adaption strategies and resource management protocols developed by the project partners. The PMC-based approach is complementary to the experimental and simulation-based analysis conducted by project partners by providing insights in the energy-utility, reliability and other performance characteristics from a global and long-run per-spective. PMC results on the quantitative behavior can guide the optimization of resource management algorithms and can be useful to predict the perfor-mance of management algorithms on future hardware, where measurements are impossible.

Besides the prominent state-explosion problem, a series of problems have to be addressed. One challenging task is to find eligible model parameters (stochastic distributions, cost values) that fit with "reality", without zooming into complex details of, e.g., hardware primitives (caches, busses, memory organization, and so on). By means of a simple spinlock protocol, we illustrated in [6,5] a stepwise refinement approach for the model generation that incorporates cache effects in the distributions and relies on the parallel use of measurements and PMC for small numbers of processes. In this way, we obtained evidence for the appropri-ateness of the model and then applied PMC – in combination with sophisticated techniques to tackle the state-explosion problem – for larger numbers of processes where measurements are no longer feasible.

Surprising to us was the observation that the operating-system experts iden-tified several quantitative measures as highly relevant that are not supported by state-of-the-art probabilistic model checkers. Some of them are reducible to standard queries that are supported by probabilistic model checkers. This, e.g., applies to the evaluation of *relativized long-run properties*, where the task is to compute the probability of a certain temporal property or the expected value of a random function, relativized by some constraint for the starting states and under the assumption that the system has reached its steady state. As an example, for an energy-aware resource management protocol, the expected amount of energy consumed by threads in a given mode or the probability to access a requested shared resource while not exceeding a given energy budget, is most interesting when the system is in equilibrium and particularities of the initialization phase have lost their influence. Theoretically more interesting were those quantitative measures demanded by the operating-system colleagues that have not or only barely been addressed by the model-checking community before. This mainly concerns requirements on the interplay of several objectives, possibly depend-ing on each other. For instance, minimizing the energy consumption of a sys-tem without regarding its productivity or utility is rather pointless, since these measures clearly depend on the energy spent during an execution of the system. One obvious possibility to include utility requirements into an energy-minimizing objective is to require a minimal step-wise utility of the system, which however may lead to a restriction towards overall inefficient executions consuming much energy. As this example shows, more involved approach need to be identified to faithfully reason over the interplay of objectives.

For Markov decision processes (MDPs), a variety of dynamic and linear-programming techniques have been proposed to find Pareto optimal or compromise solutions for multi-objective requirements given by a series of constraints on expected rewards or the probabilities for ω-regular path properties, see, e.g., [42,18,23,37,24]. Although these techniques can be very useful to reason about the interplay of energy efficiency and utility requirements, we identified three important classes of quantitative measures that serve to formalize other types of multi-objective requirements and that have obtained less attention so far in the PMC community.

The first two are *quantiles* and *conditional probabilities* resp. *conditional accumulated costs*, where the probability measure is relativized by some temporal condition. Both quantiles and conditional probabilities are standard in mathematics and statistics and both appear naturally for analyzing the interplay of two or more objectives, such as energy efficiency and utility. Quantiles, for instance, can be used to formalize the minimal cost (e.g., energy) needed to achieve a certain degree of utility with high probability. Conditional probabilities can, e.g., be used to formalize the chance to execute a list of jobs when the available energy is bounded by some energy budget, under the condition that a certain degree of utility will be achieved.

The computation of quantiles in discrete Markovian models is computationally hard (NP-hard even for discrete Markov chains as a consequence of the results in [33]). Our recent paper [40] presents computation schemes for quantiles in MDPs based on a linear program. The naïve implementation of this LP-based approach turned out to be far too slow for practical applications. However, with several heuristics to improve the performance [4], we obtained rather good results for several case studies, including an energy-aware job scheduler and an energy-aware bonding network device [22].

Extensions of branching-time logics with conditional probability operators have been studied for discrete [2,1,29] and continuous-time [25] Markov chains. The proposed model-checking procedure for computing $\mathbb{P}_s^{\mathcal{M}}(\varphi \mid \psi)$ relies on the obvious approach to compute the quotient of $\mathbb{P}_s^{\mathcal{M}}(\varphi \wedge \psi)$ and $\mathbb{P}_s^{\mathcal{M}}(\psi)$. The computation of maximal or minimal conditional probabilities in models with nondeterminism and probabilism, such as MDPs, is more difficult since there is no computation scheme that simply takes the quotient of two (standard) maximal/minimal probabilities, where extrema range over all possible resolutions of the nondeterminism (formalized by *schedulers*). An exponential-time model-checking algorithm that relies on the inspection of a finite (but possibly exponentially large) class of finite-memory schedulers taken as candidates to achieve extremal conditional probabilities has been proposed in [2,1]. The lack of tools that provide special engines for the computation of conditional probabilities has motivated our recent paper [9], where we presented transformations for discrete Markov chains and MDPs that reduce the task of computing (maximal/minimal) conditional probabilities to the task of computing unconditional probabilities in the transformed model. First experiments with a prototype implementation indicate that our transformation-based methods are feasible in practice and outperform the quotient-based method for discrete Markov chains.

The third class of quantitative measures that we identified to be significant for the interference of energy and utility constraints is the *energy-utility ratio*, where both the energy consumption and the achieved degree of utility are formalized using accumulated values of weight functions. The nonprobabilistic logical framework presented in [12] that extends temporal logics by operators for accumulated values provides an elegant approach to formalize properties imposing constraints on accumulated values and could be used to express various constraints on the energy-utility ratio. This framework is too powerful for our purposes, since various model-checking problems are shown to be undecidable. Nevertheless, there are decidability results for special formula types. We restrict here our attention to path properties of the form $\psi_\theta \wedge \varphi$, where ψ_θ imposes a condition on the energy-utility ratio with respect to a *quality threshold* θ and φ is a side-constraint, e.g., a reachability or parity condition. In our setting, ψ_θ is of the form $\psi_\theta = \Box(\frac{\text{utility}}{\text{energy}} \geqslant \theta)$. Intuitively, ψ_θ then stands for the path property that at any moment on a path the ratio between the utility achieved and the consumed energy so far is at least θ.

Examples for interesting tasks in, e.g., Markov chains are the almost-sure verification problem that asks whether $\psi_\theta \wedge \varphi$ holds with probability 1 or the quantitative analysis that aims to compute the probability for $\psi_\theta \wedge \varphi$. These tasks share some similarities with the questions that have been addressed in the context of nonprobabilistic game structures [16,17] and probabilistic models with single weight functions [16,35]. Ratio objectives for weighted MDPs have been studied for example in [43] where the goal is to synthesize a scheduler for a given weighted MDP that maximizes or minimizes the average ratio payoff of two integer weight functions.

Contribution and Outline. The relevant concepts of discrete and continuous-time Markov chains and MDPs are summarized in Section 2. Section 3 presents the concept of relativized long-run probabilities and expectations for Markov chains as introduced in our previous papers [6,7]. Sections 4 and 5 deal with conditional probabilities and quantiles, respectively, where we put the focus on Markov chains. While our previous work on conditional probabilities [9] and on quantiles [40,4] is restricted to the discrete-time case, we expand on this in this article with computation schemes for continuous-time Markov chains. Section 6 introduces energy-utility MDPs and presents decidability results for reasoning about energy-utility ratios under ω-regular side-constraints in Markov chains and MDPs. Section 7 contains some concluding remarks.

2 Theoretical Foundations

This section provides a brief summary of our notation for Markov chains and Markov decision processes and related concepts. For further details we refer to textbooks on model checking [20,8] and on probability theory and Markovian models [38,31,28,36].

The reader is assumed to be familiar with ω-automata and temporal logics. See, e.g., [20,27,8]. We often use notation of linear temporal logic (LTL) and computation tree logic (CTL), where $\Diamond$, $\Box$, $\bigcirc$ and U stand for the temporal modalities "eventually", "always", "next" and "until", while $\exists$ and $\forall$ are used as CTL-like path quantifiers. The notion *path property* is used for any language consisting of infinite words over 2^{AP} where AP is the underlying set of atomic propositions. LTL-formulas are often identified with the path property of infinite words over the alphabet 2^{AP} that are models for the formulas. Having in mind temporal logical specifications, we use the logical operators $\vee$, $\wedge$, $\neg$ for union, intersection and complementation of path properties.

Distributions. A distribution on a nonempty, countable set S is a function $\mu : S \to [0,1]$ such that $\sum_{s \in S} \mu(s) = 1$. For $U \subseteq S$, $\mu(U)$ denotes $\sum_{s \in U} \mu(s)$.

Markov Chains. A (discrete) Markov chain is a pair $\mathcal{M} = (S, P)$ where S is a finite, nonempty set of states and $P : S \times S \to [0,1]$ a function, called the transition probability function, such that $\sum_{s' \in S} P(s, s') = 1$ for each state s. Paths in $\mathcal{M}$ are finite or infinite sequences $s_0\, s_1\, s_2 \ldots$ of states built by transitions, i.e., $P(s_i, s_{i+1}) > 0$ for all $i \geqslant 0$. The length $|\pi|$ denotes the number of transitions taken in π. If $\pi = s_0\, s_1 \ldots s_n$ is a finite path, then $first(\pi) = s_0$ denotes the first state of π, and $last(\pi) = s_n$ the last state of π. We refer to the value

$$\Pr(\pi) \;=\; \prod_{0 \leqslant i < n} P(s_i, s_{i+1})$$

as the probability for π. We write $FinPaths(s)$ for the set of all finite paths π with $first(\pi) = s$. Similarly, $InfPaths(s)$ stands for the set of infinite paths starting in s. We use $FinPaths$ and $InfPaths$ to denote the sets of all finite paths, respectively infinite paths. A *weight* function for $\mathcal{M}$ is a function $wgt : S \times S \to \mathbb{Z}$ such that $wgt(s, s') = 0$ if $P(s, s') = 0$. For a finite path $\pi = s_0\, s_1 \ldots s_n$ the (accumulated) weight of π is defined by the sum of the weights of its transitions:

$$wgt(\pi) \;=\; \sum_{i=0}^{n-1} wgt(s_i, s_{i+1})$$

Occasionally, we also consider weight functions with rational values and refer to them as *rational-valued weight functions*. If wgt is nonnegative, i.e., $wgt(s, s') \geqslant 0$ for all states s, s', then we refer to wgt as a *reward function*. We also consider weight functions $wgt_{st} : S \to \mathbb{Z}$ for the states (rather than transitions). These can be encoded as weight functions for the transitions by defining

$wgt(s, s') = wgt_{st}(s)$ for all states s' with $P(s, s') > 0$. The accumulated weight of a finite path $\pi = s_0 s_1 \ldots s_n$ is defined accordingly:

$$wgt_{st}(\pi) \quad = \quad \sum_{i=0}^{n-1} wgt(s_i)$$

The *probability space* induced by Markov chains is defined using classical concepts of measure and probability theory. The basic elements of the underlying sigma-algebra are the cylinder sets spanned by the finite paths, i.e., the sets $Cyl(\pi)$ consisting of all infinite paths $\tilde{\pi}$ such that π is a prefix of $\tilde{\pi}$. Elements of this sigma-algebra are called *(measurable) path events*. The probability measure $\mathbb{P}_s^{\mathcal{M}}$ is defined on the basis of standard measure-extension theorems stating the existence of a unique probability measure $\mathbb{P}_s^{\mathcal{M}}$ with $\mathbb{P}_s^{\mathcal{M}}(\, Cyl(\pi)\,) = \Pr(\pi)$ for all $\pi \in FinPaths(s)$, whereas cylinder sets of paths π with $first(\pi) \neq s$ have measure 0 under $\mathbb{P}_s^{\mathcal{M}}$.

With these definitions one can also reason about expected values of random functions on infinite paths. We consider here the expected accumulated weight for reaching a target set $F \subseteq S$. Let $wgt[\Diamond F] : InfPaths \to \mathbb{Z}$ be the partial function such that $wgt[\Diamond F](s_0 s_1 s_2 \ldots) = wgt(s_0 s_1 \ldots s_n)$ if $\{s_0, \ldots, s_{n-1}\} \cap F = \varnothing$ and $s_n \in F$. For infinite paths $\tilde{\pi}$ that do not visit F, i.e., $\tilde{\pi} \not\models \Diamond F$, $wgt[\Diamond F](\tilde{\pi})$ is undefined. Assuming that F will be reached from state s almost surely, i.e., $\mathbb{P}_s^{\mathcal{M}}(\Diamond F) = 1$, then the expected value of $wgt[\Diamond F]$ under the probability measure $\mathbb{P}_s^{\mathcal{M}}$ is well-defined and given by:

$$\mathbb{E}_s^{\mathcal{M}}(\, wgt[\Diamond F]\,) \quad = \quad \sum_{\pi} \Pr(\pi) \cdot wgt(\pi)$$

where π ranges over all finite paths $s_0 s_1 \ldots s_n$ starting in $s_0 = s$ such that $\{s_0, \ldots, s_{n-1}\} \cap F = \varnothing$ and $s_n \in F$.

Continuous-Time Markov Chains (CTMCs). A continuous-time Markov chain (CTMC) is a pair $\mathcal{C} = (\mathcal{M}, E)$ where $\mathcal{M} = (S, P)$ is a discrete Markov chain and $E : S \to \mathbb{R}_{\geqslant 0}$ a function that specifies an exit-rate for each state s. $\mathcal{M}$ is called the *embedded discrete Markov chain* that describes the time-abstract operational behavior. Intuitively, $E(s)$ specifies the frequency of taking a transition from s. More formally, $E(s)$ is the rate of an exponential distribution and the probability to take some transitions from s within t time units is given by $1 - e^{-E(s) \cdot t}$ where e is Euler's number. The probability to take a specific transition from s to s' within t time units is then:

$$P(s, [0, t], s') \quad = \quad P(s, s') \cdot \left(1 - e^{-E(s) \cdot t}\right)$$

As a consequence, $1/E(s)$ is the average sojourn time in state s. A *trajectory* (or *timed path*) of $\mathcal{C}$ is a path in $\mathcal{M}$ augmented with the sojourn times in the states. Formally, a trajectory in $\mathcal{C}$ is an alternating sequence $s_0 t_0 s_1 t_1 s_2 t_2 \ldots$ of states s_i and nonnegative real numbers t_i such that $P(s_i, s_{i+1}) > 0$. An infinite trajectory $\tilde{\vartheta}$ is said to be *time-divergent* if $\sum_{i \geqslant 0} t_i = \infty$. In this case, if $t \in \mathbb{R}_{\geqslant 0}$ then $\tilde{\vartheta}@t$ denotes the state s_i where i is the greatest index such that $t_0 + t_1 + \ldots + t_i \leqslant t$. For finite trajectories we require that they end in a state. We write *FinTraj*(*s*) for the set of finite trajectories starting in state *s* and *InfTraj*(*s*) for the infinite time-divergent trajectories from *s*. If the starting state *s* is irrelevant, we omit *s* and write *FinTraj* and *InfTraj*. In CTMCs, state weights are understood as weights per time spend in the corresponding state, and therefore need to be scaled with the corresponding sojourn times in trajectories during accumulation. Given weight functions $wgt : S \times S \to \mathbb{Z}$ and $wgt_{st} : S \to \mathbb{Z}$, the accumulated weights of a finite trajectory $\vartheta = s_0 t_0 s_1 t_1 \ldots t_n s_n$ are given by:

$$wgt(\vartheta) = \sum_{i=0}^{n-1} wgt(s_i, s_{i+1}) \qquad wgt_{st}(\vartheta) = \sum_{i=0}^{n-1} t_i \cdot wgt_{st}(s_i)$$

To reason about probabilities for conditions on trajectories, one can again rely on standard concepts of measure and probability theory to define a sigma-algebra where the events are infinite trajectories. Let $s_0 s_1 \ldots s_n$ be a finite path in the embedded discrete Markov chain $\mathcal{M}$ and let $I_0, I_1, \ldots, I_{n-1}$ be bounded real intervals in $[0, \infty[$. We write $\mathcal{T} = s_0 I_0 s_1 I_1 \ldots I_{n-1} s_n$ for the set of all finite trajectories $s_0 t_0 s_1 t_1 \ldots t_{n-1} s_n$ with $t_i \in I_i$ for $0 \leqslant i < n$ and refer to $\mathcal{T}$ as a *symbolic finite trajectory*. The infinite trajectories $\tilde{\vartheta}$ that have a prefix in $\mathcal{T}$ constitute the cylinder set $Cyl(\mathcal{T})$. For state *s*, $\mathbb{P}_s^{\mathcal{C}}$ is the unique probability measure on the smallest sigma-algebra containing the sets $Cyl(\mathcal{T})$ for all symbolic finite trajectories $\mathcal{T}$ such that

$$\mathbb{P}_s^{\mathcal{C}}\big(\, Cyl(\mathcal{T})\,\big) = \prod_{0 \leqslant i < n} P(s_i, I_i, s_{i+1})$$

where for *s*, $s' \in S$ and real numbers t_1, t_2 with $t_1 \leqslant t_2$ we have:

$$P(s, [t_1, t_2], s') = P(s, s') \cdot \big(e^{-E(s)t_1} - e^{-E(s)t_2} \big)$$

It is well-known that under $\mathbb{P}_s^{\mathcal{C}}$ almost all trajectories are time-divergent [21]. The probability measures $\mathbb{P}_s^{\mathcal{C}}$ rely on the assumption that *s* is the initial state. If μ is a distribution, viewed as an initial distribution, then $\mathbb{P}_\mu^{\mathcal{C}} = \sum_{s \in S} \mu(s) \cdot \mathbb{P}_s^{\mathcal{C}}$.

The long-run behavior of CTMCs can be formalized using steady-state distributions formalizing the mean fraction of time spent in the states in infinite trajectories. For the initial distribution μ, the *steady-state probability* of state *s* is defined by:

$$\mathrm{StPr}_\mu^{\mathcal{C}}(s) = \lim_{t \to \infty} \mathbb{P}_\mu^{\mathcal{C}}\big\{ \tilde{\vartheta} \in InfTraj(s) : \tilde{\vartheta}@t = s \big\}$$

The above limit exists for all finite-state CTMCs, but might depend on the initial distribution μ.

Markov Decision Processes (MDPs). MDPs can be seen as a generalization of Markov chains where the operational behavior in a state *s* consists of a nondeterministic selection of an enabled action α, followed by a probabilistic choice of the successor state, given *s* and α. Formally, an MDP is a tuple

$\mathcal{M} = (S, Act, P)$ where S is a finite set of states, Act a finite set of actions and $P : S \times Act \times S \to [0,1]$ a function such that for all states $s \in S$ and $\alpha \in Act$:

$$\sum_{s' \in S} P(s, \alpha, s') \in \{0, 1\}$$

We write $Act(s)$ for the set of actions that are enabled in s, i.e., $P(s, \alpha, s') > 0$ for some $s' \in S$. For technical reasons, we require that $Act(s) \neq \varnothing$ for all states s. Obviously if the sets $Act(s)$ are singletons for all states s, then $\mathcal{M}$ can be seen as a Markov chain. *Paths* are finite or infinite alternating sequences $s_0\, \alpha_0\, s_1\, \alpha_1\, s_2\, \alpha_2 \ldots$ of states and actions such that $P(s_{i-1}, \alpha_{i-1}, s_i) > 0$ for all $i \geqslant 1$. Notation that has been introduced for Markov chains can now be adapted for MDPs, such as *first*(π), *FinPaths*(s), *InfPaths*(s).

Reasoning about probabilities for path properties in MDPs requires the selection of an initial state and the resolution of the nondeterministic choices between the possible transitions. The latter is formalized via *schedulers*, often also called policies or adversaries, which take as input a finite path and select an action to be executed. For the purposes of this paper, it suffices to consider deterministic, possibly history-dependent schedulers, i.e., partial functions $\mathfrak{S} : FinPaths \to Act$ such that $\mathfrak{S}(\pi) \in Act(\,last(\pi)\,)$ for all finite paths π. Given a scheduler $\mathfrak{S}$, an $\mathfrak{S}$-*path* is any path that might arise when the nondeterministic choices in $\mathcal{M}$ are resolved using $\mathfrak{S}$. Thus, $\pi = s_0\, \alpha_0\, s_1\, \alpha_1 \ldots \alpha_n\, s_n$ is an $\mathfrak{S}$-path iff $P\big(s_{k-1}, \mathfrak{S}(pref(\pi, k)), s_k\big) > 0$ for all $1 \leqslant k \leqslant n$. Here, $pref(\pi, k)$ denotes the prefix of π consisting of the first k steps in π and ending in state s_k. Infinite $\mathfrak{S}$-paths are defined accordingly.

For an MDP $(\mathcal{M}, s)$, i.e. an MDP as before with some distinguished initial state $s \in S$, the behavior of $(\mathcal{M}, s)$ under $\mathfrak{S}$ is purely probabilistic and can be formalized by an infinite tree-like Markov chain $\mathcal{M}_s^{\mathfrak{S}}$ where the states are the finite $\mathfrak{S}$-paths starting in s. The probability measure $\mathbb{P}_s^{\mathfrak{S}}$ for measurable sets of the infinite paths in the Markov chain $\mathcal{M}_s^{\mathfrak{S}}$, can be transferred to infinite $\mathfrak{S}$-paths in $\mathcal{M}$ starting in s. Thus, if Φ is a path event then $\mathbb{P}_{\mathcal{M},s}^{\mathfrak{S}}(\Phi)$ denotes its probability under scheduler $\mathfrak{S}$ for starting state s. For a worst-case analysis of a system modeled by an MDP $\mathcal{M}$, one ranges over all schedulers (i.e., all possible resolutions of the nondeterminism) and considers the maximal or minimal probabilities for Φ:

$$\mathbb{P}_s^{\min}(\Phi) \;=\; \inf_{\mathfrak{S}} \mathbb{P}_s^{\mathfrak{S}}(\Phi) \qquad\qquad \mathbb{P}_s^{\max}(\Phi) \;=\; \sup_{\mathfrak{S}} \mathbb{P}_s^{\mathfrak{S}}(\Phi)$$

We use weight functions of the following types: $wgt_{st} : S \to \mathbb{Z}$ for the states (as for Markov chains) and $wgt : S \times Act \times S \to \mathbb{Z}$ for state-action-state tuples. The accumulated weight of finite paths as well as the expected accumulated weight $\mathbb{E}_s^{\mathfrak{S}}\big(wgt[\lozenge F] \big)$ to reach a target set F under some scheduler $\mathfrak{S}$ with $\mathbb{P}_s^{\mathfrak{S}}(\lozenge F) = 1$ is defined in the obvious way.

Assumptions and Relaxed Notation for Path Properties. For the methods proposed in the following sections, we suppose that the transition probabilities are rational. When using LTL-like or CTL-like notation with atomic

propositions in AP we suppose the Markov chain or MDP $\mathcal{M}$ under consideration is extended by a labeling function that declares which atomic propositions
$a \in$ AP hold in which states. At several places, we use single states or sets of
states in $\mathcal{M}$ as atomic propositions with the obvious meaning. For the interpretation of LTL- or CTL-like formulas in $\mathcal{M}$, the probability annotations (as well
as the action labels in case of an MDP and the exit rates in CTMCs) are ignored
and $\mathcal{M}$ is viewed as an ordinary Kripke structure. A path property φ is said to
be measurable if the set of infinite paths $\tilde{\pi}$ in $\mathcal{M}$ satisfying φ is a path event,
i.e., an element of the induced sigma-algebra. It is well known that all ω-regular
path properties are measurable [41]. We abuse notation and identify measurable
path properties and the induced path event. Thus, if $\mathcal{M}$ is an MDP then

$$\mathbb{P}_s^{\mathfrak{S}}(\varphi) \ = \ \mathbb{P}_s^{\mathfrak{S}} \{ \tilde{\pi} \in \mathit{InfPaths}(s) \ : \ \tilde{\pi} \models \varphi \}$$

denotes the probability for φ under scheduler $\mathfrak{S}$ and starting state s.

3   Relativized Long-Run Probabilities

In the classical approach of PMC for Markov chains, the quantitative analysis
computes the probabilities $\mathbb{P}_s^{\mathcal{M}}(\varphi)$ for the given path event for all states s in
the given Markov chain. Many low-level operating-system protocols are, how-
ever, designed to run for an indefinite period of time and most relevant for the
quantitative analysis is the long-run behavior, when exceptional phenomena of
the initialization phase lose their influence. Possible sources for significant different quantitative behaviors in the long-run and in the initialization phase are,
e.g., variations in the frequency of cache misses or competitions between processes to access shared resources. This indicates that the quantitative analysis
should be carried out assuming the steady-state distribution as the initial distribution. However, for many relevant quantitative measures only some states
that have positive steady-state probability are appropriate as reference points.
For instance, to determine the average time that a process waits for a requested
shared resource in the long-run, the relevant reference points are those states
with positive steady-state probability that are entered immediately after the
considered process has performed its request operation. That is, what we need
are long-run probabilities of temporal properties or long-run expected values,
relativized by some state condition that declares the relevant reference points.
The mathematical definition of *relativized long-run probabilities* for CTMCs is as
follows. Suppose $\mathcal{C} = (\mathcal{M}, E)$ is a CTMC, μ an initial distribution as in Section 2
and U a subset of S such that the steady-state probability for U is positive, i.e.,
$\mathrm{StPr}_\mu^{\mathcal{C}}(U) > 0$. Given an event φ, then the long-run probability for φ relativized
to U is given by:

$$\mathbb{LP}_\mu^{\mathcal{C}}[U](\varphi) \ = \ \mathbb{P}_\theta^{\mathcal{C}}(\varphi)$$

where θ is the relativized steady-state distribution where the condition is given by U. Formally, θ is the distribution on S given by $\theta(s) = 0$ if $s \in S \setminus U$ and

$$\theta(s) \;=\; \frac{\text{StPr}_\mu^\mathcal{C}(s)}{\text{StPr}_\mu^\mathcal{C}(U)} \qquad \text{if } s \in U$$

Likewise, we define the relativized long-run expected value of random functions for infinite trajectories, such as the relativized long-run average accumulated costs to reach a goal set F:

$$\mathbb{LE}_\mu^\mathcal{C}\,[U]\,(wgt[\lozenge F]) \;=\; \mathbb{E}_\theta^\mathcal{C}(wgt[\lozenge F])$$

For the case of discrete Markov chains, the definition of relativized long-run probabilities is analogous, except that we have to deal with the Cesàro limit:

$$\text{StPr}_\mu^\mathcal{M}(s) \;=\; \lim_{n\to\infty} \frac{1}{n+1}\cdot\sum_{i=0}^{n}\mathbb{P}_\mu^\mathcal{M}\big\{\,s_0\,s_1\,s_2\ldots \in \mathit{InfPaths}(s)\, :\, s_i = s\,\big\}$$

for the definition of steady-state probabilities. The Cesàro limit always exists in finite Markov chains and is computable via an analysis of bottom strongly connected components and linear equation systems.

State-of-the art probabilistic model checkers support the computation of steady-state probabilities, but there is no direct support to compute the probabilities for temporal path properties or expected accumulated values under steady-state probabilities. For our case studies on low-level OS code we extended the prominent model checkers PRISM [32] and MRMC [30] to compute relativized long-run probabilities and average accumulated rewards. To illustrate the demand of relativized long-run probabilities and expectations we now detail two of our case studies.

Example 1 (Spinlock). In a case study on a locking protocol modeled as a discrete Markov chain [6,5] the concept of relativized long-run properties was crucial. We performed exhaustive experiments to compute various quantitative measures that have been identified to be most relevant. We provide here three examples. The first one is the long-run probability for a process acquiring a lock within the next time step, assuming that the process just requested the lock:

$$\mathbb{LP}_\mu^\mathcal{C}\,[\text{``lock requested''}]\,(\bigcirc\text{``get lock''})$$

Another example is the long-run probability that the same process gets the lock two times in a row:

$$\mathbb{LP}_\mu^\mathcal{C}\,[\text{``lock released''}]\,(\text{``lock free''}\ \mathrm{U}\ \text{``get lock''})$$

As an example for relativized long-run expectations one may ask for the average time or energy spent on getting access to the lock:

$$\mathbb{LE}_\mu^\mathcal{C}\,[\text{``lock requested''}]\,(wgt[\lozenge\,\text{``get lock''}]) \qquad\blacksquare$$

Example 2 (Probabilistic-Write/Copy-Select). Also in the context of the quantitative analysis [7] of a novel synchronization scheme suggested by McGuire [34], called *Probabilistic-Write/Copy-Select (PWCS)*, long-run probabilities play an important role. The idea behind PWCS is that the protocol avoids expensive atomic operations for synchronizing access to shared objects. Instead, PWCS makes inconsistencies detectable (using hashes or version numbers) and recoverable. The protocol builds on the assumption that the probability for data races is very small for typical workload scenarios. Hence, the PWCS protocol belongs to a new and interesting class of algorithms, which make use of the inherent randomness (called inherent nondeterminism in [34]) provided by modern parallel hardware architectures, which can potentially also be used as an alternative solution for breaking symmetries. Our formal analysis of a variant of the PWCS protocol with multiple writers uses a CTMC and centers around the probability of data races in the long-run for different types of workload scenarios. PWCS makes use of an indexed list of replicas of the original data object to increase the probability for a reader to find at least one consistent replica. That is, the writers operate in a write cycle in which they modify the replicas successively in an increasing order. The readers visit the replicas in a reversed order within their read cycle to find at least one consistent version of the data, i.e., a replica not damaged due to concurrent write operations. Figure 1 shows the transition system for the kth replica (left) and the CTMC for the ith writer (right), whereas the CTMC for the jth reader is depicted in Figure 2. The CTMC $\mathcal{C}$ for the PWCS protocol arises from a parallel composition of the CTMC for the writers and the readers and the ordinary transition systems for the replicas. We used here CSP-like notations for synchronous send (!) and receive (?) actions. Rates (denoted by greek letters in the figures) are only attached to send actions, while the matching receive action in the replicas is "reactive" and has no rate. (See [7] for a formal definition of the product.) Examples for relevant quantitative measures are:

(1) the probability to successfully read a replica for a reader (in the long-run):

$$\mathbb{LP}_\mu^\mathcal{C} [\text{"reading started"}] (\Diamond \text{"reading successful"})$$

(2) the probability to write at least $c \in \mathbb{N}$ consistent replicas within a write cycle

$$\mathbb{LP}_\mu^\mathcal{C} [\text{"writing started"}] (\phi),$$

where ϕ stands for a formula stating that the writing of at least c replicas is not interrupted by any other writer within one write cycle.

(3) the average time or energy costs (in the long-run) for repairing a replica:

$$\mathbb{LE}_\mu^\mathcal{C} [\text{"just damaged"}] (wgt[\Diamond \text{"consistent"}])$$

For the computation we annotate all states of $\mathcal{C}$ with reward 1 for the time and the respective energy costs to compute the above relativized long-run accumulated reward. ∎

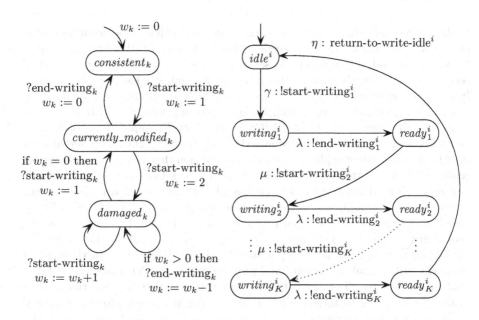

Fig. 1. Transition system for the kth replica (left) and CTMC for the ith writer (right)

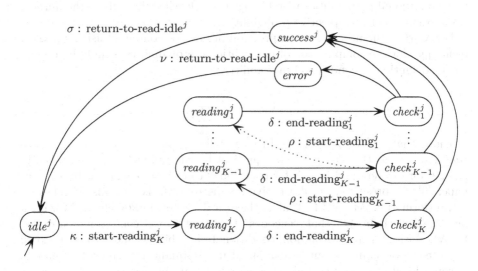

Fig. 2. CTMC for the jth reader

4   Conditional Probabilities and Rewards

Probabilities and expectations under the assumption that some additional temporal condition holds are often needed within the quantitative analysis of protocols. In general, any measurable set of paths with positive probability could serve as a condition when computing either a probability value or expected costs. In practice the conditions are often liveness properties or time or cost bounded reachability properties representing progress assumptions. For instance, in the context of the case study on the PWCS-protocol (cf. Example 2) our operating-system colleagues asked us to compute the expected time until a reader successfully reads a consistent replica, under the assumption that it will find a consistent replica without interference. When F formalizes the set of states where a consistent replica is read and G is the set of states without an interference, this amounts to computing a conditional expected value of the form $\mathbb{E}_s^{\mathcal{M}}\big(\,wgt[\Diamond F]\,|\,GUF\,\big)$. Recall that expectations of a path property are undefined if the corresponding probability is less than 1. Hence, the corresponding unconditional expected accumulated reward $\mathbb{E}_s^{\mathcal{M}}\big(\,wgt[\Diamond F]\,\big)$ is undefined if $\mathbb{P}_s^{\mathcal{M}}(\Diamond F) < 1$. In the context of energy-utility analysis, conditional probabilities or expectations are useful to analyze the energy-efficiency, while assuming that a certain condition on the achieved utility is guaranteed. Vice versa, one might ask, e.g., for the expected utility, while not exceeding a given energy budget. Conditional probabilities also offer an alternative to the standard methods for dealing with fairness assumptions. This is of particular interest for MDPs where the classical analysis ranges over fair schedulers (i.e., schedulers that generate almost surely paths where the fairness assumption holds) and excludes unfair schedulers, whereas an approach with conditional probabilities might range over all schedulers where the fairness assumptions hold with positive probability.

For Markov chains, conditional probabilities can be computed using standard techniques for the computation of unconditional probabilities, simply by relying on the definition of conditional probabilities:

$$\mathbb{P}_s^{\mathcal{M}}\big(\varphi\,|\,\psi\big) \;=\; \frac{\mathbb{P}_s^{\mathcal{M}}\big(\varphi\wedge\psi\big)}{\mathbb{P}_s^{\mathcal{M}}(\psi)}$$

provided that $\mathbb{P}_s^{\mathcal{M}}(\psi)$ is positive. In what follows, we refer to ψ as the *condition* and to φ as the *objective*. This quotient-based approach has been taken in [2,1,29] for discrete and in [25] for continuous-time Markov chains. In a recent paper [9], we presented an alternative for discrete Markov chains. It relies on a transformation $\mathcal{M} \rightsquigarrow \mathcal{M}_\psi$ that replaces a discrete Markov chain $\mathcal{M}$ with a new Markov chain $\mathcal{M}_\psi$ that "encodes" the effect of the condition ψ imposed for $\mathcal{M}$. With this transformation, conditional probabilities and also conditional expected rewards in $\mathcal{M}$ can be computed using standard methods to compute unconditional probabilities and expected rewards in $\mathcal{M}_\psi$ for an analogous event resp. an analogous random function. We recall the definition of $\mathcal{M}_\psi$ as proposed in [9] for the case where ψ is a reachability condition. Then, we show that the same transformation is also applicable for CTMCs.

Conditional Probabilities in DTMCs. Let $\mathcal{M} = (S, P)$ be a discrete Markov chain and $\psi = \Diamond F$ where $F \subseteq S$. The new Markov chain $\mathcal{M}_\psi = (S_\psi, P_\psi)$ has the state space $S_\psi = S^{bef} \cup S^{nor}$ where:

$$S^{bef} = \left\{ s^{bef} : s \in S, s \models \exists \Diamond F \right\} \qquad S^{nor} = \left\{ s^{nor} : s \in S \right\}.$$

Intuitively, s^{nor} is simply a copy of state s in $\mathcal{M}$ ("normal mode"). The states s^{bef} are only needed if F is reachable from s in $\mathcal{M}$ and the behavior of s^{bef} in $\mathcal{M}_\psi$ agrees with the behavior of s in $\mathcal{M}$ under the condition that s will indeed reach F. We refer to the states s^{bef} as a copy of s in the "before mode". Intuitively, $\mathcal{M}_\psi$ starts in the before mode. As soon as F has been reached, $\mathcal{M}_\psi$ switches to the *normal mode* where $\mathcal{M}_\psi$ behaves as $\mathcal{M}$. This is formalized by the transition probability function of $\mathcal{M}_\psi$. If $s \in S \setminus F$ and $v \in S$ with $s \models \exists \Diamond F$ and $v \models \exists \Diamond F$:

$$P_\psi\left(s^{bef}, v^{bef} \right) \;=\; P(s, v) \cdot \frac{\mathbb{P}_v^{\mathcal{M}}(\Diamond F)}{\mathbb{P}_s^{\mathcal{M}}(\Diamond F)}$$

For $s \in F$, we define $P_\psi(s^{bef}, v^{nor}) = P(s, v)$, modeling the switch from before to normal mode. For the states in normal mode, the transition probabilities are given by $P_\psi(s^{nor}, v^{nor}) = P(s, v)$. In all other cases, $P_\psi(\cdot) = 0$.

Obviously, all states in the before mode can reach $F^{bef} = \{ v^{bef} : v \in F \}$. This yields $\mathbb{P}_{s^{bef}}^{\mathcal{M}_\psi}(\Diamond F^{bef}) = 1$ for all states s of $\mathcal{M}$ with $s \models \exists \Diamond F$. Suppose now that φ is a measurable path property formalized, e.g., in LTL and using atomic propositions for the states in $\mathcal{M}$. Assuming that the copies s^{bef} and s^{nor} satisfy the same atomic propositions as s, then there is a one-to-one correspondence between the infinite paths $\tilde{\pi}$ in $\mathcal{M}$ with $\tilde{\pi} \models \varphi \wedge \Diamond F$ and the infinite paths $\tilde{\pi}_\psi$ in $\mathcal{M}_\psi$ with $\tilde{\pi}_\psi \models \varphi$. More precisely, each path $\tilde{\pi}_\psi$ in $\mathcal{M}_\psi$ induces a path $\tilde{\pi}_\psi|_\mathcal{M}$ in $\mathcal{M}$ by dropping the mode annotations. Vice versa, if $\tilde{\pi}$ is a path in $\mathcal{M}$ with $\tilde{\pi} \models \Diamond F$ or $\tilde{\pi} \models \Box \exists \Diamond F$ then $\tilde{\pi}$ can be augmented with mode annotations to obtain a path $\tilde{\pi}_\psi$ in $\mathcal{M}_\psi$ with $\tilde{\pi}_\psi|_\mathcal{M} = \tilde{\pi}$. This yields a probability preserving one-to-one correspondence between the cylinder sets in $\mathcal{M}_\psi$ and the cylinder sets in $\mathcal{M}$ spanned by finite paths $s_0 s_1 \ldots s_n$ of $\mathcal{M}$ such that for all $i \in \{0, 1, \ldots, n-1\}$: if $s_i \not\models \exists \Diamond F$ then $s_j \in F$ for some $j < i$. Hence:

$$\mathbb{P}_s^{\mathcal{M}}\left(\varphi \mid \Diamond F \right) \;=\; \mathbb{P}_{s^{bef}}^{\mathcal{M}_\psi}\left(\varphi \right)$$

for all measurable path properties φ. The analogous statement holds for the expected values of random functions.

This approach can be generalized for the case of ω-regular conditions. We have implemented the above transformation for LTL conditions within the probabilistic model checker PRISM and evaluated the implementation with a case study on a bounded retransmission protocol [9]. Among others, we used our implementation to compute the probability of observing two retries for sending a data fragment under the condition that globally no errors occur, i.e.,

$$\mathbb{P}_s^{\mathcal{M}}\left(\Diamond \text{ "second retry for fragment"} \mid \neg \Diamond \text{"finish with error"} \right).$$

As conditional expectation we considered, e.g., the expected (energy) costs for data transmission under the condition that the data was transmitted successfully:

$$\mathbb{E}_s^{\mathcal{M}}\big(\, wgt[\lozenge \text{ "finish with success"}] \mid \lozenge \text{ "finish with success"}\,\big)$$

In the context of our experiments it turned out that indeed using this implementation was more efficient than the standard quotient-based approach. For details we refer to [9].

Conditional Probabilities in CTMCs. We now adapt this transformation for CTMCs. Let $\mathcal{C} = (\mathcal{M}, E)$ where $\mathcal{M}$ is as before and $E : S \to \mathbb{R}$ is an exit rate function. We define:

$$\mathcal{C}_\psi = (\mathcal{M}_\psi, E_\psi) \quad \text{where} \quad E_\psi(s^{bef}) = E_\psi(s^{nor}) = E(s)$$

for all states s in $\mathcal{C}$. The above mentioned one-to-one correspondence between paths in $\mathcal{M}$ satisfying $\lozenge F$ or $\square \exists \lozenge F$ and paths in $\mathcal{M}_\psi$ carries over to trajectories in the CTMCs $\mathcal{C}$ and $\mathcal{C}_\psi$. If Φ is a measurable set of trajectories in $\mathcal{C}$ then:

$$\Phi_\psi = \big\{ \tilde{\vartheta}_\psi : \tilde{\vartheta} \in \Phi, \, \tilde{\pi} \models \lozenge F \vee \square \exists \lozenge F \big\}$$

where $\tilde{\vartheta}_\psi$ results from $\tilde{\vartheta}$ by adding the obvious mode annotations to the states.

Theorem 1. *With the notations as before, Φ_ψ is measurable in the sigma-algebra induced by $\mathcal{C}_\psi$ and for each s of $\mathcal{M}$ with $s \models \exists \lozenge F$:*

$$\mathbb{P}_s^{\mathcal{C}}\big(\Phi \mid \lozenge F\big) = \mathbb{P}_{s^{bef}}^{\mathcal{C}_\psi}\big(\Phi_\psi\big)$$

Proof. Measurability of Φ_ψ is clear, given the measurability of Φ and the set $\{\tilde{\vartheta} \in InfTraj : \tilde{\vartheta} \models \lozenge F \vee \square \exists \lozenge F\}$. It remains to prove the claim for cylinder sets. The above mentioned one-to-one correspondence between finite paths in $\mathcal{M}$ that do not contain a state s with $s \not\models \exists \lozenge F$, unless F has been visited before, and $\mathcal{M}_\psi$ resp. finite trajectories in $\mathcal{C}$ and $\mathcal{C}_\psi$ yields a one-to-one correspondence between the cylinder sets (symbolic finite trajectories) in $\mathcal{C}$ and the cylinder sets in $\mathcal{C}_\psi$. The above transformation $\pi \mapsto \pi_\psi$ for paths in the embedded discrete Markov chain can be adapted for (symbolic) finite trajectories. We define $\mathcal{T}_\psi$ as the symbolic finite trajectory that results from $\mathcal{T}$ by adding appropriate mode annotations. Likewise, for a symbolic trajectory $\mathcal{T}_\psi$ in $\mathcal{C}_\psi$ we define $\mathcal{T}_\psi|_{\mathcal{C}}$ as the symbolic finite trajectory obtained from $\mathcal{T}_\psi$ by dropping the mode annotations.

Claim: Let $\mathcal{T} = s_0\, I_0\, s_1\, I_1 \ldots I_{n-1}\, s_n$ be a symbolic finite trajectory in $\mathcal{C}$ such that $\{s_0, \ldots, s_n\} \cap F \neq \varnothing$ or $s_i \models \exists \lozenge F$ for all $1 \leqslant i \leqslant n$. Then,

$$\mathbb{P}_s^{\mathcal{C}}\big(\, Cyl(\mathcal{T}) \mid \lozenge F\big) = \mathbb{P}_{s^{bef}}^{\mathcal{C}_\psi}\big(\, Cyl(\mathcal{T}_\psi)\big)$$

Proof of the claim. We only address the case that $\mathcal{T}$ contains some state in F, say $s_k \in F$ and $\{s_0, \ldots, s_{k-1}\} \cap F = \varnothing$ where $0 \leqslant k \leqslant n$. In particular, $\tilde{\vartheta} \models \lozenge F$ for all $\tilde{\vartheta} \in Cyl(\mathcal{T})$ and therefore

$$\mathbb{P}_s^{\mathcal{C}}\big(\, Cyl(\mathcal{T}) \wedge \lozenge F\big) = \mathbb{P}_s^{\mathcal{C}}\big(\, Cyl(\mathcal{T})\big),$$

where $\Diamond F$ is identified here with the set of infinite trajectories that contain at least one F-state. We now show that the conditional probability of the cylinder set of $\mathcal{T}$ in $\mathcal{C}$ agrees with the unconditional probability of the cylinder set

$$\mathcal{T}_\psi = s_0^{bef} I_0 s_1^{bef} I_1 \ldots I_k s_k^{bef} I_{k+1} s_{k+1}^{nor} I_{k+1} \ldots I_{n-1} s_n^{nor}$$

in $\mathcal{C}_\psi$. Indeed, with $s = s_0$ we have:

$$\mathbb{P}_{s^{bef}}^{\mathcal{C}_\psi}\big(\, Cyl(\mathcal{T}_\psi) \, \big)$$

$$= \prod_{i=0}^{k-1} P_\psi(s_i^{bef}, I_i, s_{i+1}^{bef}) \cdot P_\psi(s_k^{bef}, I_k, s_{k+1}^{nor}) \cdot \prod_{i=k+1}^{n-1} P_\psi(s_i^{nor}, I_i, s_{i+1}^{nor})$$

$$= \prod_{i=0}^{k-1} \left(P(s_i, I_i, s_{i+1}) \cdot \frac{\mathbb{P}_{s_{i+1}}^{\mathcal{C}}(\Diamond F)}{\mathbb{P}_{s_i}^{\mathcal{C}}(\Diamond F)} \right) \cdot \prod_{i=k}^{n-1} P(s_i, I_i, s_{i+1})$$

$$= \prod_{i=0}^{n-1} P(s_i, I_i, s_{i+1}) \cdot \prod_{i=0}^{k-1} \frac{\mathbb{P}_{s_{i+1}}^{\mathcal{C}}(\Diamond F)}{\mathbb{P}_{s_i}^{\mathcal{C}}(\Diamond F)}$$

$$= \mathbb{P}_{s_0}^{\mathcal{C}}\big(\, Cyl(\mathcal{T}) \, \big) \cdot \frac{1}{\mathbb{P}_{s_0}^{\mathcal{C}}(\Diamond F)} \quad = \quad \mathbb{P}_{s_0}^{\mathcal{C}}\big(\, Cyl(\mathcal{T}) \mid \Diamond F \, \big)$$

The argument for finite trajectories in $\mathcal{C}$ that do not contain an F-state, but consist of states s with $s \models \exists \Diamond F$, is analogous. ∎

As a consequence of Theorem 1, if φ is a measurable trajectory property, e.g., formalized by a time-bounded until formula then

$$\mathbb{P}_s^{\mathcal{C}}\big(\varphi \mid \Diamond F\big) \quad = \quad \mathbb{P}_{s^{bef}}^{\mathcal{C}_\psi}\big(\varphi\big)$$

and the analogous statement for the expected values of random functions. For instance, if $rew : S \to \mathbb{N}$ is a reward function for the states in $\mathcal{M}$ then

$$\mathbb{E}_s^{\mathcal{C}}\big(\, rew[\Diamond F] \mid \Diamond F \, \big) \quad = \quad \mathbb{E}_{s^{bef}}^{\mathcal{C}_\psi}\big(\, rew_\psi[\Diamond F_\psi] \, \big),$$

where $F_\psi = \{s^{bef} : s \in S, s \models \exists \Diamond F\} \cup \{s^{nor} : s \in S\}$ and $rew_\psi(s^{bef}) = rew(s)$ if $s \models \exists \Diamond F$ and $rew_\psi(s^{nor}) = rew(s)$ for all states s in $\mathcal{M}$. Thus, the computation of conditional probabilities and rewards in CTMCs where the condition is an *untimed* reachability property can be carried out using the presented transformation $\mathcal{C} \rightsquigarrow \mathcal{C}_\psi$ and applying standard techniques for the computation of unconditional probabilities or expected rewards in $\mathcal{C}_\psi$. One advantage of this approach is that the transformed $\mathcal{C}_\psi$ only depends on the condition ψ and it can be used for different objectives. The construction of $\mathcal{C}_\psi$ is simple since it only requires the computation of (untimed) reachability probabilities.

The presented transformation $\mathcal{C} \rightsquigarrow \mathcal{C}_\psi$ can be adapted easily for (untimed) ω-regular conditions. This has been explained in [9] for discrete Markov chains, using a representation of the condition ψ by a deterministic ω-automaton $\mathcal{A}$ and the (standard) product construction $\mathcal{M} \otimes \mathcal{A}$ to reduce the probabilities of ψ in $\mathcal{M}$ to reachability conditions in $\mathcal{M} \otimes \mathcal{A}$. The adaption of this transformation-based approach for CTMCs is straightforward. Thus, our transformation yields an alternative to the quotient-based approach of [25] for the case of untimed conditions. [25] deals with conditions and objectives that are represented as path formulas of the logic CSRL (continuous stochastic reward logic) [3], and thus can handle timed conditions. It would be interesting to study whether our transformation-based approach can be extended for conditions with timing constraints, e.g., provided by timed automata specifications [19].

Conditional Probabilities in MDPs. The task to reason about conditional probabilities in Markov decision processes is more challenging. The matter is that for the computation of, e.g.,

$$\mathbb{P}_s^{\max}(\varphi \,|\, \psi) \;=\; \max_{\mathfrak{S}} \mathbb{P}_s^{\mathfrak{S}}(\varphi \,|\, \psi) \;=\; \max_{\mathfrak{S}} \frac{\mathbb{P}_s^{\mathfrak{S}}(\varphi \wedge \psi)}{\mathbb{P}_s^{\mathfrak{S}}(\psi)}$$

we cannot simply maximize the nominator and denominator. This problem has been addressed first in [2,1], where an extension of PCTL over MDPs [11] by a conditional probability operator has been presented. The presented model-checking algorithm relies on an exhaustive search (with heuristic bounding techniques) in some finite, but potentially exponentially large class of finite-memory schedulers. In [9] we improved this result by presenting a polynomial transformation $\mathcal{M} \rightsquigarrow \mathcal{M}_{\varphi|\psi}$ for reachability objectives and conditions, which has been shown to be the core problem for reasoning about ω-regular objectives and conditions by using automata representations of the objective and the condition.

5   Quantiles

Quantiles play a central role in statistics, where for a given random variable R and probability value p, the quantile is defined as the least value r such that the probability for $R \geqslant r$ is at least p. In the context of system analysis, quantiles can provide useful insights about the interplay of two objectives:

> Reduce the cost (formalized by R), while providing guarantees on
> the utility (probability for some event is larger than p).

This allows combining energy (and other cost measures) with utility aspects like throughput, latency and other QoS measures. Typical examples for quantiles are the minimal time that a thread has to wait for a requested resource with probability $\geqslant 0.98$ or the minimal amount of energy required to complete a list of tasks without missing some deadline with probability $\geqslant 0.98$.

For discrete Markov chains over a state space S, quantiles can be formalized using a reward function $rew \colon S \to \mathbb{N}$ (i.e., a nonnegative weight function for the states) to model R and a reward-bounded temporal constraint formalizing the utility objective. This formalization is more general than using time-bounded temporal constraints, since those can also be expressed by a reward function assigning reward 1 to all states. When $\varphi[r]$ denotes a path property imposing bounds on the costs formalized by the reward r, the above mentioned quantile refers to

$$\mathsf{Qu}_s^{\mathcal{M}}(\mathsf{Pr}_{\geqslant p}(\varphi[?])) \;=\; \min\{\, r \in \mathbb{N} : \mathbb{P}_s^{\mathcal{M}}(\varphi[r]) \geqslant p \,\}.$$

The bounded path property $\varphi[r]$ could be, e.g., an upper reward-bounded until formula $\varphi[r] = GU^{\leqslant r}F$ over sets of states $G, F \subseteq S$. Intuitively, $GU^{\leqslant r}F$ denotes the constraint that F will eventually be reached along some finite path π consisting of G-states with an accumulated reward of at most r (i.e., $rew(\pi) \leqslant r$). For instance, in the spinlock protocol case study (see Example 1), we used a reward function for encoding the waiting time and computed the quantile $\mathsf{Qu}_s^{\mathcal{M}}(\mathsf{Pr}_{\geqslant 0.8}(\lozenge^{\leqslant ?}\text{"get lock"}))$ for a state s that has just requested the lock. This corresponds to the minimal required time to ensure that the lock is acquired with a probability of at least 80%. Instead of $GU^{\leqslant r}F$, $\varphi[r]$ can be any other reward-bounded path property where the function $r \mapsto \mathbb{P}_s^{\mathcal{M}}(\varphi[r])$ is increasing, i.e., where $\mathbb{P}_s^{\mathcal{M}}(\varphi[r]) \leqslant \mathbb{P}_s^{\mathcal{M}}(\varphi[r+1])$ for all $r \in \mathbb{N}$.

Other examples for quantiles aim to maximize the utility formalized by the reward r by imposing constraints on the costs, e.g., by lower reward-bounded until properties $\psi[r] = GU^{>r}F$ over states $G, F \subseteq S$. The path property $GU^{>r}F$ is fulfilled by all those paths only visiting G-states until eventually reaching F with an accumulated reward exceeding r. In this case, the quantile is defined as the maximal value r such that $\psi[r]$ holds with sufficiently large probability:

$$\mathsf{Qu}_s^{\mathcal{M}}(\mathsf{Pr}_{\geqslant p}(\psi[?])) \;=\; \max\{\, r \in \mathbb{N} : \mathbb{P}_s^{\mathcal{M}}(\psi[r]) \geqslant p \,\}.$$

Note that similar as in the minimizing quantile, $\psi[r]$ can be any reward-bounded path property where $r \mapsto \mathbb{P}_s^{\mathcal{M}}(\psi[r])$ is decreasing, i.e., $\mathbb{P}_s^{\mathcal{M}}(\psi[r]) \geqslant \mathbb{P}_s^{\mathcal{M}}(\psi[r+1])$ for all $r \in \mathbb{N}$. The definition for strict probability bounds "$> p$" rather than "$\geqslant p$" is analogous. Quantiles with upper probability bounds are reducible to quantiles with lower probability bounds [40,4], e.g., for $\varphi[r] = GU^{\leqslant r}F$:

$$\mathsf{Qu}_s^{\mathcal{M}}(\mathsf{Pr}_{\leqslant p}(\varphi[?])) \;=\; \max\{\, r \in \mathbb{N} : \mathbb{P}_s^{\mathcal{M}}(\varphi[r]) \leqslant p \,\} \;=\; \mathsf{Qu}_s^{\mathcal{M}}(\mathsf{Pr}_{> p}(\varphi[?]))-1$$

Similarly, reward-bounded always and reachability properties can be transformed into each other, e.g., $\mathsf{Qu}_s^{\mathcal{M}}(\mathsf{Pr}_{\geqslant p}(\Box^{\leqslant ?}G))$ is equivalent to $\mathsf{Qu}_s^{\mathcal{M}}(\mathsf{Pr}_{\leqslant 1-p}(\lozenge^{\leqslant ?}\neg G))$.

In this paper, we only define quantiles for discrete Markov chains. However, more general quantiles for MDPs can also be considered, which require a further declaration whether the probability bound is imposed for all schedulers or only for some scheduler [40,4].

Computing Quantiles in DTMCs. In [40] a linear-programming approach for computing quantiles for MDPs has been proposed. In order to support a clean presentation, we continue presenting the approach towards computing quantiles for discrete Markov chains only. Then, the linear program of [40] boils down to a single linear-equation system. For instance, if $\varphi[r] = GU^{\leqslant r}F$, the values $x_{s,i} = \mathbb{P}_s^{\mathcal{M}}(\varphi[i])$ for $0 \leqslant i \leqslant r$ can be computed with the following equation scheme:

$$x_{s,i} = \sum_{u \in S} P(s,u) \cdot x_{u,i-rew(s)} \quad \text{if } rew(s) \leqslant i,\ s \notin F \text{ and } s \models \exists GUF,$$

and $x_{s,i} = 1$ for $s \in F$ and $x_{s,i} = 0$ if $s \not\models \exists G U F$ or $rew(s) > i$. The value $\mathrm{Qu}_s^{\mathcal{M}}(\mathrm{Pr}_{\geqslant p}(\varphi[?]))$ is then determined by computing $x_{s,i}$ for increasing i until $x_{s,i} \geqslant p$.

The approach by [40] only deals with quantiles concerning upper reward-bounded properties $\varphi[r]$ or $\psi[r]$ where the utility is specified by a lower or upper bound on the probability. In [4], this approach has been generalized towards computing lower reward-bounded properties. Furthermore, [4] introduced expectation quantiles, where the utility is given by a bound on the expected utility depending on the costs (e.g., consumed energy). In this setting, two reward functions for the states are considered:

$$rew^u : S \to \mathbb{N} \quad \text{for the utility, and}$$
$$rew^e : S \to \mathbb{N} \quad \text{for the energy}$$

The induced function $\mathsf{energy} : FinPaths \to \mathbb{N}$ for the accumulated costs along finite paths is given by $\mathsf{energy}(s_0 \ldots s_n) = \sum_{i=0}^{n-1} rew^e(s_i)$. For a given energy bound $r \in \mathbb{N}$, the random variable $\mathsf{utility}[\Diamond^{\mathsf{energy}\leqslant r} F] : InfPaths \to \mathbb{N}$ is given by:

$$\mathsf{utility}[\Diamond^{\mathsf{energy}\leqslant r} F](s_0 s_1 s_2 \ldots) = rew^u(s_0 s_1 \ldots s_n)$$
$$\text{if } \{s_0, \ldots, s_{n-1}\} \cap F = \varnothing, \; s_n \in F \text{ and } \mathsf{energy}(s_0 \ldots s_n) \leqslant r$$

For infinite paths $\tilde{\pi}$ that do not visit F within the given energy bound r, $\mathsf{utility}[\Diamond^{\mathsf{energy}\leqslant r} F](\tilde{\pi})$ is irrelevant. Assuming that F will be reached from state s almost surely within the given energy bound, i.e., $\mathbb{P}_s^{\mathcal{M}}(\Diamond^{\mathsf{energy}\leqslant r} F) = 1$, then the expected value of $\mathsf{utility}[\Diamond^{\mathsf{energy}\leqslant r} F]$ under $\mathbb{P}_s^{\mathcal{M}}$ is well-defined and given by:

$$\mathbb{E}_s^{\mathcal{M}}\big(\mathsf{utility}[\Diamond^{\mathsf{energy}\leqslant r} F]\big) = \sum_{\pi} \mathrm{Pr}(\pi) \cdot \mathsf{utility}[\Diamond^{\mathsf{energy}\leqslant r} F](\pi),$$

where π ranges over all finite paths $s_0 s_1 \ldots s_n$ starting in $s_0 = s$ such that $\{s_0, \ldots, s_{n-1}\} \cap F = \varnothing$, $s_n \in F$ and $\mathsf{energy}(s_0 \ldots s_n) \leqslant r$. We are interested in the quantile

$$q_s = \min\big\{ r \in \mathbb{N} : \mathbb{P}_s^{\mathcal{M}}(\Diamond^{\mathsf{energy}\leqslant r} F) = 1 \text{ and } \mathbb{E}_s^{\mathcal{M}}\big(\mathsf{utility}[\Diamond^{\mathsf{energy}\leqslant r} F]\big) \geqslant x \big\},$$

where $\min \varnothing = \infty$. Let

$$S' = \big\{ s \in S : \mathbb{P}_s^{\mathcal{M}}(\Diamond^{\mathsf{energy}\leqslant r} F) = 1 \text{ for some } r \in \mathbb{N} \big\}.$$

Then, for each such state $s \in S'$ the smallest energy bound such that F is visited almost surely is given by

$$r_s = \mathrm{Qu}_s^{\mathcal{M}}\big(\mathrm{Pr}_{\geqslant 1}(\Diamond^{\mathsf{energy}\leqslant ?} F)\big).$$

If $s \in S' \backslash F$, then $u \in S'$ and $r_u \leqslant r_s - rew^e(s)$ for all $u \in S$ with $P(s, u) > 0$.

The values $x_{s,i} = \mathbb{E}_s^{\mathcal{M}}\big(\mathsf{utility}[\Diamond^{\mathsf{energy}\leqslant i} F]\big)$ for $s \in S'$ and $i \geqslant r_s$ are well-defined and can be computed via the following linear-equation system:

$$x_{s,i} = rew^u(s) + \sum_{u \in S} P(s, u) \cdot x_{u, i-rew^e(s)} \quad \text{if } s \notin F \text{ and } i \geqslant r_s,$$

where $x_{s,i} = 0$ for $s \in F$ and $i \geqslant 0$. In all other cases, the values $x_{s,i}$ are irrelevant. If $\mathbb{E}_s^{\mathcal{M}}(\text{utility}[\lozenge^{\text{energy} \leqslant i} F]) > x$ for a nonnegative rational number x and $s \in S'$, then the value q_s is well-defined and can be computed by an iterative computation of the values $x_{s,i}$ for $i = r_s, r_s+1, \ldots$ until $x_{s,i} \geqslant x$.

Quantiles in CTMCs. For continuous-time Markov chains, quantiles can be defined in a similar way as for discrete Markov chains. However, in the case of CTMCs we have to consider trajectories rather than paths, in which case the quantile can be a real number (rather than an integer) and min and max in the definitions need to be replaced with inf and sup, respectively.

Let $\mathcal{C} = (\mathcal{M}, E)$ be a CTMC and *rew* a reward function possibly consisting of state and action rewards. We consider here a CSRL-like [3] reward-bounded reachability constraint $\varphi[r] = \lozenge^{\leqslant r} F$, where $F \subseteq S$. An infinite trajectory $\tilde{\vartheta} = s_0 t_0 s_1 t_1 s_2 t_2 \ldots$ satisfies $\varphi[r]$ if there is some $n \in \mathbb{N}$ such that $rew(s_0 t_0 \ldots t_{n-1} s_n) \leqslant r$ and $s_n \in F$. If $p \in [0,1]$ and s is a state in $\mathcal{C}$ such that $s \notin F$ and $\mathbb{P}_s^{\mathcal{C}}(\lozenge F) > p$, then we define:

$$\text{Qu}_s^{\mathcal{M}}(\text{Pr}_{\geqslant p}(\lozenge^{\leqslant ?} F)) \quad = \quad \inf\{r \in \mathbb{R} : \mathbb{P}_s^{\mathcal{C}}(\lozenge^{\leqslant r} F) \geqslant p\}$$

We are not aware of any method presented in the literature computing quantiles of this form. A simple approximation scheme works as follows. The first step is an *exponential search* to determine the smallest $i \in \mathbb{N}$ such that

$$\mathbb{P}_s^{\mathcal{C}}(\lozenge^{\leqslant 2^i} F) \geqslant p.$$

For this step we might use known algorithms for computing reward-bounded reachability probabilities in CTMCs [3]. The existence of such an index i is guaranteed by the assumption $\mathbb{P}_s^{\mathcal{C}}(\lozenge F) > p$. If $i \geqslant 1$, then we perform a *binary search* to determine some value $r \in [2^{i-1}, 2^i]$ such that

$$\mathbb{P}_s^{\mathcal{C}}(\lozenge^{\leqslant r - \frac{\varepsilon}{2}} F) < p \quad \text{and} \quad \mathbb{P}_s^{\mathcal{C}}(\lozenge^{\leqslant r + \frac{\varepsilon}{2}} F) \geqslant p$$

for some user-defined $\varepsilon > 0$. Then, r is indeed an ε-approximation of the quantile $\text{Qu}_s^{\mathcal{M}}(\text{Pr}_{\geqslant p}(\lozenge^{\leqslant ?} F))$.

In the case where the exponential search aborts immediately with $i = 0$, we proceed in the first step by an exponential search to the left (by considering the reward-bounds $2^0 = 1, \frac{1}{2}, \frac{1}{4}, \frac{1}{8}, \ldots$) and determine the smallest $i \in \mathbb{N}$ with

$$\mathbb{P}_s^{\mathcal{C}}(\lozenge^{\leqslant 2^{-i}} F) < p.$$

The second step then corresponds to a binary search in the interval $[2^{-i}, 2^{-i+1}]$.

6 Reasoning about the Energy-Utility Ratio

In this section, we detail another approach to combine energy and utility objectives by considering the ratio between the utility achieved during a system execution and the energy consumed, called *energy-utility ratio*. To reason about properties relying on this ratio, our aim is to apply probabilistic model-checking techniques. The undecidability results by Boker et al. [12] stated for temporal logics extended by accumulations of numeric variables interpreted over nonprobabilistic models impose limitations on reasoning about quotients of two accumulated values (such as energy-utility ratios). Nevertheless, there are decidable yet useful patterns of properties detailed in [12] and in work on energy games [16,17], which motivate a closer look at these patterns concerning the energy-utility ratio in the probabilistic model-checking setting.

In particular, we consider requirements formalized by an ω-regular condition φ, which are combined with an invariance condition on the energy-utility ratio. More precisely, given a quality threshold θ, we are interested in those executions of a probabilistic system satisfying φ where at any time the energy-utility ratio is at least θ. This pattern is rather natural, ensuring energy efficiency of the considered system as well as fulfilling the requirements on functionality.

Energy-Utility MDP (EUM). To reason about systems where the energy-utility ratio is of interest, we deal with MDPs $\mathcal{M}^{e,u} = (\mathcal{M}, wgt^e, wgt^u)$ that are equipped with two (state) weight functions: wgt^e for energy and wgt^u for utility. These weight functions formalize the current amount of energy (respectively, utility) that has been consumed (respectively, achieved) so far during the execution. Formally, the consumed energy along a finite path $\pi = s_0 \, \alpha_0 \, s_1 \, \alpha_1 \ldots \alpha_n \, s_n$, is defined by:

$$\text{energy}(\pi) \;=\; \sum_{i=0}^{n-1} wgt^e(s_i)$$

Likewise, utility(π) is defined by adding the utility weights of the states (except for the last one) along π according to wgt^u. We do not restrict the range of utility(π), but suppose that energy(π) is positive for all finite paths π starting from some distinguished (initial) state $s = s_0$. This constraint can be checked in polynomial time using standard shortest-path algorithms (see Lemma 1 below). In order to formalize the requirements involving the energy-utility ratio already detailed above, let us assume a given ω-regular condition φ and a safety condition ψ_θ expressing that the energy-utility ratio is always at least the quality threshold θ, which is assumed to be an arbitrary rational number. The energy-utility ratio is formalized by the function ratio $= \frac{\text{utility}}{\text{energy}}$ that assigns positive rational numbers to finite paths of an EUM. Then, the path properties we focus on in this section can be expressed by:

$$\psi_\theta \wedge \varphi, \quad \text{where} \quad \psi_\theta \;=\; \square(\,\text{ratio} \geqslant \theta\,).$$

The semantics of ψ_θ is as expected. If $\tilde{\pi} = s_0 \, \alpha_0 \, s_1 \, \alpha_1 \ldots$ is an infinite path in $\mathcal{M}$ and ratio a function that assigns to each finite path a rational number, then:

$$\tilde{\pi} \models \square(\,\text{ratio} \geqslant \theta\,) \quad \text{iff} \quad \text{ratio}(\,\text{pref}(\tilde{\pi}, k)\,) \geqslant \theta \text{ for all } k \in \mathbb{N}.$$

Clearly, $\tilde{\pi} \models \psi_\theta \wedge \varphi$ iff $\tilde{\pi} \models \psi_\theta$ and $\tilde{\pi} \models \varphi$. We are now interested in minimizing or maximizing the probability for $\psi_\theta \wedge \varphi$ in a given EUM. First, we show the decidability of the almost-sure model-checking problem in EUMs that asks whether

$\mathbb{P}_s^{\max}(\psi_\theta \wedge \varphi) = 1$ or $\mathbb{P}_s^{\min}(\psi_\theta \wedge \varphi) = 1$ (see Theorem 2). The corresponding problems where the task is to check whether $\mathbb{P}_s^{\max}(\psi_\theta \wedge \varphi) > 0$ or $\mathbb{P}_s^{\min}(\psi_\theta \wedge \varphi) > 0$ are more difficult as we will see in Example 3.

Theorem 2 (Almost-sure problems in EUMs). *Let $\mathcal{M}^{e,u}$ be an EUM as above, s a state in $\mathcal{M}$, φ a reachability, Büchi or parity property and $\theta \in \mathbb{Q}$ be a quality threshold. Then:*

(a)   *The problem "does $\mathbb{P}_s^{\min}(\psi_\theta \wedge \varphi) = 1$ hold?" is in P.*

(b)   *The problem "does $\mathbb{P}_s^{\max}(\psi_\theta \wedge \varphi) = 1$ hold?" is in NP ∩ coNP.*

For proving Theorem 2 we employ some technical transformations first. To reason about path properties of the form $\psi_\theta \wedge \varphi$, we switch from $\mathcal{M}^{e,u}$ to an MDP $(\mathcal{M}, wgt)$ with a single rational-valued state weight function $wgt : S \to \mathbb{Q}$ defined by:

$$wgt(s) \quad = \quad wgt^u(s) \; - \; \theta \cdot wgt^e(s)$$

We then have:

$$\frac{\mathsf{utility}(\pi)}{\mathsf{energy}(\pi)} \geqslant \theta \quad \text{iff} \quad wgt(\pi) \geqslant 0$$

The values of wgt might be rational. Since the weight functions for the energy and the utility are supposed to be integer-valued, we multiply wgt with the denominator of θ to obtain an integer-valued weight function. This permits to assume that $wgt(s) \in \mathbb{Z}$ for all states s.

In order to prove statement (a), i.e., the upper polynomial bound for the minimizing almost-sure model-checking problem, we first observe that

$$\mathbb{P}_s^{\min}(\psi_\theta \wedge \varphi) = 1 \quad \text{iff} \quad \mathbb{P}_s^{\min}(\psi_\theta) = 1 \text{ and } \mathbb{P}_s^{\min}(\varphi) = 1.$$

The condition $\mathbb{P}_s^{\min}(\varphi) = 1$ can be checked in time polynomial in the size of $\mathcal{M}^{e,u}$ using standard techniques (see, e.g., [8]). We now address the task to check whether $\mathbb{P}_s^{\min}(\psi_\theta) = 1$. Again we combine the weight functions for the energy and utility to a single weight function wgt and replace the constraint $\square(\mathsf{ratio} \geqslant \theta)$ with $\square(wgt \geqslant 0)$.

Lemma 1. *Let $\mathcal{M}$ be an MDP with a weight function $wgt : S \to \mathbb{Q}$, and s a state in $\mathcal{M}$. The task to check whether*

$$\mathbb{P}_s^{\min}\big(\square(wgt \geqslant 0)\big) \quad = \quad 1$$

is solvable in polynomial time. The same holds for any other constraint "$\geqslant \theta$" or "$> \theta$" (rather than "$\geqslant 0$") where θ is an arbitrary rational number.

Proof. The main argument is that:

$$\mathbb{P}_s^{\min}\big(\square(wgt \geqslant 0)\big) = 1 \quad \text{iff} \quad s \not\models \exists \lozenge(wgt < 0).$$

The decidability of the latter is derivable from known results for energy games [16,17] or the result of [12], but the time complexity of these algorithms designed for checking more complex properties is not polynomial. We consider the weighted graph that results from $\mathcal{M}$ by ignoring the action names and transition probabilities. That means, the nodes are the states and there is an edge from s to s' with weight w iff there is an action $\alpha \in Act(s)$ with $P(s, \alpha, s') > 0$. Using standard shortest-path algorithms (e.g., Floyd or Bellmann-Ford) we compute the length $\Delta(s, s')$ of a shortest path from s to s' for all state pairs (s, s'). Here, $\Delta(s, s') = +\infty$ if s' is not reachable from s and $\Delta(s, s') = -\infty$ if there is a cycle with negative weight containing a state $\tilde{s}$ that belongs to a path from s to s'. Then, $\mathbb{P}^{\min}_s(\square(wgt \geqslant 0)) < 1$ if and only if $\Delta(s, s') < 0$ for some state s'.

"if": Pick some finite path π from s to s' with $wgt(\pi) = \Delta(s, s')$ and some scheduler $\mathfrak{S}$ such that π is a $\mathfrak{S}$-path. Then, the cylinder set of π has positive measure under $\mathfrak{S}$ and $\tilde{\pi} \not\models \square(wgt \geqslant 0)$ for all $\tilde{\pi} \in Cyl(\pi)$.

"only if": Let $\mathfrak{S}$ be a scheduler with $\mathbb{P}^{\mathfrak{S}}_s\big(\square(wgt \geqslant 0)\big) < 1$. Then:

$$\mathbb{P}^{\mathfrak{S}}_s\big(\lozenge(wgt < 0)\big) \; > \; 0$$

Hence, there is a finite $\mathfrak{S}$-path π starting in s such that $wgt(\pi) < 0$. With $s' = last(\pi)$ we get:

$$\Delta(s, s') \; \leqslant \; wgt(\pi) \; < \; 0$$

Obviously, the constraint "$\geqslant 0$" can be replaced with "$\geqslant \theta$" or "$> \theta$".   ∎

In [16], an upper complexity bound of NP∩coNP was established for the almost-sure model-checking problem of energy parity objectives in MDPs. Due to our transformation of $\mathcal{M}^{e,u}$ towards $(\mathcal{M}, wgt)$, this result can be directly applied to our setting and leads to a proof for statement (b) in Theorem 2. In combination with Lemma 1, this completes the proof of Theorem 2.

The qualitative model-checking problem that asks whether $\mathbb{P}^{\mathfrak{S}}_s(\psi_\theta \wedge \varphi)$ is positive for some or all schedulers $\mathfrak{S}$ is much more difficult, since it depends on the concrete transition probabilities. This even holds for the case of energy-utility Markov chains (i.e., an EUM where $Act(s)$ is a singleton for all states s) as the following example shows.

Example 3. Consider the following Markov chain $\mathcal{M}$ with three states s_0, s^+ and s^-. The transition from s^+ and s^- to s_0 have probability 1. For the transition probabilities of action walk in state s_0 we deal with a parameter $p \in {]}0, 1{[}$.

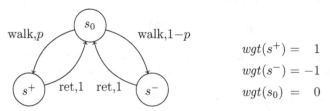

$$wgt(s^+) = 1$$
$$wgt(s^-) = -1$$
$$wgt(s_0) = 0$$

For instance, $(\mathcal{M}, wgt)$ results from the energy-utility Markov chain $\mathcal{M}^{e,u}$ when dealing with the quality threshold $\theta = 1$, the energy weights $wgt^e(s^+) = 1$ and $wgt^e(s^-) = 1$ and the utility weights $wgt^u(s^+) = 2$ and $wgt^u(s^-) = 0$, while all other energy and utility weights are 0.

The finite paths $\pi = s_0 s^1 s_0 s^2 \dots s^k s_0$ that start and end in s_0 (where $s^1, \dots, s^k \in \{s^+, s^-\}$) augmented with their accumulated weights $wgt(\pi)$ constitute a biased random walk starting at position 0 (the accumulated weight of the path starting in state s_0 of length 0) with possible steps from position $\ell \in \mathbb{Z}$ to position $\ell-1$ with probability $1-p$ (for taking the cycle $s_0 s^- s_0$) and to position $\ell+1$ with probability p (for taking the cycle $s_0 s^+ s_0$).

It is well-known that for $p > \frac{1}{2}$, with positive probability the random walk drifts to the right and never reaches position -1, while for $0 < p \leqslant \frac{1}{2}$, position -1 will be visited almost surely. Thus:

$$\mathbb{P}^{\mathcal{M}}_{s_0}\big(\Box(wgt \geqslant 0)\big) > 0 \quad \text{iff} \quad p > \tfrac{1}{2}$$

As a consequence, the answers for the qualitative model-checking problem "does $\mathbb{P}^{\mathcal{M}^{e,u}}_s\big(\Box(\text{ratio} \geqslant \theta)\big) > 0$ hold?" for a given energy-utility Markov chain $\mathcal{M}^{e,u}$ and the analogous questions "does $\mathbb{P}^{\max}_s\big(\Box(\text{ratio} \geqslant \theta)\big) > 0$ hold?" or "does $\mathbb{P}^{\min}_s\big(\Box(\text{ratio} \geqslant \theta)\big) > 0$ hold?" for a given EUM $\mathcal{M}^{e,u}$ and quality threshold θ can depend on the specific transition probabilities. This rules out simple algorithms relying on shortest-path arguments as in part (b) of Theorem 2. ∎

Despite the observation of the previous example, the positive qualitative model-checking problem "does $\mathbb{P}^{\mathcal{M}^{e,u}}_s(\psi_\theta \wedge \varphi) > 0$ hold?" and even the quantitative model-checking problem are decidable when $\mathcal{M}^{e,u}$ is a Markov chain. As we will see now, this is a consequence of decidability results for model checking probabilistic pushdown automata (pPDA) [15].

Theorem 3 (Quantitative analysis of energy-utility Markov chains). *Let $\mathcal{M}^{e,u}$ be an energy-utility Markov chain, φ an ω-regular property and $\rho \in [0,1]$ a rational threshold. Then it is decidable whether $\mathbb{P}^{\mathcal{M}^{e,u}}_s(\psi_\theta \wedge \varphi) \leqslant \rho$.*

Proof. As explained before, we switch from $\mathcal{M}^{e,u}$ to $(\mathcal{M}, wgt)$. From $(\mathcal{M}, wgt)$ we now construct a pPDA $\mathcal{P}$ with one stack symbol. The control states of $\mathcal{P}$ are the states S from $\mathcal{M}$, including additional states $init_s$ for all $s \in S$. Moreover, for each transition $s \to s'$ in $\mathcal{M}$ (i.e., each pair of states s, s' of states where $P(s, s') > 0$) with $wgt(s) \notin \{-1, 0, +1\}$ and $k = |wgt(s)|$, $\mathcal{P}$ contains $k-1$ additional control states $s_1, \dots, s_{k-1}$ with transitions $s \to s_1 \to s_2 \to \dots s_{k-1} \to s'$, all pushing (if $wgt(s)$ is positive), keeping the stack as it is (if $wgt(s) = 0$), or popping (if $wgt(s)$ is negative). The transition probabilities of the transitions $s_i \to s_{i+1}$ are 1 for $1 \leqslant i < k$ where $s_k = s'$, whereas the transition $s \to s_1$ has probability $P(s, s')$. Furthermore, $init_s$ has a single transition to s with probability 1, pushing one stack symbol. Hence, during a run in $\mathcal{P}$ starting in s with one stack symbol, the accumulated weight of $(\mathcal{M}, wgt)$ is represented by the size of the stack in $\mathcal{P}$. Since the initial states $init_s$ are never visited again, all infinite runs in $\mathcal{P}$ correspond to those in $\mathcal{M}$ satisfying ψ_θ. In [15] it was shown that the quantitative model-checking problem is decidable for any ω-regular condition expressed over the control states of $\mathcal{P}$. Dealing with a parity condition φ and assuming that state s has the same parity color as $init_s$ and the states $s_1, \dots, s_{k-1}$ that have been introduced to mimic an outgoing transition from s to s' in $\mathcal{M}$ by $|wgt(s)|$ push or pop operations, we obtain the decidability of the question whether $\mathbb{P}_s^{\mathcal{M}}(\Box(\mathsf{ratio} \geqslant 0) \wedge \varphi) \leqslant \rho$, which is equivalent to $\mathbb{P}_s^{\mathcal{M}^{e,u}}(\psi_\theta \wedge \varphi) \leqslant \rho$. If φ is an arbitrary ω-regular property, then we construct a deterministic parity automaton $\mathcal{A}$ for φ and deal with the standard product-approach (see, e.g., [8]). ∎

The dual case of this theorem deciding $\mathbb{P}_s^{\mathcal{M}^{e,u}}(\psi_\theta \wedge \varphi) \geqslant \rho$ can be established analogously. We leave the question of appropriate upper complexity bounds on solving the quantitative model-checking problem open for further research. Note that the size of the one-counter pPDA constructed in the proof above can be exponential in the size of $(\mathcal{M}, wgt)$, where the weights are supposed to be represented as decimal or binary numbers. Besides the quantitative model-checking problem detailed here, there are recent approaches approximating the probability in a one-counter pPDA satisfying an ω-regular property, possible in polynomial time [39]. It is clear that when $\mathcal{M}$ is an MDP, a construction as above can also be achieved towards a one-counter MDP [13], i.e., one-counter pPDA with actions assigned to the probabilistic transition relation. However, existing model-checking approaches for one-counter MDPs [13,14] do not allow for model-checking nontrivial properties involving ratios of accumulated weights.

7 Conclusions

In this article we reported on our experience with probabilistic model checking for evaluating low-level resource management protocols. We put the focus on the inspirations that we got for theoretical work on quantitative measures to reason about the duality of two weight functions, such as energy and utility. As sketched in this article, we partly provided solutions, but still more theoretical work on new algorithms, complexity-theoretic statements as well as tool support with sophisticated techniques to tackle the state-explosion problem, needs to be addressed in future work. Although the research field on energy games and related structures is very active, algorithms for reasoning about quotients of accumulated values, such as energy-utility ratios, are rare. The undecidability results for the model checking problem of temporal logics with accumulated values [12] impose some limitations. Nevertheless, as the work on energy games and the results in Section 6 show, there are several patterns of relevant properties where algorithmic reasoning is possible. However, the presented results are just a first step towards the theoretical foundations of energy-utility ratios in Markovian models.

Acknowledgement. The authors are grateful to Marcus Völp and Hermann Härtig for inspiration from practical problems; and to Marcus Daum and Steffen Märcker for their work on relativized long-run properties, quantiles and conditional probabilities, as well as the corresponding implementation.

References

1. Andrés, M.: Quantitative Analysis of Information Leakage in Probabilistic and Nondeterministic Systems. PhD thesis, UB Nijmegen (2011)
2. Andrés, M., van Rossum, P.: Conditional probabilities over probabilistic and nondeterministic systems. In: Ramakrishnan, C.R., Rehof, J. (eds.) TACAS 2008. LNCS, vol. 4963, pp. 157–172. Springer, Heidelberg (2008)
3. Baier, C., Cloth, L., Haverkort, B., Hermanns, H., Katoen, J.-P.: Performability assessment by model checking of Markov reward models. Formal Methods in System Design 36(1), 1–36 (2010)
4. Baier, C., Daum, M., Dubslaff, C., Klein, J., Klüppelholz, S.: Energy-utility quantiles. In: Rozier, K.Y. (ed.) NFM 2014. LNCS, vol. 8430, pp. 285–299. Springer, Heidelberg (2014)
5. Baier, C., Daum, M., Engel, B., Härtig, H., Klein, J., Klüppelholz, S., Märcker, S., Tews, H., Völp, M.: Chiefly symmetric: Results on the scalability of probabilistic model checking for operating-system code. In: 7th Conference on Systems Software Verification (SSV). Electronic Proceedings in Theoretical Computer Science, vol. 102, pp. 156–166 (2012)
6. Baier, C., Daum, M., Engel, B., Härtig, H., Klein, J., Klüppelholz, S., Märcker, S., Tews, H., Völp, M.: Waiting for locks: How long does it usually take? In: Stoelinga, M., Pinger, R. (eds.) FMICS 2012. LNCS, vol. 7437, pp. 47–62. Springer, Heidelberg (2012)
7. Baier, C., Engel, B., Klüppelholz, S., Märcker, S., Tews, H., Völp, M.: A probabilistic quantitative analysis of probabilistic-write/copy-select. In: Brat, G., Rungta, N., Venet, A. (eds.) NFM 2013. LNCS, vol. 7871, pp. 307–321. Springer, Heidelberg (2013)
8. Baier, C., Katoen, J.-P.: Principles of Model Checking. MIT Press (2008)
9. Baier, C., Klein, J., Klüppelholz, S., Märcker, S.: Computing conditional probabilities in Markovian models efficiently. In: Ábrahám, E., Havelund, K. (eds.) TACAS 2014. LNCS, vol. 8413, pp. 515–530. Springer, Heidelberg (2014)
10. Barnat, J., Brim, L., Černá, I., Češka, M., Tůmová, J.: ProbDiVinE-MC: Multicore LTL model checker for probabilistic systems. In: 5th International Conference on Quantitative Evaluation of Systems (QEST), pp. 77–78. IEEE Computer Society (2008)
11. Bianco, A., de Alfaro, L.: Model checking of probabilistic and non-deterministic systems. In: Thiagarajan, P.S. (ed.) FSTTCS 1995. LNCS, vol. 1026, pp. 499–513. Springer, Heidelberg (1995)
12. Boker, U., Chatterjee, K., Henzinger, T., Kupferman, O.: Temporal specifications with accumulative values. In: 26th Annual IEEE Symposium on Logic in Computer Science (LICS), pp. 43–52. IEEE Computer Society (2011)
13. Brázdil, T., Brozek, V., Etessami, K.: One-counter stochastic games. In: 30th IARCS Annual Conference on Foundations of Software Technology and Theoretical Computer Science (FSTTCS). Leibniz International Proceedings in Informatics (LIPIcs), vol. 8, pp. 108–119. Schloss Dagstuhl - Leibniz-Zentrum fuer Informatik (2010)
14. Brázdil, T., Brozek, V., Etessami, K., Kucera, A.: Approximating the termination value of one-counter MDPs and stochastic games. Information and Computation 222, 121–138 (2013)
15. Brázdil, T., Kučera, A., Stražovský, O.: On the Decidability of Temporal Properties of Probabilistic Pushdown Automata. In: Diekert, V., Durand, B. (eds.) STACS 2005. LNCS, vol. 3404, pp. 145–157. Springer, Heidelberg (2005)

16. Chatterjee, K., Doyen, L.: Energy and mean-payoff parity Markov decision processes. In: Murlak, F., Sankowski, P. (eds.) MFCS 2011. LNCS, vol. 6907, pp. 206–218. Springer, Heidelberg (2011)

17. Chatterjee, K., Doyen, L.: Energy parity games. Theoretical Computer Science 458, 49–60 (2012)

18. Chatterjee, K., Majumdar, R., Henzinger, T.: Markov decision processes with multiple objectives. In: Durand, B., Thomas, W. (eds.) STACS 2006. LNCS, vol. 3884, pp. 325–336. Springer, Heidelberg (2006)

19. Chen, T., Han, T., Katoen, J.-P., Mereacre, A.: Model checking of continuous-time Markov chains against timed automata specifications. Logical Methods in Computer Science 7(1) (2011)

20. Clarke, E., Grumberg, O., Peled, D.: Model Checking. MIT Press (2000)

21. Desharnais, J., Panangaden, P.: Continuous stochastic logic characterizes bisimulation of continuous-time Markov processes. Journal of Logic and Algebraic Programming 56(1-2), 99–115 (2003)

22. Dubslaff, C., Klüppelholz, S., Baier, C.: Probabilistic model checking for energy analysis in software product lines. In: 13th International Conference on Modularity, MODULARITY 2014 (to appear, 2014)

23. Etessami, K., Kwiatkowska, M., Vardi, M., Yannakakis, M.: Multi-objective model checking of Markov decision processes. Logical Methods in Computer Science 4(4) (2008)

24. Forejt, V., Kwiatkowska, M., Norman, G., Parker, D., Qu, H.: Quantitative multi-objective verification for probabilistic systems. In: Abdulla, P.A., Leino, K.R.M. (eds.) TACAS 2011. LNCS, vol. 6605, pp. 112–127. Springer, Heidelberg (2011)

25. Gao, Y., Xu, M., Zhan, N., Zhang, L.: Model checking conditional CSL for continuous-time Markov chains. Information Processing Letters 113(1-2), 44–50 (2013)

26. Garavel, H., Lang, F., Mateescu, R., Serwe, W.: CADP 2011: a toolbox for the construction and analysis of distributed processes. Software Tools and Technology Transfer (STTT) 15(2), 89–107 (2013)

27. Grädel, E., Thomas, W., Wilke, T. (eds.): Automata, Logics, and Infinite Games. LNCS, vol. 2500. Springer, Heidelberg (2002)

28. Haverkort, B.: Performance of Computer Communication Systems: A Model-Based Approach. Wiley (1998)

29. Ji, M., Wu, D., Chen, Z.: Verification method of conditional probability based on automaton. Journal of Networks 8(6), 1329–1335 (2013)

30. Katoen, J.-P., Zapreev, I., Hahn, E., Hermanns, H., Jansen, D.: The ins and outs of the probabilistic model checker MRMC. Performance Evaluation 68(2), 90–104 (2011)

31. Kulkarni, V.: Modeling and Analysis of Stochastic Systems. Chapman & Hall (1995)

32. Kwiatkowska, M., Norman, G., Parker, D.: Probabilistic symbolic model checking with PRISM: A hybrid approach. International Journal on Software Tools for Technology Transfer (STTT) 6(2), 128–142 (2004)

33. Laroussinie, F., Sproston, J.: Model checking durational probabilistic systems. In: Sassone, V. (ed.) FOSSACS 2005. LNCS, vol. 3441, pp. 140–154. Springer, Heidelberg (2005)

34. McGuire, N.: Probabilistic write copy select. In: 13th Real-Time Linux Workshop, pp. 195–206 (2011)

35. Oualhadj, Y.: The value problem in stochastic games. PhD thesis, Université Science et Technologies, Bordeaux I (2012)

36. Panangaden, P.: Measure and probability for concurrency theorists. Theoretical Computer Science 253(2), 287–309 (2001)
37. Perny, P., Weng, P.: On finding compromise solutions in multiobjective Markov decision processes. In: 19th European Conference on Artificial Intelligence (ECAI). Frontiers in Artificial Intelligence and Applications, vol. 215, pp. 969–970. IOS Press (2010)
38. Puterman, M.: Markov Decision Processes: Discrete Stochastic Dynamic Programming. John Wiley & Sons, Inc., New York (1994)
39. Stewart, A., Etessami, K., Yannakakis, M.: Upper bounds for newton's method on monotone polynomial systems, and P-time model checking of probabilistic one-counter automata. In: Sharygina, N., Veith, H. (eds.) CAV 2013. LNCS, vol. 8044, pp. 495–510. Springer, Heidelberg (2013)
40. Ummels, M., Baier, C.: Computing quantiles in Markov reward models. In: Pfenning, F. (ed.) FOSSACS 2013. LNCS, vol. 7794, pp. 353–368. Springer, Heidelberg (2013)
41. Vardi, M.: Automatic verification of probabilistic concurrent finite-state programs. In: 26th IEEE Symposium on Foundations of Computer Science (FOCS), pp. 327–338. IEEE Computer Society (1985)
42. Viswanathan, B., Aggarwal, V., Nair, K.: Multiple criteria Markov decision processes. TIMS Studies in the Management Sciences 6, 263–272 (1977)
43. von Essen, C., Jobstmann, B.: Synthesizing efficient controllers. In: Kuncak, V., Rybalchenko, A. (eds.) VMCAI 2012. LNCS, vol. 7148, pp. 428–444. Springer, Heidelberg (2012)

(Co)Algebraic Characterizations
of Signal Flow Graphs

Henning Basold[1,3], Marcello Bonsangue[2,3],
Helle Hvid Hansen[1,3], and Jan Rutten[3,1]

[1] Radboud University Nijmegen
[2] Leiden University
[3] CWI Amsterdam

Abstract One of the first publications of Prakash Panangaden is about compositional semantics of digital networks, back in 1984. Digital networks transform streams of input signals to streams of output signals. If the output streams of the components of the network are functions of their input streams, then the behavior of the entire network can be nicely characterized by a recursive stream function. In this paper we consider signal flow graphs, i.e., open synchronous digital networks with feedbacks, obtained by composing amplifiers, mergers, copiers, and delayers. We give two characterizations of the recursive stream functions computed by signal flow graphs: one algebraic in terms of localization of modules of polynomials, and another coalgebraic in terms of Mealy machines. Our main result is that the two characterizations coincide.

"Tell all the truth but tell it slant - success in circuit lies."

— Emily Dickinson

1 Introduction

Signal flow graphs are a graphical representation for the analysis, modeling and evaluation of linear systems as studied, for example, in signal processing algorithms and systems theory [6,20]. They also occur in the literature as linear flow graphs [16], linear finite state machines [17] or linear (sequential) circuits [25,18]. Here we will also refer to them simply as *circuits* since they are a special case of digital networks.

Signal flow graphs are directed graphs in which the nodes transform input signals into output signals (by amplifying, copying and merging), the arcs communicate signals without delay [12,13] (unlike in data flow graphs [11,7]), and signal delay is implemented by registers. An arc which connects only a node at its target is called an input end, similarly, an output end connects only a node at its source. We classify signal flow graphs, along two parameters: being open/closed and feedforward/feedback, where a signal flow graph is *open* if it has an input end, otherwise it is *closed*. A signal flow graph is *feedforward* if it has no cycles; otherwise it is a *feedback* circuit. Our main object of study are open, feedback circuits. All other cases are viewed as instances of them.

F. van Breugel et al. (Eds.): Panangaden Festschrift, LNCS 8464, pp. 124–145, 2014.

The behavior of a signal flow graph can be nicely characterized by a recursive stream function. This works as follows. The internal state of the circuit is its register contents. The internal state in the next step can be calculated as a linear combination of the input and the registers. That is, the dynamics of a circuit can be expressed as a system of linear of equations, one for each register. Since we consider open circuits, the corresponding linear systems may have more variables than equations where these extra (read-only) variables correspond to the input arcs of the circuit.

One way of associating a stream transformation to an open linear system is to solve it *algebraically*. Assuming that signals are elements of a unital, commutative ring R, we present a generalization of the method given in [23] for solving *closed* linear systems. The solution of a closed linear system is a fraction of polynomials (i.e., a rational stream [24]). The solution of an open linear system is also a fraction of polynomials, but its numerator consists of two or more polynomials: one represents the initial register contents, and the others represent the dependency on the input arcs. More precisely, we show that the solutions of open linear systems are characterized by the localization of free modules of polynomials over R.

Our second observation is that open linear systems (over a unital commutative ring R) can be viewed as (generally infinite) Mealy machines with input and output in R, and a free R-module as state space. Since Mealy machines are coalgebras [3,9], they come equipped with a unique behavior map associating each state of a Mealy Machine to a causal function from (input) streams to (output) streams. In this way we obtain a *coalgebraic* characterization of signal flow graphs.

Our main result is to show that the algebraic and the coalgebraic characterizations of linear systems coincide. As a consequence we obtain a novel sound and complete axiomatization of signal flow graphs, as well as an effective procedure for deciding circuit equivalence.

Related Work. A strictly smaller class of signal flow graphs and their behaviors has already been studied coalgebraically in [23,24], where the behaviors of closed feedback circuits with signals from a field are characterized as rational streams. In fact, our method for computing stream transformations for open feedback circuits is a generalization of the one in [23,24] for computing streams. Also in [23], the behaviors of open feedback circuits in which all registers are initially 0 were characterized as stream functions that multiply the input stream by a fixed rational stream, but no algebraic characterization was provided. An alternative algebraic calculus (without polynomials but using fixed points) for closed feedback circuits with signals over a field is given in [19], which yields also a sound and complete axiomatization of rational streams. An extension of the latter calculus (again without polynomials but using fixed points) to weighted automata over alphabets of arbitrary size and weights in a semiring is presented in [4]. Our method to represent stream transformations by fractions of polynomials over two or more generators is inspired by the work in [10].

Finally, we mention the classical methods of finding closed forms for linear recurrences [28], which correspond to closed systems. A well-known example is the formula for computing the Fibonacci numbers, involving the golden ratio. There is also work (e.g., in [25]) on finding closed forms for what we call an open system here, i.e., linear recurrences with a parameter. Such closed forms allow efficient computation of single values, but usually it is difficult to check for equivalence of closed forms. This differs from our approach which yields methods for effective comparison, but not so much efficient computation of single values.

Overview. In Section 2 we recall basic facts from ring theory and universal coalgebra. Signal flow graphs and their relationship with open linear systems are briefly discussed in Section 3. In Section 4 we present the relevant algebraic structures needed to solve open linear systems, and show their correspondence with subsets of causal functions. Open linear system are solved algebraically in Section 5, and coalgebraically in Section 6. The two solutions are shown to coincide in Section 7. We summarize our results, and discuss future directions in Section 8.

2   A Bit of Algebra and Coalgebra

In this section we recall the basic material from ring theory and from the coalgebraic stream calculus needed throughout the paper. A more extensive introduction to commutative ring theory can be found in [1,8]. For the stream calculus and universal coalgebra we refer to [21,23].

2.1   Rings, Modules and Algebras

Throughout this paper we let R denote a *unital commutative ring*, that is, a ring $(R, +, \cdot, 0, 1)$ in which the multiplication is commutative with neutral element 1. Furthermore, we assume R is *non-trivial*, i.e., $0 \neq 1$ in R. We call an element $a \in R$ *invertible* if it has a multiplicative inverse $a^{-1} \in R$, i.e., $a \cdot a^{-1} = 1 = a^{-1} \cdot a$. We denote by $R^\times \subseteq R$ the set of invertible elements of R. If for $a, b \in R$ the inverses a^{-1}, b^{-1} exist, then $(ab)^{-1} = b^{-1} a^{-1}$.

A *(unital) R-module* is an abelian group $(M, +, 0)$ together with a scalar multiplication, written am for $a \in R$ and $m \in M$, such that for every $a, b \in R$ and every $m, n \in M$ the following identities hold:

$$(a + b)m = am + bm \qquad a(m + n) = am + an$$
$$1m = m \qquad a(bm) = (ab)m.$$

Both rings and modules come with the usual notion of homomorphism. Module homomorphisms will also be called *linear maps*.

A map $f \colon S \to R$ is said to have *finite support* if $f(x) \neq 0$ only for finitely many elements of $x \in S$. For every set S, the *free R-module* over S exists and can be constructed as the set R^S of all maps in $S \to R$ with finite support. Addition and scalar multiplication are defined point-wise. Often we write an element $m \in R^S$ as a linear combination $m = a_1 x_1 + \cdots + a_n x_n$, where $x_1, \ldots, x_n$ are the support of m. By universality of free constructions, every function $f\colon S \to M$, where M is an R-module, can be uniquely extended to a linear map $\overline{f}\colon R^S \to M$ such that $f = \overline{f} \circ i$, where $i(x) = 1x$ is the inclusion of S into R^S. The extension of f is given by $\overline{f}(a_1 x_1 + \cdots + a_n x_n) = a_1 f(x_1) + \cdots + a_n f(x_n)$.

A subset V of a module M is *linearly independent* if whenever $\sum_{i=1}^{k} a_i v_i = 0$ for some $a_i \in R$ and $v_i \in V$, then we have $a_i = 0$ for all $1 \le i \le k$. If $f\colon S \to M$ is an injective map into an R-module M such that $f(S)$ is a linearly independent subset of M, then $\overline{f}\colon R^S \to M$ is injective.

If R is commutative and non-trivial, then finitely generated free modules behave like vector spaces over a field. In fact, every generator set of R^S has the same cardinality as S. In this case the ring R is said to have an *invariant basis number* and S is called a *basis* for the free module. Having such a basis, linear maps between free finitely generated R-modules can be seen as matrices.

An R-*algebra* is an R-module that is also a commutative ring having a multiplication that is bilinear. Equivalently, an R-algebra is a pair (A, ψ) such that A is a ring and $\psi\colon R \to A$ is a ring homomorphism.

For example, every ring R is trivially an R-algebra, and hence an R-module. A prototypical example of an R-algebra is the ring of polynomials $R[X]$ in a single variable X (with the inclusion $a \mapsto aX^0$).

2.2   Coalgebras

Given a functor $F\colon \mathbf{Set} \to \mathbf{Set}$ on the category of sets and functions, an F-*coalgebra* consists of a set S together with a *structure map* $c\colon S \to FS$. An F-coalgebra *homomorphism* $f\colon (S_1, c) \to (S_2, d)$ is a map $f\colon S_1 \to S_2$ such that $d \circ f = F(f) \circ c$. The F-coalgebras together with their homomorphisms form a category denoted by $\mathrm{Coalg}(F)$. A *subcoalgebra* of an F-coalgebra (S_1, c_1) is an F-coalgebra (S_2, c_2) if the inclusion map $S_2 \hookrightarrow S_1$ is a homomorphism.

An F-coalgebra (Z, ζ) is said to be *final* if for any F-coalgebra (S, c) there exists a unique homomorphism $\widetilde{c}\colon (S, c) \to (Z, \zeta)$. The carrier Z can be thought of as the set of all *observable behaviors* of F-coalgebras, and the unique homomorphism $\widetilde{c}\colon (S, c) \to (Z, \zeta)$ is therefore also called the *behavior map*. A final F-coalgebra, if it exists, is unique up to isomorphism. The structure map ζ of a final coalgebra is necessarily an isomorphism [15].

An F-*bisimulation* between two F-coalgebras (S_1, c_1) and (S_2, c_2) is a relation $B \subseteq S_1 \times S_2$ that can be equipped with an F-coalgebra structure b such that both projections $\pi_1\colon B \to S_1$ and $\pi_2\colon B \to S_2$ are F-coalgebra homomorphisms. Two elements $s_1 \in S_1$ and $s_2 \in S_2$ are *bisimilar* if there exists an F-bisimulation B containing the pair (s_1, s_2). We will use bisimulations as a tool for proving that two states have the same observable behavior.

Proposition 1. *Let (S_1, c_1) and (S_2, c_2) be two F-coalgebras with $s_1 \in S_1$ and $s_2 \in S_2$, and assume that a final F-coalgebra exists. If s_1 and s_2 are bisimilar, then $\widetilde{c_1}(s_1) = \widetilde{c_2}(s_2)$, i.e., they have the same behavior.*

The converse of the above proposition holds under the assumption that the functor F preserves weak pullbacks [21].

2.3 Elements of Stream Calculus

For a ring R, coalgebras for the functor $\mathcal{S} = R \times (-)$ are called *stream automata*. The final $\mathcal{S}$-coalgebra is given by the set of all *streams* $\sigma \in R^\omega = \mathbb{N} \to R$ together with the structure map ξ defined by $\xi(\sigma) = (\sigma(0), \sigma')$, where $\sigma(0)$ is the *initial value* of σ is σ' its *derivative* [22]), i.e., for all $n \in \mathbb{N}$, $\sigma'(n) = \sigma(n+1)$. The inverse of ξ is given by the *cons* map defined by $(r : \sigma)(0) = r$, $(r : \sigma)(n+1) = \sigma(n)$ for all $n \in \mathbb{N}$.

We define operations on R^ω by means of *behavioral differential equations* [22], i.e., by specifying their initial value and derivative. The following operations become relevant for the algebraic characterizations of circuit behaviors in Section 5.

Initial value	Derivative	Name
$[r](0) = r$	$[r]' = [0]$	constant
$X(0) = 0$	$X' = [1]$	shift
$(\sigma + \tau)(0) = \sigma(0) + \tau(0)$	$(\sigma + \tau)' = \sigma' + \tau'$	sum
$(\sigma \times \tau)(0) = \sigma(0) \cdot \tau(0)$	$(\sigma \times \tau)' = \sigma' \times \tau + [\sigma(0)] \times \tau'$	convolution product
$\sigma^{-1}(0) = \sigma(0)^{-1}$	$(\sigma^{-1})' = [-\sigma(0)^{-1}] \times \sigma' \times \sigma^{-1}$	convolution inverse

In the first column, $r \in R$ and the operations $+$, $\cdot$ and $(-)^{-1}$ on the right-hand side of the equations are operations on R. We note that the inverse is only defined on streams σ for which $\sigma(0) \in R^\times$. Note that $(\sigma \times X) = (0, \sigma_0, \sigma_1, \dots)$ for all $\sigma \in R^\omega$. We will use the so-called *fundamental theorem of stream calculus* [22].

Proposition 2. *For any $\sigma \in R^\omega$ we have:* $\sigma = [\sigma(0)] + (\sigma' \times X)$.

Proof. We note that $(r : \tau) = (r, \tau_0, \tau_1, \dots) = [r] + (0, \tau_0, \tau_1, \dots) = [r] + (\tau \times X)$ for any $\tau \in R^\omega$, and thus we obtain the desired result: $\sigma = (\sigma_0 : \sigma') = [\sigma_0] + (\sigma' \times X)$. $\qquad\blacksquare$

Streams over R form a unital, commutative ring $(R^\omega, +, \times, [0], [1])$, and an R-algebra via the embedding $a \mapsto [a]$ of R into R^ω. In other words R^ω is an R-module with the scalar multiplication $a\sigma = [a] \times \sigma$. We denote by $[R^\omega, R^\omega]$ the set of all *stream transformations*, i.e., all functions from R^ω to R^ω. It forms a ring under point-wise sum and (convolution) product, as well as an R^ω-algebra via the embedding of R^ω as the subring of constant maps. A stream transformation $f : R^\omega \to R^\omega$ is said to be *causal* whenever the n-th output of f depends only on the elements up to n of its input. More precisely, f is causal if for all $\sigma, \tau \in R^\omega$, if $\sigma(k) = \tau(k)$ for all $k \leq n$ then $f(\sigma)(n) = f(\tau)(n)$.

Causal stream transformations play a key role in this paper. For example, all constant maps as well as the identity are causal. We let $C(R^\omega) \subseteq [R^\omega, R^\omega]$ denote the subset of causal transformations on R^ω. Since causal functions are closed under point-wise sum and point-wise convolution product, the set $C(R^\omega)$ inherits the ring structure as well as the R^ω-algebra structure from $[R^\omega, R^\omega]$.

3   Signal Flow Graphs and Linear Systems

3.1   Signal Flow Graphs

A *signal flow graph* [20,23] is a directed graph in which arcs connect up-to two nodes. Nodes, also called gates, perform operations on signals from incoming arcs and output the result on outgoing arcs. Signals are assumed to be elements of a unital, commutative ring R. The *amplifier gate* ($\xrightarrow{\ a\ }$) performs the scalar multiplication of the incoming signal with $a \in R$. The *adder gate* ($\oplus$) outputs the sum of its inputs. A *copier* ($\bullet$) simply copies its input to two (or more) output ends. Finally, a *register* ($\rightarrow\boxed{a}\rightarrow$) is a one-element buffer which outputs its content and stores the incoming signal. The initial content of the register is thereby a. Arcs with no connecting gates at their source are *input ends*, whereas those with no connecting gates at their target are *output ends*. For clarity, input ends are marked with an input stream ι. We will also refer to signal flow graphs simply as *circuits*. For technical simplicity we will consider only circuits with at most one input ι end and one output end.

A circuit with no input end is *closed*, otherwise it is *open*. A circuit is called *feedforward* if it contains no cycles; otherwise it is a *feedback* circuit. In order for feedback circuits to have a well-defined behavior, all cycles are required to pass through at least one register. Intuitively, the reason is that otherwise we would end up with equations which may not have unique solutions. The condition will be used in the construction in Section 3.2.

By feeding signals to the input end of a circuit we observe signals on its output end. Since there is no limit on the number of signals a circuit can react to, the *behavior* of a circuit is given by a function transforming input streams to output streams. Closed circuits do not need any input, and their behavior is given by constant stream functions, or, equivalently, by streams.

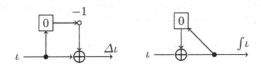

Fig. 1. A differentiator and an integrator signal flow graph

For example, the leftmost circuit in Fig. 1, implements a *discrete differential* $\Delta\iota$ where, for all $n \in \mathbb{N}$,

$$(\Delta\iota)(0) = \iota(0) \quad \text{and} \quad (\Delta\iota)(n+1) = \iota(n+1) - \iota(n).$$

It consists of a copier, a register with initial value 0, a multiplication by -1 and an adder. The rightmost circuit implements the *discrete integral* $\int\iota$ defined, for all $n \in \mathbb{N}$, by

$$(\textstyle\int\iota)(0) = \iota(0) \quad \text{and} \quad (\textstyle\int\iota)(n+1) = \iota(n+1) + (\textstyle\int\iota)(n).$$

Note that the discrete differential circuit is feedforward, whereas the integration circuit is a feedback circuit.

By composing the two circuits (i.e., by linking the output arc of one circuit with the input arc of the other), we obtain a new circuit that implements $\int(\Delta\iota)$. One can show by fairly straightforward induction that $\int(\Delta\iota)(n) = \iota(n)$ for all $n \in \mathbb{N}$. However, for more involved circuits, this may not always be easy. One of the applications of our results is an algebraic calculus for proving circuit equivalence (as opposed to point-wise, inductive reasoning).

3.2   Linear Systems

Signal flow graphs are graphical representations of linear systems, i.e., systems of linear equations [5]. An *(n-dimensional open) linear system* $L = (V, M, O, I)$ is defined as follows. The set $V = \{v_1, \ldots, v_n\}$ is a set of n variables denoting the registers, and in addition, we have a variable v_{in} denoting the input signal. Together they form the set $\overline{V} = \{v_{\text{in}}\} + V$. We use the free module R^V to model the assignment of register contents: every element $\sum_{i=1}^{n} s_i v_i \in R^V$ (written as $(s_1, \ldots, s_n)$) denotes the assignment of s_i to the register v_i. Analogously, the elements of $R^{\overline{V}}$ combine the input value for v_{in} and assignments to registers. Next, M and O are linear maps $M\colon R^{\overline{V}} \to R^V$ and $O\colon R^{\overline{V}} \to R$ that describe the circuit wiring through which new values are fed to the registers and to the output end. Since M is a linear map between free modules it can be represented as an $n \times (n+1)$-matrix over R with entries $m_{i,j}$ coming from: $M(v_{\text{in}}) = \sum_{i=1}^{n} m_{i,0} v_i$ and $M(v_j) = \sum_{i=1}^{n} m_{i,j} v_i$, for $1 \leq j \leq n$. Similarly, O is a $1 \times (n+1)$-matrix with entries o_i given by $o_0 = O(v_{\text{in}})$, and $o_i = O(v_i)$, for $1 \leq i \leq n$. Together, M and O describe a system of $n + 1$ equations in the variables $v_{\text{in}}, v_1, \ldots, v_n$. Finally, $I = (r_1, \ldots, r_n) \in R^V$ is the vector of initial register contents.

One should think of the register contents over time as streams. If we denote the current state by $(s_1, \ldots, s_n)$ (now viewed as a tuple of streams), the input stream by ι, the next state of the system by $(s'_1, \ldots, s'_n)$, and the output stream by o, then they satisfy the following system of stream differential equations:

$$
\left.
\begin{array}{ll}
s_1(0) = r_1 & s'_1 = m_{1,0}\iota + m_{1,1}s_1 + \cdots + m_{1,n}s_n \\
\quad\vdots & \quad\vdots \\
s_n(0) = r_n & s'_n = m_{n,0}\iota + m_{n,1}s_1 + \cdots + m_{n,n}s_n
\end{array}
\right\} M
$$

$$
\underbrace{}_{I}
$$

$$
o = o_0\iota + o_1 s_1 + \cdots + o_n s_n \left.\right\} O
$$

By adapting the constructions given in [23] and [19], for every signal flow graph C with one input end, one output end, finitely many gates and n registers, we define its associated n-dimensional open linear system

$$
\mathcal{L}(C) = (V, M, O, I) \tag{1}
$$

as follows. We set $V = \{v_1, \ldots, v_n\}$, and take $I = (r_1, \ldots, r_n) \in R^V$ to be the vector of the initial content of the registers of C. To define M, we build for each register v_i a linear combination over $v_{in}, v_1, \ldots, v_n$ by traversing the graph in reverse direction starting from the input arc of v_i and ending at a register or an input end. These reverse paths form a syntax tree, say T_i, with the root being the source of the arc entering the register v_i and leaves among $v_{in}, v_1, \ldots, v_n$. The tree branches at adder-gates and passes through a number of scalar multiplier gates. If a branch ends in register v_j the tree has a leaf v_j, similarly for paths ending in the input end ι, the tree has a leaf v_{in}. Since we assume all cycles pass through a register, this tree is finite. Each tree T_i gives rise to a linear combination $m_{i,0}v_{in} + m_{i,1}v_1 + \cdots + m_{i,n}v_n \in R^{\overline{V}}$ by evaluating T_i top-down in $R^{\overline{V}}$, and we define the i-th row of M (seen as a matrix) to be $M_i = (m_{i,0}, m_{i,2}, \ldots, m_{i,n})$. For the output of the circuit we get again a tree, which yields a linear combination $O \in R^{\overline{V}}$ in the same way.

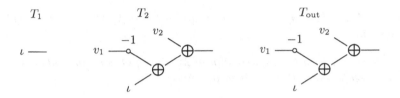

Fig. 2. Trees for flow in the composition of ciruits in Fig. 1

Example 3. Let v_1, v_2 denote the two registers from left to right in the composition of the circuits given in Fig. 1 that computes $\int \Delta \iota$. The constructed trees for the registers and the output are shown in Fig. 2. These trees result in linear combinations $\iota + 0v_1 + 0v_2$, $\iota - v_1 + v_2$ and again $\iota - v_1 + v_2$, hence the linear system given by

$$V = \{v_1, v_2\} \quad I = \begin{pmatrix} 0 \\ 0 \end{pmatrix} \quad M = \begin{pmatrix} 1 & 0 & 0 \\ 1 & -1 & 1 \end{pmatrix} \quad O = \begin{pmatrix} 1 & -1 & 1 \end{pmatrix}.$$

Conversely, we can construct from every linear system L a circuit, such that its associated linear system is L.

Proposition 4. *For all open linear systems L, there is a linear circuit $\mathcal{C}(L)$ such that $\mathcal{L}(\mathcal{C}(L)) = L$. In other words, transforming $\mathcal{C}(L)$ back into a linear system, see (1), yields the original system L again.*

Proof. Let $L = (V, M, O, I)$ be a linear system. Figure 3 sketches the linear circuit $\mathcal{C}(L)$. In order to keep the picture simple, we use arrows pointing to and originating from v_i instead of drawing the full graph. By applying the definition, we can check that $\mathcal{L}(\mathcal{C}(L)) = L$. □

The shape of the matrix M of a linear system associated to a signal flow graph gives us a precise characterization of closed and feedforward circuits.

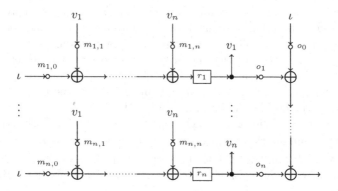

Fig. 3. Signal flow graph constructed from an open linear system

Lemma 5. *Let C be a signal flow graph, and $\mathcal{L}(C) = (V, M, O, I)$ its associated linear system.*

1. *If C is closed, then the first column of M and of O contains only 0's.*
2. *If C is feedforward, then M is of the form*

$$
M = \begin{pmatrix}
m_{1,0} & 0 & 0 & \cdots & 0 & 0 \\
m_{2,0} & m_{2,1} & 0 & \cdots & 0 & 0 \\
& & & \vdots & & \\
m_{n,0} & m_{n,1} & m_{n,2} & \cdots & m_{n,n-1} & 0
\end{pmatrix}
$$

The second part follows from the fact that, in a feedforward circuit, the register variables can be ordered such that the input of a register only depends on the registers preceding it in the order.

4   Algebraic Structures for Signal Flow Graphs

The ring structure on streams forms the basis of the algebraic structures that characterize the behaviors of signal flow graphs. Recall that R is assumed to be a unital, commutative ring. In this section we describe the relevant algebras. We begin with the *ring $R[X]$ of polynomials over R*. It consists of all streams with only finitely many non-zero elements, i.e., streams of the form $[a_0] + [a_1] \times X + \cdots + [a_n] \times X^n$ for $a_0, \ldots, a_n \in R$. The following is a well-known fact about polynomials.

Proposition 6. *The set $R[X]$ of polynomials is a subring of R^ω.*

Polynomials are not closed under inverse, but we can extend the ring $R[X]$ to fractions of polynomials using a construction called *localization* [1]. Let U be the set of all invertible polynomial streams, i.e., $U = \{p \in R[X] \mid p(0) \in R^\times\}$.

We observe that U is multiplicatively closed, a necessary condition to form the *localization of $R[X]$ (viewed as an $R[X]$-module) at U*:

$$R[X]\left[U^{-1}\right] = \{[p : u] \mid p \in R[X], u \in U\}.$$

Elements in $R[X]\left[U^{-1}\right]$ are equivalence classes with respect to the relation $\sim$ on $R[X] \times U$ defined by

$$(p_1, u_1) \sim (p_2, u_2) \text{ iff } \exists v \in U : vp_1u_2 = vp_2u_1.$$

Note that the extra v can be left out if, e.g., R is an integral domain. Using sum and convolution product of streams, we define addition and multiplication by scalars from $R[X]$ on $R[X]\left[U^{-1}\right]$ as follows:

$$[p_1 : u_1] + [p_2 : u_2] = [p_1u_2 + p_2u_1 : u_1u_2]$$
$$q[p : u] = [qp : u].$$

The above operations turn $R[X]\left[U^{-1}\right]$ into an $R[X]$-module with $[0 : 1]$ as additive identity, and such that, for all $u \in U$ and $p \in R[X]$, $u[p : uq] = [p : q]$ In fact, $R[X]\left[U^{-1}\right]$ is also a ring.

The behaviors of closed feedforward signal flow graphs are precisely the polynomial streams $R[X]$, whereas those of closed feedback signal flow graphs are the rational streams, i.e., the $R[X]$-module $R[X]\left[U^{-1}\right]$, cf. [23].

For open, feedforward signal flow graphs, we will show that the relevant algebraic structure is given by the free $R[X]$-module $R[X]^{\{\iota,1\}}$ generated by the two elements set $\{\iota, 1\}$. The intuition for these generators is that ι represents an unknown input stream, and 1 captures the initial content of the registers. Elements of $R[X]^{\{\iota,1\}}$ are of the form $p\iota + q1$ where $p, q \in R[X]$. In the sequel, we will write such an element simply as $p\iota + q$. As for closed signal flow graphs, allowing for feedback means constructing fractions. In the open case this means that we will consider the localization of $R[X]^{\{\iota,1\}}$ at the set U of invertible polynomials:

$$R[X]^{\{\iota,1\}}\left[U^{-1}\right] = \{[p\iota + q : u] \mid p, q \in R[X], u \in U\}.$$

Similar to the previous localization, fractions $[p\iota + q : u]$ are equivalence classes with respect to the relation $\sim$ on $R[X]^{\{\iota,1\}} \times U$ defined by:

$$(p_1\iota + q_1, u_1) \sim (p_2\iota + q_1, u_2) \text{ iff } \exists v \in U : v(p_1\iota + q)u_2 = v(p_2\iota + q)u_1.$$

As usual, we write $\frac{p\iota+q}{u}$ instead of $[p\iota + q : u]$. Addition and scalar multiplication are defined as expected, turning $R[X]^{\{\iota,1\}}\left[U^{-1}\right]$ into an $R[X]$-module which is free among all $R[X]$-modules M for which the assignment $\lambda_u : x \mapsto ux$ is a linear isomorphism on M for all $u \in U$.

Proposition 7. *The $R[X]$-module $R[X]^{\{\iota,1\}}\left[U^{-1}\right]$ together with the linear inclusion map $\varphi : R[X]^{\{\iota,1\}} \to R[X]^{\{\iota,1\}}\left[U^{-1}\right]$ given by $\varphi(x) = \frac{x}{1}$ satisfies the following universal property [1,8].*

If M is an R[X]-module such that for all u ∈ U, the linear maps $\lambda_u \colon M \to M$ given by $\lambda_u(x) = ux$ are isomorphisms, and if $f \colon R[X]^{\{\iota,1\}} \to M$ is a linear map, then there is a unique linear map $\overline{f} \colon R[X]^{\{\iota,1\}} [U^{-1}] \to M$ extending f such that the following diagram commutes:

$$
\begin{array}{ccc}
R[X] & \xrightarrow{\;\varphi\;} & R[X]^{\{\iota,1\}} [U^{-1}] \\
& \searrow{\scriptstyle f} & \Big\downarrow{\scriptstyle !\overline{f}} \\
& & M
\end{array}
$$

The extension of f is given by $\overline{f}([x : u]) = \lambda_u^{-1}(f(x))$. Moreover, if f is injective, then its extension $\overline{f}$ is injective as well.

The following lemma relates the algebraic constructions used so far.

Lemma 8. *There are inclusions among the constructed R[X]-modules, as indicated in the following commuting diagram*

$$
\begin{array}{ccc}
R[X] [U^{-1}] & \xhookrightarrow{\;j\;} & R[X]^{\{\iota,1\}} [U^{-1}] \\
{\scriptstyle \varphi_1}\Big\uparrow & & {\scriptstyle \varphi_2}\Big\uparrow \\
R[X] & \xhookrightarrow{\;\;i\;\;} & R[X]^{\{\iota,1\}}
\end{array}
$$

where φ_1, φ_2 are the inclusions into the localizations, i is given by $x \mapsto x1$, and j is the linear extension $j = \overline{\varphi_2 \circ i}$. All these inclusions are injective R[X]-module homomorphisms.

The elements of the four algebras above denote causal stream functions. The polynomials $R[X]$ are by definition streams, or, equivalently, constant stream transformations (which are clearly causal). For the algebras corresponding to feedforward, closed and arbitrary circuits we have semantic maps $[\![-]\!]_{\mathrm{ff}}, [\![-]\!]_{\mathrm{c}}$ and $[\![-]\!]$ as shown here in the following diagram (we give their definitions below):

$$
\begin{array}{ccccc}
R[X]^{\{\iota,1\}} & \hookrightarrow & R[X]^{\{\iota,1\}} [U^{-1}] & \hookleftarrow & R[X] [U^{-1}] \\
\Big\uparrow & \searrow{\scriptstyle [\![-]\!]_{\mathrm{ff}}} & \Big\downarrow{\scriptstyle [\![-]\!]} & \swarrow{\scriptstyle [\![-]\!]_{\mathrm{c}}} & \\
\{\iota,1\} & \xrightarrow{\;\;g\;\;} & C(R^\omega) & &
\end{array}
$$

Here $g \colon \{\iota,1\} \to C(R^\omega)$ is the map defined by $g(\iota) = \mathrm{id}_{R^\omega}$ and $g(1) = \sigma \mapsto [1]$, for all $\sigma \in R^\omega$. Since $R[X]^{\{\iota,1\}}$ is the *free* $R[X]$-module over the set $\{\iota,1\}$ the map $[\![-]\!]_{\mathrm{ff}} \colon R[X]^{\{\iota,1\}} \to C(R^\omega)$ is defined as the unique linear map extending g.

For all $u \in U$, the scalar multiplication $\lambda_u \colon C(R^\omega) \to C(R^\omega)$ on $C(R^\omega)$, sends f to $u \times f$ (point-wise convolution). Since U consists of all invertible polynomial streams, λ_u has an inverse $\lambda_u^{-1}(f) = u^{-1} \times f$, and hence each

λ_u is a linear isomorphism on $C(R^\omega)$. We can thus apply the universal property of the localization (Prop. 7) in order to uniquely define the linear map $[\![-]\!]\colon R[X]^{\{\iota,1\}}\,[U^{-1}] \to C(R^\omega)$ as the extension of $[\![-]\!]_{\mathrm{ff}}\colon R[X]^{\{\iota,1\}} \to C(R^\omega)$.

Finally, the map $[\![-]\!]_{\mathrm{c}}\colon R[X]\,[U^{-1}] \to C(R^\omega)$ is obtained by restricting $[\![-]\!]$ along the inclusion $R[X]\,[U^{-1}] \hookrightarrow R[X]^{\{\iota,1\}}\,[U^{-1}]$. Note that $[\![[p:q]]\!]_{\mathrm{c}}$ is simply the constant map that sends every stream to the rational stream $p \times q^{-1}$.

Theorem 9 (Soundness and completeness). *For all* $x, y \in R[X]^{\{\iota,1\}}\,[U^{-1}]$, $x = y$ *if and only if* $[\![x]\!] = [\![y]\!]$.

Soundness is, in fact, what allows us to define $[\![-]\!]\colon R[X]^{\{\iota,1\}}\,[U^{-1}] \to C(R^\omega)$ as (linear) map. Completeness, on the other hand, is a consequence of the fact that all the maps $[\![-]\!]_{\mathrm{ff}}$, $[\![-]\!]_{\mathrm{c}}$, and $[\![-]\!]$ are injective, since id_{R^ω} and $\sigma \mapsto [1]$ are linearly independent in the $R[X]$-module $C(R^\omega)$.

5   Solving Linear Systems, Algebraically

In this section we give a matrix-based method for computing the solution of an *open* linear system. Our method is a novel adaptation of the method presented in [23] for solving *closed* linear systems.

Let $L = (V, M, O, I)$ be an n-dimensional open linear system with (stream) variables $V = \{v_1, \ldots, v_n\}$. We use the fact that the matrix M over R can be seen as a matrix over $R[X]$ by applying the inclusion $a \mapsto [a]$ to each entry in M. This allows us to multiply M (entry-wise) with scalars from $R[X]$. Likewise, we implicitly apply the entry-wise inclusion to view I as a vector over $R[X]$. More abstractly, we are using the $R[X]$-algebra structure on the matrices.

Informally, a state solution to L is an assignment $s_\sigma\colon V \to R^\omega$, which depends on an input stream σ, and satisfies the equations of L when taking $\iota = \sigma$. By the fundamental theorem of stream calculus (Prop. 2), for any such assignment, we have $s_\sigma(v_i) = [s_\sigma(v_i)(0)] + s_\sigma(v_i)'X$ for each $v_i \in V$. Hence, as a system of constraints L is equivalent to the system expressed by:

$$s_\sigma = I + (MX)\begin{pmatrix}\iota\\s_\sigma\end{pmatrix}$$

This system is, in turn, equivalent to the "square" system:

$$\begin{pmatrix}\iota\\s_\sigma\end{pmatrix} = \begin{pmatrix}\iota\\I\end{pmatrix} + \begin{pmatrix}\mathbf{0}\\MX\end{pmatrix}\begin{pmatrix}\iota\\s_\sigma\end{pmatrix}$$

where $\mathbf{0}$ denotes a row of $n+1$ 0's. Finally, this system is equivalent to

$$\left(\mathbf{I}_{n+1} - \begin{pmatrix}\mathbf{0}\\MX\end{pmatrix}\right)\begin{pmatrix}\iota\\s_\sigma\end{pmatrix} = \begin{pmatrix}\iota\\I\end{pmatrix} \tag{2}$$

where $\mathbf{I}_{n+1}$ is the $(n+1)$-dimensional identity matrix. Formally, the system (2) can be seen as an equation over the $\overline{V}$-fold direct sum (copower) $\overline{V} \odot \Gamma = \bigoplus_{\overline{V}} \Gamma$

of Γ where $\Gamma = R[X]^{\{\iota,1\}}\left[U^{-1}\right]$, as before. Elements of $\overline{V}\odot\Gamma$ are $(n{+}1)$-vectors, and to ease readability, we introduce the notation $\left(\begin{smallmatrix}\iota\\x\end{smallmatrix}\right) = \iota v_{\text{in}} + x \in \overline{V}\odot\Gamma$ where $x = (x_1,\ldots,x_n) \in V\odot\Gamma$. Using the usual matrix-vector product, we can apply M to vectors from the $R[X]$-module $\overline{V}\odot\Gamma$, and solve the system using standard methods from linear algebra [1]. Note also that (2) does not depend on the output O of the linear system. We therefore call it a *state equation*, and a solution will be referred to as a *state solution* of L.

Definition 10. *Let $L = (V,M,O,I)$ be a linear system. A state solution to L is an element $s \in V\odot\Gamma$ such that*

$$\left(\mathbf{I}_{n+1} - \begin{pmatrix}\mathbf{0}\\MX\end{pmatrix}\right)\begin{pmatrix}\iota\\s\end{pmatrix} = \begin{pmatrix}\iota\\I\end{pmatrix}.$$

A solution s to L thus contains for each $v_i \in V$ an element s_i in the localization $\Gamma = R[X]^{\{\iota,1\}}\left[U^{-1}\right]$. Since we have seen in the previous section that localization elements denote causal functions, we get a stream assignment $s_\sigma \colon V \to R^\omega$ defined by $s_\sigma(v_i) = [\![s_i]\!](\sigma)$, for all streams $\sigma \in R^\omega$. The dependency on the input ι is thus formalized via the free module $R[X]^{\{\iota,1\}}$.

Proposition 11. *Every n-dimensional open linear system $L = (V,M,O,I)$ has a unique state solution $s \in V\odot\left(R[X]^{\{\iota,1\}}\left[U^{-1}\right]\right)$.*

If $s \in V\odot\Gamma$ is the unique state solution of L, then the *output solution of L* is the localization element obtained by applying the output map O to the $(n+1)$-vector $\iota v_{\text{in}} + s$, that is, the output solution of L is defined as

$$O(\iota v_{\text{in}} + s) \in R[X]^{\{\iota,1\}}\left[U^{-1}\right].$$

Note that the output solution is uniquely defined for L due to the existence of a unique state solution.

Definition 12. *If L is an open linear system, then we define the* algebraic solution *to L, denoted by $[\![L]\!]$, as the algebraic semantics of the output solution: $[\![L]\!] = [\![O(\iota v_{\text{in}} + s)]\!] \in C(R^\omega)$.*

In Section 7 we will show that $[\![L]\!]$ is indeed the behavior of any signal flow graph represented by L.

Example 13. We solve the system from Example 3 using the above method.

$$\begin{pmatrix}1 & 0 & 0\\-X & 1 & 0\\-X & X & 1-X\end{pmatrix}\begin{pmatrix}\iota\\s_1\\s_2\end{pmatrix} = \begin{pmatrix}\iota\\0\\0\end{pmatrix}$$

Using Gaussian elimination, we find the state solutions $s_1 = \iota X$, $s_2 = \frac{\iota X - \iota X^2}{1-X} = \iota X$ and the output solution

$$o = 1\iota - \iota X + 1\frac{\iota X - \iota X^2}{1-X} = \iota - \iota X + \iota X = \iota.$$

Hence the composition $\int(\Delta\iota)$ is the identity, as expected.

6   Solving Linear Systems, Coalgebraically

In this section we describe a coalgebraic method of associating a stream transformation to a linear system. The key observation is that each linear system can be viewed as a Mealy machine. The *Mealy machines* we will consider are coalgebras for the functor $\mathcal{M} = (R \times (-))^R$. Intuitively, for a set of states S, a structure map $c\colon S \to \mathcal{M}S$ assigns to each state $x \in S$ and input value $a \in R$ a pair $(o, y) \in R \times S$, where o is the output and y is the next state of the machine in state x on input a.

A final $\mathcal{M}$-coalgebra exists, and is given by the set $C(R^\omega)$ of causal stream transformations equipped with the structure map $\delta\colon C(R^\omega) \to (R \times C(R^\omega))^R$ defined for all $f \in C(R^\omega)$ and $a \in R$ by $\delta(f)(a) = (f[a], f^{(a)})$ where for all $\sigma \in R^\omega$,

$$
\begin{aligned}
f[a] &= f(a : \sigma)(0) &\in R, \\
f^{(a)}(\sigma) &= f(a : \sigma)' &\in R^\omega,
\end{aligned}
\tag{3}
$$

(see e.g. [9]). Note that because f is causal the definition of $f[a]$ is independent of the choice of $\sigma \in R^\omega$, and $f^{(a)}$ is causal as well.

By instantiating the notion of bisimulation to the functor $\mathcal{M}$, we obtain that a bisimulation between Mealy machines (S_1, c_1) and (S_2, c_2) is a relation $B \subseteq S_1 \times S_2$ such that for all $(s_1, s_2) \in B$ and all $a \in R$ the following holds: if $(o_1, t_1) = c_1(s_1)(a)$ and $(o_2, t_2) = c_2(s_2)(a)$ then

$$
o_1 = o_2 \qquad \text{and} \qquad (t_1, t_2) \in B.
$$

Definition 14 (Linear machines). *Let $L = (V, M, O, I)$ be an open linear system with extra input variable v_{in}. We define the* Mealy machine *associated with L as the Mealy machine (R^V, c_L), where $c_L\colon R^V \to (R \times R^V)^R$ is defined by*

$$
c_L(x)(a) = (O(av_{\text{in}} + x), M(av_{\text{in}} + x))
$$

for all $a \in R$ and $x \in R^V$.

We point out that for the linear machine (R^V, c_L) associated with some L, the actual structure map $c_L\colon R^V \to (R \times R^V)^R$ is, in general, not linear, only its uncurried form $\overline{c_L}\colon R^V \times R \to R \times R^V$ is linear.

A Mealy machine (S, c), is called a *linear machine* if S is an R-module and the uncurried structure map $\overline{c}\colon S \times R \to R \times S$ is linear. Clearly, not all Mealy machines are linear machines. In particular, the final $\mathcal{M}$-coalgebra $(C(R^\omega), \delta)$ is not a linear machine, because $\overline{\delta}$ is *not* linear, even though δ itself is linear. So the final Mealy machine is linear in the (coalgebraic) "curried form" whereas linear machines are linear in "uncurried form".

We denote by $\langle \text{LM} \rangle$ the least subcoalgebra of $(C(R^\omega), \delta)$ containing the behaviors of all linear machines. Similarly, $\langle \text{LM}_c \rangle$ and $\langle \text{LM}_{\text{ff}} \rangle$ denote the least subcoalgebras of $(C(R^\omega), \delta)$ containing the behaviors of all linear machines associated to closed, respectively feedforward, circuits.

We define the coalgebraic solution of an open linear system via the final Mealy machine.

Definition 15. *Let $L = (V, M, O, I)$ be an open linear system. We define the coalgebraic solution of L, denoted by $\langle\!\langle L \rangle\!\rangle$, to be the coalgebraic behavior of the initial state I in its Mealy machine (R^V, c_L), that is, $\langle\!\langle L \rangle\!\rangle = \widetilde{c_L}(I) \in C(R^\omega)$.*

Example 16. Taking the linear system in Example 3 (our running example), we calculate the first few outputs and states of the resulting Mealy machine with input stream $\sigma \in R^\omega$. To this end, we set $s_0 = I$ and $(o_k, s_{k+1}) = c_L(s_k)(\sigma_k)$ and compute the concrete values:

$$o_0 = O(\sigma_0 v_{\text{in}} + s_0) = \sigma_0 - 0 + 0 = \sigma_0$$
$$s_1 = M(\sigma_0 v_{\text{in}} + s_0) = \sigma_0 v_1 + (\sigma_0 - 0 + 0)v_2 = \sigma_0 v_1 + \sigma_0 v_2$$
$$o_1 = O(\sigma_1 v_{\text{in}} + s_1) = \sigma_1 - \sigma_0 + \sigma_0 = \sigma_1$$
$$s_2 = M(\sigma_1 v_{\text{in}} + s_1) = \sigma_1 v_1 + (\sigma_1 - \sigma_0 + \sigma_0)v_2 = \sigma_1 v_1 + \sigma_1 v_2$$
$$\vdots$$

Clearly, we get $\langle\!\langle L \rangle\!\rangle(\sigma) = \widetilde{c_L}(I)(\sigma) = (o_0, o_1, \dots) = \sigma$ as expected.

7   Algebraic and Coalgebraic Solutions Coincide

In the previous two sections we have seen an algebraic and a coalgebraic method for assigning a causal stream transformation to a linear system. In this section we will show that the two methods lead to the same element of $C(R^\omega)$.

To begin with, we show that the localization $R[X]^{\{\iota,1\}}[U^{-1}]$ can be given a Mealy machine structure such that the algebraic semantics $[\![-]\!]$ coincides with its coalgebraic behavior map. For a more compact notation, we define $\Gamma = R[X]^{\{\iota,1\}}[U^{-1}]$. The Mealy structure on Γ is defined by a two-step procedure that mimics the definition of the structure map of the final Mealy machine in (3): For all causal functions $f \in C(R^\omega)$,

$$f \longmapsto \lambda a \in R.f(a:-) \longmapsto \lambda a \in R.\lambda \sigma \in R^\omega.(f(a:\sigma)(0), f(a:\sigma)').$$

Let $x \in \Gamma$. To mimic the leftmost step above, we need an element $x_a \in \Gamma$ such that $[\![x_a]\!] = [\![x]\!](a{:}-)$. The idea is to obtain x_a by substituting $a{:}\iota = a + \iota X$ for ι in x. Formally, we define the *substitution* $x[y/\iota]$ of $y \in R[X]^{\{\iota,1\}}$ for ι in x as the linear extension of the map $\rho_y \colon \{\iota, 1\} \to \Gamma$ with $\rho_y(\iota) = y$ and $\rho_y(1) = 1$, i.e., $x[y/\iota] := \overline{\rho_y}(x)$. More concretely, for $x = (p\iota + q)u^{-1}$ and $y = r\iota + t$, we have

$$x[y/\iota] = (pr\iota + (pt + q))u^{-1}.$$

Lemma 17. *For all $x \in R[X]^{\{\iota,1\}}[U^{-1}]$, $a \in R$ and $\sigma \in R^\omega$,*

$$[\![x[(a + \iota X)/\iota]]\!](\sigma) = [\![x]\!](a : \sigma).$$

We now make the observation, that $x[(a + \iota X)/\iota]$ lies in the submodule $G \subseteq \Gamma$ where ι always occurs "guarded", namely in the form ιX:

$$G = \left\{ \frac{p\iota X + q}{u} \;\middle|\; p, q \in R[X], u \in U \right\}.$$

Due to the guardedness, we can define linear maps $o \colon G \to R$ and $d \colon G \to \Gamma$, which should be thought of as taking initial value and derivative of guarded stream expressions, by inductively applying the behavioral differential equations from Section 2.3 with the special case that $o(\iota X) = 0$ and $d(\iota X) = \iota$. The Mealy machine on Γ is obtained by composing the substitution with the maps $o \colon G \to R$ and $d \colon G \to \Gamma$.

Definition 18. *The localization* $\Gamma = R[X]^{\{\iota, 1\}}[U^{-1}]$ *can be turned into a Mealy machine with the structure map* $\gamma \colon \Gamma \to (R \times \Gamma)^R$ *defined by*

$$\gamma(x)(a) = (o(x[(a + \iota X)/\iota]), d(x[(a + \iota X)/\iota])).$$

Concretely, for $x = \frac{p\iota + q}{u} \in \Gamma$ *and* $a \in R$ *we have:*

$$o(x[(a + \iota X)/\iota]) = \frac{p(0)a + q(0)}{u(0)}$$

$$d(x[(a + \iota X)/\iota]) = \frac{p}{u}\iota + \frac{u(0)(p'a + q') - (p(0)a + q(0))u'}{u(0)u}.$$

Since we defined o and d inductively using the behavioral differential equations, one can show using Lem. 17 and Def. 18 that for all $x \in \Gamma$, all $a \in R$ and all $\sigma \in R^\omega$:

$$o(x[(a + \iota X)/\iota]) \qquad = [\![x]\!](a : \sigma)(0)$$
$$[\![d(x[(a + \iota X)/\iota])]\!](\sigma) = [\![x]\!](a : \sigma)'.$$

In other words, we have shown the following lemma.

Lemma 19. *The algebraic semantics* $[\![-]\!] \colon \Gamma \to C(R^\omega)$ *is a Mealy machine homomorphism from* (Γ, γ) *to* $(C(R^\omega), \delta)$. *By finality of* $(C(R^\omega), \delta)$, *the algebraic semantics coincides with the coalgebraic behavior map, that is,* $[\![x]\!] = \tilde{\gamma}(x)$ *for all* $x \in \Gamma$.

We will use the above lemma to show our main result.

Theorem 20. *For any open linear system* L, *the algebraic solution of* L *coincides with the coalgebraic solution of* L: $[\![L]\!] = \langle\!\langle L \rangle\!\rangle$.

Proof. Let $L = (V, M, O, I)$ be an n-dimensional open linear system and let $s \in V \odot \Gamma$ be the unique state solution of L. Furthermore, let (R^V, c_L) be the Mealy machine associated to L with initial state I. The proof is divided into two steps. We leave out some details.

First, we construct a Mealy machine $(V \odot \Gamma, d \colon V \odot \Gamma \to \mathcal{M}(V \odot \Gamma))$ on the solution space. The map d is obtained by applying the Mealy structure (Γ, γ)

point-wise. For $x = (x_1, ..., x_n) \in V \odot \Gamma$ and $a \in R$, let $d(x)(a) = (O(av_{\mathrm{in}} + o), x')$ where $o = \sum_{i=1}^{n} o_i v_i$ and $x' = \sum_{i=1}^{n} x'_i v_i$ with $\gamma(x_i)(a) = (o_i, x'_i)$ for $i = 1, \ldots, n$. With this definition, we can show that for all $x \in V \odot \Gamma$, $\tilde{d}(x) = \tilde{\gamma}(O(\iota v_{\mathrm{in}} + x))$. By Lem. 19 we get $[\![O(\iota v_{\mathrm{in}} + x)]\!] = \tilde{\gamma}(O(\iota v_{\mathrm{in}} + x))$, and hence $[\![O(\iota v_{\mathrm{in}} + x)]\!] = \tilde{d}(x)$.

In the second step we build a bisimulation B between $(V \odot \Gamma, d)$ and (R^V, c_L) with $(s, I) \in B$ from which $\tilde{d}(s) = \widetilde{c_L}(I)$ and hence $[\![O(\iota v_{\mathrm{in}} + s)]\!] = \widetilde{c_L}(I)$ follows, as desired. The relation B is constructed in the following way. We denote by $I_{\sigma[n]} \in R^V$ the state reached in (R^V, c_L) after reading $\sigma[n] = (\sigma(0), \ldots, \sigma(n-1))$ when starting in state I. For 0 we put $I_{\sigma[0]} = I$ for all $\sigma \in R^\omega$. Analogously, we denote by $s_{\sigma[n]} \in V \odot \Gamma$ the state reached in the Mealy machine $(V \odot \Gamma, d)$ after reading $\sigma[n]$ when starting from the state solution s, and again we put $s_{\sigma[0]} = s$.

Using the fact that s is a state solution, we can show that $s_{\sigma[n]}$ is obtained by applying M repeatedly to the inputs $\sigma(0), \ldots, \sigma(n-1)$ and s. To this end, let $\pi \colon \overline{V} \odot \Gamma \to V \odot \Gamma$ be the evident projection and $M^{[k]} = \pi \circ \left(\begin{smallmatrix} 0 \\ M \end{smallmatrix}\right)^k$ the k-fold composition of M, followed by this projection. Furthermore, we define $x_n = \sum_{i=0}^{n-1} \sigma_i X^i + \iota X^n \in \Gamma$ and observe, that for $n \geq 1$ the x_n is in fact an element of G, the submodule of Γ in which ι occurs in guarded form. With a bit of patience one arrives at the following explicit definitions

$$s_{\sigma[n]} = M^{[n]} \begin{pmatrix} 0 \\ s[x_n/\iota] \end{pmatrix} + \sum_{k=0}^{n-1} M^{[k+1]} \begin{pmatrix} x_n^{(n-k-1)} \\ 0 \end{pmatrix}$$

$$I_{\sigma[n]} = M^{[n]} \begin{pmatrix} 0 \\ I \end{pmatrix} + \sum_{k=0}^{n-1} M^{[k+1]} \begin{pmatrix} \sigma_{n-k-1} \\ 0 \end{pmatrix}.$$

It is now easy to see that, applying o, we have $o\left(x_n^{(n-k-1)}\right) = \sigma_{n-k-1}$ and $o(s[x_n/\iota]) = I$. By point-wise application of o we deduce, that the outputs of $d(s_{\sigma[n]})(a)$ and $c_L(I_{\sigma[n]})(a)$ match for all $a \in R$. Moreover, one can also show, that the next states are $s_{\tau[n+1]}$ and $I_{\tau[n+1]}$ for some $\tau \in R^\omega$ with $\tau[n] = \sigma[n]$ and $\tau(n) = a$. Thus the relation $B = \left\{ \left(s_{\sigma[n]}, I_{\sigma[n]} \right) \mid \sigma \in R^\omega, n \in \mathbb{N} \right\}$ is a bisimulation. Finally, by definition $(s, I) = \left(s_{\sigma[0]}, I_{\sigma[0]} \right) \in B$ and the claim follows.   □

We end this section with a Kleene-style theorem showing that the module $R[X]^{\{\iota, 1\}} [U^{-1}]$ characterizes precisely the behaviors of *all* open linear systems.

Theorem 21. *For every open linear system L, the unique output solution x of L is in $R[X]^{\{\iota, 1\}} [U^{-1}]$. Conversely, for every $x \in R[X]^{\{\iota, 1\}} [U^{-1}]$ there is an open linear system L such that x is the output solution of L.*

Proof. One direction of the above theorem is just Prop. 11. In order to prove the other direction, we sketch here how to construct from $x \in R[X]^{\{\iota, 1\}} [U^{-1}]$ an open linear system L with x as output solution. Assume $x = \frac{p\iota + q}{u}$ where, without loss of generality, $u_0 = 1$.

We take as L the linear system associated with the following signal flow graph:

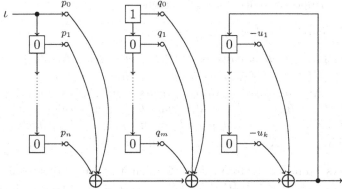

The register contents are represented as a state vector $s = \begin{pmatrix} s_p & s_q & s_u \end{pmatrix}^T \in R^{n+(m+1)+k}$ separated into $s_p \in R^n$, $s_q \in R^{m+1}$ and $s_u \in R^k$.

More concretely, $L = (V, M, O, I)$ with variables $V = V_p + V_q + V_u$ using $V_p = \{1, \dots, n\}, V_q = \{1, \dots, m+1\}$ and $V_u = \{1, \dots, k\}$ is defined as follows. The initial state is $I = (I_p, I_q, I_u)$ where

$$I_p = (0, \dots, 0) \qquad I_q = (1, 0, \dots, 0) \qquad I_u = (0, \dots, 0).$$

We describe the matrix M by letting $s^{(a)} = M(a v_{\text{in}} + s) = M(a v_{\text{in}} + s_p + s_q + s_u)$:

$$
\begin{aligned}
s_{p,1}^{(a)} &= a & s_{p,i+1}^{(a)} &= s_{p,i}, & 1 \le i \le n-1 \\
s_{q,1}^{(a)} &= 0 & s_{q,i+1}^{(a)} &= s_{q,i}, & 1 \le i \le m \\
s_{u,1}^{(a)} &= O(a v_{\text{in}} + s) & s_{u,i+1}^{(a)} &= s_{u,i}, & 1 \le i \le k-1.
\end{aligned}
$$

The output matrix O is

$$O = (p_0, p_1, \dots, p_n, q_0, \dots, q_m, -u_1, \dots, -u_k).$$

To prove that the output behavior of this linear system is x we consider the Mealy machine $(R^{n+(m+1)+k}, c_L)$ associated to L. Consider now the relation $B \subseteq R^{n+(m+1)+k} \times \Gamma$ given by

$$
B = \left\{ (s, z) \;\middle|\; z = \frac{p}{u} \iota + \frac{1}{u} \left(\sum_{i=1}^{n} s_{p,i} p^{(i)} + \sum_{i=1}^{m+1} s_{q,i} q^{(i-1)} - \sum_{i=1}^{k} s_{u,i} u^{(i)} \right) \right\}
$$

One can verify that B is a bisimulation between $(R^{n+(m+1)+k}, c_L)$ and (Γ, γ), and that $(I, x) \in B$, hence we have that $\widetilde{c_L}(I) = \widetilde{\gamma}(x) = [\![x]\!]$.

By solving the system L algebraically, we get a unique output solution $x_o \in R[X]^{\{\iota, 1\}} [U^{-1}]$ with $[\![x_o]\!] = \widetilde{c_L}(I)$ by Thm. 20. Since $[\![-]\!]$ is injective (Thm. 9) we have $x = x_o$. □

Due to Thm. 21, given an open linear system L, we can simply refer to $[\![L]\!] = \langle\!\langle L \rangle\!\rangle$ as the *behavior* of L. Given a signal flow graph C, we refer to $[\![\mathcal{L}(C)]\!] = \langle\!\langle \mathcal{L}(C) \rangle\!\rangle$ as the *behavior* of C.

We can use Thm. 21 to give a precise characterization of the behaviors of the subclasses of closed, respectively feedforward, circuits.

Corollary 22. *We have the following axiomatizations:*

a) *The behaviors* $\langle \mathrm{LM} \rangle$ *of all circuits is an* $R[X]$*-submodule of* $C(R^\omega)$*, and* $\langle \mathrm{LM} \rangle \cong R[X]^{\{\iota,1\}} [U^{-1}]$ *as* $R[X]$*-modules and as Mealy machines.*

b) *The behaviors* $\langle \mathrm{LM_c} \rangle$ *of closed, feedback circuits is an* $R[X]$*-submodule of* $C(R^\omega)$*, and* $\langle \mathrm{LM_c} \rangle \cong R[X] [U^{-1}]$ *as* $R[X]$*-modules and as Mealy machines.*

c) *The behaviors* $\langle \mathrm{LM_{ff}} \rangle$ *of open, feedforward circuits is an* $R[X]$*-submodule of* $C(R^\omega)$*, and* $\langle \mathrm{LM_{ff}} \rangle \cong R[X]^{\{\iota,1\}}$ *as* $R[X]$*-modules and as Mealy machines.*

Proof. **a)** It follows immediately from Thm. 21 that the image of the algebraic semantics map $[\![-]\!]\colon R[X]^{\{\iota,1\}} [U^{-1}] \to C(R^\omega)$ is $\langle \mathrm{LM} \rangle$, and by soundness and completeness (Thm. 9), $[\![-]\!]$ is an injective linear map of $R[X]$-modules, hence a module isomorphism from $R[X]^{\{\iota,1\}} [U^{-1}]$ to $\langle \mathrm{LM} \rangle$. From the fact that $[\![-]\!]\colon R[X]^{\{\iota,1\}} [U^{-1}] \to \langle \mathrm{LM} \rangle$ is also a bijective Mealy homomorphism, it follows that $R[X]^{\{\iota,1\}} [U^{-1}]$ and $\langle \mathrm{LM} \rangle$ are also isomorphic as Mealy machines. Since $[\![-]\!]$ is both an $R[X]$-linear map and a Mealy homomorphism which moreover is injective, it suffices for the remaining items to show that the restrictions of $[\![-]\!]$ to $R[X] [U^{-1}]$ and $R[X]^{\{\iota,1\}}$ have range $\langle \mathrm{LM_c} \rangle$ and $\langle \mathrm{LM_{ff}} \rangle$, respectively.

b) Let $x = \frac{q}{u} \in R[X] [U^{-1}]$ and $j(x) = \frac{0\iota + q}{u} \in R[X]^{\{\iota,1\}} [U^{-1}]$ its embedding. Then we construct a circuit C for x such that $[\![x]\!]_c = [\![j(x)]\!] = [\![\mathcal{L}(C)]\!]$, following the proof of Thm. 21. By construction, this circuit will be independent of the input, i.e., C is closed and hence $[\![x]\!]_c \in \langle \mathrm{LM_c} \rangle$. Conversely, if C is a closed, feedback circuit and $L = (V, M, O, I)$ its associated linear system, then the first column of O is 0 (cf. Lem. 5). Consequently, the output solution of L is of the form $j(x) = \frac{0\iota + q}{u}$ for some $x \in R[X] [U^{-1}]$ with $\langle L \rangle = [\![j(x)]\!] = [\![x]\!]_c$, thus $[\![-]\!]_c\colon R[X] [U^{-1}] \to \langle \mathrm{LM_c} \rangle$ is onto.

c) Let $x = p\iota + q \in R[X]^{\{\iota,1\}}$ and $\varphi_2(x) = \frac{p\iota + q}{1} \in R[X]^{\{\iota,1\}} [U^{-1}]$ its embedding. Then we construct again a circuit C for x such that $[\![x]\!]_{ff} = [\![\varphi_2(x)]\!] = [\![\mathcal{L}(C)]\!]$, following the proof of Thm. 21. By construction, C is feedforward (since $u_1, \dots, u_k$ will all be 0), and hence $[\![x]\!]_{ff} \in \langle \mathrm{LM_{ff}} \rangle$. Conversely, if C is an open, feedforward circuit and $L = (V, M, O, I)$ its associated linear system, then M is of the "lower-triangular form" given in Lem. 5. It follows that the output solution of L will be of the form $\varphi(x) = \frac{p\iota + q}{1}$ for some $x \in R[X]^{\{\iota,1\}}$ with $\langle L \rangle = [\![\varphi(x)]\!] = [\![x]\!]_{ff}$, so $[\![-]\!]_{ff}\colon R[X] [U^{-1}] \to \langle \mathrm{LM_{ff}} \rangle$ is onto. $\square$

Remark 23. From the coalgebraic point of view, one may wonder, whether any of our algebraic characterizations of open signal flow graphs is a fixed point for the functor $\mathcal{M}$ of Mealy machines. For *closed* feedback signal flow graphs, the localization $R[X] [U^{-1}]$ is a fixed point of the functor for streams [4,19]. In the general case of *open* signal flow graphs the result is negative, i.e., $\gamma\colon \Gamma \to (R \times \Gamma)^R$ is *not* an isomorphism, for $\Gamma = R[X]^{\{\iota,1\}} [U^{-1}]$. It is indeed easy to see that d is not surjective by taking, for example $f \in (R \times \Gamma)^R$ with $f(a) = (1, \iota) \in R \times \Gamma$

for all $a \in R$. Let us assume there is an element $x = \frac{p\iota + q}{u} \in \Gamma$ with $d(x) = f$. Then we necessarily have $h(x[(a + \iota X)/\iota]) = 1$ and we can deduce that $p_0 = 0$, for otherwise a cannot be arbitrary. It follows that x is of the following form (for new $p, q, u \in R[X]$)

$$x = \frac{b + qX + pX\iota}{b + uX}$$

From the requirement that $(x[(a+\iota X)/\iota])' = \iota$, we can derive that $paX\iota = b + uX$ and hence, by taking initial output, that $0 = b$ which is a contradiction. Thus there is no $x \in \Gamma$ with $d(x) = f$, i.e., d is not surjective and therefore Γ is not a fixed point of the functor $\mathcal{M}$.

8   Concluding Remarks

Our main contribution in this paper is the axiomatization of signal flow graphs using standard mathematical concepts and techniques, such as polynomials and module localization. In the following table we give an overview of the algebras corresponding to different classes of signal flow graphs.

type	feedforward	feedback
closed	Free R-algebra $R[X]$ of polynomials	Localization $R[X][U^{-1}]$
open	Free $R[X]$-module $R[X]^{\{\iota,1\}}$	Localization $R[X]^{\{\iota,1\}}[U^{-1}]$

Our results yield a method for deciding circuit equivalence by comparing solutions in the localization $\Gamma = R[X]^{\{\iota,1\}}[U^{-1}]$. Deciding whether $\frac{p_1\iota + q_1}{u_1} = \frac{p_2\iota + q_2}{u_2}$ boils down to finding a $v \in U$ such that $v(p_1u_2 - p_2u_1) = 0$ and $v(q_1u_2 - q_2p_1) = 0$ hold (using that $R[X]^{\{\iota,1\}}$ is freely generated). If R is an integral domain, then this problem reduces to the simple problem of deciding equivalence of polynomials: $p_1u_2 = p_2u_1$ and $q_1u_2 = q_2p_1$. If equality in R is effectively decidable, then polynomial equivalence is effectively decidable, since $R[X]$ is the free commutative R-algebra over the single generator X. Summarizing, if R is an integral domain in which equality is effectively decidable, then so is equality in Γ.

We have restricted our attention to circuits with at most one input end, and one output end. It is straightforward to extend our result to more inputs by using different generators $\iota_1, \ldots, \iota_k$ for each input end. Multiple outputs, on the other hand, can be represented by changing the underlying ring to R^m (with component-wise operations).

All the work in this paper is based on the assumption that signals are elements of a *commutative* ring. There are, however, interesting rings used in systems theory which are non-commutative, such as the ring of matrix polynomials [27]. An interesting future direction is the generalization of our results using non-commutative localization. This raises two problems: first one needs different conditions on the ring to still have an invariant basis number, so that matrices still represent linear maps. The second problem is that in the localization one generally loses the property that every element is of the form $\frac{a}{u}$, instead they will be sums of such fractions. For discussions on these issues see for example [14].

The localization Γ and the causal functions $C(R^\omega)$ both carry algebraic as well as coalgebraic structure. This suggests the presence of a more abstract description in terms of bialgebras for a distributive law [2,26]. However, it is not clear what the involved monad is, and as discussed after Def. 14 it is also not clear how open linear systems can be viewed as coalgebras over a category of algebras. We do not exclude the possibility of a bialgebraic modeling of open linear systems, but we have to leave it as future work.

References

1. Atiyah, M., MacDonald, I.G.: Introduction to Commutative Algebra. Addison-Wesley series in mathematics. Westview Press (1994)
2. Bartels, F.: On Generalised Coinduction and Probabilistic Specification Formats. Ph.D. thesis, Vrije Universiteit Amsterdam (2004)
3. Bonsangue, M.M., Rutten, J.J.M.M., Silva, A.: Coalgebraic logic and synthesis of Mealy machines. In: Amadio, R.M. (ed.) FOSSACS 2008. LNCS, vol. 4962, pp. 231–245. Springer, Heidelberg (2008)
4. Bonsangue, M.M., Milius, S., Silva, A.: Sound and complete axiomatizations of coalgebraic language equivalence. ACM Transactions on Computational Logic 14(1), 7–57 (2013)
5. Chen, W.-K.: On flow graph solutions of linear algebraic equations. SIAM Journal on Applied Mathematics 142, 136–142 (1967)
6. Crochiere, R., Oppenheim, A.: Analysis of linear digital networks. Proceedings of IEEE 4, 581–595 (1975)
7. Davis, A.L., Keller, R.M.: Dataflow program graphs. Computer 15(2), 26–41 (1982)
8. Eisenbud, D.: Commutative Algebra: With a View Toward Algebraic Geometry. Graduate Texts in Mathematics Series. Springer (1995)
9. Hansen, H.H.: Subsequential transducers: a coalgebraic perspective. Information and Computation 208(12), 1368–1397 (2010)
10. Hansen, H.H., Rutten, J.J.M.M.: Symbolic synthesis of Mealy machines from arithmetic bitstream functions. Scientific Annals of Computer Science 20, 97–130 (2010)
11. Kahn, G.: The semantics of a simple language for parallel programming. Information Processing 74, 471–475 (1974)
12. Keller, R.M., Panangaden, P.: Semantics of networks containing indeterminate operators. In: Brookes, S.D., Winskel, G., Roscoe, A.W. (eds.) Seminar on Concurrency. LNCS, vol. 197, pp. 479–496. Springer, Heidelberg (1985)
13. Keller, R.M., Panangaden, P.: Semantics of digital networks containing indeterminate modules. Distributed Computing 1(4), 235–245 (1986)
14. Lam, T.Y.: Lectures on Modules and Rings. In: Graduate Texts in Math. Springer (1999)
15. Lambek, J.: A fixpoint theorem for complete categories. Mathematische Zeitschrift 103(2), 151–161 (1968)
16. Mason, S.J.: Feedback theory: Some properties of linear flow graphs. In: Proceedings of IRE 41, pp. 1144–1156 (1953)
17. Massay, J.L., Sain, M.K.: Codes, Automata, and Continuous Systems: Explicit Interconnections. IEEE Trans. Automatic Control 12(6), 644–650 (1967)
18. Massay, J.L., Sain, M.K.: Inverses of Linear Sequential Circuits. IEEE Trans. Comput. 17, 330–337 (1968)

19. Milius, S.: A sound and complete calculus for finite stream circuits. In: Proceedings of LICS 2010, pp. 421–430. IEEE (2010)
20. Parhi, K.K., Chen, Y.: Signal flow graphs and data flow graphs. In: Handbook of Signal Processing Systems, pp. 791–816. Springer (2010)
21. Rutten, J.J.M.M.: Universal coalgebra: a theory of systems. Theoretical Computer Science 249(1), 3–80 (2000)
22. Rutten, J.J.M.M.: A coinductive calculus of streams. Mathematical Structures in Computer Science 15(1), 93–147 (2005)
23. Rutten, J.J.M.M.: A tutorial on coinductive stream calculus and signal flow graphs. Theoretical Computer Science 343(3), 443–481 (2005)
24. Rutten, J.J.M.M.: Rational streams coalgebraically. Logical Methods in Computer Science 3(9), 1–22 (2008)
25. Scherba, M.B., Roesser, R.P.: Computation of the transition matrix of a linear sequential circuit. IEEE Trans. Computers 22(4), 427–428 (1973)
26. Turi, D., Plotkin, G.: Towards a mathematical operational semantics. In: Proceedings of LICS 1997, pp. 280–291. IEEE Computer Society (1997)
27. Vidyasagar, M.: Control System Synthesis. A Factorization Approach. The MIT Press (1988)
28. Wilf, H.S.: Generating functionology, 2nd edn. Academic Press, Inc. (1992)

Fractal Sets as Final Coalgebras Obtained by Completing an Initial Algebra

Prasit Bhattacharya, Lawrence S. Moss*,
Jayampathy Ratnayake, and Robert Rose

Department of Mathematics, Indiana University,
Bloomington, IN 47405, USA

Abstract. This paper is concerned with final coalgebra representations of fractal sets. The background to our work includes Freyd's Theorem: the unit interval is a final coalgebra of a certain endofunctor on the category of bipointed sets. Leinster's far-ranging generalization of Freyd's Theorem is also a central part of the discussion, but we do not directly build on his results. Our contributions are in two different directions. First, we demonstrate the connection of final coalgebras and initial algebras; this is an alternative development to one of his central contributions, working with resolutions. Second, we are interested in the metric space aspects of fractal sets. We work mainly with two examples: the unit interval $[0, 1]$ and the Sierpiński gasket $\mathbb{S}$ as a subset of $\mathbb{R}^2$.

1 Introduction

This paper is a contribution to the presentation of fractal sets in terms of final coalgebras. The first result on this topic was Freyd's Theorem in [6]: the unit interval $[0, 1]$ is the final coalgebra of a functor $X \mapsto X \oplus X$ on the category of bipointed sets. Leinster [10] offers a sweeping generalization of this result. He is able to represent many of what would be intuitively called *self-similar* spaces using (a) bimodules (also called profunctors or distributors), (b) an examination of non-degeneracy conditions on functors of various sorts; (c) a construction of final coalgebras for the types of functors of interest using a notion of resolution. In addition to the characterization of fractal sets as sets, his seminal paper also characterizes them as topological spaces.

Our major contribution is to suggest that in many cases of interest, point (c) above on resolutions is not needed in the construction of final coalgebras. Instead, one may obtain a number of spaces of interest as the Cauchy completion of an initial algebra, and this initial algebra is the set of points in a colimit of an ω-sequence of finite metric spaces. This generalizes Hutchinson's characterization of fractal attractors in [8] as closures of the orbits of the critical points. In addition to simplifying the overall machinery, it also presents a metric space which is "computationally related" to the overall fractal. For example, when

---

* This work was partially supported by a grant from the Simons Foundation (#245591 to Lawrence Moss).

F. van Breugel et al. (Eds.): Panangaden Festschrift, LNCS 8464, pp. 146–167, 2014.

applied to Freyd's construction, our method yields the metric space of dyadic rational numbers in $[0,1]$. When applied to the Sierpiński gasket $\mathbb{S}$, it yields a countable metric space whose completion is $\mathbb{S}$ and which might be taken as the set of finitely described points in $\mathbb{S}$.

Our second contribution is not completed at this time, but it is a set of results on *metric space* characterizations of final coalgebras. This point was raised as an open issue in Hasuo, Jacobs, and Niqui [7]; this paper was important to us because it emphasized algebras in addition to coalgebras, and because it highlighted endofunctors defined using quotient metrics. Indeed, we use quotient metrics in the main results of this paper. We say that our work is not completed, and by this we mean that what we have is a fully worked out example rather than a general theory. Still, since $[0,1]$ and $\mathbb{S}$ are important in their own right, we feel working them out in detail will inform a more general theory.

Related work. We know of a few papers which discuss final coalgebras obtained as Cauchy completions of initial algebras. Perhaps the most relevant is Adámek [2], since it summarizes and extends work that came before it. All of these papers work under assumptions that are not satisfied in our setting. Specifically, [2] has results on locally finitely presentable categories; these are complete, in contrast to the main categories here. To make this point differently, our main categories, Bi (bipointed sets) and Tri (tripointed sets) do not have final objects. Most of the constructions of final coalgebras in the literature *start* from a final object. So for this reason, what we are doing is not a special case of the results in [2].

2   Freyd's Theorem on the Unit Interval

We begin our work by reviewing Peter Freyd's characterization in [6] of the unit interval $[0,1]$ in terms of bipointed sets. Our proof is a little simpler than the original. In our presentation, we include a detour, connecting the final coalgebra characterization of the unit interval to an initial algebra characterization of the dyadic rationals in $[0,1]$.

A *bipointed set* is a triple $(X, \top, \bot)$, where X is an arbitrary set, and $\top$ and $\bot$ belong to X. A bipointed set is *proper* (or *non-degenerate*) if $\bot \neq \top$. We restrict attention to the proper bipointed sets from now on. Let Bi be the category of *proper bipointed sets*; a morphism in Bi is required to preserve the constants $\top$ and $\bot$.

We are mainly interested in a certain endofunctor on bipointed sets which we'll write as $X \mapsto M \otimes X$. (The original notation for $M \otimes X$ was $X \oplus X$, but we have changed this to match Leinster [10]. The reason for the tensor product notation $\otimes$ is that we are constructing a quotient of the product set $M \times X$, as we shall see shortly.) Here M is a two-element set $\{\ell, r\}$. For a bipointed set X, $M \otimes X$ is the product $M \times X$, modulo the identification $(\ell, \top) = (r, \bot)$. Further, we take $\top_{M \otimes X} = (r, \top)$, and $\bot_{M \otimes X} = (\ell, \bot)$. We complete the definition of $M \otimes X$ as an endofunctor in the following way: if $f : X \to Y$ is a morphism of bipointed sets, then $M \otimes f : M \otimes X \to M \otimes Y$ is $(m, x) \mapsto (m, f(x))$.

Background on initial algebras and final coalgebras. We now recall some general definitions from category theory. Given an endofunctor, that is a functor from some category to itself, say $F : C \to C$, an *algebra* for F is a pair $(A, a : FA \to A)$. Dually, a *coalgebra* for F is a pair $(A, a : A \to FA)$. The difference is the direction of the arrow. In either case, the object A is called the *carrier*, and the morphism a is the *structure*. Often we suppress the structure when referring to algebras or coalgebras. Algebras for F are often called F-algebras, and similarly for coalgebras.

A morphism of algebras $f : A \to B$ is a morphism in the underlying category such that $a \circ f = b \circ Ff$, as on the left below:

$$
\begin{array}{ccc}
FA \xrightarrow{\ a\ } A & \qquad & A \xrightarrow{\ a\ } FA \\
\ \downarrow{\scriptstyle Ff} \quad\ \downarrow{\scriptstyle f} & & \ \downarrow{\scriptstyle f} \quad\ \downarrow{\scriptstyle Ff} \\
FB \xrightarrow{\ b\ } B & & B \xrightarrow{\ b\ } FB
\end{array}
$$

A morphism of coalgebras is a morphism $f : A \to B$ such that the diagram on the right commutes. An *initial algebra* is an F-algebra with the property that there is a unique algebra morphism from it to any algebra (of the same functor). A *final coalgebra* is a coalgebra with the property that for every coalgebra, there is a unique coalgebra morphism into it. Initial algebras and final coalgebras occur in many places in mathematics and theoretical computer science. Many examples of these, along with their theory, may be found in the survey volume [3]. We do need two of the central results in the theory.

Lemma 1 (Lambek's Lemma [9]). *The structure map of an initial algebra is a categorical isomorphism. That is, if (A, a) is an initial algebra, then a has an inverse $a^{-1} : FA \to A$ such that $a^{-1} \circ a = \mathrm{id}_{FA}$, and $a \circ a^{-1} = \mathrm{id}_A$. The same hold for final coalgebra structures, mutatis mutandis.*

Theorem 2 (Adámek [1]). *Let C be a category with initial object 0. Let $F : C \to C$ be an endofunctor. Consider the* initial chain

$$
0 \xrightarrow{\ !\ } F0 \xrightarrow{\ F!\ } F^2 0 \xrightarrow{\ F^2 !\ } \cdots \qquad F^n 0 \xrightarrow{\ F^n !\ } F^{n+1} 0 \qquad \cdots
$$

Suppose that the colimit

$$
A = colim_{n<\omega} F^n 0
$$

exists, and write $i_n : F^n 0 \to A$ for the cocone morphism. Suppose that F preserves this colimit. Let $a : FA \to A$ be the unique morphism so that $a \circ Fi_n = i_{n+1}$ for all n. Then (A, a) is an initial algebra of F.

We apply this with C being Bi. The category does have an initial algebra, the two-point bipointed set $\{\bot, \top\}$. For the functor F, we take $X \mapsto M \otimes X$. So $F0$ is a three-point bipointed set $\{\bot, \top, \frac{1}{2}\}$, and the map $! : 0 \to F0$ is the inclusion. The $\frac{1}{2}$ here is merely suggestive; we could use any object. But using $\frac{1}{2}$ foreshadows what we shall soon see in the metric setting. Continuing, $F^2 0$ would be $\{\bot, \frac{1}{4}, \frac{1}{2}, \frac{3}{4}, \top\}$; again, the map $F! : F0 \to F^2 0$ is the inclusion. This explains how the initial chain is constructed. Its colimit in the category is the set $\mathbb{D}$ of dyadic rationals in $[0, 1]$ with $\bot = 0$ and $\top = 1$. (The dyadic rationals are the rationals whose denominator is a power of two.) The structure $d : F\mathbb{D} \to \mathbb{D}$ is

$$d(\ell, x) = x/2$$
$$d(r, x) = (x + 1)/2$$

It is easy to check that this is well-defined. To apply Theorem 2, we must verify that F preserves the colimit of the initial chain. We omit this detail, and merely state the result which we are after; see [3] for more on this result.

Theorem 3 ([3]). $(\mathbb{D}, d)$ is an initial algebra of $F : \mathsf{Bi} \to \mathsf{Bi}$.

We now turn to the final coalgebra of F. There is a dual of Theorem 2, first stated by Barr [5]. However, Barr's result requires that the underlying category C have a final object, and the category Bi of bipointed sets does not have a final object. [To see this, suppose towards a contradiction that A were final. Let B be the three-element Bi-object $\{\top, \bot, *\}$. There are at least two Bi-morphisms $i : B \to A$. One is $\top \mapsto \top$, $\bot \mapsto \bot$, and $* \mapsto \top$. The other is $\top \mapsto \top$, $\bot \mapsto \bot$, and $* \mapsto \bot$. This shows that A is not final.] Instead, we use a different approach. One way is to basically "guess" the final coalgebra and verify the guess. This is what we shall do in Theorem 4 below.

Let I be the unit interval $[0, 1]$, with $\bot = 0$ and $\top = 1$. So I is an object in Bi. We have a coalgebra structure $i : I \to FI$ quite closely related to the inverse of d above:

$$i(x) = \begin{cases} (\ell, 2x) & \text{if } x \leq 1/2 \\ (r, 1 - 2x) & \text{if } x \geq 1/2 \end{cases}$$

(If $x = 1/2$, then both cases agree because $(\ell, 1) = (r, 0)$.) Note that i is a bijection, and indeed an isomorphism in Bi. In particular, i^{-1} exists in Bi.

Theorem 4 (Freyd [6]). (I, i) is a final coalgebra of $F : \mathsf{Bi} \to \mathsf{Bi}$.

Proof. Let $(X, e : X \to FX)$ be a coalgebra, where X is any biointed set. The homset $\mathsf{Set}(X, I)$ is a complete metric space with the metric given by the supremum of pointwise distances. And the homset $\mathsf{Bi}(X, I)$ is a closed subset of this space. Thus $\mathsf{Bi}(X, I)$ is a complete metric space. It is also non-empty. Consider the function

$$\varphi : \mathsf{Bi}(X, I) \to \mathsf{Bi}(X, I)$$

given by

$$\varphi(f) = i^{-1} \circ (Ff) \circ e.$$

In other words, $\varphi(f)$ makes the diagram below commute:

$$
\begin{array}{ccc}
X & \xrightarrow{\ e\ } & FX \\
{\scriptstyle \varphi(f)} \downarrow & & \downarrow {\scriptstyle Ff} \\
I & \xrightarrow[\ i\ ]{} & FI
\end{array}
$$

Then φ is a contraction of a non-empty complete metric space. (The contractive property means that in the natural metric on $\mathsf{Bi}(X, I)$,

$$d(\varphi(f), \varphi(g)) < \tfrac{1}{2} d(f, g).$$

This property is easy to check; it uses the fact that in our definition of FX, we scaled the metric of X by $\frac{1}{2}$.) Thus φ has a unique fixed point. A fixed point φ^* of φ is exactly a coalgebra morphism from X to I, and so we are done.

We mentioned before that this proof of Theorem 4 is simpler than the original argument. Note also that the proof does not really use Theorem 3. So it is not clear that there is a connection between these results at all. The central point in this paper is to draw such a connection, based on the observation that the unit interval is the Cauchy completion of the dyadic rationals in $[0, 1]$. Our way of making the connection is via a metric version of Bi.

MS is the category of metric spaces, with all distances bounded by 1. The morphisms in MS are the *short maps* $f : X \to Y$ of metric spaces: for all $x, y \in X$,

$$d_Y(f(x), f(y)) \leq d_X(x, y)$$

These are also called *non-expanding maps*. (Incidentally, we choose the short morphisms as the morphisms in this category because it is the most standard choice in the literature on coalgebras. However, at the very end of this paper we shall see a reason to perhaps re-consider this choice.)

Let the category BiMS of *bipointed metric spaces* be the category whose objects are bipointed sets which are also metric spaces, and with the property that $d(\bot, \top) = 1$. The morphisms are short maps which preserve $\bot$ and $\top$. The endofunctor $X \to M \otimes X$ works in the metric setting.

Theorem 5. *The initial algebra of $X \to M \otimes X$ on BiMS is the bipointed metric space of dyadic rationals $(\mathbb{D}, d)$. The final coalgebra is the unit interval (I, i).*

The initiality result is similar to what we saw in Theorem 3. The finality result is proved in nearly the same way as Theorem 4, except that we must show one extra point: if X is a bipointed set, then the set of short maps from X to I is non-empty. (We need this because we want to apply the contraction mapping theorem to get a fixed point, but to do this we must know that the space is actually non-empty.) Fortunately, in this case, the result is easy: $x \mapsto d(x, \bot)$ is a BiMS-morphism of any space into $[0, 1]$. The analogous result for the Sierpiński gasket will be much more difficult to obtain.

3   The Sierpiński Gasket

Freyd's Theorem suggests that many interesting topological or metric spaces might have final coalgebra characterizations. This point has been substantiated by Leinster [10], especially on topics related to the topology. We are after something similar for the metric aspects. This paper begins this development by an extended discussion of the *Sierpiński gasket* (or *Sierpiński triangle*) $\mathbb{S} \subseteq \mathbb{R}^2$. This is sometimes taken to be the prototypical fractal set, and so it is a good place to begin. As we shall see, not all of the results on $\mathbb{S}$ are straightforward generalizations of what we have seen for the unit interval.

We first recall the characterization of $\mathbb{S}$ using *iterated function systems*. Consider the maps $\sigma_a, \sigma_b, \sigma_c : \mathbb{R}^2 \to \mathbb{R}^2$ given by

$$\begin{aligned}
\sigma_a(x, y) &= (x/2, y/2) + (1/4, \sqrt{3}/4) \\
\sigma_b(x, y) &= (x/2, y/2) \\
\sigma_c(x, y) &= (x/2, y/2) + (1/2, 0)
\end{aligned} \tag{1}$$

The idea is that σ_a takes a point to the midpoint of the segment determined by it and $(1/2, \sqrt{3}/2)$; σ_b acts similarly using $(0, 0)$, and σ_c uses $(1, 0)$. These maps σ_i extend to subsets of $\mathbb{R}^2$ by taking images.

Definition 1. *The Sierpiński gasket is the unique non-empty compact subset* $\mathbb{S} \subseteq \mathbb{R}^2$ *such that*

$$\mathbb{S} = \sigma_a(\mathbb{S}) \cup \sigma_b(\mathbb{S}) \cup \sigma_c(\mathbb{S}). \tag{2}$$

$\mathbb{S}$ is shown in Figure 1.

Later on, we shall need some notation for certain subsets of the plane. Let

$$M = \{a, b, c\}$$

We use the letter m as a variable over M in the sequel.

Let M^* be the set of finite words from M. We define triangles tr_w for $w \in M^*$ by recursion:

$$\begin{aligned}
tr_\varepsilon \ &= \text{the triangle with vertices} \\
&\quad \text{top} = (1/2, \sqrt{3}/2), \text{ left} = (0, 0), \text{ and right} = (1, 0) \\
tr_{mw} &= \sigma_m(tr_w)
\end{aligned} \tag{3}$$

Let

$$R_n = \bigcup \{tr_w : w \text{ is a word of length } n\}.$$

Also, for an infinite sequence $\alpha = \alpha_0 \alpha_1 \cdots \alpha_n \cdots \in M^\omega$, let

$$p_\alpha = \text{the unique point in } \bigcap_n tr_{\alpha_0 \alpha_1 \cdots \alpha_n}.$$

That is, $\bigcap_n tr_{\alpha_0 \alpha_1 \cdots \alpha_n}$ is a nested intersection of a family of non-empty compact sets, and the diameter of the sets tends to 0. By the Cantor intersection theorem, this intersection is a singleton.

The following is a special case of the classical result of Hutchinson [8] on fractal attractors.

Proposition 6. $\mathbb{S}$ *has the following characterizations:*

$$\begin{aligned}
\mathbb{S} &= \bigcap_n R_n \\
&= \{p_\alpha : \alpha \in M^\omega\} \\
&= \text{the closure of } \bigcup_{w \in M^*} \{x : x \text{ is a vertex of } t_w\}.
\end{aligned}$$

Fig. 1. The Sierpiński gasket $\mathbb{S} \subseteq \mathbb{R}^2$

Different sequences in M^ω might well be associated with the same point via the operation p. For example, let $\alpha = baccc\cdots$, and let $\beta = bcaaa\cdots$. Then $p_\alpha = p_\beta = (3/8, \sqrt{3}/8)$. However, the sequence α is a *resolution* of p_α. And so what we have is the familiar phenomenon that a given point might have more than one resolution.

In addition, Proposition 6 gives three different ways to think about $\mathbb{S}$. We just saw that writing $\mathbb{S} = \{p_\alpha : \alpha \in M^\omega\}$ connects $\mathbb{S}$ with resolutions. But the third way is closest to what we do in this paper.

4   Tripointed Sets

The main categories involved in our rendering of the Sierpiński gasket as a final coalgebra are *tripointed sets* and *tripointed metric spaces*.

Definition 2. *A* tripointed set *is a set X together with distinguished elements* top, left, *and* right. *X is* proper *(or* non-degenerate*) if the distinguished elements are distinct. (As with bipointed sets, we are only really interested in proper tripointed sets.)*

We let Tri *be the category of proper tripointed sets, taking as morphisms the functions respecting the distinguished points.*

Example 1. The initial object I of Tri is {top, left, right}. But Tri has no final object. To see this, suppose that X were final. Let Y contain top, left, and right, and also a fourth point $*$. There are at least three short maps from Y to X, since $*$ can be sent to any of the distinguished points.

Example 2. The Sierpiński gasket is an object of Tri. For the structure, we take top $= (1/2, \sqrt{3}/2)$, left $= (0,0)$, and right $= (1,0)$.

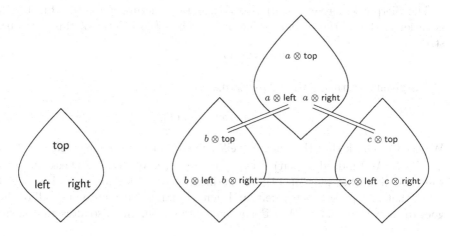

Fig. 2. On the left we have a Tri-object X. On the right we show $M \otimes X$. It is the cartesian product of $M = \{a, b, c\}$ with X, with three pairs of points identified. The identifications are shown with the doubled lines. The top of $M \otimes X$ is $a \otimes \mathsf{top}$, and similarly for left and right. In Section 5, we shall study a version of this functor for TriMS-objects. In that setting the metric multiplies the distances in the copies of M by $\frac{1}{2}$, and then takes the quotient metric associated with the identifications shown above. Pairs of points identified by the doubled lines count as distance 0 apart.

The functor $X \mapsto M \otimes X$ on Tri. Recall the Set endofunctor $X \mapsto M \times X$, where M is $\{a, b, c\}$. For a function $f : X \to Y$, $M \times f : M \times X \to M \times Y$ is $(m, x) \mapsto (m, fx)$. We are interested in a version of this functor for Tri.

We define $\sim$ on $M \times X$ to be the relation

$$(a, \mathsf{left}) \sim (b, \mathsf{top}) \qquad (a, \mathsf{right}) \sim (c, \mathsf{top}) \qquad (b, \mathsf{right}) \sim (c, \mathsf{left}) \qquad (4)$$

Let $\approx$ be the reflexive and symmetric closure of $\sim$. This relation $\approx$ is an equivalence relation. We take $M \otimes X$ to be the quotient of $M \times X$ by the relation $\approx$. See Figure 2 for a picture. We write $m \otimes x$ instead of the equivalence class $[(m, x)]$. So we have equalities

$$a \otimes \mathsf{left} = b \otimes \mathsf{top} \qquad a \otimes \mathsf{right} = c \otimes \mathsf{top} \qquad b \otimes \mathsf{right} = c \otimes \mathsf{left}$$

We call these points the *connection points of $M \otimes X$.*

Returning to our tripointed set $M \otimes X$, we now see that this set is a tripointed space with

$$\begin{aligned}
\mathsf{top}_{M \otimes X} &= a \otimes \mathsf{top} \\
\mathsf{left}_{M \otimes X} &= b \otimes \mathsf{left} \\
\mathsf{right}_{M \otimes X} &= c \otimes \mathsf{right}
\end{aligned}$$

If X is a proper tripointed set, so is $M \otimes X$. Moreover, the operation $X \mapsto M \otimes X$ extends to a functor Tri $\to$ Tri.

The Sierpiński gasket $\mathbb{S}$ carries a coalgebra structure σ for this functor. It is easier to define the algebra structure which is the inverse of this coalgebra structure,

$$\sigma^{-1} : M \otimes \mathbb{S} \to \mathbb{S}$$

σ^{-1} is given in terms of the maps σ_a, σ_b, and σ_c in (1):

$$\sigma^{-1}(m \otimes x) = \sigma_m(x). \tag{5}$$

We need to check that this map is well-defined. For example, $a \otimes \text{left} = b \otimes \text{top}$ in $M \otimes \mathbb{S}$. And indeed $\sigma_a(\text{left}) = (\frac{1}{4}, \frac{\sqrt{3}}{2}) = \sigma_b(\text{top})$. We have to check the same thing for the other pairs in (4); the verifications are similar. It is also easy to check that σ^{-1} preserves top, left, and right. Finally, it is one-to-one. All of this goes to show that $\sigma : \mathbb{S} \to M \otimes \mathbb{S}$ is a coalgebra on Tri, and also an isomorphism.

5   Tripointed Metric Spaces

The functor $X \mapsto M \times X$ on MS. The operation $X \mapsto M \times X$ extends to an endofunctor on MS in the following way. We take

$$d((m, x), (n, y)) = \begin{cases} \frac{1}{2}d(x, y) & \text{if } m = n \\ 1 & \text{if } m \neq n \end{cases}$$

If $f : X \to Y$ is a morphism in MS, then so is $M \times f : M \times X \to M \times Y$.
    It is easy to show that if X is complete, then $M \times X$ is also complete.

Tripointed metric spaces. An object of TriMS is an object of MS (a metric space with distances bounded by 1) with three distinguished points top, left, and right which are required to be of pairwise distance 1. Morphisms are short maps which preserve the distinguished points.

Example 3. The initial object I of TriMS is $I = \{\text{top}, \text{left}, \text{right}\}$ from Example 1 with pairwise distances 1.

Example 4. Another example is the unit triangle $\mathbb{T}$, where

$$\mathbb{T} = \{(x, y) \in \mathbb{R}^2 : y \geq 0 \text{ and } y \leq x\sqrt{3} \text{ and } y \leq \sqrt{3}(1 - x)\}.$$

We take $\text{top} = (\sqrt{3}/2, 1)$, $\text{left} = (0, 0)$, and $\text{right} = (1, 0)$.

The functor $X \mapsto M \otimes X$ on TriMS. We turn $X \mapsto M \otimes X$ into an endofunctor on TriMS by taking the set $M \otimes X$ and then using the quotient metric. This means that the distance from $m \otimes x$ to $n \otimes y$ is the infimum over all finite paths in $M \times X$ of the *score*, where the score is the sum of the distances (in $M \times X$) along the path, but where we count 0 for pairs in the relation $\sim$.

Example 5. Let X be the initial object of TriMS (see Example 3) together with another point x such that

$$d(\mathsf{top}, x) = 1 = d(\mathsf{left}, x) \qquad d(\mathsf{right}, x) = 1/3.$$

There are essentially two paths from $a \otimes x$ to $b \otimes x$: via the path

$$(a, x) \xrightarrow{\quad\frac{1}{2}\quad} (a, \mathsf{left}) \approx (b, \mathsf{top}) \xrightarrow{\quad\quad\frac{1}{2}\quad\quad} (b, x)\ .$$

$$(a, x) \xrightarrow{\quad\frac{1}{6}\quad} (a, \mathsf{right}) \approx (c, \mathsf{top}) \xrightarrow{\quad\frac{1}{2}\quad} (c, \mathsf{left}) \approx (b, \mathsf{right}) \xrightarrow{\quad\frac{1}{6}\quad} (b, x)$$

In terms of the score, the second path is shorter, and its score is $\frac{5}{6}$. Indeed, $d(a \otimes x, b \otimes x) = \frac{5}{6}$.

Generalizing what we saw in Example 5 just above, we have the following result.

Lemma 7. *For all $x, y \in X$, distances in $M \otimes X$ are calculated as follows:*

$$d(a \otimes x, a \otimes y) = \tfrac{1}{2}d(x, y)$$
$$d(a \otimes x, b \otimes y) = \min\left(\tfrac{1}{2}(d(x, \mathsf{left}) + d(\mathsf{top}, y)), \tfrac{1}{2}(d(x, \mathsf{right}) + 1 + d(\mathsf{right}, y)) \right)$$

and similarly for the other distance calculations.

Proof. We check the first assertion, concerning two pairs in the same copy of X, say (a, x) and (a, y).

Consider a finite sequence of pairs in $M \times X$, and the sum of the distances between pairs which are not related by $\sim$. Clearly we have

$$d(a \otimes x, a \otimes y) \leq \tfrac{1}{2}d(x, y)$$

The main work is to show the reverse inequalities by a detailed examination of the quotient metric.

If our sequence consists only of pairs in the a-copy, then by the triangle inequality of X, the score has to be at most half the distance from x to y in the original space X. If the sequence has two points in the b-copy, we may assume that those points are different connection points; if they are the same point, then we could get a smaller score by omitting everything between them. But if the sequence has two connection points, then its score is at least $1/2$. The last case is when the sequence has two connection points of different types other than a, say (c, top) and (b, right). In this case, again we have a score of at least $\frac{1}{2}$.

The second assertion is proved similarly. The point is that no sequence of minimal score may use all three connection points. The shortest path between $a \otimes x$ and $b \otimes y$ could either go through $(a, \mathsf{left}) = (b, \mathsf{top})$; or it could go through $(a, \mathsf{right}) = (c, \mathsf{top})$, and then to $(c, \mathsf{left}) = (b, \mathsf{right})$. (This last option occurs in Example 5.)

Proposition 8. $X \mapsto M \otimes X$ *is an endofunctor on* TriMS.

Proof. Lemma 7 implies that the distances in $M \otimes X$ between distinct corner points is 1. For example, consider top and left.

$$d(\mathsf{top}_{M \otimes X}, \mathsf{left}_{M \otimes X}) = d(a \otimes \mathsf{top}, b \otimes \mathsf{left})$$
$$= \min(\tfrac{1}{2}(1+1), \tfrac{1}{2}(1+1+1))$$
$$= 1$$

This checks that $M \otimes X$ is an object in $M \otimes X$. We next check the parallel assertion for morphisms.

Let $f : X \to Y$ be a short map. We show that $M \otimes f : M \otimes X \to M \otimes Y$. One example case concerns $x_1, x_2 \in X$ and asks us that show that

$$d_{M \otimes Y}(a \otimes f(x_1), b \otimes f(x_2)) \leq d_{M \otimes X}(a \otimes x_1, b \otimes x_2).$$

We sketch the proof. Both of the distances above are minima, and so it is sufficient to show that the following two inequalities hold:

$$d_Y(f(x_1), \mathsf{left}) + d_Y(\mathsf{top}, f(x_2)) \leq d_X(x_1, \mathsf{left}) + d_X(\mathsf{top}, x_2)$$
$$d_Y(f(x_1), \mathsf{right}) + d_Y(\mathsf{top}, f(x_2)) \leq d_X(x_1, \mathsf{right}) + d_X(\mathsf{top}, x_2)$$

Now these inequalities hold because f preserves top, left, and right and is itself a short map. For example,

$$d_Y(f(x_1), \mathsf{left}) = d_Y(f(x_1), f(\mathsf{left})) \leq d_X(x_1, \mathsf{left}).$$

The full details of the proof are based on these observations and similar ones.

Example 6. With $\mathbb{S}$ the Sierpiński gasket in the usual metric, we have an algebra $\sigma^{-1} : M \otimes \mathbb{S} \to \mathbb{S}$ (see (5)). This map σ^{-1} is a short map, as is easily verified. On the other hand, its inverse $\sigma : \mathbb{S} \to M \otimes \mathbb{S}$ is not a short map. For example, $\sigma((\tfrac{1}{2}, \tfrac{\sqrt{3}}{2})) = (a, \mathsf{top})$, $\sigma((\tfrac{1}{2}, 0)) = (b, \mathsf{right})$, and

$$d_{\mathbb{S}}((\tfrac{1}{2}, \tfrac{\sqrt{3}}{2}), (\tfrac{1}{2}, 0)) \quad = \tfrac{\sqrt{3}}{2}$$
$$d_{M \otimes \mathbb{S}}((a, \mathsf{top}), (b, \mathsf{left})) = 1$$

The theorem below is about tripointed *sets*, but the proof makes a detour into tripointed *metric spaces*.

Theorem 9. $\mathbb{S} \to M \otimes \mathbb{S}$ *is a final coalgebra of* $M \otimes X$ *on* Tri.

Proof. The proof is almost the same as that of Theorem 4, so we merely sketch this one and point out where it differs from the earlier proof. Let $(X, e : X \to M \otimes X)$ be a coalgebra, where X is any tripointed set. The main point is to consider the homset $\mathsf{Tri}(X, \mathbb{S})$. This is easily seen to be a complete metric space.

It is also non-empty: we can give X the discrete metric, and then take any function preserving top, left, and right. Consider the function

$$\varphi : \mathsf{Tri}(X, \mathbb{S}) \to \mathsf{Tri}(X, \mathbb{S})$$

given by

$$\varphi(f) = \sigma^{-1} \circ (M \otimes f) \circ e.$$

(Here σ^{-1} is from (5), and we also saw it in Example 6.) In other words, $\varphi(f)$ makes the diagram below commute:

$$
\begin{array}{ccc}
X & \xrightarrow{\ e\ } & M \otimes X \\
{\scriptstyle \varphi(f)} \downarrow & & \downarrow {\scriptstyle M \otimes f} \\
\mathbb{S} & \xleftarrow[\ \sigma^{-1}\ ]{} & M \otimes \mathbb{S}
\end{array}
$$

(We use σ^{-1} rather than σ because σ^{-1} is a short map.) Then φ is a contraction of a non-empty complete metric space. Thus φ has a unique fixed point φ^*, and the fixed points of φ are exactly the coalgebra morphisms from X to $\mathbb{S}$. Note also that there is no reason to expect that φ^* would be a short map, even if X carried a metric and e were itself short, since our proof starts by throwing away the metric on X.

6   Isometric Embeddings and the Initial Chain

As with any morphism of metric spaces, we say that a TriMS morphism $\eta : X \to Y$ is an isometric embedding if it preserves all distances.

Proposition 10. *Concerning isometric embeddings:*

1. *The unique map $\eta : I \to M \otimes I$ is an isometric embedding, where I is the initial object of* TriMS *from Example 3.*
2. *If $\eta : X \to Y$ is an isometric embedding, so is $M \otimes \eta : M \otimes X \to M \otimes Y$.*

Proof. The first part follows from the fact that $M \otimes I$ is an object in TriMS. The second follows easily from Lemma 7; see the proof of Proposition 8 for a similar argument.

As a result, we have a chain of isometric embeddings

$$I \xrightarrow{\ \ \eta\ \ } M \otimes I \xrightarrow{\ M \otimes \eta\ } M \otimes M \otimes I \xrightarrow{\ M \otimes M \otimes \eta\ } \cdots \qquad (6)$$

In this situation, the colimit exists and is preserved by the functor $M \otimes -$.

Theorem 11. *The colimit of the chain in (6) exists. Call this space G. G carries the structure of an initial algebra $\eta : M \otimes G \to G$ for $M \otimes X$ on* TriMS.

Proof. We are going to use Adámek's Theorem 2. The category TriMS has an initial object. Given a chain of isometric embeddings as in (6), the colimit exists in the category of metric spaces, and is the set-theoretic union of the spaces. This colimit is an object of Tri, since the initial object is isometrically embedded in it. This TriMS object is G in our theorem. The map $i_n : M^n \otimes I \to G$ is the inclusion. This space G is a fixed point: $M \otimes G \to G$. Moreover, the mediating morphism $g : M \otimes G \to G$ given by the colimit is

$$m \otimes (n_1 \otimes n_2 \otimes \cdots \otimes n_k \otimes d) \quad \mapsto \quad m \otimes n_1 \otimes n_2 \otimes \cdots \otimes n_k \otimes d.$$

By Adámek's Theorem 2, (G, g) is an initial algebra.

6.1   Concrete Presentation of G

It might be useful to see a presentation of G in very concrete terms. We have in mind readers not familiar with Adámek's Theorem, but who might be familiar with the basics of universal algebra or abstract data types.

Consider a signature Σ of terms which takes symbols top, left, and right as constants, and symbols a, b, and c as unary function symbols. Then we would get terms such as $a(b(c(\text{right})))$ and $b(b(\text{left}))$. In order to match our earlier notation, we are going to write these as $a \otimes b \otimes c \otimes \text{right}$ and $b \otimes b \otimes \text{left}$, respectively. Consider the free Σ-algebra modulo the equations E

$$
\begin{aligned}
\text{top} &= a \otimes \text{top} & a \otimes \text{left} &= b \otimes \text{top} \\
\text{left} &= b \otimes \text{left} & a \otimes \text{right} &= c \otimes \text{top} \\
\text{right} &= c \otimes \text{right} & b \otimes \text{right} &= c \otimes \text{left}
\end{aligned}
\tag{7}
$$

This is the set G of all terms on the signature Σ modulo the smallest equivalence relation containing the pairs in E above and closed under the three function symbols. Technically, the elements of G are not terms but rather equivalence classes of terms. (This is often written as $T_{\Sigma,E}$.) In fact, an $(M \otimes X)$-algebra on Tri is exactly a Σ-algebra satisfying E with the property that top, left, and right are distinct. This correspondence extends to Σ-algebra morphisms and $M \otimes X$-algebra morphisms. Then it is a standard result that $G = T_{\Sigma,E}$ is the initial Σ-algebra satisfying E.

We take G and impose the unique metric such that $d(\text{top}, \text{left}) = d(\text{top}, \text{right}) = d(\text{left}, \text{right}) = 1$, and also such that the equations in Lemma 7 hold. This concludes our digression on the concrete presentation of G as a metric space.

Proposition 12. *As a metric space, G is totally bounded.*

Proof. Let $M^n \otimes I$ be the nth term in the chain in (6), starting with $M^0 \otimes I = I$, and $M^1 \otimes I = M \otimes I$. One checks easily that for all n, each point in $M^{n+1} \otimes I$ is of distance $\leq 2^{-(n+1)}$ from some point in M^n. Then using the triangle inequality, it follows that the distance of each point in $M^{n+k} \otimes I$ to some point in $M^n \otimes I$ is at most $\sum_{i=1}^{k} 2^{-(n+1+i)} < 2^{-n}$. Thus, $M^n \otimes I$ is a finite set of points in G with the property that every point in G is within 2^{-n} of some point in it.

6.2   The Final Coalgebra of $X \to M \otimes X$

This section constructs the final coalgebra of $X \to M \otimes X$ on TriMS. We need a few preliminaries.

Complete tripointed spaces. We let TriCMS be the tripointed metric spaces which are *complete*: every Cauchy sequence has a limit. Every metric space is isometrically embedded in its Cauchy completion, and so we have a Cauchy completion functor

$$C : \mathsf{TriMS} \to \mathsf{TriCMS}.$$

For a tripointed metric space X, consider

$$M \otimes C(X) \xrightarrow{\beta_X} C(M \otimes X) \xrightarrow{\gamma_X} M \otimes C(X)$$

given by

$$\beta_X(m \otimes (x_0, x_1, \dots)) = (m \otimes x_0, m \otimes x_1, \dots)$$
$$\gamma_X((m_i \otimes x_i)_i) = m^* \otimes (x_{j_1}, x_{j_2}, \dots,)$$

In the definition of γ_X, for each sequence of points in $M \otimes X$, some $m \in M$ must occur infinitely many times as the first coordinate m_i, since M is finite. Let m^* be "first" in some pre-chosen order on M. Let $(x_{j_i})_i$ be the subsequence of $(x_i)_i$ such that the corresponding first coordinates are all m^*.

Lemma 13. $\beta : C(M \otimes -) \to M \otimes C(-)$ *and* $\gamma : M \otimes C(-) \to C(M \otimes -)$ *are natural isomorphisms.*

Proof. One checks that for all X, β_X and γ_X well-defined, that they are inverses (modulo equivalence of Cauchy sequences), that they are short maps (hence they are isometries), that they preserve the distinguished points, and that they are the components of natural transformations. All of these verifications are elementary.

The final coalgebra of $X \to M \otimes X$. We have seen the initial algebra

$$(G, \eta : M \otimes G \to G)$$

of $X \to M \otimes X$ on TriMS. By Lambek's Lemma 1, η is an isomorphism. Applying the Cauchy completion functor C, we have another algebra

$$(CG, C\eta \circ \beta_G : M \otimes CG \to C(M \otimes G) \to CG).$$

Moreover, $C\eta \circ \beta_G$ is an isomorphism. Let us shorten our notation by writing S for CG and s for $(C\eta \circ \beta_G)^{-1}$. So we have a coalgebra $(S, s : S \to M \otimes S)$ on TriMS. S is the completion of a totally bounded space (Proposition 12) and is therefore compact. Forgetting the metric structure, S a coalgebra for $M \otimes X$ on Tri.

Theorem 14 (Leinster [10]). $(S, s : S \to M \otimes S)$ *is a final coalgebra for* $M \otimes X$ *on* Tri.

Proof. We have seen the argument twice before, in Theorems 4 and 9. The main points are that S is complete, so therefore $\mathsf{Tri}(X, S)$ is complete; also $\mathsf{Tri}(X, S)$ is non-empty, and i is an isometry.

Theorem 14 is due to Leinster, but as with Freyd's Theorem, our proof is different. We feel that going through the initial algebra G offers an advantage of sorts: G is the countable metric space of approximations to points in S. It is a little like what is called the set of finite addresses of points in $\mathbb{S}$, but it is obtained as the colimit of the initial sequence of an endofunctor on TriMS.

7   Finality of S for $M \otimes X$ on TriMS

Theorem 14 shows that S is the final coalgebra of $M \otimes X$ on Tri. In this section, we present a related result. To begin, recall from Proposition 8 that $M \otimes X$ is an endofunctor on TriMS. As S is an object of TriMS and its structure s is a short map, we thus can ask whether S is a final coalgebra of $M \otimes X$ on TriMS. It is important to see that this is not an immediate consequence of Theorem 14. Given a short map of tripointed metric spaces $e : X \to M \otimes X$, we do have a unique coalgebra morphism $f : X \to S$. We must show that f is a short map.

Notation. Let $\boldsymbol{m} = m_1, m_2, \ldots, m_k \in M^k$. For $x \in X$, write $\boldsymbol{m} \otimes x$ for $m_1 \otimes \cdots \otimes m_k \otimes x$. Further, $M^i \otimes X$ denotes the space

$$\overbrace{M \otimes M \otimes \cdots \otimes M}^{i} \otimes X.$$

Finally, we use a notation like $M^j \otimes e$ as a shorthand for the obvious map.

Lemma 15. *Let X be an object in* TriMS. *For all natural numbers i, all $\boldsymbol{m} \in M^i$ and all $x, y \in X$ the distance in $M^i \otimes X$ between*

$$d(\boldsymbol{m} \otimes x, \boldsymbol{m} \otimes y) \leq 2^{-i}.$$

Thus for all $x, x', y, y' \in X$

$$|d(\boldsymbol{m} \otimes x, \boldsymbol{n} \otimes y) - d(\boldsymbol{m} \otimes x', \boldsymbol{n} \otimes y')| \leq 2^{1-i}. \tag{8}$$

Proof. The first assertion is proved by an easy induction on i. For the second, we use the triangle inequality and two applications of the first assertion:

$$d(\boldsymbol{m} \otimes x, \boldsymbol{n} \otimes y) - d(\boldsymbol{m} \otimes x', \boldsymbol{n} \otimes y')| \leq d(\boldsymbol{m} \otimes x, \boldsymbol{m} \otimes x') + d(\boldsymbol{n} \otimes y, \boldsymbol{n} \otimes y')$$
$$\leq 2^{-i} + 2^{-i}$$

Theorem 16. (S, s) *is a final coalgebra of $M \otimes X$ on* TriMS.

Proof. Let $(X, e : X \to M \otimes X)$ be a coalgebra of $M \otimes -$ on TriMS. Forgetting the metric structure and the shortness of e, we know from Theorem 9 that there is a unique coalgebra morphism $f : X \to S$. This map preserves top, left, and right, and the main task is to show that f is a short map. We do this by showing that for $\varepsilon > 0$, and all $x, y \in X$,

$$d(f(x), f(y)) \leq d(x, y) + 2^{2-i}.$$

An easy induction on i shows that the composite below is a short map:

$$X \xrightarrow{\;e\;} M \otimes X \xrightarrow{\;M \otimes e\;} M \otimes M \otimes X \xrightarrow{\;M \otimes M \otimes e\;} \cdots \xrightarrow{\;M^{i-1} \otimes e\;} M^i \otimes X \; .$$
$$(9)$$

Let $x, y \in X$. Denote the image of x and y under the composite above by $\boldsymbol{m} \otimes x'$ and $\boldsymbol{m} \otimes y'$ respectively, where $\boldsymbol{m} = m_0, \ldots, m_{k-1} \in M^i$, $\boldsymbol{n} = n_0, \ldots, n_{k-1} \in M^i$ and $x', y' \in X$ Using the shortness of the map in (9), we see that

$$d(\boldsymbol{m} \otimes x', \boldsymbol{n} \otimes y') \leq d(x, y) \tag{10}$$

Since f is a coalgebra morphism to S we get:

$$(M^{i-1} \otimes s \circ \ldots \circ s)(f(x)) = \boldsymbol{m} \otimes f(x')$$
$$(M^{i-1} \otimes s \circ \ldots \circ s)(f(y)) = \boldsymbol{n} \otimes f(y')$$

Now s is an isometry, as are $M \otimes s$, $\ldots$, $M^{k-1} \otimes s$. Using the version of (9) that starts with S and s, we see that

$$d(f(x), f(y)) = d(\boldsymbol{m} \otimes f(x'), \boldsymbol{n} \otimes f(y')). \tag{11}$$

(Note the contrast to (10), where we have an inequality instead.)

Recall that the category TriMS has an initial object $I = \{\mathsf{top}, \mathsf{left}, \mathsf{right}\}$, and for any object in Z of TriMS, the initial map

$$i_Z : I \to Z$$

is an isometric embedding and so are the maps

$$M^i \otimes i_Z : M^i \otimes Z \to M^i \otimes Z$$

for all i (see Proposition 10). In our case we use the following diagram with both arrows being isometric embeddings

$$M^i \otimes S \longleftarrow M^i \otimes I \longrightarrow M^i \otimes X$$

to conclude that

$$d_{M^i \otimes S}(\boldsymbol{m} \otimes \mathsf{top}_S, \boldsymbol{n} \otimes \mathsf{top}_S) = d_{M^i \otimes X}(\boldsymbol{m} \otimes \mathsf{top}_X, \boldsymbol{n} \otimes \mathsf{top}_X). \tag{12}$$

where top_X and top_S denote the images of $\mathsf{top} \in I$ under the two evident initial maps. Therefore

$$
\begin{aligned}
d_{M^i \otimes S}(f(x), f(y)) &= d_{M^i \otimes S}(\boldsymbol{m} \otimes f(x'), \boldsymbol{n} \otimes f(y')) && \text{by (11)} \\
&\leq d_{M^i \otimes S}(\boldsymbol{m} \otimes \mathsf{top}_S, \boldsymbol{n} \otimes \mathsf{top}_S) + 2^{1-i} && \text{by Lemma 15} \\
&= d_{M^i \otimes X}(\boldsymbol{m} \otimes \mathsf{top}_X, \boldsymbol{n} \otimes \mathsf{top}_X) + 2^{1-i} && \text{by (12)} \\
&\leq d_{M^i \otimes X}(\boldsymbol{m} \otimes x', \boldsymbol{m} \otimes y') + 2^{1-i} + 2^{1-i} && \text{by Lemma 15} \\
&\leq d_{M^i \otimes X}(x, y) + 2^{2-i} && \text{by (10)}
\end{aligned}
$$

This for all i completes the proof.

8   The Relation between $\mathbb{S}$ and S

Let us summarize what we know about the spaces $\mathbb{S}$, G, and S:

1. $\mathbb{S}$ is the Sierpiński gasket, the unique non-empty compact subset of $\mathbb{R}^2$ satisfying

$$\mathbb{S} = \sigma_a(\mathbb{S}) \cup \sigma_b(\mathbb{S}) \cup \sigma_c(\mathbb{S}).$$

   We know that as a tripointed set, it is a final coalgebra of $X \mapsto M \otimes X$.
2. As explained in Section 6.1, G is the set of expressions $m_1 \otimes m_2 \otimes \cdots \otimes m_k \otimes x$, where $m_i \in M$ and $x \in \{\mathsf{top}, \mathsf{left}, \mathsf{right}\}$; we also quotient by a certain set E of equations. G carries a metric structure, and as such it is an initial algebra of $X \mapsto M \otimes X$, considered as an endofunctor on TriMS. It also is the case that forgetting the metric structure, G is an initial algebra for this endofunctor on Tri.
3. S is the Cauchy completion of G. S is a final coalgebra of $X \mapsto M \otimes X$ on Tri, and also of this endofunctor on TriMS.

As tripointed sets, S and $\mathbb{S}$ are final coalgebras for the same endofunctor. Thus there are inverse coalgebra morphisms i and j:

$$
\begin{array}{ccc}
S \xrightarrow{\;s\;} M \otimes S & \qquad & \mathbb{S} \xrightarrow{\;\sigma\;} M \otimes \mathbb{S} \\
\downarrow{i} \qquad \downarrow{M \otimes i} & \qquad & \downarrow{j} \qquad \downarrow{M \otimes j} \\
\mathbb{S} \xrightarrow[\;\sigma\;]{} M \otimes \mathbb{S} & \qquad & S \xrightarrow[\;s\;]{} M \otimes S
\end{array}
$$

Both i and j are Tri-morphisms, and they are the topic of this section.

Recall from Example 6 that σ^{-1} is short. By initiality in the metric setting, there is a unique short map $h : G \to \mathbb{S}$ such that the diagram below commutes:

$$
\begin{array}{ccc}
M \otimes G & \xrightarrow{\;\eta\;} & G \\
\downarrow{M \otimes h} & & \downarrow{h} \\
M \otimes \mathbb{S} & \xrightarrow[\;\sigma^{-1}\;]{} & \mathbb{S}
\end{array}
\qquad (13)
$$

As an illustration of how h works, $h(a \otimes b \otimes \mathsf{top}) = \sigma_a(\sigma_b(\mathsf{top}))$.

Example 7. The map h is not an isometry. For example, $d_G(b \otimes \text{left}, a \otimes \text{right}) = 1$. But $h(b \otimes \text{left}) = h(\text{left}) = (0,0)$, $h(a \otimes \text{right}) = (3/4, \sqrt{3}/4)$, and

$$d_\mathbb{S}((0,0), (3/4, \sqrt{3}/4)) = \sqrt{3}/2.$$

We need an observation concerning the map i.

Lemma 17. $i = Ch$.

Proof. Consider the diagram below:

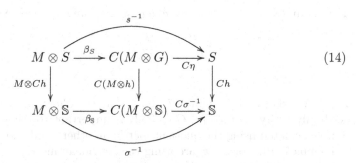

$$(14)$$

The square on the right comes from the diagram in (13), applying C. The square on the left is the naturality noted in Lemma 13. The top region commutes by the definition of s. It is easy to check that the bottom part of the diagram also commutes. It follows that $h : S \to \mathbb{S}$ is an algebra morphism. We take the inverses of s^{-1} and σ^{-1}, and we forget the metric structure. (We need to forget this metric structure, since σ is not a short map.) As a map of tripointed sets, h is a coalgebra morphism. By finality, $Ch = i$.

One last preliminary is needed before our main result in this section. It is an addendum to the triangle inequality.

Lemma 18. *Let $\triangle ABC$ be a triangle in the plane. Then*

$$\overline{AB} + \overline{AC} \leq \overline{BC}\sqrt{\frac{2}{1 - \cos A}}$$

Proof. Let us write K for $\sqrt{2/(1 - \cos A)}$. Since $\angle A$ is part of a triangle, $-1 < \cos A < 1$. Thus $K > 1$. Also, $2K^2 \cos A + 2 = 2K^2 - 2$. For all x and y,

$$(K^2 - 1)x^2 - 2(K^2 \cos A + 2)xy + (K^2 - 1)y^2$$
$$= (K^2 - 1)x^2 - 2(K^2 - 1)xy + (K^2 - 1)y^2$$
$$= (K^2 - 1)(x - y)^2$$
$$\geq 0$$

Thus the first of the four lines above is ≥ 0. Rearranging things,

$$x^2 + 2xy + y^2 \leq K^2(x^2 + y^2 - 2xy \cos A).$$

$$(15)$$

We apply (15) to the sides of our triangle, taking x to be $\overline{AB}$, and y to be $\overline{AC}$:

$$(\overline{AB} + \overline{AC})^2 \leq K^2(\overline{AB}^2 + \overline{AC}^2 - 2\overline{AB} \cdot \overline{AC} \cos A) = K^2(\overline{BC})^2.$$

In the line just above we used the law of cosines from trigonometry. Taking square roots gives the assertion in our lemma.

Two cases of Lemma 18 will be of interest. When $\angle A = 60^\circ$, we get $\overline{AB} + \overline{AC} \leq 2\overline{BC}$. When $\angle A = 120^\circ$, we get $\overline{AB} + \overline{AC} \leq \frac{2\sqrt{3}}{3}\overline{BC}$.

Theorem 19. $i : S \to \mathbb{S}$ *is a bilipschitz isomorphism. Specifically, for all* $x, y \in S$,

$$\frac{\sqrt{3}}{4} d(x, y) \leq d(i(x), i(y)) \leq d(x, y) . \tag{16}$$

Proof. We showed in Lemma 17 above that $i = Ch$. Since G is dense in S, we need only verify (16) for $h : G \to \mathbb{S}$. It is important to recall that distances in S are calculated using the quotient metric; in other words, we use the equations in Lemma 7. But for $\mathbb{S}$, we are using the Euclidean metric.

It is convenient to picture the points in $M^n \otimes I$ as arranged in a graph (see some pictures below). For each n, we take the distance of a segment in $M^n \otimes I$ to be 2^{-n}. Then the distance between two points in $M^n \otimes I$ (as we have been considering it) is exactly the length of the shortest path in the graph. Thus, we need to show that the graph-theoretic distance between two points is at most the Euclidean distance multiplied by $4\sqrt{3}/3$.

Lemma 20. *For all* $x \in M^n \otimes I$, *there is a path from the* top *of* $M^n \otimes I$ *to* x *whose graph-theoretic distance is at most* $2\sqrt{3}/3$ *times the Euclidean distance.*

Proof. For each x, we first construct a *principal path* top $= p_0, p_1, \ldots, p_n = x$ from the top to x. Each p_i will belong to $M^i \otimes I$. We take $p_0 =$ top.

If $x = p_0$, then the path is simply p_0.

If x belongs to the top main triangle tr_a, then it is of the form $a \otimes y$. And by induction hypothesis, we already have principal path for y, say top $= p_0, p_1, \ldots, p_n = y$. We take the principal path for x to be top $= a \otimes p_0, a \otimes p_1, \ldots, a \otimes p_n = x$. (In other words, we take the principal path for y and then apply σ_a to each point.)

If x belongs to the left main triangle tr_b, then it is $b \otimes y$. Again by induction hypothesis, we already have principal path for y, say top $= p_0, p_1, \ldots, p_n = y$. We have three subcases. If p_1 is to the left of p_0, then we take the principal path for x to be top $= p_0, b \otimes p_2, \ldots, b \otimes p_n = b \otimes y = x$. Note that we omit $b \otimes p_1$ in this subcase. The second subcase is when p_1 is to the right of p_0. In this case, we take the principal path for x to be top $= p_0, b \otimes p_1, b \otimes p_2, \ldots, b \otimes p_n = b \otimes y = x$. In the last subcase, x is directly under $p_0 =$ top. In this case, $x = b \otimes$ right $= c \otimes$ left, and we may take the principal path for x to be either top, $a \times$ left, x, or else top, $a \times$ right, $\otimes$

If x belongs to the right main triangle tr_c, then the definitions work similarly. Here are some illustrations of points x and the principal paths for them.

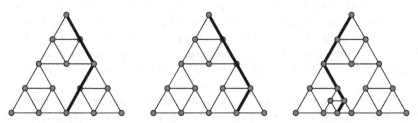

The path in the first figure is used in the second, and the path in the second is used in the third. In fact, the path in the first figure could go the other way, left instead of right. This possibility is important, as we shall see shortly.

We might have to modify the end of the path, because we would like to arrange that no point on the principal path for x is directly above x, except when x is of the form $a \otimes a \otimes \cdots \otimes a \otimes b \otimes \mathsf{left} = a \otimes a \otimes \cdots \otimes a \otimes c \otimes \mathsf{right}$. For all other x, we can indeed arrange that the principal path not have any point directly above x. To do this, we would only need to take the last two points on the path, and change the direction.

An easy induction on n now shows that the path zig-zags across the vertical line through x, as shown on the left below.

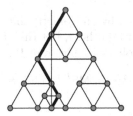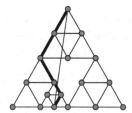

Next, it follows easily that if we incline the line so that it intersects top, the principal path zig-zags across it. This uses the assumption that no point on the principal path is directly above x. For example, if x is in the left main triangle (as in our pictures above), then the inclined line on the right still intersects all segments determined by the principal path.

To complete the proof, we use Lemma 18 in each of the triangles formed by the principal path and the line from x to top.

Now we complete the proof of Theorem 19. We are done when one of the points is top; and similarly we are done when one of the points is left or right. In both of those cases, we have a better bound than the general bound stated in Theorem 19: $\sqrt{3}/2 \cdot d(x,y) \le d(i(x), i(y))$. But in general, we combine paths to one of the connection points, and again use Lemma 18; we use the special case of $60°$-triangles. We would get $\sqrt{3}/4 \cdot d(x,y) \le d(i(x), i(y))$.

This completes the proof of Theorem 19.

Remark 1. The bound of $\frac{2}{3}\sqrt{3}$ in Lemma 20 is sharp. The lower bound of $\sqrt{3}/4 \approx$ .433 in Theorem 19 may be improved to $1/2$. However, the argument in this case is much more involved than what we have seen in Lemma 20. In effect, one has to show that all lines in $\mathbb{R}^2$ connecting pairs of points in $M^n \otimes I$ may be "zig-zagged" by paths in $M^n \otimes I$ itself. (One would like to use results like those in Aichholzer [4], but our graphs are not triangulations; they involve colinear triples, for example.) Of course, it is sufficient in Theorem 19 to have *any* bound.

Metric spaces with lipschitz maps. The reader might wonder about the significance of the bilipschitz result, Theorem 19. This points to the subtle question of the maps in the category of "metric spaces." We have been working with metric spaces and short maps. But here are settings where one makes other choices. Another choice would be the continuous maps. An even smaller choice would be the Lipschtiz continuous functions. These are the maps $f : X \to Y$ of metric spaces with the property that there is a constant K such that $d(f(x), f(y)) \leq K \cdot d(x, y)$ for all $x, y \in X$. Let MS^{Lip} be the category of 1-bounded metric spaces and Lipschitz continuous functions. An isomorphism in MS^{Lip} is exactly a bilipschtiz bijection. Frequently in the literature on metric spaces one does indeed identify spaces which are bilipschitz isomorphism. In comparison with other notions of equivalence for metric spaces, we have

$$\text{isometry} \Longrightarrow \text{bilipschitz isomorphism} \Longrightarrow \text{homeomorphism}$$

We had hoped to find $\mathbb{S}$ and S to be related by an isometry, but this was not to be. (The parallel result for the unit interval did hold, but this seems like a very special result.) We have the following corollary to Theorem 19:

Corollary 21. $\sigma : M \otimes \mathbb{S} \to \mathbb{S}$ *is a coalgebra in* MS^{Lip}*, and it is isomorphic to* (S, s).

Having said this, a good next question in this line of work would be: in MS^{Lip}, is $\mathbb{S}$ a final coalgebra of $M \otimes -$? We have not yet pursued this question.

9   Conclusion

This paper presented some specific results on the Sierpiński gasket $\mathbb{S}$. We considered the categories Tri of tripointed sets and TriMS of tripointed metric spaces. Both categories carry an endofunctor $X \to M \otimes X$ defined in a natural way. The initial algebra of this functor on TriMS is the set of finite addresses (tensor expressions) of points in $\mathbb{S}$. Its Cauchy completion, S, is the final coalgebra of the endofunctor. This finality result goes through in the metric setting, and this is one of our main results. As tripointed sets, S and $\mathbb{S}$ are isomorphic. But the isomorphism is not an isometry of the associated metric spaces. Indeed, the natural metric structure on $\mathbb{S}$ as a coalgebra is not an isometry. The best we can say is that as metric spaces, $\mathbb{S}$ and S are bilipschitz equivalent: there is a bijection between these spaces with the property that distances in one space are uniformly bounded by multiples of the corresponding distances in the other space. This bilipschitz equivalence involves work on paths in the finite approximations to the Sierpiński gasket. It also extends to show that the coalgebras (S, s) and $(\mathbb{S}, \sigma)$ are isomorphic in MS^{Lip}.

Naturally, the main open question in all of this work is whether the results on the Sierpiński gasket extend to all of the other fractal sets in $\mathbb{R}^n$. That is, it is possible to consider specific self-similar sets, such as the Sierpiński carpet or the unit interval, and then render them as finial coalgebras, the same way as we have seen for the Sierpiński gasket. The initial algebras in these cases are also interesting. However, we do not yet have a general theory that offers an insightful generalization of these examples.

Acknowledgment. We thank anonymous referees for comments and suggestions which have improved our presentation.

References

1. Adámek, J.: Free algebras and automata realizations in the language of categories. Comment. Math. Univ. Carolinae 14, 589–602 (1974)
2. Adámek, J.: On final coalgebras of continuous functors. Theoret. Comput. Sci. 294(1-2), 3–29 (2003)
3. Adámek, J., Milius, S., Moss, L.S.: Initial Algebras and Terminal Coalgebras: a Survey (2011) (unpublished ms.)
4. Aichholzer, O.: The path of a triangulation. In: Proc. 15th Ann. ACM Symp. Computational Geometry, Miami Beach, Florida, USA, pp. 14–23 (1999)
5. Barr, M.: Terminal coalgebras in well-founded set theory. Theoret. Comput. Sci. 114(2), 299–315 (1993); Additions and corrections in Theoret. Comput. Sci. 124 (1), 189–192 (1994)
6. Freyd, P.: Real coalgebra, post on the Categories mailing list (December 22, 1999), www.mta.ca/~cat-dist
7. Hasuo, I., Jacobs, B., Niqui, M.: Coalgebraic representation theory of fractals (Extended Abstract). In: Proc. Mathematical Foundations of Programming Semantics (MFPS XXVI), Electr. Notes Comp. Sci., vol. 265, pp. 351–368 (2010)
8. Hutchinson, J.: Fractals and self-similarity. Indiana Univ. Math. Journal 30(5), 713–747 (1981)
9. Lambek, J.: A Fixpoint Theorem for Complete Categories. Math. Z. 103, 151–161 (1968)
10. Leinster, T.: A general theory of self-similarity. Advances in Mathematics 226, 2935–3017 (2011)

Leaving Traces:
A Note on a Sound and Complete Trace Logic for Concurrent Constraint Programs

Frank S. de Boer[1] and Maurizio Gabbrielli[2]

[1] Centrum Wiskunde & Informatica
[2] University of Bologna

Abstract. Concurrent constraint programs operate on data which is represented by a constraint of a given cylindric constraint system. Such a system provides an algebraic representation of first-order logic. It features a (binary) entailment relation, a binary union operation for adding information (formally defined as the least upper bound of two constraints with respect to the entailment relation) and, and, finally, existential quantification of variables.

The main contribution of this paper is a *sound and complete* proof theory based on traces of input/output constraints for reasoning about the correctness of concurrent constraint programs.

1 Introduction

One of the many influential papers co-authored by Prakash Panangaden is on the *semantic foundations of concurrent constraint programming* [8]. That paper also forms the very basis of this contribution to this "festschrift" dedicated to the 60th anniversary of Prakash. By this contribution we hope to provide further evidence of the lasting relevance of the work of Prakash in general, and of his work on concurrent constraint programming in particular.

In this paper we address the main challenge of a *complete* proof theory for reasoning about the correctness of concurrent constraint programs (*ccp* programs, for short). A sound and complete proof theory for *confluent ccp* programs has been introduced in [2]. However in [2] it is also shown that the theory is sound but incomplete for the general class of *non-determinstic ccp* programs. This is due to the fact that in [2] the correctness of a *ccp* program is described in terms of properties of its final results, that is, its *resting points*. However, these observables do not provide sufficient information for a compositional semantics of *ccp*, for which additional information is needed about *how* these final results are obtained. Interestingly, this additional information is needed because of the interplay between local variables in *ccp* and non-deterministic choice. In fact, both deterministic *ccp* and *ccp* without local variables allow a compositional semantics in terms of the final results.

In [3] a compositional and fully abstract model of *ccp* has been introduced based on *traces* of input/output constraints, where an output constraint is generated by the program itself and the input constraint by its environment. We show

F. van Breugel et al. (Eds.): Panangaden Festschrift, LNCS 8464, pp. 168–179, 2014.
© Springer International Publishing Switzerland 2014

in this paper how this trace semantics forms the basis of a corresponding sound and complete logic for the verification of *ccp* programs, following the approach of trace logics for CSP (Communicating Sequential Processes), as presented for example in [9].

Related work. Other approaches to proving correctness of *ccp* programs are based on a straightforward translation in intuitionistic linear logic, see for example [5,7]. However, this approach is applicable only to a restricted class of liveness and safety properties, quoting [4]:

> The logical characterization of other observable properties of CC (Concurrent Constraint) computations, such as the set of terminal stores, i.e. constraints of terminal configurations with or without suspended agents, is more delicate.

2   Getting Started

Concurrent constraint programs operate on data which is represented by an element, a constraint c, of a given cylindric constraint system $\mathbf{C}$. Such a system consists of an algebraic representation of first-order logic. It provides a (binary) entailment relation $\vdash$, a binary union operation $\sqcup$ for adding information (formally defined as the least upper bound of two constraints with respect to the entailment relation) and, for each variable x, an unary operation $\exists_x$, for hiding information about x. Since the operator $\sqcup$ corresponds to logical conjunction we will often refer to it with this terminology.

Assuming a given cylindric constraint system $\mathbf{C}$ the syntax of *finite ccp* agents is given by the following grammar:

$$A ::= \mathbf{tell}(c) \mid \sum_{i=1}^{n} \mathbf{ask}(c_i) \to A_i \mid A \parallel B \mid \exists x A$$

where the c, c_i are supposed to be *finite constraints* (i.e. algebraic elements) in $\mathcal{C}$.

Action prefixing is denoted by $\to$, non-determinism is introduced via the guarded choice construct $\sum_{i=1}^{n} \mathbf{ask}(c_i) \to A_i$, parallel composition is denoted by $\parallel$, and a notion of locality is introduced by the agent $\exists x A$ which behaves like A with x considered local to A.

For technical convenience only we restrict in this paper to finite *ccp* agents. In fact, the counterexample to completeness of the proof theory in [2] is finite. In general a sound proof theory for finite *ccp* agents can be extended to recursion by the inclusion of an appropriate induction rule in the logic. Completeness can be extended assuming the logic is expressive enough (see [2]).

In the next subsection we describe formally the operational semantics of *ccp*.

3   Getting Fully Operational

We define a semantic basis of our trace logic in terms of the transition system of *ccp* introduced in [8]. Let us briefly discuss the rules in Figure 1, where computation is described in terms of (transitions on) configurations consisting of a process and a constraint representing the store (in which the process is evaluated). In order to represent successful termination we introduce the auxiliary agent **stop**: it cannot make any transition. Rule **R1** shows that we are considering here the so called "eventual" tell: the agent **tell**(c) adds c to the store d without checking for consistency of $c \sqcup d$ and then stops. According to rule **R2** the guarded choice operator gives rise to global non-determinism: the external environment can affect the choice since **ask**(c_j) is enabled iff the store d entails c_j, and d can be modified by other agents. Rules **R3** models the parallel composition operator in terms of *interleaving*. The agent $\exists x A$ behaves like A, with x considered *local* to A, i.e. the information on x provided by the external environment is hidden from A and, conversely, the information on x produced locally by A is hidden from its external environment. To describe locality the syntax has been extended by an agent $\exists^d x A$ where d is a local store of A containing information on x which is hidden in the external store. Initially the local store is empty, i.e. $\exists x A = \exists^{true} x A$.

R1 $\langle \mathbf{tell}(c), d \rangle \rightarrow \langle \mathbf{stop}, c \sqcup d \rangle$

R2 $\langle \sum_{i=1}^{n} \mathbf{ask}(c_i) \rightarrow A_i, d \rangle \rightarrow \langle A_j, d \rangle$ $j \in [1, n]$ *and* $d \vdash c_j$

R3 $\dfrac{\langle A, c \rangle \rightarrow \langle A', d \rangle}{\langle A \parallel B, c \rangle \rightarrow \langle A' \parallel B, d \rangle}$

R4 $\dfrac{\langle A, d \sqcup \exists_x c \rangle \rightarrow \langle B, d' \rangle}{\langle \exists^d x A, c \rangle \rightarrow \langle \exists^{d'} x B, c \sqcup \exists_x d' \rangle}$

Fig. 1. The transition system for *ccp*

Using the transition system described by (the rules in) Figure 1 we can now define our compositional trace semantics (along the lines of [3]), where a trace θ is a (possibly empty) *monotonically increasing* (with respect to the entailment relation) sequence of input constraints $in(c)$ and output constraints $out(c)$, ending in a final resulting constraint. Since the constraints arising from the syntax are finite, we also assume that a trace contains only finite constraints[1].

Operationally, the traces $\mathcal{T}(A)(c)$ of an agent A starting from an initial store c can be generated recursively as follows:

$$\mathcal{T}(A)(c) = \{ out(d) \cdot \theta \mid \langle A, c \rangle \rightarrow \langle B, d \rangle \text{ and } \theta \in \mathcal{T}(B)(d) \}$$
$$\cup$$
$$\{ in(d) \cdot \theta \mid d \vdash c \text{ and } \theta \in \mathcal{T}(A)(d) \}$$
$$\cup$$
$$\{ \theta \mid \langle A, c \rangle \rightarrow \langle B, c \rangle \text{ and } \theta \in \mathcal{T}(B)(c) \}$$
$$\cup$$
$$\{ c \mid \langle A, c \rangle \nrightarrow \}.$$

---

[1] Note that here we implicitly assume that if c is a finite element then also $\exists_x c$ is finite.

This recursive definition can be justified by a straightforward least fixed-point construction. Note that in this *input-enabled* trace semantics the inputs describe arbitrary assumptions of its environment (as described by the second clause above). A trace $\theta \in T(A)(c)$ without any input constraints corresponds with an execution of A in isolation starting from the initial constraint c and terminating with the final constraint of θ.

4   Specifying Traces

We propose a *typed* specification language for traces which is based on a many-sorted signature consisting of the sort B of Boolean values, a sort C which represents the set of constraints of the given constraint system (each constraint c is represented by a constant in the language), and a sort T of traces of input/output constraints followed by a final constraint. The sort T contains the sort C of constraints as a subsort so that each constraint itself is a trace, denoting the final result.

Formally, given a set of (many-sorted) operators $op : S_1 \times \cdots \times S_n \to S$ and for each sort mutually disjoint sets of variables, every variable x of sort S is an expression of sort S in the specification language and for every operator $op : S_1 \times \cdots \times S_n \to S$, expressions t_i of sort S_i, for $i = 1, \ldots, n$, $op(t_1, \ldots, t_n)$ is an expression of sort S. Formulas are constructed from Boolean expressions by means of the logical operations of conjunction, negation, implication and quantification over variables of any sort (including the *second-order* quantification over variables of the sort T of traces). Apart from the entailment relation on constraints, represented by a Boolean binary operation $C \times C \to B$, and the usual Boolean operators, the logic further includes the following operators.

Undefined Trace. The constant

$$\bot : T$$

denotes the "undefined" trace.

Input/output. The strict trace constructors

$$in, out : C \times T \to T$$

add an input/output constraint to a trace which is only defined in case the monotonicity of the entailment relation between the constraints of the resulting trace is preserved and the final result entails the added constraint. For example,

$$in(x = a, true) = \bot.$$

On the other hand,

$$in(x = a, x = a \sqcup y = b) \neq \bot.$$

For notational convenience, in the sequel we denote $in(c, t)$ and $out(c, t)$ by $in(c) \cdot t$ and $out(c) \cdot t$, for any trace expression t. Note that since C is as a subsort of T, all expressions in the initial algebra denoting traces consist of a sequence of input/output constraints followed by the undefined trace $\perp$ or a constraint which represents a possible final result, i.e., a constraint from which no further transition is possible. In general the final constraint of a trace entails all its input/output constraints. However, the conjunction of all the input/output constraints does not necessarily entail the final result. A trace is *complete* if the conjunction of all its input/output constraints is equivalent to its final resulting constraint. In our trace logic incomplete traces are mainly used for the *under-specification* of the environment, e.g., such traces allow to abstract from when exactly the inputs of the environment are generated. For example, the trace expression $out(c) \cdot d$, where c and d are variables of sort C, denotes an incomplete trace which consists of a single output constraint and abstracts from the inputs.

Final constraint. The operation

$$\gamma : T \to C$$

extracts from a trace its final constraint ($\gamma(\perp)$ equals an arbitrary constraint). Its definition (inductive on the length of the trace) is straightforward and therefore omitted.

Parallel Composition. The (commutative) parallel composition

$$\| : T \times T \to T$$

satisfies the following axioms (here and in the sequel the variables c and d are of sort C, and the variables t and t' are of sort T):

- $c \parallel c = c$
- $(in(c) \cdot t) \parallel (in(c) \cdot t') = in(c) \cdot (t \parallel t')$
- $(out(c) \cdot t) \parallel (in(c) \cdot t') = out(c) \cdot (t \parallel t')$

In all other cases $t \parallel t' = \perp$. Note that thus every output must match with a corresponding input. Two matching inputs which coincide represent an assumption about the environment of the parallel composition.

Local variables. For each variable x we introduce a relation

$$\sqsubseteq_x : (T \times T) \to B$$

such that $\theta \sqsubseteq_x \theta'$ if and only if θ and θ' differ at most with respect to the information about x such that the input constraints of θ and the output constraints of θ' do not contain any information about x. Formally, this relation is the smallest relation satisfying the following axioms.

- $\exists_x d = \exists_x d' \to d \sqsubseteq_x d'$
- $t \sqsubseteq_x t' \to out(d) \cdot t \sqsubseteq_x out(\exists_x d) \cdot t'$
- $(\exists_x d = \exists_x d' \wedge t \sqsubseteq_x t') \to in(\exists_x d) \cdot t \sqsubseteq_x in(d') \cdot t'$

Trace equivalence. We introduce next the following equivalence relation $\simeq$ on traces which respects the *in/out* operations and satisfies the following axioms:

   Constraint equivalence

- $c = d \rightarrow c \simeq d$

   Empty input/output

- $in(true) \cdot t \simeq t$
- $out(true) \cdot t \simeq t$

   Input/output (de)composition

- $in(c \sqcup d) \cdot t \simeq in(c) \cdot in(c \sqcup d) \cdot t$
- $out(c \sqcup d) \cdot t \simeq out(c) \cdot out(c \sqcup d) \cdot t$

These axioms can be justified by a fully abstract semantics as discussed in [3]. Note that they allow reasoning about traces modulo a *normal* form consisting of an alternating sequence of input/output constraints different from *true*.

Input enabledness. To describe logically in a concise manner the input enabledness of ccp programs we introduce in the trace logic a preorder $\preceq$ on traces, defined as the smallest relation which subsumes the above trace equivalence (i.e., $\simeq \subseteq \preceq$), respects the *in/out* operations (e.g., $t \preceq t'$ implies $in(c) \cdot t \preceq in(c) \cdot t'$) and satisfies the following axioms:

- $t \neq \perp \rightarrow \neg(\perp \preceq t \lor t \preceq \perp)$
- $t \preceq in(c) \cdot t$

Note that thus $t \preceq in(c) \cdot t$ implies that c is entailed by the final constraint $\gamma(t)$ of t (the *in* operation being undefined otherwise). This preorder allows to add input constraints as long as they are entailed by the final result. In other words, $t \preceq t'$ if and only if t' provides more information about the environment. For example, instead of having to specify *explicitly* that a trace t consists of a single output of a constraint c preceded and followed by arbitrary inputs, this now can be simply expressed by $out(c) \cdot \gamma(t) \preceq \gamma(t)$.

Validity. The intended (many-sorted) model $M(\mathcal{C})$ of the above specification language includes the Boolean values "true" and "false", the given constraint system $\mathcal{C}$ and the (finte) traces of input/output constraints in $\mathcal{C}$ followed by a final constraint, and is defined in terms of the standard interpretation of the operators. Given a trace specification ϕ, we denote by

$$M(C), \sigma \models \phi$$

that ϕ holds in $M(C)$ under the variable assignment σ (of values in $M(C)$). Its definition is standard and proceeds by induction on ϕ. This induction relies on a semantics $Val(e)(M(C), \sigma)$ of expressions. Its definition also proceeeds by a straightforward induction on the structure of e. As a special case we show the following clause

$$Val(in(c) \cdot t)(M(C), \sigma) = in(\sigma(c)) \cdot_s \sigma(t)$$

Here c and t are variables ranging over constraints and traces, respectively, and the semantic operation "$\cdot_s$" adds the input constraint $in(\sigma(c))$ to $\sigma(t)$, in case the resulting sequence is montonically increasing and the final constraint of $\sigma(t)$ entails $\sigma(c)$, otherwise the undefined trace $\bot$ results.

By $\models \phi$ we denote that $M(C), \sigma \models \phi$, for all variable assignments σ.

5   Verifying Trace Spec's

Given a trace specification $\phi(t)$, with a (distinguished) trace variable t, by

$$A \text{ sat } \phi(t)$$

we denote that A satisfies the trace specification $\phi(t)$. In the theory below for proving correctness of ccp programs we implicitly assume that $t \neq \bot$ and that t is a complete trace (which can be easily expressed in the logic). It consists of an axiom for the tell action, for each operator of the programming language a single rule, and it includes a weakening rule.

Tell

$$tell(c) \text{ sat } out(c) \cdot \gamma(t) \preceq t$$

Note that $t \neq \bot$ implies that $\gamma(t) \vdash c$ (because the *out* operation is only defined if the output constraint is entailed by the final result[2]). Further, it is worthwhile to observe the use of the preorder $\preceq$ in the implicit specification of input enabledness.

Parallel Composition

$$\frac{A_i \text{ sat } \phi_i(t), \ i = 1, 2}{A_1 \parallel A_2 \text{ sat } \exists t_1, t_2 : t \simeq t_1 \parallel t_2 \ \wedge \phi_1(t_1) \wedge \phi_2(t_2)}$$

Here $\phi_i(t_i)$ denotes the result of replacing in ϕ_i the trace variable t by t_i. Note that $t \neq \bot$ implies that t_1 and t_2 are indeed compatible.

Hiding

$$\frac{A \text{ sat } \phi(t)}{\exists x A \text{ sat } \exists t' \sqsubseteq_x t : \phi(t')}$$

For notational convenience we use here a simple form of bounded quantification. The above specification of $\exists x P$ expresses that a trace t of $\exists x P$ can be obtained from a trace t' which does not contain information about the local variable x in an input constraint and which satisfies $\phi(t')$, by eliminating information about the local variable x in the output constraints of t' and replacing the information about the local variable x in its input constraints by new information about the *global* variable x (according the definition of $\sqsubseteq_x$).

---

[2] Recalla that $\gamma(t)$ denotes the final constraint of the trace t.

Choice

$$\frac{A_i \ sat \ \phi_i(t)}{\Sigma_i \mathbf{ask}(c_i) \to A_i \ sat \ (\gamma(t) \preceq t \wedge \bigwedge_i \gamma(t) \not\vdash c_i) \vee \bigvee_i in(c_i) \cdot t \simeq t \wedge \phi_i(t)}$$

So either t is a trace consisting only of inputs such that its final result does not entail any of the asked constraints or t is equivalent to a trace which starts with an input of one of the asked constraints and t satisfies the corresponding trace specification.

Weakening

$$\frac{A \ sat \ \phi(t) \qquad \models \phi(t) \to \psi(t)}{A \ sat \ \psi(t)}$$

Let us illustrate the above trace logics by the following two simple examples.

Example 1. Consider the ccp program

$$\mathbf{tell}(c_1) \parallel \mathbf{tell}(c_2)$$

We have that $\mathbf{tell}(c_i)$, $i = 1, 2$, satisfies the trace specification

$$out(c_i) \cdot \gamma(t) \preceq t$$

The parallel composition thus satisfies

$$\exists t_1, t_2 : t_1 \parallel t_2 \simeq t \wedge out(c_1) \cdot \gamma(t_1) \preceq t_1 \wedge out(c_2) \cdot \gamma(t_2) \preceq t_2$$

From the definition of parallel composition and the preorder $\preceq$ it follows that

$$out(c_1 \sqcup c_2) \cdot \gamma(t) \preceq t$$

By the weakening rule we thus obtain

$$\mathbf{tell}(c_1) \parallel \mathbf{tell}(c_2) \ sat \ out(c_1 \sqcup c_2) \cdot \gamma(t) \preceq t$$

Example 2. Next we consider proving correctness of the counter example to completeness to the proof theory based on resting points ([2]): $\exists x A$ where A denotes

$$\mathbf{ask}(x = a) \to \mathbf{tell}(y = b) + \mathbf{ask}(true) \to \mathbf{tell}(z = c)$$

By the tell axiom and the choice rule we derive

$$A \ sat \ \begin{pmatrix} (\gamma(t) \preceq t \wedge \gamma(t) \not\vdash x = a \wedge \gamma(t) \not\vdash true) \\ \vee \\ (in(x = a) \cdot t \simeq t \wedge out(y = b) \cdot \gamma(t) \preceq t) \\ \vee \\ (in(true) \cdot t \simeq t \wedge out(z = c) \cdot \gamma(t) \preceq t) \end{pmatrix}$$

By the weakening rule ($\gamma(t) \vdash true$ and $in(true) \cdot t \simeq t$):

$$A \; sat \left(\begin{array}{c} (in(x = a) \cdot t \simeq t \land out(y = b) \cdot \gamma(t) \preceq t) \\ \lor \\ out(z = c) \cdot \gamma(t) \preceq t \end{array} \right)$$

Applying next the hiding rule we obtain

$$\exists x A \; sat \; \exists t' \sqsubseteq_x t \left(\begin{array}{c} (in(x = a) \cdot t' \simeq t' \land out(y = b) \cdot \gamma(t') \preceq t') \\ \lor \\ out(z = c) \cdot \gamma(t') \preceq t' \end{array} \right)$$

By definition of $\simeq_x$ we have

$$\neg(in(x = a) \cdot t' \sqsubseteq_x t)$$

So we derive

$$\exists x A \; sat \; \exists t' \sqsubseteq_x t : out(z = c) \cdot \gamma(t') \preceq t'$$

by a trivial application of the weakening rule. A final application of the weakening rule then gives the desired result

$$\exists x A \; sat \; out(z = c) \cdot \gamma(t) \preceq t$$

6 Wrapping Up: Soundness and Completeness

Let

$$\vdash A \; sat \; \phi(t)$$

denote that the trace specification $A \; sat \; \phi(t)$ is derivable from the above proof system. Further, let

$$\mathcal{T}_\simeq(A)(true) = \{\theta \mid \theta \simeq \theta', \text{ for some } \theta' \text{ such that } \theta' \in \mathcal{T}(A)(true\}$$

Then

$$\models A \; sat \; \phi(t)$$

denotes that $M(C), \sigma \models \phi(t)$, for every variable assignment σ such that $\sigma(t) \in \mathcal{T}_\simeq(A)(true)$.

We have the following main soundness and completeness theorem.

Theorem 1

$$\vdash A \; sat \; \phi(t) \text{ if and only if } \models A \; sat \; \phi(t)$$

To sketch its proof, we first introduce the following abbreviations for the trace specification schema describing the *ccp* operators in the above proof theory:

$PAR(\phi_1(t), \phi_2(t)) = \exists t_1, t_2 : t \simeq t_1 \parallel t_2 \land \phi_1(t_1) \land \phi_2(t_2)$
$HIDE(\phi(t)) = \exists t' \sqsubseteq_x t : \phi(t')$

$$CHOICE_n^{\bar{c}}(\phi_1(t),\ldots,\phi_n(t))=$$
$$(\gamma(t) \preceq t \wedge \bigwedge_i d \nvdash c_i) \vee \bigvee_i in(c_i)\cdot t \simeq t \wedge \phi_i(t)$$
where $\bar{c}=c_1,\ldots,c_n$

The following lemma states the monotonicity of the above trace specification schema.

Lemma 1. *We have*

- $\models (\phi_i(t) \to \psi_i(t)) \to (PAR(\phi_1(t),\phi_2(t)) \to PAR(\psi_1(t),\psi_2(t)))$
  for $i=1,2$, *where* $\psi_j = \phi_j$, *for* $i \neq j$.
- $\models (\phi(t) \to \psi(t)) \to (HIDE(\phi(t)) \to HIDE(\psi(t)))$
- $\models (\phi_i(t) \to \psi_i(t)) \to (CHOICE_n^{\bar{c}}(\ldots,\phi_i(t),\ldots) \to CHOICE_n^{\bar{c}}(\ldots,\psi_i(t),\ldots))$
  where $i=1,\ldots,n$

Let for every agent A the trace specification $\phi_A(t)$ be defined inductively as follows:

$A = \mathbf{tell}(c)$: $\phi_A(t) \leftrightarrow out(c) \cdot \gamma(t) \preceq t$
$A = A_1 \parallel A_2$: $\phi_A(t) \leftrightarrow PAR(\phi_{A_1}(t),\phi_{A_2}(t))$
$A = \exists x B$: $\phi_A(t) \leftrightarrow HIDE(\phi_B(t))$
$\Sigma_i^n \mathbf{ask}(c_i) \to A_i$: $\phi_A(t) \leftrightarrow CHOICE_n^{\bar{c}}(\phi_{A_1}(t),\ldots,\phi_{A_n}(t))$
   where $\bar{c}=c_1,\ldots,c_n$

The following lemma follows by a straightfoward induction of $\phi_A(t)$.

Lemma 2. *For every (finite) ccp agent A we have*

$$\vdash A \ sat \ \phi_A(t)$$

The next lemma shows that $\phi_A(t)$ describes the trace semantics of *ccp* agent A.

Lemma 3. *For every agent A the trace specification $\phi_A(t)$ describes its trace semantics $\mathcal{T}_\simeq$. That is,*

$$\{\sigma(t) \mid M(C),\sigma \models \phi_A(t)\} = \mathcal{T}_\simeq(A)(true)$$

The proof of this lemma proceeds by a simultaneous induction on the *ccp* agent A and the length of traces.

The following corollary follows immediately from Lemma 3 and the above definition of $\models A \ sat \ \phi(t)$.

Corollary 1. *For every (finite) ccp agent A we have*

$$\models A \ sat \ \phi(t) \text{ if and only if } \models \phi_A(t) \to \phi(t)$$

Soundness. As a characteristic example we prove soundness of the parallel composition rule. Let

$$\models A_i \ sat \ \phi_i(t),$$

for $i = 1, 2$, and $A = A_1 \parallel A_2$. From these assumptions and Corollary 1 it follows that

$$\models \phi_{A_i}(t) \rightarrow \phi_i(t)$$

By the monotonicity of the PAR scheme (Lemma 1) we thus derive that

$$\models PAR(\phi_{A_1}(t), \phi_{A_2}(t)) \rightarrow PAR(\phi_1(t), \phi_2(t))$$

By Lemma 3

$$\models \phi_A(t) \leftrightarrow PAR(\phi_{A_1}(t), \phi_{A_2}(t)))$$

So we have

$$\models \phi_A(t) \rightarrow PAR(\phi_1(t), \phi_2(t))$$

Thus we conclude from Lemma 3 again that

$$\models A \ sat \ PAR(\phi_1(t), \phi_2(t))$$

Completeness. To prove completeness, let

$$\models A \ sat \ \phi(t)$$

From our assumption and Corollary 1 we thus derive that

$$\models \phi_A(t) \rightarrow \phi(t)$$

By lemma 2

$$\vdash A \ sat \ \phi_A(t)$$

So an application of the weakening rule gives us the desired result

$$\vdash A \ sat \ \phi(t)$$

7   What Next?

Courcelle's famous theorem from 1990 [1] states that any property of graphs definable in monadic second-order logic (MSO) can be decided in linear time on any class of graphs of bounded treewidth. Since traces are sequences of monotonically increasing *finite* (input/output) constraints we want to apply the above result to constraint systems which can be modeled as graphs of bounded treewidth (with respect to the entailment relation) by embedding our trace logic which involves second-order quantification over sequences into monadic second-order logic, representing traces by monadic predicates (on input/output constraints).

Of particular interest is an extension of our work to the specification and verification of the infinite behavior of *ccp* programs. A promising approach to this problem involves an application of co-induction, as discussed in [6] for a simple class of imperative programs.

Acknowledgement. We thank the reviewers for their valuable comments.

References

1. Courcelle, B.: The monadic second-order logic of graphs, ii: Infinite graphs of bounded width. Mathematical Systems Theory 21(1), 187–221 (1988)
2. De Boer, F.S., Gabbrielli, M., Marchiori, E., Palamidessi, C.: Proving concurrent constraint programs correct. ACM Transactions on Programming Languages and Systems (TOPLAS) 19(5), 685–725 (1997)
3. De Boer, F.S., Palamidessi, C.: A fully abstract model for concurrent constraint programming. In: TAPSOFT 1991, pp. 296–319. Springer, Heidelberg (1991)
4. Fages, F.: Concurrent constraint programming and linear logic (abstract). In: Proceedings of the 2nd ACM SIGPLAN International Conference on Principles and Practice of Declarative Programming, PPDP 2000, p. 44. ACM, New York (2000)
5. Fages, F., Ruet, P., Soliman, S.: Phase semantics and verification of concurrent constraint programs. In: LICS, pp. 141–152 (1998)
6. Nakata, K., Uustalu, T.: A hoare logic for the coinductive trace-based big-step semantics of while. In: Gordon, A.D. (ed.) ESOP 2010. LNCS, vol. 6012, pp. 488–506. Springer, Heidelberg (2010)
7. Ruet, P., Fages, F.: Concurrent constraint programming and non-commutative logic. In: Nielsen, M. (ed.) CSL 1997. LNCS, vol. 1414, pp. 406–423. Springer, Heidelberg (1998)
8. Saraswat, V.A., Rinard, M., Panangaden, P.: The semantic foundations of concurrent constraint programming. In: Proceedings of the 18th ACM SIGPLAN-SIGACT Symposium on Principles of Programming Languages, pp. 333–352. ACM (1991)
9. Zwiers, J., de Roever, W.P., van Emde Boas, P.: Compositionality and concurrent networks: Soundness and completeness of a proofsystem. In: Brauer, W. (ed.) ICALP 1985. LNCS, vol. 194, pp. 509–519. Springer, Heidelberg (1985)

Privacy from Accelerating Eavesdroppers: The Impact of Losses

Adam Bognat[1] and Patrick Hayden[2]

[1] Department of Physics, McGill University,
Montréal, Québec, Canada
[2] Department of Physics, Stanford University, CA, USA

Abstract. We investigate a communication scenario in which two inertial observers attempt to securely communicate quantum information via a noisy channel in the presence of a uniformly accelerating eavesdropper. Due to her acceleration, the eavesdropper is subject to Unruh noise which can potentially be exploited to design a secure communication protocol. This problem had previously been studied by Panangaden and co-authors for the special case in which the channel between the inertial observers is noiseless. In this article, we consider noise in the form of a lossy bosonic channel. Our calculations demonstrate that for a fixed acceleration, there is a secure communication protocol provided the noise is below a threshold value.

Keywords: relativistic quantum information, private quantum capacity.

1 Introduction

While most quantum information theory assumes that the computing and communication of the participants all take place in a shared inertial frame, there are many conceptual challenges and practical reasons to extend the theory beyond this special case. Perhaps most obviously, protocols designed to distribute entanglement or teleport quantum information must be modified if they are to perform properly when the participants do not share an inertial frame [1–3]. More intriguingly, relativistic causality can be exploited to provide solutions to cryptographic problems [4]. In addition, the interplay between metrology and quantum information has led to proposals for observing effects in relativistic quantum field theory that had previously seemed inaccessible to experiment [5]. It has even been discovered that the no-cloning theorem admits a generalization that characterizes exactly how quantum information can be replicated in space and time [6, 7].

Modelling a physically realistic relativistic quantum information protocol is a challenging problem, and only recently have tenable, explicit models been proposed for even relatively simple communication tasks. The challenge is to model the exchange of information between two localized agents in spacetime, whereby the agents communicate by encoding their information in specific states of an appropriate quantum field. [8] has considered the problem of an inertial sender

F. van Breugel et al. (Eds.): Panangaden Festschrift, LNCS 8464, pp. 180–190, 2014.

communicating with a uniformly accelerating receiver via the coherent states of a massless scalar field and find that, due to the Unruh effect, the secret key rate of continuous variable quantum key distribution is reduced compared to the inertial case, while [9] models how general relativistic effects impact satellite-based quantum communication and derives similar conclusions. While these studies are done in very explicit terms and consider very specific implementations of certain quantum communication protocols, they do not much shed light on the information-theoretical limits of such protocols, and the complexity of the analysis leaves it unamenable to such considerations. Indeed, much work remains to be done to marry information-theoretical techniques with realistic models of relativistic communication protocols.

In this article, we take a step in this direction by revisiting the problem of two *relativistic* inertial observers, Alice and Bob, who wish to securely transmit quantum information to each other in the presence of a uniformly accelerating eavesdropper, Eve; henceforth A, B, and E will respectively denote Alice, Bob, and Eve's Hilbert spaces. Alice encodes her message in the excited states of a quantum scalar field and then transmits to Bob over a channel $\mathcal{N} : A \to B$. Due to the Unruh effect [10–13], whereby Eve perceives the vacuum state of the quantum field in Alice's frame as thermal, the message that Eve intercepts will be subject to noise, the effect of which can be modelled by a quantum channel $\mathcal{E} : A \to E$, the so-called *Unruh channel*. One hopes that this noise can be exploited to ensure private quantum communication between Alice and Bob, at least for some range of accelerations. This question is formalized by defining the *private quantum capacity*, which is the optimal rate at which qubits can be transmitted through repeated uses of $\mathcal{N}$ while simultaneously ensuring that any eavesdropping using $\mathcal{E}$ is completely thwarted.

In the case that the channel from Alice to Bob is noiseless, Brádler, Hayden and Panangaden [14, 15] have shown that that the private quantum capacity $Q_p(\mathrm{id}, \mathcal{E})$ is non-zero for all accelerations. In that case, the private quantum capacity admits a single-letter formula, *viz.*

$$Q_p(\mathrm{id}, \mathcal{E}) = \max \frac{1}{2} I(A'; E_C)_\tau, \tag{1}$$

where the maximization is over all states of the form $\tau = (\mathrm{id} \otimes \mathcal{N}_C)\psi_{A'A}$ and $\mathcal{N}_C$ denotes the channel complementary to $\mathcal{N}$, that is, the channel to Eve's environment. In words, the capacity is obtained by maximizing one-half the mutual information between Alice's purifying system and Eve's environment over pure states $\psi_{A'A}$. Because the Unruh channel is covariant and $I(A'; E_C)_\tau$ is a concave function of ψ_A [16], the expression is maximized for the maximally mixed input state for dual-rail qubit [14, 15]. In that case, (1) evaluates to

$$Q_p(\mathcal{N}, \mathcal{E}) = \frac{1}{2}\left(1 - \sum_{k=1}^{\infty}(1 - z)^2 z^{k-1}\frac{k(k+1)}{2}\log\frac{k+1}{k}\right), \tag{2}$$

where z parametrizes Eve's proper acceleration [14, 15].

In what follows, we will consider the case in which the channel from Alice to Bob is itself noisy. In that scenario, there is no single-letter formula known for the private quantum capacity, nor can we use the covariance of the channel to prove that the maximally mixed state is optimal because the modified capacity formula is not a concave function of the input density operator. Instead, we will have to explicitly optimize the expression for the private quantum capacity over Alice's input states. We make some progress towards this end by demonstrating that a lower bound on the capacity is maximized by the maximally mixed state, and find that, unlike in this noiseless case, this expression vanishes for non-zero acceleration parameter z.

2   Noise Models

2.1   The Unruh Channel

A detailed description of the Unruh effect and the corresponding quantum channel can be found in [15]. For our purposes, a conceptual description motivating the definition of the channel will suffice.

We will assume that Alice encodes her quantum information in the single-particle excitations of one of two Unruh modes of her quantum field; to wit, she transmits to Bob qubit states of the form $|\psi\rangle = \alpha\hat{a}_1^\dagger|0,0\rangle + \beta\hat{a}_2^\dagger|0,0\rangle$, where $\hat{a}_i$ is the annihilation operator corresponding to the i'th mode. States of this form are called *dual-rail qubits* and the states $|1,0\rangle = \hat{a}_1^\dagger|0,0\rangle$ and $|0,1\rangle = \hat{a}_2^\dagger|0,0\rangle$ comprise the *dual-rail basis*.

When quantizing a field, the usual prescription is to separate the classical solutions of the field into positive and negative frequency modes and assign annihilation operators to the positive-frequency modes and creation operators to the negative-frequency modes and demand that the canonical commutation relations hold. Because the separation into positive and negative frequency modes requires specifying a time-like Killing vector field, different observers will, in general, make different assignments, resulting in the non-uniqueness of the definition of the vacuum state of a quantum field. The creation and annihilation operators of different observers are related by so-called *Bogoliubov transformations*, which are, in general, linear transformations that preserve the canonical commutation relations [17].

Thus, Eve's motion will induce a transformation of Alice's transmitted state. Because the spacetime seen by a uniformly accelerating observer admits a horizon, one must trace out the causally disconnected components of Alice's transformed state to obtain the state that is accessible to Eve. This procedure leaves Eve with a mixed state, and so from Eve's point of view, the intercepted state has been subject to some kind of noise. Quantitatively, the transformation induced on Alice's i'th mode is equivalent to the action of the unitary operator

$$U_{A_iC_i} = \exp[r(\hat{a}_i^\dagger\hat{c}_i^\dagger - \hat{a}_i\hat{c}_i)]. \tag{3}$$

In the above expression, C_i plays the role of Eve's i'th environment mode, corresponding to the Fock space of the causally disconnected modes of her field, and

r is related to Eve's acceleration as follows: if Eve's proper acceleration is a and the frequency of the i'th mode is ω, then $\tanh r = \exp(-\pi\omega/a)$.

Strictly peaking, (3) is valid only if Alice encodes her information in her Unruh modes, given in terms of the Minkowski modes in a rather non-trivial way. While the Unruh modes do not readily admit a clean physical interpretation, defining the problem this way makes it amenable to information-theoretical considerations, though ultimately, the additional complications of frequency mixing between Minkowski and Unruh modes needs to be taken into account [18]. In addition, the Unruh modes are globally extended objects, and do not give a realistic model of directed communication between two localized observers, but this holds for the Minkowski modes as well.

As described in [15], the effect of this transformation on the dual-rail qubit encoding is as follows.

Definition 1. *The qubit Unruh channel $\mathcal{E} : A \to E$ is the quantum channel defined by*

$$\mathcal{E}(\psi_A) = \mathrm{Tr}_C \left[(U_{A_1 C_1} \otimes U_{A_2 C_2}) \psi_A \otimes |0,0\rangle\langle 0,0|_C (U^\dagger_{A_1 C_1} \otimes U^\dagger_{A_2 C_2}) \right].$$

The action of the channel on an input qubit state can be written

$$\mathcal{E}(\psi_A) = (1-z)^3 \bigoplus_{k=0}^{\infty} z^k \sigma_A^{(k)} \tag{4}$$

for matrices $\sigma_A^{(k)}$, where $z = \tanh^2 r$. If $J_i^{(k)}$ represents the i'th generator of the irreducible k-dimensional representation of $SU(2)$ and ψ_A has Bloch vector $\boldsymbol{n}$, then

$$\sigma_A^{(k)} = \frac{k}{2} \mathrm{id}^{(k+1)} + \sum_{i=1}^{3} n_i J_i^{(k+1)} \tag{5}$$

In the formula, $\mathrm{id}^{(j)}$ is the $j \times j$ identity matrix. Subscripts such as A_i and C_j are subsystem labels chosen to match the corresponding annihilation operators $\hat{a}_i$ and $\hat{c}_j$. The eavesdropper system here is $E = A = A_1 A_2$. An important role is also played by the channel complementary to the Unruh channel in our investigation of the private quantum capacity. The complementary channel is the channel to the eavesdropper's environment $E_C = C_1 C_2$.

Definition 2. *The channel complementary to the qubit Unruh channel $\mathcal{E}_C : A \to E_C$ is the quantum channel defined by*

$$\mathcal{E}_C(\psi_A) = \mathrm{Tr}_A \left[(U_{A_1 C_1} \otimes U_{A_2 C_2}) \psi_A \otimes |0,0\rangle\langle 0,0|_C (U^\dagger_{A_1 C_1} \otimes U^\dagger_{A_2 C_2}) \right].$$

As detailed in [15], the action of the complementary channel on an input qubit state can be written in terms of Unruh channel itself. Specifically,

$$\mathcal{E}_C(\psi_A) = z\overline{\mathcal{E}(\psi_A)} + (1-z)\omega_0 \tag{6}$$

where $\omega_0 = (1 - z)^2 \left[\mathrm{id}^{(1)} \oplus z\,\mathrm{id}^{(2)} \oplus z^2\,\mathrm{id}^{(3)} \oplus \cdots \right]$. Therefore, up to complex conjugation, the output of the complementary channel is a degraded version of the output of the Unruh channel. The Unruh channel is therefore *conjugate degradable* [19].

2.2   The Lossy Bosonic Channel

The simplest noise model for our purposes is the *lossy bosonic channel* (LBC), which models the transfer of excitations into the environment. Mathematically, the LBC affords a compact description in terms of the transformation it induces on Alice's annihilation operators, given by

$$\hat{a}_i \to \sqrt{\eta}\hat{a}_i + \sqrt{1 - \eta}\hat{f}_i \tag{7}$$

$$\hat{f}_i \to -\sqrt{1 - \eta}\hat{a}_i + \eta\hat{f}_i. \tag{8}$$

Here, $\hat{f}_i$ is the annihilation operator of the associated environmental mode. This transformation is equivalent to conjugation of the annihilation operators by the unitary operator

$$S_i = \exp(\theta(\hat{a}_i^\dagger \hat{f}_i - \hat{a}_i \hat{f}_i^\dagger)) \tag{9}$$

with $\theta = \arctan\sqrt{\frac{1-\eta}{\eta}}$. Assuming that the channel acts independently on each of Alice's input modes, we may explicitly write down the action of the channel on Alice's input state, as follows:

Definition 3. *The lossy bosonic channel* $\mathcal{N} : A \to B$ *is the quantum channel defined by* $\mathcal{N}(\psi_A) = \mathrm{Tr}_F \left[S_1^\dagger \otimes S_2^\dagger (\psi_A \otimes |0,0\rangle\langle 0,0|_F) S_1 \otimes S_2 \right]$. *The action of the channel on an input* ψ_A *is given by*

$$\mathcal{N}(\psi_A) = \eta\psi_A + (1 - \eta)|0,0\rangle\langle 0,0|_A \tag{10}$$

Thus, we have that with probability η the state is transmitted unaltered, and with probability $1 - \eta$ the qubit is absorbed by the environment and the state is projected onto the zero-photon subspace. Henceforth η will be called the *noise parameter*. It should be mentioned that the channel represented by (10) is known as an *erasure* channel in the quantum information literature [20], while in the context of quantum optics, the operator (9) represents the transformation induced by a beam splitter [17].

3   Private Quantum Capacity: The Noisy Case

Different communication capacities for bosonic channels have been previously considered in the literature [21, 22]. Herein, we consider the private quantum capacity $Q_p(\mathcal{N}, \mathcal{E})$, which is, intuitively, the maximum rate at which qubits can be reliably transmitted through $\mathcal{N}$ in such a way that if they were intercepted en route and examined via $\mathcal{E}$, the eavesdropped output would be essentially independent of the qubit transmitted. More formally, let $\Phi^{(j)}$ represent a Schmidt rank j maximally entangled state and $\pi^{(j)}$ the rank j maximally mixed state.

Definition 4. *An (n, k, δ, ϵ) private entanglement transmission code from Alice to Bob consists of an encoding channel $\mathcal{A}$ taking a k-qubit system R' into the input of $\mathcal{N}^{\otimes n}$ and a decoding channel $\mathcal{B}$ taking the output of $\mathcal{N}^{\otimes n}$ to a k-qubit system $C \cong R'$ satisfying*

1. *Transmission:* $\left\| (\mathrm{id} \otimes \mathcal{B} \circ \mathcal{N}^{\otimes n} \circ \mathcal{A})(\Phi^{(2^k)}) - \Phi^{(2^k)} \right\|_1 \leq \delta.$

2. *Privacy:* $\left\| (\mathrm{id} \otimes \mathcal{E}^{\otimes n} \circ \mathcal{A})(\Phi^{(2^k)}) - \pi^{(2^k)} \otimes (\mathcal{E}^{\otimes n} \circ \mathcal{A})(\pi^{(2^k)}) \right\|_1 \leq \epsilon.$

A rate Q is an *achievable rate for private entanglement transmission if for all $\delta, \epsilon > 0$ and sufficiently large n there exist $(n, \lfloor nQ \rfloor, \delta, \epsilon)$ private entanglement transmission codes. The* private quantum capacity $Q_p(\mathcal{N}, \mathcal{E})$ *is the supremum of all the achievable rates.*

As mentioned earlier, in the case that the channel from Alice to Bob is noisy, there is no known single-letter formula for the private quantum capacity. Denoting by $I(X\rangle Y)_\rho$ the *coherent information* $H(Y)_\rho - H(XY)_\rho$ between X and Y, we may write the capacity as

$$Q_p(\mathcal{N}, \mathcal{E}) = \lim_{n \to \infty} \max \frac{1}{2n} [I(A'\rangle B^n)_\rho - I(A'\rangle E^n)_\tau] \tag{11}$$

where $\rho = (\mathrm{id} \otimes \mathcal{N}) \psi_{A'A}$ and $\tau = (\mathrm{id} \otimes \mathcal{E}) \psi_{A'A}$, and the maximization is taken over all pure states $\psi_{A'A}$ [15]. A' can be taken to be isomorphic to A. Fixing $n = 1$ and evaluating the expression above for any state $\psi_{A'A}$ yields a lower bound on $Q_p(\mathcal{N}, \mathcal{E})$, that is, an achievable rate for private quantum communication:

$$L_{Q_p}(\mathcal{N}, \mathcal{E}, \psi_A) = \frac{1}{2} (I(A'\rangle B)_\rho - I(A'\rangle E)_\tau). \tag{12}$$

(L_{Q_p} is written as a function of ψ_A rather than $\psi_{A'A}$ because its value is independent of the choice of purification to A'.)

The rest of the paper will be devoted to computing the lower bound for appropriately chosen input density matrices ψ_A, which will provide achievable rates of private quantum communication. We will find that ψ_A maximally mixed gives the best lower bound, which will allow us to quantitatively explore the competition between noise in the channel between the inertial observers and noise induced by the eavesdropper's acceleration.

4   Computing $L_{Q_p}(\mathcal{N}, \mathcal{E}, \psi_A)$

The coherent information terms appearing in the expression for $L_{Q_p}(\mathcal{N}, \mathcal{E}, \psi_A)$ are difficult to calculate for arbitrary inputs ψ, but special properties of the erasure and Unruh channels will simplify our computation by allowing us to consider a restricted class of inputs. To begin, the coherent information between Alice's reference system and Bob evaluates to

$$I(A'\rangle B)_\rho = (2\eta - 1)H(A)_\psi. \tag{13}$$

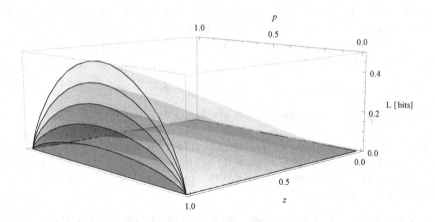

Fig. 1. Achievable rates of private quantum communication. The lower bound $\max(L_{Q_p}, 0)$ on the private quantum capacity is plotted as a function of the input parameter p and the acceleration parameter z, plotted for 6 different values of the noise parameter η, ranging from $\eta = 1.0$, which gives the highest communication rates, down in steps of 0.1 to $\eta = 0.5$, for which no private quantum communication is possible. The lower bound has a single global maximum at $p = 1/2$ for all values of z and η, unless the function is everywhere nonpositive for $p \in [0, 1]$.

Moreover, since both the erasure channel and the qubit Unruh channel are $SU(2)$ covariant, the coherent information between Alice's reference and Eve depends only on the eigenvalues of the input and not on the basis in which it is diagonal. Thus, we may without loss of generality restrict ourselves to input states of the form $\psi_A = p|1, 0\rangle\langle 1, 0| + (1 - p)|0, 1\rangle\langle 0, 1|$ and just maximize $L_{Q_p}(\mathcal{N}, \mathcal{E}, \psi_A)$ over the *input parameter* p.

The coherent information between Alice's reference and Eve can be written

$$I(A'\rangle E)_\tau = H(E)_\tau - H(E_C)_\tau \tag{14}$$

where E_C is the output $\mathcal{E}_C(\psi_{A'A})$ of the channel complementary to the Unruh channel. Writing the action of the Unruh channel and its complementary channel on Alice's input as

$$\mathcal{E}(\psi_{A'A}) = \bigoplus_{k=1}^{\infty} \beta_k(z)\sigma_A^{(k)} \tag{15}$$

$$\mathcal{E}_C(\psi_{A'A}) = \bigoplus_{k=1}^{\infty} \beta_k(z)\phi_C^{(k)} \tag{16}$$

with $\beta_k(z) = (1 - z)^3 z^{k-1}$, we have that

$$L_{Q_p}(\mathcal{N}, \mathcal{E}, \psi_A) = (2\eta - 1)H(A)_\psi + \sum_{k=1}^{\infty} \beta_k(z)[H(\phi_C^{(k)}) - H(\sigma_A^{(k)})], \tag{17}$$

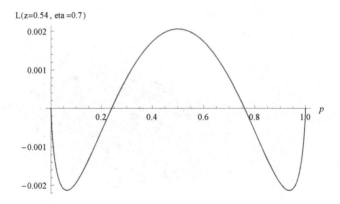

Fig. 2. The lower bound L_{Q_p} as a function of p for $z = 0.54$ and $\eta = 0.7$. The function is clearly neither concave nor convex. Deviations from concavity tend to occur only when η is tuned to give only a very small private quantum communication rate, in this case on the order of only 0.002 qubits per channel use.

where the operators $\sigma_A^{(k)}$ and $\phi_C^{(k)}$ were introduced in Section 2.1. One can write the difference of entropies above explicitly in terms of their eigenvalues, but the resulting expression is neither elegant nor particularly enlightening. Indeed, a plot of the expression is far more illuminating; Figure 1 shows the lower bound on the private quantum capacity as a function of the input parameter p and acceleration parameter z for a range of values of the noise parameter. We have verified numerically that the maximally mixed input state, corresponding to $p = 1/2$, maximizes $L_{Q_p}(\mathcal{N}, \mathcal{E}, \psi_A)$ whenever this quantity is non-negative.

Note, however, that the usual argument justifying the optimality of the maximally mixed state fails: L_{Q_p} is a difference of two coherent informations, the first for the erasure channel and the second for the Unruh channel. Because the channels are degradable [23] and conjugate degradable [15, 19], respectively, the individual coherent informations are concave. However, the difference of concave functions L_{Q_p} is not concave itself, as illustrated in Figure 2.

In the absence of a proof of additivity for (11), which would allow us to ignore the limit and consider the $n = 1$ case to evaluate the capacity, we cannot conclude that $L_{Q_p}(\mathcal{N}, \mathcal{E})$ evaluated at the maximally mixed state gives an exact expression for the capacity. Nevertheless, this quantity remains an informative lower bound and achievable rate for private quantum communication; explicitly, it is given by

$$L_{Q_p}(\mathcal{N}, \mathcal{E}, \mathrm{id}_A/2) = \frac{1}{2}\left((2\eta - 1) - \sum_{k=1}^{\infty}(1-z)^2 z^{k-1}\frac{k(k+1)}{2}\log\frac{k+1}{k} \right) \quad (18)$$

which reduces to (2) when $\eta = 1$, as it should. Indeed, the mutual information in (1) can be written as

$$I(A'; E_C)_\tau = H(A)_\psi + H(E_C)_\tau - H(E)_\tau, \quad (19)$$

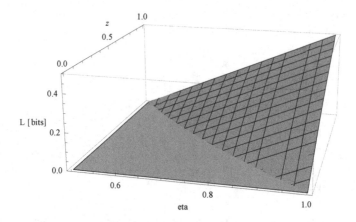

Fig. 3. Achievable rates for private quantum communication, calculated using the lower bound $\max(0, L_{Q_p})$ evaluated for the maximally mixed state, as a function of the noise and acceleration parameters. There is a clear trade-off between the amount of loss in the channel between the inertial observers, parametrized by $1 - \eta$, and the minimum acceleration required to ensure the possibility of private quantum communication.

while $L_{Q_p}(\mathcal{N}, \mathcal{E}, \psi_A)$ takes the form

$$(2\eta - 1)H(A)_\psi + H(E_C)_\tau - H(E)_\tau \qquad (20)$$

when $\mathcal{N}$ is the erasure channel; the effect of the lossy bosonic channel on the first term is merely to multiply it by a constant. Figure 3 shows a plot of $L_{Q_p}(\mathcal{N}, \mathcal{E}, \mathrm{id}_A/2)$ as a function of the noise and acceleration parameters. We see that for values of η away from 1, the rate vanishes for some non-zero value of the acceleration parameter. This precisely characterizes the trade-off between the noise that Eve experiences, which can be appropriately exploited to design a secure communication protocol, and the noise between Alice and Bob, which must be corrected against to have a robust communication protocol at all. In particular, for $z = 1$, corresponding to infinite acceleration, in which case Eve can glean no information about Alice's state, we see that the rate vanishes for $\eta \leq 1/2$, as it should in accordance with the no-cloning theorem; for if information could be transmitted when $\eta \leq 1/2$, it could equally well be recovered from the environment, thereby producing two copies of the same input state.

5   Conclusions

We have studied the problem of private communication of quantum information between two inertial observers, Alice and Bob, in the presence of an accelerating eavesdropper, when the channel between Alice and Bob is noisy. We considered the simplest model of a noisy communication channel – the lossy bosonic channel – and found that, while a single-letter formula for the private quantum capacity could not be obtained, a useful lower bound could be calculated. This quantity suggests that, for a given noise parameter, the private quantum capacity should vanish for some *finite* value of Eve's acceleration, whereas in the noiseless case, the capacity was non-zero for all accelerations. This can be understood as a trade-off between the rate at which Alice can encode her information so that Bob receives them with high fidelity, and the cost of securing that information from a potential eavesdropper. However, the communication model used is not particularly realistic for directed communication between two localized relativistic observers, and that a full information-theoretical analysis of the explicit communication models considered in the literature remains to be done. With respect to the calculation done in this article, the remaining open question is whether taking the limit in (11) results in a significantly improved capacity. We do not expect this to be the case, but examples of non-additive channel capacities do occur in quantum information [24]. Finally, more realistic modelling of the communication protocol could change the capacity in an unexpected way.

Acknowledgments. Writing this paper was an opportunity to revisit a problem one of us first had the pleasure of exploring with Prakash Panangaden. We are both deeply indebted to Prakash for his insights into relativistic information theory, of course, but more so for his mentorship, advice and generosity.

This work was supported by the Canada Research Chairs program, the Perimeter Institute, CIFAR, NSERC, FQXi and ONR through grant N000140811249. The Perimeter Institute is supported by Industry Canada and Ontario's Ministry of Economic Development & Innovation.

References

1. Peres, A., Terno, D.R.: Quantum information and relativity theory. Reviews of Modern Physics 76, 93–123 (2004)
2. Gingrich, R.M., Adami, C.: Quantum entanglement of moving bodies. Physical Review Letters 89(27), 270402 (2002)
3. Alsing, P.M., Milburn, G.J.: Teleportation with a uniformly accelerated partner. Physical Review Letters 91(18), 180404 (2003)
4. Kent, A.: Unconditionally secure bit commitment. Physical Review Letters 83, 1447–1450 (1999)
5. Martin-Martinez, E., Fuentes, I., Mann, R.B.: Using Berry's phase to detect the Unruh effect at lower accelerations. Physical Review Letters 1067, 131301 (2011)
6. Hayden, P., May, A.: Summoning information in spacetime, or where and when can a qubit be? arXiv preprint arXiv:1210.0913 (2012)
7. Kent, A.: A no-summoning theorem in relativistic quantum theory. Quantum Information Processing 12(2), 1023–1032 (2013)
8. Downes, T., Ralph, T., Walk, N.: Quantum communication with an accelerated partner. Physical Review A 87(1), 012327 (2013)
9. Bruschi, D.E., Ralph, T., Fuentes, I., Jennewein, T., Razavi, M.: Spacetime effects on satellite-based quantum communications. arXiv preprint arXiv:1309.3088 (2013)

10. Fulling, S.A.: Nonuniqueness of canonical field quantization in Riemannian space-time. Physical Review D 7(10), 2850–2862 (1973)
11. Davies, P.C.W.: Scalar production in Schwarzschild and Rindler metrics. Journal of Physics A: Mathematical and General 8, 609 (1975)
12. Unruh, W.G.: Notes on black-hole evaporation. Physical Review D 14(4), 870–892 (1976)
13. Unruh, W.G., Wald, R.M.: What happens when an accelerating observer detects a Rindler particle. Physical Review D 29(6), 1047–1056 (1984)
14. Brádler, K., Hayden, P., Panangaden, P.: Private information via the unruh effect. Journal of High Energy Physics 8, 074 (2009)
15. Brádler, K., Hayden, P., Panangaden, P.: Quantum communication in Rindler spacetime. Communications in Mathematical Physics 312(2), 361–398 (2012)
16. Bennett, C.H., Shor, P.W., Smolin, J.A., Thapliyal, A.V.: Entanglement-assisted capacity of a quantum channel and the reverse Shannon theorem. IEEE Transactions on Information Theory 48(10), 2637–2655 (2002)
17. Agarwal, G.S.: Quantum optics. Cambridge University Press (2013)
18. Bruschi, D.E., Louko, J., Martín-Martínez, E., Dragan, A., Fuentes, I.: Unruh effect in quantum information beyond the single-mode approximation. Physical Review A 82(4), 042332 (2010)
19. Brádler, K., Dutil, N., Hayden, P., Muhammad, A.: Conjugate Degradability and the Quantum Capacity of Cloning Channels. Journal of Mathematical Physics 51, 072201 (2010)
20. Nielsen, M.A., Chuang, I.L.: Quantum computation and quantum information. Cambridge University Press (2000)
21. Holevo, A.S., Werner, R.F.: Evaluating capacities of bosonic gaussian channels. Physical Review A 63(3), 032312 (2001)
22. Giovannetti, V., Guha, S., Lloyd, S., Maccone, L., Shapiro, J.H., Yuen, H.P.: Classical capacity of the lossy bosonic channel: The exact solution. arXiv preprint quant-ph/0308012 (2003)
23. Devetak, I., Shor, P.W.: The capacity of a quantum channel for simultaneous transmission of classical and quantum information. Communications in Mathematical Physics 256(2), 287–303 (2005)
24. Hayden, P., Winter, A.: Counterexamples to the maximal p-norm multiplicativity conjecture for all $p > 1$. Communications in Mathematical Physics 284(1), 263–280 (2008)

The Complexity of Computing a Bisimilarity Pseudometric on Probabilistic Automata

Franck van Breugel[1] and James Worrell[2]

[1] Department of Electrical Engineering and Computer Science,
York University 4700 Keele Street, Toronto, M3J 1P3, Canada
[2] Department of Computer Science, University of Oxford
Parks Road, Oxford, OX1 3QD, UK

In honour of Prakash Panangaden on the occasion of his 60th birthday

Abstract. We consider the problem of approximating and computing a bisimilarity pseudometric on the state space of a probabilistic automaton. We show that the distances are rational and that both approximation and exact computation of distances are in **PPAD**. In the proofs, a generalization of the classical game-theoretic characterization of bisimilarity, given in terms of simple stochastic games, plays a central role.

1 Some History

In the mid 1990's Panangaden became interested in probabilistic systems with continuous state spaces. For these systems, he coined the term *labelled Markov processes*. A web search for this term returns more than twenty-five thousand results, showing that it has been widely adopted. This term is also the title of Panangaden's most recent book [46].

Initially, Panangaden was interested in defining a notion of *behavioural equivalence* for labelled Markov processes. Such an equivalence relation on the state space captures which states behave the same. Together with Blute, Desharnais and Edalat [7], he generalized *probabilistic bisimilarity*, introduced by Larsen and Skou [42], from finite to continuous state spaces. Later Desharnais, Edalat and Panangaden provided a *logical characterization* of probabilistic bisimilarity in [23]. They considered a fragment of the probabilistic modal logic introduced by Larsen and Skou in [42] and showed that states satisfy the same logical formulae if and only if they are probabilistic bisimilar.

During a meeting in February 1998, Panangaden pointed out to Desharnais (at the time his PhD student) and the first author of this paper (at the time his postdoc) that his notion of probabilistic bisimilarity, or any other equivalence relation based on the probabilities of the labelled Markov process for that matter, was not *robust*. Small changes to any of those probabilities may cause equivalent states to become inequivalent or vice versa. This started the quest for a robust notion of behavioural equivalence for labelled Markov processes.

F. van Breugel et al. (Eds.): Panangaden Festschrift, LNCS 8464, pp. 191–213, 2014.

Almost a decade earlier, Giacalone, Jou and Smolka [33] had already observed that Larsen and Skou's notion of probabilistic bisimilarity was not robust. They proposed to use a *pseudometric*, rather than an equivalence relation, to capture the behavioural similarity of states. The distance between two states, a real number in the unit interval, captures their behavioural similarity. The smaller their distance, the more alike they behave. Distance zero captures that states are behaviourally indistinguishable. The pseudometric put forward by Giacalone *et al.* was only defined for what they call deterministic probabilistic processes. They close their paper with an open problem: generalize their pseudometric to the nondeterministic setting.

In May 1998, Edalat discussed metrics on probability measures with the first author and suggested to look at the Hutchinson metric [35]. As it turned out, over a span of four decades this metric was reinvented several times and is originally due to Kantorovich [39] and should, therefore, be called the *Kantorovich metric*. In October 1998, the first author sketched in a grant proposal how the Kantorovich metric can be used to define a behavioural pseudometric on a class of labelled Markov processes using the theory of *coalgebra*. For a detailed discussion of this theory we refer the reader to, for example, Jacobs' textbook [36]. Some of these ideas later appeared in [8].

While the first author was trying to exploit coalgebras to define a pseudometric on the state space of a labelled Markov process, Panangaden and Desharnais were attempting to adapt their *logic* to a *quantitative setting*. In January 1999, in collaboration with Gupta and Jagadeesan, they proved that their pseudometric is a quantitative generalization of probabilistic bisimilarity by showing that distance zero coincides with probabilistic bisimilarity, a key result that was missing from the first author's work. Their behavioural pseudometric, the proof that it generalizes probabilistic bisimilarity and several other results were published later that year in [24]. This solved the problem posed by Giacalone *et al.* almost a decade earlier.

In September 1999, the second author of this paper came up with a general notion of similarity. This notion appeared later in [59]. Worrell was looking for applications of this notion in a quantitative setting. In August 2000, Rutten visited Oxford and met with the second author. At the time, Rutten was also working on quantitative notions of bisimilarity, in particular metric bisimilarity [48]. Rutten told Worrell about the work on the Kantorovich metric by the first author. The latter visited Rutten in Amsterdam in October 2000. Rutten told Van Breugel about Worrell's work. This resulted in the first collaboration of the authors of this paper. In the next couple of months, we showed that the behavioural pseudometric defined by means of the theory of coalgebra coincides with the bisimilarity pseudometric defined in terms of a logic by Desharnais *et al.* As a consequence, also this behavioural pseudometric generalizes probabilistic bisimilarity. This and some other results appeared in [12].

2   The Problem, Related Work, and Our Results

In [24], Desharnais *et al.* wrote "The extension of the methods of this paper to systems which have both probability and traditional nondeterminism remains open and will be the object of future study." They also wrote "In future work, we will explore efficient algorithms and complexity results for our metrics." In this paper, we address both, that is, we consider a behavioural pseudometric for probabilistic automata, which contain probability as well as nondeterminism, and we focus on complexity results.

The paper [24] revived interest in behavioural pseudometrics, which were first proposed by Giacalone *et al.* in [33]. Over the last 15 years, more than 100 papers on behavioural pseudometrics have been published. We will not attempt to provide an overview of that vast body of work, but only discuss those papers that are most relevant to this paper.

In this paper, we will restrict ourselves to systems with *finitely many states*. Most papers on behavioural pseudometrics consider finite state systems. A few papers, including a paper by Ferns, Panangaden and Precup [30], allow for systems with infinite state spaces.

In most papers, either the transitions or the states are labelled. The former type of labelling usually captures interaction of the system with its environment. The latter type is generally used to express properties of interest that hold in particular states. Although the interpretations of these two types of labelling are quite different, they generally have the same expressive power. That is, a system with labellled transitions can be encoded as a system with labelled states such that the encoding preserves the behavioural equivalences or distances, and vice versa. Such an encoding can be found, for example, in the textbook [3, Section 7.1.2] of Baier and Katoen. In this paper, we restrict our attention to systems the *states* of which are *labelled*. In our examples, we use different colours to distinguish states that are labelled differently.

The systems, called *labelled Markov systems*, considered by Desharnais *et al.* in [24] only consider probabilistic choices. That is, for each state of the system, its outgoing transitions form a probability distribution on the state space. Such a collection of outgoing transitions we will call a *probabilistic transition*. An example of a labelled Markov system is presented below. In this example, state 2 goes to state 3 with probability $\frac{1}{4}$ and with probability $\frac{3}{4}$ to state 4. To avoid clutter, transitions with probability one are not labelled. For example, state 3 goes to state 3 with probability one. The bisimilarity pseudometric of Desharnais *et al.* gives rise to the following distances for this example.

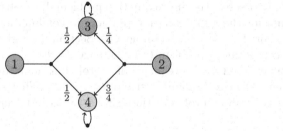

	1	2	3	4
1	0	$\frac{1}{4}$	$\frac{1}{2}$	1
2	$\frac{1}{4}$	0	$\frac{3}{4}$	1
3	$\frac{1}{2}$	$\frac{3}{4}$	0	1
4	1	1	1	0

We will not discuss here how these distances are computed. For algorithms to compute these, we refer the reader to, for example, [2,15]. Instead we will argue informally that these distances capture the similarity of the behaviour of the states. The distance from state 4 to the other states is one, since state 4 is blue whereas the other states are red. Let us, for example, consider the probability that a state can reach a blue state. For the states 1, 2, 3 and 4 this probability is $\frac{1}{2}$, $\frac{3}{4}$, 0, and 1, respectively. Note that with respect to this particular property state 2 behaves more like state 1 than state 3. This is reflected in the distance function.

The systems that we consider in this paper, called *probabilistic automata*, contain both probabilistic and nondeterministic choices. Behavioural equivalences for these systems were first studied by Segala and Lynch in [52]. These automata are also known as Segala automata (see, for example, [6]). Recall that the outgoing transitions of a state of a labelled Markov system form a probabilistic transition. Each state of a probabilistic automaton has a set of such probabilistic transitions eminating from it. An example of a probabilistic automaton is presented below. In this example, state 2 has two nondeterministic alternatives. The one takes the automaton to state 3 with probability 1. The other goes to state 3 with probability $\frac{1}{4}$ and to state 4 with probability $\frac{3}{4}$.

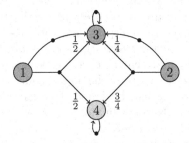

Below, we will consider algorithms and their complexity for three different problems related to a bisimilarity pseudometric, say d, on the state space of a probabilistic automaton. The *decision problem*, given states s_1 and s_2 of a probabilistic automaton and a rational q, raises the question whether $d(s_1, s_2) < q$. The *approximation problem*, given states s_1 and s_2 of a probabilistic automaton and a rational ϵ, consists of finding a rational q such that $|d(s_1, s_2) - q| < \epsilon$. The *computation problem*, given states s_1 and s_2 of a probabilistic automaton, simply aims at computing $d(s_1, s_2)$. These three problems are closely related. An algorithm that solves the decision problem in combination with binary search gives rise to an algorithm that solves the approximation problem. If the distances are rational, an algorithm for the approximation problem in combination with the continued fraction algorithm (see, for example, [50, Section 6.1]) gives rise to an algorithm for the computation problem (see [29, page 2540]).

A bisimilarity pseudometric on the state space of a probabilistic automaton was first presented by Deng, Chothia, Palamidessi and Pang in [22]. They showed that their pseudometric generalizes probabilistic bisimilarity as introduced by Segala and Lynch in [52].

De Alfaro, Majumdar, Raman and Stoelinga considered a slightly different pseudometric in [21]. We will discuss the difference in Section 4. They showed that the approximation problem for their pseudometric can be expressed in the first order theory over the reals. The latter theory is decidable, as shown by Tarski in [56]. His algorithm has nonelementary complexity. The problem was shown to be double exponential by Collins in [18].

Chatterjee, De Alfaro, Majumdar and Raman [14] considered the decision problem and the approximation problem. In their paper, they first transformed a quadratic optimization problem into a linear optimization problem that can be solved by linear programming. Next, they showed that this linear optimization characterization, capturing the approximation problem, can be expressed in term of the existential fragment of the first order theory over the reals. Since this theory is decidable in **PSPACE**, as shown by Cann in [13], they concluded that the decision problem is in **PSPACE**. As a consequence, the approximation problem is also in **PSPACE**.

Fu [31] showed that the bisimilarity distances for probabilistic automata are rational. We will provide an alternative proof of that result for a slightly different pseudometric in Section 7 of this paper. Furthermore, he proved that the decision problem is in **NP ∩ coNP**. Similar to our definition in Section 4, he defined his pseudometric as a least fixed point of some functional Δ. Fu's proof consists of two main steps. He presented a refinement algorithm which shows that the problem of deciding whether a given rational fixed point of Δ is the least fixed point is in **P**. Furthermore, he showed how to guess a rational fixed point of Δ. The proof can be adapted to show that the decision problem is in **UP ∩ coUP** [32]. Recall that **UP** contains those problems in **NP** with a unique accepting computation.

Desharnais, Laviolette and Tracol [26] and Tracol, Desharnais and Zhioua [58] also introduced pseudometrics for labelled Markov systems and probabilistic automata, respectively. Their pseudometrics generalize probabilistic bisimilarity as well. They are different from the pseudometrics on labelled Markov systems and probabilistic automata discussed above. Examples showing the difference can be found in [26, Example 7] and [58, Example 5]. To solve the computation problem for their pseudometrics, they developed iterative algorithms. In each iteration, a maximum flow problem needs to be solved. The resulting algorithms show that the computation problem for their pseudometrics is in **P**.

The complexity class **PPAD**, which is short for *polynomial parity argument in a directed graph*, was introduced by Papadimitriou in [47]. It lies between the search problem versions of **P** and **NP**. The class captures the basic principles of path-following algorithms like those of Lemke and Howson [43] and Scarf [49]. This complexity class **PPAD** is defined by one of its complete problems, called *end of the line*. This problem is defined as follows. Let G be a possibly exponentially large directed graph with no isolated vertices and with every vertex having at most one predecessor and at most one successor. The graph G is given by a polynomial-time computable function $f(v)$, polynomial in the size of v, which returns the predecessor and successor of the vertex v, if these exist.

Given a vertex v in the graph G with no predecessor, the end of the line problem is to find a vertex different from vertex v with no predecessor or no successor. Note that such a vertex exists due to the parity argument that the graph G has an even number of vertices with no predecessor or no successor. Finding Nash equilibria of two player games is **PPAD**-complete, as shown by Chen and Deng in [16]. Kintali *et al.* [41] present several other **PPAD**-complete problems. Etessami and Yannakakis [28] have shown that computing the value of a simple stochastic game is in **PPAD**.

In this paper, we consider the bisimilarity pseudometric on probabilistic automata introduced by Deng *et al.* [22]. We show that the approximation problem and the computation problem for this pseudometric are in **PPAD**. To prove these results, we exploit simple stochastic games.

Stochastic games were introduced by Shapley [54]. Condon [19] was the first to study simple stochastic games from a complexity theory point of view. A *simple stochastic game* consists of a directed graph whose vertices are partitioned into sets of *max vertices*, *min vertices*, and *average vertices* and two special vertices called *0-sink* and *1-sink*. One of the vertices is the *start vertex*. Each vertex has two outgoing edges, apart from the 0-sink and 1-sink, which have none.

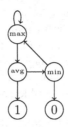

A simple stochastic game is played by two players, *Player 0* and *Player 1*, with a single token. Initially, the token is on the start vertex. At each step of the game, the token is moved from a vertex to one of its successors. At a min vertex Player 0 chooses the successor, at a max vertex Player 1 chooses the successor, and at an average vertex the successor is chosen randomly. Player 1 wins a play of the game if the token reaches a 1-sink; Player 0 wins if play reaches a 0-sink or continues forever without reaching a sink.

The *value* of a simple stochastic game is the probability that Player 1 wins the game when both players play optimally. The value of the above simple stochastic game, where the max vertex is the start vertex, is $\frac{1}{2}$. The *decision problem*, given a simple stochastic game, asks whether its value is smaller than $\frac{1}{2}$. The *approximation problem*, given a simple stochastic game and a rational ϵ, consists of finding a rational q such that $|v - q| < \epsilon$, where v is the value of the game. The *computation problem*, given a simple stochastic game, simply aims at computing the value of the game.

The decision problem for simple stochastic games was shown to be in **NP** ∩ **coNP** by Condon [19] and, in fact, is known to be in **UP** ∩ **coUP** (see, for example, the paper [61] by Yannakakis). However, after 20 years, the exact complexity remains unknown. The computation problem was shown to be in **PLS** (*polynomial local search*) by Yannakakis [60] via a strategy improvement algorithm. The same problem has also been shown to be in **PPAD** by Juba [38] and Etessami and Yannakakis [28], ultimately relying on Scarf's algorithm for computing approximate fixed points of continuous functionals. More recently the problem has been placed in the classes **CLS** (*continuous local search*) and **CCLS** (*convex continuous local search*) by Daskalakis and Papadimitriou [20]. It is still not known whether this problem is complete for any complexity class.

As promulgated by Stirling, ordinary bisimilarity can be characterized in term of a *two player game*. We will not discuss this game here, but refer the reader to [55, Section 3.2]. In [26], Desharnais *et al.* introduce for each $\varepsilon \in [0,1]$ a notion of ε-bisimilarity, which generalizes probabilistic bisimilarity for labelled Markov systems. They also present a two player game that characterizes ε-bisimilarity and generalizes the bisimilarity game. In Section 6, we will show that the bisimilarity pseudometric for probabilistic automata can also be characterized as a two player game. As we will see, our game is a simple stochastic game that also generalizes the bisimilarity game.

Etessami and Yannakakis [28] showed that computing a fixed point of a *polynomial piecewise linear* functional is in **PPAD**. We will introduce and discuss this notion in Section 7. Etessami and Yannakakis used their result to show that **PPAD** contains a variety of problems, including computing the values of simple stochastic games, finding Nash equilibria of two player games, and computing fixed points of discretized Brouwer functions. In this paper, we will show that their result is also applicable to our setting.

We reduce the approximation problem of the bisimilarity pseudometric on a probabilistic automaton to the approximation problem of the simple stochastic game that generalizes the bisimilarity game. The size of this game depends exponentially on the branching degree of the automaton and so our reduction is not polynomial-time in general. Nevertheless, we are still able to inherit the **PPAD** complexity from simple stochastic games. Here, we use the above mentioned result of Etessami and Yannakakis.

3   Metrics and Orders

Distance functions on the states of a probabilistic automaton, that is, functions that map each pair of states to an element of the unit interval, carry a natural order and metric. These will play a key role in our technical development. Next, we will collect definitions and results from the literature that we will use later in this paper.

Let X be a set. The relation $\sqsubseteq \; \subseteq [0,1]^{X \times X} \times [0,1]^{X \times X}$ is defined by

$$d_1 \sqsubseteq d_2 \text{ iff } d_1(x_1, x_2) \le d_2(x_1, x_2) \text{ for all } x_1, x_2 \in X.$$

One can verify that $\langle [0,1]^{X \times X}, \sqsubseteq \rangle$ is a complete lattice (see, for example, [25, Lemma 3.2]).

The function $\| \cdot - \cdot \| : [0, 1]^{X \times X} \times [0, 1]^{X \times X} \rightarrow [0, 1]$ is defined by

$$\|d_1 - d_2\| = \sup_{x_1, x_2 \in X} |d_1(x_1, x_2) - d_2(x_1, x_2)|.$$

One can also check that $\langle [0, 1]^{X \times X}, \| \cdot - \cdot \| \rangle$ is a nonempty complete metric space (see, for example, [4, Section 1.1.2]).

To define the bisimilarity pseudometric on the states of a probabilistic automaton, we will use two key ingredients: a distance function on nonempty and finite sets and a distance function on probability distributions. The former captures the nondeterministic choices in a probabilistic automaton, whereas the latter deals with the probabilistic choices in the automaton.

We denote the set of *nonempty and finite* subsets of a set X by $\mathcal{P}(X)$. We lift a distance function on X to a distance function on $\mathcal{P}(X)$ as follows.

Definition 1. *Let X be a set. The function $\mathcal{P} : [0, 1]^{X \times X} \rightarrow [0, 1]^{\mathcal{P}(X) \times \mathcal{P}(X)}$ is defined by*

$$\mathcal{P}(d)(A_1, A_2) = \max \left\{ \max_{x_1 \in A_1} \min_{x_2 \in A_2} d(x_1, x_2), \max_{x_2 \in A_2} \min_{x_1 \in A_1} d(x_2, x_1) \right\}.$$

The above is known as the *Hausdorff metric* (in case d is a metric). One can show that $\mathcal{P}$ preserves both the order and the metric: for all $d_1, d_2 \in [0, 1]^{X \times X}$, if $d_1 \sqsubseteq d_2$ then $\mathcal{P}(d_1) \sqsubseteq \mathcal{P}(d_2)$, that is, $\mathcal{P}$ is monotone, and $\|\mathcal{P}(d_1) - \mathcal{P}(d_2)\| \leq \|d_1 - d_2\|$, that is, $\mathcal{P}$ is nonexpansive.

Recall that $d \in [0, 1]^{X \times X}$ is a pseudometric if for all $x_1, x_2, x_3 \in X$, $d(x_1, x_1) = 0$, $d(x_1, x_2) = d(x_2, x_1)$, and $d(x_1, x_3) \leq d(x_1, x_2) + d(x_2, x_3)$. If d is a pseudometric then $\mathcal{P}(d)$ is a pseudometric as well (see, for example, [9, Proposition A.25]).

We denote the set of *rational* probability distributions on a set X by $\mathcal{D}(X)$. To lift a distance function on X to a distance function on $\mathcal{D}(X)$, we use the set of nonexpansive functions from the set X endowed with the distance function d to the unit interval, which we denote by $(X, d) \twoheadrightarrow [0, 1]$. Recall that a function $f : X \rightarrow [0, 1]$ is *nonexpansive* if $|f(x_1) - f(x_2)| \leq d(x_1, x_2)$ for all $x_1, x_2 \in X$.

Definition 2. *Let X be a set. The function $\mathcal{D} : [0, 1]^{X \times X} \rightarrow [0, 1]^{\mathcal{D}(X) \times \mathcal{D}(X)}$ is defined by*

$$\mathcal{D}(d)(\mu_1, \mu_2) = \sup \left\{ \sum_{x \in X} f(x)(\mu_1(x) - \mu_2(x)) \; \middle| \; f \in (X, d) \twoheadrightarrow [0, 1] \right\}.$$

The above is known as the *Kantorovich metric* (in case d is a metric). One can show that $\mathcal{D}$ is monotone (see, for example, [11, Proposition 38]) and nonexpansive (see, for example, [10, Section 3]). One can also prove that $\mathcal{D}(d)$ is a pseudometric if d is a pseudometric (see, for example, [27, Proposition 2.5.14]).

In Corollary 5, we present a dual characterization of the above definition of the Kantorovich metric. It is this characterization that we will use in our proofs.

This dual characterization is captured by means of couplings, a notion used for bounding the rate of convergence of Markov chains (see, for example, [45, Chapter 11]) and introduced next.

Definition 3. *Let X be a set. Let μ_1, $\mu_2 \in \mathcal{D}(X)$. Then $\omega \in \mathcal{D}(X \times X)$ is a coupling of μ_1 and μ_2 if for all x_1, $x_2 \in X$,*

$$\sum_{x_2 \in X} \omega(x_1, x_2) = \mu_1(x_1) \ and \ \sum_{x_1 \in X} \omega(x_1, x_2) = \mu_2(x_2).$$

In other words, ω is a joint probability distribution whose marginals are μ_1 and μ_2. We denote the set of couplings of μ_1 and μ_2 by Ω_{μ_1,μ_2}. Using the duality theorem of linear programming (see, for example, [17, Theorem 5.1]) we can characterise $\mathcal{D}$ as follows.

Theorem 4. *Let X be a finite set. For all $d \in [0, 1]^{X \times X}$ and μ_1, $\mu_2 \in \mathcal{D}(X)$,*

$$\mathcal{D}(d)(\mu_1, \mu_2) = \min \left\{ \sum_{x_1, x_2 \in X} \omega(x_1, x_2) d(x_1, x_2) \ \middle| \ \omega \in \Omega_{\mu_1,\mu_2} \right\}.$$

The above result is a special case of the Kantorovich-Rubinstein duality theorem [40]. The set of couplings Ω_{μ_1,μ_2} is a convex polytope. We denote its set of vertices by $V(\Omega_{\mu_1,\mu_2})$. Since a linear function on a convex polytope attains its minimum at a vertex (see, for example, [50, Chapter 8]), we obtain the following characterisation of $\mathcal{D}$.

Corollary 5. *Let X be a finite set. For all $d \in [0, 1]^{X \times X}$ and μ_1, $\mu_2 \in \mathcal{D}(X)$,*

$$\mathcal{D}(d)(\mu_1, \mu_2) = \min \left\{ \sum_{x_1, x_2 \in X} \omega(x_1, x_2) d(x_1, x_2) \ \middle| \ \omega \in V(\Omega_{\mu_1,\mu_2}) \right\}.$$

4   Probabilistic Automata

The topic of this paper is a bisimilarity pseudometric on a probabilistic automaton and the complexity to compute it. Next, we will introduce probabilistic automata. Furthermore, we will define a pseudometric on the state space of a probabilistic automaton.

Definition 6. *A probabilistic automaton is a tuple $(S, L, \rightarrow, \ell)$ consisting of*

- *a nonempty finite set S of states,*
- *a nonempty finite set L of labels,*
- *a finite total transition relation $\rightarrow \subseteq S \times \mathcal{D}(S)$, and*
- *a labelling function $\ell : S \rightarrow L$.*

Note that for simplicity we suppose the transition relation $\to$ to be total, that is, for each $s \in S$ there exists a $\mu \in \mathcal{D}(S)$ such that $(s, \mu) \in \to$.

For the remainder of this section, we fix a probabilistic automaton $(S, L, \to, \ell)$. Instead of $(s, \mu) \in \to$, we will write $s \to \mu$, where $s \in S$ and $\mu \in \mathcal{D}(S)$. Next, we introduce the notion of probabilistic bisimilarity for probabilistic automata due to Segala and Lynch [52]. To define this notion, we first show how to lift a relation on states to a relation on probability distributions on states.

Definition 7. *The lifting of a relation $R \subseteq S \times S$ is the relation $\bar{R} \subseteq \mathcal{D}(S) \times \mathcal{D}(S)$ defined by $\mu_1 \bar{R} \mu_2$ if there exists a coupling $\omega \in \Omega_{\mu_1,\mu_2}$ such that $\omega(s_1, s_2) > 0$ implies $s_1 R s_2$ for all $s_1, s_2 \in S$.*

This notion of lifting can be found, for example, in [37, Definition 4.3]. It can be used to define probabilistic bisimilarity as follows.

Definition 8. *A relation $R \subseteq S \times S$ is a probabilistic bisimulation if $s_1 R s_2$ implies*

- $\ell(s_1) = \ell(s_2)$,
- *if $s_1 \to \mu_1$ then there exists $s_2 \to \mu_2$ such that $\mu_1 \bar{R} \mu_2$,*
- *if $s_2 \to \mu_2$ then there exists $s_1 \to \mu_1$ such that $\mu_1 \bar{R} \mu_2$.*

States s_1 and s_2 are probabilistic bisimilar, denoted $s_1 \sim s_2$, if $s_1 R s_2$ for some probabilistic bisimulation R.

Segala and Lynch also introduced another notion of probabilistic bisimilarity for probabilistic automata. That notion is obtained from the above definition by replacing probabilistic transitions with convex combinations of probabilistic transitions. These transitions are also known as combined transitions [52] and mixed transitions [14,21,31]. These convex combinations of probabilistic transitions correspond to randomized schedulers, whereas probabilistic transitions correspond to deterministic schedulers. For a detailed discussion of both notions of probabilistic bisimilarity we refer the reader to Segala's thesis [51, Chapter 8].

As we already mentioned in Section 1, a behavioural pseudometric can be defined by means of a real valued interpretation of a logic and also in terms of a terminal coalgebra. As shown by Desharnais, Gupta, Jagadeesan and Panangaden in [25], a behavioural pseudometric can also be defined as a fixed point. In [11], the authors in collaboration with Hermida and Makkai showed that all three approaches give rise to the same pseudometric for labelled Markov systems. Here, we will use the fixed point approach. As we will see below, the bisimilarity pseudometric is defined as the least fixed point of the following functional.

Definition 9. *The function $\Delta : [0,1]^{S \times S} \to [0,1]^{S \times S}$ is defined as follows. If $\ell(s_1) \neq \ell(s_2)$ then $\Delta(d)(s_1, s_2) = 1$. Otherwise*

$$\Delta(d)(s_1, s_2) = \mathcal{P}(\mathcal{D}(d))(\{\, \mu_1 \mid s_1 \to \mu_1 \,\}, \{\, \mu_2 \mid s_2 \to \mu_2 \,\}).$$

From the facts that $\mathcal{P}$ and $\mathcal{D}$ are monotone, we can conclude that Δ is monotone as well. According to Tarski's fixed point theorem [57], a monotone function on a complete lattice has a least fixed point. Hence, Δ has a least fixed point, which we denote by δ. This is the bisimilarity pseudometric introduced by Deng *et al.* in [22].

De Alfaro *et al.* [21], Chatterjee *et al.* [14] and Fu [31] consider a behavioural pseudometric for probabilistic automata that generalizes the notion of probabilistic bisimilarity defined in terms of combined transitions. Their pseudometric is defined as the least fixed point of the functional obtained from the one defined in Definition 9 by replacing the sets of transitions with their convex closures. To illustrate the difference between this pseudometric and the one we study, we consider the probabilistic automaton below. In our pseudometric, the states 1 and 2 are $\frac{1}{2}$ apart, whereas they have distance zero if we consider combined transitions.

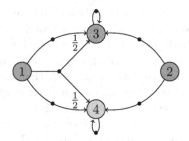

From the facts that $\mathcal{P}$ and $\mathcal{D}$ are nonexpansive, we can conclude that Δ is nonexpansive as well. Since Δ is monotone and nonexpansive, we can conclude from [10, Corollary 1] that the closure ordinal of Δ is ω, that is, δ is the least upper bound of $\{\,\Delta^n(\mathbf{0}) \mid n \in \mathbb{N}\,\}$, where the distance function $\mathbf{0}$ maps every pair of states to zero. The latter characterization of δ allows for inductive proofs. For example, to conclude that δ is a pseudometric, it suffices to prove by induction on n that $\Delta^n(\mathbf{0})$ is a pseudometric.

As shown by Deng *et al.* [22, Corollary 2.14], the pseudometric δ is a quantitative generalization of probabilistic bisimilarity, since distance zero coincides with probabilistic bisimilarity.

Theorem 10. *For all s_1, $s_2 \in S$, $\delta(s_1, s_2) = 0$ if and only if $s_1 \sim s_2$.*

5   Simple Stochastic Games

In this section we present background on simple stochastic games in preparation for our main results in Section 6 and 7.

Definition 11. *A simple stochastic game is a tuple (V, E, P) consisting of*

- *a finite directed graph (V, E) such that*
  - *V is partitioned into the sets*

  * $V_{\max}$ *of max vertices,*
  * $V_{\min}$ *of min vertices,*
  * V_{rnd} *of random vertices,*
  * V_0 *of 0-sinks, and*
  * V_1 *of 1-sinks,*
- *the vertices in V_0 and V_1 have outdegree zero and all other vertices have outdegree at least one,*
- *a function $P : V_{\mathrm{rnd}} \to \mathcal{D}(V)$ such that for each vertex $v \in V_{\mathrm{rnd}}$, $P(v)(w) > 0$ iff $(v, w) \in E$.*

The above definition is slightly more general than the one given by Condon in [19] and described in Section 2. Note that the outdegree of min, max and random vertices is at least one (instead of exactly two), there may be multiple 0-sinks and 1-sinks (rather than exactly one), and the outgoing edges of a random vertex are labelled with rationals (rather than $\frac{1}{2}$). However, a simple stochastic game as defined above can be transformed in polynomial-time into a simple stochastic game as defined in Section 2, as shown by Zwick and Paterson [62, Section 6].

Let $\mathcal{G}$ be a simple stochastic game. A *strategy* for Player 0 is a function $\sigma_0 : V_{\min} \to E$ that assigns an outgoing edge to each min vertex. Likewise a strategy for Player 1 is a function $\sigma_1 : V_{\max} \to E$ that assigns an outgoing edge to each max vertex. These strategies are known as pure stationary strategies. We can restrict ourselves to these strategies since both players of a simple stochastic game have optimal strategies of this type (see, for example, [44]).

Such strategies determine a sub-game, denoted $\mathcal{G}_{\sigma_0,\sigma_1}$, in which each max vertex and each min vertex has outdegree one (see [19, Section 2] for details). Such a game can naturally be viewed as a Markov chain. We write $\phi_{\sigma_0,\sigma_1} : V \to [0,1]$ for the function that gives the probability of a vertex in this Markov chain to reach a 1-sink.

The *value function* $\phi : V \to [0,1]$ of a simple stochastic game is defined as $\min_{\sigma_0} \max_{\sigma_1} \phi_{\sigma_0,\sigma_1}$. It is folklore that the value function of a simple stochastic game can be characterised as the least fixed point of the following monotone function (see, for example, [38, Section 2.2 and 2.3]).

Definition 12. *The function $\Phi : [0,1]^V \to [0,1]^V$ is defined by*

$$\Phi(f)(v) = \begin{cases} 0 & \text{if } v \in V_0 \\ 1 & \text{if } v \in V_1 \\ \max\{ f(w) \mid w \in V \wedge (v,w) \in E \} & \text{if } v \in V_{\max} \\ \min\{ f(w) \mid w \in V \wedge (v,w) \in E \} & \text{if } v \in V_{\min} \\ \sum_{w:(v,w)\in E} P(v)(w)\, f(w) & \text{if } v \in V_{\mathrm{rnd}} \end{cases}$$

One can prove that Φ is nonexpansive. Since Φ is also monotone, we can conclude from [10, Corollary 1] that the closure ordinal of Φ is ω, that is, ϕ is the least upper bound of $\{ \Phi^n(\mathbf{0}) \mid n \in \mathbb{N} \}$, where the function $\mathbf{0}$ maps every vertex to zero. Again, this characterisation allows for inductive proofs.

We say that a simple stochastic game $\mathcal{G}$ is a *stopping game* if for every Player 0 strategy σ_0 and Player 1 strategy σ_1, each vertex in $\mathcal{G}_{\sigma_0,\sigma_1}$ can reach a sink.

That is, $\mathcal{G}$ terminates with probability one. While Φ need not have a unique fixed point in general, it does in case $\mathcal{G}$ is a stopping game (see [19, Lemma 1]). Stopping games play a key role in Section 7.

The following result is due to Etessami and Yannakakis [28]. This result can already be found in Juba's thesis [38, Theorem 5], although it should be mentioned that there is an oversight in his proof.

Theorem 13. *The problem of computing the value function of a simple stochastic game is in* **PPAD**.

As we will explain in Section 7, the proof of the **PPAD** bound for computing the bisimilarity pseudometric involves an appropriate generalisation of the proof of Theorem 13, as given in [29, Section 5].

6   The Bisimilarity Game

Fix a probabilistic automaton $\mathcal{A}$ and write δ for the bisimilarity pseudometric on $\mathcal{A}$. We will characterise δ in terms of a certain simple stochastic game $\mathcal{G}$, which we call the *bisimilarity game*. This properly generalises the well-known game-theoretic characterisation of bisimilarity in the non-probabilistic setting (see, for example, [55, Section 3.2] for details). The latter can be recovered in the special case that all probabilistic transitions in $\mathcal{A}$ are Dirac distributions.

In the bisimilarity game $\mathcal{G}$ we think of Player 0 as trying to show that two states are probabilistic bisimilar, while Player 1 tries to prove that they are not probabilistic bisimilar. There is a vertex v_{s_1,s_2} for each pair of automaton states s_1 and s_2. If $\ell(s_1) \neq \ell(s_2)$ then the vertex is a 1-sink. Otherwise, Player 1 selects a transition, say $s_1 \to \mu_1$, and Player 0 responds by selecting a matching transition $s_2 \to \mu_2$ and a vertex $\omega \in V(\Omega_{\mu_1,\mu_2})$. Then play proceeds to a new vertex $v_{s_1',s_2'}$ chosen according to the distribution ω.

In the following, we use $M = \{ \mu \in \mathcal{D}(S) \mid \exists s \in S : s \to \mu \}$ for the set of probabilistic transitions occurring in automaton $\mathcal{A}$. The formal definition of the bisimilarity game is as follows.

Vertices. For all s_1, $s_2 \in S$, there is a vertex v_{s_1,s_2}. If $\ell(s_1) \neq \ell(s_2)$ then this vertex is a 1-sink; otherwise it is a max vertex. For all $s \in S$ and $\mu \in M$, there is a min vertex $v_{\mu,s}$. For all μ_1, $\mu_2 \in M$, there is a min vertex v_{μ_1,μ_2}. Furthermore, for all $\omega \in V(\Omega_{\mu_1,\mu_2})$, there is a random vertex v_ω.

Edges. There is an edge from each max vertex v_{s_1,s_2} to each min vertex v_{μ_1,s_2} such that $s_1 \to \mu_1$ and an edge to each min vertex v_{μ_2,s_1} such that $s_2 \to \mu_2$. There is an edge from each min vertex v_{μ_1,s_2} to each min vertex v_{μ_1,μ_2} such that $s_2 \to \mu_2$. There is an edge from each min vertex v_{μ_1,μ_2} to each random vertex v_ω such that $\omega \in V(\Omega_{\mu_1,\mu_2})$. Finally, there is an edge from each random vertex v_ω to each vertex v_{s_1,s_2} such that (s_1, s_2) is in the support of ω, that is, $\omega(s_1, s_2) > 0$. The probability of this edge is $\omega(s_1, s_2)$.

By construction of bisimilarity game $\mathcal{G}$, there is a direct correspondence between the function Φ from Definition 12 associated to $\mathcal{G}$ and the function Δ from Definition 9 associated to the automaton $\mathcal{A}$. From this correspondence it is straightforward that the respective least fixed points of Φ and Δ agree.

Theorem 14. *For all s_1, $s_2 \in S$, $\phi(v_{s_1,s_2}) = \delta(s_1, s_2)$.*

Proof. Since ϕ and δ are the least upper bounds of $\{\Phi^n(\mathbf{0}) \mid n \in \mathbb{N}\}$ and $\{\Delta^n(\mathbf{0}) \mid n \in \mathbb{N}\}$, respectively, it suffices to show that for all s_1, $s_2 \in S$ and $n \in \mathbb{N}$,

$$\Phi^{4n}(\mathbf{0})(v_{s_1,s_2}) = \Delta^n(\mathbf{0})(s_1, s_2)$$

by induction on n. Obviously, the above holds if $n = 0$. Let $n > 0$. We distinguish the following cases.

- If $\ell(s_1) \neq \ell(s_2)$ then the vertex v_{s_1,s_2} is a 1-sink and, hence,

$$\Phi^{4n}(\mathbf{0})(v_{s_1,s_2}) = 1 = \Delta^n(\mathbf{0})(s_1, s_2).$$

- If $\ell(s_1) = \ell(s_2)$ then

$$\Phi^{4n}(\mathbf{0})(v_{s_1,s_2})$$

$$= \max \left\{ \max_{s_1 \to \mu_1} \Phi^{4n-1}(\mathbf{0})(v_{\mu_1,s_2}), \max_{s_2 \to \mu_2} \Phi^{4n-1}(\mathbf{0})(v_{\mu_2,s_1}) \right\}$$

$$= \max \left\{ \max_{s_1 \to \mu_1} \min_{s_2 \to \mu_2} \Phi^{4n-2}(\mathbf{0})(v_{\mu_1,\mu_2}), \max_{s_2 \to \mu_2} \min_{s_1 \to \mu_1} \Phi^{4n-2}(\mathbf{0})(v_{\mu_2,\mu_1}) \right\}$$

$$= \max \left\{ \max_{s_1 \to \mu_1} \min_{s_2 \to \mu_2} \min_{w \in V(\Omega_{\mu_1,\mu_2})} \Phi^{4n-3}(\mathbf{0})(v_w), \right.$$

$$\left. \max_{s_2 \to \mu_2} \min_{s_1 \to \mu_1} \min_{w \in V(\Omega_{\mu_2,\mu_1})} \Phi^{4n-3}(\mathbf{0})(v_w) \right\}$$

$$= \max \left\{ \max_{s_1 \to \mu_1} \min_{s_2 \to \mu_2} \min_{w \in V(\Omega_{\mu_1,\mu_2})} \sum_{s_1',s_2' \in S} w(s_1', s_2')\, \Phi^{4n-4}(\mathbf{0})(v_{s_1',s_2'}), \right.$$

$$\left. \max_{s_2 \to \mu_2} \min_{s_1 \to \mu_1} \min_{w \in V(\Omega_{\mu_2,\mu_1})} \sum_{s_1',s_2' \in S} w(s_1', s_2')\, \Phi^{4n-4}(\mathbf{0})(v_{s_1',s_2'}) \right\}$$

$$= \max \left\{ \max_{s_1 \to \mu_1} \min_{s_2 \to \mu_2} \min_{w \in V(\Omega_{\mu_1,\mu_2})} \sum_{s_1',s_2' \in S} w(s_1', s_2')\, \Delta^{n-1}(\mathbf{0})(s_1', s_2'), \right.$$

$$\left. \max_{s_2 \to \mu_2} \min_{s_1 \to \mu_1} \min_{w \in V(\Omega_{\mu_2,\mu_1})} \sum_{s_1',s_2' \in S} w(s_1', s_2')\, \Delta^{n-1}(\mathbf{0})(s_1', s_2') \right\}$$

[induction hypothesis]

$$= \max \left\{ \max_{s_1 \to \mu_1} \min_{s_2 \to \mu_2} \mathcal{D}(\Delta^{n-1}(\mathbf{0}))(\mu_1, \mu_2), \max_{s_2 \to \mu_2} \min_{s_1 \to \mu_1} \mathcal{D}(\Delta^{n-1}(\mathbf{0}))(\mu_2, \mu_1) \right\}$$

[Corollary 5]

$$= \Delta^n(\mathbf{0})(s_1, s_2).$$

□

Thus the distance between automaton states s_1 and s_2 in the bisimilarity pseudometric is the value of the vertex v_{s_1,s_2} in the bisimilarity game.

We remark that the translation from the probabilistic automaton $\mathcal{A}$ to the bisimilarity game $\mathcal{G}$ incurs an exponential blow-up in general. This is because the convex polytope Ω_{μ_1,μ_2} for two given distributions μ_1, $\mu_2 \in M$ can have up to $n!$ vertices, where n is the number of states in $\mathcal{A}$. For example, if μ_1 and μ_2 are both the uniform distribution on S, then the set of couplings Ω_{μ_1,μ_2} is (up to scaling by $\frac{1}{n}$) the set of doubly stochastic $n \times n$ matrices. Here it is well-known that the vertices are the permutation matrices (see, for example, [53, Theorem 8.4]).

In general the bisimilarity game $\mathcal{G}$ need not be a stopping game. For each $\varepsilon \in [0,1)$ we define a *discounted version* $\mathcal{G}^\varepsilon$ by adding a new 0-sink, scaling all probabilities out of the random vertices by ε, and sending the remaining mass $1 - \varepsilon$ to the new 0-sink. Clearly, $\mathcal{G}^\varepsilon$ is a stopping game. We will use this game in the next section.

7   Computing the Bisimilarity Pseudometric Is in PPAD

Let $\mathcal{A} = (S, L, \rightarrow, \ell)$ be a probabilistic automaton. In this section we show that for each pair of states s_1, $s_2 \in S$ the distance $\delta(s_1, s_2)$ is a rational number that can moreover be computed in **PPAD**. To prove this we make use of the characterisation of δ in terms of the bisimilarity game $\mathcal{G}$. Notwithstanding the exponential blow-up in going from $\mathcal{A}$ to $\mathcal{G}$, with some effort we are able to adapt the technique used to obtain a **PPAD** bound for computing the value of simple stochastic games in [28] to computing the bisimilarity pseudometric.

Our complexity bounds assume that the rational transition probabilities in $\mathcal{A}$ are encoded as pairs of integers, with each integer given in binary. We write $||\mathcal{A}||$ for the length of the representation of $\mathcal{A}$.

Theorem 15. *For all s_1, $s_2 \in S$, $\delta(s_1, s_2)$ is a rational number whose size is bounded by a polynomial in $||\mathcal{A}||$.*

Proof. As we have observed in Section 5, the number of vertices of the bisimilarity game $\mathcal{G}$ need not be bounded by a polynomial in $||\mathcal{A}||$. Note, however, that for each Player 0 strategy σ_0 and Player 1 strategy σ_1, the subgame $\mathcal{G}_{\sigma_0,\sigma_1}$ has a number of vertices polynomial in $||\mathcal{A}||$. In particular, each min vertex of the form v_{μ_1,μ_2} has a single successor v_ω in the sub-game $\mathcal{G}_{\sigma_0,\sigma_1}$ rather than potentially exponentially many successors as in the overall game $\mathcal{G}$. Moreover, all transition probabilities appearing in $\mathcal{G}$ are of size bounded by a polynomial in $||\mathcal{A}||$. Indeed for each vertex $v_\omega \in V_{\text{rnd}}$, ω is a vertex of a polytope defined by a system of linear equations whose coefficients are transition probabilities in $\mathcal{A}$. Thus the size of ω is bounded by a polynomial in $||\mathcal{A}||$ (see, for example, [15, Proposition 12]).

Let σ_0^* and σ_1^* be strategies that realise the value of the bisimilarity game in each vertex. The value of each vertex can be expressed as the probability to reach a 1-sink in the sub-game $\mathcal{G}_{\sigma_0^*,\sigma_1^*}$. As shown in [19, Lemma 2] this value is rational and has size bounded by a polynomial in $\mathcal{G}_{\sigma_0^*,\sigma_1^*}$ (and therefore also $\|\mathcal{A}\|$ by the above considerations).

Finally, recall from Theorem 14 that the bisimilarity distance $\delta(s_1, s_2)$ of any pair of automaton states $s_1, s_2 \in S$ is the value of the vertex v_{s_1,s_2} in the bisimilarity game. The result immediately follows. $\qquad\square$

Recall from Section 6 the bisimilarity game $\mathcal{G}$ derived from the probabilistic automaton $\mathcal{A}$. The value of this game was defined as the least fixed point of the associated functional Φ. Following the proof of Theorem 14 note that for a max vertex v_{s_1,s_2} we have

$$\Phi^2(f)(v_{s_1,s_2}) = \max \left\{ \max_{s_1 \to \mu_1} \min_{s_2 \to \mu_2} f(v_{\mu_1,\mu_2}), \max_{s_2 \to \mu_2} \min_{s_1 \to \mu_1} f(v_{\mu_2,\mu_1}) \right\} \quad (1)$$

and for a min vertex v_{μ_1,μ_2} we have

$$\Phi^2(f)(v_{\mu_1,\mu_2}) = \min_{\omega \in V(\Omega_{\mu_1,\mu_2})} L_\omega(f) \quad (2)$$

where $L_\omega(f) = \displaystyle\sum_{s_1',s_2' \in S} \omega(s_1', s_2') f(s_1', s_2')$.

We introduce

$$U_1 = \{ v_{s_1,s_2} \mid v_{s_1,s_2} \in V_1 \}$$
$$U_{\max} = \{ v_{s_1,s_2} \mid v_{s_1,s_2} \in V_{\max} \}$$
$$U_{\min} = \{ v_{\mu_1,\mu_2} \mid v_{\mu_1,\mu_2} \in V_{\min} \}$$
$$U = U_1 \cup U_{\max} \cup U_{\min}$$

By a slight abuse of notation, in the rest of this section we consider Φ^2 as a self-map of $[0,1]^U$ defined by Equation (1) and (2) and $\Phi^2(f)(v_{s_1,s_2}) = 1$ for all $v_{s_1,s_2} \in U_1$.

We will show that Φ^2 is *polynomial piecewise linear* in the sense of [28]. In our context, Φ^2 is *polynomial piecewise linear* if for each rational vector $f \in [0,1]^U$ we can compute in polynomial time (in the representation of f and $\mathcal{A}$)

- a set I of linear inequalities with rational coefficients defining a cell C containing f, where the size of I and the coefficients of the inequalities in I are of size bounded by a polynomial in $\|\mathcal{A}\|$, and
- a rational matrix $A \in [0,1]^{U \times U}$ and a rational vector $b \in [0,1]^U$, where the rationals are bounded by a polynomial in $\|\mathcal{A}\|$, such that $\Phi^2(g) = Ag + b$ for all $g \in C$.

Etessami and Yannakakis [28] have shown that the fixed-point functional associated with a simple stochastic game is polynomial piecewise linear. Recall that in general we experience an exponential blow-up going from $\mathcal{A}$ to $\mathcal{G}$. Hence, we cannot allude to this result of Etessami and Yannakakis to conclude that Φ is polynomial piecewise linear. Instead, we exploit the structure of the game $\mathcal{G}$ to get polynomial complexity of Φ^2 in the original automaton $\mathcal{A}$, which may be exponentially smaller than $\mathcal{G}$.

Proposition 16. Φ^2 *is polynomial piecewise linear.*

Proof. First, we will show that it suffices to prove that $\Phi^2(-)(v) : [0,1]^U \to [0,1]$ is polynomial piecewise linear for each vertex $v \in U$. Fix a rational $f \in [0,1]^U$. Assume that for each $v \in U$ we can compute in polynomial time a set I_v of linear inequalities defining a cell $C_v \subseteq [0,1]^U$ containing f and a linear function coinciding with $\Phi^2(-)(v)$ on the cell C_v, that is, a rational vector $a_v \in [0,1]^U$ and a rational $b_v \in [0,1]$ such that $\Phi^2(g)(v) = a_v g + b_v$ for all $g \in C_v$. Suppose that the size of I_v and the rational coefficients of the linear inequalities in I_v and the rationals in a_v and the rational b_v are all of size bounded by a polynomial in $\|\mathcal{A}\|$. Now define the cell $C = \bigcap_{v \in U} C_v$. Then we can compute in polynomial time the set $\bigcup_{v \in U} I_v$ of linear inequalities defining the cell C. Obviously, C contains f. Furthermore, we can combine the vectors a_v into a matrix A and the rationals b_v into a vector b such that $\Phi^2(g) = Ag + b$ for all $g \in C$. Clearly, the size of I and the rational coefficients of the linear inequalities in I and the rationals in A and b are all of size bounded by a polynomial in $\|\mathcal{A}\|$. Hence, Φ^2 is polynomial piecewise linear.

Let $v_{s_1,s_2} \in U_1$. In this case, for all $g \in [0,1]^U$, $\Phi^2(g)(v_{s_1,s_2}) = 1$. Obviously, $\Phi^2(-)(v_{s_1,s_2})$ is polynomial piecewise linear.

Fix $v_{s_1,s_2} \in U_{\max}$ and consider the set of linear inequalities $I_{v_1,v_2} = \{f \in [0,1]^U \mid f(v_1) \leq f(v_2)\}$ for $v_1, v_2 \in U_{\min}$. There are polynomially many such inequalities (in the number of states of $\mathcal{A}$) and these inequalities define cells. Fix a rational $f \in [0,1]^U$. From Equation (1) we have that $\Phi^2(f)(v_{s_1,s_2}) = f(v)$ for some $v \in U_{\min}$. Let C be the cell containing f. For each $g \in C$, we have that for all $v_1, v_2 \in U_{\min}$,

$$f(v_1) \leq f(v_2) \text{ iff } g(v_1) \leq g(v_2).$$

As a consequence, $\Phi^2(g)(v_{s_1,s_2}) = g(v)$. Hence, $\Phi^2(-)(v_{s_1,s_2})$ is polynomial piecewise linear.

Next, fix $v_{\mu_1,\mu_2} \in U_{\min}$ and consider the set of linear inequalities $I_{\omega_1,\omega_2} = \{f \in [0,1]^U \mid L_{\omega_1}(f) \leq L_{\omega_2}(f)\}$, where $\omega_1, \omega_2 \in V(\Omega_{\mu_1,\mu_2})$. Similar to the previous case, these inequalities define cells. Again, fix a rational $f \in [0,1]^U$. Clearly, the vertex ω that minimises the term $L_{(-)}(f)$ on the right-hand side of Equation (2) is determined by the cell C that contains f. Note that the size of I_{ω_1,ω_2} for $\omega_1, \omega_2 \in V(\Omega_{\mu_1,\mu_2})$ is potentially exponential in $\|\mathcal{A}\|$. However, we will show next that each cell C can be defined by only polynomialy many inequalities, and these can be computed in polynomial time.

Using the network simplex algorithm (see, for example, [1, Chapter 11]) we can compute in polynomial time (in the representation of f, μ_1, and μ_2) a vertex $\omega \in V(\Omega_{\mu_1,\mu_2})$ that minimises the linear function $L_{(-)}(f) : V(\Omega_{\mu_1,\mu_2}) \to [0,1]$. According to, for example, [15, Proposition 12], ω consists of rationals of size bounded by a polynomial in $\|\mathcal{A}\|$. Let $\omega_1, \ldots, \omega_k$ be the adjacent vertices of ω in $V(\Omega_{\mu_1,\mu_2})$. The number of such vertices is at most the number of linear constraints defining Ω_{μ_1,μ_2}; moreover we can compute each such vertex in polynomial time from ω by performing one pivoting step of the simplex algorithm.

Now define the cell

$$C = \{\, g \in [0,1]^U \mid \forall 1 \leq i \leq k : L_\omega(g) \leq L_{\omega_i}(g) \,\}.$$

That is, C contains those vectors g for which the vertex ω is a local optimum when minimizing the linear function $L_{(-)}(g)$. Since a local optimum of a linear function on a convex polytope is a global optimum (see, for example, [34, Theorem 16.25]), for $g \in C$ we have that $L_\omega(g) \leq L_{\omega'}(g)$ for all $\omega' \in V(\Omega_{\mu_1,\mu_2})$. Hence, $\Phi^2(g)(v_{\mu_1,\mu_2}) = L_\omega(g)$ for all $g \in C$ and, therefore, $\Phi^2(-)(v_{\mu_1,\mu_2})$ is polynomial piecewise linear.   □

Theorem 17. *The problem of approximating δ is in **PPAD**.*

Proof. Recall the value function $\phi : V \to [0,1]$ of the bisimilarity game $\mathcal{G}$. By Theorem 14 we have that $\delta(s_1, s_2) = \phi(v_{s_1,s_2})$ for all $s_1, s_2 \in S$, so it will suffice to show that approximating ϕ is in **PPAD**.

Given an integer N in binary, for each pair of states $s_1, s_2 \in S$ we wish to compute a value q such that $|\phi(v_{s_1,s_2}) - q| < \frac{1}{N}$. Let us fix $s_1, s_2 \in S$ for the remainder of this proof.

The value function $\phi^\varepsilon : V \to [0,1]$ of the discounted version $\mathcal{G}^\varepsilon$ of the bisimilarity game is the least fixed point of the function $\Phi^\varepsilon : [0,1]^V \to [0,1]^V$ defined by

$$\Phi^\varepsilon(f)(v) = \begin{cases} \varepsilon \sum_{w:(v,w)\in E} P(v)(w)\, f(w) & \text{if } v \in V_{\mathrm{rnd}} \\ \Phi(f)(v) & \text{otherwise} \end{cases}$$

By [19, Lemma 8] there is some $\varepsilon \in [0,1)$ such that the value of each vertex of the bisimilarity game $\mathcal{G}$ is within $\frac{1}{N}$ of the value of the same vertex in the stopping game $\mathcal{G}^\varepsilon$. That is, $|\phi(v_{s_1,s_2}) - \phi^\varepsilon(v_{s_1,s_2})| < \frac{1}{N}$. In fact, exploiting again the fact that $\mathcal{G}_{\sigma_0,\sigma_1}$ has size polynomial in $\|\mathcal{A}\|$ for each pair of strategies σ_0 and σ_1, the argument of [19, Lemma 8] shows that ε can be chosen to have bit-length polynomial in $\|\mathcal{A}\|$ and the bit-length of N. We will take $\phi^\varepsilon(v_{s_1,s_2})$ as our approximation to $\phi(v_{s_1,s_2})$, so it remains to show that the former can be computed in **PPAD**.

From Proposition 16 and the fact that ε can be chosen to have bit-length polynomial in $\|\mathcal{A}\|$ and the bit-length of N, we can conclude that $(\Phi^\varepsilon)^2$ is polynomial piecewise linear. Thus, by [28, Theorem 23], *some* fixed point of $(\Phi^\varepsilon)^2$ can be computed in **PPAD**. But, expanding definitions (as in the proof of Theorem 14), we have $(\Phi^\varepsilon)^4(f) = \Phi^4(\varepsilon \cdot f)$. Since Φ is non-expansive it follows that $(\Phi^\varepsilon)^4$ is contractive. From Banach's fixed point theorem [5] it follows that the fixed point of $(\Phi^\varepsilon)^4$ is unique. Since any fixed point of $(\Phi^\varepsilon)^2$ is also a fixed point of $(\Phi^\varepsilon)^4$, we can conclude that $(\Phi^\varepsilon)^2$ has a unique fixed point. Because ϕ^ε is a fixed point of Φ^ε, it is also a fixed point of $(\Phi^\varepsilon)^2$. Therefore, ϕ^ε is the *unique* fixed point of $(\Phi^\varepsilon)^2$. Hence, we can conclude that ϕ^ε can be computed in **PPAD**.   □

Corollary 18. *The problem of computing δ is in **PPAD**.*

Proof. We use here the approach described by Etessami and Yannakakis [29, page 2540]. Let $N \in \mathbb{N}$ be an upper bound on the denominators of the distances $\delta(s_1, s_2)$ for $s_1, s_2 \in S$. By Theorem 15, N has bit-length polynomial in $\|\mathcal{A}\|$.

By Theorem 17, we can compute an approximation of δ to within $\frac{1}{2N}$ in **PPAD**. If we round all distances in this approximation to the nearest rational with denominator at most N then we recover δ. This rounding can be done in polynomial time using the continued fractions method (see, for example, [50, Section 6.1]). □

8   Conclusion

We have reduced the problem of computing the bisimilarity metric on probabilistic automata to that of computing the value of an associated simple stochastic game: the *bisimilarity game*. Although this reduction is exponential, we have shown that the **PPAD** complexity of simple stochastic games carries over to the bisimilarity metric. Our results rely in particular on [28, Theorem 23]. From the proof of this theorem in [28] one sees that ultimately the reduction of the computation problem for the bisimilarity metric to an end-of-line problem proceeds via Scarf's algorithm for computing approximate fixed points of continuous functionals. The latter can be seen as a *path-following* algorithm (see, e.g., [28, Section 2]).

In future work we would like to further compare the complexity of the respective computation problems for the bisimilarity metric and simple stochastic games. It is natural to ask whether there is a *polynomial* reduction of the former to the latter. In the absence of such a reduction, the complexity of computing the bisimilarity metric can naturally be studied in terms of searching for local optima, for example, via strategy improvement algorithms. Such an approach has been pursued for labelled Markov systems in [2]. Recall from Section 2 that the computation problem for simple stochastic games is known to belong to the classes **PLS** [60], **CLS** [20], and **CCLS** [20], which are all defined in terms of local-search algorithms where each step of the search can be done in polynomial time. It is thus natural to wonder if the computation problem for the bisimilarity metric also belongs in these classes.

Acknowledgements. We are thankful to Di Chen for his contributions to our collaborative research on which some of this work is based. We would also like to thank Taolue Chen, Hongfei Fu, and Paul Goldberg for discussions related to this research. We are grateful to Josée Desharnais and Jan Rutten for filling in some of the blanks in Section 1. Furthermore, we would like to thank the referees for their constructive feedback. Last but not least, we would like to thank Prakash Panangaden. He challenged us with research questions such as the ones addressed in this paper. He also provided us with feedback on numerous occasions. Furthermore, we thank him for his advise and friendship.

References

1. Ahuja, R.K., Magnanti, T.L., Orlin, J.B.: Network Flows: Theory, Algorithms and Applications. Prentice Hall, Upper Saddle River (1993)
2. Bacci, G., Bacci, G., Larsen, K.G., Mardare, R.: On-the-fly exact computation of bisimilarity distances. In: Piterman, N., Smolka, S.A. (eds.) TACAS 2013. LNCS, vol. 7795, pp. 1–15. Springer, Heidelberg (2013)
3. Baier, C., Katoen, J.-P.: Principles of Model Checking. MIT Press, Cambridge (2008)
4. de Bakker, J.W., de Vink, E.P.: Control Flow Semantics. MIT Press, Cambridge (1996)
5. Banach, S.: Sur les opérations dans les ensembles abstraits et leurs applications aux equations intégrales. Fundamenta Mathematicae 3, 133–181 (1922)
6. Bartels, F., Sokolova, A., de Vink, E.P.: A hierarchy of probabilistic system types. In: Gumm, H.P. (ed.) CMCS 2003, Warsaw, Poland. Electronic Notes in Theoretical Computer Science, vol. 82(1), pp. 57–75. Elsevier (April 2003)
7. Blute, R., Desharnais, J., Edalat, A., Panangaden, P.: Bisimulation for labelled Markov processes. In: LICS 1997, Warsaw, Poland, pp. 149–158. IEEE (June/July 1997)
8. van Breugel, F.: A note on the Hutchinson metric (July 1999)
9. van Breugel, F.: An introduction to metric semantics: Operational and denotational models for programming and specification languages. Theoretical Computer Science 258(1/2), 1–98 (2001)
10. van Breugel, F.: On behavioural pseudometrics and closure ordinals. Information Processing Letters 112(18), 715–718 (2012)
11. van Breugel, F., Hermida, C., Makkai, M., Worrell, J.: Recursively defined metric spaces without contraction. Theoretical Computer Science 380(1/2), 143–163 (2007)
12. van Breugel, F., Worrell, J.B.: Towards quantitative verification of probabilistic systems. In: Orejas, F., Spirakis, P.G., van Leeuwen, J. (eds.) ICALP 2001. LNCS, vol. 2076, pp. 421–432. Springer, Heidelberg (2001)
13. Canny, J.: Some algebraic and geometric computations in PSPACE. In: Simon, J. (ed.) STOC 1988, Chicago, IL, USA, pp. 460–467. ACM (May 1988)
14. Chatterjee, K., de Alfaro, L., Majumdar, R., Raman, V.: Algorithms for game metrics. In: Hariharan, R., Mukund, M., Vinay, V. (eds.) FSTTCS 2008, Bangalore, India. Leibniz International Proceedings in Informatics, vol. 2, pp. 107–118. Schloss Dagstuhl – Leibniz-Zentrum fuer Informatik (December 2008)
15. Chen, D., van Breugel, F., Worrell, J.: On the complexity of computing probabilistic bisimilarity. In: Birkedal, L. (ed.) FOSSACS 2012. LNCS, vol. 7213, pp. 437–451. Springer, Heidelberg (2012)
16. Chen, X., Deng, X.: Settling the complexity of two-player Nash equilibrium. In: FOCS 2006, Berkeley, CA, USA, pp. 261–272. IEEE (October 2006)
17. Chvátal, V.: Linear Programming. W.H. Freeman and Company, New York (1983)
18. Collins, G.E.: Quantifier elimination for real closed fields by cylindrical algebraic decomposition. In: Brakhage, H. (ed.) GI-Fachtagung 1975. LNCS, vol. 33, pp. 134–183. Springer, Heidelberg (1975)
19. Condon, A.: The complexity of stochastic games. Information and Computation 96(2), 203–224 (1992)

20. Daskalakis, C., Papadimitriou, C.: Continuous local search. In: Choffrut, C., Lengauer, T. (eds.) SODA 2011, San Francisco, CA, USA, pp. 790–804. SIAM (January 2011)
21. de Alfaro, L., Majumdar, R., Raman, V., Stoelinga, M.: Game relations and metrics. In: LICS 2007, Wroclaw, Poland, pp. 99–108. IEEE (July 2007)
22. Deng, Y., Chothia, T., Palamidessi, C., Pang, J.: Metrics for action-labelled quantitative transition systems. In: Cerone, A., Wiklicky, H. (eds.) QAPL 2005, Edinburgh, Scotland. Electronic Notes in Theoretical Computer Science, vol. 153(2), pp. 79–96. Elsevier (March 2005)
23. Desharnais, J., Edalat, A., Panangaden, P.: A logical characterization of bisimulation for labelled Markov processes. In: LICS 1998, Indianapolis, IN, USA, pp. 478–487. IEEE (July 1998)
24. Desharnais, J., Gupta, V., Jagadeesan, R., Panangaden, P.: Metrics for labeled Markov systems. In: Baeten, J.C.M., Mauw, S. (eds.) CONCUR 1999. LNCS, vol. 1664, pp. 258–273. Springer, Heidelberg (1999)
25. Desharnais, J., Gupta, V., Jagadeesan, R., Panangaden, P.: The metric analogue of weak bisimulation for probabilistic processes. In: LICS 2002, Copenhagen, Denmark, pp. 413–422. IEEE (July 2002)
26. Desharnais, J., Laviolette, F., Tracol, M.: Approximate analysis of probabilistic processes: logic, simulation and games. In: QEST 2008, Saint-Malo, France, pp. 264–273. IEEE (September 2008)
27. Edgar, G.A.: Integral, Probability, and Fractal Measures. Springer, New York (1998)
28. Etessami, K., Yannakakis, M.: On the complexity of Nash equilibria and other fixed points (extended abstract). In: FOCS 2007, Providence, RI, USA, pp. 113–123. IEEE (October 2007)
29. Etessami, K., Yannakakis, M.: On the complexity of Nash equilibria and other fixed points. SIAM Journal on Computing 39(6), 2531–2597 (2010)
30. Ferns, N., Panangaden, P., Precup, D.: Metrics for Markov decision processes with infinite state spaces. In: UAI 2005, Edinburgh, Scotland, pp. 201–208. AUAI Press (July 2005)
31. Fu, H.: Computing game metrics on Markov decision processes. In: Czumaj, A., Mehlhorn, K., Pitts, A., Wattenhofer, R. (eds.) ICALP 2012, Part II. LNCS, vol. 7392, pp. 227–238. Springer, Heidelberg (2012)
32. Fu, H.: Personal communication (January 2013)
33. Giacalone, A., Jou, C.-C., Smolka, S.: Algebraic reasoning for probabilistic concurrent systems. In: PROCOMET 1990, Sea of Gallilee, Israel, pp. 443–458. North-Holland (April 1990)
34. Har-Peled, S.: Geometric Approximation Algorithms. American Mathematical Society, Providence (2011)
35. Hutchinson, J.: Fractals and self similarity. Indiana University Mathematics Journal 30(5), 713–747 (1981)
36. Jacobs, B.: Introduction to Coalgebra (2012)
37. Jonsson, B., Larsen, K.G.: Specification and refinement of probabilistic processes. In: LICS 1991, Amsterdam, The Netherlands, pp. 266–277. IEEE (July 1991)
38. Juba, B.: On the hardness of simple stochastic games. Master's thesis, Carnegie Mellon University, Pittsburgh, PA, USA (May 2005)
39. Kantorovich, L.: On the transfer of masses. Doklady Akademii Nauk 5(1), 1–4 (1942) (in Russian); Translated in Management Science 5(1), 1–4 (October 1958)

40. Kantorovich, L.V., Rubinstein, G.S.: On the space of completely additive functions. Vestnik Leningradskogo Universiteta 3(2), 52–59 (1958) (in Russian)

41. Kintali, S., Poplawski, L.J., Rajaraman, R., Sundaram, R., Teng, S.-H.: Reducibility among fractional stability problems. In: FOCS 2009, Atlanta, GA, USA, pp. 283–292. IEEE (October 2009)

42. Larsen, K., Skou, A.: Bisimulation through probabilistic testing. In: POPL 1989, Austin, TX, USA, pp. 344–352. ACM (January 1989)

43. Lemke, C., Howson, J.J.T.: Equilibrium points of bimatrix games. Journal of the Society for Industrial and Applied Mathematics 12(2), 413–423 (1964)

44. Liggett, T.M., Lippman, S.A.: Stochastic games with perfect information and time average payoff. SIAM Review 11(4), 604–607 (1969)

45. Mitzenmacher, M., Upfal, E.: Probability and Computing: Randomized Algorithms and Probabilitic Analysis. Cambridge University Press, Cambridge (2005)

46. Panangaden, P.: Labelled Markov Processes. Imperial College Press, London (2009)

47. Papadimitriou, C.H.: On the complexity of the parity argument and other inefficient proofs of existence. Journal of Computer and System Sciences 48(3), 498–532 (1994)

48. Rutten, J.: Relators and metric bisimulations. In: Jacobs, B., Moss, L., Reichel, H., Rutten, J. (eds.) CMCS 1998, Lisbon, Portugal. Electronic Notes in Theoretical Computer Science, vol. 11, pp. 252–258. Elsevier (March 1998)

49. Scarf, H.: The approximation of fixed points of a continuous mapping. Journal of the Society for Industrial and Applied Mathematics 15(5), 1328–1343 (1967)

50. Schrijver, A.: Theory of Linear and Integer Programming. John Wiley & Sons, Chichester (1986)

51. Segala, R.: Modeling and Verification of Randomized Distributed Real-Time Systems. PhD thesis, Massachusetts Institute of Technology, Cambridge, MA, USA (June 1995)

52. Segala, R., Lynch, N.: Probabilistic simulations for probabilistic processes. In: Jonsson, B., Parrow, J. (eds.) CONCUR 1994. LNCS, vol. 836, pp. 481–496. Springer, Heidelberg (1994)

53. Serre, D.: Matrices: Theory and Applications. Springer, New York (2010)

54. Shapley, L.S.: Stochastic games. Proceedings of the Academy of Sciences 39(10), 1095–1100 (1953)

55. Stirling, C.: Modal and Temporal Properties of Processes. Springer, New York (2001)

56. Tarski, A.: A decision method for elementary algebra and geometry. University of California Press, Berkeley (1951)

57. Tarski, A.: A lattice-theoretic fixed point theorem and its applications. Pacific Journal of Mathematics 5(2), 285–309 (1955)

58. Tracol, M., Desharnais, J., Zhioua, A.: Computing distances between probabilistic automata. In: Massink, M., Norman, G. (eds.) QAPL 2011, Saarbrücken, Germany. Electronic Proceedings in Theoretical Computer Science, vol. 57, pp. 148–162. Elsevier (April 2011)

59. Worrell, J.: Coinduction for recursive data types: partial orders, metric spaces and Omega-categories. In: Reichel, H. (ed.) CMCS 2000, Berlin, Germany. Electronic Notes in Theoretical Computer Science, vol. 33, pp. 337–356. Elsevier (March 2000)

60. Yannakakis, M.: The analysis of local search problems and their heuristics. In: Chof-frut, C., Lengauer, T. (eds.) STACS 1990. LNCS, vol. 415, pp. 298–311. Springer, Heidelberg (1990)
61. Yannakakis, M.: Computational Aspects of Equilibria. In: Mavronicolas, M., Papadopoulou, V.G. (eds.) SAGT 2009. LNCS, vol. 5814, pp. 2–13. Springer, Heidelberg (2009)
62. Zwick, U., Paterson, M.: The complexity of mean payoff games. In: Li, M., Du, D.-Z. (eds.) COCOON 1995. LNCS, vol. 959, pp. 1–10. Springer, Heidelberg (1995)

From Haar to Lebesgue via Domain Theory

Will Brian and Michael Mislove

Department of Mathematics,
Tulane University,
New Orleans, LA 70118
{wbrian,mislove}@tulane.edu

Abstract. If $C \simeq 2^{\mathbb{N}}$ is the Cantor set realized as the infinite product of two-point groups, then a folklore result says the Cantor map from C into $[0, 1]$ sends Haar measure to Lebesgue measure on the interval. In fact, C admits many distinct topological group structures. In this note, we show that the Haar measures induced by these distinct group structures are all the same. We prove this by showing that Haar measure for any group structure is the same as Haar measure induced by a related abelian group structure. Moreover, each abelian group structure on C supports a natural total order that determines a map onto the unit interval that is monotone, and hence sends intervals in C to subintervals of the unit interval. Using techniques from domain theory, we show this implies this map sends Haar measure on C to Lebesgue measure on the interval, and we then use this to prove any two group structures on C have the same Haar measure.

Keywords: Cantor set, Cantor map, compact group, Haar measure, Lebesgue measure, Stone duality.

1 Introduction

The discovery of the middle-third Cantor set in the late 1800s led to the first construction of a continuous map of the unit interval onto itself whose derivative is zero almost everywhere. Another remarkable – in fact, folklore – result about the Cantor set is that the restriction of the same map to the Cantor set sends Haar measure on the compact group $2^{\mathbb{N}}$ to Lebesgue measure on the interval [14]. In this note we generalize this result to any compact totally disconnected second countable infinite group. Any topological group structure on the Cantor set is the strict projective limit of finite groups, and conversely, the limit of a countable projective system of finite groups is a topological group on the Cantor set. In fact, any compact totally disconnected second countable group is either finite or a strict projective limit of a countable family of finite groups.

Any locally compact group admits a unique (up to scalar factor) translation-invariant Borel measure called *Haar measure*, and Haar measure is finite (and hence normalized to be a probability measure) iff the group is compact. For example, Haar measure on $(\mathbb{R}, +)$ is Lebesgue measure, and Haar measure on any discrete group is counting measure. There are two main results in this paper:

F. van Breugel et al. (Eds.): Panangaden Festschrift, LNCS 8464, pp. 214–228, 2014.
© Springer International Publishing Switzerland 2014

the first is that any two topological group structures on the Cantor set have the same Haar measure, and the second is that the natural map from the Cantor set to the unit interval sends Haar measure to Lebesgue measure. To prove the first of these results, we first show that any strict projective system of finite groups can be replaced by a system of finite abelian groups, so that each of the replacement groups has the same cardinality as the corresponding group in the original projective system. Since the probability functor is continuous on compact Hausdorff spaces, Haar measure on the limit of a projective system of finite groups is the limit of the Haar measures on the finite groups. Any two finite groups of the same cardinality have the same Haar measure, so this implies Haar measure on the limit of the projective system of finite abelian groups is the same as Haar measure on the limit of the original projective system.

The advantage of a projective system of finite abelian groups is that each is a product of cyclic groups, which allows us to define a total order on each of these groups relative to which the projection maps from larger to smaller groups are monotone. This implies these total orders induce a complete total order on the limit, the Cantor set C, and from this it follows that the natural map from C onto the unit interval is monotone and Lawson continuous, if we view C as a continuous lattice. Using domain theory, we then show that Haar measure on the Cantor set assigns the same length to each interval in C as Lebesgue measure assigns to the image of the interval under the map, which implies that the map sends Haar measure on the Cantor set to Lebesgue measure on the unit interval.

1.1 Outline of the Results

Our focus is on the Cantor set C, which can be defined abstractly as a second countable perfect Stone space, i.e., a compact Hausdorff perfect zero-dimensional space that has a countable base for its topology. Here *perfect* means every point is a limit point; second countability implies C is the projective limit of a countable family of finite sets. We will study two additional structures with which C can be endowed:

(1) The structure of a topological group – the leading example is $C \simeq 2^{\mathbb{N}}$, the infinite product of two-point groups, but like $2^{\mathbb{N}}$, any topological group structure on C can be realized as the strict projective limit of a countable system of finite groups and group homomorphisms, and

(2) A total order relative to which C is complete lattice.

Because the probability functor on compact Hausdorff spaces is continuous, viewing the Cantor set C as a compact group that is the strict projective limit of finite groups, C_n implies that Haar measure on C is the limit of the Haar measures on the C_ns, where Haar measure on each C_n has the uniform distribution. We show we can replace any topological group structure on C_n with an "equivalent" abelian group structure, in the sense that the Haar measure is the same for both groups. As a finite abelian group, the replacement group structure is isomorphic to a finite product of cyclic groups, and we show that we can construct the replacement group $\mathcal{C}_n$ so that it satisfies $\mathcal{C}_n \simeq \bigoplus_{k \leq n} \mathbb{Z}_{a_k}$ is a direct product of n finite cyclic groups.

Since $\mathbb{Z}_k$ admits a natural total order for each k, this allows us to define the lexicographic order on $\mathcal{C}_n \simeq \bigoplus_{k \leq n} \mathbb{Z}_{a_k}$ for each n, and then the quotient mapping $\mathcal{C}_m \to \mathcal{C}_n$ is monotone for each $n \leq m$. These total orders therefore induce a complete total order on $\mathcal{C}$, which means $\mathcal{C}$ is a complete lattice in this order. Then the topology on $\mathcal{C}$ is the Lawson topology from the theory of continuous lattices.

Applying Stone duality allows us to interpret each finite quotient $\mathcal{C}_n$ as a partition of $\mathcal{C}$ into subintervals, and then Haar measure on $\mathcal{C}_n$ assigns equal lengths to each of these intervals. Next, we show that there is a natural map from $\mathcal{C}$ to $[0, 1]$ that is monotone and Lawson continuous. We show this assigns the same length to each subinterval of $\mathcal{C}$ determined by $\mathcal{C}_n$ as Lebesgue measure assigns to its image in $[0, 1]$.

The final piece of the puzzle relies on verifying that the length Haar measure on $\mathcal{C}$ assigns to each closed subinterval is the same as the length that Lebesgue measure assigns to its image in $[0, 1]$. Since both measures are continuous (i.e., they assign measure 0 to points), and the clopen (= closed and open) intervals in $\mathcal{C}$ map to the closed intervals in $[0, 1]$, inner regularity implies these measures assign the same measure to open intervals, and it follows that the image of Haar measure on $\mathcal{C}$ under the natural map is Lebesgue measure on the interval.

1.2   The Plan of the Paper

In the next section, we review some background material from domain theory and from the theory of compact abelian groups. Most of the latter is well-known, but we include some proofs for completeness sake. The treatment of domain theory includes a version of Stone duality. The following section constitutes the main part of the paper, where we analyze the Cantor set when it is equipped with an arbitrary abelian topological group structure making it a topological group.

2   Background

In this section we present the background material we need for our main results.

2.1   Domains

Our results rely fundamentally on domain theory. Most of the results that we quote below can be found in [2] or [4]; we give specific references for those that are not found there.

To start, a *poset* is a partially ordered set. A poset is *directed complete* if each of its directed subsets has a least upper bound; here a subset S is *directed* if each finite subset of S has an upper bound in S. A directed complete partial order is called a *dcpo*. The relevant maps between dcpos are the monotone maps that also preserve suprema of directed sets; these maps are usually called *Scott continuous*.

These notions can be presented from a purely topological perspective: a subset $U \subseteq P$ of a poset is *Scott open* if (i) $U = \uparrow U \equiv \{x \in P \mid (\exists u \in U)\, u \leq x\}$ is an upper set, and (ii) if $\sup S \in U$ implies $S \cap U \neq \emptyset$ for each directed subset $S \subseteq P$. It is routine to show that the family of Scott-open sets forms a topology on any poset; this topology satisfies $\downarrow x \equiv \{y \in P \mid y \leq x\} = \overline{\{x\}}$ is the closure of a point, so the Scott topology is always T_0, but it is T_1 iff P is a flat poset.[1] A mapping between dcpos is Scott continuous in the order-theoretic sense iff it is a monotone map that is continuous with respect to the Scott topologies on its domain and range.

If P is a dcpo, and $x, y \in P$, then x *approximates* y iff for every directed set $S \subseteq P$, if $y \leq \sup S$, then there is some $s \in S$ with $x \leq s$. In this case, we write $x \ll y$ and we let $\downarrow y = \{x \in P \mid x \ll y\}$. A *basis* for a poset P is a family $B \subseteq P$ satisfying $\downarrow y \cap B$ is directed and $y = \sup(\downarrow y \cap B)$ for each $y \in P$. A *continuous poset* is one that has a basis, and a dcpo P is a *domain* if P is a continuous dcpo. An element $k \in P$ is *compact* if $x \ll x$, and P is *algebraic* if $KP = \{k \in P \mid k \ll k\}$ forms a basis. Domains are sober spaces in the Scott topology.

Domains also have a Hausdorff refinement of the Scott topology which will play a role in our work. The *weak lower topology* on P has the sets of the form if $O = P \setminus \uparrow F$ as a basis, where $F \subset P$ is a finite subset. The *Lawson topology* on a domain P is the common refinement of the Scott- and weak lower topologies on P. This topology has the family

$$\{U \setminus \uparrow F \mid U \text{ Scott open \& } F \subseteq P \text{ finite}\}$$

as a basis. The Lawson topology on a domain is always Hausdorff.

A domain is *coherent* if its Lawson topology is compact. We denote the closure of a subset $X \subseteq P$ of a coherent domain in the Lawson topology by $\overline{X}^\Lambda$.

Example 1. A basic example of a domain is the unit interval; here $x \ll y$ iff $x = 0$ or $x < y$. The Scott topology on the $[0,1]$ has open sets $[0,1]$ together with $\uparrow x = (x, 1]$ for $x \in (0,1]$. Since domains are closed under finite products, $[0,1]^n$ is a domain in the product order, where $x \ll y$ iff $x_i \ll y_i$ for each i; a basis of Scott-open sets is formed by the sets $\uparrow x$ for $x \in [0,1]^n$ (this last is true in any domain).

The Lawson topology on $[0,1]$ has basic open sets $(x, 1] \setminus [y, 1]$ for $x < y$ – i.e., sets of the form (x, y) for $x < y$, which is the usual topology. Then, the Lawson topology on $[0,1]^n$ is the product topology from the usual topology on $[0,1]$. This shows $[0,1]$ is a coherent domain.

Since $[0,1]$ has a least element, the same results apply for any power of $[0,1]$, where $x \ll y$ in $[0,1]^J$ iff $x_j = 0$ for almost all $j \in J$, and $x_j \ll y_j$ for all $j \in J$. Thus, every power of $[0,1]$ is a coherent domain.

Similarly, the middle-third Cantor set $\mathcal{C} \subseteq [0,1]$ is a domain in the order it inherits from $[0,1]$. But while $K[0,1] = \{0\}$, the compact elements of $\mathcal{C}$ consist

---

[1] A space X is T_0 if given any pair of points, there is an open set containing exactly one of the points; X is T_1 if $\{x\}$ is a closed set for each $x \in X$.

of the least upper bounds of the open intervals that are deleted from $[0,1]$ to form $\mathcal{C} - \frac{2}{3}, \frac{2}{9}, \frac{8}{9}, \ldots$. Thus, $y = \sup K\mathcal{C} \cap \downarrow y$ for each $y \in \mathcal{C}$, so $\mathcal{C}$ is an *algebraic* domain, in fact a complete algebraic lattice.

A more interesting example of a coherent domain is $\mathsf{Prob}(D)$, the family of probability measures on a coherent domain D, where $\mu \leq \nu$ iff $\mu(U) \leq \nu(U)$ for every Scott-open subset $U \subseteq D$. For example, $\mathsf{Prob}([0,1])$ is a coherent domain. In fact, the category COH of coherent domains and Scott continuous maps is closed under the application of the functor Prob [10].

Embedding-Projection Pairs. One of the features of domain theory is its ability to provide solutions to *domain equations* – these are abstract domains that satisfy structural requirements, most often ones needed in defining models for programming language constructs. Of course, the most famous domain equation is $D \simeq [D \to D]$, which can be solved in any of the number of Cartesian closed categories of domains. We don't need anything so sophisticated, but we can use the basic approach to solving domain equations to realize Stone spaces as algebraic lattices.

Definition 1. *Let P and Q be posets. An* embedding–projection pair *between P and Q is a pair of monotone mappings $e\colon Q \to P$ and $p\colon P \to Q$ satisfying $p \circ e = 1_Q$ and $p \circ e \leq 1_P$, where the order on functions is pointwise.*

The main result we need is the following:

Theorem 1. *Let $(P_i, e_{i,j}, p_{i,j})_{i \leq j \in I}$ be an indexed family of domains P_i and Scott-continuous e–p pairs $e_{i,j}\colon P_i \to P_j$, $p_{i,j}\colon P_j \to P_i$ for $i \leq j$. Then $P = \{(x_i)_{i \in I} \mid p_{i,j}(x_j) = x_i\}$ is a domain, and the projection maps $\pi_i\colon P \to P_i$ together with the mappings $e_i\colon P_i \to P$ by $(e_i(x))_j = \begin{cases} p_{i,j}(x) & \text{if } i \leq j \\ e_{i,j}(x_j) & \text{if } j \leq i \end{cases}$ form Scott-continuous e–p pairs. Moreover, if each P_i is algebraic, then so is P, and $KP = \bigcup_i e_i(KP_i)$.*

2.2   The Prob Monad on Comp

It is well known that the family of probability measures on a compact Hausdorff space is the object level of a functor which defines a monad on Comp, the category of compact Hausdorff spaces and continuous maps. As outlined in [7], this monad gives rise to several related monads:

- On Comp, it associates to a compact Hausdorff space X the free *barycentric algebra* over X, the name deriving from the counit $\epsilon\colon \mathsf{Prob}(S) \to S$ which assigns to each measure μ on a probabilistic algebra S its barycenter $\epsilon(\mu)$.
- On the category $\mathsf{CompMon}$ of compact monoids and continuous monoid homomorphisms, Prob gives rise to a monad that assigns to a compact monoid S the free compact affine monoid over S.

- On the category CompGrp of compact groups and continuous homomor-phisms, Prob assigns to a compact group G the free compact affine monoid over G; in this case the right adjoint sends a compact affine monoid to its group of units, as opposed to the inclusion functor, which is the right adjoint in the first two cases.

If we let SProb(X) denote the family of subprobability measures on a compact Hausdorff space X, then it's routine to show that SProb defines monads in each of the cases just described, where the only change is that the objects now have a 0 (i.e., they are affine structures with 0-element, allowing one to define scalar multiples $r \cdot x$ for $r \in [0,1]$ and $x \in$ SProb(X), as well as affine combinations).

There is a further result we need about Prob which relates to its role as an endofunctor on Comp and its subcategories. The following result is due to Fedorchuk:

Theorem 2 (Fedorchuk [3]). *The functor* Prob: Comp $\to$ Comp *is normal; in particular,* Prob *preserves inverse limits.*

2.3   Stone Duality

In modern parlance, Marshall Stone's seminal result states that the category of Stone spaces – compact Hausdorff totally disconnected spaces – and continuous maps is dually equivalent to the category of Boolean algebras and Boolean algebra maps. The dual equivalence sends a Stone space to the Boolean algebra of its compact-open subsets; dually, a Boolean algebra is sent to the set of prime ideals, endowed with the hull-kernel topology. This dual equivalence was used to great effect by Abramsky [1] where he showed how to extract a logic from a domain constructed using Moggi's monadic approach, so that the logic was tailor-made for the domain used to build it.

Our approach to Stone duality is somewhat unconventional, but one that also has been utilized in recent work by Gehrke [5,6]. The idea is to realize a Stone space as a projective limit of finite spaces, a result which follows from Stone duality, as we now demonstrate.

Theorem 3 (Stone Duality). *Each Stone space X can be represented as a projective limit $X \simeq \varprojlim_{\alpha \in A} X_\alpha$, where X_α is a finite space. In fact, each X_α is a partition of X into a finite cover by clopen subsets, and the projection $X \twoheadrightarrow X_\alpha$ maps each point of X to the element of X_α containing it.*

Proof. If X is a Stone space, then $\mathcal{B}(X)$, the family of compact-open subsets of X is a Boolean algebra. Clearly $\mathcal{B}(X) \simeq \varinjlim_{\alpha \in A} \mathcal{B}_\alpha$ is the injective limit of its family $\{\mathcal{B}_\alpha \mid \alpha \in A\}$ of finite Boolean subalgebras. For a given $\alpha \in A$, we let X_α denote the finite set of atoms of $\mathcal{B}_\alpha$. Then $\mathcal{B}_\alpha \hookrightarrow \mathcal{B}(X)$ implies $\mathcal{B}_\alpha$ is a family of clopen subsets of X, and the set of atoms of $\mathcal{B}_\alpha$ are pairwise disjoint, and their sup – i.e., union – is all of X, so X_α forms a partition of X into clopen subsets, Thus there is a continuous surmorphism $X \twoheadrightarrow X_\alpha$ sending each element of X to the unique atom in X_α containing it. The family $\{\mathcal{B}_\alpha \mid \alpha \in A\}$ is an injective system, since given $\mathcal{B}_\alpha$ and $\mathcal{B}_\beta$, the Boolean subalgebra they generate is again finite. Dually the family $\{X_\alpha \mid \alpha \in A\}$ is a projective system, and since $\mathcal{B}(X) \simeq \varinjlim_{\alpha \in A} \mathcal{B}_\alpha$, it follows that $X \simeq \varprojlim_{\alpha \in A} X_\alpha$.

We note that a corollary of this result says that it is enough to have a basis for the family of finite Boolean subalgebras of $\mathcal{B}(X)$ in order to realize X as a projective limit of finite spaces, where by a *basis*, we mean a directed family whose union generates all of $\mathcal{B}(X)$.

2.4   Compact Groups

We now recall some results about compact topological groups. We include proofs of some results that are well-known in the interest of completeness. A standard reference for group theory is [12], and an excellent reference for the theory of compact groups is [8]

To begin, a *topological group* is a T_1-topological space G that is also a group, and for which the multiplication $\cdot\colon G \times G \to G$ and inversion $x \mapsto x^{-1}\colon G \to G$ mappings are continuous. A basic result is that all topological groups are Hausdorff spaces. A *compact group* is a topological group whose topology is compact.

We are interested in group structures on the Cantor set, which can be characterized as a metrizable perfect Stone space. That is, a *Cantor set* is a compact Hausdorff zero-dimensional space that has a countable base for its topology, and in which every point is a limit point. It is well-known that any Cantor set has a base of clopen subsets. We prove a stronger result for groups on the Cantor set.

Proposition 1. *If G is a compact group whose underlying space it zero dimensional, then G admits a neighborhood base of the identity consisting of clopen normal subgroups.*

Proof. We start with a basis $\mathcal{O}$ of clopen neighborhoods of the identity, which exists in any Stone space. Since inversion is a homeomorphism (being its own inverse), each $O \in \mathcal{O}$ satisfies $O^{-1} \in \mathcal{O}$, so $O \cap O^{-1} \in \mathcal{O}$, which implies it is no loss of generality to assume that $O = O^{-1}$ for each $O \in \mathcal{O}$.

Now, since multiplication is continuous and O is both compact and open, $O = e \cdot O \subseteq O$ implies there is a $U \in \mathcal{O}$ with $U \cdot O \subseteq O$. But then $U \subseteq O$, and so $U^2 \subseteq O$, and by induction, $U^n \subseteq O$ for each $n > 0$. Since U is symmetric, this implies the subgroup H_U that U generates is a subset of O. And since U is open, so is H_U (which also implies H_U is closed).

For the claim about normal subgroups, we first recall that the family of conjugates $\mathcal{H} = \{xHx^{-1} \mid x \in G\}$ of a closed subgroup $H < G$ is closed in the space of closed subsets of G endowed with the Vietoris topology, which is compact since G is compact. Moreover, G acts continuously on $\mathcal{H}$ by $(x, H) \mapsto xHx^{-1}\colon G \times \mathcal{H} \to \mathcal{H}$. The kernel $K = \{x \in G \mid xHx^{-1} = H\}$ of this action is then a normal subgroup of G, and if H is clopen, then K is clopen as well. But since G acts transitively on this family of conjugates, it follows that $|G/K| = |\mathcal{H}|$. Since K is open and G is compact, G/K is finite, and so there are only finitely many cosets

xHx^{-1}. Then their intersection $\bigcap_{x \in G} xHx^{-1} \subseteq H$ is a clopen normal subgroup of G inside H. Since G has a basis of clopen subgroups H around e by the first part, and we can refine each of these with a clopen normal subgroup by taking $\bigcap_{x \in G} xHx^{-1}$, it follows that G has a basis of clopen normal subgroups around e.

Corollary 1. *Any compact zero-dimensional group is the strict projective limit* [2] *of finite groups.*

Proof. If G is compact and zero-dimensional, then e has a basis $\mathcal{N}$ of clopen normal subgroups by the Proposition. If $N \in \mathcal{N}$, then since G is compact, G/N also is compact and the quotient map $\pi_N \colon G \to G/N$ is open. But $N \in \mathcal{N}$ is open, so G/N is discrete, which implies there are only finitely many cosets in G/N, i.e., G/N is finite. The family $\mathcal{N}$ is directed by reverse set inclusion, and for $M \subseteq N \in \mathcal{N}$, we let $\pi_{N,M} \colon G/M \to G/N$ be the natural projection. Then the family $(G/N, \pi_{N,M})_{M \leq N \in \mathcal{N}}$ forms a strict projective system of finite groups which satisfies $G \simeq \varprojlim_{\mathcal{N}} G/N$.

Remark 1. We also note that since any topological group is homogeneous, a topological group must satisfy the property that either every point is a limit point, or else the group is discrete. Thus, the underlying space of a compact group is either perfect or the group is finite. In particular, a topological group on a Stone space forces the space to be finite or perfect. By a *Cantor group*, we mean a topological group structure on a Cantor set (which we also assume is metrizable).

2.5   Haar Measure on Cantor Groups

Definition 2. *A Borel measure μ on a topological group G is* left translation invariant *if $\mu(xA) = \mu(A)$ for all $x \in G$ and all measurable sets $A \subseteq G$.*

A fundamental result of topological group theory is that each locally compact group admits a left translation invariant Borel measure which is unique up to scalar constant; i.e., if μ and ν are left translation invariant measures on the locally compact group G, then there is a constant $c > 0$ such that $\mu(A) = c \cdot \nu(A)$ for every measurable set A. Any such measure is called a *Haar measure*. If G is compact, the measure μ is assumed to satisfy $\mu(G) = 1$, which means this measure is unique. Notice in particular that Haar measure on any discrete group is counting measure, and on a finite group, it is normalized counting measure.

   We now establish an important result we need for the main result of this section.

Proposition 2. *Let G and H be compact groups and let $\phi \colon G \to H$ be a continuous surmorphism. Then $\phi(\mu_G) = \mu_H$, where μ_G and μ_H are Haar measure on G and H, respectively.*

---

[2] A projective system is *strict* if the projection maps all are surjections.

Proof. Let $K = \ker \phi$, and let $A \subseteq G/K$ be measurable and $x \in G/K$. Since ϕ is a surmorphism, there is $x_0 \in G$ with $\phi(x_0) = x$. Then

$$\pi_K(\mu_G)(xA) = \mu_G(\phi^{-1}(xA)) = \mu_G(\phi^{-1}(x) \cdot \phi^{-1}(A)) = \mu_G(x_0 K \cdot \phi^{-1}(A))$$
$$\stackrel{*}{=} \mu_G(x_0 \phi^{-1}(A)) = \mu_G(\phi^{-1}(A)) = \pi_K(\mu_G)(A),$$

where $\stackrel{*}{=}$ follows from the normality of K and the fact that $\phi^{-1}(A)$ is saturated with respect to K, and the next equality follows because μ_G is Haar measure on G. Thus $\phi(\mu_G)$ is a Haar measure on H. The result then follows by the uniqueness of Haar measure on a compact group.

The main result of this section is the following:

Theorem 4. *If G is a topological group whose underlying space is a Cantor set, then there is an abelian topological group structure on G that has the same Haar measure as the original topological group structure.*

Proof. Since G is a Cantor set, Corollary 1 implies $G \simeq \varprojlim_k G/N_k$ of a countable chain of finite groups, where $k \leq k'$ implies $N_{k'} \subseteq N_k$. For each k, we define groups G_k as follows:

1. $G_1 = \mathbb{Z}_{n_1}$, where $n_1 = |G/N_1|$, and
2. for $k > 1$, $G_k = G_{k-1} \oplus \mathbb{Z}_{n_k}$, where $n_k = |\ker \pi_{G/N_k, G/N_{k-1}}|$.

In short, $G_k = \oplus_{l \leq k} \mathbb{Z}_{n_l}$, where $n_1 = |G/N_1|$, and $n_k = |\ker \pi_{G/N_k, G/N_{k-1}}|$ for $k > 1$. Thus, G_k is a direct product of cyclic groups, and $|G_k| = |G/N_k|$ for each k. Since G_k and G/N_k are both finite, this last implies Haar measure on G/N_k is the same as Haar measure on G_k for each k.

Clearly there is a canonical projection $\pi_{k,k'}: G_k \to G_{k'}$ whenever $k' \leq k$. So we have a second strict projective system $(G_k, \pi_{k,k'})_{k' \leq k}$, and since $G/N_k \simeq G_k$ for each k *qua* topological spaces, it follows that $G \simeq \varprojlim_{k' \leq k}(G_k, \pi_{k',k})$, again *qua* topological spaces.

Next, Theorem 2 implies that the limit of the sequence $\{\mu_{G_k}\}_k$ is a Borel measure μ on $G \simeq \varprojlim_{k' \leq k}(G_k, \pi_{k',k})$ whose image under the quotient map $G \to G_k$ is μ_k, Haar measure on G_k. But if G_A denotes the limit of the projective system $(G_k, \pi_{k',k})_{k' \leq k}$ *qua* compact abelian groups, then Proposition 2 implies Haar measure on G_A also has image μ_{G_k} under the quotient map $G_A \to G_k$. Since limits are unique, this implies $\mu_{G_A} = \mu$.

Now the Haar measures $\mu_{G/N_k} = \mu_k$ on G/N_k and on G_k are equal by design, and Proposition 2 implies Haar measure μ_G on G with its original compact group structure maps to μ_{G/N_k} under the quotient map $G \to G/N_k$. Again limits are unique, so we conclude that $\mu_G = \mu_{G_A}$, the Haar measure induced on $G_A \simeq \varprojlim_{k' \leq k}(G_k, \pi_{k',k})$ *qua* compact abelian group.

Remark 2. We note that the same result holds for general (ie., nonmetrizable) compact group structures on Stone spaces. The only thing that changes is that the group may require a directed family of finite quotients that may be uncountable.

3   Defining an Bialgebraic Lattice Structure on $\mathcal{C}$

According to the proof of Theorem 4 we can assume we are given a Cantor group $\mathcal{C} \simeq \varprojlim(\mathcal{C}_n, \pi_{m,n})_{n \leq m}$ where each $\mathcal{C}_n = \oplus_{i \leq n}\mathbb{Z}_{n_i}$ is a product of n cyclic groups, and the mapping $\pi_{m,n}\colon \mathcal{C}_m \to \mathcal{C}_n$ is the projection map onto the first n factors of $\mathcal{C}_m$ for $n \leq m$. In particular, this representation relies on a fixed sequence of finite cyclic groups $\{\mathbb{Z}_{n_i} \mid i > 0\}$ satisfying $\mathcal{C}_n = \oplus_{i \leq n}\mathbb{Z}_{n_i}$, and without loss of generality, we can assume that $n_i > 1$ for each i – this follows from the fact that $\mathcal{C}$ is a perfect (hence uncountable) Stone space and each quotient group $\mathcal{C}_n$ is finite.

Theorem 5. $\mathcal{C}$ *admits a total order relative to which it is a complete bialgebraic lattice*[3] *endowed with the Lawson topology.*

Proof. We first note that we can define a total order on $\mathcal{C}_n = \oplus_{n_i \leq n}\mathbb{Z}_{n_i}$ to be the lexicographic order, where we endow $\mathbb{Z}_{n_i}$ with its total order from $\mathbb{N}$.

Next, the projection mapping $\pi_{m,n}\colon \mathcal{C}_n \to \mathcal{C}_m$ is monotone and clearly Scott continuous, for $n \leq m$, and we can define embeddings $\iota_{m,n}\colon \mathcal{C}_m \to \mathcal{C}_n$ by
$$\iota_{m,n}(x)_i = \begin{cases} x_i & \text{if } j \leq m \\ 0 & \text{if } m < j \end{cases}, \text{ and clearly } \iota_{m.n} \text{ is monotone and Scott continuous.}$$
Moreover, it is clear that $\pi_{m,n} \circ \iota_{m,n} = 1_{\mathcal{C}_n}$ and $\iota_{m,n} \circ \pi_{m,n} \leq 1_{\mathcal{C}_m}$ for $n \leq m$.

So, we have a system $((\mathcal{C}_n, \leq_n), \iota_{m,n}, \pi_{m,n})_{n \leq m}$ of e–p pairs in the category of algebraic lattices and Scott-continuous maps. By Theorem 1, $\varprojlim((\mathcal{C}_n), \leq_n), \pi_{n,m})_{n \leq m}$ is an algebraic lattice whose compact elements are the union of the images of the $\mathcal{C}_n$s under the natural embeddings, $\iota_n(x)_j = \begin{cases} x_j & \text{if } j \leq n \\ 0 & \text{if } n < j \end{cases}$. But this is the same family of finite sets and projection maps that define the original projective system, which implies $\mathcal{C}$ has a total order relative to which it is an algebraic lattice.

To see that $\mathcal{C}^{op}$ also is algebraic, we note that since $\mathcal{C}$ is totally ordered and complete, each $x \in K\mathcal{C}$ has a corresponding $x' = \sup(\downarrow x \setminus \{x\}) \in K\mathcal{C}^{op}$. If $y \notin K\mathcal{C}$ and $y < z \in \mathcal{C}$, then since $\mathcal{C}$ is algebraic, $z = \sup(\downarrow z \cap K\mathcal{C})$, so there is some $x \in K\mathcal{C}$ with $y < x \leq z$. But then $y \leq x' \in K\mathcal{C}^{op}$. It follows that $y = \inf(\uparrow y \cap K\mathcal{C}^{op})$ for $y \in \mathcal{C}$, so $\mathcal{C}^{op}$ also is algebraic.

Finally, the Lawson topology on an algebraic lattice is compact and Hausdorff, and it is refined by the original topology on $\mathcal{C}$, so the two topologies agree.

Remark 3. We note that $K\mathcal{C} = \bigcup_n \iota_n(\mathcal{C}_n)$, and the mappings $\iota_n\colon \mathcal{C}_n \to K\mathcal{C}$ are injections, so we often elide the injections ι_n and simply regard $\mathcal{C}_n$ as a subset of $\mathcal{C}$. Note as well that ι_n is a group homomorphism for each n, so this identification applies both order-theoretically and group theoretically.

Theorem 5 allows us to define the natural map $\phi\colon \mathcal{C} \to [0, 1]$: For each n, $\mathcal{C}_n = \oplus_{i \leq n}\mathbb{Z}_{n_i}$, endowed with the lexicographic order. For $x \in \mathcal{C}_n$, we define $\phi_n(x) =$

---

[3] A lattice L is *bialgebraic* if L and L^{op} are both algebraic lattices.

$\sum_{i \leq n} \frac{x_i}{n_1 \cdot n_2 \cdots n_i}$.[4] Then ϕ_n is monotone, and $n \leq m$ implies $\phi_m \circ \iota_{m,n} = \phi_n$. Thus we have a monotone mapping $\phi \colon KC \to [0,1]$. The fundamental theorem of domain theory implies ϕ admits a Scott-continuous extension $\widehat{\phi} \colon C \to [0,1]$.

In fact, note that $\phi \colon KC \to [0,1]$ is stictly monotone: if $x < y$, then $\phi(x) < \phi(y)$. This implies ϕ is one-to-one on KC, and clearly its image is dense in $[0,1]$. Now, for any $s \in (0,1]$, if $x \in C$ satisfies $\widehat{\phi}(x) < s$, then $s - \widehat{\phi}(x) > 0$, so we can choose $n > 0$ large enough so there are $x_n, y_n \in C_n \subseteq KC$ satisfying $x_n \leq x < y_n$ and $\phi_n(y_n) < s$. Hence $C \setminus \widehat{\phi}^{-1}([s,1])$ is weak-lower open in C, from which it follows that $\widehat{\phi}^{-1}([s,1])$ is weak-lower closed, which is to say $\widehat{\phi}^{-1}([s,1]) = \uparrow z$ for some $z \in C$. But this implies that $\widehat{\phi}$ is Lawson continuous. Since C is compact in the Lawson topology, this implies $\widehat{\phi}(C) = [0,1]$.

Moreover, since $[0,1]$ is connected and $\widehat{\phi}$ is monotone, it follows that $\widehat{\phi}(x') = \widehat{\phi}(x)$ for each $x \in KC$. We summarize this discussion as

Corollary 2. *The mapping $\widehat{\phi} \colon KC \to [0,1]$ by $\widehat{\phi}(x) = \sum_{i \leq n} \frac{x_i}{n_1 \cdot n_2 \cdots n_i}$ is strictly monotone (hence injective), and it has a Lawson-continuous, monotone and surjective extension defined by $\widehat{\phi}(x) = \sup \widehat{\phi}(\downarrow x \cap KC)$. Moreover, for each $x \in KC$, $\widehat{\phi}(x') = \widehat{\phi}(x)$, where $x' = \sup(\downarrow x \setminus \{x\}) \in KC^{op}$.*

4   Mapping Haar Measure to Lebesgue Measure

We now come to the main result of the paper. Our goal is to show that there is a natural map from any Cantor group onto the unit interval that sends Haar measure to Lebesgue measure. According to Theorem 4, any compact group structure on a Cantor set has the same Haar measure as a group structure realized as the strict projective limit of a sequence of finite abelian groups, and Theorem 5 and Corollary 2 show there is a Lawson continuous monotone mapping of C onto the unit interval for such a group structure. We now show that this map sends Haar measure on C as an abelian group to Lebesgue measure.

Recall that the abelian group structure satisfies $C = \varprojlim_{n>0} (\bigoplus_{i \leq n} \mathbb{Z}_{k_i}, \pi_{m,n})$, where $k_i > 1$ for each i, and $\pi_{m,n} \colon C_m \to C_n$ is the projection on the first n factors, for $n \leq m$. Theorem 5 says C has a total order relative to which it is a complete bialgebraic lattice, and it is this order structure we exploit in our proof.

Remark 4. Recall that $b \in KC$ implies $b' = \sup(\downarrow b \setminus \{b\}) \in KC^{op}$ and $\widehat{\phi}(b) = \widehat{\phi}(b')$. We need this fact because the clopen intervals in C all have the form $[a, b']$ for some $a \leq b \in KC$. Indeed, according to Stone duality (Theorem 3), in a representation of a Stone space as a strict projective limit of finite spaces, each finite quotient space corresponds to a partition of the space into clopen sets.

---

[4] An intuitive way to understand ϕ_n for each n is that $\mathbb{Z}_{n_1}$ divides the interval into n_1 subintervals, $\mathbb{Z}_{n_2}$ divides each of those into n_2 subintervals, and so on. So ϕ_n maps the elements of C_n to those those rationals in $[0,1]$ that can be expressed precisely in an expansion using $n_1, n_2, \ldots$ as successive denominators.

If the Stone space is totally ordered and the representation is via monotone maps, then the elements of each partition are clopen intervals. In particular, if $\pi_n \colon \mathcal{C} \to \mathcal{C}_n$ is a projection map, then $\pi_n^{-1}(x) = [a, b']$ for some $a, b \in K\mathcal{C}$, for each $x \in \mathcal{C}_n$

Throughout the following, we let $\mu_{\mathcal{C}}$ denote Haar measure on $\mathcal{C}$, and let λ denote Lebesgue measure on $[0, 1]$.

Proposition 3. *If $a \leq b \in \mathcal{C}_n$, then $\lambda(\widehat{\phi}([a, b'])) = \mu_{\mathcal{C}_n}([a, b]_{\mathcal{C}_n})$.*

Proof. On one hand, $\lambda(\widehat{\phi}([a, b'])) = \widehat{\phi}(b') - \widehat{\phi}(a) = \widehat{\phi}(b) - \widehat{\phi}(a)$. On the other, $\mu_{\mathcal{C}_n}([a, b]_{\mathcal{C}_n}) = \frac{|[a, b]_{\mathcal{C}_n}|}{|\mathcal{C}_n|}$ since $\mathcal{C}_n$ is finite. Now $\mathcal{C}_n = \bigoplus_{i \leq n} \mathbb{Z}_{k_i}$ in the lexicographic order, and we show these two values agree by induction on n. Indeed, since $a \leq b \in \mathcal{C}_n$, we have $a = (a_1, \ldots, a_i)$ and $b = (b_1, \ldots, b_j)$ for some $i, j \leq n$, and then $\widehat{\phi}(a) = \sum_{l \leq i} \frac{a_l}{k_1 \cdots k_l}$ and $\widehat{\phi}(b) = \sum_{l \leq j} \frac{b_l}{k_1 \cdots b_l}$. By padding a and b with 0s, we can assume $i = j = n$. Then

$$\lambda(\widehat{\phi}([a, b'])) = \widehat{\phi}(b') - \widehat{\phi}(a) = \widehat{\phi}(b) - \widehat{\phi}(a)$$

$$= \sum_{l \leq n} \frac{b_l}{k_1 \cdots b_l} - \sum_{l \leq n} \frac{a_l}{k_1 \cdots k_j}$$

$$= \left(\sum_{l \leq n-1} \frac{b_l}{k_1 \cdots b_l} - \sum_{l \leq n-1} \frac{a_l}{k_1 \cdots k_i} \right)$$

$$+ \left(\left| \frac{b_n}{k_1 \cdots k_n} - \frac{a_n}{k_1 \cdots k_n} \right| \right)$$

$$\overset{\dagger}{=} \frac{|[a^*, b^*]_{\mathcal{C}_{n-1}}|}{|\mathcal{C}_{n-1}|} + \left(\left| \frac{b_n}{k_1 \cdots k_n} - \frac{a_n}{k_1 \cdots k_n} \right| \right)$$

$$= \frac{|[a, b]_{\mathcal{C}_n}|}{|\mathcal{C}_n|} = \mu_{\mathcal{C}}([a, b]_{\mathcal{C}_n}),$$

where $a^* = (a_1, \ldots, a_{n-1}) \leq b^* = (b_1, \ldots, b_{n-1}) \in \mathcal{C}_{n-1}$ so that $\overset{\dagger}{=}$ follows by induction.

Theorem 6. *Let $\mathcal{O}([0, 1])$ denote the family of open subsets of $[0, 1]$. Then $\lambda \colon \mathcal{O}([0, 1]) \to [0, 1]$ and $\mu_{\mathcal{C}} \circ \widehat{\phi}^{-1} \colon \mathcal{O}([0, 1]) \to [0, 1]$ are the same mapping.*

Proof. Let $U \in \mathcal{O}([0, 1])$ be an open set. Since $\widehat{\phi}$ is Lawson continuous, $\widehat{\phi}^{-1}(U)$ is open in $\mathcal{C}$, and since $\mathcal{C}$ is a Stone space, it follows that $\widehat{\phi}^{-1}(U) = \bigcup\{K \mid K \subseteq \widehat{\phi}^{-1}(U) \text{ clopen}\}$. Now, $\widehat{\phi}$ is a continuous surjection, so $\widehat{\phi}(K)$ is compact and

$$U = \widehat{\phi}(\widehat{\phi}^{-1}(U)) = \widehat{\phi}\left(\bigcup\{K \mid K \subseteq \widehat{\phi}^{-1}(U) \text{ clopen}\}\right)$$

$$= \bigcup\{\widehat{\phi}(K) \mid K \subseteq \widehat{\phi}^{-1}(U) \text{ clopen}\}.$$

Next, any clopen $K \subseteq \mathcal{C}$ is compact, and because $K\mathcal{C}$ is dense, $(\exists n > 0)(\exists a_i < b_i \in \mathcal{C}_i)$ $K = \bigcup_{i \leq n}[a_i, b_i']$. Moreover, we can assume $[a_i, b_i'] \cap [a_j, b_j'] = \emptyset$ for $i \neq j$.

Then

$$\mu_C(K) = \sum_i \mu_C([a_i, b_i']) = \sum_i \lambda(\widehat{\phi}([a_i, b_i'])),$$

the last equality following from Proposition 3. Since the intervals $[a_i, b_i']$ are pairwise disjoint, if $\widehat{\phi}([a_i, b_i']) \cap \widehat{\phi}([a_j, b_j']) \neq \emptyset$ then either $b_i' = a_j'$ or $b_j' = a_i'$. In either case, $\widehat{\phi}([a_i, b_i']) \cap \widehat{\phi}([a_j, b_j'])$ is singleton, and then since λ is continuous,

$$\mu_C(K) = \sum_i \lambda(\widehat{\phi}([a_i, b_i'])) = \lambda(\bigcup_i \widehat{\phi}([a_i, b_i'])) = \lambda(K). \qquad (1)$$

Finally, since μ_C and λ are both inner regular, we have

$$\begin{aligned}
\lambda(U) &= \lambda(\widehat{\phi}(\widehat{\phi}^{-1}(U))) \\
&= \lambda\left(\bigcup \{\widehat{\phi}(K) \mid K \subseteq \widehat{\phi}^{-1}(U) \text{ clopen}\}\right) \\
&\overset{\dagger}{=} \bigcup \{\lambda(\widehat{\phi}(K)) \mid K \subseteq \widehat{\phi}^{-1}(U) \text{ clopen}\} \\
&\overset{\ddagger}{=} \bigcup \{\mu_C(K) \mid K \subseteq \widehat{\phi}^{-1}(U) \text{ clopen}\} \\
&\overset{\#}{=} \mu_C\left(\bigcup \{K \mid K \subseteq \widehat{\phi}^{-1}(U) \text{ clopen}\}\right) \\
&= \mu_C(\widehat{\phi}^{-1}(U)).
\end{aligned}$$

where $\overset{\dagger}{=}$ follows by the inner regularity of λ,[5] $\overset{\ddagger}{=}$ follows from Equation 1, and $\overset{\#}{=}$ follows from the inner regularity of μ_C.

Corollary 3. *If we endow C with the structure of topological group with Haar measure μ_C, then there is a continuous mapping $\widehat{\phi}\colon C \to [0, 1]$ satisfying $\widehat{\phi}(\mu_C) = \lambda$.*

Proof. If $A \subseteq [0, 1]$ is Borel measurable, then $\widehat{\phi}(\mu_C)(A) = \mu_C(\widehat{\phi}^{-1}(A))$. We have shown $\widehat{\phi}(\mu_C)(A) = \lambda(A)$ in case A is open. But since the open sets generate the Borel σ-algebra the result follows.

Theorem 7. *Let C_1 and C_2 be Cantor sets with topological group structures with Haar measures μ_{C_1} and μ_{C_2}, respectively. Then $\mu_{C_1} = \mu_{C_2}$.*

Proof. By Theorem 4, we can assume that the group structures on C_1 and C_2 are both abelian, and then Theorem 5 and Corollary 2 show there are Lawson-continuous monotone mappings of $\widehat{\phi}_1\colon C_1 \to [0, 1]$ and $\widehat{\phi}_2\colon C_2 \to [0, 1]$ both onto the unit interval. Since KC_i' are both countable, $\widehat{\phi}_1\colon C_1 \setminus KC_1' \to [0, 1]$ is a Borel isomorphism onto its image, as is $\widehat{\phi}_2\colon C_2 \setminus KC_2' \to [0, 1]$. Then the composition $\widehat{\phi}_2^{-1} \circ \widehat{\phi}_1\colon C_1 \setminus KC_1' \to C_2 \setminus KC_2'$ is a Borel isomorphism onto its image (that also is an order isomorphism). Then, for any measurable set $A \subseteq C_1$, $\mu_{C_1}(A) = \lambda(\widehat{\phi}_1(A)) = \mu_{C_2}(A)$, proving the claim.

---

[5] It is straightforward to argue that *any* compact set $C \subseteq U$ is contained in $\bigcup\{\widehat{\phi}(K) \mid K \subseteq \widehat{\phi}^{-1}(U) \text{ clopen}\}$.

Remark 5. In the last proof, we could have restricted the mappings to $\mathcal{C}_i \setminus (K\mathcal{C}_i \cup K\mathcal{C}_i')$ for $i = 1, 2$. Then the induced map $\widehat{\phi_2}^{-1} \circ \widehat{\phi_1}$ is a homeomorphism as well as an order isomorphism. On the other hand, the mappings we did use map are one-to-one, in particular on the elements of $[0, 1]$ that are expressible as fractional representations using the families $\{\mathbb{Z}_{n_i} \mid i > 0\}$ and $\{\mathbb{Z}_{n_i'} \mid i > 0\}$.

5 Summary

We have studied the topological groups structures on the Cantor set $\mathcal{C}$ and shown that any such structure has an "equivalent" abelian group structure, in the sense that the Haar measures are the same. We also showed any representation of $\mathcal{C}$ as an abelian group admits a continuous mapping onto the unit interval sending Haar measure to Lebesgue measure. Finally, we showed that Haar measure on $\mathcal{C}$ is the same, regardless of the group structure over which it is defined.

This work is the outgrowth of a talk by the second author at a Dagstuhl seminar in 2012. A final comment in that talk sketched a domain-theoretic approach to showing that Haar measure on $\mathcal{C} \simeq 2^{\mathbb{N}}$ maps to Lebesgue measure. We were inspired to look more closely at this issue because of the enthusiasm Prakash Panangaden expressed for that result. So, as a 60th birthday present, we offer this paper, and hope the recipient enjoys this presentation as well.

Happy Birthday, Prakash!!

Acknowledgements. Both authors gratefully acknowledge the support of the US NSF during the preparation of this paper; the second author also thanks the US AFOSR for its support during the preparation of this paper.

References

1. Abramsky, S.: Domain theory in logical form. Annals of Pure and Applied Logic 51, 1–77 (1991)
2. Abramsky, S., Jung, A.: Domain Theory. In: Handbook of Logic in Computer Science, pp. 1–168. Clarendon Press (1994)
3. Fedorchuk, V.: Probability measures in topology. Russ. Math. Surv. 46, 45–93 (1991)
4. Gierz, G., Hofmann, K.H., Lawson, J.D., Mislove, M., Scott, D.: Continuous Lattices and Domains. Cambridge University Press (2003)
5. Gehrke, M., Grigorieff, S., Pin, J.-É.: Duality and equational theory of regular languages. In: Aceto, L., Damgård, I., Goldberg, L.A., Halldórsson, M.M., Ingólfsdóttir, A., Walukiewicz, I. (eds.) ICALP 2008, Part II. LNCS, vol. 5126, pp. 246–257. Springer, Heidelberg (2008)
6. Gehrke, M.: Stone duality and the recognisable languages over an algebra. In: Kurz, A., Lenisa, M., Tarlecki, A. (eds.) CALCO 2009. LNCS, vol. 5728, pp. 236–250. Springer, Heidelberg (2009)
7. Hofmann, K.H., Mislove, M.: Compact affine monoids, harmonic analysis and information theory, in: Mathematical Foundations of Information Flow. AMS Symposia on Applied Mathematics 71, 125–182 (2012)

8. Hofmann, K.H., Morris, S.: The Structure Theory of Compact Groups, de Gruyter Studies in Mathematics, 2nd edn., vol. 25, p. 858. de Gruyter Publishers (2008)
9. Jones, C.: Probabilistic Nondeterminism, PhD Thesis, University of Edinburgh (1988)
10. Jung, A., Tix, R.: The troublesome probabilistic powerdomain. ENTCS 13, 70–91 (1998)
11. Mislove, M.: Topology. domain theory and theoretical computer science. Topology and Its Applications 89, 3–59 (1998)
12. Rotman, J.: An Introduction to the Theory of Groups, Graduate Studies in Mathematics, 4th edn. Springer (1999)
13. Saheb-Djarhomi, N.: CPOs of measures for nondeterminism. Theoretical Computer Science 12, 19–37 (1980)
14. http://en.wikipedia.org/wiki/Cantor_set

Word Order Alternation in Sanskrit via Precyclicity in Pregroup Grammars

Claudia Casadio[1] and Mehrnoosh Sadrzadeh[2,*]

[1] Dept. of Philosophy, Chieti University, IT
`casadio@unich.it`
[2] Queen Mary University of London, UK
`mehrs@eecs.qmul.ac.uk`

Abstract. We develop a pregroup grammar for a basic fragment of Sanskrit, similar to the fragment used by Lambek in his treatment of English in a number of papers. We make reference to recent work of the authors involving the characterization of cyclic rules in pregroup grammars to treat word order alternation (e. g. in English, Persian, Italian, and Latin) and analyse this phenomena in Sanskrit. Pregroups are introduced by Lambek; they are partially ordered compact closed categories. The latter have been invoked to analyze quantum protocols in the work of Abramsky and Coecke. Sanskrit was the ancient official language of India and remains one of its main religious and literary languages.

Keywords: Type Grammar, Pregroup, Cyclic Rules, Movement.

Dedicated to Prakash Panangaden, on his 60th Birthday.

1   Introduction

Pregroups are partially ordered monoids in which every element has a left and a right adjoint. They have been introduced by Jim Lambek as a simplification of the structure of residuated monoids [13, 24, 26]. Category theoretically, pregroups are partial order versions of compact closed categories, in contrast to residuated moniods, which are partial order versions of monoidal closed categories [28, 30].

Similarly to Lambek's Syntactic Calculus [23], which was based on residuated monoids, pregroups were developed to reason about grammatical structures in natural language. They also come equipped with a cut-free sequent calculus [10]. Because of their simple nature, pregroups have been applied, in a short span of time, to reason about the structure of a wide range of languages from English and French to Arabic and Persian (see [6, 12, 15, 27, 31]). What is missing from the

* The authors are grateful to Jim Lambek for stimulating suggestions given in the course of a preliminary presentation of this work. Thanks are also due to an anonymous referee for comments and corrections, and for suggesting to use the word *alternation*. Support by EPSRC (grant EP/F042728/1) is gratefully acknowledged by M. Sadrzadeh; support by MIUR (PRIN 2012, 60% 2013) by C. Casadio.

F. van Breugel et al. (Eds.): Panangaden Festschrift, LNCS 8464, pp. 229–249, 2014.

list is the *mother* of all languages, Sanskrit, which is moreover claimed to have a perfect grammatical structure. In this paper, we use pregroups to analyse a basic fragment of Sanskrit, similar to the fragment used by Lambek in his treatment of English in a number of papers, and consecutively by various other authors in their analysis, such as the work of current authors on Persian, Hungarian, and Latin [14, 31, 32].

We invoke recent work of the authors about precyclic properties in pregroups [17], and show that a weaker version of Yetter's cyclic axiom [35] hold in pregroup algebras, via the translation between residuated monoids and pregorups [13] and Abrusci's cyclic sequent calculus rules for a non-commutative linear logic [1, 2]. Then we use ideas from [16] on word order alternation in language and develop corresponding permutations and transformations to analyse word order alternations in Sanskrit and study its movement patterns. We type the Sanskrit sentence assuming the canonical word order suggested in Apte [5]. We also type a number of Sanskrit compounds based on the classification of Gillon [20]. The focus of the current paper, however, is reasoning about alternation of word order in Sanskrit. To achieve this, we use the rules of word order and movement in Sanskrit according to the analysis of Gillon [18] and most of the examples are also taken from the work of Gillon [18–20]. The present work is intended as the basis for a further more detailed analysis, such as extending the work of [7] from production rules to a pregroup grammar.

The connection to Prakash's work is that pregorups are compact closed categories; these have been used to analyze quantum protocols in the work of Abramsky and Coecke. They have also been applied to reason about vector space semantics of natural language, in the work of Clark, Coecke, and Sadrzadeh. This latter work develops a primary syntactic analysis of the language under consideration in a pregroup grammar. So our analysis of Sanskrit can be used to provide a vector space semantics, based on the texts written in Sanskrit. There is also a connection to Prakash himself, since Sanskrit was the ancient official language of India and remains one of its main religious and literary languages.

2   Pregroup Algebras

Lambek developed the calculus of pregroups as an alternative to his Syntactic Calculus [23], usually known as Lambek Calculus [29]. While the latter is an intuitionistic system, based on the operation of implication, the former is a classical system, based on the operation of multiplicative conjunction (for details see [9, 11, 13, 26]). The mathematical and logical properties of pregroups are studied in [9, 10, 24, 28]. In a short span of time, pregroups have been applied to the grammatical analysis of fragments of a wide range of languages: English [25], French [6, 27], Italian [12], Arabic [8], Polish [22], Persian [31] and others [15].

In a nutshell, a pregroup P, denoted by the structure $(P, \cdot \, , 1, \leq, ()^l, ()^r)$, is a partially ordered monoid (formally defined in the next section), '$\cdot$' is a multiplicative operation, 1 is the unit of this multiplication, and each element $p \in P$ has both a left adjoint p^l and a right adjoint p^r, so that the following *adjunction inequalities* hold:

$$p^l \cdot p \leq 1 \leq p \cdot p^l \qquad\qquad p \cdot p^r \leq 1 \leq p^r \cdot p$$

The two inequalities on the left side of 1 are referred to as *contractions*, while the two at the right side are called *expansions*; adjoints are unique and contravariant:

$$p \leq q \quad\Longrightarrow\quad q^l \leq p^l \text{ and } q^r \leq p^r$$

As a consequence of the *compact* property of the monoidal operation (which gives rise to the adjunction inequalities), the unit 1 and the multiplication are *self dual* [9, 24, 26, 28], that is:

$$1^l = 1 = 1^r \qquad (p \cdot q)^l = q^l \cdot p^l \qquad (p \cdot q)^r = q^r \cdot p^r$$

Some other properties of pregroups are as follows:

1- The adjoint of multiplication is the multiplication of adjoints but in the reverse order, that is:

$$(p \cdot q)^l = q^l \cdot p^l \qquad (p \cdot q)^r = q^r \cdot p^r$$

2- The adjoint operation is order reversing, that is:

$$p \leq q \Longrightarrow q^r \leq p^r \qquad \text{and} \qquad p \leq q \Longrightarrow q^l \leq p^l$$

3- Composition of the opposite adjoints is identity, that is:

$$(p^l)^r = (p^r)^l = p$$

4- Composition of the same adjoints is not identity, that is:

$$p^{ll} = (p^l)^l \neq p, \qquad p^{rr} = (p^r)^r \neq p$$

This leads to the existence of iterated adjoints [24], so that each element of a pregroup can have countably many iterated adjoints, for instance we have:

$$\cdots, p^{ll}, p^l, p, p^r, p^{rr}, \cdots$$

A group is a pregroup where $p^r = p^l$ for all $p \in P$. Another example of a pregroup is the set of all monotone unbounded maps on integers $f \colon Z \to Z$. In this pregroup, function composition is the monoid multiplication and the identity map is its unit; the underlying order on integers lifts to an order on the maps whose Galois adjoints are their pregroup adjoints, defined as follows:

$$f^l(x) = min\{y \in Z \mid x \leq f(y)\} \qquad f^r(x) = max\{y \in Z \mid f(y) \leq x\}$$

A residuated monoid $(M, \leq, \cdot, 1, /, \backslash)$ is a partially ordered monoid, where the monoid multiplication has a right $- \backslash -$ and a left $-/-$ adjoint, that is, for $a, b, e \in M$ we have

$$b \leq a \backslash e \quad\Leftrightarrow\quad a \cdot b \leq e \quad\Leftrightarrow\quad a \leq e/b$$

The passage from residuated monoids (on which the Lambek Calculus is based) to pregroups can be thought of as replacing the two adjoints of the monoid multiplication with the two adjoints of the elements. If a residuated monoid has a dualizing object, i.e. an object $0 \in M$ satisfying $(0/p) \setminus 0 = p = 0/(p \setminus 0)$ for $p \in M$, then one can define for each element a left and a right negation as $p^0 := p \setminus 0$ and $^0p := 0/p$. It would then be tempting to think of these negations as the two pregroup adjoints, i.e. to define $p^0 = p^r$ and $^0p = p^l$. The problem with this definition is that the operation involved in $a \setminus b$ (or b/a) - the linear logic "par" - is different from the operation in $(a \cdot b)$ - the tensor product. One can however translate, on this basis, Lambek Calculus expressions into pregroups, provided that these two operations are identified with the pregroup unique operation: then all the $a \setminus b$ (or b/a) types will become $a^r b$ (or $b\, a^l$). In thise sense, pregroups are non conservative extensions of the Lambek Calculus [24, 26].

3   Precyclic Properties

According to Yetter [35], an element c of a partially ordered monoid M is said to be *cyclic* whenever, for all $a, b \in M$, we have:

$$a \cdot b \leq c \implies b \cdot a \leq c$$

Although this definitions was first used in the setting of residuated monoids, one can as well use it for partially ordered monoids that are not necessarily residuated, since obviously the definition does not involve the adjoints to the multiplication. For the reader more familiar with Yetter's original definition, note that whenever a monoid is residuated, the cyclic condition becomes equivalent to one involving the adjoints, as follows:

$$c/a = a \setminus c$$

We say that a partially ordered monoid (residuated or not) is *cyclic* whenever it has a cyclic element. Residuated monoids admit the notion of *dualization*. An element d of a residuated monoid is *dualizing* whenever for all $a \in M$ we have:

$$(d/a) \setminus d = a = d/(a \setminus d)$$

If the dualizing element of M is furthermore cyclic, we obtain:

$$d/(d/a) = a = (a \setminus d) \setminus d$$

Using the usual translation between residuated monoids and pregroups [13], we can investigate whether and how the above notions may hold in a pregroup. In particular, we can show that 1 is a dualizing element which is not necessarily cyclic.

Definition 1. *Given an element x of a residuated monoid M, we denote its translation into a pregroup by $t(x)$. For all $a, b \in M$, this translation is defined as follows:*

$$t(1) = 1, \quad t(a \cdot b) = t(a) \cdot t(b), \quad t(a \setminus b) = t(a)^r \cdot t(b), \quad t(a/b) = t(a) \cdot t(b)^l$$

Proposition 1. *In a pregroup P, the element 1 is dualizing; if P is proper, that is $a^l \neq a^r$ [9], then 1 is not cyclic.*

Proof. To verify the first part, we have to show that for all $a \in P$, the translation of $(1/a)\backslash 1 = a = 1/(a\backslash 1)$ holds in a pregroup. That is we have to show:

$$t((1/a)\backslash 1) = t(a) = t(1/(a\backslash 1))$$

For the left hand side we have

$$t((1/a)\backslash 1) = (t(1/a))^r \cdot t(1) = (1 \cdot t(a)^l)^r \cdot 1 = ((t(a)^l)^r \cdot 1) \cdot 1 = (t(a)^l)^r = t(a)$$

For the right hand side we have

$$t(1/(a\backslash 1)) = t(1) \cdot t(a\backslash 1)^l = 1 \cdot (t(a)^r \cdot t(1))^l = 1 \cdot (1 \cdot (t(a)^r)^l) = (t(a)^l)^r = t(a)$$

To verify the second part, we have to show that for all $a \in P$, the translation of $1/a \neq a\backslash 1$ holds in a pregroup. This is true since we have:

$$t(1/a) = t(1) \cdot t(a)^l = t(a)^l \quad \neq \quad t(a\backslash 1) = t(a)^r \cdot 1 = t(a)^r$$

However, pregroups do admit a weak form of cyclicity, which we refer to by using the term *precyclicity*, described below:

Proposition 2. *The following hold in any pregroup P, for any $p, q, r \in P$*

$(i)\, pq \leq r \implies q \leq p^r r$ $\qquad\qquad$ $(ii)\, q \leq rp \implies qp^r \leq r$

$(iii)\, qp \leq r \implies q \leq rp^l$ $\qquad\qquad$ $(iv)\, q \leq pr \implies p^l q \leq r$

Proof. Consider the first case. Suppose $pq \leq r$, since the multiplication operation of a pregroup is order preserving, we multiply both sides by p^r from the left and obtain $(*)$ $p^r pq \leq p^r r$. Now from the axioms of a pregroup it follows that $q \leq p^r pq$, by order preservation of multiplication, unity of 1, and from the two validities $q \leq q$ and $1 \leq p^r p$. From this and $(*)$, by transitivity it follows that $q \leq p^r r$. Consider the second case now. Suppose $q \leq rp$, multiply both sides from the right with p^r and obtain $qp^r \leq rpp^r$, by a similar argument as before we have that $rpp^r \leq r$ in any pregroup, hence by transitivity it follows that $qp^r \leq r$. The proofs of the other two cases are similar.

As a consequence we obtain:

Corollary 1. *The following hold in any pregroup P, for any $a, b \in P$:*

$(1)\, 1 \leq ab \overset{(ll)}{\implies} 1 \leq ba^{ll}$ $\qquad\qquad$ $(2)\, 1 \leq ab \overset{(rr)}{\implies} 1 \leq b^{rr} a$

Proof. Suppose $1 \leq ab$; by case (iv) of Proposition 2 we have that $a^l 1 \leq b$, since 1 is the unit of P this is equivalent to $1a^l \leq b$, from which by case (iii) of Proposition 2 it follows that $1 \leq ba^{ll}$. The proof of the second case is similar.

Informally, case (1) of the above corollary says that whenever a juxtaposition of types, e.g. ab, is above the monoidal unit, then so is a permuted version of it, where a moves from the left of b to the right of it, but as a result of this movement, a gets marked with double adjoints ll to register the fact that it came from the *left*. That is why this property is annotated with (and we thus refer to it by) ll. Case (2) is similar, except that in this case it is b that moves from the right of a to its left, hence it is marked with rr.

A more direct connection between these properties and cyclicity is highlighted by the sequent calculus for linear logic. In this context, the cyclic properties were originally represented via the exchange rule, first introduced by Girard [21]:

$$\frac{\vdash \Gamma, A}{\vdash A, \Gamma} \; CycExch$$

Later, Abrusci generalised this rule in the following way referring to its logic as Pure Non-Commutative Classical Linear Logic (**SPNCL′**)[1]:

$$\frac{\vdash \Gamma, A}{\vdash \neg^r\neg^r A, \Gamma} \; Cyc^{+2} \qquad\qquad \frac{\vdash A, \Gamma}{\vdash \Gamma, \neg^l\neg^l A} \; Cyc^{-2}$$

Using the translations $\neg^r\neg^r A := A^{rr}$ and $\neg^l\neg^l A := A^{ll}$, and Buzskowsi's interpretation map of compact bi-linear logic into pregroups [10], one can easily see that the properties of Corollary 1 are the semantic counterparts of Abrusci's cyclic rules (the empty sequent on the left is the unit of multiplication).

4   Pregroup Grammars

To analyse a natural language we use a *pregroup grammar*. On analogy with other type logical or categorial grammars, a pregroup grammar consists in a free pregroup generated over a set of basic grammatical types together with the assignment of the pregroup types to the vocabulary of the language. To exemplify consider the set of basic types $\{\pi, o, p, n, s\}$, representing five basic grammatical roles as follows:

π: subject o: object p: predicate s: sentence n: noun phrase

The linguistic reading of partial orders on basic types is as follows: whenever we have $a \leq b$ we read it as 'a word of type a can also have type b'. Examples of these partial orders are $n \leq \pi$ and $n \leq o$. The free pregroup generated over the above basic types includes *simple* types such as π^l, π^r, o^l, o^r, and *compound* types such as $oo^l, \pi\pi^l, s^r s, \pi^r s, \pi^r s o^l$.

A sentence of type s is defined to be grammatical whenever the multiplication - corresponding to syntactic composition - of the types of its constituents is less than or equal to the type s. This means that the type of a sentences is derivable from the types assigned to its constituents. The computations that lead to deciding about this matter are referred to as *grammatical reductions*. For example, the type assignments to the words of the declarative sentence 'John saw Mary.' are the following:

$$\text{John} \quad \text{saw} \quad \text{Mary.}$$
$$\pi \quad (\pi^r s \, o^l) \quad o$$

The grammatical reduction corresponding to this example is as follows:

$$\pi \, (\pi^r s \, o^l) \, o \leq 1 \, s \, 1 = s$$

In this computation, the subject is inputted to the verb via the inequality $\pi\pi^r \leq 1$, similarly, the object is inputted via the inequality $o^l o \leq 1$, and since 1 is the unit of juxtaposition the result is the type of the sentence. Considering some more complex examples involving adjectives, adverbials and the predicative copula 'to be', one can perform the following computations by assigning $(\pi^r s \, o^l)$ to the transitive verb 'admires', π to the subject role of the noun 'girl', $(o o^l)$ to 'beautiful' in its adjectival role and p to its predicative role (as the complement of the copula), $(\pi\pi^l)$ to the definite article 'the' in its subject position and $o o^l$ to its object position, $(\pi^r s \, p^l)$ to the predicative copula 'is', and $(s^r s)$ to the adverbial 'sincerely'[1]:

$$\text{John admires} \quad \text{the} \quad \text{beautiful girl.}$$
$$\pi \quad (\pi^r s \, o^l) \quad (o o^l) \quad (o o^l) \quad o \qquad \qquad \leq s$$

$$\text{John admires} \quad \text{the} \quad \text{beautiful girl sincerely.}$$
$$\pi \quad (\pi^r s \, o^l) \quad (o o^l) \quad (o o^l) \quad o \quad (s^r s) \leq s$$

$$\text{The} \quad \text{girl} \quad \text{is} \quad \text{beautiful.}$$
$$(\pi\pi^l) \quad \pi \quad (\pi^r s \, p^l) \quad p \qquad \qquad \leq s$$

A significantly large set of natural languages have been analysed by using pregroup grammars (see e.g. [15, 26]). The computations that lead to type reductions can be depicted by the under-link diagrams, like in the example above, that are reminescent of the planar proof nets of non-commutative linear logic, as shown in the calculi developed in [2, 9], allowing for a geometrical representation of the possible connections between words and expressions of the sentences of a given language.

5   The Grammar of a Basic Fragment of Sanskrit

Arguably, Sanskrit is the language that has one of oldest grammar books. These were first written down by Panini, who lived in the period between 2nd to 4th century BC and later commented on by Mahabhashya, in 2nd century BC. Sanskrit itself, is said to have originated in Hindi hymns such as the Vedic verses, which go back to 2000 BC. These religious writings later led to Barahamanas

---

[1] Following [26], we separate the subject and object roles, instead of using n for both.

which played a major role in the institutional language of India. The great legend of Mahdbharata from 6th century BC and the philosophical poem of Bhagavad-Glta are the two most important texts written in Sanskrit. It was on the basis on these texts that Panini presented the grammar of Sanskrit. His work is a set of algebraic formula-like rules that produce what is greatly known as *Sanskrit Compounds*. These are mainly formed based on euphonic rules, rarely seen in any other language. This is the reason why Sanskrit is claimed to have a "perfect grammar". However, it has also been argued that these rules describe an artificial language, one that has been used in religious and highly literary texts and not one that is spoken by the people in the street [34].

The formation of Sanskrit compounds is complex and usually relies on enriched versions of the context-free rules [20]. In this paper we focus on a small set of compounds, as well as the general rules of sentence formation. It is argued that since the grammar of Sanskrit was based on hymns and poems, sentences did not play a major role in its grammar, as originally put down by scholars. In this paper we work with the invaluable exposition of Sanskrit grammar in the style of European grammars by Apte [5] and a modern take of it by using generative means by Gillon [18–20].

According to Apte, Sanskrit sentence can be of three kind: *simple, complex,* and *compound.* Simple sentences are the ones that have one subject and one verb. Complex sentences have one subject, one principal verb, and two or more secondary verbs. Compound sentences consist of two or more principal sentences. Verbs can be classified into copular and non-copular. Copulars are sentences where a subject is connected to a predicate, also referred to by a complement. Like in most languages, the best representative of a copular case is the verb *to be;* '*aste*' in Sanskrit. Similarly to Latin and certain European languages such as Spanish, the verbs of the copular sentences are sometimes dropped, but for now we disregard this case. Sanskrit also uses '*aste*' for existential purposes, e.g. in the sentence '*There is a man.*' ('*The man is.*') in Sanskrit. In most of these cases the copula is not dropped. The non-copular sentences are those in which the verb is reporting the occurrence of an action, such as *loving* or *going.* These verbs, may have objects, hence called *transitive,* or not, hence called *intransitive.*

The canonical pattern of a simple Sanskrit sentence is as follows:

Subject - Subject-Enlargement - Verb - Object - Object-Enlargement - Adverb

Enlargements of the subject and the object are ways of qualifying nouns; they consist of adjectives and other nouns and compounds. The subject and object can be nouns, noun compounds, or pronouns. If the verb needs more than one object, there will be an order on the objects, e.g. preliminary, secondary etc; this order is kept in the sentence.

The simplest type of compounds are *aluk* compounds. In these compounds the leftmost part of the compound is inflected and the rest of the compound is not. The remaining compounds are of the *luk* type, where the *avyayibhava* compounds are the easiest; they are the inflected adverbs, for instance obtained from a preposition followed by a noun. We also deal with the *nan-tatpurusa* compounds, which are prefixed with the bound morpheme *a* or *an* and the *upapada-tatpurusa* ones, which end with bound morphemes derived from a verbal root, such as *bhida*, *jna*, and *ghna*.

Sanskrit is a highly inflected language, with three genders (masculine, feminine, neutral), three numbers (singular, dual, plural), and eight cases (nominative, accusative, dative, ablative, locative, vocative, instrumental). For the purposes of demonstration, we deal with the singular nominative and accusative inflections. The former has ending 'h' and the latter has ending 'm'. So for instance, the name *Rama* will have the form *Ramah* in singular nominative case and the form *Ramam* in singular accusative case.

5.1 A Pregroup Grammar for Sanskrit

We start with a set of basic types $\{\rho, \pi, o, p, n, s\}$. Apart from ρ, these types are the same as in English: ρ denotes the subject of a predicative copular[2]. The enlargements are treated as modifiers of the subject, the object or the predicate and, since they are placed to the right of the unit they are modifying, they will have a right adjoint type. Consequently, the subject-enlargement will be assigned the type $\pi^r \pi$, the object-enlargement the type $o^r o$, and the predicate enlargement the type $p^r p$.

An intransitive verb has type $\pi^r s$ and a transitive verb has type $\pi^r s o^l$. The Sanskrit copula '*aste*' (*is*) in its existential form has type $\pi^r s$ and in its predicative form has type $\rho^r s p^l$. The adverb is treated as a sentence modifier: it takes a sentence and modifies it, and as it occurs to the right of the verb, it will also have the type of a right adjoint of the sentence: $s^r s$. On this basis, the pregroup type assignment table corresponding to the general Sanskrit sentence structure is as follows:

	Subject	Subject-Enl	Verb	Object	Object-Enl	Adverb
	π	$\pi^r \pi$		o	$o^r o$	$s^r s$
Intransitive			$\pi^r s$			
Existential Copular			$\pi^r s$			
Transitive			$\pi^r s o^l$			
Predicative Copular			$\rho^r s p^l$			

and sentences with a transitive verb or a predicative copular are typed as follows:

Subject	Subject-Enlargement	Verb	Object	Object-Enlargement	Adverb.
π	$(\pi^r \pi)$	$(\pi^r s o^l)$	o	$o^r o$	$s^r s$

Subject	Subject-Enlargement	Verb	Predicate	Predicate-Enlargement	Adverb.
ρ	$(\rho^r \rho)$	$(\rho^r s p^l)$	p	$p^r p$	$s^r s$

---

[2] As we will see in Section 7.2, copulars play a more complex role in word order alternation in Sanskrit than in English. That is why we assign a different type to the predicative copular subjects.

The intransitive verb and the existential copulars are special cases of the examples below, where the object and predicate (correspondingly for each case) are dropped.

As an example of a transitive sentence, consider the sentence 'Rama from the old city saw Súbhadrā (a) beautiful woman'[3], which types as follows:

Ramah   from the old city   saw   Súbhadrā   (a) beautiful woman.
Ramah   kapiJjalArma   apasyat   Súbhadrā   maJjunAzI
π $(\pi^r \pi)$ $(\pi^r s\, o^l)$ o $(o^r o)$

As an example of a copular sentence, consider the sentence 'Rama from the old city [is] asked for instruction (as a teacher)', in which 'adhISTa' is an adjective in predicative role, and types as follows:

Rama   from the old city   [is]   asked for instruction.
Ramah   kapiJjalArma   [aste]   adhISTa
ρ $(\rho^r \rho)$ $(\rho^r s\, p^l)$ p

As for the compounds, we only describe here the case of transitive sentences; the case of copulars is obtained by replacing o with p and π with ρ or ϕ, where appropriate. The internal structure of a compound is set in such a way that its resulting type, after the internal cancellations have taken place, is either $\pi^r \pi$ or $o^r o$. We develop a procedure for two-word $\pi^r \pi$ compounds, trusting that the two-word $o^r o$ compounds can be treated in a similar fashion, basically by substituting the type π with o and any of the π adjoints with the corresponding o adjoint. The treatment for k-word compounds follows shortly below. For *aluke* compounds, we type the first word of the compound as $\pi^r \pi \pi^l$ and the second word as π, so that the first word inputs the second word and outputs a subject enlargement. The same methodology is used for the *avyayibhava* and *nan-tatpurusa* compounds, where the preposition of the first case and the morphemes of the second case are assumed to input the noun and output a subject enlargement. As for the *upapada-tatpurusa* compounds, the methodology is basically the same, but because the morpheme occurs at the end of the compound, it has to have type $\pi^r \pi^r \pi$. These assignments are expressed in the following table:

	Type of Compound	Word 1	Word 2
1	*aluke*	$\pi^r \pi \pi^l$	π
	luke		
2	*avyayibhava*	$\pi^r \pi \pi^l$	π
3	*nan-tatpurusa*	$\pi^r \pi \pi^l$	π
4	*upapada-tatpurusa*	π	$\pi^r \pi^r \pi$

---

[3] Súbhadrā in Mahaabharata in the wife of Arjuna, one of the heroes of the poem.

The above procedure extends to compounds that have more than two words as follows. Suppose there are k words in a compound, then the first word of the compounds of type 1 to 3 will be typed as follows:

$$\pi^r \pi \underbrace{\pi^l \cdots \pi^l}_{k-1}$$

The compounds of type 4 will be typed as follows:

$$\underbrace{\pi^r \cdots \pi^r}_{k-1} \pi^r \pi$$

Here are three examples; the first is *aluke*, the second is *avyayibhava* and the third is *upapada-tatpurusa*. Following [18], the inflected word (i.e. the first one of the *aluke* compound) is denoted by the label $N1$, the inflected adverb by $D1$, the non-inflected noun by N, and the non-inflected adjective by A.

$$[[_{N1} \ \text{atamane}] \ [_N \ \text{pada}]] \quad \to \ [_N \ \text{atamanepada}]$$
$$\pi^r \ \pi \ \pi^l \qquad\qquad \pi \qquad \leq \qquad \pi^r \ \pi$$
$$\text{oneself} \qquad\qquad \text{voice} \qquad\quad \text{voice for oneself}$$

$$[[_{N1} \ \text{upari}] \ [_N \ \text{bhumi}]] \quad \to \ [_N \ \text{uparibhumi}]$$
$$\pi^r \ \pi \ \pi^l \qquad\quad \pi \qquad \leq \qquad \pi^r \ \pi$$
$$\text{above} \qquad\quad \text{earth} \qquad \text{above the ground}$$

$$[_N [_N \ \text{sarva}] \quad [_N \ \text{-jna}]] \to [_N \ \text{sarvajna}]$$
$$\pi \qquad\qquad \pi^r \pi^r \pi \quad \leq \quad \pi^r \pi$$
$$\text{all} \qquad\qquad \text{knowing} \quad \text{omniscient}$$

As for inflections, we assign the type $n^r \pi$ to the nominative case morpheme and the type $n^r o$ to the accusative case morpheme. When these attach themselves to the end of a noun, which is of type n, the cancellation between the type n of the noun and the type n^r of these morphemes, will produce a subject π and an object o, respectively. The singular nominative and accusative cases are typed in the following table:

Case	Type of Morpheme	Reduction
Nominative	$n^r \pi$	$n(n^r \pi) \leq \pi$
Accusative	$n^r o$	$n(n^r o) \leq o$

So, for instance, we assign the type n to *Rama* and the type $n^r \pi$ to the nominative morpheme 'h', hence *Rama-h* will have type π. Similarly, we assign the type $n^r o$ to the accusative morpheme 'm', hence obtaining the type o for *Rama-m*.

$$\text{Rama} \quad \text{h} \ \to \text{Ramah} \qquad \text{Rama} \quad \text{m} \ \to \text{Ramam}$$
$$n \quad (n^r \pi) \ \leq \ \pi \qquad\qquad n \ (n^r o) \ \leq \ o$$

6   Cyclic Rules and Alternation of Word Order

By alternation of word order, we mean that certain language units within a sentence move from *after* the verb to *before* it, or from *before* the verb to *after* it, such that the resulting composition of words is still a grammatical sentence, although they may convey a different meaning. Non-inflectional languages such as English seem to have a fixed word order and inflectional ones, such as Latin, a free word order. However, the former cases do allow for alternations in word order as a result of, for instance, putting emphasis on a part of speech. At the same time, the latter cases do not allow for all possible permutations without changing the meaning.

Pregroups were not able to reason about alternations of word order in a general way and we offer a solution here. We propose to enrich the pregroup grammar of a language with a set of *precyclic transformations* that allow for substituting certain type combinations with their *precyclic permutations*. These transformations differ from language to language and express different, language specific, movement patterns. Within each language, they are restricted to a specific set so that not all word orders become permissible. More formally, we define:

Definition 2. *In a pregroup P, whenever $1 \leq ab \implies 1 \leq ba^{ll}$ or $1 \leq b^{rr}a$, then we refer to ba^{ll} and $b^{rr}a$ as precyclic permutations of ab and denote this relationship by $ab \overset{\sigma(ll)}{\leadsto} ba^{ll}$ and $ab \overset{\sigma(rr)}{\leadsto} b^{rr}a$.*

Definition 3. *In a pregroup P, for ba^{ll} and $b^{rr}a$ precyclic permutations of ab, and any $A, B, C \in P$, we define the following precyclic transformations[4]:*

$$
\begin{aligned}
(ll)\text{-\textbf{transformation}} \quad & A \leq B(ab)C \overset{(ll)}{\leadsto} A \leq B(ba^{ll})C \\
(rr)\text{-\textbf{transformation}} \quad & A \leq B(ab)C \overset{(rr)}{\leadsto} A \leq B(b^{rr}a)C
\end{aligned}
$$

Definition 4. *A precyclic pregroup grammar is a pregroup grammar with a set of precyclic transformations.*

The intuitions behind the (ll) and (rr) annotations of the σ denotations and the transformation rules are as described after Corollary 1. For instance, the (ll) permutation rule allows us to substitute the type ab with a permuted version of it, obtained by moving a from the left of the compound to the right of it, but as a result of this movement, we have to annotate a with ll, hence it becomes a^{ll}. This annotation registers the fact that a has moved to its current position from the left. The case for the (rr) rule is the same, here we substitute ab with $b^{rr}a$ and the annotation registers the fact that b has moved to its current position from the right of the compound.

Assuming these definitions, we formalise patterns of movement by providing a procedure on the basis of which one can derive the type of the verb after the movement, from its type before the movement. Hence, given the reduction of the

---

[4] These transformations prevent us from making isolated assumptions such as $1 \leq so^l$ and block the generation of meaningless inequalities such as $1o \leq (so^l)o \leq s$.

original sentence and this procedure, one is able to obtain the reduction of the sentence in which the alternation of word order has happened. The movement is captured by the idea that arguments of certain words and phrases with complex types, e.g. adjectives and verb phrases, can be moved before or after them, as an effect of stress or other semantic vs. pragmatic intentions. The procedure to extend the existing pregroup grammar of a language to include the word order changes resulting from movement has two main steps:

1. Decide which word or phrase w allows which forms of movement and encode this information about movement in the precyclic permutations of the type of each such word or phrase w.
2. Form a precyclic pregroup grammar from the pregroup grammar of a language by turning the above permutations into precyclic transformations.

The permutations are encoded as follows:

1.(i) If w is of type $p^r q$, i.e. it requires an argument of type p *before* it, and p can be moved after w, then allow for the cyclic permutation $p^r q \overset{\sigma(ll)}{\leadsto} q p^l$.
1.(ii) Else, if w is of type $q p^l$, i.e. it requires an argument of type p *after* it, and p can be moved before w, then allow for the cyclic permutation $q p^l \overset{\sigma(rr)}{\leadsto} p^r q$.

The transformations are encoded as follows:

2.(i) If w is from step 1(i), add an (ll)-transformation by taking $a = p^r$ and $b = q$ and computing $b a^{ll} = q(p^r)^{ll} = q p^l$.
2.(ii) Else, if w is from step 1(ii), add an (rr)-transformation by taking $a = q$ and $b = p^l$ and computing $b^{rr} a = (p^l)^{rr} q = p^r q$.

Finally, one says that a string of words is a grammatical sentence, whenever either the types of its words, as assigned by the pregroup grammar, reduce to s, or their transformed versions do.

English has different word order patterns, as discussed in detail by [4]. The basic English word order is SVO (Subject-Verb-Object), but this order may change as a result of object topicalisation or VP-preposing. Topicalisation allows for the object to move from after the verb phrase to before it. VP-preposing allows for the infinitive verb phrase to move from after the auxiliary or modal verb to before it. These permissible movements are encoded in the following precyclic transformations:

Moving Unit	Permutation	Transformation
Object	$s o^l \overset{\sigma(rr)}{\leadsto} o^r s$	$A \leq B(s o^l)C \overset{(rr)}{\leadsto} A \leq B(o^r s)C$
Infinitive	$s i^l \overset{\sigma(ll)}{\leadsto} i^r s$	$A \leq B(s i^l)C \overset{(rr)}{\leadsto} A \leq B(i^r s)C$

As an example of topicalization, consider the simple transitive sentence 'I saw him', and its topicalized form 'Him I saw', which are typed as follows:

$$\text{I saw him. :} \quad \pi(\pi^r s o^l) o \leq 1(s o^l) o \leq 1 s 1 \leq s$$

$$\text{Him I saw. :} \quad o \, \pi(\pi^r s o^l) \leq o(s o^l) \overset{(rr)}{\leadsto} o(o^r s) \leq s$$

We are not allowed to reduce the other possible four orderings (*him saw I, saw I him, saw him I, I him saw*) to s, since for obtaining similar permutations we need either the subject to move to after the verb, or subject and object invert their relative position; in both cases the consequence is that the subject and the verb occur in configurations like verb-subject (inversion) or subject-object-verb (separate) not admitted by the English grammar, as pointed out in [4]. Hence, we have not included the unlawful permutations that lead to these cases; examples of these are $(*_1)$ $\pi^r s \overset{\sigma(ll)}{\leadsto} s\pi^l$ and $(*_2)$ $\pi^r so^l \overset{\sigma(ll)}{\leadsto} so^l\pi^l$. As another example, consider the sentence 'He must love her', as typed below:

$$\text{He must love her.}$$
$$\pi \ (\pi^r si^l) \ (io^l) \ \ o$$

Here we can have both topicalisation (case (1) below) and VP-preposing (case (2) below). The type assignments and reductions of these cases are as follows:

(1) Her he must love. : $o\,\pi(\pi^r s\,i^l)(i\,o^l) \le o\,(so^l) \overset{(rr)}{\leadsto} o\,(o^r\,s) \le s$

(2) Love her he must. : $(i\,o^l)\,o\,\pi\,(\pi^r s\,i^l) \le i\,(s\,i^l) \overset{(rr)}{\leadsto} i\,(i^r\,s) \le s$

Non-permissible combinations like 'must love her he' or 'must love he her' cannot be derived, because they require, as before, a transformation corresponding to the precyclic permutation $\pi^r so^l \overset{\sigma}{\leadsto} so^l\pi^l$, in which the subject is expected to occur after the verb, that has not been included into the pregroup grammar.

7 Alternation of Word Order in Sanskrit

Like Latin, Sanskrit is a case-sensitive or inflectional language. That is, it has morphemes that attach themselves to the end of the words and specify their grammatical role in the sentence. For instance a subject may be marked with the morpheme 'h', for the nominative case, and an object with the morpheme 'm', for the accusative case, as in the following transitive sentence:

Ramah   apasyat   Govindam.
Rama   saw   Govinda

which is typed as follows:

$$\text{Ramah apasyat Govindam.}$$
$$\pi \qquad (\pi^r \ s \ o^l) \qquad o \qquad \le s$$

One might think that, no matter where these words appear in the sentence, one can be certain of their role and it need not be the case, as it is in English, that the word order tells you which word is the subject and which is the object. For instance the above Sanskrit sentence may as well be as follows and there might be no doubt that still '*Ramah*' is its subject and '*Govindam*' is its object:

Govindam   apasyat   Ramah.
Govindam   saw   Ramah

Whereas if the order of the words of a sentence changes in English, the roles change as well. The sentence 'Rama saw Govinda' has a very different meaning with respect to 'Govinda saw Rama': subject and object exchange their roles. Because of these matters, it is argued that Sanskrit has a free word order. However, a completely free word order for Sanskrit has been debated, e.g. by Staal [33] and later by Gillon [18] and even the original work of Apte [5] expresses concern for this presupposition. Apte insists that certain word orders may not be ungrammatical, but they are certainly awkward. In the following, we first present Staal's view and his constraints on movement, and then review Gillon's view and give some examples for each case.

7.1   Staal's Constraints

According to Staal [33] word order is free among the branches (sisters) of one and the same constituent. So for instance in the above example, *Ramah* is one of the branches and *apasyat Govindam* is the other one, which itself consists of two branches: *apasyat* and *Govindam*. Consequently, *apasyat* and *Govindam* can change their order, and then *Ramah* can change its order with regard to these two possible orders in *apasyat Govindam*.

These alternations are formalised via just two permutations:

Moving Unit	Permutation	Transformation
Subject	$\pi^r s \overset{\sigma(ll)}{\leadsto} s\pi^l$	$A \leq B(\pi^r s)C \overset{(ll)}{\leadsto} A \leq B(s\pi^l)C$
Object	$\pi^r s o^l \overset{\sigma(rr)}{\leadsto} o^r \pi^r s$	$A \leq B(\pi^r so^l)C \overset{(rr)}{\leadsto} A \leq B(o^r \pi^r s)C$

We have the following sentences and reductions:

– When *apasyat* and *Govindam* swap order

$$\text{Ramah Govindam apasyat.}$$

$$\pi \qquad o \qquad (\pi^r so^l) \overset{(rr)}{\leadsto} \pi\, o\, (o^r \pi^r s) \leq s$$

– When *Ramah* swaps order with *apaysat Govindam*

$$\text{apasyat Govindam Ramah.}$$

$$(\pi^r\, s\, o^l) \qquad o \qquad \pi \;\leq\; \underline{(\pi^r s)}\, \pi \overset{(ll)}{\leadsto} (s\, \pi^l)\, \pi \leq s$$

– When *Ramah* swaps order with *Govindam apaysat*

$$\text{Govindam apasyat Ramah.}$$

$$o \qquad (\underline{\pi^r so^l}) \qquad \pi \overset{(rr)}{\leadsto} o(o^r \pi^r s)\pi \leq (\underline{\pi^r s})\pi \overset{(ll)}{\leadsto} (s\pi^l)\pi \leq\; s$$

The two other possibilities, which according to Staal are not allowed, are as follows

$(*_1)$ *Govindam Ramah apasyat* $(*_2)$ *apasyat Ramah Govindam*

These are not derivable using the permutations of the above table. The first case needs the permutation $(*_1)$ $so^l \overset{\sigma(rr)}{\leadsto} o^r s$ and the second one needs the permutation $(*_2)$ $\pi^r so^l \overset{\sigma(ll)}{\leadsto} so^l \pi^l$, which we have not allowed.

7.2   Gillon's Conjectures

In [18], after reviewing Staal's constraints, Gillon observes that certain word orders that Staal's theory considered unlawful, do occur in his reference Sanskrit corpus. This corpus consists of examples from Apte and some older texts. His observation shows that Staal's constraints might be too rigid. Gillon goes on to find a pattern within these occurrences and develop appropriate constraints for word order alternation in Sanskrit. The constraints he discovers show themselves when working with sentences in which at least the subject and the verb have modifiers. In this paper, we study some of these constraints using precyclic transformations.

Although Gillon suggests that simple transitive sentences, such as the example of the previous section, allow for a fairly free word order, none of his witness sentences are simple. That is, in all the sentences he presents and studies the subject and/or object have modifiers, whereas in a simple transitive sentence subject and object do not have modifiers. Hence, in this section we let the invalid permutations of the previous section remain invalid and allow for new predictive permutations including the following ones:

$$(o^r o) \stackrel{\sigma(ll)}{\rightsquigarrow} (oo^l) \qquad (\pi^r \pi) \stackrel{\sigma(ll)}{\rightsquigarrow} (\pi\pi^l) \qquad (\rho^r \rho) \stackrel{\sigma(ll)}{\rightsquigarrow} (\rho\rho^l) \qquad (s^r s) \stackrel{\sigma(ll)}{\rightsquigarrow} (ss^l)$$

These permutations, enable the subject and object change order with their enlargements, and the verb swap place with adverbs (which can be seen as verb enlargements and some times even referred to as verb complements). At the same time, we will not allow for a fully free word order within a simple transitive sentence, that is a subject and object that do not have enlargements will obey Staal's constraints. If at any point, one wants to allow for these alternations as well (hence allowing for all of the six permutations of words in a three-word transitive sentence), one can include $(*_1)$ and $(*_2)$ of the previous section in the set of lawful permutations.

Extraposition from Subject. Here the suggestion is that the subject modifier can be separated from the subject, despite the fact that they form a constituent and according to Staal should not be separated. As an example consider the sentence "Rama from the old city saw Govinda", where the subject "Rama" has the enlargement "from the old city":

Rama	from the old city	apasyat	Govinda.
Ramah	kapiJjalArma	saw	Govindam.
π	$(\pi^r \pi)$	$(\pi^r s\, o^l)$	o

Here, "Rama" and "from the old city" form one constituent and, although they can change position, allowing for phrases such as "from the old city Rama", according to Staal, they should never be separated from each other in a sentence. However, according to Gillon's witness sentences, it should be possible to form the following sentence, in which this separation does indeed occur[5]:

---

[5] We do not give exactly the same example as Gillon, since his examples are retrieved from a real corpus and are rather complex, e.g. the example for this case is in question form, and we have not discussed question forms here.

from the old city   saw   Govinda   Rama.
kapiJjalArma   apasyat   Govindam   Ramah.
$(\pi^r\pi)$ $(\pi^r so^l)$ o π

This sentence types in the following way: first the subject modifier's type will change from $(\pi^r\pi)$ to $(\pi\pi^l)$ via an (ll)-permutation; then the subject swaps places with the verb phrase "saw Govinda" of type $(\pi^r s)$ via the following (ll)-permutation:

$$(\pi^r s)\pi \overset{\sigma(ll)}{\rightsquigarrow} \pi(\pi^r s)^{ll}$$

As a result of these movements, the subject modifier type $(\pi\pi^l)$ and the subject type π will cancel out. Now it remains to input the subject to the verb; this is done by swapping the type of the verb with the subject and permuting it to $(s\pi^l)$. The corresponding permutations are as follows:

$$\pi(\pi^r s)^{ll} \overset{\sigma(rr)}{\rightsquigarrow} (\pi^r s)\pi \qquad (\pi^r s) \overset{\sigma(ll)}{\rightsquigarrow} (s\pi^l)$$

The full reduction is shown below:

from the old city   saw   Govinda   Rama.
kapiJjalArma   apasyat   Govindam   Ramah.

$(\pi^r\pi)$ $(\pi^r so^l)$ o π $\leq$ $(\pi^r\pi)(\pi^r s)\pi \overset{(ll)}{\rightsquigarrow} (\pi\pi^l)(\pi^r s)\pi$

$\overset{(ll)}{\rightsquigarrow} \underline{(\pi\pi^l)}\pi(\pi^r s)^{ll} \leq \pi(\pi^r s)^{ll}$

$\overset{(rr)}{\rightsquigarrow} (\underline{\pi^r s})\pi \overset{(ll)}{\rightsquigarrow} (s\pi^l)\pi \leq s$

Extraposition from Verb Phrase. Here, Gillon observes that the object can be separated from its verb and move (on its own and leaving the verb behind) to the beginning of the sentence, which is exactly the $(*_1)$ case that was discarded by Staal. However, Gillon argues that this probably only happens in sentences where the object has an enlargement and that this enlargement will remain in situ, keeping the (perviously maintained by the object) connectivity to the verb. An example is the sentence "Ramah saw Govindam from the old city":

Ramah   apasyat   Govindam   kapiJjalArma
π $(\pi^r so^l)$ o $(o^r o)$ $\leq s$

A verb phrase extraposition of the above sentence results in the following:

Govindam   Ramah   apasyat   kapiJjalArma.
o π $(\pi^r so^l)$ $(o^r o)$

We type this sentence by first inputting the subject to the verb and reducing the $\pi(\pi^r s o^l)$ phrase to $(s o^l)$. Then we apply a (ll)-permutation to this phrase and the object modifier and swap their places; this permutation is as follows:

$$(s o^l)(o^r o) \overset{\sigma(ll)}{\rightsquigarrow} (o^r o)(s o^l)^{ll}$$

Now it becomes possible to input the object to its enlargement and obtain the modified object via the reduction $o(o^r o) \leq o$. At this stage, we send the verb back via the following (rr)-transformation:

$$o\,(s\,o^l)^{ll} \overset{(rr)}{\rightsquigarrow} (s\,o^l)\,o$$

All that remains to be done is to input the object to the verb via the reduction $(so^l)o \leq o$ and obtain a sentence. The full reduction is shown below:

Govindam Ramah apasyat kapiJjalArma.

$$o \qquad \pi \qquad (\pi^r s\,o^l) \qquad (o^r o) \qquad \leq o(so^l)(o^r o) \overset{(ll)}{\rightsquigarrow} o(o^r o)(so^l)^{ll} \leq$$
$$o(so^l)^{ll} \overset{(rr)}{\rightsquigarrow} (s\,o^l)o \leq s$$

Extraposition from Verb Complement. In such cases, adverbial phrases can move to the beginning of the sentence. For instance, a sentence such as

Ramah apasyat Govindam suzevya.
Rama   saw   Govinda   dearly
π $(\pi^r s\,o^l)$ o $(s^r s)$

can be turned into the following one

suzevya Ramah apasyat Govindam.
$(s^r s)$ π $(\pi^r s\,o^l)$ o

To type the above, we first input the subject and the object to the verb to obtain $(s^r s)\,s$, and then permute the adverbial type $(s^r s)$ to $(s\,s^l)$. The full reduction is as follows:

suzevya Ramah apasyat Govindam.
$(s^r s)$ π $(\pi^r s\,o^l)$ o $\leq (s^r s)s \overset{(ll)}{\rightsquigarrow} (ss^l)s \leq s$

Copular Sentences. A special case, upon which both Staal and Gillon agree, is the word order in copular sentences. According to [19], in the existential copular sentences, the subject can move from before the verb to after it, regardless of the explicit or implicit presence of the verb *aste*. However, in the predicative copulars, when *aste* is present, the subject always stays before the verb. We deal with this case by using the different type of the subject of the predicative copular. Because of this difference in type (ρ rather than π), the permutation that allowed for the subject of other sentences to move to after the verb, that is $\pi^r s \overset{\sigma(ll)}{\rightsquigarrow} s\pi^l$, is not applicable to predicative copular case. In other words, since we did not include the following unlawful permutation, we bar the movement of the subject of a predicative copular to after *aste*.

$$(*_3)\ \rho^r s \overset{\sigma(ll)}{\rightsquigarrow} s\rho^l$$

All the other plausible copular movements and their corresponding lawful movements can be reasoned about, similarly to the cases presented above. For instance, we can have the following sentence:

From the old city Súbhadrā   is   a lady.
  kapiJjalArma   Súbhadrā   aste   adhISTa.
  $(\rho^r \rho)$ ρ $(\rho^r s\, p^l)$ p

We type the above by applying just one transformation, namely $(\rho^r \rho) \overset{\sigma(ll)}{\rightsquigarrow} (\rho \rho^l)$ and then proceed the cancelations as usual. That is, the ρ in *Súbhadrā* will be cancelled out with the ρ^l in the new type of *kapiJjalArma*, the result will cancel out with the ρ^r in the type of *aste* and so on. On the contrary, the following alternation will be unlawful and not derivable:

(*)   is   a lady   from the old city Súbhadrā.
   aste   adhISTa   kapiJjalArma   Súbhadrā.
  $(\rho^r s\, p^l)$ p $(\rho^r \rho)$ ρ $\not\leqslant$ s

8   Conclusions

In this paper, we applied the calculus of pregroups to the analysis of a basic fragment of the mother of Indo-European languages: Sanskrit. In particular we have used the recent development of precyclic transformations (work based on [16, 17]) to treat word order alternations in this language. In [17], we provided a preliminary analysis of word order in Persian and Latin using cyclic permutations and transformations in pregroups. The alternation of word order in Persian, at least in simple sentences, is closer to that in English, induced by topicalisation and preposing. Perhaps it is not surprising that the change of word order in Sanskrit is more similar to Latin, rather than to Persian, despite the geographical closeness of the regions in which these languages were spoken. In both Latin and Sanskrit, simple sentences can in principle enjoy a free word order (although Staal disagrees), but when the constituents have complements, there are some restrictions governing the discontinuities that lead the complements to be separated from the parts of speech they are modifying.

A proper treatment of the full lexicon (see, for instance, [7] for a formalisation using generative rewriting rules) and syntactic rules of Sanskrit is far from having been achieved in this paper. We offer some results both from the point of view of morphological properties and of syntactical organization, with special attention to word order, in particular, as a result of extraposition from the subject, verb and object clauses. We distinguish and study a property of pregroups, called *precyclicty*, and develop corresponding transformation rules to apply this property within types in compounds. These rules allow us to derive the new types of the words, caused by alternations, from their original types, in the canonical sentence structure. It remains for future work to study the expressive power of such transformation rules and expand the current work to a larger fragment of the language.

References

1. Abrusci, M.: Phase Semantics and Sequent Calculus for Pure Noncommutative Classical Linear Propositional Logic. Journal of Symbolic Logic 56, 1403–1451 (1991)
2. Abrusci, V.M.: Classical Conservative Extensions of Lambek Calculus. Studia Logica 71, 277–314 (2002)
3. Abrusci, V.M., Ruet, P.: Non-commutative Logic I: the Multiplicative Fragment. Annals of Pure and Applied Logic 101, 29–64 (1999)
4. Ades, A.E., Steedman, M.J.: On the Order of Words. Linguistics and Philosophy 4, 517–558 (1982)
5. Apte, V.: The Student's Guide to Sanskrit Composition, A Treatise on Sanskrit Syntax for Use of Schools and Colleges, Poona, India, 24th edn. Lokasamgraha Press (1960)
6. Bargelli, D., Lambek, J.: An algebraic approach to french sentence structure. In: de Groote, P., Morrill, G., Retoré, C. (eds.) LACL 2001. LNCS (LNAI), vol. 2099, pp. 62–78. Springer, Heidelberg (2001)
7. Bhargava, M., Lambek, J.: A Production Grammar for Sanskrit Kinship Terminology. Theoretical Linguistics 18, 45–60 (1992)
8. Bargelli, D., Lambek, J.: An Algebraic Approach to Arabic Sentence Structure. Linguistic Analysis 31, 301–315 (2001)
9. Buszkowski, W.: Lambek Grammars Based on Pregroups. In: de Groote, P., Morrill, G., Retoré, C. (eds.) LACL 2001. LNCS (LNAI), vol. 2099, pp. 95–109. Springer, Heidelberg (2001)
10. Buszkowski, W.: Cut elimination for Lambek calculus of adjoints. In: Abrusci, V.M., Casadio, C. (eds.) New Perspectives in Logic and Formal Linguistics, Proceedings of the 5th Roma Workshop, Rome, pp. 85–93 (2002)
11. Buszkowski, W.: Type Logics and Pregroups. Studia Logica 87, 145–169 (2007)
12. Casadio, C., Lambek, J.: An Algebraic Analysis of Clitic Pronouns in Italian. In: de Groote, P., Morrill, G., Retoré, C. (eds.) LACL 2001. LNCS (LNAI), vol. 2099, pp. 110–124. Springer, Heidelberg (2001)
13. Casadio, C., Lambek, J.: A Tale of Four Grammars. Studia Logica 71, 315–329 (2002)
14. Casadio, C., Lambek, J.: A Computational Algebraic Approach to Latin Grammar. Research on Language and Computation 3, 45–60 (2005)
15. Casadio, C., Lambek, J. (eds.): Recent Computational Algebraic Approaches to Morphology and Syntax, Polimetrica, Milan (2008)
16. Casadio, C., Sadrzadeh, M.: Clitic movement in pregroup grammar: A cross-linguistic approach. In: Bezhanishvili, N., Löbner, S., Schwabe, K., Spada, L. (eds.) TbiLLC 2009. LNCS, vol. 6618, pp. 197–214. Springer, Heidelberg (2011)
17. Casadio, C., Sadrzadeh, M.: Cyclic Properties from Linear Logic to Pregroups. In: Presented at: From Categories to Logic, Linguistics and Physics: A tribute for the 90th birthday of Joachim Lambek, CRM, Montreal (September 21, 2013)
18. Gillon, B.: Word Order in Classical Sanskrit. Indian Linguistics 57, 1–35 (1996)
19. Gillon, B.: Subject Predicate Order in Classical Sanskrit, Language and Grammar, Studies in Mathematical Linguistics and Natural Language. In: Casadio, C., Scott, P.J., Seely, R.A.G. (eds.) Center for the Study of Language and Information (2005)
20. Gillon, B.: Tagging Classical Sanskrit Compounds. In: Kulkarni, A., Huet, G. (eds.) Sanskrit Computational Linguistics. LNCS, vol. 5406, pp. 98–105. Springer, Heidelberg (2009)

21. Girard, J.Y.: Linear Logic. Theoretical Computer Science 50, 1–102 (1987)
22. Kiślak - Malinowska, A.: Polish Language in Terms of Pregroups. In: Casadio, C., Lambek, J. (eds.) Recent Computational Algebraic Approaches to Morphology and Syntax, Polimetrica, Milan, pp. 145–172 (2008)
23. Lambek, J.: The Mathematics of Sentence Structure. American Mathematics Monthly 65, 154–169 (1958)
24. Lambek, J.: Type Grammar Revisited. In: Lecomte, A., Perrier, G., Lamarche, F. (eds.) LACL 1997. LNCS (LNAI), vol. 1582, pp. 1–27. Springer, Heidelberg (1999)
25. Lambek, J.: A Computational Algebraic Approach to English Grammar. Syntax 7, 128–147 (2004)
26. Lambek, J.: From Word to Sentence, A Computational Algebraic Approach to Grammar, Polimetrica, Monza, Milan (2008)
27. Lambek, J.: Exploring Feature Agreement in French with Parallel Pregroup Computations. Journal of Logic, Language and Information 19, 75–88 (2010)
28. Lambek, J.: Compact Monoidal Categories from Linguistics to Physics. In: Coecke, B. (ed.) New Structures for Physics. Lecture Notes in Physics, vol. 813, pp. 467–487. Springer (2011)
29. Moortgat, M.: Categorical Type Logics. In: van Benthem, J., ter Meulen, A. (eds.) Handbook of Logic and Language, pp. 93–177. Elsevier, Amsterdam (1997)
30. Preller, A., Lambek, J.: Free Compact 2-Categories. Mathematical Structures in Computer Science 17, 309–340 (2007)
31. Sadrzadeh, M.: Pregroup Analysis of Persian Sentences. In: [15]
32. Sadrzadeh, M.: An Adventure into Hungarian Word Order with Cyclic Pregroups. In: Models, Logics, and Higher-Dimensional Categories, a tribute to the work of Michael Makkai, Centre de Recherches Mathématiques. Proceedings and Lecture Notes, vol. 53, pp. 263–275. American Mathematical Society (2011)
33. Staal, J.F.: Word Order in Sanskrit and Universal Grammar. D. Reidel Publishing Co., Dordrecht (1967)
34. Whitney, W.D.: Sanskrit Grammar: including both the classical language, and the older dialects, of Veda and Brahamana, 2nd edn. Harvard University Press, Cambridge (1889)
35. Yetter, D.N.: Quantales and (non-Commutative) Linear Logic. Journal of Symbolic Logic 55, 41–64 (1990)

The Logic of Entanglement

Bob Coecke

Department of Computer Science,
University of Oxford
coecke@cs.ox.ac.uk

Abstract. We expose the information flow capabilities of pure bipartite entanglement as a theorem — which embodies the exact statement on the 'seemingly acausal flow of information' in protocols such as teleportation. We use this theorem to re-design and analyze known protocols (e.g. logic gate teleportation and entanglement swapping) and show how to produce some new ones (e.g. parallel composition of logic gates). We also show how our results extend to the multipartite case and how they indicate that entanglement can be measured in terms of 'information flow capabilities'. Ultimately, we propose a scheme for automated design of protocols involving measurements, local unitary transformations and classical communication.

Prakash! Ok, here we go. In 2001 I found myself being in Montreal without any further academic job prospects, having done a PhD in the at that time career-suicidal area of quantum foundations, followed by work in the already buried area of quantum logic. An angel fell from the sky in the form of Prakash who put me on a plane to Oxford, to work with Samson. The ticket turned out to be a one-way one.

Besides this direct personal help Prakash has also been instrumental in raising a new community which draws from modern computer science mathematics and feeds this not only in quantum computer science but also in the foundations of physics as well. Key to these where that Barbados meeting which Prakash has been holding on this area since 2004, the Quantum Physics and Logic workshop series, and there were several one-off events too.

All of this only didn't consolidate my own career, but also the careers of many who followed, maybe not all represented here in this volume, but I am sure that I am speaking for all of them when praising Prakash's guidance and support to create an entire new generation of scientists at the interface of logic, computer science, and foundational physics.

This Paper. My fate in particular changed with the paper presented here, which for reasons of many top journals editors being either complete idiots or bitter frustrated failed scientists (and usually both), never got published. Ten years later this is the ideal opportunity to finally get this stuff a legitimate status. Below is the slightly adapted 2004 arXiv version [6].

F. van Breugel et al. (Eds.): Panangaden Festschrift, LNCS 8464, pp. 250–267, 2014.

This work first appeared in 2003 as a 160 page research report [5]. Before that it was presented at the 1st QPL in Ottawa where it endangered my life when I ended up being attacked by a member of the audience who got really upset by the "as if time goes backward" statement. The joke is that exactly this aspect of the paper, via Svetlichny whom only mentioned it in a footnote [11], made it into the New York Times as a "new model for quantum time-travel" when it ended up being experimentally implemented, ... but it was in fact already implemented long before in [8], by some decent people who knew how to cite.

More importantly, this work was the starting point of a diagrammatic reformulation of quantum theory, on which a textbook jointly written with Aleks Kissinger is forthcoming, and the categorical axiomatization of this work with Abramsky yielded categorical quantum mechanics [1].

1   Introduction

Entanglement has always been a primal ingredient of fundamental research in quantum theory, and more recently, quantum computing. By studying it we aim at understanding the operational/physical significance of the use of the Hilbert space tensor product for the description of compound quantum systems. Many typical quantum phenomena are indeed due to compound quantum systems being described within the tensor product $\mathcal{H}_1 \otimes \mathcal{H}_2$ and not within a direct sum $\mathcal{H}_1 \oplus \mathcal{H}_2$.

In this paper we reveal a new structural ingredient of the supposedly well-understood *pure bipartite entanglement*, that is, we present a new theorem about the tensor product of Hilbert spaces. It identifies a 'virtual flow of information' in so-called *entanglement specification networks*. For example, it is exactly this flow of information which embodies *teleporting* [3] an unknown state from one physical carrier to another. Furthermore, our theorem (nontrivially) extends to *multipartite entanglement*. We also argue that it provides a new way of conceiving entanglement itself and hence of *measuring entanglement*:

$$\text{entanglement} \equiv \text{information flow capabilities}$$

Indeed, our result enables *reasoning about quantum information flow* without explicitly considering classical information flow — this despite the impossibility of transmitting quantum information through entanglement without the use of a classical channel.

Using our theorem we can fairly trivially reconstruct protocols such as *logic gate teleportation* [7] and *entanglement swapping* [12]. It moreover allows smooth generation of new protocols, of which we provide an example, namely the conversion of accumulation of inaccuracies causing 'sequential composition' into *fault-tolerant* 'parallel composition' [10]. Indeed, when combing our new insights on the flow of information through entanglement with a model for the flow of classical information we obtain a powerful tool for designing protocols involving entanglement.

An earlier and extended version of this paper is available as a research report [5]. It contains details of proofs, other/larger pictures, other references, other applications and some indications of connections with logic, proof theory and functional programming.

2   Classical Information Flow

By the spectral theorem any non-degenerated measurement on a quantum system described in a n-dimensional complex Hilbert space $\mathcal{H}$ has the shape

$$M = x_1 \cdot P_1 + \ldots + x_n \cdot P_n.$$

Since the values $x_1, \ldots, x_n$ can be conceived as merely being tokens distinguishing the projectors $P_1, \ldots, P_n$ in the above sum we can abstract over them and conceive such a measurement as a set

$$M \simeq \{P_1, \ldots, P_n\}$$

of n mutually orthogonal projectors which each project on a one-dimensional subspace of $\mathcal{H}$. Hence, by von Neumann's projection postulate, a measurement can be conceived as the system being subjected to an *action* P_i and the observer being informed about which action happened (e.g. by receiving the token x_i).

In most quantum information protocols the indeterminism of measurements necessitates a flow of classical information e.g. the 2-bit classical channel required for teleportation [3]. We want to separate this *classical information flow* from what we aim to identify as the *quantum information flow*. Consider a protocol involving local unitary operations, (non-local) measurements and classical communication e.g. teleportation:

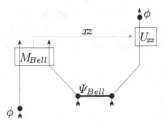

We can decompose such a protocol in

1. a *tree* with the consecutive operations as nodes, and, in case of a measurement, the emerging branches being labeled by tokens representing the projectors;
2. the configuration of the operations in terms of the time when they are applied and the subsystem to which they apply.

Hence we abstract over spatial dynamics. The nodes in the tree are connected to the boxes in the *configuration picture* by their temporal coincidence. For teleportation we thus obtain

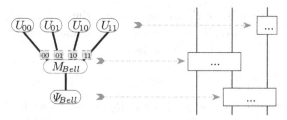

Classical communication is encoded in the tree as the dependency of operations on the labels on the branches below it e.g. the dependency of the operation U_{xz} on the variable xz stands for the 2-bit classical channel required for teleportation. We will also replace any initial state Ψ by the projector P_Ψ on it, which can be conceived as its *preparation* e.g. P_{EPR} is the preparation of an EPR-pair. It should be clear that for each path from the root of the tree to a leaf, by 'filling in the operations on the included nodes in the corresponding boxes of the configuration picture', we obtain a network involving only local unitary operations and (non-local) projectors e.g. one network

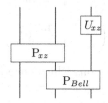

for each of the four values xz takes. It will be these networks (from which we extracted the classical information flow) for which we will reveal the quantum information flow. Hence each projector in it which is not a preparation is to be conceived *conditionally*.

3   Bipartite Entanglement

Let $\mathcal{H}_1$ and $\mathcal{H}_2$ be two finite dimensional complex Hilbert spaces. The elements of $\mathcal{H}_1 \otimes \mathcal{H}_2$ are in bijective correspondence with those of $\mathcal{H}_1 \to \mathcal{H}_2$, the vector space of *linear maps* with domain $\mathcal{H}_1$ and codomain $\mathcal{H}_2$, and also with those of $\mathcal{H}_1 \rightarrowtail \mathcal{H}_2$, the vector space of *anti-linear maps* with domain $\mathcal{H}_1$ and codomain $\mathcal{H}_2$. Given a base $\{e_\alpha^{(1)}\}_\alpha$ of $\mathcal{H}_1$ and a base $\{e_\beta^{(2)}\}_\beta$ of $\mathcal{H}_2$ this can easily be seen through the correspondences

$$\sum_{\alpha\beta} m_{\alpha\beta} \langle e_\alpha^{(1)} \mid - \rangle \cdot e_\beta^{(2)} \stackrel{\text{L}}{\simeq} \sum_{\alpha\beta} m_{\alpha\beta} \cdot e_\alpha^{(1)} \otimes e_\beta^{(2)}$$

$$\sum_{\alpha\beta} m_{\alpha\beta} \langle - \mid e_\alpha^{(1)} \rangle \cdot e_\beta^{(2)} \stackrel{\text{aL}}{\simeq} \sum_{\alpha\beta} m_{\alpha\beta} \cdot e_\alpha^{(1)} \otimes e_\beta^{(2)}$$

where $(m_{\alpha\beta})_{\alpha\beta}$ is the *matrix* of the corresponding function in bases $\{e_\alpha^{(1)}\}_\alpha$ and $\{e_\alpha^{(2)}\}_\beta$ and where by

$$\langle e_\alpha^{(1)} \mid - \rangle : \mathcal{H}_1 \to \mathcal{H}_2 \quad \text{and} \quad \langle - \mid e_\alpha^{(1)} \rangle : \mathcal{H}_1 \rightarrowtail \mathcal{H}_2$$

we denote the functionals which are respectively the linear and the anti-linear duals to the vector $e_\alpha^{(1)}$. While the second correspondence does not depend on the choice of $\{e_\alpha^{(1)}\}_\alpha$ the first one does since

$$\langle c \cdot e_\alpha^{(1)} | - \rangle = \bar{c} \cdot \langle e_\alpha^{(1)} | - \rangle \quad \text{and} \quad \langle - | c \cdot e_\alpha^{(1)} \rangle = c \cdot \langle - | e_\alpha^{(1)} \rangle.$$

We can now represent the *states* of $\mathcal{H}_1 \otimes \mathcal{H}_2$ by functions in $\mathcal{H}_1 \to \mathcal{H}_2$ or in $\mathcal{H}_1 \looparrowright \mathcal{H}_2$, and vice versa, these functions represent states of $\mathcal{H}_1 \otimes \mathcal{H}_2$. Omitting normalization constants, an attitude we will abide by throughout this paper, examples of linear maps encoding states are:

$$\text{id} := \begin{pmatrix} 1 & 0 \\ 0 & 1 \end{pmatrix} \overset{\text{L}}{\simeq} |00\rangle + |11\rangle$$

$$\pi := \begin{pmatrix} 0 & 1 \\ 1 & 0 \end{pmatrix} \overset{\text{L}}{\simeq} |01\rangle + |10\rangle$$

$$\text{id}^* := \begin{pmatrix} 1 & 0 \\ 0 & -1 \end{pmatrix} \overset{\text{L}}{\simeq} |00\rangle - |11\rangle$$

$$\pi^* := \begin{pmatrix} 0 & -1 \\ 1 & 0 \end{pmatrix} \overset{\text{L}}{\simeq} |01\rangle - |10\rangle$$

The last three of these four functions which encode the *Bell-base* states are (up to a scalar multiple) the *Pauli matrices*

$$\sigma_x \equiv X := \pi \qquad \sigma_y \equiv Y := i\pi^* \qquad \sigma_z \equiv Z := \text{id}^*$$

plus the *identity* which encodes the *Bell-state*. We can also encode each projector

$$P_\Psi : \mathcal{H}_1 \otimes \mathcal{H}_2 \to \mathcal{H}_1 \otimes \mathcal{H}_2 :: \Phi \mapsto \langle \Psi \mid \Phi \rangle \cdot \Psi$$

with $\Psi \in \mathcal{H}_1 \otimes \mathcal{H}_2$ by a function either in $\mathcal{H}_1 \to \mathcal{H}_2$ or $\mathcal{H}_1 \looparrowright \mathcal{H}_2$. Hence we can use these (linear or anti-linear) *functional labels* both to denote the states of $\mathcal{H}_1 \otimes \mathcal{H}_2$ and the projectors on elements of $\mathcal{H}_1 \otimes \mathcal{H}_2$. We introduce a graphical notation which incarnates this.

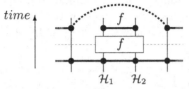

The box $\boxed{\;f\;}$ depicts the projector which projects on the bipartite state labeled by the (anti-)linear function f and the barbell $\bullet\!\!-\!\!^{f}\!\!-\!\!\bullet$ depicts that state itself. Hence the projector $\boxed{\;f\;}$ acts on the multipartite state represented by $\bullet\!\!-\!\!\bullet\;\;\bullet\!\!-\!\!\bullet$ and produces a pure tensor consisting of (up to a normalization constant) $\bullet\!\!-\!\!^{f}\!\!-\!\!\bullet$ and some remainder. Hence this picture portrays 'preparation of the f-labeled state'.

By an *entanglement specification network* we mean a collection of bipartite projectors $\boxed{\;f\;}$ 'configured in space and time' e.g.

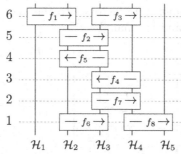

The arrows indicate which of the two Hilbert spaces in $\mathcal{H}_1 \otimes \mathcal{H}_j$ is the domain and which is the codomain of the labeling function.

Such a network can also contain local unitary operations — which we will represent by a grey square box $\boxed{U}$. We will refer to the lines labeled by some Hilbert space $\mathcal{H}_i$ ($\simeq$ time-lines) as *tracks*.

Definition 1. A *path* is a line which progresses along the tracks either forward or backward with respect to the actual physical time, and, which:

(i) respects the four possibilities

for entering and leaving a bipartite projector;

(ii) passes local unitary operations unaltered, that is

(iii) does not end at a time before any other time which it covers.

An example of a path is the grey line below.

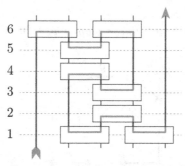

The notion of a path allows us to make certain predictions about the output $\Psi_{\mathbf{out}}$ of a network, that is, the state of the whole system after all projectors have been effectuated. Before stating the theorem we illustrate it on our example. Let

$$\Psi_{\mathbf{in}} := \phi_{in} \otimes \sum_{\alpha_2 \ldots \alpha_5} \Phi^{in}_{\alpha_2 \ldots \alpha_5} \cdot e^{(2)}_{\alpha_2} \otimes e^{(3)}_{\alpha_3} \otimes e^{(4)}_{\alpha_4} \otimes e^{(5)}_{\alpha_5}$$

be its input state. This input state factors into the *pure factor* ϕ_{in}, which we call the *input of the path*, and a remainder.

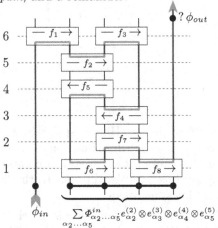

It should be clear that after effectuating all projectors we end up with an output which factors in the bipartite state labeled by f_1, the bipartite state labeled by f_2 and a remaining pure factor ϕ_{out} — which we call the *output of the path*. Our theorem (below) predicts that

$$\phi_{out} = (f_8 \circ f_7 \circ f_6 \circ f_5 \circ f_4 \circ f_3 \circ f_2 \circ f_1)(\phi_{in}). \tag{1}$$

Be aware of the fact that the functions $f_1, \ldots, f_8$ are not physical operations but labels obtained via a purely mathematical isomorphism. Moreover, the order in which they appear in the composite (1) has no obvious relation to the temporal order of the corresponding projectors. Their order in the composite (1) is:

the order in which the path passes through them

— this despite the fact that the path goes both forward and backward in physical time. Here's the theorem.

Lemma 1. *For f, g and h anti-linear maps and U and V unitary operations we have*

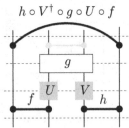

Proof. Straightforward verification or see [5] §5.1.   □

Theorem 1. (i) *Given are an entanglement specification network and a path. Assume that:*

1. *The order in which the path passes through the projectors is*

$$f_1 \Rightarrow f_2 \Rightarrow \ldots \Rightarrow f_{k-1} \Rightarrow f_k.$$

2. *The input of the path is a pure factor ϕ_{in}.*
3. *Ψ_{out} has a non-zero amplitude.*

Then the output of the path is (indeed) a pure factor ϕ_{out} which is explicitly given by

$$\phi_{out} = (f_k \circ f_{k-1} \circ \ldots \circ f_2 \circ f_1)(\phi_{in}). \tag{2}$$

(ii) *If the path passes forwardly through U then U will be part of the composite (2); if it passes backwardly through U then $U^\dagger$ will be part of the composite (2).*

Proof. Lemma 1 is the crucial lemma for the proof. For a full proof see [5] §5. □

It might surprise the reader that in the formulation of Theorem 1 we didn't specify whether $f_1, \ldots, f_k$ are either linear or anti-linear, and indeed, we slightly cheated. The theorem is only valid for $f_1, \ldots, f_k$ anti-linear. However, in the case that $f_1, \ldots, f_k$ are linear, in order to make the theorem hold it suffices to conjugate the matrix elements of those functional labels for which the path enters (and leaves) the corresponding projector 'from below' (see [5] §4.1):

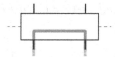

In most practical examples these matrix elements are real (see below) and hence the above theorem also holds for linear functional labels. One also verifies that if a path passes though a projector in the opposite direction of the direction of an anti-linear functional label f, then we have to use the *adjoint* $f^\dagger$ of the anti-linear map f in the composite (2) — the matrix of the adjoint of an anti-linear map $f^\dagger$ is the transposed of the matrix of f (see [5] §4.2). Finally note that we did not specify that at its input a path should be directed forwardly in physical time, and indeed, the theorem also holds for paths such as

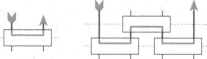

We discuss this in Section 5.

4   Re-designing Teleportation

By Theorem 1 we have

$$\tag{3}$$

due to

$$(\mathrm{id} \circ \mathrm{id})(\phi) = \phi.$$

When conceiving the first projector as the preparation of a Bell-pair while tilting the tracks we indeed obtain 'a' teleportation protocol.

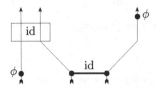

However, the other projector has to 'belong to a measurement' e.g.

$$M_{Bell} := \{\mathrm{P}_{\mathrm{id}}, \mathrm{P}_\pi, \mathrm{P}_{\mathrm{id}^*}, \mathrm{P}_{\pi^*}\}\,.$$

Hence the above introduced protocol is a *conditional* one. We want to make it *unconditional*.

Definition 2. Paths are *equivalent* iff for each input ϕ_{in} they produce the same output ϕ_{out}.

Corollary 1. *For U unitary and $g \circ U = U \circ g$ we have that*

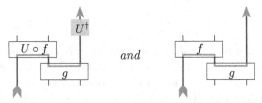

and

are equivalent paths.

Proof. Since $U^\dagger \circ g \circ (U \circ f) = g \circ f$ the result follows by Theorem 1.   □

Intuitively, one can move the box $\boxed{U^\dagger}$ along the path and permute it with projectors whose functional labels commute with U ($=$ commute with $U^\dagger$) until it gets annihilated by the U-factor of $\boxed{U \circ f}$. Applying Corollary 1 to

$$f, g := \mathrm{id} \quad \text{and} \quad U \in \{\mathrm{id}, \pi, \mathrm{id}^*, \pi^*\},$$

since $\pi^\dagger = \pi$, $(\mathrm{id}^*)^\dagger = \mathrm{id}^*$ and $(\pi^*)^\dagger = -\pi^*$, we obtain four conditional teleportation protocols

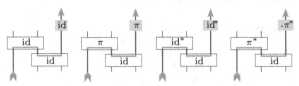

of which the one with $U := \text{id}$ coincides with (3). These four together constitute an unconditional teleportation protocol since they correspond to the four paths 'from root to leaf' of the tree discussed in Section 2, from which then also the 2-bit classical channel emerges.

In order to obtain the teleportation protocol as it is found in the literature, observe that $\pi^* = \pi \circ \text{id}^*$, hence

$$
\begin{array}{c|c|c}
\circ & \text{id} & \text{id}^* \\
\hline
\text{id} & \text{id} & \text{id}^* \\
\hline
\pi & \pi & \pi^*
\end{array}
$$

and thus we can factor — with respect to composition of functional labels — the 2-bit Bell-base measurement in two 1-bit 'virtual' measurements ($\vee$ stands for 'or'):

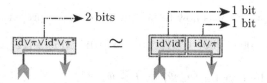

Note that such a decomposition of M_{Bell} does not exist with respect to $\otimes$ nor does it exist with respect to composition of projector actions. All this results in

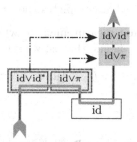

which is the standard teleportation protocol [3].

The aim of *logic gate teleportation* [7] is to teleport a state and at the same time subject it to the action of a gate f. By Theorem 1 we evidently have

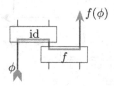

We make this protocol unconditional analogously as we did it for ordinary teleportation.

Corollary 2. *For U and V unitary and $g \circ V = U \circ g$ we have that*

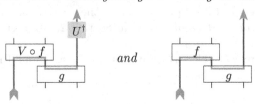

are equivalent paths.

Proof. Analogous to that of Corollary 2.   □

We apply the above to the case

$$f := \mathrm{id} \otimes \mathrm{id} \qquad \text{and} \qquad g := \mathrm{CNOT}$$

that is, the first projector is now to be conceived as the preparation of the state

$$\Psi_{\mathrm{CNOT}} = |00\rangle \otimes |00\rangle + |01\rangle \otimes |01\rangle + |10\rangle \otimes |11\rangle + |11\rangle \otimes |10\rangle .$$

Let Ψ_f be defined either by $f \overset{\mathrm{L}}{\simeq} \Psi_f$ or $f \overset{\mathrm{aL}}{\simeq} \Psi_f$.

Proposition 1. $\Psi_{f \otimes g} = \Psi_f \otimes \Psi_g$; $\mathrm{P}_{f \otimes g} = \mathrm{P}_f \otimes \mathrm{P}_g$.

Proof. The first claim is verified straightforwardly. Hence $\mathrm{P}_{f \otimes g} \equiv \mathrm{P}_{\Psi_{f \otimes g}} = \mathrm{P}_{\Psi_f \otimes \Psi_g} = \mathrm{P}_{\Psi_f} \otimes \mathrm{P}_{\Psi_g} \equiv \mathrm{P}_f \otimes \mathrm{P}_g$ what completes the proof.   □

Hence we can factor the 4-qubit measurement to which the second projector belongs in two Bell-base measurements, that is, we set

$$V \in \{U_1 \otimes U_2 \mid U_1, U_2 \in \{\mathrm{id}, \pi, \mathrm{id}^*, \pi^*\}\} .$$

The resulting protocol

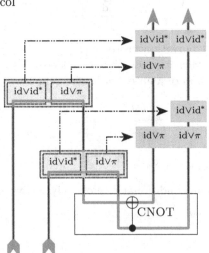

is the one to be found in [7] — recall that $U^\dagger$ factors as a tensor since CNOT is a member of the Clifford group.

Our last example in this section involves the passage *from sequential to parallel composition of logic gates*. Due to the accumulation of inaccuracies in sequential composition [10] it would be desirable to have a fault-tolerant parallel alternative. This would for example be useful if we have a limited set of available gates from which we want to generate more general ones e.g. generating all Clifford group gates from CNOT gates, Hadamard gates and phase gates via tensor and composition. By Theorem 1 the network

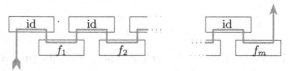

realizes the composite $f_m \circ \ldots \circ f_1$ conditionally. Again this protocol can be made unconditional — an algorithm which captures the general case can be found in [5] §3.4. Note that by Theorem 1 it suffices to make unitary corrections only at the end of the path [5] §3.4.

5   Entanglement Swapping

By Theorem 1 we have

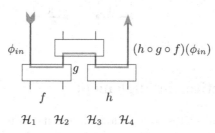

However, Theorem 1 assumes ϕ_{in} to be a pure factor while it is part of the output $\Psi_{\mathbf{out}}$ of the network. This fact constraints the network by requiring that

$$h \circ g \circ f \overset{\mathrm{aL}}{\simeq} \phi_{in} \otimes \phi_{out}$$

for some ϕ_{in} and ϕ_{out} i.e. the state labeled by $h \circ g \circ f$ has to be *disentangled* — which is equivalent to the range of $h \circ g \circ f$ being one-dimensional [5] §5.3.

Using Lemma 1 this pathology can be overcome by conceiving the output state of the bipartite subsystem described in $\mathcal{H}_1 \otimes \mathcal{H}_4$ not as a pair (ϕ_{in}, ϕ_{out}) but as a *function* $\varphi : \mathcal{H}_1 \hookrightarrow \mathcal{H}_4$ which relates any input $\phi_{in} \in \mathcal{H}_1$ to an output $\phi_{out} := \varphi(\phi_{in}) \in \mathcal{H}_4$. Hence we conceive the above network as producing a function

$$\varphi := h \circ g \circ f \overset{\mathrm{aL}}{\simeq} \Psi_\varphi$$

where $\Psi_{\mathbf{out}} = \Psi_\varphi \otimes \Psi_g$ with

$$\Psi_\varphi \in \mathcal{H}_1 \otimes \mathcal{H}_4 \quad \text{and} \quad g \overset{\mathrm{aL}}{\simeq} \Psi_g \in \mathcal{H}_2 \otimes \mathcal{H}_3.$$

To such a function produced by a network we can provide an input via a unipartite projector. The generic example (which can be easily verified) is

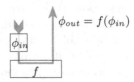

One can then conceive $\boxed{\;f\;}$ as a λ-*term* $\lambda\phi.f\phi$ [2] and the process of providing it with an input via a unipartite projector embodies the β-*reduction* [2]

$$(\lambda\phi.f\phi)\phi_{in} \overset{\beta}{=} f(\phi_{in}).$$

As we will see below we can 'feed' such a function at its turn as an input of *function type* in another network. This view carries over to the interpretation of multipartite entanglement where it becomes crucial.

The entanglement swapping protocol [12] can now be derived analogously as the teleportation protocol by setting $f = g = h := \mathrm{id}$ in the above. For this particular case Lemma 1 becomes

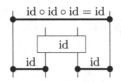

Details can be found in [5] §6.2.

6   Multipartite Entanglement

The passage from states to functions as inputs and outputs enables to extend our functional interpretation of bipartite entanglement to one for multipartite entanglement. In general this involves *higher order functions* and hence the use of denotational tools from modern logic and proof theory such as λ-calculus [2].

Whereas (due to commutativity of $- \otimes -$) a bipartite tensor $\mathcal{H}_1 \otimes \mathcal{H}_2$ admits interpretation as a function either of type $\mathcal{H}_1 \rightsquigarrow \mathcal{H}_2$ or of type $\mathcal{H}_1 \multimap \mathcal{H}_2$, a tripartite tensor (due to associativity of $- \otimes -$) admits interpretation as a function of a type within the union of two (qualitatively different) families of types namely

$$\mathcal{H}_i \rightsquigarrow (\mathcal{H}_j \rightsquigarrow \mathcal{H}_k) \quad \text{and} \quad (\mathcal{H}_i \rightsquigarrow \mathcal{H}_j) \rightsquigarrow \mathcal{H}_k.$$

Explicitly, given

$$\sum_{\alpha\beta} M_{\alpha\beta\gamma} \cdot e_\alpha^{(1)} \otimes e_\beta^{(2)} \otimes e_\gamma^{(3)} \in \mathcal{H}_1 \otimes \mathcal{H}_2 \otimes \mathcal{H}_3$$

we respectively obtain

$$f_1 : \mathcal{H}_1 \leftrightsquigarrow (\mathcal{H}_2 \leftrightsquigarrow \mathcal{H}_3)$$

$$:: \sum_\alpha \psi_\alpha \cdot e_\alpha^{(1)} \mapsto \sum_{\beta\gamma} \left(\sum_\alpha \bar{\psi}_\alpha M_{\alpha\beta\gamma} \right) \langle - \mid e_\beta^{(2)} \rangle \cdot e_\gamma^{(3)}$$

and

$$f_2 : (\mathcal{H}_1 \leftrightsquigarrow \mathcal{H}_2) \leftrightsquigarrow \mathcal{H}_3$$

$$:: \sum_{\alpha\beta} m_{\alpha\beta} \langle - \mid e_\alpha^{(1)} \rangle \cdot e_\beta^{(2)} \mapsto \sum_\gamma \left(\sum_{\alpha\beta} \bar{m}_{\alpha\beta} M_{\alpha\beta\gamma} \right) \cdot e_\gamma^{(3)}$$

as the corresponding functions — the complex conjugation of the coefficients $\bar{\psi}_\alpha$ and $\bar{m}_{\alpha\beta}$ is due to the anti-linearity of the maps. The appropriate choice of an interpretation for a tripartite projector depends on the *context* i.e. the configuration of the whole network to which it belongs. A *first order function* f_1 enables interpretation in a configuration such as

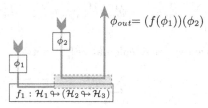

One can think of this tripartite projector as producing a bipartite one at its 'output'. A *second order function* f_2 — recall that a definite integral is an example of a second order function — enables interpretation in the configuration

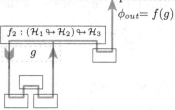

We illustrate this in an example — we will not provide an analogue to Theorem 1 for the multipartite case since even its formulation requires advanced denotational tools. Consider the following configuration.

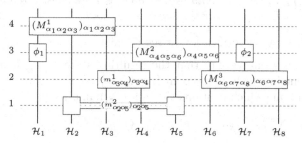

For 'good' types we can draw a 'compound' path.

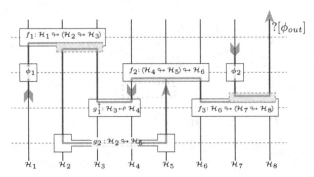

If a multipartite analogue to Theorem 1 truly holds one would obtain

$$\phi_{out} = (f_3 \circ f_2)(g_2 \circ (f_1(\phi_1))^\dagger \circ g_1^\dagger)(\phi_2).$$

Hence in terms of matrices we predict $\phi_{\alpha_8}^{out}$ to be

$$\sum_{\alpha_1 \ldots \alpha_7} \bar{\phi}_{\alpha_7}^2 m_{\alpha_3 \alpha_4}^1 \phi_{\alpha_1}^1 \bar{M}_{\alpha_1 \alpha_2 \alpha_3}^1 m_{\alpha_2 \alpha_5}^2 \bar{M}_{\alpha_4 \alpha_5 \alpha_6}^2 M_{\alpha_6 \alpha_7 \alpha_8}^3.$$

To verify this we explicitly calculate $\phi_{\alpha_8}^{out}$. Set

$$\Psi^\tau = \sum_{i_1 \ldots i_8} \Psi_{i_1 \ldots i_8}^\tau \cdot e_{i_1}^{(1)} \otimes \ldots \otimes e_{i_8}^{(8)}$$

where Ψ^0 is the (essentially arbitrary) input of the network and Ψ^τ for $\tau \in \{1, 2, 3, 4\}$ is the state at time $\tau + \epsilon$. For $I \subseteq \{1, \ldots, 8\}$ and $I^c := \{1, \ldots, 8\} \setminus I$ let P_Φ^I stipulate that this projector projects on the subspace

$$\Phi \otimes \bigotimes_{i \in I^c} \mathcal{H}_i \quad \text{for some} \quad \Phi \in \bigotimes_{i \in I} \mathcal{H}_i.$$

Lemma 2. *If* $\Psi^\tau = P_\Phi^I(\Psi^{\tau-1})$ *then*

$$\Psi_{i_1 \ldots i_8}^\tau = \sum_{j_\alpha | \alpha \in I} \Psi_{i_1 \ldots i_8 [j_\alpha / i_\alpha | \alpha \in I]}^{\tau-1} \bar{\Phi}_{(j_\alpha | \alpha \in I)} \Phi_{(i_\alpha | \alpha \in I)}$$

where $i_1 \ldots i_8 [j_\alpha / i_\alpha \mid \alpha \in I]$ denotes that for $\alpha \in I$ we substitute the index i_α by the index j_α which ranges over the same values as i_α.

Proof. Straightforward verification or see [5] §6.4.   □

Using Lemma 2 one verifies that the resulting state $\Psi_{i_1 \ldots i_8}^4$ factors into five components, one in which no index in $\{i_1, \ldots, i_8\}$ appears, three with indices in $\{i_1, \ldots, i_7\}$ and one which contains the index i_8 namely

$$\sum_{\substack{l_4 l_5 l_6 l_7 \\ m_1 m_2 m_3}} m_{m_2 l_5}^2 m_{m_3 l_4}^1 M_{l_6 l_7 i_8}^3 \phi_{m_1}^1 \bar{M}_{l_4 l_5 l_6}^2 \bar{\phi}_{l_7}^2 \bar{M}_{m_1 m_2 m_3}^1.$$

Substituting the indices m_1, m_2, m_3, l_4, l_5, l_6, l_7, i_8 by $\alpha_1, \ldots, \alpha_8$ we exactly obtain our prediction for $\phi_{\alpha_8}^{out}$.

It should be clear from our discussion of multipartite entanglement that, provided we have an appropriate entangled state involving a sufficient number of qubits, we can implement arbitrary (linear) λ-terms.

7   Discussion

For a unitary operation $U : \mathcal{H} \to \mathcal{H}$ there is a *flow of information* from the input to the output of U in the sense that for an input state ϕ the output $U(\phi)$ fully depends on ϕ.

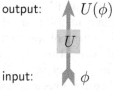

output: $U(\phi)$

input: ϕ

How does a projector P_ψ act on states? After renormalization and provided that $\langle \phi \,|\, \psi \rangle \neq 0$ the input state ϕ is not present anymore in the output $\psi = \mathrm{P}_\psi(\phi)$. At first sight this seems to indicate that through projectors on one-dimensional subspaces there cannot be a flow of information cfr. the 'wall' in the picture below.

Theorem 1 provides a way around this obstacle.

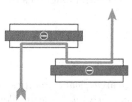

While there cannot be a flow from the input to the output, there is a 'virtual flow' between the two inputs and the two outputs of a bipartite projector whenever it is configured within an appropriate context. And such a bipartite projector on a state in $\mathcal{H}_1 \otimes \mathcal{H}_2$ can act on this flow as any (anti-)linear function f with domain in $\mathcal{H}_1$ and codomain in $\mathcal{H}_2$ — which is definitely more general than unitary operations and also more general than actions by (completely) *positive maps*. This behavioral interpretation extends to *multipartite entanglement*, and, as is shown in [5] §6.6, it also enables interpretation of *non-local unitary operations*.

The wall within a projector incarnates the fact that

$$\mathrm{P}_\psi \overset{\mathrm{L}}{\simeq} \psi \otimes \psi.$$

Indeed, one verifies that *disentangled states* $\psi \otimes \phi$ are in bijective correspondence with those linear maps which have a one-dimensional range [5] §5.3, that is, since states correspond to one-dimensional subspaces, disentangled states correspond to (partial) constant maps on states. Since constant maps incarnate the absence of information flow (cfr. 'the wall' mentioned above):

$$\frac{\text{entangled}}{\text{disentangled}} \simeq \frac{\text{information flow}}{\text{no information flow}}.$$

Pursuing this line of thought of conceiving entanglement in terms of its *information flow capabilities* yields a proposal for *measuring pure multipartite entanglement* [5] §7.5 — given a measure for pure bipartite entanglement e.g. majorization [9].

The use of Theorem 1 in Sections 4 and 5 hints towards *automated design* of general protocols involving entanglement. We started with a simple configuration which conditionally incarnates the protocol we want to implement. Conceiving this conditional protocol as a pair consisting of (i) a single path 'from root to leaf' in a tree, and, (ii) a configuration picture, we can extend the tree and the configuration picture with unitary corrections in order to obtain an unconditional protocol. It constitutes an interesting challenge to produce an explicit *algorithm* which realizes this given an appropriate front-end design language.

Recent proposals for fault-tolerant quantum computers of which the architecture is manifestly different from the circuit model require a different mathematical setting for programming them and reasoning about them [4]. We are convinced that the insights obtained in this paper provide the appropriate tool for doing so.

Acknowledgments. We thank Samson Abramsky, Howard Barnum, Sam Braunstein, Ross Duncan, Peter Hines, Radha Jagadeesan and Prakash Panangaden for useful input.

References

1. Abramsky, S., Coecke, B.: A categorical semantics of quantum protocols. In: Proceedings of 19th IEEE Conference on Logic in Computer Science, pp. 415–425. IEEE Press (2004), arXiv:quant-ph/0402130; Extended revised version: Categorical quantum mechanics. In: Engesser, K., Gabbay, D.M., Lehmann, D., (eds.): Handbook of Quantum Logic and Quantum Structures, pp. 261–323. Elsevier (2009), arXiv:0808.1023
2. Barendrecht, H.P.: The Lambda Calculus — Its Syntax and Semantics. North-Holland (1994)
3. Bennett, C.H., Brassard, G., Crépeau, C., Jozsa, R., Peres, A., Wooters, W.K.: Teleporting an unknown quantum state via dual classical and Einstein-Podolsky-Rosen channels. Physical Review Letters 70, 1895–1899 (1993)
4. Raussendorf, R., Browne, D.E., Briegel, H.-J.: Measurement-based quantum computation on cluster states. Physical Review A 68, 022312 (2003), arXiv:quant-ph/0301052

5. Coecke, B.: The Logic of entanglement. An invitation. Research Report PRG-RR-03-12 Oxford University Computing Laboratory (2003), http://www.cs.ox.ac.uk/techreports/oucl/RR-03-12.ps
6. Coecke, B.: The Logic of entanglement (2004), arXiv:quant-ph/0402014
7. Gottesman, D., Chuang, I.L.: Quantum teleportation is a universal computational primitive. Nature 402, 390–393 (1999), arXiv:quant-ph/9908010
8. Laforest, M., Laflamme, R., Baugh, J.: Time-reversal formalism applied to maximal bipartite entanglement: theoretical and experimental exploration. Physical Review A 73, 032323 (2006), arXiv:quant-ph/0510048
9. Nielsen, M.A.: Conditions for a class of entanglement transformations. Physical Review Letters 83, 436–439 (1999)
10. Preskill, J.: Reliable quantum computers. Proceedings of the Royal Society of London A454, 385–410 (1997), quant-ph/9705031
11. Svetlichny, G.: Effective quantum time travel (2009), arXiv 0902.4898
12. Żukowski, M., Zeilinger, A., Horne, M.A., Ekert, A.K.: 'Event-ready-detectors' Bell experiment via entanglement swapping. Physical Review Letters 71, 4287–4290 (1993)

Free Energy of Petri Nets

Vincent Danos and Ilias Garnier

LFCS, School of Informatics, University of Edinburgh, UK

1 Introduction

One can often extract valuable properties of Markov chains (MCs) by "compressing" them using various notions of bisimulation (exact or approximate, strong or weak, etc) [10,3,2,4]. Typically a bisimulation will lead from a concrete and perhaps overly detailed system to a simpler and more abstract one [7]. In this paper, we will go the opposite way! We will show that for the subclass of continuous-time MCs (ctMCs) corresponding to thermodynamically consistent stochastic mass action Petri nets (the standard model for chemical reactions for fast diffusing chemicals), one can construct in a systematic fashion concrete versions of the dynamics. These concrete MCs are functionally bisimilar to their abstract counterpart and admit a simpler description of their invariant probability (equivalently, of their energy function). This can sometimes reveal interesting equilibrium properties of the original chain.

To see how the benefits of the construction come about, and fix a few notations which we will re-use in the main development, we start with a simple class of 'urn models'. These are traditionally introduced as discrete models of diffusion.

1.1 Continuous Time Ehrenfest Urns

Transitions correspond to one particle moving from one urn to another. One has a constant number N of (non-interacting) particles, a set k of urns, and the state of the system is modeled by an occupancy map $\sigma : k \to \mathbb{N}$. For each urn i, $\sigma(i)$ specifies the number of particles in i. Thus, a state σ can be seen as a vector in $\mathbb{R}^k$. We will write $\mathbf{e_i}$s for the canonical basis vectors.

As the total number of particles is constant, one has a global invariant $\sum_i \sigma(i) = N$. In particular, for any (deterministic) initial condition, the set of reachable states is strongly connected and finite. The rate at which one particle travels from urn to urn is described by a $k \times k$ matrix R with real-valued coefficients (non-negative for the off-diagonal ones). Hence, we obtain a global ctMC with rate function Q_a whose off-diagonal values are:

$$Q_a(\sigma, \sigma - \mathbf{e_i} + \mathbf{e_j}) = \sigma(i)R_{ij} \qquad for\ i \neq j \tag{1}$$

Say we are interested in computing the fixed point of Q_a. To do this, we introduce a new chain Q_c which tracks each particle individually. The intuition is that, as the particles do not interact, their final distribution in the concrete Q_c model will be a simple product of each individual ones. The fixed point of the original chain Q_a will just be the projection of this simple product distribution.

F. van Breugel et al. (Eds.): Panangaden Festschrift, LNCS 8464, pp. 268–289, 2014.

A state of the concrete system is now a map $\tau : N \to k$ from invidual particles to the urn where they sit. We write $\tau[n \mapsto i]$ for the map identical to τ save for particle n sitting in i.

There is an evident projection map π from concrete states to abstract ones:

$$\pi(\tau)(i) = |\tau^{-1}(i)|$$

Each individual particle n in N travels over k following the rate matrix R defined above. Hence the concrete chain Q_c is the asynchronous parallel composition of N copies of R:

$$Q_c(\tau[n \mapsto i], \tau[n \mapsto j]) = R_{ij} \qquad for \ i \neq j \qquad (2)$$

Let us now assume that R is irreducible -i.e. has one strongly connected component, and has a fixed point probability. By classical MC theory [13], this implies that the fixed point $\mathcal{E}$ is unique and a global attractor, meaning:

$$\mathcal{E} = \lim_{t \to +\infty} p_n^T(0) \, e^{tR}$$

for any choice of the initial probability $p_n(0)$.

We want to prove that the equilibrium of the compound chain Q_c: 1) is the product over N of the individual (and identical) $\mathcal{E}$, 2) is unique, 3) and is also a global attractor for Q_c. To see this, we remark that the asynchronous parallel composition of ctMCs Q_1, Q_2 can be written as a Kronecker sum:

$$Q_1 \oplus Q_2 := Q_1 \otimes I_2 + I_1 \otimes Q_2$$

where $\otimes$ is the tensor product, and I_1, I_2 identity matrices of same dimensions as Q_1, Q_2. So, by definition, $Q_c = \bigoplus_N R$. Besides, it is easy to see that $e^{A \oplus B} = e^A \otimes e^B$. This gives the dynamics of the concrete chain in terms of R:

$$p_c(t) = (\bigotimes_N p_n(0))^T e^{tQ_c} = \bigotimes_N p_n(0)^T e^{tR}$$

where we assume the initial probability is a tensor (i.e. a product probability). Hence $p_c(t)$ stays always a product, and in the limit:

$$\lim_{t \to +\infty} p_c(t) = \bigotimes_N \mathcal{E}$$

Therefore, with $\mathcal{E}_c := \bigotimes_N \mathcal{E}$, this implies point 1) and point 3) for tensor initial states. Furthermore, Q_c being a product of irreducibles is clearly irreducible, hence points 2) and (full) 3) follow.

It remains to lift this concrete fixed point to the abstract state space. As we will prove shortly, it so happens that the projection π is a (forward functional stochastic) bisimulation. Intuitively, the abstract flow of probability is preserved in the concrete system. Formally, this means that for all σ, σ', and τ in $\pi^{-1}(\sigma)$:

$$Q_a(\sigma, \sigma') = \sum_{\tau' \in \pi^{-1}(\sigma')} Q_c(\tau, \tau') \qquad (3)$$

It is easy to verify that π is a bisimulation. Suppose one has an abstract transition $Q_a(\sigma, \sigma')$, then one must have $\sigma' = \sigma - e_i + e_j$ for some $i \neq j$. To match this transition on the concrete side, one needs to pick an n in i and move it to j. That is to say:

$$\sum_{\tau' \in \pi^{-1}(\sigma')} Q_c(\tau, \tau') = \sum_{n \in \tau^{-1}(i)} Q_c(\tau[n \mapsto i], \tau[n \mapsto j]) = \sigma(i)R_{ij}$$

which is equal to $Q_a(\sigma, \sigma')$ (Eqn. 1). This establishes that π is a bisimulation as defined in (Eqn. 3).

1.2   Categorical Aside

The fact that π is a bisimulation (Eqn. 1) can be neatly captured as a commutative diagram in the category of, say, finite sets and real-valued matrices. In this setting (Eqn. 1) simply says that $\pi Q_a = Q_c \pi$ (note the inversion of the composition due to the convention that $Q(x, y)$ is the rate to jump from x to y, which is the opposite of the usual linear algebraic convention) where π is now recast as a rectangular matrix in the obvious way: $\pi(\tau, \sigma) = 1$ iff $\pi(\tau) = \sigma$. This in turn immediately implies (and is in fact equivalent to) that $\pi e^{tQ_a} = e^{tQ_c} \pi$.

Again, we can express this commutation as a diagram in the category of probabilistic relations [14,15].[1] This diagram is shown in Fig. 1 in the specific case of our example (but is completely general). Therein, the function π being deterministic, is now seen as taking Dirac values $\delta_{\pi(\tau)}$. This category has a final object, the one point (measurable) space, written 1. An arrow p from 1 to X is just a probability p on X. Hence the existence of a steady state can be expressed as a commuting triangle (see Fig. 1). Also the composition of p with a (measurable) function $X \to Y$ is no other than the image probability $f(p)$. It follows, diagramatically, that the image of a steady state probability through a functional bisimulation π is a steady state of the target of π.

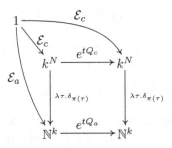

Fig. 1. Forward stochastic bisimulation as a diagram in the category of stochastic relations; the fact that $\mathcal{E}_c$ is a steady state of Q_c can be expressed as the commutation of the upper triangle

---

[1] This category is the Kleisli category of Giry's probabilistic monad [8], also known as the category of probabilistic mappings [11], or as the category of Markov kernels.

1.3   The Ehrenfest Steady State

We can readily use this general result to obtain a steady state of Q_a simply by aggregating the steady state probability of Q_c in the pre-image of each abstract state:

$$
\begin{aligned}
\mathcal{E}_a(\sigma) &= \sum_{\tau \in \pi^{-1}(\sigma)} \mathcal{E}_c(\tau) \\
&= \sum_{\tau \in \pi^{-1}(\sigma)} \prod_N \mathcal{E}(\tau(n)) \\
&= \sum_{\tau \in \pi^{-1}(\sigma)} \prod_k \mathcal{E}(i)^{\sigma(i)} = |\pi^{-1}(\sigma)| \prod_k \mathcal{E}(i)^{\sigma(i)}
\end{aligned}
$$

The cardinality of $\pi^{-1}(\sigma)$ is by definition a multinomial coefficient; hence the general Ehrenfest steady state is:

$$
\mathcal{E}_a(\sigma) = \binom{N}{\sigma(1) \cdots \sigma(k)} \prod_k \mathcal{E}(i)^{\sigma(i)} \tag{4}
$$

The same diagram guarantees that Q_a inherits the irreducibility of Q_c, and therefore $\mathcal{E}_a$ is unique and a global attractor.

Thus we see that a more concrete and bisimilar description of a process can sometimes be more amenable to calculations and reveal some of its properties. Specifically, moving over from a multiset semantics to a word-based semantics (with distinguishable particles) can provide insight in the dynamics. As we will show in this paper, and perhaps surprisingly, a similar 'concretization' can be made made to work for a vastly larger class of ctMCs, namely that of stochastic and thermodynamically consistent (to be defined below) Petri nets. This can be thought of as a vast generalisation of our opening example (although, strictly speaking, the classes are incomparable, as in the latter and simpler case we do not assume thermodynamic consistency).

1.4   Plan of the Paper

This paper is structured as follows: in §2, Petri nets are introduced as well as their qualitative and stochastic semantics. A proof is provided that under some weak restrictions there exists an equilibrium distribution from which the free energy function of the net can be derived – entailing the thermodynamic consistency of these restricted Petri nets. Word-based rewriting systems are considered and the shape of their equilibrium probability is derived under some general hypotheses on the properties of the embedding of Petri nets into those systems. In §3 Petri nets are rephrased in categorical terms. The state-space of word-based rewriting systems is described. The fact that multisets arise as a quotient of words is made precise by exhibiting an equivalence between the category of words and the category of multisets. In §4, two distinct attempts at concretising Petri nets are described: in the first instance, a concrete dynamics is described that preserves an invariant on the size of words. In the second instance, this rewriting system is generalised to non size-preserving Petri nets. In both cases, results of bisimulation and convergence to thermodynamically consistent equilibria are given. The paper concludes on some general remarks on the results achieved in this paper and discusses how this approach could be applied to more general systems.

2   Mass-Action Petri Nets

2.1   General Definitions

Petri nets. We recall the qualitative semantics of Petri nets (see Ref. [12] for a more detailed overview). A Petri net $\mathcal{N}$ is given by a finite set of places P and of transitions T equipped with a pair of input $i : T \to P \to \mathbb{N}$ and output $o : T \to P \to \mathbb{N}$ functions, noted $\mathcal{N} = \langle P, T, i, o \rangle$. It is insightful to interpret Petri nets as chemical reaction networks, where places correspond to chemical species and transitions correspond to chemical reactions. The input function i then encodes the number of molecules of each species consumed by a given transition, whereas the output function o encodes the amount of molecules that are produced.

The state space of a net is the set of *markings* $\mathbb{N}^P$. Markings are multisets, that is P-indexed vectors of non-negative integers, and as such they are endowed with an obvious additive structure. If $x \in \mathbb{N}^P$ is a marking and $A \in P$ is a place, let $x(A)$ be the number of elements (called *tokens*) in that place for the particular marking x. The chemical interpretation of a marking is simply a mixture of species.

The moves on the state graph correspond to firing of transitions. Given a transition $r \in T$ and a state $x \in \mathbb{N}^P$, there exists an r-labelled move:

$$x \longrightarrow_r x - i(r) + o(r) \quad \text{if} \quad x \geq i(r)$$

The guard $x \geq i(r)$ intuitively ensures that there are enough reactants for r to activate. This dynamics can be expressed in a vectorial form. Let us write $p = |P|$ and $q = |T|$. The functions i and o define the stoichiometric matrix C with p lines, q columns and elements $C_{j,k} = o(k)(j) - i(k)(j)$. When a net in vectorial form is considered, it will be noted $\mathcal{N} = \langle P, T, C \rangle$.

In the canonical basis $\{\mathbf{e}_1, \ldots, \mathbf{e}_q\}$ of $\mathbb{R}^q$, the outcome of $C\mathbf{e}_r$ is the net effect of firing transition r once. This lifts to arbitrary linear combinations of firings. Conversely, any vector $\mathbf{p} \in \mathbb{R}^p$ can be read as the per-place "cost" of a token. Given $\mathbf{r} \in \mathbb{R}^q$, the p-cost of firing $\mathbf{r}$ is hence $\mathbf{p}^t C \mathbf{r}$. This view of the dynamics overlooks the restriction that markings should always be point-wise non-negative, but embeds enough information to reason on asymptotic properties of the stochastic semantics.

Transition invariants are vectors $\mathbf{r} \in \mathbb{N}^q$ such that $C\mathbf{r} = \mathbf{0}$. In the light of the previous paragraph, these vectors are combinations of firings that leave markings (i.e. reactants) invariant. Transition invariants correspond to *cycles* in the transition graph. Place invariants are vectors $\mathbf{p} \in \mathbb{R}^p$ so that $\mathbf{p}^t C = \mathbf{0}$, i.e. costs for reactants that make any firing of a transition cost 0.

Simple and symmetric nets. The rest of the paper will concentrate on the study of *simple* and *symmetric* Petri nets (see e.g. [5] for more details); we say that a Petri net $\mathcal{N} = \langle P, T, C \rangle$ is:

  - *simple* iff no two transitions have identical jumps, i.e. identical columns in C;
  - *symmetric* iff every transition r has an inverse transition r^* with $i(r^*) = o(r)$ and $i(r) = o(r^*)$.

Example 1. Consider the finite set of places $P = \{A; B\}$ and a net $\mathcal{N}$ with a cyclic set of reactions $t_1 = A \mapsto B$ and $t_2 = B \mapsto A$, with an initial marking A. The semantics gives rise to an infinite run $t_1 t_2 t_1 \ldots$ The stoichiometric matrix C is $\left[\begin{smallmatrix} -1 & +1 \\ +1 & -1 \end{smallmatrix}\right]$. But the same stoichiometric matrix is compatible with the reaction $t_1' = 2A \mapsto A + B$, which gives rise to the empty run if started from the same marking. In a *simple* network, one *cannot* have both transitions t_1 and t_1'.

Mass action stochastic semantics. Petri nets can be endowed with *mass action* semantics, emphasising the chemical interpretation of their dynamics. The state space of a Petri net is countably infinite and by finiteness of T the branching factor is bounded, hence any assignment of rates to its transitions will determine a ctMC. Let us recall the the definition of a ctMC:

Definition 1. *(Continuous-Time Markov Chain) A ctMC on a countable set Ω is a family of random variables $\{X_t\}_{t \in [0;+\infty)}$ with values in Ω which obey the Markov property:*

$$P(X_{s+t} = j | X_s = i, X_{s_1} = i_1, \ldots, X_{s_n} = i_n) = P(X_t = j | X_0 = i)$$

$\forall s, t > 0, \forall i, i_{1\ldots n}, j \in \Omega.$

Equivalently, the time evolution of the probabilistic state of a ctMC (or a more general Markov process) is described by the *master equation* $p_t = e^{tQ} p_0$ where $p_t : \Omega \times \Omega \to [0; 1]$ describes the time evolution of the jumping probability, p_0 is the initial probabilistic state and Q is called the *rate matrix* or *infinitesimal generator*. It is Q that specifies how fast the process jumps from a state to another. Let us recall that the definition of stochastic forward bisimulation given in the introduction (Eqn. 3) applies perfectly to this definition of ctMCs.

Going forward with our chemical interpretation of Petri nets, we use *mass action* semantics: the rate of a given transition is defined to be proportional to the number of ways in which it can be fired, i.e. proportional to the number of place-preserving *injections* from the input of the transition to the current state. Let us note $[X; Y]$ the injections from a finite set X to a finite set Y. The number of such injections is given by $|[X; Y]| = \frac{|Y|!}{(|Y|-|X|)!}$.

Let r be a transition, A a place, x a marking, and $i(r)$ the input multiset of r, the number of place-preserving injections from $i(r)$ to x is:

$$\prod_{A \in P} |[i(r)(A); x(A)]| = \frac{x!}{(x - i(r))!}$$

where we have written $x!$ for the multiset exponential $\prod_{A \in P} x(A)!$.

Definition 2. *(Mass action Petri net) For any simple and symmetric Petri net $\mathcal{N} = \langle P, T, C \rangle$, given reaction rate constants $k : T \to \mathbb{R}^+$, the associated mass action Petri net is a ctMC $\mathcal{N}_{ma} = \langle \mathcal{N}, Q, p_0 \rangle$ on the state space $\Omega = \mathbb{N}^P$ with initial probability p_0 and rate matrix:*

$$Q(x, x') = \begin{cases} k(r) \dfrac{x!}{(x - i(r))!} & \text{if } x \longrightarrow_r x' \\ 0 & \text{otherwise} \end{cases} \tag{5}$$

The simplicity assumption ensures that this is a well-defined specification: any move $x \to x'$ can only correspond to one transition $r \in T$. Also, the initial distribution is rescaled by dividing it by the size of the fibre its argument belongs to. Also, observe that we allow self-loops: we do not enforce $Q(x, x) = -\sum_{y \neq x} Q(x, y)$. This does not alter the transient nor the steady-state probabilities [7]. This completes the ctMC structure associated to a simple, symmetric and mass-action Petri net (a sisma net for short).

2.2   Equilibrium and Entropy of Mass-Action Petri Nets

We now study the existence of an equilibrium (to be defined shortly) for a ctMC built as in the previous paragraph. With the notations introduced above, a probability $\mathcal{E}$ on Ω is a steady state for Q iff $\mathcal{E}^T Q = 0$. A sufficient (but far from necessary!) condition for this to happen is the *detailed balance* condition:

$$\mathcal{E}(x)Q(x, x') = \mathcal{E}(x')Q(x', x) \tag{6}$$

for all x, x' in Ω. When this condition is satisfied, we say that $\mathcal{E}$ is an *equilibrium* for Q. Just to be clear, $\mathcal{E}$ can be a steady state without being an equilibrium. Indeed for an equilibrium to exist, some form of reversibility must hold in the sense that $Q(x, x') \neq 0$ iff $Q(x', x) \neq 0$.

Given a sisma Petri net $\langle P, T, C \rangle$ with rate map $k : T \to \mathbb{R}^+$, we set the *transitional energy vector* $K : T \to \mathbb{R}$ to be:

$$K(r) = \log k(r^*) - \log k(r)$$

Theorem 1 ([5]). *A sisma net $\langle P, T, C \rangle$ has an equilibrium iff $K \in ker(C)^\perp$.*

Proof. Suppose $\mathcal{E}$ is an equilibrium, and $x \longrightarrow x'$; using equations (5) and (6), we get:

$$\frac{\mathcal{E}(x)}{\mathcal{E}(x')} = \frac{Q(x', x)}{Q(x, x')} = e^{K(r)} \frac{x'!}{x!}. \tag{7}$$

This relation extends to any path $\phi : x_1 \longrightarrow_{\mathbf{r}}^* x_n$, with $\mathbf{r}$ the multiset of reactions used by ϕ:

$$\frac{\mathcal{E}(x_1)}{\mathcal{E}(x_n)} = \prod_i \frac{Q(x_{i+1}, x_i)}{Q(x_i, q_{i+1})} = e^{\langle K, \mathbf{r} \rangle} \frac{x_n!}{x_1!}. \tag{8}$$

In particular, if ϕ is a loop, then $e^{\langle K, \mathbf{r} \rangle} = 1$, equivalently $\langle K, \mathbf{r} \rangle = 0$. Suppose now $\mathbf{r}$ in $\mathbb{N}^T$ is a transition invariant. It is easy to see that there is a loop ϕ (in fact countably many) which visits the multiset $\mathbf{r}$. Besides, $ker(C)$ has a basis with rational coordinates in the canonical basis as C has integer coefficients. In fact, a $ker(C)$ basis can be chosen with integer coefficients by scaling, and even positive by exploiting the fact that $r^* = -r$ as a basis vector. It follows that, K will be orthogonal to such a basis of $ker(C)$ (hence to any), hence $K \in ker(C)^\perp$.

Conversely, if $K \in ker(C)^\perp = Im(C^T)$, there exists a pricing of species ϵ such that $K = C^T \epsilon$. Therefore, for a path $\phi : x \longrightarrow_{\mathbf{r}}^* x'$, $\langle K, \mathbf{r} \rangle = \langle C^T \epsilon, \mathbf{r} \rangle = \langle \epsilon, C\mathbf{r} \rangle =$

$\langle \epsilon, x' - x \rangle$ depends only of its endpoints. Hence (Eqn. 8) can be used to define an $\mathcal{E}$ which will be an equilibrium (which will be unique up to an additive constant, per connected component of the underlying transition graph). One can prove easily that the per component normalization factor is always finite (see Ref. [5]).

Thermodynamic interpretation. The existence of an equilibrium can be expressed as the existence of an *energy* function $E : \Omega \to \mathbb{R}^+$ with $\mathcal{E}(x) = e^{-E(x)}/Z$ and $Z = \sum_x e^{-E(x)}$. The energy is only defined up to an additive constant. Taking the logarithm of the equilibrium condition on paths (Eqn. 8), we get:

$$E(x_1) - E(x_n) = \log \mathcal{E}(x_n) - \log \mathcal{E}(x_1)$$
$$= -\langle C^T \epsilon, \mathbf{r} \rangle + \log x_1! - \log x_n! = \langle \epsilon, x_1 \rangle - \langle \epsilon, x_n \rangle + \log x_1! - \log x_n!$$

Hence we have simply $E(x) := \langle \epsilon, x \rangle + \log x!$.

Note that the ϵ function associated to K in the derivation above is not unique in general. If two distinct energies E_1 and E_2 can be defined in this way, their difference on a given state x, in the connected component of a reference state x_0, is:

$$E_1(x) - E_2(x) = \langle \epsilon_1 - \epsilon_2, x \rangle$$
$$= \langle \epsilon_1 - \epsilon_2, x_0 + C\mathbf{r} \rangle \qquad \textit{for any path from } x_0 \textit{ to } x$$
$$= \langle \epsilon_1 - \epsilon_2, x_0 \rangle \qquad \epsilon_1 - \epsilon_2 \in Ker(C^T)$$
$$= E_1(x_0) - E_2(x_0)$$

Hence versions only differ by a constant on each connected component, and any choice uniquely defines an equilibrium per component. The first term of E entails cost minimisation – as such, it is an order term, while the second term tends to the uniformisation of the amount of each species in the state and is hence a disorder – or *entropic* – term.

Seeing this decomposition, one could be tempted to interpret E as a free energy function, identifying $\langle \epsilon, x \rangle$ to an internal energy term and $-log(x!)$ to an entropy – but this negative entropy would make little sense from a physical standpoint. In general, entropy corresponds to the *degeneracy factor* of the energy, i.e. it quantifies its non-injectivity. Indeed, one can rewrite the partition function $Z = \sum_x e^{-E(x)} = \sum_{E_i} N(E_i)e^{-E_i}$ where E_i is a particular value taken by the energy (in the case of Petri nets, there are only countably many) and $N(E_i)$ is the number of states with energy E_i in the reachability class of interest. For any fixed state x with energy E_i, the degeneracy factor of E_i corresponds to the number of distinct solutions y to the equation:

$$\langle \epsilon, x \rangle + log(x!) = \langle \epsilon, y \rangle + log(y!)$$

Let us first study the order and disorder terms of $E(x)$ separately. The following points are easily proved.

- One has $log(x!) = log(y!)$ if and only if y is obtained from x by renaming species, i.e. $y = (A \in P) \mapsto x(\tau A)$ for some $\tau \in Sym(P)$.
- One has $\langle \epsilon, x \rangle = \langle \epsilon, y \rangle$ exactly for all ys on the hyperplane normal to ϵ and passing by x.

Therefore, the states fixing both the order and disorder terms simultaneously will be those obtained from x by renamings of species that preserve energy.

It remains to study the degeneracies arising from correlations between the order and disorder term, i.e. symmetries that fix $E(x)$ without fixing both $\epsilon(x)$ and $log(x!)$. Such an energy-preserving transformation can only arise from a transfer of energy between the order and disorder components. In general, for fixed x and ϵ we seek states y s.t.

$$\langle \epsilon, y \rangle = \langle \epsilon, x \rangle + log(x!) - log(y!)$$
$$= C - log(y!)$$

Clearly, one can craft cases where this equation has non-trivial solutions. It is not trivial to determine even whether there is an asymptotic bound on the number of solutions as the energy grows. However, the non-linearity of the right-hand side versus the linearity of the left hand of the equation side hints that these symmetries are unlikely.

As an aside, we can compute the asymptotic variation of energy when the state undergoes an "infinitesimal" transition, which in our case corresponds to adding a token, say A, to the multiset state x. Let μ_A be this variation:

$$\mu_A = \frac{\partial E}{\partial x(A)} = \epsilon(A) + \frac{\partial \log(x(A)!)}{\partial x(A)} \sim \epsilon(A) + \log(x(A))$$

The cost of adding an A to x decomposes into a constant cost $\epsilon(A)$ which one can think of as the internal energy of A, and the logarithm of the number of As already present. No matter how energetically cheap A is, the second entropic term $\log(x(A))$ will eventually dominate and make the creation of more As increasingly unlikely.

2.3   Concretisation of the State-Space and Concrete Equilibrium

The natural state space of Petri nets is the set of multisets over places. The purpose of this paper is to study how the dynamics and the equilibrium can be explained in terms of another, more concrete state space of *words* over places. The objective of this subsection is to build a plausible concrete equilibrium probability, with the idea that the concrete ctMC still to be defined will have to converge to this equilibrium. Assume given a thermodynamically consistent sisma net $\mathcal{N} = \langle P, T, C \rangle$ with a cost function $\epsilon : \mathbb{N}^P \to \mathbb{R}$ and an energy function $E : \Omega \to \mathbb{R}^+$ as defined in Subsec. 2.2. Let P^* be the set of all finite words over P and let $P^n \subset P^*$ be the set of words of length n. Consider again the mapping $\pi : P^n \to \mathbb{N}^P$ of Sec. 1. One seek concrete counterparts ϵ', E', defined on *words* and inducing a concrete equilibrium $\mathcal{E}'(w) \propto exp(-E'(w))$. Let us make the two following fundamental hypotheses:

1. *(Energy conservation)* the concrete cost is the same as the abstract one: $\epsilon' = \epsilon$,
2. *(Equipartition)* the equilibrium probability is divided equally among all elements in the fibre of a multiset: $\mathcal{E}'(w) = \mathcal{E}(x)/|\pi^{-1}(x)|$.

These two hypotheses can be intuitively justified as follows. The first one states that since the cost of a multiset is only a function of its contents, it should be the same for words. The second hypothesis states that the concrete dynamics is not biased by the structure of the words, i.e. is invariant under the symmetries generated by permuting symbols. The transformation we propose has the deliberate effect of multiplying the degeneracy factor by the size of the fibre of π.

Lemma 1. *(Size of the inverse image of a multiset) Let P^n be the set of words of length n and $x \in \mathbb{N}^P$ a multiset with $\sum_A x(A) = n$. Then one has:*

$$|\pi^{-1}(x)| = \frac{n!}{x!}$$

Proof. This is a trivial application of the orbit-stabilizer lemma: $x!$ is the order of the stabilizer subgroup of any word in $\pi^{-1}(x)$, under the natural action of S_n. The ratio yields the size of the orbit, i.e. the number of words in the fibre of x.

With the lemma in place and under the assumptions of energy conservation and equipartition, we can compute E':

$$\begin{aligned}
E'(w) &= -\log \mathcal{E}'(w) \\
&= -\log \mathcal{E}(\pi(w)) + \log |\pi^{-1}\pi(w))| \\
&= \langle \epsilon, \pi(w) \rangle \rangle + \log(\pi(w)!) + \log |\pi^{-1}\pi(w))| \\
&= \langle \epsilon, w \rangle + \log |w|!
\end{aligned}$$

So, the *concrete energy* is $E'(w) = \langle \epsilon, w \rangle + \log |w|!$.

As an aside, one can easily show that the free energy F is preserved by the transformation. The free energy corresponding to an energy shell (and by extension to any state x in this shell) is by definition $F(E_i) = E_i - log(N_i)$. For a state, $F(x) = E(x) - log(N(x))$. After concretisation, one has $F'(x) = E'(x) - log(N(x) \cdot |\pi^{-1}(x)|) = F(x)$, hence the free energy is conserved by concretisation.

3   Categorical Presentation of Nets

In order to facilitate the construction of the concrete transition system, we recast nets in a categorical setting. In Sec. 4, we will ground word rewriting in this framework. The plan of this section is as follows:

1. we provide a categorical view of multisets and words,
2. the operational semantics of Petri nets is presented using double pushout (DPO) rewriting,
3. we exhibit an equivalence of categories between the categories of words and multisets.

For the rest of this section, let us fix an arbitrary sisma net $\mathcal{N} = \langle P, T, C \rangle$.

3.1   A Category of Multisets and Inclusions

The category $\mathbb{M}$ of multisets and functions over a fixed alphabet P has elements of $\mathbb{N}^P$ as objects and colour-preserving (i.e. place-preserving) functions as morphisms. This corresponds to the intuitive concept of multiset inclusion. More formally, if $x, y \in \mathbb{N}^P$ are two objects then a multiset morphism $m : x \to y$ is a colour-indexed family of set-theoretic functions $(m_A : x(A) \to y(A))_{A \in P}$ between the objects $x(A)$ and $y(A)$ seen as finite ordinals. Identities and composition of morphisms are that of **Set**.

The operational semantics of Petri nets can be given as a rewriting system expressed in double pushout (DPO) terms [16,6,9]. One can associate to any transition $r \in T$ a span $i(r) \twoheadleftarrow_g \lhd k \rhd_h \twoheadrightarrow o(r)$ in $\mathbb{M}$, where the component k is called the "glueing object" and represents the part of the state conserved by the rewriting. Note that the span is taken in the subcategory of injective morphisms and multisets. Given a total and injective matching morphism $m : i(r) \rightarrow x$ representing the application point of the rule on the multiset x, a rewrite is encoded as a direct derivation diagram composed of two pushout diagrams, as pictured below:

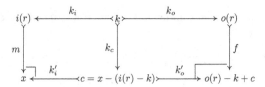

The rewrite is performed in two steps: the left-hand side of the rule $i(r)$ is removed from the target except for the conserved part k, and the right-hand side $o(r)$ is added – still modulo the conserved part k. The next figure is an example of such a DPO rewrite step, where the two-line notation on the arrows embodies the actual mapping from source to target tokens.

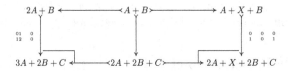

This example illustrates a transition $AA+B \rightarrow A+X+B$ whose firing will conserve the first A, erase the second, create an X and conserve the B. Let us take advantage of this simple example to draw the attention of the reader to the fact that many variations of this diagram would yield an identical resulting multiset. The degrees of freedom are the combinations of the choice of conserved symbols and the symmetries of the matching morphism.

3.2   A Category of Words and Inclusions

Given the same fixed alphabet P as for multisets, the category $\mathbb{W}$ of words has elements of P^* as objects and functions as morphisms (symbol order is not preserved). Given two words w_1, w_2, a morphism $m : w_1 \rightarrow w_2$ is a set-theoretic function $u(m) : |w_1| \rightarrow |w_2|$ between the integers $|w_i|$ seen as finite ordinals, such that symbols are preserved. The identities as well as associativity are provided by **Set**. If $u(m)$ is a bijection, w_1 and w_2 are isomorphic words, i.e. permutations of one another. The i-th symbol of a word w will be noted $w[i]$, not to be confused with $w(A)$, the number of symbols A in w. Words can be concatenated, endowing $\mathbb{W}$ with the obvious structure. Importantly, this structure lifts on morphisms as well: given any pair of morphisms $f : v_1 \rightarrow v_2$, $g : w_1 \rightarrow w_2$, one can build the pairing $\langle f, g \rangle : v_1 \cdot w_1 \rightarrow v_2 \cdot w_2$, where $\cdot$ is word concatenation.

3.3   Functors between the Category of Words and the Category of Multisets

From words to multisets. The category of multisets $\mathbb{M}$ is obtained from $\mathbb{W}$ by a quotient of the words under the relation of being isomorphic in $\mathbb{W}$. This yields a functor $\pi : \mathbb{W} \to \mathbb{M}$, defined on words $w : n \to P$ as:

$$\pi(w) = A \mapsto |w^{-1}(A)|,$$

i.e. to each w is associated the functor mapping any place $A \in P$ to its number of occurrences in w. To any $\mathbb{W}$-morphism $m : w_1 \to w_2$ is associated a morphism $\pi(m) : \pi(w_1) \to \pi(w_2)$ defined as a family $\{m_A : \pi(w_1, A) \to \pi(w_2, A)\}_{A \in P}$ of injections indexed by each colour (i.e. each place). This family arises as the projections of m along each colour. In particular, the injections are preserved by π. This projection process is illustrated by the following example.

$$m : AAB \overset{\substack{012\\132}}{\rightarrowtail} AABACB \qquad\qquad m_A : 2A \overset{\substack{01\\12}}{\rightarrowtail} 3A$$

This process is repeated for each place, yielding the image morphism. It is straightforward to see that identities are mapped to identities, and preservation of composition corresponds to the preservation of each colour.

From multisets to words. Given a total order $\prec$ on P a functor π^* in the opposite way to π can be defined, mapping any multiset to its $\prec$-sorted word. Concretely, it is defined on objects as $\pi^* = x \mapsto \prod_{A \in P_\prec} A^{x(A)}$, effectively *sorting* the multiset. It can readily be seen that any $\mathbb{M}$-morphism $h : x \to y$ can be decomposed into a family of morphisms $h_A : x(A) \to y(A)$. On morphisms, define $\pi^*(h) = \prod_{A \in P_\prec} h_A$ where the product is interpreted as the pairing defined in Sec. 3.2. What follows is an example of the action of both π and π^* on objects and morphisms.

$$AAB \overset{\pi}{\dashrightarrow} 2A + B \overset{\pi^*}{\dashrightarrow} AAB$$

$$AABACB \dashrightarrow 3A + 2B + C \dashrightarrow AAABBC$$

(left column) $\substack{012\\132}$ (middle) $\substack{01\ 0\\12\ 0}$ (right) $\substack{012\\123}$

It appears that π^* is only a right inverse to π: sorting modifies the orders of tokens according to $\prec$, the arbitrarily specified order on places. Notice that π is locally stable: the injections are preserved in each component. More importantly, π verifies the following property.

Lemma 2. π *is an equivalence of categories.*

Proof. Essentially surjectivity is trivial. It remains to prove fullness and faithfulness. Let $w_1, w_2 \in Obj(\mathbb{W})$ be two words and $\pi(w_1), \pi(w_2)$ the corresponding multisets. For any colour-preserving multiset inclusion $m : \pi(w_1) \to \pi(w_2)$, one has a corresponding $\pi^*(m) : \pi^*(\pi(w_1)) \to \pi^*(\pi(w_1))$. By construction, for any w, there exists an isomorphism $\alpha_w : \pi^*(\pi(w)) \to w$ (indeed, $\pi^*(\pi(w))$ is w after stable sorting). Hence, there exists a morphism $(\alpha_{w_1}^{-1}; \pi^*(m); \alpha_{w_2}) : w_1 \to w_2$ whose image is m. For the other direction, one also resorts to local stability: two different colour-preserving inclusions $m_1, m_2 : w_1 \to w_2$ must yield multiset inclusions $\pi(m_1), \pi(m_2)$ which disagree on some colour.

4   Concrete Transition Systems

We can now proceed to the definition of transition systems implementing concrete versions of the stochastic Petri net dynamics exposed in Sec. 2. The transitions systems we expose in this section are designed to respect the following criteria: they should be functionally bisimilar to the Petri net dynamics while respecting the steady state constraints set in Sec. 2.3, namely *energy conservation* and *equipartition*. Taking advantage of the definitions of Sec. 3, the concrete transition systems will be presented as DPO-based word rewriting systems set in the category $\mathbb{W}$. For the rest of this section, let us assume fixed an arbitrary sisma Petri net $\mathcal{N} = \langle P, T, C \rangle$ and also that an arbitrary ordering of places has been chosen so as to satisfy the definition of the functor π^* (Sec. 3.3). This defines a *concrete rewriting system*.

Definition 3. *(Concrete rewriting system) The concrete rewrite system $\mathcal{N}_c$ associated to $\mathcal{N}$ is the tuple $\langle P, T_c \rangle$ where:*

- *P is the same set of places as in $\mathcal{N}$ considered as a finite alphabet;*
- *$T_c \subseteq P^* \times P^*$ is a set of word rewrite rules s.t. $\hat{r} = (\pi^*(i(r)), \pi^*(o(r))) \in T_C \Leftrightarrow r \in T$. We will note $i(\hat{r}) = \pi^*(i(r))$ and $o(\hat{r}) = \pi^*(o(r))$.*

Without loss of generality, any rule $\hat{r} \in T_c$ can be noted as a $\mathbb{W}$ span $i(\hat{r}) \longleftarrow k \longrightarrow o(\hat{r})$, with k some conserved symbols.

In plain terms, each concrete rewrite rule $\hat{r} \in T_c$ is built as an arbitrary ordering of the input and output of some Petri net transition in T. We restrict our matches to multiset and word injections, as in standard Petri nets. Note that all matchings from the original Petri net are preserved.

Example 2. Consider the second net in example 1, with transitions $2A \mapsto A + B$ and $B \mapsto A$. A possible concrete rewriting system for the order $A \prec B$ is the set of concrete transitions $\{(AA, AB); (B, A)\}$. Another one for the order $B \prec A$ is $\{(AA, BA); (B, A)\}$.

Rules can be applied at so-called *redexes*, which correspond exactly to injective matches from left-hand sides of rules onto words.

Definition 4. *(Redex) A redex is a combination of a rule $\hat{r} \in T_c$ and of a word w such that there exists a total injection $m : i(\hat{r}) \rightarrowtail w$. The set of redexes corresponding to a word w is defined as $\mathcal{R}_{\mathcal{N}_c}(w) = \{(m, \hat{r}) \mid \hat{r} \in T_c \wedge m : i(\hat{r}) \rightarrowtail w\}$. The restriction of $\mathcal{R}_{\mathcal{N}_c}(w)$ to a particular rule $\hat{r}$ is the restriction $\mathcal{R}_{\mathcal{N}_c, \hat{r}}(w)$.*

One can define the *erasure* part of a DPO rewrite using word complementation. Word complementation corresponds to the removal of all symbols in a word w_2 that are in the image of some redex $m : w_1 \rightarrowtail w_2$.

Definition 5. *(Word complement) For any two words w_1, w_2 and any total injection $m : w_1 \rightarrowtail w_2$, a partial re-indexing function $f : |w_2| \rightarrow |w_2| - |w_1|$ can be defined as $f(i) = i - |\{j < i \mid j \in codom(m)\}|$ if $i \notin codom(m)$, $f(i) = \bot$ otherwise. Injectivity of f follows the verification of the inequality $|\{j \mid i_1 < j < i_2\}| < i_2 - i_1$ for all $i_1 < i_2 \in dom(f)$. The outcome of complementation is the morphism $w_2 - w_1 @ m \triangleq w : |w_2| - |w_1| \rightarrow P$ s.t. $w[i] = w_2[f^{-1}(i)]$, well-defined by injectivity of f.*

Erasure is defined in terms of complementation: for any word w, any redex $m : i(\hat{r}) \rightarrow w$ and any conserved symbols specified by $g : k \rightarrow i(\hat{r})$, define $c = w - (i(\hat{r}) - k @ k_i) @ m$. The span $i(\hat{r}) \twoheadleftarrow_{k_i} \prec k \succ_{k_c} \twoheadrightarrow c$ has w as a pushout object: any

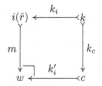

cospan $i(\hat{r}) \longrightarrow w' \longleftarrow c$ making the span into a commutative diagram factors uniquely up-to isomorphism through a morphism $u : w \rightarrow w'$ (not necessarily injective), constructed by inverting each branch of the cospan and building the injection u in the obvious way. Note that with our definition, the morphism $k_i' : c \rightarrowtail w$ has the property of being both injective and *monotonic*: symbol order is preserved on conserved symbols. It remains to define the second part of the DPO rewrite, concerned with symbol insertion. We simply say that for a diagram as above and a morphism $k_o : k \rightarrowtail o(\hat{r})$, any cospan $c \succ_{k_o'} \twoheadrightarrow w' \twoheadleftarrow_f \prec o(\hat{r})$ making the span $c \twoheadleftarrow_{k_c} \prec k \succ_{k_o} \twoheadrightarrow o(\hat{r})$ into a commutative diagram and w' a pushout object defines a valid insertion. This general scheme will be used in two distinct solutions in Sec. 4.1 and 4.2. We first investigate the case of a dynamics constrained to stay in a shell where words have an arbitrary fixed length and follow by showing how to extend this to the general case.

4.1 Size-Preserving Dynamics

Let us assume given a Petri net $\mathcal{N}$ and its concrete counterpart $\mathcal{N}_c$ as above, and suppose further that the dynamics is size preserving, i.e. that the total number of tokens is an invariant of the dynamics. This last condition of size preservation can be stated as the existence of constants $0 < k \leq n$ where $\forall r \in T, |i(r)| = |o(r)| = k$ and with an initial state w s.t. $|x| = n$. Given a redex (in span notation) $m : i(\hat{r}) \rightarrow w$, a rewrite can be interpreted as *in-place* substitution in w of the symbols of the input $i(\hat{r})$ by the symbols of the output $o(\hat{r})$ at the position specified by m. Parts of w outside the image of m are left untouched. We choose to not conserve symbols: the glueing object is the empty word.

Definition 6. *(Size-preserving rewrite) For any word w and any redex $m : i(\hat{r}) \rightarrowtail w$ in $\mathcal{R}_{\mathcal{N}_c}(w)$, the word $w' = w\{i(\hat{r})\backslash o(\hat{r})@m\}$ defined below makes the diagram a DPO rewrite.*

$$i \notin codom(m) \rightarrow w'[i] = w[i]$$
$$i \in codom(m) \rightarrow w'[i] = o(\hat{r})[m^{-1}(i)]$$

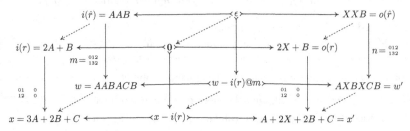

Example 3. What follows is an example of how a multiset rewrite can be mapped down to the word rewriting system in the size-preserving case. The back face is a concrete rewrite, with redex $AAB \rightarrowtail AABACB$.

$$
\begin{array}{ccc}
i(\hat{r}) = AAB & \xleftarrow{\quad} \epsilon \xrightarrow{\quad} & XXB = o(\hat{r}) \\
i(r) = 2A+B & \xleftarrow{} 0 \xrightarrow{} & 2X+B = o(r) \quad n = \substack{012\\132} \\
m = \substack{012\\132} & & \\
w = AABACB & \xleftarrow{} w-i(r)@m \xrightarrow{} & AXBXCB = w' \\
x = 3A+2B+C & \xleftarrow{} x-i(r) \xrightarrow{} & A+2X+2B+C = x'
\end{array}
$$

Observe that the resulting word stays identical for any automorphism of w that is preserved by the rewrite. In the previous example, one easily checks that $m' = \substack{012\\312}$ is a matching s.t. $w\{i(\hat{r})\backslash o(\hat{r})@m\} = w\{i(\hat{r})\backslash o(\hat{r})@m'\}$. The number of such matchings is called the *thickness* of the transition $w \longrightarrow_{\hat{r}} w'$.

Definition 7. *(Thickness) Let w, w' be words and $\hat{r}$ be a concrete rule. Let us note:*

$$\mathcal{C}_{\mathcal{N}_c, \hat{r}}(w, w') \triangleq \{m : i(\hat{r}) \rightarrow w \mid w' = w\{i(\hat{r})\backslash o(\hat{r})@m\}\}$$

the set of redexes inducing a rewrite from w to w'. The thickness of the transition $w \longrightarrow_{\hat{r}} w'$ is defined as $|\mathcal{C}_{\mathcal{N}_c, \hat{r}}(w, w')|$.

An important property required to prove thermodynamic consistency is that the thickness of the forward $\hat{r}$ and backward $\hat{r}^*$ transitions should match.

Lemma 3. *(Forward and backward thickness are equal) If $w \longrightarrow_r w'$, then one has the equality:*

$$|\mathcal{C}_{\mathcal{N}_c, \hat{r}}(w, w')| = |\mathcal{C}_{\mathcal{N}_c, \hat{r}}(w', w)|$$

Proof. This is a direct consequence of the symmetry of DPO rewrites: any forward diagram is a backward diagram when read right-to-left, and conversely. See [9], Sec. 2, p. 33.

Concrete ctMC. This size-preserving dynamics can be given stochastic semantics. As will be shown, it turns out that the corresponding ctMC is *forward bisimilar* to the original (Petri net-based) one. Let us proceed by giving a ctMC corresponding to $\mathcal{N}_c$ and prove that π is indeed a functional forward bisimulation in the size-preserving case.

A property of the size-preserving transition system is that it enjoys a version of the simplicity property of abstract nets. This allows us to stay in the format of standard ctMCs:

Proposition 1. *(Simplicity for the concrete transition system) Given any two words* w, w', *there is at most one transition* $\hat{r}$ *s.t.* $w' = w\{i(\hat{r})\backslash o(\hat{r})@m\}$ *for some matching* m.

Proof. The existence of $\hat{r}_1 \neq \hat{r}_2$ s.t. $w\{i(\hat{r}_1)\backslash o(\hat{r}_1)@m_1\} = w\{i(\hat{r}_2)\backslash o(\hat{r}_2)@m_2\}$ would imply the existence of $r_1, r_2 \in T$ with $\pi(w) \longrightarrow_{r_1 \neq r_2} \pi(w')$, violating the simplicity hypothesis. ∎

The definition of the size-preserving concrete ctMC is derived from the concrete rewriting system $\langle P, T_c \rangle$ (Def. 3).

Definition 8. *(Size-preserving concrete ctMC) Consider a mass-action net* $\langle \mathcal{N}, Q, p_0 \rangle$ *such that the support of the initial distribution* p_0 *lies in the multisets of size* n. *The concrete ctMC* $\langle Q', p'_0 \rangle$ *is defined on the state space* $\Omega' = P^n$ *of words of length* n *and has components defined as follows.*

- *Q' is the rate matrix s.t. for any word w, w' and rule $\hat{r} \in T_c$*

$$Q'(w, w') = \begin{cases} k(r)|\mathcal{C}_{\mathcal{N}_c, \hat{r}}(w, w')| & \text{if } \exists m. w' = w\{i(\hat{r})\backslash o(\hat{r})@m\} \\ 0 & \text{otherwise} \end{cases}$$

- *The initial distribution p'_0 is such that for all w, $p'_0(w) = p_0(\pi(w))/|\pi^{-1}(\pi(w))|$.*

To put it in words, the concrete rate corresponds to the thickness of the transition.

Bisimulation and equilibrium correctness. Proving that our concrete dynamics is bisimilar (cf. Eq. 3) to the abstact one is now a simple matter of unfolding definitions.

Proposition 2. π *is a bisimulation between* Q *and* Q'.

Proof. Let x and x' be multisets such that $x \longrightarrow_r x'$. By simplicity, this transition is unique. Let $w \in \pi^{-1}(x)$. One has:

$$\sum_{w' \in \pi^{-1}(x')} Q'(w, w') = \sum_{w' \in \pi^{-1}(x')} k(r)|\mathcal{C}_{\mathcal{N}_c, \hat{r}}(w, w')|$$
$$= k(r)|\mathbb{W}(i(\hat{r}), w)| \qquad \text{\textit{(Sum of all thicknesses on the fibre)}}$$
$$= k(r)|\mathbb{M}(i(r), x)| \qquad \text{\textit{(fullness \& faithfulness of } }\pi\text{\textit{)}}$$
$$= Q(x, x')$$

However, this result alone does not ensure that the image equilibrium respects the requirements stated in Sec. 2.3. i.e. that it is of the shape $\mathcal{E}'(w) \propto \exp -E'(w)$. A sufficient condition is for $\mathcal{E}'$ to verify the detailed balance conditions (equation 6) for any pair of related words w, w', which it does, as witnessed by this theorem.

Theorem 2. $\mathcal{E}'(w)$ *verifies* $\mathcal{E}'(w) \propto \exp\left(-(\langle \epsilon, w \rangle + \log(|w|!))\right)$

Proof.

$$\log\left(\frac{q(w', w)}{q(w, w')}\right) = K(r) + \log\left(\frac{|\mathcal{C}_{\mathcal{N}_c, \hat{r}}(w', w)|}{|\mathcal{C}_{\mathcal{N}_c, \hat{r}}(w, w')|}\right)$$
$$= K(r) \qquad \text{\textit{(Lemma 3)}}$$
$$= E'(w') - E'(w)$$
$$= \log\left(\frac{\mathcal{E}'(w)}{\mathcal{E}'(w')}\right)$$

Notice that the entropic term vanishes. Indeed, the term $\log(|w|!)$ stays by definition constant in size-preserving dynamics, and can thus be dropped of the energy term.

4.2 Unconstrained Dynamics

The size-preservation constraint allowed for a straightforward definition of word rewriting, by replacing symbols in-place. This obviated the need to consider the problem of symbol insertion in its relation to the equipartition hypothesis. As the following example shows, in unconstrained word rewriting systems some insertion strategies yield incorrect dynamics.

Failure of a naïve insertion strategy. Let us consider a net restricted to a unique transition $\hat{r} = \epsilon \rightarrow A$ creating a symbol A out of nothing. The question arises as to where this symbol should be inserted in the word on which $\hat{r}$ is applied. Let us arbitrarily choose to insert the symbol at the end of the target word. As an example, consider the concrete rewrite $w = A^n B A^m \longrightarrow_{\hat{r}} A^n B A^{m+1} = w'$. Computing the forward and backward rates, one obtains:

$$Q(w, w') = k(r) \cdot 1$$
$$Q(w', w) = k(r^*)(m + 1)$$

Indeed, there are $m + 1$ ways to erase a symbol in w' yielding w as a result. Observe that this naïve dynamics is biased towards states with a short tail of As. However, in the long run it will converge to states with $n = 0$. Computing the log ratio of the backward and forward rates, one has:

$$\log \left(\tfrac{Q(w',w)}{Q(w,w')} \right) = \epsilon(A) + \log(m + 1)$$
$$\neq E'(w') - E'(w) = \epsilon(A) + \log(n + m + 2)$$

The deterministic insertion strategy that was chosen is incompatible with the equipartition hypothesis it is supposed to satisfy.

Insertions uniformly at random. The previous paragraph showed the bias induced by selecting a deterministic strategy to perform insertions. A plausible solution is hence to perform them uniformly at random, i.e. to attribute an equal probability to each possible insertion. Considering the previous example, one observes that there are $m + 1$ ways of inserting to the right of B, yielding a uniform insertion probability of $(m + 1)/(m + n + 2)$. This results in the following forward and backward rates:

$$Q(w, w') = k(r)(m + 1)/(m + n + 2)$$
$$Q(w', w) = k(r^*)(m + 1)$$

The log ratio yields the expected energy difference:

$$\log \left(\tfrac{Q(w',w)}{Q(w,w')} \right) = K(r) + \log \tfrac{m+1}{(m+1)/(m+n+2)} = K(r) + \log(|w'|)$$
$$= E'(w') - E'(w) = \epsilon(A) + \log(n + m + 2)$$

In the general situation, one has to consider the case of an arbitrary rule $\hat{r} : i(\hat{r}) \rightarrow o(\hat{r})$. In this setting, many distinct possibilities arise as to how to perform the rewrite, depending on whether symbols are conserved or erased and reinserted at random. The cardinality of the hom-set between two words is a function of this choice, inducing different dynamics (but the same equilibrium!). The choice here is quite arbitrary. In our definition of the concrete transition graph, the chosen option is to erase all symbols matched by the left-hand side and insert uniformly at random all symbols in the right-hand side.

In order to proceed constructively one needs to build the multiset of all possible insertions of a finite set into a word. For the uniform insertion probability to be properly normalised, it is also required to compute the cardinality of the multiset of insertions. One easily guesses that it corresponds to some rising factorial, noted $a^{(b)} = \prod_{i=1}^{b} a + i = (a+b)!/a!$. Let us proceed in order and start with the definition of the insertion operation, and more precisely by the insertion of an arbitrary symbol in a word.

Definition 9. *(Symbol insertion) Let σ be a finite set and $w = a_1 \cdot a_n \in \sigma$ an arbitrary word. The insertion of a symbol x in w is a multiset of words noted $x \,/\!\!\Lambda\, w$ and defined by induction:*

$$x \,/\!\!\Lambda\, \epsilon = x \mapsto 1 \quad x \,/\!\!\Lambda\, a \cdot w = (x \cdot a \cdot w) + (a \cdot (x \,/\!\!\Lambda\, w))$$

Using multisets allows us counting multiple occurrences of the same word. Here, one has as a trivial fact that $|x \,/\!\!\Lambda\, w| = |w| + 1$. Let us continue with the unordered insertion of a whole word into a word.

Definition 10. *(Unordered word insertion) Let σ_1 and σ_2 be two finite sets and let $v = a_1 \cdots a_m$ and $w = b_1 \cdots b_n \in \sigma_2^*$ be two finite words. The multiset of all insertions of v into w is defined inductively as:*

$$\epsilon \,/\!\!\Lambda\, w = w \mapsto 1 \quad a \cdot v \,/\!\!\Lambda\, w = \sum_{w' \in a /\!\!\Lambda w} v \,/\!\!\Lambda\, w'$$

An important but easy property is that the multiset $v \,/\!\!\Lambda\, w$ is independent of the actual order of symbols in v.

Lemma 4. *(Order independence of insertion) Let v and w be any words as in Def. 10. Then for any permutation ρ, one has that $v \,/\!\!\Lambda\, w = \rho(v) \,/\!\!\Lambda\, w$.*

Proof. It is enough to prove the case for ρ a simple transposition exchanging elements of indices $i, i + 1$. This further reduces to proving the commutation result $a_1 \,/\!\!\Lambda\, (a_2 \,/\!\!\Lambda\, w) = a_2 \,/\!\!\Lambda\, (a_1 \,/\!\!\Lambda\, w)$, easily proven by induction on w.

The following lemma gives the size of the multiset of all possible insertions.

Lemma 5. *(Number of insertions) Let σ_1 and σ_2 be two finite sets, and let $w = b_1 \cdots b_n \in \sigma_2^*$ be a finite word. The number of insertions $\sigma_1 \,/\!\!\Lambda\, w$ of the symbols of σ_1 in w is:*

$$|\sigma_1 \,/\!\!\Lambda\, w| = |w|^{(|\sigma_1|)}$$

Proof. Let us proceed by induction on $|\sigma_1|$. For $|\sigma_1| = 0$, one has $|\sigma_1 \,/\!\!\Lambda\, w| = n^{(0)} = 1$. For $|\sigma_1 + \{a\}| = i + 1$ the induction hypothesis yields $|\sigma_1 \,/\!\!\Lambda\, w| = n^{(|\sigma_1|)}$ words. For each of these words, there are $n + |\sigma_1| + 1$ insertion possibilities for a, yielding the expected result.

Concrete ctMC. These combinatorial results enable us to define a concrete transition system implementing uniform insertion. Given a Petri net $\mathcal{N}$ and its concrete counterpart $\mathcal{N}_c$ as defined in Def. 3, recall that the concrete transition system is built by associating to each redex (Def. 4) a *multiset* of outcomes, each outcome built by erasing the left-hand side of the rule and inserting randomly the right-hand side.

Definition 11. *(Multiset of outcomes) Given a redex $i(\hat{r}) \overset{m}{\rightarrowtail} w$, the multiset of outcomes is $\mathcal{O}(w, \hat{r}, m) = o(\hat{r}) \barwedge (w - i(\hat{r})@m)$.*

Given all that, the definition of the concrete operational semantics can finally be stated explicitly as a ctMC. The transition system enjoys the simplicity property inherited from abstract nets:

Proposition 3. *(Simplicity of the concrete dynamics) Given any two words w_1, w_2, there is at most one transition $\hat{r}$ s.t. $W_c(w_1, \hat{r}, w_2) > 0$.*

Proof. The proof proceeds similarly as in the size-preserving case, by showing that the existence of two concrete transitions between the same words would lift to the Petri net, violating simplicity.

Definition 12. *(Concrete ctMC) Let us consider the same input data as in Def. 8. The concrete ctMC $\langle Q', p_0' \rangle$ is defined on the state space $\Omega' = P^*$. Its components are defined below.*

- Q' *is the rate matrix s.t. for any word w and rule $\hat{r}$,*

$$Q'(w, w') = \begin{cases} k(r) \sum_m \frac{\mathcal{O}(w, \hat{r}, m)(w')}{|\mathcal{O}(w, \hat{r})|} & \text{if } \exists m.w' = w\{i(\hat{r})\backslash o(\hat{r})@m\} \\ 0 & \text{otherwise} \end{cases}$$

where m ranges in $\mathbb{W}(i(\hat{r}), w)$.
- *The initial distribution p_0' is such that for all w, $p_0'(w) = p_0(\pi(w))/|\pi^{-1}(\pi(w))|$.*

Notice how the rate of a transition is rescaled by its multiplicity. An easy property of our rewrite system is that the number of distinct ways of performing $w \longrightarrow_r w'$ is equal to its inverse $w' \longrightarrow_{r^*} w$. This also proceeds from the symmetry of DPO rewrites. This intuition is made precise by the following result.

Lemma 6. *For any words w, w' and reaction $\hat{r}$, one has the equality:*

$$\sum_m \mathcal{O}(w, \hat{r}, m)(w') = \sum_m \mathcal{O}(w', \hat{r}^*, m)(w)$$

Proof. Any w' in $\sum_m \mathcal{O}(w, \hat{r}, m)$ arises as the outcome of a DPO rewrite (in our case, with glueing object ϵ), which by symmetry of DPOs corresponds to a DPO rewrite from w' to w.

Bisimulation and equilibrium correctness. The concrete dynamics is built so as to properly reflect the Petri net dynamics. The following theorem states as expected that the mapping from words to multisets is a bisimulation, as in the size-preserving case.

Theorem 3. π *is a functional forward bisimulation between Q and Q'.*

Proof. Let x and x' be multisets such that $x \longrightarrow_r x'$. By simplicity, this transition is unique. Let $w \in \pi^{-1}(x)$. One has:

$$
\sum_{w' \in \pi^{-1}(x')} Q'(w, w') = \frac{k(r)}{|\mathcal{O}(w, \hat{r})|} \sum_{w'} \sum_m \mathcal{O}(w, \hat{r}, m)(w')
$$

$$
= \frac{k(r)}{|\mathcal{O}(w, \hat{r})|} |\mathcal{O}(w, \hat{r})| \cdot |[i(\hat{r}); w]| \qquad \text{(Summing on all outcomes)}
$$

$$
= k(r)|\mathbb{W}(i(\hat{r}), w)|
$$
$$
= k(r)|\mathbb{M}(i(r), x)| \qquad \text{(fullness \& faithfulness of } \pi\text{)}
$$
$$
= Q(x, x')
$$

Much less obvious is the property of reaching a concrete equilibrium whose image through π matches the abstract equilibrium. The random insertion strategy is precisely what one needs to ensure reaching that good equilibrium. As in the case of size-preserving dynamics, a sufficient condition is for the log-ratio of the rates to yield an energy difference corresponding to the definition of E' in Sec. 2.3.

Theorem 4. *(Equilibrium correctness) The unconstrained dynamics converges to an equilibrium respecting energy conservation and equipartition.*

Proof. Unfolding the detailed balance equations, one has:

$$
\log\left(\frac{q(w', w)}{q(w, w')}\right) = K(r) + \log\left(\frac{\sum_m \mathcal{O}(w', \hat{r}^*, m)(w)}{\sum_m \mathcal{O}(w, \hat{r}, m)(w')} \frac{|\mathcal{O}(w, \hat{r})|}{|\mathcal{O}(w', \hat{r}^*)|}\right)
$$

$$
= K(r) + \log\left(\frac{|w - i(\hat{r})|^{(|o(\hat{r})|)}}{|w' - o(\hat{r})|^{(|i(\hat{r})|)}}\right)
$$

One has by definition of the transition graph $c \triangleq w' - o(\hat{r}) = w - i(\hat{r})$. Also, recall that $a^{(b)} = (a + b)!/a!$. Altogether, this yields the desired result:

$$
\log\left(\frac{q(w', w)}{q(w, w')}\right) = K(r) + \log\left(\frac{(|c| + |o(\hat{r})|)!}{(|c| + |i(\hat{r})|)!}\right)
$$

$$
= K(r) + \log\frac{|w'|!}{|w|!}
$$

$$
= E'(w') - E'(w)
$$

This achieves the construction of a word-based transition system respecting equipartition and energy preservation.

4.3 Analysis of the Construction

The energy function E' is precisely the one needed to counterbalance the entropy increase induced by concretisation, ensuring free energy preservation. A physical intuition is that holding onto a word in the concrete system does not give any way of extracting any more "work" than in the original system: that is in essence a consequence of the equipartition constraint (and not of bisimulation).

In order for the net to converge to the good equilibrium E', a crucial step was to use an uniformly random insertion strategy. An open question is whether this is the only solution. The equilibrium is a direct outcome of the detailed balance equations, i.e. of the ratio of the backward to forward rates between any two words. In our construction, the rate of a jump from w to w' through $\hat{r}$ is defined to be proportional to the "density" of w' in the multiset of outcomes of rewriting w with $\hat{r}$. The shape of E' is directly given by the cardinality of this multiset of outcomes together with the cancellation of forward and backward thicknesses. Bisimulation only requires the global flow of probability to match the abstract one. This gives a set of sufficient conditions to build a correct concrete ctMC:

- enforce forward bisimulation,
- ensure that forward and backward thicknesses are equal,
- normalize the rates of outgoing jumps (i.e. the thicknesses) from e.g. w through $\hat{r}$ by $|\mathcal{O}(w, \hat{r})|$.

An crucial observation is that the concretisation has simplified the (species-dependent) entropic term $log(x!)$ into a term $log(|w|!)$. In other terms, concretisation allows to push irrelevant part of the combinatorics into the entropy and simplify the expression of the probabilistic fixed point. This is also what happens in the Ehrenfest case: the multinomial term in the equilibrium of the abstract system disappears when one goes to the concrete system.

5 Conclusion and Future Work

We have successfully embedded stochastic Petri nets into a word-based rewriting system. A first result is that the energy function is simplified by moving to a more concrete transition system: in the concrete system, the mass-action entropic potential term is a function of the size of the word, independently of its contents, whereas in the abstract case entropy also takes into account the diversity of symbols. This echoes what we observed in the case of Ehrenfest urns, as explained previously. We also derived results hinting at the fact that the degeneracy factor of stochastic Petri nets is low, implying that the energy is a good proxy to the free energy: we plan to substantiate our heuristic arguments in future work. It should be noted that our qualitative word-based transition system is similar to the pre-net model of [1]: each pre-net corresponds to a fixed choice of what we call a matching in our setting.

An unexpected hurdle in the construction of the concrete transition systems was handling *creation of tokens* – this prompted the main ingredient of the paper, namely uniformly random insertion. It stands as a particular solution to a more general problem, that of deriving a concrete stochastic process from an abstract one, such that the concrete equilibrium is in some functional relation with the abstract one. In our case, we enforced *equipartition*, but one could easily imagine more involved settings. We also plan to study more structured state spaces, such as trees, graphs or generalisations thereof, where we expect similar phenomena to arise. Uniformly random insertion corresponds to enumerating all the possible structure-preserving embeddings of an object of size n into an object of size $n + 1$. This suggests using a formalism such as Joyal's species of structures as a generic setting where to apply our ideas.

References

1. Baldan, P., Bruni, R., Montanari, U.: Pre-nets, read arcs and unfolding: a functorial presentation (2002)
2. Blute, R., Desharnais, J., Edalat, A., Panangaden, P.: Bisimulation for labelled markov processes. In: 12th Annual IEEE Symposium on Logic in Computer Science, LICS 1997 Proceedings, pp. 149–158 (1997)
3. Buchholz, P.: Exact and ordinary lumpability in finite markov chains. Journal of Applied Probability 31(1), 59–75 (1994)
4. Buchholz, P.: Bisimulation relations for weighted automata. Theoretical Computer Science 393(1-3), 109–123 (2008)
5. Danos, V., Oury, N.: Equilibrium and termination ii: the case of petri nets. Mathematical Structures in Computer Science 23, 290–307 (2013)
6. Ehrig, H., Pfender, M., Schneider, H.J.: Graph-grammars: An algebraic approach. In: SWAT (FOCS), pp. 167–180 (1973)
7. Feret, J., Henzinger, T., Koeppl, H., Petrov, T.: Lumpability abstractions of rule-based systems. Theoretical Computer Science 431(0), 137–164 (2012)
8. Giry, M.: A categorical approach to probability theory. In: Mosses, P.D., Nielsen, M. (eds.) CAAP 1995, FASE 1995, and TAPSOFT 1995. LNCS, vol. 915, pp. 68–85. Springer, Heidelberg (1995)
9. Heindel, T.: A category theoretical approach to the concurrent semantics of rewriting: adhesive categories and related concepts. PhD thesis (2009)
10. Larsen, K.G., Skou, A.: Bisimulation through probabilistic testing. In: Conference Record of the 16th ACM Symposium on Principles of Programming Languages (POPL), pp. 344–352 (1989)
11. Lawvere, F.W.: The category of probabilistic mappings - with applications to stochastic processes, statistics, and pattern recognition. Seminar Handout Notes (1962)
12. Murata, T.: Petri Nets: Properties, Analysis and Applications. Proceedings of the IEEE 77, 541–580 (1989)
13. Norris, J.R.: Markov chains. Cambridge series in statistical and probabilistic mathematics. Cambridge University Press (1998)
14. Panangaden, P.: Probabilistic relations. In: School of Computer Science, McGill University, Montreal, pp. 59–74 (1998)
15. Panangaden, P.: Labelled Markov Processes. Imperial College Press (2009)
16. Rozenberg, G. (ed.): Handbook of Graph Grammars and Computing by Graph Transformations. Foundations, vol. 1. World Scientific (1997)

Laudatio for Prakash Panangaden

Josée Desharnais

Université Laval, Québec, Canada
josee.desharnais@ift.ulaval.ca

It is with great emotion that I write this tribute to Prakash. The minute I set foot into his office was one of those rare pivotal moments where one's future changes. The path he led me on was exceptional and fascinating. From a somewhat purposeless dreamy math student, he smoothly transformed me into a critical part of what seemed and would become an exciting and impactful research project.

Prakash's started his scientific career in physics. He obtained an MSc from the Indian Institute of Technology, Kanpur, an M. Sc. from the University of Chicago and a PhD from the University of Wisconsin-Milwaukee, all in the area of quantum field theory. He then switched to computer science, obtaining an MSc from the University of Utah, where he had done postdoctoral research in quantum field theory.

At this point I must retell a savoury anecdote – after all, Prakash is a wonderful story teller. After Utah, Prakash followed his partner to Cornell University where he decided to start studying category theory. What better way to study a topic than to teach the subject? Doubly so when you are such a good performer. He stuck a note to the door of a random classroom: "Lecture on Category Theory, every Monday at 10." From week to week, the number of attendants grew in such a way that after a short while he found himself in the chair's office, being offered an assistant professor position. Prakash likes to recount how he suffered all weekend long, being forced by the chair to wait and ponder the offer until Monday before accepting it. How nervous he was on the Monday morning, waiting for the chair to come into the office, afraid he would have changed his mind! Of course, he had not.

At Cornell University, he worked on semantics, concurrency, type theory and logic. With his wife Laurie, he went on to Canada in 1990 and has been at McGill University ever since. Over the years, he has been an invited professor in Aarhus, Edinburgh, Oxford and Paris, as well as visiting scientist in Amsterdam, Cambridge, Toronto and Sydney. He has just been elected fellow of the Academy of Science of the Royal Society of Canada. His work has been and will remain influential in many fields of mathematics, computer science and physics, reflecting his varied interests. The papers in the present book bear witness to this.

Yes, the facts are there, his CV is impressive, but the most essential is, as Saint-Exupéry's little prince puts it, "invisible to the eye." Two of Prakash's most impressive qualities are his exceptional intuition and his broad vision of computer science and science in general. In a discussion, even when he knows epsilon about a subject, he is often able to formulate a conjecture or an idea that will become a crucial part to his interlocutor's progress. Imagine when he talks about something he actually knows well! He can put things into perspective like no other.

F. van Breugel et al. (Eds.): Panangaden Festschrift, LNCS 8464, pp. 290–291, 2014.

Prakash has an incredible memory for a large variety of facts: of course in mathematics, computer science and physics, but you can also ask him who has won track and field events at the Olympic Games and the results of tennis or cricket tournaments from any year. He will know. He also knows intricate details from Tolkien's Lord of the Rings and from Harry Potter; he has read the latter with his daughter Jane in a never-ending loop, only interrupted by the arrival of a new book of the series!

Prakash is a very nice man, who nonetheless has been known to occasionally exhibit a somewhat boiling side. What a contrast between how he acted with a research partner, and how he treated a wandering student who would knock just below the "do not disturb" piece of paper stuck to his door, to ask him some futile question. Akin to the Incredible Hulk, he would change from a agreeable companion to a bearish and scary bundle of anger. At some point, perhaps to spare his energy, he changed the "do not disturb" note to "knock and die." He would lose faith in humanity when a student knocked anyway to see what would happen.

While preparing this laudatio, I consulted the acknowledgement section of my Ph. D. thesis. I find revealing that, in the source file, I hid the sentence "I am honoured to have worked with him." He would not have liked this official-sounding phrase, as he does not like any notion of rank. As far as he was concerned, I was a colleague from the start. He wanted me to use the familiar "tu" form to address him in French. We spoke French for about a year, as he insisted he wanted to improve his command of the language. He likes to tell that it is only when we switched to English that the research really took off.

I would like to complete this personal portrait of Prakash with a few things he said: a few strong statements, as he likes to make, a few anecdotal phrases that marked me or our relationship:

— Let's make a list! (That's how one starts without too much commitment.)
— Never apologize (typically in a talk). Your slides have been prepared in a rush? Do not admit it, just be brilliant.
— Why would we remove negation in the logic? For sure it is not possible and even if we could, nobody would be interested in that.
— Let's prove Borel's theorem! (after some dinner)
— Say something, just say something!
— One needs a "ten-year" to celebrate a tenure! (A ten-year-old bottle of Dom Pérignon)
— What? Olive oil?! We will miss the plane! They're closing the gate! (In Italy)
— Are we there yet? (driving towards Christiana)

In addition to all these qualities that we all have benefited from, thank you Prakash for your unswerving support and for being my most faithful ambassador. You are the most precious advisor, sending your students to top conferences, introducing them to dozens of people, from bright students to computer science stars. Your will to put them up front, to fit them into stimulating teams of researchers is exemplary. Thank you for your legendary enthusiasm and your commitment to advances in Mathematics and Computer Science. It is really no surprise soooooo many people want to work with you.

Generalized Differential Privacy: Regions of Priors That Admit Robust Optimal Mechanisms[*]

Ehab ElSalamouny[1,2], Konstantinos Chatzikokolakis[1], and Catuscia Palamidessi[1]

[1] INRIA, CNRS and LIX, Ecole Polytechnique, France
[2] Faculty of Computers and Informatics, Suez Canal University, Egypt

Abstract. Differential privacy is a notion of privacy that was initially designed for statistical databases, and has been recently extended to a more general class of domains. Both differential privacy and its generalized version can be achieved by adding random noise to the reported data. Thus, privacy is obtained at the cost of reducing the data's accuracy, and therefore their *utility*.

In this paper we consider the problem of identifying *optimal* mechanisms for generalized differential privacy, i.e. mechanisms that maximize the utility for a given level of privacy. The utility usually depends on a prior distribution of the data, and naturally it would be desirable to design mechanisms that are *universally optimal*, i.e., optimal for all priors. However it is already known that such mechanisms do not exist in general. We then characterize maximal *classes of priors* for which a mechanism which is optimal for all the priors of the class *does exist*. We show that such classes can be defined as convex polytopes in the priors space.

As an application, we consider the problem of privacy that arises when using, for instance, location-based services, and we show how to define mechanisms that maximize the quality of service while preserving the desired level of geo-indistinguishability.

1 Prologue

Privacy is an instance of the general problem of information protection, which constitutes one of the main topics of the research of our team Cométe. The history of our interest for this topic has an important milestone in the visit of Prakash to Cométe in 2006, in the context of our équipe associée Printemps. We had been working for a while on a probabilistic approach to anonymity, and when Prakash arrived, he suggested to consider an information-theoretic approach instead. This was the beginning of a very fruitful collaboration between Prakash and our team, and two of the papers that originated from this collaboration became the backbone of the PhD thesis of Konstantinos Chatzikokolakis. Furthermore, the collaboration with Prakash influences, still today, our research on information protection, in the sense that our research is characterized by the paradigmatic view of a system as a noisy channel – the central concept of information theory. The present paper, which explores the properties of the channel matrix in the

---

[*] This work is partially funded by the Inria large scale initiative CAPPRIS, the EU FP7 grant no. 295261 (MEALS), the INRIA Equipe Associée PRINCESS, and by the project ANR-12-IS02-001 PACE.

F. van Breugel et al. (Eds.): Panangaden Festschrift, LNCS 8464, pp. 292–318, 2014.

context of differential privacy, is a tribute to the fundamental role that Prakash has had in Cométe's scientific life and evolution.

2 Introduction

It is often the case that a privacy threat arises not because of direct access to sensitive data by unauthorized agents, but rather because of the information they can infer from correlated public data. This phenomenon, known as *information leakage*, is quite general and it has been studied in several different domains, including programming languages, anonymity protocols, and statistical databases (see, for instance, [1–3]). Naturally, the settings and the approaches vary from domain to domain, but the principles are the same.

In the case of statistical databases, the public information is typically defined by the kind of queries we are allowed to ask, and the concerns for privacy focus on the consequences that the participation in the databases may have for the confidential data of a *single individual*. Differential privacy [4, 5] was designed to control these consequences. Since it has been recognized that the deterministic methods offer little resistance to composition attacks (i.e. to the combination of information inferred from different databases, see for instance [6, 7]), differential privacy targets probabilistic mechanisms, i.e. mechanisms that answer the query in a probabilistic fashion. Typically, they generate the output by adding random noise to the true answer, according to some probabilistic distribution. The aim of differential privacy is to guarantee that the participation of a single individual in the database will not affect too much the probability of each reported answer. More precisely, (the log of) the ratio between the likelihoods of obtaining a certain answer, from any two *adjacent* databases (i.e., differing only for the presence of an individual), must not exceed a given parameter ϵ. The rationale of this notion comes from the fact that it is equivalent to the property that the reported answer does not change significantly the probabilistic knowledge of the individual data. Differential privacy has become very popular thanks to the fact that it is easy to implement: it is sufficient to add Laplacian noise to the true answer. Furthermore, the notion and the implementation are independent from the side knowledge of the adversary about the underlying database (represented as a prior probability distiribution over possible databases). Finally, it is compositional, in the sense that the privacy loss caused by the combination of attacks is the sum of the single privacy losses.

There have been several studies aimed at applying differential privacy to other areas. In this work, we focus on the approach proposed in [8], which introduced the concept of $d_{\mathcal{X}}$-privacy, suitable for any domain $\mathcal{X}$ equipped with a notion of distance $d_{\mathcal{X}}$. Given a mechanism K from the set of secrets $\mathcal{X}$ to distribution over some set of outputs $\mathcal{Z}$, we say that K satisfies $d_{\mathcal{X}}$-privacy if for any two secrets x_1 and x_2, and any output z, the log of the ratio between $K(x_1)$ and $K(x_2)$ does not exceed $d_{\mathcal{X}}(x_1, x_2)$. Note that $d_{\mathcal{X}}$-privacy is an extension of differential privacy: the latter can be obtained by setting $\mathcal{X}$ to be the set of databases (seen as tuples of individual records) and $d_{\mathcal{X}}$ to be the Hamming distance between these tuples, scaled by ϵ. Furthermore, it is a *conservative extension*, in the sense that it preserves the implementability by means of Laplacian noise, the independence from the prior probability, the interpretation in terms of probabilistic knowledge, and the compositionality properties. From the practical point of view, $d_{\mathcal{X}}$-privacy is particularly suitable to protect the accuracy of the values, like in the case of smart-meter signatures [8] and the precise geographical position in location-based services [9]. Similar extensions of differential privacy obtained by generalizing the distance or the adjacency relation have been considered in [10–12].

Besides guaranteeing privacy, a mechanism should of course provide an answer which is "useful" enough for the service it has been designed. This second goal is measured in terms of *utility*, which represents the average gain that a rational user obtains from the reported answer. More precisely, let y be the true answer and let z be the output reported by the mechanism. On the basis of the latter, the user tries to make a guess y' (remapping) about the (hidden) true answer y. His gain $g(y, y')$ is determined by a given function g. The utility is then defined as *the expected gain under the best possible remapping*. While the gain function can take various forms, in this paper we restrict our analysis to the *binary* gain function, which evaluates to 1 when the user's guess is the same as the query result ($y = y'$) and evaluates to 0 otherwise.

Obviously, there is a trade-off between privacy and utility, and we are interested in mechanisms that offer maximal utility for the desired level of $d_{\mathcal{X}}$-privacy. Such mechanisms are called *optimal*. Naturally, we are also interested in mechanisms that are *universally* optimal, i.e., optimal under any prior[1], as we don't want to design a different mechanism for each user[2]. A famous result by Gosh et al. [13] states that this is possible for the *counting queries*, namely the queries of the form "how many records in the database have the property p", for some p. Unfortunately Brenner and Nissim showed that in differential privacy universally optimal mechanisms do not exist for any other kind of query [14]. However, one can still hope that it is possible to design mechanisms that are optimal for a significant class of users. These are exactly the main objectives of this paper: identify regions of priors which admit a *robust* optimal mechanism, i.e. a mechanism whose optimality is not affected by changes in the prior (within the region), and provide a method to construct such mechanism.

A related issue that we consider in this paper is the amount of information leaked by a mechanism, a central concept in the area of *quantitative information flow* . There have been various proposals for quantifying the information leakage, we consider here an *information-theoretic approach* based on Rényi min-entropy [15, 16], which is suitable for one-try attacks. A main difference between the min-entropy leakage and $d_{\mathcal{X}}$-privacy is that the former measures the *expected* risk of disclosure of sensitive information, while the latter focuses on the worst case, i.e., it considers catastrophic any such disclosure, no matter how unlikely it is.

Recently, researchers have investigated the relation between differential privacy and min-entropy leakage [17–19], and in particular it has been proved in [18] that differential privacy induces a bound on the min-entropy leakage, which is met by a certain mechanism for the uniform prior (for which min-entropy leakage is always maximum). In this paper, we extend the above result to provide a more accurate bound for any prior in the special regions described above. More precisely, we provide a bound to the leakage specific to the prior and that can be met, under a certain condition, by a suitable mechanism.

---

[1] Note that, in contrast to $d_{\mathcal{X}}$-privacy, utility *does* depend on the prior.

[2] We recall that the prior represents the side knowledge of the user.

Contributions

- We identify, for an arbitrary metric space $(\mathcal{Y}, d_{\mathcal{Y}})$, the class of the $d_{\mathcal{Y}}$-regular distributions of $\mathcal{Y}$. The interest of this class is that for each prior distribution in it we are able to provide a specific upper bound to the utility of any $d_{\mathcal{Y}}$-private mechanism. We characterize this class as a geometric region, and we study its properties.
- We describe a $d_{\mathcal{Y}}$-private mechanism, called "tight-constraints mechanism", which meets the upper bound for every $d_{\mathcal{Y}}$-regular prior, and is therefore robustly optimal in that region. We provide necessary and sufficient conditions for the existence of such mechanism, and an effective method to test the conditions and to construct the mechanism.
- We consider the domain of databases $(\mathcal{X}, d_{\mathcal{X}})$, where $d_{\mathcal{X}}$ is the Hamming distance, and we recast the above definitions and results in terms of min-entropy leakage. We are able to improve a result from the literature which says that differential privacy induces a bound on the min-entropy leakage for the uniform prior: We provide more accurate bounds, and show that these bounds are valid for all the $d_{\mathcal{X}}$-regular priors (not just for the uniform one). A construction similar to the one in the previous point yields the tight-constraints mechanism which reaches those upper bounds.

A preliminary version of this paper, restricted to standard differential privacy, and without proofs, appeared in POST 2013.

Plan of the paper. In the next section we recall the basic definitions of generalized differential privacy, utility, and min-entropy mutual information. Section 4 introduces the notion of $d_{\mathcal{Y}}$-regular prior, investigates the properties of these priors, and gives a geometric characterization of their region. Section 5 shows that for all $d_{\mathcal{Y}}$-regular priors on the true answers (resp. databases), $d_{\mathcal{Y}}$-privacy induces an upper bound on the utility (resp. on the min-entropy leakage). Section 6 identifies a mechanism which reaches the above bounds for every $d_{\mathcal{Y}}$-regular prior, and that is therefore the universally optimal mechanism (resp. the maximally leaking mechanism) in the region. Section 7 illustrates our methodology and results using the example of the sum queries and location privacy. Section 8 concludes and proposes some directions for future research.

3 Preliminaries

In this section we recall the generalized variant of differential privacy from [8], considering an arbitrary set of secrets $\mathcal{X}$, equipped with a metric $d_{\mathcal{X}}$. We then discuss two instantiations of the general definition: first, *standard differential privacy* is defined on databases under the Hamming distance. Second, *geo-indistinguishability* [9], a notion of location privacy, is obtained by using geographical locations as secrets, under the Euclidean distance. Finally, we recall a standard way for measuring the utility of a mechanism, and the notion of min-mutual information.

3.1 Generalized Privacy

As discussed in the introduction, a generalized variant of differential privacy can be defined on an arbitrary set of secrets $\mathcal{X}$, equipped with a metric $d_{\mathcal{X}}$. Intuitively, $d_{\mathcal{X}}(x, x')$

gives the "distinguishability level" between secrets x, x', based on the privacy semantics that we wish to obtain. The smaller the distinguishability level is, the harder it should be for the adversary to distinguish the two secrets, hence offering privacy, while secrets at great distance are allowed to be distinguished, giving the possibility to obtain some controlled knowledge about the secret.

A *mechanism* from $\mathcal{X}$ to $\mathcal{Z}$ is a function $K : \mathcal{X} \to \mathcal{P}(\mathcal{Z})$, where $\mathcal{P}(\mathcal{Z})$ denotes the set of probability distributions over some set of outputs $\mathcal{Z}$. In this paper we consider $\mathcal{X}, \mathcal{Z}$ to be finite, hence the involved distributions to be discrete. The mechanism's outcome $K(x)$ is then a probability distribution, and $K(x)(z)$ is the probability of an output $z \in \mathcal{Z}$ when running the mechanism on $x \in \mathcal{X}$. For simplicity we write $K : \mathcal{X} \to \mathcal{Z}$ to denote a machanism from $\mathcal{X}$ to $\mathcal{Z}$ (omitting $\mathcal{P}$).

The multiplicative distance $d_{\mathcal{P}}$ between probability distributions $\mu_1, \mu_2 \in \mathcal{P}(\mathcal{Z})$ is defined as $d_{\mathcal{P}}(\mu_1, \mu_2) = \sup_{z \in \mathcal{Z}} |\ln \frac{\mu_1(z)}{\mu_2(z)}|$ with the convention that $|\ln \frac{\mu_1(z)}{\mu_2(z)}| = 0$ if both $\mu_1(z), \mu_2(z)$ are zero and ∞ if only one of them is zero.

We are now ready to give the definition of $d_{\mathcal{X}}$-privacy:

Definition 1. *A mechanism* $K : \mathcal{X} \to \mathcal{Z}$ *satisfies* $d_{\mathcal{X}}$-*privacy, iff* $\forall x, x' \in \mathcal{X}$:

$$d_{\mathcal{P}}(K(x), K(x')) \le d_{\mathcal{X}}(x, x')$$

or equivalently:

$$K(x)(z) \le e^{d_{\mathcal{X}}(x,x')} K(x')(z) \quad \forall z \in \mathcal{Z}$$

The intuition behind this definition is that the attacker's ability to distinguish two secrets should depend on their distinguishability level $d_{\mathcal{X}}(x, x')$. The closer two secrets are, the more similar the mechanism's output on those secrets should be, making it harder for the adversary to distinguish them. Depending on the choice of $d_{\mathcal{X}}$, the definition can be adapted to the application at hand, giving rise to different notions of privacy.

In [8], two alternative characterizations of $d_{\mathcal{X}}$-privacy are also given, in which the attacker's knowledge is explicitly quantified, which makes it easier to understand the privacy guarantees obtained by a particular choice of $d_{\mathcal{X}}$.

Answering queries. In practice, we often want to learn some information about our secret, that is we want to obtain the answer to a query $f : \mathcal{X} \to \mathcal{Y}$. To do so privately, we can compose f with a "noise" mechanism $H : \mathcal{Y} \to \mathcal{Z}$, thus obtaining an "oblivious" mechanism $H \circ f : \mathcal{X} \to \mathcal{Z}$, called oblivious since the answer depends only on $f(x)$ and not on x itself. The role of H is to add random noise to the true query result $f(x)$ and produce a "noisy" reported output $z \in \mathcal{Z}$.

Since we assume all sets to be finite, the mechanism H can be described by a stochastic matrix $H = (h_{yz})$, called the *noise matrix*, whose rows are indexed by the elements of $\mathcal{Y}$ and whose columns are indexed by the elements of $\mathcal{Z}$. Hence, h_{yz} is the probability of reporting z when the true query result is y.

Given a metric $d_{\mathcal{Y}}$ on $\mathcal{Y}$, the generalized definition of privacy allows us to directly talk about the privacy of H, without involving f at all. Using matrix notation, $d_{\mathcal{Y}}$-privacy for H (Definition 1) can be written as

$$h_{yz} \le e^{d_{\mathcal{Y}}(y,y')} h_{y'z} \qquad \forall y, y' \in \mathcal{Y}, z \in \mathcal{Z} \tag{1}$$

A natural question, then, is how d_χ-privacy of the composed mechanism $H \circ f$ relates to d_y-privacy of H. The connection between the two comes from the concept of *uniform Δ-sensitivity.*

Definition 2. *A sequence $y_1, \ldots, y_n$ is called a* chain *from y_1 to y_n. We say that such chain is tight if $d_y(y_1, y_n) = \sum_i d_y(y_i, y_{i+1})$. Two elements $y, y' \in \mathcal{Y}$ are called Δ-expansive iff $d_y(y, y') = \Delta d_\chi(x, x')$ for some $x \in f^{-1}(y), x' \in f^{-1}(y')$. A chain is Δ-expansive iff all steps y_i, y_{i+1} are Δ-expansive.*
  Finally, f is uniformly Δ-sensitive wrt d_χ, d_y iff:

- *for all $x, x' \in \mathcal{X}$: $d_y(f(x), f(x')) \leq \Delta d_\chi(x, x')$, and*
- *for all $y, y' \in \mathcal{Y}$: there exists a tight and Δ-expansive chain from y to y'.*

The intuition behind this definition is that f expands distances by at most Δ, and there are no answers that are always the results of a smaller expansion: all $y, y' \in \mathcal{Y}$ can be linked by a chain in which the expansion is exactly Δ. Under this condition, it has been shown in [8] that the privacy of H characterizes that of $H \circ f$.

Theorem 1 ([8]). *Assume that f is uniformly Δ-sensitive wrt d_χ, d_y. Then H satisfies d_y-privacy if and only if $H \circ f$ satisfies Δd_χ-privacy.*

In the remaining of the paper, we give results about d_y-privacy for H, for an arbitrary metric d_y, independently from any function f. The results can be used either to talk about the privacy of H itself, or – given the above theorem – about the privacy of oblivious mechanisms of the form $H \circ f$, for some function f for which uniform sensitivity can be established. A typical case of uniform sensitivity arises in standard differential privacy when d_y is the metric obtained from the *induced graph* of f, as discussed in the next section. But uniform sensitivity can be established for other types of metrics; some examples are given in [8].

3.2 Differential Privacy

The notion of differential privacy, introduced by Dwork in [4], imposes constraints on data reporting mechanisms so that the outputs produced by two databases differing only for one record are almost indistinguishable. Let V be a universe of values and u the number of individuals. The set of all possible databases (u-tuples of values from V) is $\mathcal{V} = V^u$. Two databases $x, x' \in \mathcal{V}$ are called *adjacent*, written $x \sim x'$, iff they differ in the value of exactly one individual. The adjacency relation $\sim$ defines a graph, and the length of the shortest path between two databases x, x' in the graph, written $d_h(x, x')$, defines a metric called the Hamming distance. In other words, $d_h(x, x')$ is the number of individuals in which x and x' differ.

The property of ϵ-differential privacy requires that, for any two adjacent databases, the ratio of the probabilities of producing a certain output is bound by e^ϵ. It is easy to see that this property is equivalent to ϵd_h-privacy, under the Hamming distance d_h.

Given a query $f : \mathcal{V} \to \mathcal{Y}$, the adjacency relation $\sim$ can be extended to $\mathcal{Y}$, giving rise to the *induced graph* $\sim_f$ of f [14, 19], defined as:

$$y \sim_f y' \quad \text{iff} \quad x \sim x' \text{ for some } x \in f^{-1}(y), x' \in f^{-1}(y')$$

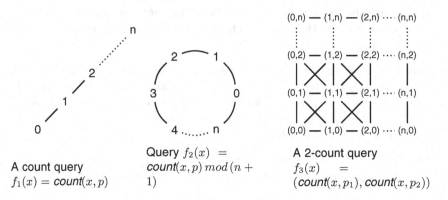

A count query
$f_1(x) = count(x, p)$

Query $f_2(x) =$
$count(x, p) \, mod \, (n + 1)$

A 2-count query
$f_3(x) =$
$(count(x, p_1), count(x, p_2))$

Fig. 1. The induced graph of different queries

Figure 1 shows the induced graph of three different queries. In these examples $count(x, p)$ refers to a counting query which returns the number of records in the database x which satisfy a certain property p. Other queries in the figure are expressed using the *count* function.

Furthermore, let $d_{\sim_f}(y, y')$ be the metric on $\mathcal{Y}$ defined as the shortest $\sim_f$-path from y to y'. It has then been shown in [8] that any function f is uniformly 1-sensitive wrt $d_h, d_{\sim_f}$. As a consequence of this, and of Theorem 1, ϵ-differential privacy of an oblivious mechanism $H \circ f$ can be characterized by the $\epsilon d_{\sim_f}$-privacy privacy of H.

Corollary 1. *For any query $f : \mathcal{V} \to \mathcal{Y}$, H satisfies $\epsilon d_{\sim_f}$-privacy if and only if $H \circ f$ satisfies ϵd_h-privacy.*

3.3   Geo-Indistinguishability

An advantage of the generalized definition of privacy is that it can be applied in cases when there is a single individual involved – hence the notion of adjacency is inadequate – by using a metric that gives a meaningful notion of privacy for the application at hand. An example of such a notion is *geo-indistinguishability* [9], proposed as a formal notion of location privacy in the context of Location Based Services (LBSs).

Consider a mobile user, typically using a GPS-enabled hand-held device, who wishes to obtain information related to his current location, for instance restaurants close to him. To do so, he can query an LBS provider, providing his actual location x as part of the query. However, location information is not only inherently sensitive itself, but also correlated to a variety of other sensitive information, such as political and religious beliefs, medical information, etc. Hence, the user would like to perform the LBS query privately, that is without disclosing his exact location to the provider. Note that protecting the user's *identity* is not the goal here; in fact, the user might wish to be authenticated to the service provider in order to obtain personalized recommendations. What he is interested in, instead, is hiding his *location*.

A possible solution is to use a *location obfuscation* mechanism [20], producing a noisy location z which is reported to the service provider. A natural goal then is to formalize the privacy guarantees provided by such a mechanism, for which various approaches have been proposed in the literature [21].

Geo-indistinguishability provides such a formal definition of location privacy, and can be expressed as an instance of $d_{\mathcal{X}}$-privacy. Secrets $\mathcal{X}$ are now locations (a subset of $\mathbb{R}^2$), and ϵ-geo-indistinguishability is ϵd_2-privacy, where d_2 is the Euclidean distance between locations.[3] Intuitively, $d_{\mathcal{P}}(K(x), K(x')) \leq \epsilon d_2(x, x')$ requires that the closer (geographically) two locations x, x' are, the more likely to produce the same reported location z they should be. This allows the provider to get some approximate information necessary to provide the service (e.g. distinguish locations in Paris from those in London), but prevents him from learning x with high accuracy (since locations x' close to x produce the same z with similar probabilities).

The results of this paper refer to an arbitrary metric between secrets, hence they are directly applicable to geo-indistinguishability. A case-study in the context of location privacy is given in Section 7.2.

3.4 Utility Model

The main role of a noise mechanism $H : \mathcal{Y} \to \mathcal{Z}$ is to guarantee $d_{\mathcal{Y}}$-privacy while providing useful information about the true query result, i.e. to satisfy a trade-off between the privacy and utility. For quantifying the utility of H we follow a standard model from [13]. Let $y \in \mathcal{Y}$ be the result of executing a query f. The mechanism $H : \mathcal{Y} \to \mathcal{Z}$ processes y and produces an output z in some domain $\mathcal{Z}$ to the user. Based on the reported output z and prior knowledge about the likely results of f, she applies a remapping function $R : \mathcal{Z} \to \mathcal{Y}$ to z to produce a guess $y' \in \mathcal{Y}$ for the real query result. Note that the composite mechanism $R \circ H : \mathcal{Y} \to \mathcal{Y}$ is a mechanism whose output domain is the query results domain $\mathcal{Y}$. We say that H is remapped to $R \circ H$ by the remap R. Now, with the user's guessed value y', a real-valued *gain function* $g : \mathcal{Y} \times \mathcal{Y} \to \mathbb{R}$ quantifies how informative y' is compared to the real query result y. In this paper we restrict our analysis to the binary gain function g_b which is defined as $g_b(y, y') = 1$ iff $y' = y$ and 0 otherwise. The choice of this gain corresponds to the preference of a user to guess the true query result.

In practice, the user usually bases her guess y' about the real query result on prior knowledge about the underlying secret and the underlying query. This knowledge is modeled by a probability distribution π (called *prior*) over the domain $\mathcal{Y}$ of query results. Now the utility of a mechanism $H : \mathcal{Y} \to \mathcal{Z}$ with respect to a prior π and a remap $R : \mathcal{Z} \to \mathcal{Y}$ is the expected value of the underlying gain function g_b, and is therefore expressed as

$$\mathcal{U}(H, \pi, R) = \sum_{y,y'} \pi_y \, (HR)_{yy'} \, g_b(y, y'). \tag{2}$$

Using the definition of g_b, the above expression reduces to a convex combination of the diagonal elements of HR as follows.

$$\mathcal{U}(H, \pi, R) = \sum_y \pi_y \, (HR)_{yy}. \tag{3}$$

---

[3] Note that any other meaningful geographical distance could also be used, such as the Manhattan or a map-based distance.

Accordingly, we say that a d_y-private mechanism H is d_y-optimal for a prior π if there is a remap R such that $\mathcal{U}(H, \pi, R)$ is maximal for all d_y-private mechanisms and all remaps.[4] In general the optimality of a mechanism depends on the prior (related to the user). That is a mechanism that is optimal for a prior may not be optimal for another one. In the setting of differential privacy, it has been proven [14] that for any query, other than a single counting one, there is no mechanism that is optimal for all priors simultaneously. Nevertheless, we identify in Section 4 a region of priors, where it is possible to find a single mechanism which is optimal to all of them.

3.5 Min-mutual Information

In this section we recall the use of an information-theoretic notion, namely mutual information, to quantify the amount of information conveyed by a mechanism H : $\mathcal{Y} \rightarrow \mathcal{Z}$ as an information theoretic channel.

Following recent works in the area of quantitative information flow ([15–17]), we adopt *Rényi's min-entropy* ([22]) as our measure of uncertainly. The min-entropy $\mathcal{H}_\infty(\pi)$ of a prior π, defined as $\mathcal{H}_\infty(\pi) = -\log_2 \max_i \pi_i$, measures the user's uncertainty about the query result. Then, the corresponding notion of *conditional* min-entropy, defined as $\mathcal{H}_\infty(H, \pi) = -\log_2 \sum_{z \in \mathcal{Z}} \max_y \pi_y h_{yz}$, measures the uncertainty about the query result after observing an output $z \in \mathcal{Z}$. Finally, subtracting the latter from the former brings us to the notion of min-mutual information:

$$\mathcal{L}(H, \pi) = \mathcal{H}_\infty(\pi) - \mathcal{H}_\infty(H, \pi)$$

which measures the amount of information about the query result conveyed by the mechanism H. In the area of quantitative information flow this quantity is known as *min-entropy leakage*; the reader is referred to [15] for more details about this notion.

4   Regular Priors

In this section we describe a region of priors, called 'd_y-regular'. These priors are determined by the metric d_y on the domain $\mathcal{Y}$. Recall that the d_y-privacy constraints for H can be written as $h_{yz}/h_{y'z} \geq e^{-d_y(y,y')}$ for all $y, y' \in \mathcal{Y}$. Since every lower bound $e^{-d_y(y,y')}$ depends only on y, y', the constraints can be described altogether by a square matrix Φ formed by such lower bounds. We refer to this matrix as the *privacy-constraints* matrix.

Definition 3 (privacy-constraints matrix). *The* privacy-constraints matrix Φ *of a metric d_y is a square matrix, indexed by $\mathcal{Y} \times \mathcal{Y}$, where $\phi_{yy'} = e^{-d_y(y,y')}$ for all $y, y' \in \mathcal{Y}$.*

Note that Φ is symmetric ($\phi_{yy'} = \phi_{y'y}$) due to the symmetry of d_y. Recall that d_y describes the privacy restrictions imposed on the domain $\mathcal{Y}$. In particular these restrictions become vacuous if $d_y(y, y') \rightarrow \infty$ for all $y, y' : y \neq y'$. In this extreme case the privacy-constraints matrix Φ converges to the identity matrix where each diagonal

---

[4] Note that there may exist many optimal mechanisms for a given prior.

entry is 1 and all other entries are 0. We now define the d_y-regular priors, in terms of the privacy-constraints matrix of d_y. For a vector μ having cardinality $|\mathcal{Y}|$, we use $\mu \geq 0$ to denote $\forall y : \mu_y \geq 0$.

Definition 4 (d_y-**regular prior**). *A prior π is called d_y-regular iff there exists a row vector $\mu \geq 0$ such that $\pi = \mu \Phi$.*

In the following we describe the common properties of these priors and also give a geometric characterization for their region comparing it to the whole prior space. As a first observation, this region converges to the entire prior space when the privacy constraints on $\mathcal{Y}$ become vacuous. This is because, as described above, Φ approaches the identity matrix where the vector μ exists for each prior π (just define $\mu = \pi$).

An important property of any d_y-regular prior is that the ratio between any two of its entries $\pi_y, \pi_{y'}$ is always bound by $e^{d_y(y,y')}$. Because of this property, such a prior is called d_y-regular.

Proposition 1. *For every d_y-regular prior π and for all $y, y' \in \mathcal{Y}$ we have that $\pi_y / \pi_{y'} \leq e^{d_y(y,y')}$.*

Proof. By Definition 4, the ratio $\pi_y / \pi_{y'}$ is given by

$$\pi_y / \pi_{y'} = \frac{\sum_{y''} \mu_{y''} \phi_{y''y}}{\sum_{y''} \mu_{y''} \phi_{y''y'}}. \tag{4}$$

By the definitions of $\phi_{y''y'}, \phi_{y''y}$ we also have that

$$\phi_{y''y'} = e^{-d_y(y'',y')} \geq e^{-(d_y(y'',y)+d_y(y,y'))} = e^{-d_y(y,y')} \phi_{y''y}.$$

The above inequality is implied by the triangle inequality, $d_y(y'', y') \leq d_y(y'', y) + d_y(y, y')$ and the fact that $e^{-1} < 1$. Since $\mu_{y''} \geq 0$ for all y'', we have

$$\sum_{y''} \mu_{y''} \phi_{y''y'} \geq e^{-d_y(y,y')} \sum_{y''} \mu_{y''} \phi_{y''y}$$

Substituting the above inequality in Eq. (4) completes the proof.   □

The above property restricts nearby elements of $\mathcal{Y}$ (with respect to the metric d_y) to have 'similar' probabilities. In practice, this property holds for a large class of users who have no sharp information that discriminates between nearby elements of $\mathcal{Y}$. Note that the above property is not equivalent to Definition 4. Namely, it is not true that all priors having such a property are d_y-regular.

A consequence of the above proposition is that for any d_y-regular prior π, the probability π_y associated with $y \in \mathcal{Y}$ is restricted by upper and lower bounds as follows.

Proposition 2. *For every d_y-regular prior π and for every $y \in \mathcal{Y}$ we have that*

$$1 \Big/ \sum_{y' \in \mathcal{Y}} e^{d_y(y,y')} \leq \pi_y \leq 1 \Big/ \sum_{y' \in \mathcal{Y}} e^{-d_y(y,y')}.$$

Proof. By Proposition 1, it holds for every pair of entries $\pi_y, \pi_{y'}$ that

$$\pi_{y'} \le e^{d_\mathcal{Y}(y,y')} \pi_y \quad \text{and} \quad e^{-d_\mathcal{Y}(y,y')} \pi_y \le \pi_{y'}.$$

Summing the above inequalities over y', we get

$$\sum_{y' \in \mathcal{Y}} \pi_{y'} \le \pi_y \sum_{y' \in \mathcal{Y}} e^{d_\mathcal{Y}(y,y')} \quad \text{and} \quad \pi_y \sum_{y' \in \mathcal{Y}} e^{-d_\mathcal{Y}(y,y')} \le \sum_{y' \in \mathcal{Y}} \pi_{y'}.$$

Since $\sum_{y' \in \mathcal{Y}} \pi_{y'} = 1$, the above inequalities imply the upper and lower bounds for π_y. $\qquad\square$

One obvious implication is that any $d_\mathcal{Y}$-regular prior must have full support, that is $\pi_y > 0$ for all $y \in \mathcal{Y}$. In the following we describe the set of $d_\mathcal{Y}$-regular priors as a region in the prior space. For doing so, we first define in the following set of priors which we refer to as the corner priors.

Definition 5 (corner priors). *For every $y \in \mathcal{Y}$, a corresponding corner prior, denoted by c^y, is defined as*

$$c^y_{y'} = \frac{\phi_{yy'}}{\sum_{y'' \in \mathcal{Y}} \phi_{yy''}} \qquad \forall y' \in \mathcal{Y}.$$

Note that the above definition is sound, i.e. c^y is a probability distribution for all $y \in \mathcal{Y}$. Note also that there are $|\mathcal{Y}|$ corner priors; each one corresponds to an element $y \in \mathcal{Y}$. By inspecting the entries of c^y, observe that c^y_y has the maximum value compared to other entries, and moreover this value is exactly the upper bound specified by Proposition 2. We can therefore interpret this observation informally as c^y is 'maximally biased' to y. It can be also seen that each corner prior is $d_\mathcal{Y}$-regular. In fact for any corner c^y, there is a row vector μ that satisfies the condition in Def. 4; this vector is obtained by setting $\mu_y = 1/\sum_{y' \in \mathcal{Y}} \phi_{yy'}$ and $\mu_{y'} = 0$ for all $y' \ne y$. Here it is easy to verify that $c^y = \mu \Phi$.

Now we can describe the region of the $d_\mathcal{Y}$-regular priors using the corner priors. Precisely, this region consists of all convex combinations of the corner priors.

Proposition 3 (convexity). *A prior π is $d_\mathcal{Y}$-regular iff it is a convex combination of the corner priors, i.e. there exist real numbers $\gamma_y \ge 0$, $y \in \mathcal{Y}$ such that*

$$\pi = \sum_{y \in \mathcal{Y}} \gamma_y \, c^y \quad \text{and} \quad \sum_{y \in \mathcal{Y}} \gamma_y = 1.$$

Proof. By Definition 4, a prior π is $d_\mathcal{Y}$-regular iff there exists vector $\mu \ge 0$ such that $\pi = \mu \Phi$; that is iff there are reals $\mu_y \ge 0$ for all $y \in \mathcal{Y}$, such that π can be written as a linear combination of Φ's rows as follows.

$$\pi = \sum_{y \in \mathcal{Y}} \mu_y \Phi_y, \tag{5}$$

where Φ_y is the row of Φ corresponding to the element $y \in \mathcal{Y}$. From Def. 5, observe that each row Φ_y is equal to $\left(\sum_{y' \in \mathcal{Y}} \phi_{yy'} \right) c^y$. By substituting Φ_y in Eq. (5), we get that π is $d_\mathcal{Y}$-regular iff $\pi = \sum_{y \in \mathcal{Y}} \gamma_y c^y$ where $\gamma_y = \mu_y \left(\sum_{y' \in \mathcal{Y}} \phi_{yy'} \right)$. Note from the latter relation between γ_y and μ_y (for every $y \in \mathcal{Y}$) that the existence of the vector $\mu \ge 0$ is equivalent to the existence of the coefficients $\gamma_y \ge 0$. Finally observe that $\sum_{y \in \mathcal{Y}} \gamma_y = \sum_{y \in \mathcal{Y}} (\mu \Phi)_y = \sum_{y \in \mathcal{Y}} \pi_y = 1$. $\qquad\square$

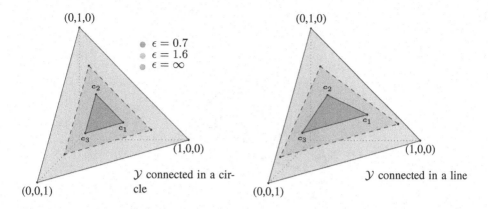

Fig. 2. Regions of $d_{\mathcal{Y}}$-regular priors for Example 1

From Proposition 3 the region of $d_{\mathcal{Y}}$-regular priors is a convex set, where each point (prior) in this region is a convex combination of the corner priors. This region is therefore geometrically regarded as a convex polytope in the prior space. Since the corner points always exists, this region is never empty. For a prior π in this region, the coefficients γ_y model the 'proximity' of π to each corner prior c^y. In particular, note that $0 \leq \gamma_y \leq 1$, and $\gamma_y = 1$ iff $\pi = c^y$. We demonstrate this geometric interpretation using the following examples.

Example 1. Consider a simple domain $\mathcal{Y}$ consisting of 3 elements organized in a graph structure where $d_g(y, y')$ is the graph distance between y, y'. Now for an arbitrary scaling number $\epsilon > 0$, we can define the metric $d_{\mathcal{Y}}$ as $d_{\mathcal{Y}}(y, y') = \epsilon\, d_g(y, y')$. Since every prior on $\mathcal{Y}$ has 3 entries (specifying the probability of every element $y \in \mathcal{Y}$), the prior space for $\mathcal{Y}$ can be represented by the 3-dimensional Euclidean space. Figure 2 visualizes the region of $d_{\mathcal{Y}}$-regular priors in two cases: when the graph structure of $\mathcal{Y}$ is a line, and when it is a circle. Note that in both cases, we have 3 corner priors c^1, c^2, c^3. In each case, the region is depicted for $\epsilon = 0.7$ and $\epsilon = 1.6$. Note in this example that ϵ controls the privacy constraints imposed by $d_{\mathcal{Y}}$-privacy, which in turn determine the size of the region of $d_{\mathcal{Y}}$-regular priors. In particular with $\epsilon = 1.6$ (less privacy), the region is larger than the one with $\epsilon = 0.7$. In general the region expands as ϵ increases and converges to the entire region of priors defined by the corner points $\{(0, 0, 1), (0, 1, 0), (0, 0, 1)\}$ when $\epsilon \to \infty$.

Example 2. Suppose that $\mathcal{Y}$ contains 4 elements, and $d_{\mathcal{Y}}$ is defined as $d_{\mathcal{Y}}(y, y') = D$ for all $y, y' : y \neq y'$. In this case every prior contains 4 entries and therefore is not possible to be plotted in the 3-dimensional space. However, using the fact that the fourth component is redundant ($\sum_i \pi_i = 1$), every prior is fully described by its 'projection' onto the 3-dimensional subspace. Figure 3 shows the projection of the $d_{\mathcal{Y}}$-regular prior region for different values of D. Again the privacy constraints enforced by $d_{\mathcal{Y}}$-privacy are determined by D. The less restricted is D (i.e. having a higher value), the bigger the region is; and eventually coincides with the entire space when $D \to \infty$.

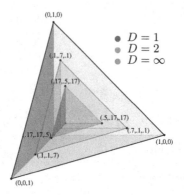

Fig. 3. Regions of d_y-regular priors for Example 2

5   Upper Bounds for Utility and Min-mutual Information

In this section, we further describe the d_y-regular priors on the domain $\mathcal{Y}$ in terms of the utility that can be achieved for these priors by a mechanism $H : \mathcal{Y} \to \mathcal{Z}$ satisfying d_y-privacy. We also describe the amount of information that can be conveyed by H to users with such priors. More precisely, we identify for any d_y-regular prior π upper bounds for the utility and min-mutual information, considering all d_y-private mechanisms and all possible remaps. These bounds are indeed induced by the privacy constraints defined by the metric d_y.

5.1   Utility

For a given domain $\mathcal{Y}$ equipped with the metric d_y, consider a d_y-private mechanism $H : \mathcal{Y} \to \mathcal{Z}$ producing observables in some domain $\mathcal{Z}$. In the following analysis we derive a linear algebraic expression for $\mathcal{U}(H, \pi, R)$, the utility of H for a prior π using the remap $R : \mathcal{Z} \to \mathcal{Y}$. Such an expression will play the main role in the subsequent results. We start by observing that the matrix product of H and the remap R describes an d_y-private mechanism $HR : \mathcal{Y} \to \mathcal{Y}$. Therefore the entries of HR satisfy the following subset of constraints.

$$e^{-d_y(y,y')} (HR)_{y'y'} \leq (HR)_{yy'}$$

for all $y, y' \in \mathcal{Y}$. Using Definition 3 of the privacy-constraints matrix Φ, and taking into account that $\sum_{y' \in \mathcal{Y}} (HR)_{yy'} = 1$ for all y (as both H and R are stochastic), we get the following inequalities.

$$\sum_{y' \in \mathcal{Y}} \phi_{yy'} (HR)_{y'y'} \leq 1, \quad \forall y \in \mathcal{Y}.$$

The inequality operators can be replaced by equalities while introducing *slack* variables $s_y : 0 \leq s_y \leq 1$ for all $y \in \mathcal{Y}$. The above inequalities can therefore be written as follows.

$$\sum_{y' \in \mathcal{Y}} \phi_{yy'} (HR)_{y'y'} + s_y = 1, \quad \forall y \in \mathcal{Y}.$$

Let the slack variables s_y form a column vector s indexed by $\mathcal{Y}$. Let also 1 denote a column vector of the same size and having all entries equal to 1. Using these vectors and the privacy-constraints matrix Φ (for the given metric d_y), the above equations can be rewritten in the following matrix form.

$$\Phi \, \mathrm{diag}(HR) + s = 1, \tag{6}$$

where $\mathrm{diag}(HR)$ is the column vector consisting of the diagonal entries of HR. Now, for any mechanism $H : \mathcal{Y} \to \mathcal{Z}$ and a remap $R : \mathcal{Z} \to \mathcal{Y}$ satisfying Eq. (6), and for a prior π, we want to refine the generic expression (3) of the utility by taking Eq. (6) into account. We start by rewriting Eq. (3) in the following matrix form.

$$\mathcal{U}(H, \pi, R) = \pi \, \mathrm{diag}(HR). \tag{7}$$

Now, let μ be a row vector such that

$$\pi = \mu \Phi. \tag{8}$$

Note that, the above matrix equation is in fact a system of $|\mathcal{Y}|$ linear equations. The yth equation in this system is formed by the yth column of Φ, and the yth entry of π as follows.

$$\mu \Phi_y = \pi_y \qquad \forall y \in \mathcal{Y}.$$

Solving this system of equations for the row vector μ has the following possible outcomes: If the matrix Φ is invertible, then, for any prior π, Eq. (8) has exactly one solution. If Φ is not invertible (i.e. it contains linearly dependent columns), then there are either 0 or an infinite number of solutions, depending on the prior π: If the entries of π respect the linear dependence relation then there are infinitely many solutions. Otherwise, the equations are 'inconsistent', in which case there are no solutions.

Whether Φ is invertible or not, we consider here only the priors where the matrix equation (8) has at least one solution μ. Note that, by definition, all the d_y-regular priors have this property, but there can be others for which the solution μ has some negative components. In some of the results below (in particular in Lemma 1) we consider this larger class of priors for the sake of generality.

Multiplying Equation (6) by μ yields

$$\mu \Phi \, \mathrm{diag}(HR) + \mu \, s = \mu 1. \tag{9}$$

Substituting Equations (8) and (7) in the above equation consecutively provides the required expression for the utility and therefore proves the following lemma.

Lemma 1. *For a metric space $(\mathcal{Y}, d_y)$ let π be any prior on $\mathcal{Y}$. Then for every row vector μ satisfying $\pi = \mu \Phi$, the utility of any d_y-private mechanism H for π using a remap R is given by*

$$\mathcal{U}(H, \pi, R) = \mu 1 - \mu s, \tag{10}$$

for a vector s satisfying $0 \le s_y \le 1$ for all $y \in \mathcal{Y}$.

Lemma 1 expresses the utility function for any d_y-private mechanism H, for a prior π satisfying $\pi = \mu \Phi$, and using a remap R. This utility is expressed as a function of the vector μ and the slack vector s. Although the matrix H and the remap R do not explicitly appear on the right side of Equation (10), the utility still depends on them indirectly through the vector s. Namely, according to Equation (6), the choice of H and R determines the slack vector s. The utility function depends also on the prior π, because the choice of π determines the set of vectors μ satisfying Eq. (8). Substituting any of these vectors in Eq. (10) yields the same value for $\mathcal{U}(H, \pi, R)$.

Now recall from Definition 4 that for every d_y-regular prior π there is μ satisfying $\pi = \mu \Phi$ and $\mu \geq 0$. This characteristic together with Lemma 1 implies an upper bound on the utility of any d_y-private mechanism H for π.

Theorem 2 (utility upper bound). *Let π be a d_y-regular prior and $H : \mathcal{Y} \to \mathcal{Z}$ be a d_y-private mechanism. Then for all row vectors $\mu \geq 0$ satisfying $\mu \Phi = \pi$, and any remap R, it holds that*

$$\mathcal{U}(H, \pi, R) \leq \sum_{y \in \mathcal{Y}} \mu_y. \tag{11}$$

Furthermore the mechanism H and remap R satisfy the equality in (11) for every d_y-regular prior iff $\Phi \operatorname{diag}(HR) = 1$.

Proof. Since π is d_y-regular, we have $\pi = \mu \Phi$ for a vector $\mu \geq 0$. Applying Lemma 1 and noting that $s_y \geq 0$ for all $y \in \mathcal{Y}$, we observe that $\mu s \geq 0$ and hence the utility is upper-bounded by $\mu 1 = \sum_{y \in \mathcal{Y}} \mu_y$.

It remains to show that this bound is attained *for every d_y-regular prior* if and only if $\Phi \operatorname{diag}(HR) = 1$, which is equivalent (according to Eq. (6)) to $s = 0$: Clearly, if $s = 0$, then applying Lemma 1 yields the equality in (11) for every d_y-regular prior. For the 'only if' direction, it is sufficient to find a regular prior for which $s = 0$ must hold to satisfy the equality in (11). For this purpose we recall that every corner prior c^y satisfies $\mu^y \Phi = c^y$ where $\mu^y_y > 0$. Now consider the prior $\bar{\pi} = (1/|\mathcal{Y}|) \sum_{y \in \mathcal{Y}} c^y$, which is d_y-regular by Proposition 3. It is easy to see that it holds $\bar{\mu} \Phi = \bar{\pi}$ where $\bar{\mu} = (1/|\mathcal{Y}|) \sum_{y \in \mathcal{Y}} \mu^y$. Observe here that $\bar{\mu}_y > 0$ for all $y \in \mathcal{Y}$. Suppose now that the equality in (11) holds for $\bar{\mu}$. Therefore it must hold, by Lemma 1, that $\bar{\mu} s = 0$. Since $\bar{\mu}_y > 0$ for all $y \in \mathcal{Y}$, it must hold that $s = 0$. This completes the proof. □

The above result can be also seen from the geometric perspective. As shown by Proposition 3, each member in the region of d_y-regular priors is described as a convex combination of the corner priors. That is there are coefficients $\gamma_y \geq 0$ for $y \in \mathcal{Y}$ which form this combination. It can be shown (as in the proof of Proposition 3) that $\gamma_y = \mu_y \left(\sum_{y' \in \mathcal{Y}} \phi_{yy'} \right)$. Hence, the upper bound given by Theorem 2 can be written as follows using the coefficients γ_y.

$$\mathcal{U}(H, \pi, R) \leq \sum_{y \in \mathcal{Y}} \frac{\gamma_y}{\sum_{y' \in \mathcal{Y}} \phi_{yy'}}.$$

Inspecting the above result for corner priors, recall that for a corner c^y, $\gamma_{y'}$ is 1 for $y' = y$ and is 0 otherwise; thus, the utility upper bound for c^y is therefore $1/\sum_{y'} \phi_{yy'}$. Moreover, the upper bound for each d_y-regular prior π can be regarded (according to the above equation) as a convex combination of the upper bounds for the corner priors. That is, from the geometric perspective, the utility upper bound for π linearly depends on its proximity to the corner priors.

5.2   Min-mutual Information

In this paper we use the information-theoretic notion of min-mutual information in two distinct ways: first, we use it to measure the information conveyed about the result of a specific query, similarly to the use of "utility" in the previous section. Mutual information and utility are indeed closely related, which allows us to transfer the bound obtained in the previous section to the information-theoretic setting.

Second, we use it to quantify the information about the secret itself, thus obtaining what is known in the area of quantitative information flow as *min-entropy leakage* [15]. The above bound can therefore be interpreted as a bound on the information leaked by any mechanism, even non-oblivious ones, independently from the actual query. For arbitrary priors, we obtain in a more natural way the bound conjectured in [17] and proven in [19]. Moreover, if we restrict to specific (d_y-regular) priors, then we are able to provide more accurate bounds.

The following result from [19] shows that min-mutual information corresponds to the notion of utility under the binary gain function and using an *optimal* remap, i.e., a remap that gives the best utility among all possible remaps, for the given prior.

Proposition 4 ([19]). *Given a mechanism* $H : \mathcal{Y} \to \mathcal{Z}$ *and a prior* π*, let* $\hat{R}$ *be an optimal remap for* π, H*. Then, we have*

$$\mathcal{L}(H, \pi) = \log_2 \frac{\mathcal{U}(H, \pi, \hat{R})}{\max_y \pi_y}$$

This connection allows us to transfer the upper-bound given by Theorem 2 to min-mutual information.

Proposition 5 (min-mutual information upper bound). *Let* π *be a* d_y*-regular prior and* $H : \mathcal{Y} \to \mathcal{Z}$ *be a* d_y*-private mechanism. Then for all row vectors* $\mu \geq 0$ *satisfying* $\mu \Phi = \pi$*, we have:*

$$\mathcal{L}(H, \pi) \leq \log_2 \frac{\sum_{y \in \mathcal{Y}} \mu_y}{\max_y \pi_y}. \tag{12}$$

Furthermore, H satisfies the equality for every d_y*-regular prior iff there is a remap R such that* $\Phi \operatorname{diag}(HR) = 1$*.*

Proof. By Proposition 4, the leakage $\mathcal{L}(H, \pi)$ is monotonically increasing with the utility $\mathcal{U}(H, \pi, \hat{R})$. By Theorem 2, this utility is upper-bounded by $\sum_{y \in \mathcal{Y}} \mu_y$. Substituting this upper bound in Proposition 4 yields the inequality (12) where the equality holds iff it holds in Theorem 2 for H and and an optimal remap $\hat{R}$. That is iff $\Phi \operatorname{diag}(H\hat{R}) = 1$. This condition is equivalent to the condition of equality in Proposition 5, because if a remap R satisfies this latter condition then it must be optimal because the utility with R (by Theorem 2) is globally maximum, that is no other remap can achieve higher utility. $\square$

The above bound holds only for d_y-regular priors. However, it is well-known ([16]) that min-mutual information is maximized by the uniform prior u, i.e. $\mathcal{L}(H, \pi) \leq \mathcal{L}(H, u)$ for all H, π. Thus, in cases when u is d_y-regular, we can extend the above bound to *any* prior.

Corollary 2. *Suppose that the uniform prior u is d_y-regular, and let $H : \mathcal{Y} \to \mathcal{Z}$ be any d_y-private mechanism. Then for all row vectors $\mu \geq 0$ satisfying $\mu \Phi = u$, and for all priors π, we have that*

$$\mathcal{L}(H, \pi) \leq \log_2(|\mathcal{Y}| \sum_{y \in \mathcal{Y}} \mu_y)$$

5.3   Quantifying the Leakage about the Database

In the previous section we considered the information about the query result that is revealed by a mechanism H. This information was measured by the min-mutual information $\mathcal{L}(H, \pi)$.

We now turn our attention to the case of standard differential privacy, with the goal of quantifying the information about the *database* that is conveyed by a differentially private mechanism K (not necessarily oblivious). Intuitively, we wish to minimize this information to protect the privacy of the users, contrary to the utility which we aim at maximizing. We can apply the results of the previous section by considering the full mechanism K, mapping databases $\mathcal{V} = V^u$ to outputs (recall that u is the number of individuals in the database and V the universe of values). Differential privacy corresponds to ϵd_h-privacy, where d_h is the *Hamming distance* on the domain $\mathcal{V}$ of databases. Correspondingly ϵd_h-regularity will concern priors π on databases $\mathcal{V}$.

In this case, $\mathcal{L}(K, \pi)$ measures the information about the database conveyed by the mechanism, which we refer to as "min-entropy leakage", and the bounds from the previous section can be directly applied. However, since we now work on a specific metric space $(\mathcal{V}, \epsilon d_h)$, we can obtain a closed expression for the bound of Corollary 2. We start by observing that due to the symmetry of the graph, the uniform prior u is ϵd_h-regular for all $\epsilon > 0$. More precisely, we can show that the vector μ of size $\mathcal{V}$ having all elements equal to

$$\left(\frac{e^\epsilon}{|V|(|V| - 1 + e^\epsilon)} \right)^u$$

satisfies $\mu \Phi = u$ and $\mu \geq 0$. Thus, applying Corollary 2 we get the following result.

Theorem 3 (min-entropy leakage upper bound). *Let $\mathcal{V} = V^u$ be a set of databases, let $\epsilon > 0$, and let K be an ϵ-differentially private mechanism. Then for all priors π on $\mathcal{V}$, we have:*

$$\mathcal{L}(K, \pi) \leq u \log_2 \frac{|V| e^\epsilon}{|V| - 1 + e^\epsilon}$$

This bound determines the maximum amount of information that *any* ϵ-differentially privacy mechanism can leak about the database (independently from the underlying query). The bound was first conjectured in [17] and independently proven in [19]; our technique gives an alternative and arguably more intuitive proof of this result.

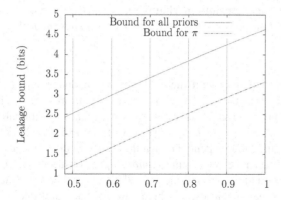

Fig. 4. Leakage bounds for various values of ϵ

Note that the above bound holds for *all priors*. If we restrict to a *specific ϵd_h-regular prior π*, then we can get better results by using the bound of Proposition 5 which depends on the actual prior. This is demonstrated in the following example.

Example 3. Consider a database of 5 individuals, each having one of 4 possible values, i.e. $\mathcal{V} = V^u$ with $V = \{1, 2, 3, 4\}$ and $u = 5$. Assume that each individual selects a value independently from the others, but not all values are equally probable; in particular the probabilities of values $1, 2, 3, 4$ are $0.3, 0.27, 0.23, 0.2$ respectively. Let π be the corresponding prior on $\mathcal{V}$ that models this information. We have numerically verified that for all $0.48 \leq \epsilon \leq 1$ (with step 0.01) π is ϵd_h-regular. Thus we can apply Proposition 5 to get an upper bound of $\mathcal{L}(K, \pi)$ for this prior.

The resulting bound, together with the general bound for all priors from Theorem 3, are shown in Figure 4. We see that restricting to a specific prior provides a significantly better bound for all values of ϵ. For instance, for $\epsilon = 0.5$ we get that $\mathcal{L}(K, \pi) \leq 1.2$ for this π, while $\mathcal{L}(K, \pi) \leq 2.5$ for all priors π.

6   Tight-Constraints Mechanisms

In general, the bounds for the utility (Theorem 2) and the min-mutual information (Proposition 5) are not tight. That is for a given metric d_y on a domain $\mathcal{Y}$, there may be no d_y-private mechanism H that meets these bounds. Nevertheless, they provide ultimate limits, induced by the d_y-privacy constraints, for all d_y-private mechanisms and d_y-regular priors. These bounds are simultaneously tight if the condition $\Phi \operatorname{diag}(HR) = 1$ is satisfied (note that this condition is independent of the underlying prior). In this section we exploit this 'tightness' condition and investigate the mechanisms that, whenever exist, satisfy this condition and are therefore optimal for the entire region of d_y-regular priors. We call these mechanisms *tight-constraints* mechanisms.

Definition 6 (A tight-constraints mechanism). *For a metric d_y, a mechanism H :
$\mathcal{Y} \to \mathcal{Y}$ is called a* tight-constraints *mechanism iff it satisfies the following conditions
for all $y, y' \in \mathcal{Y}$.*

$$e^{-d_y(y,y')} h_{y'y'} = h_{yy'}. \tag{13}$$

It is important to note that, in general, there may exist zero, one or more tight-constraints
mechanisms for a given metric d_y. The above definition enforces $|\mathcal{Y}| (|\mathcal{Y}| - 1)$ linearly
independent equations, referred to as the '*tight constraints*'. Additionally it must also
hold that $\sum_{y' \in \mathcal{Y}} h_{yy'} = 1$ for all $y \in \mathcal{Y}$. Thus we have, in total, $|\mathcal{Y}| |\mathcal{Y}|$ equations. If
these equations are linearly independent, then they solve to unique values. If these val-
ues are non-negative, then they determine a *unique* tight-constraints mechanism. On the
other hand, if these equations are not linearly independent, then there may be multiple
solutions with non-negative entries, in which case we have multiple tight-constraints
mechanisms for d_y.

6.1   Properties

The first feature that follows immediately from the definition of tight-constraints mechanisms, for a metric d_y, is that they satisfy d_y-privacy:

Proposition 6 (d_y-privacy). *For a given metric d_y, every tight-constraints mechanism
is d_y-private.*

Proof. For a tight-constraints mechanism $\hat{H}$, we want to show that for every pair of
query results y, y' and every output z, we have

$$\hat{h}_{yz} \le e^{d_y(y,y')} \cdot \hat{h}_{y'z}. \tag{14}$$

By Definition 6, for every pair of elements y, y' and every output z, we have

$$\hat{h}_{y'z} = e^{-d_y(y',z)} \cdot \hat{h}_{zz} \quad \text{and} \quad \hat{h}_{yz} = e^{-d_y(y,z)} \cdot \hat{h}_{zz}. \tag{15}$$

If $\hat{h}_{zz} = 0$ then $\hat{h}_{y'z} = \hat{h}_{yz} = 0$. In this case, Condition (14) is satisfied. Otherwise
(i.e. if $\hat{h}_{zz} \ne 0$), both $\hat{h}_{y'z}$ and $\hat{h}_{yz}$ are non-zero, and it follows from Equations (15)
that, for all inputs y and y', and every output z,

$$\hat{h}_{y'z} / \hat{h}_{yz} = e^{-(d_y(y',z) - d_y(y,z))}.$$

By the triangle inequality, we have that $d_y(y', z) - d_y(y, z) \le d_y(y, y')$. Knowing also
that $e^{-1} < 1$, it follows from the above inequality that

$$\hat{h}_{y'z} / \hat{h}_{yz} \ge e^{-d_y(y,y')}.$$

The above inequality is equivalent to Condition (14) of d_y-privacy.   □

Thanks to the above property, we can give a further useful characteristic for the tight-
constraints mechanisms distinguishing them from other d_y-private mechanisms. More
precisely, the following proposition identifies a linear algebraic condition that is satisfied *only by* the tight-constraints mechanisms for the given metric d_y:

Proposition 7 (diagonal characterization). *For a metric d_y, a d_y-private mechanism $H : \mathcal{Y} \to \mathcal{Y}$ is a tight-constraints mechanism iff*

$$\Phi \, \text{diag}(H) = 1. \tag{16}$$

Proof. If H is a tight-constraints mechanism, then by Definition 6 we have that $h_{yy'} = e^{-d_y(y,y')} h_{y'y'}$ for all $y, y' \in \mathcal{Y}$. It also holds that $\sum_{y' \in \mathcal{Y}} h_{yy'} = 1$ for all $y \in \mathcal{Y}$. Combining these equations yields

$$\sum_{y' \in \mathcal{Y}} e^{-d_y(y,y')} h_{y'y'} = 1, \qquad \forall y \in \mathcal{Y}. \tag{17}$$

Using the privacy-constraints matrix Φ, the above equations can be written in the matrix form (16). Now we prove the other direction of implication as follows. Suppose that Eq. (17) (which is equivalent to Eq. (16)) is satisfied by a d_y-private mechanism H. Then it holds for all $y, y' \in \mathcal{Y}$ that $h_{yy'} \geq e^{-d_y(y,y')} h_{y'y'}$. Suppose for a contradiction that this inequality is *strict* for some $y, y' \in \mathcal{Y}$, i.e. $h_{yy'} > e^{-d_y(y,y')} h_{y'y'}$. Then $\sum_{y' \in \mathcal{Y}} h_{yy'} > \sum_{y' \in \mathcal{Y}} e^{-d_y(y,y')} h_{y'y'} = 1$, where the last equality holds by Eq. (17). That is, the sum of the entries of a row in H is strictly greater than 1 which violates the validity of H. $\qquad\square$

The above proposition provides a way to check the existence of, and also compute, the tight-constraints mechanisms for a given metric d_y. Since Condition (16) is satisfied only by these mechanisms, there is at least one tight-constraints mechanism if there is a vector z, with non-negative entries, that satisfies the equation $\Phi \, z = 1$. In this case a tight-constraints mechanism is obtained by setting its diagonal to z, and evaluating the non-diagonal entries from the diagonal using Eqs. (13).

Now we turn our attention to the region of d_y-regular priors and identify the mechanisms that are optimal with respect to both utility and min-mutual information in this region. Precisely, we show that the set of these optimal mechanism consists exactly of all mechanisms that can be *mapped to* a tight-constraints one using some remap R.

Theorem 4 (Optimality). *Let d_y be a metric for which at least one tight-constraints mechanism exists. Then a d_y-private mechanism $H : \mathcal{Y} \to \mathcal{Z}$ is d_y-optimal (wrt both utility and min-mutual information) for every d_y-regular prior π iff there is a remap $R : \mathcal{Z} \to \mathcal{Y}$ such that HR is a tight-constraints mechanism for d_y.*

Proof. If there exists a tight-constraints mechanism H' for a given metric d_y, then H' must satisfy Eq. (16). This implies that the upper-bound in Theorem 2 is reachable by H' and the identity remap. Thus the upper-bound, in this case, is tight. Now consider a d_y-private mechanism $H : \mathcal{Y} \to \mathcal{Z}$. By Theorem 2, H meets that upper bound for the utility (and therefore is d_y-optimal) iff it satisfies the condition $\Phi \, \text{diag}(HR) = 1$, with some remap R. Since H is d_y-private, HR is also d_y-private. Now by Proposition 7, satisfying the condition $\Phi \, \text{diag}(HR) = 1$ (meaning that H is optimal) is equivalent to that HR is a tight-constraints mechanism (for d_y). Using the relation, given by Proposition 4, between utility and min-mutual information, the same argument holds for the latter. $\qquad\square$

Observe that tight-constraints mechanisms are optimal because they are mapped to themselves by the identity remap. In the light of Theorem 4, we consider the special case of the uniform prior, denoted by u, where all results in $\mathcal{Y}$ are equally likely. Note that this prior corresponds to users having unbiased knowledge about the query results, i.e. they assume that all the true results $\mathcal{Y}$ are yielded, by executing the query, with the same probability. Firstly, the following lemma proves an equivalence between the existence of at least one tight-constraints mechanism on one hand and the uniform prior u being d_y-regular on the other hand.

Proposition 8. *For a given metric d_y, there exists at least one tight-constraints mechanism iff the uniform prior u is d_y-regular.*

Proof. By Proposition 7, if there is at least a tight-constraints mechanism $\hat{H}$, then Eq. (16) must hold for this mechanism. Taking the transpose of both sides in this equation, and noting that $\Phi^t = \Phi$ (because Φ is symmetric), then we get that

$$(\mathrm{diag}(\hat{H}))^t \cdot \Phi = \mathbf{1}^t.$$

Scaling the above equation by $1/|\mathcal{Y}|$ yields the row vector u, the uniform prior, on the right hand side. Thus if a tight-constraints mechanism $\hat{H}$, exists then

$$(1/|\mathcal{Y}|)\,(\mathrm{diag}(\hat{H}))^t \cdot \Phi = u.$$

which means (By Def. 4) that u is d_y-regular, because the row vector $(\mathrm{diag}(\hat{H}))^t$ has only non-negative entries. For the opposite implication, assume that u is d_y-regular. Then by the definition there is a row vector μ with non-negative entries such that $\mu\,\Phi = u$. Taking the transpose of both sides, and multiplying by $|\mathcal{Y}|$, yields that Eq. (16) is satisfied for H, whose diagonal is given by $\mathrm{diag}(H) = |\mathcal{Y}| \cdot \mu^t$ (non-negative). Thus there exists a tight-constraints mechanism for d_y. $\qquad\square$

It is worth noticing that in general the region of d_y-regular priors may or may not include the uniform prior. However, as shown earlier in Section 4, this region is enlarged and converges to the entire prior space as the distances $d_y(y,y') \to \infty$ for all $y \neq y'$. In particular the d_y-regular priors accommodate the uniform prior u if d_y is scaled up by an appropriate factor.

In the case of ϵ-differential privacy it holds that $d_y = \epsilon\, d_h$ where d_h is the Hamming distance on databases. Thus there is always a threshold ϵ^*, above which the uniform prior u is $\epsilon\, d_h$-regular. This can provide a design criteria to *select* a setting for ϵ such that, according to Proposition 8, there is a tight-constraints mechanism that is optimal for all $\epsilon\, d_h$-regular priors.

Using Proposition 8, we can describe the optimal mechanisms for the uniform prior as a corollary of Theorem 4.

Corollary 3. *Let d_y be a metric for which there exists at least one tight-constraints mechanism. Then a mechanism H is d_y-optimal for the uniform prior on $\mathcal{Y}$ iff HR is a tight-constraints mechanism for some remap $R : \mathcal{Z} \to \mathcal{Y}$.*

In summary, the existence of tight-constraints mechanisms and their structures depend on the given metric. The choice of such metric corresponds to the required privacy guarantee. Consider in particular the conventional ϵ-differential privacy, where any two *adjacent* elements in a domain $\mathcal{Y}$ are required to be indistinguishable relative to ϵ. In this case, the domain $\mathcal{Y}$ and its adjacency relation $\sim_f$ are modeled by the graph $G = (\mathcal{Y}, \sim_f)$; and the requirement of satisfying ϵ-differential privacy for $\mathcal{Y}$ translates in our general model to the metric $d_\mathcal{Y}(y, y') = \epsilon\, d_{\sim_f}(y, y')$, where $d_{\sim_f}(y, y')$ is the graph distance between y, y'. With this metric, we find that tight-constraints mechanisms capture other known differentially-private mechanisms. For example, if we set $\mathcal{Y}$ to be the output domain of a counting query executed on a database, we find that the tight-constraints mechanism for $\mathcal{Y}$ is exactly the *truncated-geometric mechanism*, which was shown by [13] to be optimal for every prior. Also, we instantiate, in the following, the tight-constraints mechanism when the metric space $(\mathcal{Y}, d_\mathcal{Y})$ satisfies a certain symmetry. This symmetry captures, in particular, the graphs for which an optimal mechanism is constructed in [19] for the uniform prior u. Once again this mechanism is precisely a tight-constraints one. Note that an additional conclusion which we add here is that this mechanism is optimal not only for u but also for all $d_\mathcal{Y}$-regular priors.

6.2   Tight-Constraints Mechanism for Symmetric Metric Spaces

We consider the mechanisms that satisfy $d_\mathcal{Y}$-privacy for a given domain $\mathcal{Y}$. We focus here on the metric spaces $(\mathcal{Y}, d_\mathcal{Y})$ that satisfy a certain symmetry which we call *ball-size symmetry*. To describe this property, we recall the standard notion of balls in metric spaces: a *ball* of radius r around a point $y \in \mathcal{Y}$ is the set $B_r^{d_\mathcal{Y}}(y) = \{y' \in \mathcal{Y} : d_\mathcal{Y}(y, y') \le r\}$. Now we define the ball-size symmetry as follows.

Definition 7 (ball-size symmetry). *A metric space $(\mathcal{Y}, d_\mathcal{Y})$ is said to be ball-size symmetric if for all $y, y' \in \mathcal{Y}$, and all radii r, we have $|B_r^{d_\mathcal{Y}}(y)| = |B_r^{d_\mathcal{Y}}(y')|$.*

Note that the above condition is equivalent to saying that for any $y \in \mathcal{Y}$, the number of elements that are at distance r from y depends only on r, allowing us to write this number as n_r. Inspecting the privacy-constraints matrix Φ in this case, we observe that the row sum $\sum_{y'} \phi_{yy'}$ for every $y \in \mathcal{Y}$ is the same and equal to $\sum_r n_r\, e^{-r}$. This means that the column vector z, of which every element is equal to $1/\sum_r n_r\, e^{-r}$, satisfies $\Phi z = 1$ and therefore yields (by Proposition 7) the diagonal of a tight-constraints mechanism H. The other (non-diagonal) entries of H follow from the diagonal as in Definition 6. Thus we conclude the following result.

Proposition 9 (tight-constraints mechanism for symmetric metric spaces). *For any metric space $(\mathcal{Y}, d_\mathcal{Y})$ satisfying ball-size symmetry there is a tight-constraints mechanism $H : \mathcal{Y} \to \mathcal{Y}$ which is given as $h_{yy'} = e^{d_\mathcal{Y}(y,y')} / \sum_r n_r\, e^{-r}$.*

The main consequence of the above proposition is that the mechanism H is optimal for every $d_\mathcal{Y}$-regular prior including the uniform prior u.

The above result generalizes and extends a result by [19] in the context of differential privacy. The authors of [19] considered two types of graphs: distance-regular and vertex-transitive graphs. They constructed for these graphs an ϵ-differentially private mechanism optimal for the uniform prior. As shown earlier ϵ-differential privacy for a

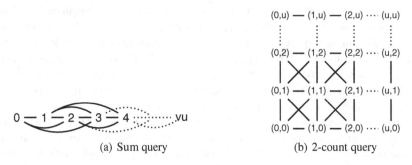

(a) Sum query   (b) 2-count query

Fig. 5. Adjacency graphs

graph $(\mathcal{Y}, \sim_f)$ translates in our setting to the metric space $(\mathcal{Y}, \epsilon\, d_{\sim_f})$. It can be easily seen that if $(\mathcal{Y}, \sim_f)$ is either distance-regular or vertex-transitive, the corresponding metric space $(\mathcal{Y}, \epsilon\, d_{\sim_f})$ is ball-size symmetric. Therefore, we can instantiate the
tight-constraints mechanism of Proposition 9 to $\epsilon d_{\sim_f}$, which gives exactly the optimal
mechanism constructed in [19]. Hence, we directly obtain the same optimality results,
and moreover our analysis shows that this mechanism is optimal on the entire region of
$\epsilon d_{\sim_f}$-regular priors, instead of only the uniform one.

7   Case-Studies

In this section we show the usefulness of the tight-constraints mechanism by applying
it to two contexts: standard differential privacy and geo-indistinguishability.

7.1   Differential Privacy: Sum and 2-Count Queries

We evaluate the tight constraints mechanism for two families of queries, namely sum
and 2-count queries. For each family, we apply the mechanism on databases consisting
of u individuals each having an integer value between 0 and v, and we compare its
utility to the geometric mechanism.

It is well-known that no universally optimal mechanism exists for these families; in
particular, the geometric mechanism, known to be optimal for a single counting query, is
not guaranteed to be optimal for sum queries or multiple counting queries. On the other
hand, as discussed in the previous section, tight-constraints mechanisms, whenever they
exist, are guaranteed to be optimal within the region of regular priors.

The comparison is made as follows: for each query, we numerically compute the
smallest ϵ (using a step of 0.01) for which a tight-constraints mechanism exists (i.e. for
which the uniform prior u is $\epsilon d_{\sim_f}$-regular, see Proposition 8). Then we compute the
utility (using an optimal remap) of both the tight constraints and the geometric mechanisms, for a range of ϵ starting from the minimum one. Note that the tight constraint
mechanism exists for any ϵ greater than the minimum one.

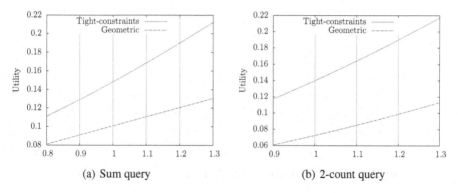

Fig. 6. Utility for various values of ϵ

Sum query. Let f be the query returning the sum of the values for all individuals, thus it has range $\mathcal{Y} = \{0, \ldots, vu\}$. By modifying the value of a single individual, the outcome of the query can be altered by at most v (when changing the value from 0 to v), thus two elements $i, j \in \mathcal{Y}$ are adjacent iff $|i - j| \leq v$. The induced graph structure on $\mathcal{Y}$ is shown in Figure 5(a) (for the case $v = 3$).

For our case-study we numerically evaluate this query for $u = 150, v = 5$ and for the uniform prior. We found that the minimum ϵ for which a tight-constraints mechanism exists (and is in fact unique since Φ is invertible) is 0.8. Figure 6(a) shows the utility of the tight-constraint mechanism, as well as that of the geometric mechanism, for values of ϵ between 0.8 and 1.3, the uniform prior and using and optimal remap. We see that the tight-constraints mechanism provides significantly higher utility than the geometric mechanism in this case.

2-count query. Consider now the query f consisting of 2 counting queries (i.e. reporting the number of users satisfying properties p_1 and p_2), thus it has range $\mathcal{Y} = \{0, \ldots, u\} \times \{0, \ldots, u\}$. By modifying the value of a single individual, the outcome of each counting query can be altered by at most 1, thus two answers $(i_1, i_2), (j_1, j_2) \in \mathcal{Y}$ are adjacent iff $|i_1 - j_1| \leq 1$ and $|i_2 - j_2| \leq 1$. The induced graph structure on $\mathcal{Y}$ is shown in Figure 5(b).

We evaluate this query for $u = 30$ and for the uniform prior. We found that the minimum ϵ for which a tight-constraints mechanism exists is 0.9. Figure 6(b) shows the utility of the two mechanisms (with the geometric being applied independently to each counting query) for values of ϵ between 0.9 and 1.3 and the uniform prior. Similarly to the sum query, we see that the tight-constraints mechanism provides significantly higher utility than the geometric mechanism in this case.

7.2   Geo-indistinguishability

As discussed in Section 3.3, geo-indistinguishability is a notion of location privacy obtained by taking $d_{\mathcal{X}} = \epsilon d_2$, where d_2 is the Euclidean distance between locations.

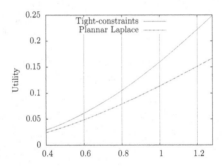

Fig. 7. Utility of location privacy mechanisms for various values of ϵ

In [9] it is shown that a planar version of the Laplace mechanism satisfies ϵ-geo-indistinguishability. The Planar Laplace mechanism is continuous, having as input and output the full $\mathbb{R}^2$, but in the case of a finite number of locations it can be discretized and truncated while still satisfying geo-indistinguishability (for a slightly adjusted ϵ).

Although the Planar Laplace mechanism is simple, efficient and easy to implement, it provides no optimality guarantees. On the other hand, for any finite number of locations, the tight-constraints mechanism, if it exists, is guaranteed to be optimal for ϵd_2-regular priors. In this section we compare the two mechanisms on a grid of 100×100 locations, with step size 1 km.

Note that constructing the tight-constraints mechanism involves inverting the matrix Φ, which can be done in time $O(|\mathcal{X}|^{2.376})$ using the Coppersmith-Winograd algorithm. This complexity is much lower than that of recent methods for computing optimal location obfuscation mechanisms. For instance, the well-known method of Shokri et al. [23] – which uses the adversary's expected error as the metric of privacy – involves solving large linear optimization problems and was evaluated to a grid of only 30 locations (compared to the 10,000 locations in our grid).

Figure 7 shows the utility of the two mechanisms for ϵ ranging from 0.4 to 1.3 and for a uniform prior. As expected, the tight-constraints mechanism offers significantly higher utility than the Planar Laplace mechanism for the same ϵ.

It should be emphasized, however, that our optimality results hold for the binary gain function, which corresponds to an attacker trying to guess the true location of the user (the utility being the probability of a correct guess). This might often be meaningful, especially when the grid size is big: guessing any incorrect cell could be considered equally bad. But it is also common to consider gain functions taking the distance between locations into account, with respect to which the tight-constraints mechanism is not guaranteed to be optimal.

8   Conclusion and Future Work

In this paper we have continued the line of research initiated by [13, 14] about the existence of differentially-private mechanisms that are universally optimal, i.e., optimal for all priors. While the positive result of [13] (for counting queries) and the negative one of [14] (for essentially all other queries) answer the question completely, the latter sets a rather dissatisfactory scenario, since counting queries are a very specific kind of queries, and in general users can be interested in very different queries. We have then considered the question whether we can achieve optimality with the same mechanism for a restricted class of priors. Fortunately the answer is positive: we have identified a region of priors, called d_y-regular, and a mechanism, called tight-constraints, which is optimal for all the priors in this region. We have also provided a complete and effectively checkable characterization of the conditions under which such mechanism exists, and an effective method to construct it. As a side result, we have improved on the existing bounds for the min-entropy leakage induced by differential privacy. More precisely, we have been able to give specific and tight bounds for each d_y-regular prior, in general smaller than the bound existing in the literature for the worst-case leakage (achieved by the uniform prior [18]).

So far we have been studying only the case of utility for binary gain functions. In the future we aim at lifting this limitation, i.e. we would like to consider also other kinds of gain. Furthermore, we intend to study how the utility decreases when we use a tight-constraints mechanism outside the class of d_y-regular priors. In particular, we aim at identifying a class of priors, larger than the d_y-regular ones, for which the tight-constraints mechanism is close to be optimal.

References

1. Sabelfeld, A., Myers, A.C.: Language-based information-flow security. IEEE Journal on Selected Areas in Communications 21(1), 5–19 (2003)
2. Chatzikokolakis, K., Palamidessi, C., Panangaden, P.: Anonymity protocols as noisy channels. Inf. and Comp. 206(2-4), 378–401 (2008)
3. Dwork, C.: A firm foundation for private data analysis. Communications of the ACM 54(1), 86–96 (2011)
4. Dwork, C.: Differential privacy. In: Bugliesi, M., Preneel, B., Sassone, V., Wegener, I. (eds.) ICALP 2006. LNCS, vol. 4052, pp. 1–12. Springer, Heidelberg (2006)
5. Dwork, C., Mcsherry, F., Nissim, K., Smith, A.: Calibrating noise to sensitivity in private data analysis. In: Halevi, S., Rabin, T. (eds.) TCC 2006. LNCS, vol. 3876, pp. 265–284. Springer, Heidelberg (2006)
6. Narayanan, A., Shmatikov, V.: Robust de-anonymization of large sparse datasets. In: Proc. of S&P, pp. 111–125 (2008)
7. Narayanan, A., Shmatikov, V.: De-anonymizing social networks. In: Proc. of S&P, pp. 173–187. IEEE (2009)
8. Chatzikokolakis, K., Andrés, M.E., Bordenabe, N.E., Palamidessi, C.: Broadening the scope of Differential Privacy using metrics. In: De Cristofaro, E., Wright, M. (eds.) PETS 2013. LNCS, vol. 7981, pp. 82–102. Springer, Heidelberg (2013)
9. Andrés, M.E., Bordenabe, N.E., Chatzikokolakis, K., Palamidessi, C.: Geo-indistinguishability: differential privacy for location-based systems. In: Proc. of CCS, pp. 901–914. ACM (2013)
10. Gaboardi, M., Haeberlen, A., Hsu, J., Narayan, A., Pierce, B.C.: Linear dependent types for differential privacy. In: Proceedings of the 40th Annual ACM SIGPLAN-SIGACT Symposium on Principles of Programming Languages, POPL 2013, pp. 357–370. ACM, New York (2013)

11. Barthe, G., Olmedo, F.: Beyond differential privacy: Composition theorems and relational logic for f-divergences between probabilistic programs. In: Fomin, F.V., Freivalds, R., Kwiatkowska, M., Peleg, D. (eds.) ICALP 2013, Part II. LNCS, vol. 7966, pp. 49–60. Springer, Heidelberg (2013)

12. Barthe, G., Köpf, B., Olmedo, F., Béguelin, S.Z.: Probabilistic relational reasoning for differential privacy. ACM Trans. Program. Lang. Syst. 35(3), 9 (2013)

13. Ghosh, A., Roughgarden, T., Sundararajan, M.: Universally utility-maximizing privacy mechanisms. In: Proc. of STOC, pp. 351–360. ACM (2009)

14. Brenner, H., Nissim, K.: Impossibility of differentially private universally optimal mechanisms. In: Proc. of FOCS, pp. 71–80. IEEE (2010)

15. Smith, G.: On the foundations of quantitative information flow. In: de Alfaro, L. (ed.) FOSSACS 2009. LNCS, vol. 5504, pp. 288–302. Springer, Heidelberg (2009)

16. Braun, C., Chatzikokolakis, K., Palamidessi, C.: Quantitative notions of leakage for one-try attacks. In: Proc. of MFPS. ENTCS, vol. 249, pp. 75–91. Elsevier (2009)

17. Barthe, G., Köpf, B.: Information-theoretic bounds for differentially private mechanisms. In: Proc. of CSF, pp. 191–204. IEEE (2011)

18. Alvim, M.S., Andrés, M.E., Chatzikokolakis, K., Degano, P., Palamidessi, C.: Differential Privacy: on the trade-off between Utility and Information Leakage. In: Barthe, G., Datta, A., Etalle, S. (eds.) FAST 2011. LNCS, vol. 7140, pp. 39–54. Springer, Heidelberg (2012)

19. Alvim, M.S., Andrés, M.E., Chatzikokolakis, K., Palamidessi, C.: On the relation between Differential Privacy and Quantitative Information Flow. In: Aceto, L., Henzinger, M., Sgall, J. (eds.) ICALP 2011, Part II. LNCS, vol. 6756, pp. 60–76. Springer, Heidelberg (2011)

20. Ardagna, C.A., Cremonini, M., Damiani, E., De Capitani di Vimercati, S., Samarati, P.: Location privacy protection through obfuscation-based techniques. In: Barker, S., Ahn, G.-J. (eds.) Data and Applications Security 2007. LNCS, vol. 4602, pp. 47–60. Springer, Heidelberg (2007)

21. Shokri, R., Theodorakopoulos, G., Boudec, J.Y.L., Hubaux, J.P.: Quantifying location privacy. In: Proc. of S&P, pp. 247–262. IEEE (2011)

22. Rényi, A.: On Measures of Entropy and Information. In: Proceedings of the 4th Berkeley Symposium on Mathematics, Statistics, and Probability, pp. 547–561 (1961)

23. Shokri, R., Theodorakopoulos, G., Troncoso, C., Hubaux, J.P., Boudec, J.Y.L.: Protecting location privacy: optimal strategy against localization attacks. In: Proc. of CCS, pp. 617–627. ACM (2012)

Bisimulation for Markov Decision Processes through Families of Functional Expressions

Norm Ferns[1,*], Doina Precup[2], and Sophia Knight[3]

[1] Département d'Informatique
École Normale Supérieure
45 rue d'Ulm, F-75230 Paris Cedex 05, France
`ferns@di.ens.fr`
[2] School of Computer Science
McGill University
Montréal, Canada, H3A 2A7
`dprecup@cs.mcgill.ca`
[3] CNRS, LORIA
Université de Lorraine
Nancy, France
`sophia.knight@gmail.com`

Abstract. We transfer a notion of quantitative bisimilarity for labelled Markov processes [1] to Markov decision processes with continuous state spaces. This notion takes the form of a pseudometric on the system states, cast in terms of the equivalence of a family of functional expressions evaluated on those states and interpreted as a real-valued modal logic. Our proof amounts to a slight modification of previous techniques [2,3] used to prove equivalence with a fixed-point pseudometric on the state-space of a labelled Markov process and making heavy use of the Kantorovich probability metric. Indeed, we again demonstrate equivalence with a fixed-point pseudometric defined on Markov decision processes [4]; what is novel is that we recast this proof in terms of integral probability metrics [5] defined through the family of functional expressions, shifting emphasis back to properties of such families. The hope is that a judicious choice of family might lead to something more computationally tractable than bisimilarity whilst maintaining its pleasing theoretical guarantees. Moreover, we use a trick from descriptive set theory to extend our results to MDPs with bounded measurable reward functions, dropping a previous continuity constraint on rewards and Markov kernels.

1 Introduction

Probabilistic bisimulation is a notion of state-equivalence for Markov transition systems, first introduced by Larsen and Skou [6] based upon bisimulation for nondeterministic systems by Park and Milner [7,8]. Roughly, states are deemed

---

* Norm Ferns' contribution was partially supported by the AbstractCell ANR-Chair of Excellence.

F. van Breugel et al. (Eds.): Panangaden Festschrift, LNCS 8464, pp. 319–342, 2014.

equivalent if they transition with the same probability to classes of equivalent states.

In the context of labelled Markov processes (LMPs), a robust quantitative notion of probabilistic bisimilarity has been devised in the form of a class of behavioural pseudometrics, or *bisimilarity metrics*, defined on the state space of a given process [9,1,10,2,11]. The defining characteristic of these metrics is that the kernel of each is probabilistic bisimilarity, and otherwise each assigns a distance between 0 and 1 that measures the degree to which two states are bisimilar.

Bisimilarity metrics were initially defined in terms of a family of functional expressions interpreted as a real-valued logic on the states of an LMP by Desharnais et al. [9], building on ideas of Kozen [12] that logic could be generalized to handle probabilistic phenomena. Subsequently, van Breugel and Worrell [2] used category theory to define another class of bisimilarity metrics, and showed that their definition was equivalent to a slightly modified version of the metrics of Desharnais et al. in terms of the family of functional expressions considered. Desharnais et al. [1] in turn reformulated this latter version solely in terms of order-theoretic fixed-point theory. A crucial component in these formulations was the recognition that the initial version, when lifted to a metric on distributions, could be expressed as a Kantorovich metric.

For finite systems, the various formulations readily admit a variety of algorithms to estimate the distances [9,3,13]. In particular, the initial formulation in terms of a family of functional expressions led to an exponential-time algorithm based on choosing a suitable finite sub-family of functionals. This was vastly improved in [3] wherein an iterative polynomial-time algorithm exploited the fact that the Kantorovich metric is a specialized linear program.

In [14,15,4], the fixed-point presentation of the LMP bisimilarity metric was adapted to finite Markov decision processes (MDPs) and MDPs with continuous state spaces, bounded continuous rewards, and (weakly) continuous Markov kernels. Insofar as finite systems are concerned, the addition of the reward parameter is minor; in fact, the iterative polynomial-time algorithm applies more or less directly [14]. Unfortunately, even for very simple toy systems the experimental space and time required is too great to be of practical use. This issue was explored in [16] where a Monte Carlo approach to estimating the Kantorovich metric, and the bisimilarity metric for MDPs in general, was devised and shown to outperform other approaches in an experimental setting. The Monte Carlo approach was even extended to MDPs in which the state space is a compact metric space [4]. However, this line of investigation is still very preliminary.

In this work, we seek to complete further the picture of bisimilarity metrics for MDPs by presenting a family of functional expressions that induce the fixed-point bisimilarity metric, in analogy with the results of [2] for LMPs. We aim to shift the study of equivalences on MDP states to the study of such families and their properties. The right choice of family might lead to a more easily computable equivalence whilst maintaining some important theoretical guarantees. Additionally, we hope further to investigate Monte Carlo approaches to estimating similarity, for example, by sampling from a given family of functions. More specifically, we carry out the following.

1. We adapt Proposition 2 of [2] to MDPs, showing that a class of functional expressions interpreted as a real-valued logic on the states of a given MDP is dense in the class of Lipschitz functions with respect to the pseudometric induced by the family. It is important to note that the proof here appears almost unchanged from [2]; what is important is that we recast the result in the terminology and conceptual framework of integral probability metrics and their generating classes of functions.
2. We remove the continuity constraints of Theorem 7, which establishes the fixed-point bisimilarity metric for continuous MDPs, using techniques from descriptive set theory. This is, to the best of our knowledge, an original result.
3. We propose a preliminary Monte Carlo technique for estimating the bisimilarity metric by sampling from the family of functional expressions that encodes bisimilarity for MDPs. This too, appears to be an original result, but is based on a heuristic method and experimental evidence presented in [17].

The paper is organized as follows. In Section 2, we provide a brief summary on the relevant development of bisimilarity metrics and related results for LMPs and MDPs. In Section 3, we establish a family of functional expressions that induces a metric equal to a previously-defined bisimilarity metric for MDPs, and then generalize the applicability of this result by removing previous continuity constraints. Finally, in Section 4 and Section 5, we propose a Monte Carlo method for approximating the bisimilarity metric by sampling from a family of functional expressions, and conclude with suggestions for future work.

2   Background

The purpose of this section is to recall the development of pseudometrics capturing bisimilarity for labelled Markov processes, and set down what has already been carried over to Markov decision processes. In doing so, we fix some basic terminology and notation for probabilistic systems. Unless otherwise stated, all the results of this section can be found in Prakash's book on labelled Markov processes [18].

2.1   Probability Measures on Polish Metric Spaces

Since we deal primarily with uncountably infinite state spaces, we must take into account the tension involved in imposing the right amount of structure on a space for general theoretical utility and imposing the right amount of structure for practical application. Much of the work on labelled Markov processes has been cast in the setting of Polish spaces and analytic spaces, which are general enough to include most practical systems of interest while structured enough to admit many useful theorems.

Definition 1.

1. *A* Polish metric space *is a complete, separable metric space.*
2. *A* Polish space *is a topological space that is homeomorphic to a Polish metric space.*
3. *An* analytic set *is a subset of a Polish space that is itself the continuous image of a Polish space.*
4. *A* standard Borel space *is a measurable space that is Borel isomorphic to a Polish space.*

If (X, τ) is a topological space, then $\mathbb{C}^b(X)$ is the set of continuous bounded real-valued functions on X. If $(X, \mathscr{B}_X)$ is a standard Borel space then we let $\mathbb{S}(X)$ and $\mathbb{P}(X)$ denote the sets of subprobability measures and probability measures on X respectively, and remark that each is also a standard Borel space [19]. We will also assume that the reader is familiar with the theory of integration. If μ is a finite measure and f is an integrable function both defined on $(X, \mathscr{B}_X)$ then we denote the integral of f with respect to μ by $\mu(f)$.

Working at the level of standard Borel spaces allows us to use the rich structure of Polish metric spaces without necessarily having to fix a metric beforehand. For example, when examining probability measures on such spaces, we can sometimes restrict to compact metric spaces, which in turn provide finite substructure for estimation schemes or over which convergence of certain functions can be made to be uniform. The following can be found in [20] and [21] and will be used to establish Theorem 9 in Section 3.

Definition 2. *Let* $\mathcal{P}$ *be a family of Borel probability measures on a metric space* (X, d).

1. $\mathcal{P}$ *is said to be* uniformly tight *iff for every* $\epsilon > 0$ *there exists a compact subset* K *of* X *such that* $P(X \backslash K) < \epsilon$ *for every* $P \in \mathcal{P}$.
2. $\mathcal{P}$ *is said to be* relatively compact *iff every sequence of elements in* $\mathcal{P}$ *contains a weakly convergent subsequence.*

Theorem 1 (Prohorov's Theorem). *Suppose* (X, d) *is a Polish metric space. Let* $\mathcal{P} \subseteq \mathbb{P}(X)$. *Then* $\mathcal{P}$ *is relatively compact if and only if it is uniformly tight.*

Theorem 2 (Dini's Theorem). *Suppose* (K, τ) *is a compact topological space,* $(f_n)_{n \in \mathbb{N}}$ *is a sequence of continuous real-valued functions on* K, *monotonically decreasing and converging pointwise to a continuous function* f. *Then* $(f_n)_{n \in \mathbb{N}}$ *converges to* f *uniformly on* K.

2.2   Stochastic Kernels and Markov Processes

Definition 3. *Let* $(X, \mathscr{B}_X)$ *and* $(Y, \mathscr{B}_Y)$ *be standard Borel spaces. A sub-Markov kernel*[1] *is a Borel measurable map from* $(X, \mathscr{B}_X)$ *to* $(\mathbb{S}(Y), \mathscr{B}_{\mathbb{S}(Y)})$. *A Markov kernel is a Borel measurable map from* $(X, \mathscr{B}_X)$ *to* $(\mathbb{P}(Y), \mathscr{B}_{\mathbb{P}(Y)})$.

---

[1] This is also known as a stochastic relation, a stochastic transition kernel, or simply a stochastic kernel.

Equivalently, K is a sub-Markov (Markov) kernel from X to Y iff

1. $K(x)$ is a sub-probability (probability) measure on $(Y, \mathscr{B}_Y)$ for each $x \in X$, and
2. $x \mapsto K(x)(B)$ is a measurable map for each $B \in \mathscr{B}_Y$.

We will simply write "K is a sub-Markov (Markov) kernel on X" when it is implicitly assumed that $Y = X$. Such stochastic kernels play the role of a transition relation in stochastic transition systems with continuous state spaces. The two Markov processes that we examine in detail are the labelled Markov process and the Markov decision process.

Definition 4. *A labelled Markov process (LMP) is a tuple $(S, \mathscr{B}_S, A, \{K_a : a \in A\})$, where $(S, \mathscr{B}_S)$ is a standard Borel space, A is a finite set of labels, and for $a \in A$, K_a is a sub-Markov kernel on S.*

Definition 5. *A Markov decision process (MDP) is a tuple $(S, \mathscr{B}_S, A, \{P_a : a \in A\}, r)$, where $(S, \mathscr{B}_S)$ is a standard Borel space, A is a finite set of actions, $r : A \times S \to \mathbb{R}$ is a bounded measurable reward function, and for $a \in A$, P_a is a Markov kernel on S.*

For each $a \in A$ we will denote by $r_a : S \to \mathbb{R}$ the function defined by $r_a(s) = r(a, s)$.

Remark 1. In [9,22,23,10], Desharnais, Panangaden, et al. consider LMPs in which the state spaces are analytic sets; this is largely because the quotient of a Polish space may fail to be Polish but is always guaranteed to be analytic. We will not consider analytic sets in this work, but the interested reader should keep this in mind.

2.3 Bisimulation

We present bisimilarity for LMPs and MDPs as outlined in [10] and [4] and note that the latter amounts to little more than a mild extension through the addition of rewards to the definition of the former.

Definition 6. *Given a relation R on a set S, a subset X of S is said to be R-closed if and only if the collection of all those elements of S that it is related to by R, $R(X) = \{s' \in S | \exists s \in X, \ sRs'\}$, is itself contained in X.*

Definition 7. *Given a relation R on a measurable space (S, Σ), we write $\Sigma(R)$ for the set of those Σ-measurable sets that are also R-closed, $\{X \in \Sigma | R(X) \subseteq X\}$.*

When R is an equivalence relation then to say that a set X is R-closed is equivalent to saying that X is a union of R-equivalence classes. In this case $\Sigma(R)$ consists of those measurable sets that can be partitioned into R-equivalence classes.

Definition 8. *Let* $(S, \mathscr{B}_S, A, \{K_a : a \in A\})$ *be an LMP. An equivalence relation* R *on* S *is a* bisimulation relation *if and only if it satisfies*

$$sRs' \Leftrightarrow \text{ for every } a \in A \text{ and for every } X \in \Sigma(R), \ K_a(s)(X) = K_a(s')(X).$$

Bisimilarity *is the largest of the bisimulation relations.*

Definition 9. *Let* $(S, \mathscr{B}_S, A, \{P_a : a \in A\}, r)$ *be an MDP. An equivalence relation* R *on* S *is a* bisimulation relation *if and only if it satisfies*

$$sRs' \Leftrightarrow \text{ for every } a \in A, \ r_a(s) = r_a(s') \text{ and for every } X \in \Sigma(R), \ P_a(s)(X) = P_a(s')(X).$$

Bisimilarity *is the largest of the bisimulation relations.*

It turns out that bisimulation for LMPS and MDPs can be equivalently cast as the maximal fixed-point of a monotone functional on a complete lattice. We present this here only in the context of MDPs; the statement for LMPs is analogous.

Theorem 3. *Let* $(S, \mathscr{B}_S, A, \{P_a : a \in A\}, r)$ *be an MDP,* τ *a Polish topology on* S *generating* $\mathscr{B}_S$ *and such that for each* a *in* A, r_a *and* P_a *are continuous with respect to* τ, $\mathbb{P}(S)$ *being endowed with the topology of weak convergence induced by* τ. *Assume that the image of* r *is contained in* $[0,1]$. *Define* $\mathcal{F} : \mathfrak{Equ} \to \mathfrak{Equ}$ *by*

$$s\mathcal{F}(R)s' \Leftrightarrow \forall a \in A, \ r_a(s) = r_a(s') \text{ and } \forall X \in \Sigma(R), \ P_a(s)(X) = P_a(s')(X),$$

where $\mathfrak{Equ}$ *is the set of equivalence relations on* S *equipped with subset ordering. Then the greatest fixed point of* $\mathcal{F}$ *is bisimilarity.*

Lastly, we remark that bisimulation for LMPs has a logical characterization, and in turn, a characterization in terms of a real-valued modal logic. We omit the details for lack of space, but return to the latter idea in subsequent sections.

2.4 Probability Metrics

Metrizing bisimilarity for Markov processes essentially involves assigning a distance to their Markov kernels via a suitable probability metric. Gibbs and Su [24] survey a variety of such metrics. LMP bisimilarity was initially defined in terms of an integral probability metric in [10], and later recast in terms of the Kantorovich metric in [2]. In order to present the Kantorovich metric, we first recall the definition of lower semicontinuity.

Definition 10. *Let* (X, τ) *be a topological space and let* $f : X \to \mathbb{R} \cup \{-\infty, \infty\}$. *Then* f *is* lower semicontinuous *if for each half-open interval of the form* (r, ∞), *the preimage* $f^{-1}(r, \infty) \in \tau$.

The Kantorovich Metric

Definition 11. *Let S be a Polish space, h a bounded pseudometric on S that is lower semicontinuous on $S \times S$ with respect to the product topology, and $Lip(h)$ be the set of all bounded functions $f : S \to \mathbb{R}$ that are measurable with respect to the Borel σ-algebra on S and that satisfy the Lipschitz condition $f(x) - f(y) \le h(x, y)$ for every $x, y \in S$. Let $P, Q \in \mathbb{P}(S)$. Then the* Kantorovich metric $\mathfrak{K}(h)$ *is defined by*

$$\mathfrak{K}(h)(P, Q) = \sup_{f \in Lip(h)} (P(f) - Q(f)).$$

Lemma 1. *Let S, h, P, and Q be as in Definition 11. Let $\Lambda(P, Q)$ consist of all measures on the product space $S \times S$ with marginals P and Q, i.e.,*

$$\Lambda(P, Q) = \{\lambda \in \mathbb{P}(S \times S) : \lambda(E \times S) = P(E) \text{ and } \lambda(S \times E) = Q(E) \text{ for all } E \in \mathscr{B}_S\}. \quad (1)$$

Then the Kantorovich metric $\mathfrak{K}(h)$ satisfies the inequality:

$$\sup_{f \in Lip(h, \mathbb{C}^b(S))} (P(f) - Q(f)) \le \mathfrak{K}(h)(P, Q) \le \inf_{\lambda \in \Lambda(P, Q)} \lambda(h) \quad (2)$$

where $Lip(h, \mathbb{C}^b(S))$ denotes functions on S that are continuous and bounded, 1-Lipschitz with respect to h, and have range $[0, \|h\|]$.

Note that h need not generate the topology on S, and so Lipschitz continuity with respect to h does not immediately imply continuity on S.

The leftmost and rightmost terms in inequality 2 are examples of infinite linear programs in duality. It is a highly nontrivial result that there is no duality gap in this case (see for example Theorem 1.3 and the proof of Theorem 1.14 in [25]).

Theorem 4 (Kantorovich-Rubinstein Duality Theorem). *Assume the conditions of Definition 11 and Lemma 1. Then there is no duality gap in equation 2, that is,*

$$\mathfrak{K}(h)(P, Q) = \sup_{f \in Lip(h, \mathbb{C}^b(S))} (P(f) - Q(f)) = \inf_{\lambda \in \Lambda(P, Q)} \lambda(h) \quad (3)$$

Note that for any point masses δ_x, δ_y, we have $\mathfrak{K}(h)(\delta_x, \delta_y) = h(x, y)$ since $\delta_{(x,y)}$ is the only measure with marginals δ_x and δ_y on the right-hand side of Equation 3. As a result, we obtain that any bounded lower semicontinuous pseudometric h can be expressed as $h(x, y) = \sup_{f \in \mathscr{F}} (f(x) - f(y))$ for some family of continuous functions $\mathscr{F}$.

Integral Probability Metrics. The intuition behind the Kantorovich metric is that the quantitive difference between two probability measures can be measured in terms of the maximal difference between the expected values with respect to the two measures, of a class of test functions - in this case, the class of Lipschitz functions. For an arbitrary class of test functions, the induced metric is known as the integral probability metric generated by that class. All definitions and results of this subsection are taken from [5].

Definition 12. *Let $\mathscr{F}$ be a subset of bounded measurable real-valued functions on a Polish metric space (S,d). Then the* integral probabilty metric *associated with $\mathscr{F}$ is the probability metric $IPM(\mathscr{F})$ on $\mathbb{P}(S)$ defined by*

$$IPM(\mathscr{F})(P,Q) = \sup_{f \in \mathscr{F}} |P(f) - Q(f)|$$

for probability measures P and Q.

For convenience, we will simply denote $IPM(\mathscr{F})$ by $\mathscr{F}$. Remark that in general $\mathscr{F}$ is allowed to take on infinite values, though we will work with bounded sets of functions to avoid this. Additionally, we remark that this probability metric in turn induces a pseudometric on S via

$$\mathscr{F}(x,y) := \mathscr{F}(\delta_x, \delta_y).$$

Thus, as an abuse of notation we will use $\mathscr{F}$ to refer to a family of functions, the associated probability metric, and the induced pseudometric, with the intended meaning clear from the context.

Definition 13. *Let $\mathscr{F}$ be a subset of bounded measurable real-valued functions on a Polish metric space (S,d). The* maximal generator *of the integral probability metric associated to $\mathscr{F}$ is the set $\mathscr{R}_{\mathscr{F}}$ of all bounded measurable real-valued functions on (S,d), each of which satisfies the following: $g \in \mathscr{R}_{\mathscr{F}}$ if and only if*

$$|P(g) - Q(g)| \leq \mathscr{F}(P,Q)$$

for every P and Q in $\mathbb{P}(S)$.

It follows that $\mathscr{R}_{\mathscr{F}}$ is the largest such family, and that $\mathscr{R}_{\mathscr{F}}(P,Q) = \mathscr{F}(P,Q)$.
  The following is Theorem 3.3 of [5].

Theorem 5. *Let $\mathscr{F}$ be an arbitrary generator of $\mathscr{R}_{\mathscr{F}}$. Then*

1. *$\mathscr{R}_{\mathscr{F}}$ contains the convex hull of $\mathscr{F}$;*
2. *$f \in \mathscr{R}_{\mathscr{F}}$ implies $\alpha f + \beta \in \mathscr{R}_{\mathscr{F}}$ for all $\alpha \in [0,1]$ and $\beta \in \mathbb{R}$;*
3. *If the sequence $(f_n)_{n \in \mathbb{N}}$ in $\mathscr{R}_{\mathscr{F}}$ converges uniformly to f, then $f \in \mathscr{R}_{\mathscr{F}}$.*

  In particular, for a given $\mathscr{F}$, $\mathscr{R}_{\mathscr{F}}$ is closed under uniform convergence.

2.5   Bisimulation Metrics

The metric analogue of bisimulation for LMPs was initially cast in terms of a family of functional expressions, interpreted as a real-valued logic over the states of a given Markov process [9]. A slightly modified version was then shown to be equivalent to a bisimultaion metric developed in the context of category theory in [2]. In [1], the authors in turn recast this latter metric fully using order-theoretic fixed-point theory for discrete systems. Finally, this method was generalized to develop a bisimulation metric for MDPs with continuous state spaces in [4].
  We present here the logic of [2] and the fixed-point results of [4], as these are the results we will attempt to join in the subsequent sections.

Definition 14. *For each $c \in (0,1]$, $\mathscr{F}_c$ represents the family of functional expressions generated by the following grammar.*

$$f ::= 1 \mid 1 - f \mid \langle a \rangle f \mid \max(f, f) \mid f \ominus q \tag{4}$$

where $q \in \mathbb{Q} \cap [0,1]$ and a belongs to a fixed set of labels A.

Let $\mathscr{M} = (S, \mathscr{B}_S, A, \{K_a : a \in A\})$ be an LMP. The interpretation of $f \in \mathscr{F}_c$, $f_{\mathscr{M}} : S \to [0,1]$, is defined inductively. Let $s \in S$. Then

$$\mathbf{1}_{\mathscr{M}}(s) = 1$$
$$(1 - f)_{\mathscr{M}}(s) = 1 - f_{\mathscr{M}}(s)$$
$$(\langle a \rangle f)_{\mathscr{M}}(s) = c \cdot K_a(s)(f_{\mathscr{M}})$$
$$\max(f_1, f_2)_{\mathscr{M}}(s) = \max((f_1)_{\mathscr{M}}(s), (f_2)_{\mathscr{M}}(s))$$
$$(f \ominus q)_{\mathscr{M}}(s) = \max(f_{\mathscr{M}}(s) - q, 0),$$

Henceforth, we shall omit the subscript $\mathscr{M}$ and use f to refer both to an expression and its interpretation, with the difference clear from the context.

Remark 2. We may also add the expressions $\min(f, f)$ and $f \oplus q$ as shorthand for the expressions $1 - \max(1 - f, 1 - f))$ and $1 - ((1 - f) \ominus q)$. The operations $\ominus$ and $\oplus$ denoted truncated subtraction in the unit interval and truncated addition in the unit interval, respectively.

The relevance of such a formulation arises via a behavioural pseudometric. The following is Theorem 3 of [2] and Theorem 8.2 of [18].

Theorem 6. *Let $\mathscr{M} = (S, \mathscr{B}_S, A, \{K_a : a \in A\})$ be an LMP and for $c \in (0,1]$, let $\mathscr{F}_c$ be the family of functional expressions defined in Definition 14. Define the map d_c on $S \times S$ as follows:*

$$d_c(x, y) = \sup_{\mathscr{F}_c} |f(x) - f(y)|. \tag{5}$$

Then d_c is a pseudometric on S whose kernel is bisimilarity.

As previously mentioned, the metric d_c can be formulated in terms of fixed-point theory, and indeed this construction has been carried over to MDPs, with the minor addition of taking into account reward differences. The following is Theorem 3.12 of [4].

Theorem 7. *Let $M = (S, \mathscr{B}_S, A, \{P_a : a \in A\}, r)$ be an MDP and let τ be a Polish topology on S that generates $\mathscr{B}_S$. Assume that the image of r is contained in $[0,1]$, and that for each a in A, r_a and P_a are continuous, $\mathbb{P}(S)$ endowed with the weak topology induced by τ. Let $c \in (0,1)$ be a discount factor, and $\mathfrak{lsc}_m$ be be the set of bounded pseudometrics on S that are lower semicontinuous on $S \times S$ endowed with the product topology induced by τ. Define $F_c : \mathfrak{lsc}_m \to \mathfrak{lsc}_m$ by*

$$F_c(h)(s, s') = \max_{a \in A}((1 - c)|r_a(s) - r_a(s')| + c \cdot \mathfrak{K}(h)(P_a(s), P_a(s')))$$

where $\mathfrak{K}(h)$ is the Kantorovich metric induced by $h \in \mathfrak{lsc}_m$. Then

1. F_c has a unique fixed point $\rho_c^* : S \times S \to [0,1]$,
2. The kernel of ρ_c^* is bisimilarity,
3. for any $h_0 \in \mathfrak{lsc}_m$, $\lim_{n \to \infty} F_c^n(h_0) = \rho_c^*$,
4. ρ_c^* is continuous on $S \times S$,
5. ρ_c^* is continuous in r and P, and
6. If MDP $M' = (S, \mathscr{B}_S, A, \{P_a : a \in A\}, k \cdot r)$ for some $k \in [0,1]$ then $\rho_{c,M'}^* = k \cdot \rho_{c,M}^*$.

Whereas the interest in finding small bisimilar systems for LMPs lies in being able to test properties of a system specified in a given logic, the interest in finding small bisimilar systems for MDPs concerns finding optimal planning strategies in terms of value functions. Given a discount factor $\gamma \in [0,1)$, the optimal value function is the unique solution to the following Bellman optimality fixed-point equation.

$$v^*(s) = \max_{a \in A} (r_a(s) + \gamma P_a(s)(v^*)) \text{ for each } s \in S.$$

In general, such a v^* need not exist. Even if it does, there may not be a well-behaved, that is to say measurable, policy that is captured by it. Fortunately, there are several mild restrictions under which this is not the case. According to Theorem 8.3.6 and its preceding remarks in [26], if the state space is Polish and the reward function is uniformly bounded then there exists a unique solution v^* to the Bellman optimality equation and there exists a measurable optimal policy for it as well.

The following is Theorem 3.20 in [4].

Theorem 8. Let $\mathscr{M} = (S, \mathscr{B}_S, A, \{P_a : a \in A\}, r)$ be an MDP and let τ be a Polish topology on S that generates $\mathscr{B}_S$. Assume that the image of r is contained in $[0,1]$, and that for each a in A, r_a and P_a are continuous, $\mathbb{P}(S)$ endowed with the weak topology induced by τ. Let $c \in (0,1)$ be a discount factor. Let v_γ^* be the optimal value function for the expected total discounted reward associated with $\mathscr{M}$ and discount factor $\gamma \in [0,1)$. Suppose $\gamma \leq c$. Then v_γ^* is Lipschitz continuous with respect to ρ_c^* with Lipschitz constant $\frac{1}{1-c}$, i.e., $|v_\gamma^*(x) - v_\gamma^*(y)| \leq (1-c)^{-1} \rho_c^*(x,y)$.

3   Bisimulation Metrics for MDPs Revisited

The goal of this section is two-fold. First, we establish a family of functional expressions as in Definition 14 that captures bisimulation for MDPs as defined in Theorem 7. This amounts to little more than Proposition 2 of [2] but using the terminology of generating classes for integral probability metrics. Second, we generalize the applicability of these results for MDPs by removing the continuity constraints in Theorem 7.

3.1   When Is the Integral Probability Metric the Kantorovich Metric?

In this section we will show that under some very mild conditions, the maximal generator of a family of functional expressions is in fact the class of Lipschitz functions with respect to the distance induced by that family. In this case, the integral probability metric and the Kantorovich metric induced by the family coincide.

The following result is Lemma 4.6 of [1], itself adapted from Proposition 2 of [2], presented almost verbatim. The imposed Lipschitz condition makes measurability concerns almost an afterthought. What really matters here is the reframing of the result in terms of the integral probability metric and its maximal generator. Doing so will allow us to examine simpler grammars for bisimulation, as well as ways of approximating these.

Theorem 9. *Suppose (S, d) is a Polish metric space and $\mathscr{F}$ is a family of real-valued functions on S that take values in the unit interval and are 1-Lipschitz continuous with respect to d. Suppose further that $\mathscr{F}$ contains the constant zero function and is closed under truncated addition with rationals in the unit interval, subtraction from the constant function 1, and taking the pointwise maximum of two functions. Let $\mathscr{R}_{\mathscr{F}}$ be the maximal generator of $\mathscr{F}$ and $Lip(\mathscr{F})$ be the set of real-valued measurable functions on S that are 1-Lipschitz with respect to the metric induced by $\mathscr{F}$. Then $\mathscr{R}_{\mathscr{F}} = Lip(\mathscr{F}) \subseteq \mathbb{C}^b(S)$.*

Proof. Firstly note that since by assumption all members of the family $\mathscr{F}$ are 1-Lipschitz continuous with respect to d, the induced pseudometric $\mathscr{F} \leq d$. Thus, $Lip(\mathscr{F}) \subseteq Lip(d) \subseteq \mathbb{C}^b(S)$. From the definition of $\mathscr{R}_{\mathscr{F}}$ applied to Dirac measures, it immediately follows that each of its members is 1-Lipschitz with respect to the pseudometric induced by $\mathscr{F}$. Thus, $\mathscr{R}_{\mathscr{F}} \subseteq Lip(\mathscr{F})$. In particular, every member of the maximal generator belongs to $\mathbb{C}^b(S)$.

The reverse inclusion $Lip(\mathscr{F}) \subseteq \mathscr{R}_{\mathscr{F}}$ is somewhat more complicated to establish. By Theorem 5, we have that $\mathscr{R}_{\mathscr{F}}$ is closed with respect to uniform convergence, and thus is also generated by $\overline{\mathscr{F}}$, the closure of $\mathscr{F}$ with respect to uniform convergence. In fact, we will show that $\mathscr{F}$ is dense in $Lip(\mathscr{F})$ in the metric of uniform convergence; for then it follows that $Lip(\mathscr{F}) = \overline{\mathscr{F}} \subseteq \mathscr{R}_{\mathscr{F}}$. We do so in two steps. First we establish the result in the case where (S, d) is a compact metric space, as this allows us to replace pointwise convergence by uniform convergence at a certain point in the proof. Finally, we extend this result to the general case of a Polish metric space by approximating it from within by a suitable compact subset.

Assume that (S, d) is a compact metric space. It is easily seen that $\overline{\mathscr{F}}$ contains the constant zero function and remains closed under truncated addition with *all* constants in the unit interval, subtraction from 1, and taking maxima; in fact, it now follows that $\overline{\mathscr{F}}$ is closed under countable suprema. To see this, suppose $(f_n)_{n \in \mathbb{N}}$ is a sequence in $\overline{\mathscr{F}}$. Since $\overline{\mathscr{F}}$ is uniformly bounded by 1 it follows that $f = \sup_{n \in \mathbb{N}} f_n$ exists and moreover it is continuous, as each f_n is 1-Lipschitz continuous with respect to d. Define $(h_n)_{n \in \mathbb{N}}$ in $\overline{\mathscr{F}}$ by $h_n = \max_{1 \leq i \leq n} f_i$. Then $(h_n)_{n \in \mathbb{N}}$ is monotonically increasing and converges pointwise to f. By Theorem 2, $(h_n)_{n \in \mathbb{N}}$ converges uniformly to f, and hence f belongs to $\overline{\mathscr{F}}$. It now also follows that $\overline{\mathscr{F}}$ is closed under truncated subtraction with constants in the unit interval, taking minima, and taking infima.

Let $g \in Lip(\mathscr{F})$. Without loss of generality, we assume the image of g belongs to $[0,1]$; for the Lipschitz property with respect to $\mathscr{F}$ implies that $\sup g - \inf g \leq 1$ and we may replace g by $g' := g - \inf g$. It follows that if g' belongs to $\mathscr{R}_{\mathscr{F}}$ then so does $g = g' + \inf g$.

Let $\epsilon > 0$. Then there exists $f_{xy\epsilon} \in \mathscr{F}$ such that

$$g(x) - g(y) \leq \mathscr{F}(x,y) \leq f_{xy\epsilon}(x) - f_{xy\epsilon}(y) + \epsilon \tag{6}$$

Define $w_{xy\epsilon} \in \overline{\mathscr{F}}$ as follows:

$$w_{xy\epsilon}(z) = \begin{cases} f_{xy\epsilon}(z) & \text{if } f_{xy\epsilon}(x) = g(x) \\ f_{xy\epsilon}(z) \ominus_{[0,1]} (f_{xy\epsilon}(x) - g(x)) & \text{if } f_{xy\epsilon}(x) > g(x) \\ f_{xy\epsilon}(z) \oplus_{[0,1]} (g(x) - f_{xy\epsilon}(x)) & \text{if } f_{xy\epsilon}(x) < g(x) \end{cases} \tag{7}$$

Then $w_{xy\epsilon}(x) = g(x)$ and $w_{xy\epsilon}(y) \leq g(y) + \epsilon$.

Let $(u_n)_{n\in\mathbb{N}}$ be a dense sequence in (S,d). Define $(\phi_{nm\epsilon})_{n,m\in\mathbb{N}} \subseteq \overline{\mathscr{F}}$ by $\phi_{nm\epsilon} = w_{u_n u_m \epsilon}$ and define $(\phi_{n\epsilon})_{n\in\mathbb{N}}$ by $\phi_{n\epsilon} = \inf_{m\in\mathbb{N}} \phi_{nm\epsilon}$. It follows that $(\phi_{n\epsilon})_{n\in\mathbb{N}} \subseteq \overline{\mathscr{F}}$. Moreover,

$$\phi_{n\epsilon}(u_n) = g(u_n) \leq g(u_n) + \epsilon \text{ and for each } m \neq n, \phi_{n\epsilon}(u_m) \leq g(u_m) + \epsilon.$$

Define $\psi_\epsilon = \sup_{n\in\mathbb{N}} \phi_{n\epsilon} \in \overline{\mathscr{F}}$. Then for any $n \in \mathbb{N}$,

$$g(u_n) \leq \psi_\epsilon(u_n) \leq g(u_n) + \epsilon. \tag{8}$$

Let $x \in S$. Then as the inequalities in line 8 hold for any subsequence of $(u_n)_{n\in\mathbb{N}}$ converging to x, and as both g and ψ_ϵ are continuous, it follows by taking limits that for any $x \in S$,

$$g(x) \leq \psi_\epsilon(x) \leq g(x) + \epsilon, \text{ or equivalently} \|\psi_\epsilon - g\| < \epsilon.$$

Define the sequence $(g_n)_{n\in\mathbb{N}}$ in $\overline{\mathscr{F}}$ by $g_n = \psi_{\frac{1}{n}}$. Then $(g_n)_{n\in\mathbb{N}}$ converges uniformly to g. Therefore, g belongs to $\overline{\mathscr{F}} \subseteq \mathscr{R}_{\mathscr{F}}$, i.e. $Lip(\mathscr{F}) \subseteq \mathscr{R}_{\mathscr{F}}$.

Now suppose (S,d) is a general Polish metric space. Let $P, Q \in \mathbb{P}(S)$. Then $\mathcal{P} = \{P, Q\}$ is finite, hence relatively compact. By Theorem 1, $\mathcal{P}$ is uniformly tight. Let $0 < \epsilon < \frac{1}{2}$. Then there exists a compact subset K of S such that $P(S\backslash K) < \epsilon$ and $Q(S\backslash K) < \epsilon$.

Let $\mathscr{G}$ denote the functions of $\mathscr{F}$ restricted to K; for $f \in \mathscr{F}$, we will write $f_K \in \mathscr{G}$. Then as $\mathscr{G}$ still contains the constant zero function, and is closed under the same operations as $\mathscr{F}$, and as (K, d) is a compact metric space, we have $\mathscr{R}_{\mathscr{G}} = Lip(\mathscr{G})$. Let $g \in Lip(\mathscr{F})$; as before, we assume without loss of generality that the image of g is contained in $[0, 1]$. Moreover, let g_K be g restricted to K and remark that $g_K \in Lip(\mathscr{G})$. Next we define $P_K, Q_K \in \mathbb{P}(K)$ by

$$P_K(E) = \frac{P(E \cap K)}{P(K)} \text{ and } Q_K(E) = \frac{Q(E \cap K)}{Q(K)}.$$

Remark that $P(K) > 1 - \epsilon > \frac{1}{2}$, and similarly for $Q(K)$, so that each is well-defined. Then as $g_K \in \mathscr{R}_\mathscr{G}$,

$$|P_K(g_K) - Q_K(g_K)| \leq \mathscr{G}(P_K, Q_K).$$

Next for any 1-bounded measurable function u on S and it's restriction u_K to K, we have

$$|P(u) - P_K(u_K)| = |P(u \cdot \delta_K) + P(u \cdot \delta_{S \setminus K}) - P_K(u_K)| = |(P(u \cdot \delta_K) - P_K(u_K)) + P(u \cdot \delta_{S \setminus K})|$$

$$\leq |1 - \frac{1}{P(K)}| \cdot P(u \cdot \delta_K) + 1 \cdot P(S \setminus K) \leq \frac{1}{1 - \epsilon} - 1 + \epsilon \leq \frac{\epsilon(2 - \epsilon)}{1 - \epsilon} \leq 3\epsilon,$$

where δ_K is the indicator function on K. Similarly $|Q(u) - Q_K(u_K)| \leq 3\epsilon$. Finally,

$$|P(g) - Q(g)| \leq |P(g) - P_K(g_K)| + |P_K(g_K) - Q_K(g_K)| + |Q_K(g_K) - Q(g)|$$

$$\leq 3\epsilon + \mathscr{G}(P_K, Q_K) + 3\epsilon \leq 6\epsilon + \sup_{f \in \mathscr{F}} |P_K(f_K) - Q_K(f_K)|$$

$$\leq 6\epsilon + \sup_{f \in \mathscr{F}} (|P_K(f_K) - P(f)| + |P(f) - Q(f)| + |Q(f) - Q_K(f_K)|)$$

$$\leq 12\epsilon + \sup_{f \in \mathscr{F}} (|P(f) - Q(f)|) \leq 12\epsilon + \mathscr{F}(P, Q).$$

As ϵ is arbitrary, it follows that $|P(g) - Q(g)| \leq \mathscr{F}(P, Q)$ and $g \in \mathscr{R}_\mathscr{F}$.

$\square$

Corollary 1. *Suppose (S, d) is a Polish metric space and $\mathscr{F}$ is a family of real-valued functions on S that take values in the unit interval and are 1-Lipschitz continuous with respect to d. Suppose further that $\mathscr{F}$ contains the constant zero function and is closed under truncated addition with rationals in the unit interval, subtraction from the constant function 1, and taking the pointwise maximum of two functions. Then the integral probability metric associated to $\mathscr{F}$ is the Kantorovich metric of the pseudometric induced by $\mathscr{F}$, i.e. $\mathscr{F}(P, Q) = \mathfrak{K}(\mathscr{F})(P, Q)$ for any $P, Q \in \mathbb{P}(S)$.*

3.2   A Family of Functional Expressions for MDP Bisimulation

We now define a family of functional expressions as in Definition 14 that when evaluated on a given MDP, capture bisimilarity.

Definition 15. *For each $c \in (0, 1]$, $\mathscr{F}_c$ represents the family of functional expressions generated by the following grammar.*

$$f ::= \mathbf{0} \,|\, \mathbf{1} - f \,|\, \langle a \rangle f \,|\, \max(f, f) \,|\, f \oplus q \qquad (9)$$

where $q \in \mathbb{Q} \cap [0, 1]$ and a belongs to a fixed set of labels A.

Let $\mathcal{M} = (S, \mathcal{B}_S, A, \{P_a : a \in A\}, r)$ be an MDP. The interpretation of $f \in \mathcal{F}_c$, $f_{\mathcal{M}} : S \to [0,1]$, is defined inductively. Let $s \in S$. Then

$$\mathbf{0}_{\mathcal{M}}(s) = 0$$
$$(\mathbf{1} - f)_{\mathcal{M}}(s) = 1 - f_{\mathcal{M}}(s)$$
$$(\langle a \rangle f)_{\mathcal{M}}(s) = r_a(s) + c \cdot P_a(s)(f_{\mathcal{M}})$$
$$max(f_1, f_2)_{\mathcal{M}}(s) = \max((f_1)_{\mathcal{M}}(s), (f_2)_{\mathcal{M}}(s))$$
$$(f \oplus q)_{\mathcal{M}}(s) = \min(f_{\mathcal{M}}(s) + q, 1).$$

As before, we shall omit the subscript $\mathcal{M}$ when it is clear from the context, and remark that the family also contains the expressions $\min(f, f)$ and $f \ominus q$.

We now show that the integral probability metric generated by $\mathcal{F}_c$ agrees with the Kantorovich metric induced by the fixed-point bisimulation metric for MDPs. This is essentially the proof method in all of Section 4 of [1].

Theorem 10. *Suppose $\mathcal{M} = (S, \mathcal{B}_S, A, \{P_a : a \in A\}, r)$ is an MDP and let τ be a Polish topology on S that generates $\mathcal{B}_S$. Assume that the image of r is contained in $[0,1]$, and that for each a in A, r_a and P_a are continuous, $\mathbb{P}(S)$ endowed with the weak topology induced by τ. Let $c \in (0,1)$ be a discount factor, and $\mathcal{F}_c$ be the family of functional expressions defined in Definition 15. Let $\mathcal{G}$ be a family of functional expressions such that $\mathcal{F}_c \subseteq \mathcal{G} \subseteq Lip(\mathcal{F}_c)$. Then the pseudometric induced by $\mathcal{G}$ coincides with the fixed-point metric ρ_c^* given by Theorem 7.*

Proof. Let (S, d) be a Polish metric space such that d generates τ. Since ρ_c^* is continuous, we can assume without loss of generality that $\rho_c^* \leq d$, as we can simply replace d by the equivalent metric $d + \rho_c^*$. By structural induction, $\mathcal{F}_c \leq \rho_c^* \leq d$, and the range of each member of $\mathcal{F}_c$ is $[0,1]$. Therefore by Corollary 1, the integral probability metric and the kantorovich metric induced by $\mathcal{F}_c$ agree.

Notice that since $\mathcal{F}_c$ is closed under subtraction from the constant function 1, we have that for any $P, Q \in \mathbb{P}(S)$

$$\begin{aligned}
\mathcal{F}_c(P, Q) &= \sup_{f \in \mathcal{F}_c} |P(f) - Q(f)| = \sup_{f \in \mathcal{F}_c} \max(P(f) - Q(f), Q(f) - P(f)) \\
&= \max(\sup_{f \in \mathcal{F}_c} P(f) - Q(f), \sup_{f \in \mathcal{F}_c} Q(f) - P(f)) \\
&= \max(\sup_{f \in \mathcal{F}_c} P(f) - Q(f), \sup_{f \in \mathcal{F}_c} P(1 - f) - Q(1 - f)) \\
&= \max(\sup_{f \in \mathcal{F}_c} P(f) - Q(f), \sup_{f \in \mathcal{F}_c} P(f) - Q(f)) \\
&= \sup_{f \in \mathcal{F}_c} P(f) - Q(f)
\end{aligned}$$

which is not necessarily the case otherwise. A simple structural induction next shows that

$$\mathcal{F}_c(x, y) = \sup_{a \in A, f \in \mathcal{F}_c} |\langle a \rangle f(x) - \langle a \rangle f(y)|.$$

Therefore,

$$\mathscr{F}_c(x,y) = \sup_{a \in A, f \in \mathscr{F}_c} \max(\langle a \rangle f(x) - \langle a \rangle f(y), \langle a \rangle f(y) - \langle a \rangle f(x))$$

$$= \sup_{a \in A, f \in \mathscr{F}_c} \max\Big((1-c)(r_a(x) - r_a(y)) + c(P_a(x)(f) - P_a(y)(f)),$$
$$(1-c)(r_a(y) - r_a(x)) + c(P_a(y)(f) - P_a(x)(f))\Big)$$

$$= \max_{a \in A} \max\Big((1-c)(r_a(x) - r_a(y)) + c \sup_{f \in \mathscr{F}_c} (P_a(x)(f) - P_a(y)(f)),$$
$$(1-c)(r_a(y) - r_a(x)) + c \sup_{f \in \mathscr{F}_c} (P_a(y)(f) - P_a(x)(f))\Big)$$

$$= \max_{a \in A} \max\Big((1-c)(r_a(x) - r_a(y)) + c \cdot \mathscr{F}_c(P_a(x), P_a(y)),$$
$$(1-c)(r_a(y) - r_a(x)) + c \cdot \mathscr{F}_c(P_a(x), P_a(y))\Big)$$

$$= \max_{a \in A}(1-c) \max\Big((r_a(x) - r_a(y)), (r_a(y) - r_a(x))\Big) + c \cdot \mathscr{F}_c(P_a(x), P_a(y))$$

$$= \max_{a \in A}(1-c)|r_a(x) - r_a(y)| + c \cdot \mathscr{F}_c(P_a(x), P_a(y))$$

$$= \max_{a \in A}(1-c)|r_a(x) - r_a(y)| + c \cdot \mathfrak{K}(\mathscr{F}_c)(P_a(x), P_a(y))$$

$$= F_c(\mathscr{F}_c)(x,y).$$

The penultimate line follows from Corollary 1. Therefore, $\mathscr{F}_c$ is a fixed-point of the functional F_c defined in Theorem 7. As the fixed-point is unique, it follows that $\rho_c^* = \mathscr{F}_c$. Finally, it follows from Theorem 9 and the definition of maximal generator that $\mathscr{G} = \mathscr{F}_c = \rho_c^*$. □

Remark 3. Theorem 10 provides another proof of Theorem 8. Consider the family $\mathscr{G}$ with the expression for the Bellman operator $\mathcal{B}_\gamma$ for the MDP $\mathcal{M} = (S, \mathscr{B}_S, A, \{P_a : a \in A\}, (1-c)r)$ and discount factor $\gamma \leq c$ in $[0,1)$. Since $Lip(\mathscr{F}_c)$ is closed under $\mathcal{B}_\gamma$ and the optimal value function scales with rewards, the result follows immediately. Otherwise, we obtain the result only for V_c^* since $\mathcal{B}_c(f) = \max_{a \in A} \langle a \rangle f$.

The usefulness of this theorem derives from our choice of $\mathscr{G}$. On the one hand, we might attempt to see what is the minimal family, if one exists, that captures bisimilarity. On the other hand, we might consider explicitly adding operators, like the Bellman operator, that could help an estimation scheme converge faster. We will explore this further in Section 4.

Practical application is still hindered by the continuity constraints on the rewards and Markov kernels, as many interesting problems model discontinuous phenomena. In the next section, we will work to remove these constraints.

3.3   The General Case: Continuity from Measurability

We conclude this section with a neat little result from descriptive set theory that was first pointed out to the authors by Ernst-Erich Doberkat at the 2012

Bellairs Workshop on Probabilistic Systems organized by Prakash. In the most interesting reinforcement learning applications, continuity of the reward process cannot be guaranteed. Amazingly, we may remove the explicit assumption of continuity in [4] and the result still holds! We seek to establish the following.

Theorem 11. *Let (X, τ) be a Polish space and $(\mathbb{P}(X), \tau_{\mathbb{P}(X)})$ be the space of probability measures on X equipped with the topology of weak convergence with respect to τ. Let $K : (X, \mathscr{B}_X) \to (\mathbb{P}(X), \mathscr{B}_{\mathbb{P}(X)})$ be a stochastic kernel. Then there exists a finer Polish topology τ' on X such that $\sigma(\tau') = \mathscr{B}_X$, $\sigma(\tau'_{\mathbb{P}(X)}) = \mathscr{B}_{\mathbb{P}(X)}$, and $K : (X, \tau') \to (\mathbb{P}(X), \tau'_{\mathbb{P}(X)})$ is continuous.*

This result is a minor reworking of the following well-known measurability-to-continuity theorem, which is Corollary 3.2.6 in [27].

Theorem 12. *Suppose (X, τ) is a Polish space, Y a separable metric space, and $f : X \to Y$ a Borel map. Then there is a finer Polish topology τ' on X generating the same Borel σ-algebra such that $f : (X, \tau') \to Y$ is continuous.*

We will also make use of this characterization of Borel σ-algebra on the set of probability measures, which is Proposition 7.25 in [28].

Proposition 1. *Let X be a separable metrizable space and $\mathbb{E}$ a collection of subsets of X which generates $\mathscr{B}_X$ and is closed under finite intersections. Then $\mathscr{B}_{P(X)}$ is the smallest σ-algebra with respect to which all functions of the form $\Theta_E(p) = p(E)$, for $E \in \mathbb{E}$, are measurable from $P(X)$ to $[0,1]$, i.e.,*

$$\mathscr{B}_{P(X)} = \sigma[\cup_{E \in \mathbb{E}} \Theta_E^{-1}(\mathscr{B}_R)].$$

For ease of exposition, we will divide the result into the following series of steps.

Lemma 2. *Let (X, τ) and K be as in Theorem 11. Then there exists an increasing sequence $(\tau_n)_{n \in \mathbb{N}}$ of Polish topologies on X finer that τ such that $\sigma(\tau_n) = \mathscr{B}_X$ and $K : (X, \tau_{n+1}) \to (\mathbb{P}(X), (\tau_n)_{\mathbb{P}(X)})$ is continuous for all $n \in \mathbb{N}$.*

Proof. By Proposition 1 for any Polish topology τ' generating $\mathscr{B}_X$, $\tau'_{\mathbb{P}(X)}$ generates $\mathscr{B}_{\mathbb{P}(X)}$. It is well known [29] that $\tau'_{\mathbb{P}(X)}$ is also a Polish topology. Therefore, $K : (X, \tau) \to (\mathbb{P}(X), \tau_{\mathbb{P}(X)})$ is a Borel map. By Theorem 12, there exists a finer Polish topology τ_0 such that $\sigma(\tau_0) = \mathscr{B}_X$ and $K : (X, \tau_0) \to (\mathbb{P}(X), \tau_{\mathbb{P}(X)})$ is continuous; but then $K : (X, \tau_0) \to (\mathbb{P}(X), (\tau_0)_{\mathbb{P}(X)})$ is Borel. Repeating this argument, there exists a finer topology τ_1 on X such that $\sigma(\tau_1) = \mathscr{B}_X$ and $K : (X, \tau_1) \to (\mathbb{P}(X), (\tau_0)_{\mathbb{P}(X)})$ is continuous. The result now easily follows for all $n \in \mathbb{N}$ by induction.   □

Lemma 3. *Let (X, τ), K, and $(\tau_n)_{n \in \mathbb{N}}$ be as in Lemma 2. Then the least upper bound topology $\tau_\infty = \vee_{n \in \mathbb{N}} \tau_n$ exists and is Polish, $\sigma(\tau_\infty) = \mathscr{B}_X$, and $K : (X, \tau_\infty) \to (\mathbb{P}(X), (\tau_n)_{\mathbb{P}(X)})$ is continuous for all $n \in \mathbb{N}$.*

Remark 4 ([27] Observation 2, pg. 93). Let $(\tau_n)_{n\in\mathbb{N}}$ be a sequence of Polish topologies on X such that for any two distinct elements x, y of X, there exist disjoint sets $U, V \in \cap_{n\in\mathbb{N}}\tau_n$ such that $x \in U$ and $y \in V$. Then the topology τ_∞ generated by $\cup_{n\in\mathbb{N}}\tau_n$ is Polish.

Proof. By Remark 4, τ_∞ exists, is Polish, and is generated by $\cup_{n\in\mathbb{N}}\tau_n$. So $\cup_{n\in\mathbb{N}}\tau_n$ is a subbasis for τ_∞. Let $O \in \tau_\infty$. Then O is an arbitrary union of finite intersections of elements of $\cup_{n\in\mathbb{N}}\tau_n$. So $O = \cup_{j\in J}(O_{j,1} \cap O_{j,2} \cap \cdots \cap O_{j,n_j})$ for some index set J. Let $i(j,k) = \min\{n \in \mathbb{N}|O_{j,k} \in \tau_n\}$ and $i(j) = \max\{i(j,k)|1 \leq k \leq n_j\}$. Then $O_j = \cap_{1\leq k\leq n_j}O_{j,k} \in \tau_{i(j)}$ because $(\tau_n)_{n\in\mathbb{N}}$ is increasing. So $O = \cup_{j\in J}O_j = \cup_{n\in\mathbb{N}}(\cup_{\{j|i(j)=n\}}O_j) = \cup_{n\in\mathbb{N}}O'_n$ where $O'_n = \cup_{\{j|i(j)=n\}}O_j \in \tau_n$. Therefore, each τ_∞-open set is a countable union of open sets in $\cup_{n\in\mathbb{N}}\tau_n$. Since each $O'_n \in \sigma(\tau_n) = \mathscr{B}_X$, $\tau_\infty \subseteq \mathscr{B}_X$ and $\sigma(\tau_\infty) \subseteq \mathscr{B}_X$. On the other hand, $\tau_0 \subseteq \tau_\infty$ implies $\mathscr{B}_X = \sigma(\tau_0) \subseteq \sigma(\tau_\infty)$. Therefore, $\sigma(\tau_\infty) = \mathscr{B}_X$.

Finally, continuity of $K : (X, \tau_\infty) \to (\mathbb{P}(X), (\tau_n)_{\mathbb{P}(X)})$ follows from that of $K : (X, \tau_{n+1}) \to (\mathbb{P}(X), (\tau_n)_{\mathbb{P}(X)})$, for all $n \in \mathbb{N}$.   □

For the next result, we will need to appeal to the famous Portmanteau Theorem, as found for example in [20].

Theorem 13 (Portmanteau Theorem). *Let P and $(P_n)_{n\in\mathbb{N}}$ be a sequence of probability measures on (X, Σ), a metric space with its Borel σ-algebra. Then the following five conditions are equivalent:*

1. $P_n \Rightarrow P$.
2. $\liminf_n \int f dP_n = \int f dP$ for all bounded, uniformly continuous real f.
3. $\limsup_n P_n(F) \leq P(F)$ for all closed F.
4. $\liminf_n P_n(G) \geq P(G)$ for all open G.
5. $\lim_n P_n(A) = P(A)$ for all P-continuity sets A.

Lemma 4. *The least upper bound of the weak topologies $\vee_{n\in\mathbb{N}}(\tau_n)_{\mathbb{P}(X)}$ exists and*

$$\vee_{n\in\mathbb{N}}(\tau_n)_{\mathbb{P}(X)} = (\tau_\infty)_{\mathbb{P}(X)}.$$

Proof. Again, $((\tau_n)_{\mathbb{P}(X)})_{n\in\mathbb{N}}$ is an increasing sequence of Polish spaces, and so $\vee_{n\in\mathbb{N}}(\tau_n)_{\mathbb{P}(X)}$ exists. Clearly, $\vee_{n\in\mathbb{N}}(\tau_n)_{\mathbb{P}(X)} \subseteq (\tau_\infty)_{\mathbb{P}(X)}$.

Suppose $P_n \Rightarrow P$ in $\vee_{n\in\mathbb{N}}(\tau_n)_{\mathbb{P}(X)}$. Then $P_n \Rightarrow P$ in $(\tau_n)_{\mathbb{P}(X)}$ for all $n \in \mathbb{N}$. Let O be a τ_∞-open set. Then as in the proof of Lemma 3, $O = \cup_{n\in\mathbb{N}}O_n$ where each $O_n \in \tau_n$. Let $G_j = \cup_{n=1}^{j}O_n \in \tau_j \subseteq \mathscr{B}_X$. Then $(G_j)_{j\in\mathbb{N}}$ increases to O. So $P_n(O) \geq P_n(G_j)$ for all $n, j \in \mathbb{N}$. So $\liminf_n P_n(O) \geq \liminf P_n(G_j) \geq P(G_j)$ for all $j \in \mathbb{N}$ by Theorem 13 in $(\tau_j)_{\mathbb{P}(X)}$. So $\liminf_n P_n(O) \geq \lim_j P(G_j) = P(O)$ by continuity from below. So $P_n \Rightarrow P$ in $(\tau_\infty)_{\mathbb{P}(X)}$ by Theorem 13 in $(\tau_\infty)_{\mathbb{P}(X)}$. Therefore, $(\tau_\infty)_{\mathbb{P}(X)} \subseteq \vee_{n\in\mathbb{N}}(\tau_n)_{\mathbb{P}(X)}$ whence equality follows.   □

We are now able to prove the main theorem of this section.

Proof (Theorem 11). By Lemmas 2 and 3, there exist Polish topologies $(\tau_n)_{n\in\mathbb{N}}$ and τ_∞ on X, finer than τ, such that $\sigma(\tau_\infty) = \mathscr{B}_X$ and $K : (X, \tau_\infty) \to$

$(\mathbb{P}(X), (\tau_n)_{\mathbb{P}(X)})$ is continuous for all $n \in \mathbb{N}$. This is equivalent to continu-ity of $K : (X, \tau_\infty) \to (\mathbb{P}(X), \vee_{n \in \mathbb{N}}(\tau_n)_{\mathbb{P}(X)})$ since convergence in $\vee_{n \in \mathbb{N}}(\tau_n)_{\mathbb{P}(X)}$ is equivalent to convergence in $(\tau_n)_{\mathbb{P}(X)}$ for all $n \in \mathbb{N}$ (again this follows from $\vee_{n \in \mathbb{N}}(\tau_n)_{\mathbb{P}(X)}$-open sets being unions of $(\tau_n)_{\mathbb{P}(X)}$-open sets). But then $K : (X, \tau_\infty) \to (\mathbb{P}(X), (\tau_\infty)_{\mathbb{P}(X)})$ is continuous by Lemma 4.   □

This argument easily extends to a countable family of stochastic kernels, so that we have the following.

Corollary 2. *Let (X, τ) be a Polish space and $(\mathbb{P}(X), \tau_{\mathbb{P}(X)})$ be the space of probability measures on X equipped with the topology of weak convergence with respect to τ. Let $(K_n)_{n \in \mathbb{N}}$ be a sequence of stochastic kernels on X. Then there exists a finer Polish topology τ' on X such that $\sigma(\tau') = \mathscr{B}_X$, $\sigma(\tau'_{\mathbb{P}(X)}) = \mathscr{B}_{\mathbb{P}(X)}$, and each $K_n : (X, \tau') \to (\mathbb{P}(X), \tau'_{\mathbb{P}(X)})$ is continuous.*

Corollary 3. *Let $\mathscr{M} = (S, \mathscr{B}_S, A, \{P_a : a \in A\}, r)$ be an MDP. Then there ex-ists a Polish topology τ on S that generates $\mathscr{B}_S$ and makes r_a and P_a continuous for each a in A, where $\mathbb{P}(S)$ is endowed with the topology of weak convergence induced by τ.*

Thus, if r is bounded and measurable we may apply Theorem 7 to obtain a quantitative form of bisimilarity. It is important to keep in mind what is going on here from a practical point of view: if we begin with a modelling scenario in which the rewards are discontinuous with *a given metric* then this amounts to changing that metric to one with respect to which rewards *are* continuous. Therefore the usefulness of this result is contingent on the modelling problem at hand not crucially being dependent on any specific metric.

With that caveat in mind, we now come to the main result of this paper, a general version of Theorem 10.

Corollary 4. *Suppose $\mathscr{M} = (S, \mathscr{B}_S, A, \{P_a : a \in A\}, r)$ is an MDP and that the image of r is contained in $[0, 1]$. Let $c \in (0, 1)$ be a discount factor, $\mathscr{F}_c$ be the family of functional expressions defined in Definition 15, and $\mathscr{G}$ be a family of functional expressions such that $\mathscr{F}_c \subseteq \mathscr{G} \subseteq Lip(\mathscr{F}_c)$. Then the pseudometric induced by $\mathscr{G}$ is the unique fixed-point ρ_c^* satisfying the equation*

$$\rho_c^*(x, y) = \max_{a \in A}((1 - c)|r_a(x) - r_a(y)| + c \cdot \mathfrak{K}(\rho_c^*)(P_a(x), P_a(y))) \text{ for all } x, y \in S$$

and whose kernel is bisimilarity.

4   Estimating Bisimulation

In this section, we discuss how focusing on families of functional expressions may make estimating bisimilarity more amenable in practice. Assume we are given an MDP $\mathscr{M} = (S, \mathscr{B}_S, A, \{P_a : a \in A\}, r)$ where the image of r is contained in $[0, 1]$. Computing a bisimilarity metric for a finite $\mathscr{M}$ has encompassed estimating the integral probability metric $\mathscr{F}_c(P, Q)$, yielding an algorithm with exponential complexity [9], computing the Kantorovich metric, $\mathfrak{K}(\cdot)(P, Q)$, yielding an algorithm with polynomial complexity [3], and solving a linear program [30].

The major issue is that although computing the linear programming formulations of bisimilarity in the ideal case can be done in polynomial time, to do so in practice is highly inefficient; to understand why, one may remark that the linear programs for a given MDP are more complex than solving for the discounted value function for that MDP; although the latter is also known to be solvable in polynomial time by linear programming [31], in practice Monte Carlo techniques have been found to be much more successful. In fact, in [4], we focused on estimating the Kantorovich metric by replacing each P and Q by empirical measures; this idea is studied in better depth in [32]. We will not focus on that approach here.

Instead we focus on a heuristic approach implicitly used in [17] and Monte Carlo techniques used in [33]. In the former, the problem at hand is, given a distribution over MDPs with a common state space, to try to find a policy that optimizes the expected total geometrically-discounted sum of rewards achieved at each state, where the average is taken over a number of sample runs performed on a number of MDPs drawn according to the given MDP distribution. The authors attack this problem by generating a family of functional expressions according to some distribution, and using these to estimate optimal planning strategies - the so-called *formula-based exploration / exploitation strategies*. In [33], the authors solve the problem of trying to compute the infimum over a given set by instead sampling and then estimating the essential infimum. Since in our case we are interested in suprema, let us recall the definition of essential supremum.

Definition 16. *Let $(X, \mathscr{B}_X, \mu)$ be a measure space. The* essential supremum *of a bounded measurable function $f : (X, \mathscr{B}_X) \to (\mathbb{R}, \mathscr{B}_{\mathbb{R}})$ is given by the following.*

$$\operatorname{ess\,sup} f = \inf\{\alpha \in \mathbb{R} : \mu(\{x \in X : f(x) > \alpha\}) = 0\}.$$

In other words, $\operatorname{ess\,sup} f$ is the least real number that is an upper bound on f except for a set of μ-measure zero. It follows that in general, $\operatorname{ess\,sup} f \leq \sup f$. Suppose further that $\mathscr{B}_X$ is a Borel σ-algebra, f is continuous, and μ is a *strictly positive measure*, i.e. every non-empty open subset of X has strictly positive μ-measure. Then since $\{x \in X : f(x) > \alpha\} = f^{-1}(\alpha, \infty)$ is open, it follows that it has μ-measure zero if and only if it is the empty set; in this case, the essential supremum and the supremum agree. We will use this in conjunction with Lemma 2 from [33], restated here in terms of the essential supremum in place of the essential infimum.

Lemma 5. *Let $(\Omega, \Sigma, \mathbb{P})$ and $(X, \mathscr{B}_X, \mu)$ be probability spaces and assume that we can sample random variables $X_1, X_2, \ldots, X_n$ mapping Ω to X, independently and identically distributed according to μ. Then if $f : X \to \mathbb{R}$ is bounded and measurable we have*

$$\max_{1 \leq i \leq n} f(X_i) \to \operatorname{ess\,sup} f \text{ in } \mu\text{-probability as } n \to \infty. \tag{10}$$

This allows for another Monte Carlo technique for (under)approximating the Kantorovich metric for bisimilarity in an MDP.

Proposition 2. *Let $\mathcal{M} = (S, \mathcal{B}_S, A, \{P_a : a \in A\}, r)$ be an MDP where the image of r is contained in $[0, 1]$. Let $c \in (0, 1)$ be a discount factor and $\mathcal{F}_c$ be the family of functional expressions defined in Definition 15 and interpreted over $\mathcal{M}$. Let $\overline{\mathcal{F}_c}$ be the closure of $\mathcal{F}_c$ with respect to uniform convergence. Let $\mu \in \mathbb{P}(\overline{\mathcal{F}_c})$ be strictly positive. Suppose $f_1, f_2, \ldots, f_n$ are independent, identically distributed samples drawn according to μ. Then*

$$\max_{1 \leq i \leq n} |P(f_i) - Q(f_i)| \to \mathcal{F}_c(P, Q) \text{ in } \mu\text{-probability as } n \to \infty. \qquad (11)$$

Proof. Since S is Hausdorff, $\mathbb{C}^b(S)$ with the uniform norm is a Banach space. Therefore, $\overline{\mathcal{F}_c}$, as a closed subset of $\mathbb{C}^b(S)$, is itself a measurable subspace when equipped with the Borel sets given by the uniform norm. For a given $P, Q \in \mathbb{P}(S)$ let $g : \overline{\mathcal{F}_c} \to \mathbb{R}$ be defined by $g(f) = |P(f) - Q(f)|$. Then g is continuous and bounded by 1. The result now follows from Lemma 5 and the preceding remarks, and Corollary 4. $\qquad \square$

Remark in particular that

$$\max_{1 \leq i \leq n} |f_i(x) - f_i(y)| \to \rho_c^*(x, y) \text{ in } \mu\text{-probability as } n \to \infty. \qquad (12)$$

To turn this into a proper algorithm is beyond the scope of this work - one needs to fix a particular measure and provide sample complexity results, among other things. However, we remark that being able to sample from a much smaller class than the class of all Lipschitz functions should improve performance regardless of how other parameters are set.

5   Conclusions

We have shown, with slight modification, that the family of functional expressions developed in [9,2] to capture quantitative bisimilarity for LMPs does the same for MDPs with continuous state spaces and bounded measurable reward functions. We have used the same techniques as in these previous works - in particular, a density result in Proposition 2 of [2] - reworded in the terminology of generating classes for integral probability metrics. The hope is that by focusing on these generating classes of functions, we may find better practical algorithms for assessing equivalence between states in a Markov process - either by under or over-approximating a particular class, or by sampling from it in some manner.

Moreover, we have used a trick from descriptive set theory to remove a previous continuity constraint on the rewards and Markov kernels in Theorem 7, thereby widening its applicability.

5.1   Related Work

The notion of bisimilarity metrics, both in terms of logical expressions and in terms of how to compute them using linear programming formulations, really derives from the work of [9] and [2] for LMPs. In [9], the emphasis was on developing a robust theoretical notion of quantitative bisimilarity and establishing a decision procedure for it, albeit with exponential complexity. In [2], the emphasis was again on establishing a robust notion of quantitative bisimilarity while at the same time yielding a theoretical polynomial complexity bound by means of the Kantorovich metric. Complexity results in general are discussed in [30]. However, in none of these are more than a few toy examples worked through, and the idea of Monte Carlo techniques for more efficient practical implementations is not broached.

The idea of examining the relationship between probability measures by studying generating classes of functions was explored in [5,34] for integral probability metrics and stochastic orders. Müller takes the point of view of looking at maximal generators for such orders, and demonstrates that in general, minimal orders may not exist.

To the best of our knowledge, the only practical work to exploit optimality based on functional expression occurs in [17]. Here, the goal is to determine an optimal planning strategy on average, when one is acting on an unknown MDP but given a distribution over its reward and transition parameters. The advantage of the functional expression approach here is that it is independent of the particulars of a given model.

5.2   Future Work

The point of view of this work is that one should focus on families of functional expressions for quantitative bisimilarity as we suspect this may be more advantageous in practice. Thus, an immediate concern is to turn Proposition 2 into a full-fledged Monte Carlo algorithm. Among the necessities are choosing the right class of functional expressions from which to sample, as small as possible a subset of $\mathscr{F}_c$, constructing a strictly positive probability measure with which to sample the class of functionals, and most importantly, a sample complexity bound to inform us of how many samples should be required for a given level of confidence.

From the theoretical side, we are interested in finding minimal classes that generate the same bisimilarity metric, and equivalences obtained from using other classes. In both cases, it might be fruitful to consider only non-empty closed subsets of $\mathbb{C}^b(S)$ with the uniform norm. We can order this space, and add in the empty set, to get a complete lattice; moreover, we can equip it with the Hausdorff metric, and the resulting Borel σ-algebra, known as the *Effros Borel space*, will be a standard Borel space provided $(S, \mathscr{B}_S)$ is as well ([27], pg. 97). Doing so may allow us to relate the differences between the equivalences induced by two families of functional expressions in terms of their quantitative difference in Effros Borel space. In particular, we are interested in coarser more easily computable equivalences, and how to relate these to the theoretical guarantees given by bisimilarity.

In statistical parlance, the interpreted class of functional expressions is just a family of random variables; and testing whether or not two states are bisimilar amounts to testing how their Markov kernels differ on this test set of random variables. Conceptually, this fits in with Prakash's view that Markov processes should be viewed as transformers of random variables [35]. As (real-valued) stochastic kernels subsume both random variables and subprobability measures, we may complete this conceptual picture by viewing a Markov process - itself a family of kernels - as a transformer of families of kernels. It remains to be seen if this point of view in general can lead to better algorithms in practice.

Acknowledgements. This work is dedicated with love to Prakash Panangaden. Absolutely none of it would have been possible without him. The authors would also like to thank Ernst-Erich Doberkat and Jean Goubault-Larrecq for suggesting Theorem 11, as well as Igor Khavkine and Jérôme Feret for verifying the results of Subsection 3.3.

References

1. Desharnais, J., Jagadeesan, R., Gupta, V., Panangaden, P.: The Metric Analogue of Weak Bisimulation for Probabilistic Processes. In: LICS 2002: Proceedings of the 17th Annual IEEE Symposium on Logic in Computer Science, July 22-25, pp. 413–422. IEEE Computer Society, Washington, DC (2002)
2. van Breugel, F., Worrell, J.: Towards Quantitative Verification of Probabilistic Transition Systems. In: Orejas, F., Spirakis, P.G., van Leeuwen, J. (eds.) ICALP 2001. LNCS, vol. 2076, pp. 421–432. Springer, Heidelberg (2001a)
3. van Breugel, F., Worrell, J.: An Algorithm for Quantitative Verification of Probabilistic Transition Systems. In: Larsen, K.G., Nielsen, M. (eds.) CONCUR 2001. LNCS, vol. 2154, pp. 336–350. Springer, Heidelberg (2001b)
4. Ferns, N., Panangaden, P., Precup, D.: Bisimulation Metrics for Continuous Markov Decision Processes. SIAM Journal on Computing 40(6), 1662–1714 (2011)
5. Müller, A.: Integral Probability Metrics and Their Generating Classes of Functions. Advances in Applied Probability 29, 429–443 (1997)
6. Larsen, K.G., Skou, A.: Bisimulation Through Probabilistic Testing. Information and Computation 94(1), 1–28 (1991)
7. Milner, R.: A Calculus of Communication Systems. LNCS, vol. 92. Springer, New York (1980)
8. Park, D.: Concurrency and Automata on Infinite Sequences. In: Proceedings of the 5th GI-Conference on Theoretical Computer Science, pp. 167–183. Springer, London (1981)
9. Desharnais, J., Gupta, V., Jagadeesan, R., Panangaden, P.: Metrics for Labeled Markov Systems. In: Baeten, J.C.M., Mauw, S. (eds.) CONCUR 1999. LNCS, vol. 1664, pp. 258–273. Springer, Heidelberg (1999)
10. Desharnais, J., Gupta, V., Jagadeesan, R., Panangaden, P.: Metrics for Labelled Markov Processes. Theor. Comput. Sci. 318(3), 323–354 (2004)
11. van Breugel, F., Hermida, C., Makkai, M., Worrell, J.: Recursively Defined Metric Spaces Without Contraction. Theoretical Computer Science 380(1-2), 143–163 (2007)

12. Kozen, D.: A Probabilistic PDL. In: STOC 1983: Proceedings of the Fifteenth Annual ACM Symposium on Theory of Computing, pp. 291–297. ACM, New York (1983)
13. van Breugel, F., Sharma, B., Worrell, J.B.: Approximating a Behavioural Pseudometric Without Discount for Probabilistic Systems. In: Seidl, H. (ed.) FOSSACS 2007. LNCS, vol. 4423, pp. 123–137. Springer, Heidelberg (2007)
14. Ferns, N., Panangaden, P., Precup, D.: Metrics for Finite Markov Decision Processes. In: AUAI 2004: Proceedings of the 20th Annual Conference on Uncertainty in Artificial Intelligence, Arlington, Virginia, United States, pp. 162–169. AUAI Press (2004)
15. Ferns, N., Panangaden, P., Precup, D.: Metrics for Markov Decision Processes with Infinite State Spaces. In: Proceedings of the 21 Annual Conference on Uncertainty in Artificial Intelligence (UAI 2005), Arlington, Virginia, pp. 201–208. AUAI Press (2005)
16. Ferns, N., Castro, P.S., Precup, D., Panangaden, P.: Methods for Computing State Similarity in Markov Decision Processes. In: Proceedings of the 22nd Annual Conference on Uncertainty in Artificial Intelligence (UAI 2006), Arlington, Virginia. AUAI Press, Arlington (2006)
17. Castronovo, M., Maes, F., Ernst., R.F.,, D.: Learning Exploration/Exploitation Strategies for Single Trajectory Reinforcement Learning. In: Proceedings of the 10th European Workshop on Reinforcement Learning (EWRL 2012), Edinburgh, Scotland, June 30-July 1, vol. 24, pp. 1–10 (2012)
18. Panangaden, P.: Labelled Markov Processes. Imperial College Press (2009)
19. Giry, M.: A Categorical Approach to Probability Theory. Categorical Aspects of Topology and Analysis, pp. 68–85 (1982)
20. Billingsley, P.: Convergence of Probability Measures. Wiley (1968)
21. Dudley, R.M.: Real Analysis and Probability. Cambridge University Press (August 2002)
22. Desharnais, J.: Labelled Markov Processes. PhD thesis, McGill University (2000)
23. Desharnais, J., Edalat, A., Panangaden, P.: Bisimulation for Labeled Markov Processes. Information and Computation 179(2), 163–193 (2002)
24. Gibbs, A.L., Su, F.E.: On Choosing and Bounding Probability Metrics. International Statistical Review 70, 419–435 (2002)
25. Villani, C.: Topics in Optimal Transportation. Graduate Studies in Mathematics, vol. 58. American Mathematical Society (2003)
26. Hernández-Lerma, O., Lasserre, J.B.: Further Topics on Discrete-Time Markov Control Processes. Applications of Mathematics. Springer, New York (1999)
27. Srivastava, S.M.: A Course on Borel Sets. Graduate texts in mathematics, vol. 180. Springer (2008)
28. Bertsekas, D.P., Shreve, S.E.: Stochastic Optimal Control: The Discrete-Time Case. Athena Scientific (2007)
29. Parthasarathy, K.R.: Probability Measures on Metric Spaces. Academic, New York (1967)
30. Chen, D., van Breugel, F., Worrell, J.: On the Complexity of Computing Probabilistic Bisimilarity. In: Birkedal, L. (ed.) FOSSACS 2012. LNCS, vol. 7213, pp. 437–451. Springer, Heidelberg (2012)
31. Puterman, M.L.: Markov Decision Processes: Discrete Stochastic Dynamic Programming. John Wiley & Sons, Inc., New York (1994)
32. Sriperumbudur, B.K., Fukumizu, K., Gretton, A., Schölkopf, B., Lanckriet, G.R.G.: On the Empirical Estimation of Integral Probability Metrics. Electronic Journal of Statistics 6, 1550–1599 (2012)

33. Bouchard-Côté, A., Ferns, N., Panangaden, P., Precup, D.: An Approximation Algorithm for Labelled Markov Processes: Towards Realistic Approximation. In: QEST 2005: Proceedings of the Second International Conference on the Quantitative Evaluation of Systems (QEST 2005) on The Quantitative Evaluation of Systems, pp. 54–61. IEEE Computer Society, Washington, DC (2005)
34. Müller, A.: Stochastic Orders Generated by Integrals: A Unified Study. Advances in Applied Probability 29, 414–428 (1997)
35. Chaput, P., Danos, V., Panangaden, P., Plotkin, G.: Approximating Markov Processes by Averaging. In: Albers, S., Marchetti-Spaccamela, A., Matias, Y., Nikoletseas, S., Thomas, W. (eds.) ICALP 2009, Part II. LNCS, vol. 5556, pp. 127–138. Springer, Heidelberg (2009)

Random Measurable Selections

Jean Goubault-Larrecq[1] and Roberto Segala[2]

[1] ENS Cachan
goubault@lsv.ens-cachan.fr
[2] Università di Verona
roberto.segala@univr.it

Abstract. We make the first steps towards showing a general "randomness for free" theorem for stochastic automata. The goal of such theorems is to replace randomized schedulers by averages of pure schedulers. Here, we explore the case of measurable multifunctions and their measurable selections. This involves constructing probability measures on the measurable space of measurable selections of a given measurable multifunction, which seems to be a fairly novel problem. We then extend this to the case of IT automata, namely, non-deterministic (infinite) automata with a history-dependent transition relation. Throughout, we strive to make our assumptions minimal.

1 Introduction

This paper grew out of an attempt at proving a "randomness for free" type theorem [5] for stochastic automata [4]. We present the first steps in this direction.

A stochastic automaton is a transition system on a measurable space Q of *states*. When in state $q \in Q$, we have access to a set $\theta(q)$ of fireable transitions, from which we choose non-deterministically. A transition is a pair (a, μ) of an *action* a, from a fixed measurable space L, and of a probability measure μ on Q. Once we have chosen such a transition, we pick the next state q' at random with respect to μ, and proceed. There are two ways to resolve the non-deterministic choice of a transition $(a, \mu) \in \theta(q)$. A *pure scheduler* σ is a function that maps each path $w = q_0 a_1 q_1 \cdots a_n q_n$ of states and actions seen so far to an element $\sigma(w)$ of $\theta(q_n)$ (or to a special termination constant $\bot$). A *randomized scheduler* η instead maps w to a (sub)probability measure concentrated on $\theta(q_n)$, thereby drawing the transition at random as well. In each case, given a scheduler η, the stochastic automaton behaves as a purely probabilistic transition system, and one can define the probability $P_\eta(\mathcal{E})$ that the automaton will follow a path that lies in the measurable set $\mathcal{E}$.

The "randomness for free" question we envision is as follows: given a randomized scheduler η, and a measurable set $\mathcal{E}$ of paths, can we find two *pure schedulers* σ^-, σ^+ such that $P_{\sigma^-}(\mathcal{E}) \leq P_\eta(\mathcal{E}) \leq P_{\sigma^+}(\mathcal{E})$? This has a number of important applications, and was solved positively by Chatterjee, Doyen *et al.* [5, Section 4], in the case that Q and L are finite. In general, we consider the following more general question: given a randomized scheduler η, and a measurable *payoff* function h from paths to $\mathbb{R}^+$, can we find two pure schedulers

F. van Breugel et al. (Eds.): Panangaden Festschrift, LNCS 8464, pp. 343–362, 2014.

σ^-, σ^+ such that $\int_\omega h(\omega)dP_{\sigma^-} \leq \int_\omega h(\omega)dP_\eta \leq \int_\omega h(\omega)dP_{\sigma^+}$? This includes the previous question, by taking the characteristic function $\chi_\mathcal{E}$ of $\mathcal{E}$ for h.

Beware that one cannot reasonably ask for the existence of a pure scheduler σ such that $P_\sigma(\mathcal{E}) = P_\eta(\mathcal{E})$. For example, let $Q = \{0, 1, 2\}$, $L = \{*\}$, $\theta(0) = \{t_1 = (*, \delta_1), t_2 = (*, \delta_2)\}$ (where δ_x is the Dirac mass at x), $\theta(1) = \theta(2) = \emptyset$. There are only two pure schedulers, one that reaches state 1 with probability 1, while the other reaches 2 with probability 1. But one can reach 1 with *arbitrary probability* p using the randomized scheduler $\eta(0) = p\delta_{t_1} + (1 - p)\delta_{t_2}$.

While Chatterjee, Doyen *et al.* are concerned with finite state and action spaces, infinite spaces are useful as well: in modeling timed probabilistic and non-deterministic transition systems [7, Example 1.1], or devices that interact with the physical world [6,1], where each state includes information about time, position, impulse and other real-valued data, for example. Considering his publication record (see for example [16] on labeled Markov processes, or [10] on Markov decision processes), Prakash would probably be the first to approve us.

Our initial aim was to prove such "randomness for free" theorems for general stochastic automata. During the year 2008, we found very complex arguments that proved only very weak versions of what we were after. We were on the wrong path. On the opposite, Chatterjee, Doyen *et al.* [5, Section 4] used a simple idea: draw pure schedulers σ themselves at random, with respect to some measure ϖ, designed so that $P_\eta(\mathcal{E}) = \int_\sigma P_\sigma(\mathcal{E})d\varpi$. The claim then follows by standard integration arguments (Fact 1 below).

The probability measure ϖ on pure schedulers has an intuitive description, too: we merely choose the transition $\sigma(w)$ at random with respect to probability $\eta(w)$, where η is our given randomized scheduler, for each finite path w, independently. Now this is the source of typical measure-theoretic complications. First, we must force our pure schedulers to be measurable. Otherwise, $\int_\sigma P_\sigma(\mathcal{E})d\varpi$, and in fact already P_σ, makes no sense. Second, we need to make clear what the intended σ-algebra is on the space of all pure schedulers. If we don't have any, nothing of the above makes any sense either. Third, what does it mean to draw $\sigma(w)$ for each w *independently*? The sheer fact that σ is measurable must enforce at least some moderate amount of non-independence.

Chatterjee, Doyen *et al.* did not need to address these issues: on finite state and action spaces, *all* pure schedulers are measurable, and the problems above disappear. Going to infinite spaces of states and actions requires extra effort.

While we were writing this paper, we soon realized that we would have to choose between: (1) solving the full question, by applying all the required measure-theoretic clout if necessary, possibly making it so complex that nobody would understand it; or (2) solving a few restricted cases, showing a few elegant proof tricks along the way. It should be clear that (2) was a better choice. We hope that Prakash will appreciate the techniques, if not the results.

That is, we shall be content to solve the problem in the special case of *IT automata*, namely, stochastic automata with a trivial action space, no option for termination, and no random choice at all: randomness will be induced by the randomized scheduler only. We shall deal with the general case in another paper.

The plan of the paper is as follows. We recapitulate some required basic measure theory in Section 2. In Section 3, we define a σ-algebra on the set $Sel(F)$ of *measurable selections* of a given multifunction F—these are just what we have called pure schedulers, for the transition relation F of a stochastic automaton without probabilistic choice; and we show that any randomized scheduler g defines a canonical probability measure ϖ_g on $Sel(F)$ such that, for every x, drawing a point at random with probability $g(x)$ gives the same result as drawing a measurable selection f with probability ϖ_g and computing $f(x)$ (Proposition 1). This is the most important construction of the paper. In Section 4, we extend this result from random measurable selections to random measurable pure schedulers of IT automata. Although the setting looks extremely similar, there is no hope of reusing the previous result. Instead, we use similar proof techniques, but with a more complex implementation. We conclude in Section 5.

2   Basics on Measure Theory

A *σ-algebra* on a set X is a family of subsets that is closed under complement and countable unions. We shall write $\overline{A}$ for the complement of A in X. A pair $X = (|X|, \Sigma_X)$ of a set $|X|$ and a σ-algebra Σ_X on $|X|$ is a *measurable space*, and the elements of Σ_X are called the *measurable subsets* of X. We shall sometimes write X instead of $|X|$ to avoid pedantry.

Given any family $\mathcal{F}$ of subsets of a set A, there is a smallest σ-algebra that contains the elements of $\mathcal{F}$. This is called the σ-algebra *generated by $\mathcal{F}$*. We shall sometimes refer to the elements of $\mathcal{F}$ as the *basic* measurable subsets of this σ-algebra, despite some ambiguity. One example is $\mathbb{R}$ with its *Borel σ-algebra*, generated by intervals. Another one is the *product* $X_1 \times X_2$ of two measurable spaces is $(|X_1| \times |X_2|, \Sigma_{X_1} \otimes \Sigma_{X_2})$, whose basic measurable subsets are the *rectangles* $E_1 \times E_2$, $E_1 \in \Sigma_{X_1}$, $E_2 \in \Sigma_{X_2}$. In general, the σ-algebra on the product of an arbitrary family of measurable spaces $(X_i)_{i \in I}$ is the one generated by the subsets $\pi_i^{-1}(E)$ where $E \in \Sigma_{X_i}$, $i \in I$, and $\pi_i \colon \prod_{i \in I} |X_i| \to |X_i|$ is the usual projection onto coordinate i.

The *coproduct* $X_1 + X_2$ of two measurable spaces X_1, X_2 is simpler: $|X_1 + X_2|$ is the disjoint union of $|X_1|$ and $|X_2|$, and $\Sigma_{X_1+X_2}$ consists of unions $E_1 \cup E_2$ of a measurable subset E_1 of X_1 and of a measurable subset E_2 of X_2. This construction generalizes to countable coproducts $\sum_{n \in \mathbb{N}} X_n$ in the obvious way.

A *measurable* map $f \colon X \to Y$ is one such that $f^{-1}(E) \in \Sigma_X$ for every $E \in \Sigma_Y$. If $\mathcal{F}$ generates Σ_Y, it is enough to check that $f^{-1}(E) \in \Sigma_X$ for every E in $\mathcal{F}$ to establish the measurability of f.

A *measure* μ on X is a map from Σ_X to $\mathbb{R}^+ \cup \{+\infty\}$ that is countably additive ($\mu(\emptyset) = 0$, and $\mu(\bigcup_{n \in \mathbb{N}} E_n) = \sum_{n \in \mathbb{N}} \mu(E_n)$ for every countable family of disjoint measurable subsets E_n). A *probability* measure is one such that $\mu(X) = 1$. The *Dirac mass* at x, δ_x, is the probability measure defined by $\delta_x(E) = 1$ if $x \in E$, 0 otherwise.

A measure μ is *concentrated on* a measurable subset A of X if and only if $\mu(X \setminus A) = 0$. For example, if X is finite and $\Sigma_X = \mathbb{P}(|X|)$, then $\mu = \sum_{x \in X} a_x \delta_x$

is concentrated on $\{x \in X \mid a_x \neq 0\}$. Any subset A (even non-measurable) of $|X|$ gives rise to a measurable *subspace*, again written A, with $\Sigma_A = \{A \cap B \mid B \in \Sigma_X\}$. If A is measurable, one can define the *restriction* $\mu_{|A}$ of μ to the subspace A, by $\mu_{|A}(B) = \mu(B)$ for every $B \in \Sigma_A$. If μ is a probability measure that is concentrated on A, then $\mu_{|A}$ is also a probability measure.

There is a standard notion of integral of measurable maps $h \colon X \to \mathbb{R}^+$ with respect to a measure μ on X, which we write $\int_{x \in X} h(x)d\mu$. Other notations found in the literature are $\int_{x \in X} h(x)\mu(dx)$ or $\langle h, \mu \rangle$. We shall also use the notation $\int_{x \in X} d\mu\, h(x)$, especially when $h(x)$ is a long formula, as in $\int_{x_1 \in E_1} d\mu_1 \int_{x_2 \in E_2} d\mu_2\, h(x_1, x_2)$. Writing χ_E for the characteristic map of a measurable subset E of X, $\int_{x \in X} \chi_E(x)d\mu$ equals the measure $\mu(E)$ of E.

Given a map $f \colon A \to |X|$ (not necessarily measurable) where X is a measurable space and A is a set, the family $(f^{-1}(E))_{E \in \Sigma_X}$ is a σ-algebra on A, called the σ-algebra *induced by* f on A. When f is the inclusion map of a subset A of $|X|$, we retrieve the subspace σ-algebra Σ_A.

If μ is concentrated on a measurable subset A of X, then $\int_{x \in X} g(x)d\mu = \int_{x \in A} g(x)d\mu_{|A}$, where A is considered as a subspace of X on the right-hand side.

We write $\mathcal{P}(X)$ for the space of all probability measures on X, with the *weak σ-algebra*, generated by the subsets $[E > r] = \{\mu \in |\mathcal{P}(X)| \mid \mu(E) > r\}$. One can equate $\mu \in \mathcal{P}(X)$ with a vector of real numbers $(\mu(E))_{E \in \Sigma_X}$, i.e., with an element of the measurable space $\mathbb{R}^{\Sigma_X}$. The weak σ-algebra is nothing else than the σ-algebra induced by the inclusion of $|\mathcal{P}(X)|$ into the product $\mathbb{R}^{\Sigma_X}$.

Given a measurable map $f \colon X \to Y$, and a measure μ on X, the formula $f[\mu](E') = \mu(f^{-1}(E'))$ defines a measure $f[\mu]$ on Y, called the *image measure* of μ by f. For any measurable $h \colon Y \to \mathbb{R}^+$, the following *change of variables* formula holds [2, Theorem 16.13]:

$$\int_{y \in Y} h(y)df[\mu] = \int_{x \in X} (h \circ f)(x)d\mu. \tag{1}$$

More trivially, the function $\mathbf{e}_X \colon X \to \mathcal{P}(X)$ that sends x to the Dirac mass δ_x is measurable. These facts assemble to define the so-called Giry monad [11, Section 1], of which $\mathbf{e}$ is the unit. (More precisely, one of the two Giry monads.) Prakash stressed the importance of this monad in [14]—probably the one paper that popularized it.

Finally, we shall use the following well-known fact near the end of the paper.

Fact 1. *For every integrable map $h \colon X \to \mathbb{R}$ on a measurable space X, for every $a \in \mathbb{R}$, if $\int_{x \in X} h(x)d\mu = a$ for some probability measure μ on X, then there are points $x^-, x^+ \in |X|$ such that $h(x^-) \le a \le h(x^+)$.*

Indeed, if x^- did not exist, say, then $h(x) > a$ for every $x \in |X|$. Let $A_n = h^{-1}(a + 1/n, +\infty)$ for every non-zero natural number n: so $|X| = \bigcup_{n \ge 1} A_n$. Since $h(x) \ge a$ for every $x \in |X|$ and $h(x) \ge a + 1/n$ if additionally $x \in A_n$, $a = \int_{x \in X} h(x)d\mu \ge a + 1/n\, \mu(A_n)$, so $\mu(A_n) = 0$. A consequence of σ-additivity is that the measure of the union of a countable chain of measurable subsets is the sup of their measures, so $1 = \mu(|X|) = \sup_{n=1}^{+\infty} \mu(A_n) = 0$: contradiction.

Carathéodory's measure extension theorem. The following measure extension theorem, due to Carathéodory, was singled out as "a very useful type of theorem" by Prakash [15, Theorem 18]; see also [2, Theorem 11.3].

A *semiring* $\mathcal{A}$ on a set Ω is a collection of subsets of Ω that contains the empty set, is closed under binary intersections, and such that the difference $A \smallsetminus B$ of any two sets $A, B \in \Omega$ can be written as a finite union of elements of $\mathcal{A}$. A map μ is *countably subadditive* on $\mathcal{A}$ if and only if for every countable disjoint family of elements A_n of $\mathcal{A}$ whose union A is in $\mathcal{A}$, $\mu(A) \leq \sum_{n \in \mathbb{N}} \mu(A_n)$.

Theorem 1 (Carathéodory). *Let Ω be a set, and $\mathcal{A}$ be a semiring on Ω. Every function $\mu \colon \mathcal{A} \to [0, +\infty]$ such that $\mu(\emptyset) = 0$, that is finitely additive and countably subadditive on $\mathcal{A}$, extends to a measure on the σ-algebra generated by $\mathcal{A}$. In particular, this is so if $\mu(\emptyset) = 0$ and μ is countably additive on $\mathcal{A}$.*

A typical application is $\Omega = \mathbb{R}$, $\mathcal{A}$ is the semiring of all half-closed intervals $[a, b)$, and $\mu[a, b) = b - a$, leading to Lebesgue measure.

The Ionescu-Tulcea Theorem. Let Q_n, $n \in \mathbb{N}$, be countably many measurable spaces, and assume countably many measurable maps $g_n \colon \prod_{i=0}^{n-1} Q_i \to \mathcal{P}(Q_n)$. One can think of Q_n as the space of all possible states of a probabilistic transition system at (discrete) time $n \in \mathbb{N}$. Given that at time n we have gone through states $q_0 \in Q_0, q_1 \in Q_1, \ldots, q_{n-1} \in Q_{n-1}$, $g_n(q_0, q_1, \cdots, q_{n-1})$ is a probability distribution along which we draw the next state q_n. The following Ionescu-Tulcea Theorem states that these data define a unique probability measure on infinite paths $q_0, q_1, \cdots, q_{n-1}, \cdots$:

Theorem 2 (Ionescu-Tulcea). *Let Q_n, $n \in \mathbb{N}$, be measurable spaces, and $g_* = (g_n \colon \prod_{i=0}^{n-1} Q_i \to \mathcal{P}(Q_n))_{n \geq 1}$ be measurable maps. For every $q_0 \in Q_0$, there is a unique probability measure $P_{g_*}(q_0)$ on $\prod_{i=0}^{+\infty} Q_i$ such that:*

$$P_{g_*}(q_0)(\prod_{i=0}^{n} E_i \times \prod_{i=n+1}^{+\infty} |Q_i|) = \chi_{E_0}(q_0) \int_{q_1 \in E_1} dg_1(q_0) \int_{q_2 \in E_2} dg_2(q_0 q_1) \cdots \int_{q_n \in E_n} dg_n(q_0 q_1 \cdots q_{n-1}).$$

(2)

Moreover, P_{g_*} defines a measurable map from Q_0 to $\mathcal{P}(\prod_{i=0}^{+\infty} Q_i)$.

We consider tuples $(q_0, q_1, q_2, \cdots, q_{n-1})$ as words, and accordingly write them as $q_0 q_1 q_2 \cdots q_{n-1}$. The notation $\int_{q_i \in E_i} d\mu_i \, h(q_i)$ (where $\mu_i = g_i(q_0 q_1 \cdots q_{i-1})$ above) stands for $\int_{q_i \in Q_i} d\mu_i \, \chi_{E_i}(q_i) h(q_i)$, and the rightmost integral in Theorem 2 is an integral of the constant 1, which is standardly omitted—i.e., the rightmost integral is $\int_{q_n \in Q_n} \chi_{E_n}(q_n) dg_n(q_0 q_1 q_2 \cdots q_{n-1}) = g_n(q_0 q_1 q_2 \cdots q_{n-1})(E_n)$.

There are several small variations on the Ionescu-Tulcea Theorem. Our version is Giry's [11, Theorem 3], except for the fact that Giry considers more general ordinal-indexed sequences of measurable spaces. We will not require that.

The following is needed in the proof of Lemma 2. Lemma 2 looks perfectly obvious, yet requires some effort to prove. Measure theory is full of these.

Lemma 1. *Under the assumptions of Theorem 2, for every $n \in \mathbb{N}$, and for every measurable map $h \colon \prod_{i=0}^{n} Q_i \to \mathbb{R}^+$, for every $q \in Q_0$,*

$$\int_{q_0 q_1 \cdots \in \prod_{i=0}^{+\infty} Q_i} h(q_0 q_1 \cdots q_n) dP_{g_*}(q) \tag{3}$$

$$= \int_{q_1 \in Q_1} dg_1(q) \int_{q_2 \in Q_2} dg_2(qq_1) \cdots \int_{q_n \in Q_n} dg_n(qq_1 q_2 \cdots q_{n-1}) h(qq_1 q_2 \cdots q_{n-1} q_n).$$

Proof. This is true for functions h of the form $\chi_{\prod_{i=0}^{n} E_i}$, $E_i \in \Sigma_{Q_i}$, as one can check by using (2). Let S be the set of measurable subsets $\mathcal{E}$ of $\prod_{i=0}^{n} Q_i$ such that (3) holds for $h = \chi_{\mathcal{E}}$, i.e., such that $P_{g_*}(q)(\mathcal{E}) = \int_{q_1 \in Q_1} dg_1(q) \int_{q_2 \in Q_2} dg_2(qq_1) \cdots$ $\int_{q_n \in Q_n} dg_n(qq_1 q_2 \cdots q_{n-1}) \chi_{\mathcal{E}}(qq_1 q_2 \cdots q_{n-1} q_n)$. S contains all the rectangles, is closed under complements (using $\chi_{\overline{\mathcal{E}}} = 1 - \chi_{\mathcal{E}}$), and under countable disjoint unions. For the latter, consider countably disjoint elements $\mathcal{E}_m$, $m \in \mathbb{N}$, of S, let $\mathcal{E} = \bigcup_{m \in \mathbb{N}} \mathcal{E}_m$, and realize that $\chi_{\mathcal{E}} = \sup_{m \in \mathbb{N}} \sum_{k=0}^{m} \chi_{\mathcal{E}_k}$. The Monotone Convergence Theorem [2, Theorem 16.2] states that integrals of non-negative real functions commute with pointwise suprema of countable chains, so $P_{g_*}(q)(\mathcal{E}) = \sup_{m \in \mathbb{N}} \sum_{k=0}^{m} \int_{q_1 \in Q_1} dg_1(q) \int_{q_2 \in Q_2} dg_2(qq_1) \cdots \int_{q_n \in Q_n} dg_n(qq_1 q_2 \cdots q_{n-1}) \chi_{\mathcal{E}_k}(qq_1 q_2 \cdots q_{n-1} q_n) = \sum_{m=0}^{+\infty} P_{g_*}(q)(\mathcal{E}_m)$. It follows that S is a σ-algebra containing the rectangles, and therefore contains $\Sigma_{\prod_{i=0}^{n} Q_i}$.

It follows easily that (3) holds for step functions h, i.e., when h is of the form $\sum_{k=0}^{m} a_k \chi_{\mathcal{E}_k}$, $m \in \mathbb{N}$, $a_k \in \mathbb{R}^+$, $\mathcal{E}_k$ measurable. Since every measurable map $h \colon \prod_{i=0}^{n} Q_i \to \mathbb{R}^+$ is the pointwise supremum of a countable chain of step functions (namely $h_m = \sum_{k=1}^{m2^m} k/2^m \chi_{h^{-1}(k/2^m, +\infty)}$, $m \in \mathbb{N}$), (3) follows by the Monotone Convergence Theorem. □

Consider now any family of measurable subsets $\mathcal{E}_n$ of $\prod_{i=0}^{n-1} Q_i \times Q_n$, $n \geq 1$, and assume that for all $q_0, q_1, \ldots, q_{n-1}$, $g_n(q_0 q_1 \cdots q_{n-1})$ draws q_n at random so that $q_0 q_1 \cdots q_{n-1} q_n$ is in $\mathcal{E}_n$. It seems obvious that what we shall get in the end is an infinite path $q_0 q_1 \cdots q_n \cdots$ such that every finite prefix $q_0 q_1 \cdots q_n$ is in $\mathcal{E}_n$. This actually needs a bit of proof. Given a measurable subset E of a product $A \times B$, and $a \in |A|$, the *vertical cut* $E_{|a}$ is the set $\{b \in |B| \mid (a, b) \in E\}$. This is measurable as soon as E is [2, Theorem 18.1 (i)].

Lemma 2. *Under the assumptions of Theorem 2, let $\mathcal{E}_n$ be measurable subsets of $\prod_{i=0}^{n-1} Q_i \times Q_n$, $n \geq 1$, and assume that for all $q_0 q_1 \cdots q_{n-1} \in \prod_{i=0}^{n-1} Q_i$, $g_n(q_0 q_1 \cdots q_{n-1})$ is concentrated on $(\mathcal{E}_n)_{|q_0 q_1 \cdots q_{n-1}}$.*

For every $q \in Q_0$, $P_{g_*}(q)$ is concentrated on the set $Path_{\mathcal{E}}$ of infinite paths whose finites prefixes $q_0 q_1 \cdots q_n$ are in $\mathcal{E}_n$ for every $n \geq 1$. If additionally $\{q\}$ is measurable in Q_0, then $P_{g_*}(q)$ is concentrated on the set $Path_{\mathcal{E}}(q)$ of those infinite paths in $Path_{\mathcal{E}}$ such that $q_0 = q$.

Proof. First, $Path_{\mathcal{E}}$ is measurable, as a countable intersection of measurable subsets $\mathcal{E}_n \times \prod_{i=n+1}^{+\infty} |Q_i|$. Since $g_n(q_0 q_1 \cdots q_{n-1})$ is concentrated on $(\mathcal{E}_n)_{|q_0 q_1 \cdots q_{n-1}}$, and the complement of the latter in $|Q_n|$ is $(\overline{\mathcal{E}_n})_{|q_0 q_1 \cdots q_{n-1}}$, $\int_{q_n \in Q_n} \chi_{(\overline{\mathcal{E}_n})_{|q_0 q_1 \cdots q_{n-1}}}$ $dg_n(q_0 q_1 \cdots q_{n-1}) = g_n(q_0 q_1 \cdots q_{n-1})((\overline{\mathcal{E}_n})_{|q_0 q_1 \cdots q_{n-1}}) = 0$. By taking $h = \chi_{\overline{\mathcal{E}_n}}$ in

(3), and realizing that $\chi_{\overline{\mathcal{E}_n}}(q_0 q_1 \cdots q_{n-1} q_n) = \chi_{(\overline{\mathcal{E}_n})_{|q_0 q_1 \cdots q_{n-1}}}(q_n)$, we obtain that $\int_{q_0 q_1 \cdots \in \prod_{i=0}^{+\infty} Q_i} \chi_{\overline{\mathcal{E}_n}}(q_0 q_1 \cdots q_n) dP_{g_*}(q) = 0$. In other words, the $P_{g_*}(q)$-measure of the complement of $\mathcal{E}_n \times \prod_{i=n+1}^{+\infty} |Q_i|$ is zero. As a consequence of σ-additivity, the union of these complements when n ranges over $\mathbb{N}$ has measure that is bounded by the sum of their measures, namely 0. So the complement of $Path_{\mathcal{E}}$ has $P_{g_*}(q)$-measure 0.

For the second claim, if $\{q\}$ is measurable, then $P_{g_*}(q)((|Q_0| \smallsetminus \{q\}) \times \prod_{i=1}^{+\infty} |Q_i|)$ is equal to 0 by (2). The measure of the complement of $Path_{\mathcal{E}}(q) = Path_{\mathcal{E}} \cap (\{q\} \times \prod_{i=1}^{+\infty} |Q_i|)$ therefore also has $P_{g_*}(q)$-measure 0. $\qquad\square$

3   Drawing Measurable Selections at Random

Before we go to the more complicated case of schedulers, we illustrate our basic technique on random choice of *measurable selections* of a multifunction. We believe this has independent interest.

A *multifunction* from a set A to a set B is a map F from A to $\mathbb{P}^*(B)$, the non-empty powerset of B. We say that F is *measurable* if and only if its graph $\text{Gr } F = \{(x, y) \mid y \in F(x)\}$ is a measurable subset of $X \times Y$. This is one of the many possible notions of measurability for relations, see [12]. The set $F(x)$ is exactly the vertical cut $(\text{Gr } F)_{|x}$, showing that for a measurable multifunction, $F(x)$ is a (non-empty) measurable subset of B (see comment before Lemma 2).

A *selection* for a multifunction F is a map $f \colon A \to B$ such that $f(x) \in F(x)$ for every $x \in A$. Every multifunction has a selection: this is the Axiom of Choice. In measure theory, we would like f to be measurable as well. Theorems guaranteeing the existence of measurable selections for certain multifunctions are called *measurable selection theorems*. There are many of them (see Wagner [18], or Doberkat [8]), but one should remember that measurable multifunctions do not have measurable selections in general: Blackwell showed that there is a multifunction from $[0, 1]$ to Baire space $\mathbb{N}^{\mathbb{N}}$ whose graph is closed (hence measurable) but has no measurable selection [3] (see also Example 5.1.7 of [17]).

Given two measurable spaces X and Y, let us write $\langle X \to Y \rangle$ for the space of all measurable maps from X to Y, with the *weak σ-algebra*. The latter is by definition the subspace σ-algebra, induced by the inclusion of $|\langle X \to Y \rangle|$ into the product space $Y^{|X|}$. In other words, this is the smallest σ-algebra that makes the maps $\varphi \in \langle X \to Y \rangle \mapsto \varphi(x)$ measurable, for every $x \in |X|$.

More generally, given a multifunction $F \colon |X| \to \mathbb{P}^*(|Y|)$, we also consider the subspace $Sel(F)$ of $\langle X \to Y \rangle$ of all measurable selections of F, with the induced σ-algebra. (Beware that $Sel(F)$ need not be a measurable subset of $\langle X \to Y \rangle$.) We again call the latter the *weak σ-algebra*, on this subset. In each case, the weak σ-algebra is generated by subsets that we write $[x \to E]$, with $x \in |X|$ and $E \in \Sigma_Y$, and defined as those measurable functions, resp. those measurable selections of F, that map x into E.

Assume now a measurable map $g \colon X \to \mathcal{P}(Y)$ such that, for every $x \in |X|$, $g(x)$ is concentrated on $F(x)$. For each $x \in |X|$, pick an element $f(x)$ in $F(x)$ with probability $g(x)$. The function f is a selection of F, but will not be measurable in general. Can we pick f at random so that f *is measurable* and $f(x)$ is drawn with probability $g(x)$? This is the question we answer in the affirmative here.

The problem looks similar to the construction of Wiener measure, a model of Brownian motion, where we would like to draw a map from $\mathbb{R}$ to some topological space at random, and this map should be continuous [2, Section 37]; or to the construction of Skorokhod's J1 topology, which allows one to make sense of random càdlàg functions. Our solution will be simpler, though: measurability is easier to enforce than continuity or being càdlàg.

One can explain the problem in terms of *independence* [2, Section 5]. Let us remind the reader that independence is not *pairwise* independence. Consider for example two independent random bits b_1 and b_2, and the random variable $b_3 = b_1 \oplus b_2$, where $\oplus$ is exclusive-or. These random variables are pairwise independent, meaning that any pair among them is formed of independent random variables. However, they are *not* independent, since given the value of any two, one obtains the third one in a deterministic way. In our case, if we are to draw a *measurable* map f at random, then the random infinite tuple $(f(x))_{x \in |X|}$ cannot be a collection of independent random variables. However, the results below essentially say that we can choose f *measurable* at random, in such a way that all *countable* sequences $(f(x_n))_{n \in \mathbb{N}}$ are independent.

A general way to draw several values at random, independently, is to draw them with respect to a product measure. The following says that product measures exists not only for finite products but also for countable products of *probability* measures. This is well-known, and can even be extended to uncountable products: this is the Łomnick-Ulam Theorem [13, Corollary 5.18].

Lemma 3. *Let μ_n be probability measures on the measurable spaces X_n, $n \in \mathbb{N}$. There is a unique probability measure μ, written $\bigotimes_{n \in \mathbb{N}} \mu_n$, on $\prod_{n \in \mathbb{N}} X_n$ such that $\mu(\bigcap_{i \in I} \pi_i^{-1}(E_i)) = \prod_{i \in I} \mu_i(E_i)$ for every finite subset I of $\mathbb{N}$, and all measurable subsets E_i of X_i, $i \in I$.*

Proof. Apply Ionescu-Tulcea's Theorem 2 to $Q_0 = \{*\}$, $Q_{n+1} = X_n$, let g_n be the constant map $g_n(q_0 q_1 \cdots q_{n-1}) = \mu_n$, and note that for every finite set I, $\bigcap_{i \in I} \pi_i^{-1}(E_i)$ is just the product $\prod_{i=0}^{n} E_i \times \prod_{i=n+1}^{+\infty} |Q_i|$, for some n large enough, and where we extend the notation E_i to denote $|Q_i|$ for $i \notin I$.   □

We shall use the following general technique to construct measurable maps.

Lemma 4 (Patching). *Let X, Y be measurable spaces, $(E_i)_{i \in I}$ be a countable partition of $|X|$ in measurable subsets ($I \subseteq \mathbb{N}$), and $(f_i)_{i \in I}$ be a matching family of measurable maps from E_i to Y. The patch $f \colon X \to Y$, defined as mapping every $x \in E_i$ to $f_i(x)$, is a measurable map.*

Proof. Categorically, this follows from the fact that X is the coproduct $\sum_{i \in I} E_i$. Alternatively, $f^{-1}(E) = \bigcup_{i \in I}(f_i^{-1}(E) \cap E_i)$ is measurable as soon as E is.   □

As an application, we show that a measurable multifunction F that has a measurable selection must have *plenty* of measurable selections. Precisely, we can fix their values, arbitrarily, at countably many arguments:

Lemma 5. *Let X, Y be measurable spaces, $F\colon X \to \mathbb{P}^*(Y)$ be a measurable multifunction with a measurable selection f, $(x_i)_{i \in I}$ be countably many points in X ($I \subseteq \mathbb{N}$), and y_i be an element of $F(x_i)$ for every $i \in I$. Write $\vec{y}$ for $(y_i)_{i \in I}$.*

There is a measurable selection $f_{\vec{y}}$ of F such that $f_{\vec{y}}(x_i) = y_i$ for every $i \in I$. Moreover, we can choose $f_{\vec{y}}$ in such a way that the mapping $\vec{y} \mapsto f_{\vec{y}}$ is itself a measurable map from $\prod_{i \in I} F(x_i)$ to $Sel(F)$.

Proof. Without loss of generality, assume that I is $\mathbb{N}$, or an initial segment $\{0, 1, \ldots, n - 1\}$ of $\mathbb{N}$. Similarly to vertical cuts, we may define horizontal cuts of $\mathrm{Gr}(F)$ at y, namely $\{x \in X \mid y \in F(x)\}$, and they are measurable as well. Write $F^{-1}(y)$ for such a vertical cut. Define E_i, for each $i \in I$, as $F^{-1}(y_i) \setminus \bigcup_{j=0}^{i-1} F^{-1}(y_j)$. Together with $E_\infty = |X| \setminus \bigcup_{i \in I} E_i$, they form a partition of $|X|$ in measurable subsets. Define f_i as the constant map on E_i equal to y_i for $i \in I$, and f_∞ as the restriction of f to E_∞, then form their patch $f_{\vec{y}}$, using Lemma 4. It is plain to see that $f_{\vec{y}}$ is a selection of F, and $f_{\vec{y}}$ is measurable.

To show that $\vec{y} \mapsto f_{\vec{y}}$ is itself measurable, we must show that the set A of tuples $\vec{y}$ such that $f_{\vec{y}} \in [x \to E]$ is measurable, for $x \in |X|$ and $E \in \Sigma_Y$. For convenience, write $E_i(\vec{y})$ for the set we called E_i above, and similarly with E_∞. Let E_i' be the set of tuples $\vec{y} \in \prod_{i \in I} F(x_i)$ such that $y_i \in F(x)$ and $y_j \notin F(x)$ for every j, $0 \le j < i$. E_i' is measurable since $F(x)$ is measurable: E_i' is just the rectangle $\prod_{j=0}^{i-1}(F(x_j) \setminus F(x)) \times (F(x_i) \cap F(x)) \times \prod_{j \in I, j > i} F(x_j)$. Also, $\vec{y} \in E_i'$ if and only if $x \in E_i(\vec{y})$. Write π_i for ith projection. Since $f_{\vec{y}}(x) \in E$ if and only if there is an $i \in I$ such that $x \in E_i(\vec{y})$ and $y_i \in E$, or for every $i \in I$, $x \notin E_i(\vec{y})$ and $f(x) \in E$, it follows that $A = (\bigcup_{i \in I} E_i' \cap \pi_i^{-1}(E))$ if $f(x) \notin E$, and $A = (\bigcup_{i \in I} E_i' \cap \pi_i^{-1}(E)) \cup (\bigcap_{i \in I} \overline{E_i'})$ otherwise. In any case, A is measurable. $\square$

Theorem 3 is the keystone of our construction, and allows us to provide foundations to the notion of a random measurable selection.

Theorem 3. *Let X, Y be two measurable spaces, and $F\colon X \to \mathbb{P}^*(Y)$ be a measurable multifunction with a measurable selection. Let also $g\colon X \to \mathcal{P}(Y)$ be a measurable map such that, for every $x \in X$, $g(x)$ is concentrated on $F(x)$.*

There is a unique probability measure ϖ_g on the space $Sel(F)$ of measurable selections of F such that $\varpi_g(\bigcap_{i=1}^n [x_i \to E_i]) = \prod_{i=1}^n g(x_i)(E_i)$ for every finite collection of pairwise distinct points $(x_i)_{1 \le i \le n}$ of X and of measurable subsets $(E_i)_{1 \le i \le n}$ of Y.

Before we prove it, we note the following consequence. Proposition 1 can also be seen as a partial implementation of the Chatterjee-Doyen-Gimbert-Henzinger idea of the introduction in the Markovian case: given a (Markovian) randomized scheduler g, draw (Markovian) pure schedulers f at random so that $f(x)$ is drawn with probability $g(x)$.

Proposition 1. *Under the assumptions of Theorem 3, let $h\colon Y \to \mathbb{R}^+$ be a measurable map, and x be a point of X, then:*

$$\int_{f \in Sel(F)} h(f(x)) d\varpi_g = \int_{y \in Y} h(y) dg(x).$$

Proof. Let α_x be the measurable map $f \in Sel(F) \mapsto f(x)$. By the change of variables formula (1), $\int_{f \in Sel(F)} h(f(x)) d\varpi_g = \int_{y \in Y} h(y) d\alpha_x[\varpi_g]$. Now note that $\alpha_x[\varpi_g](E) = \varpi_g(\alpha_x^{-1}(E)) = \varpi_g([x \to E]) = g(x)(E)$, so $\alpha_x[\varpi_g] = g(x)$.   □

Proof (of Theorem 3). We use Carathéodory's measure extension Theorem 1. Let $\mathcal{A}$ be the semiring of subsets of the form $\bigcap_{i=1}^{n}[x_i \to E_i]$ given in the statement of the Lemma. To check that this is a semiring, consider any two sets A and B of this form. We must show that $A \setminus B$ is a finite union of elements of $\mathcal{A}$. We may write A as $\bigcap_{i=1}^{n}[x_i \to E_i]$, B as $\bigcap_{j=1}^{m}[x'_j \to E'_j]$. Then $A \setminus B$ is the finite union of the sets A_j, $1 \leq j \leq m$, defined by: if $x'_j = x_k$ for some (unique) k, $1 \leq k \leq n$, then $A_j = \bigcap_{\substack{i=1 \\ i \neq j}}^{n}[x_i \to E_i] \cap [x_k \to E_i \setminus E'_j]$, else $A_j = \bigcap_{i=1}^{n}[x_i \to E_i] \cap [x'_j \to \Omega \setminus E'_j]$.

Note that $\mathcal{A}$ generates the weak σ-algebra on $Sel(F)$, by definition. Assume there is a map $\mu \colon \mathcal{A} \to [0, +\infty]$ that satisfies the formula given in the statement of the Lemma: $\mu(\bigcap_{i=1}^{n}[x_i \to E_i]) = \prod_{i=1}^{n} g(x_i)(E_i)$. For now, this is an assumption, not a definition. For it to be a definition, we would need to check that this is unambiguous: if $\bigcap_{i=1}^{n}[x_i \to E_i] = \bigcap_{j=1}^{m}[x'_j \to E'_j]$, we should verify that $\prod_{i=1}^{n} g(x_i)(E_i) = \prod_{j=1}^{m}[x'_j \to E'_j]$. This will be easier to prove later. Until then, we concentrate on the more interesting question of σ-additivity.

Let $A_k = \bigcap_{i \in I_k}[x_{ki} \to E_{ki}]$, $k \in \mathbb{N}$, be a countable family of disjoint elements of $\mathcal{A}$ whose union is some element A of $\mathcal{A}$ again, where each index set I_k is finite. We must show that $\mu(A) = \sum_{k=0}^{+\infty} \mu(A_k)$.

There is a simple trick to prove this: we exhibit a measure ($s_J[\mu_J]$ below) that coincides with μ on A and each A_k, $k \in \mathbb{N}$. We shall call this the *Łomnick-Ulam trick*, since this also subtends the classical proof of the Łomnick-Ulam theorem.

Observe that the set of points $(x_{ki})_{k \in \mathbb{N}, i \in I_k}$ is countable. For each countable set J of points of X, let $s_J \colon \prod_{x \in J} F(x) \to Sel(F)$ be the map $\vec{y} \mapsto f_{\vec{y}}$ given in Lemma 5. Let μ_J be the product probability measure $\bigotimes_{x \in J} g(x)$ on $\prod_{x \in J} F(x)$, as given in Lemma 3.

By definition, $s_J[\mu_J](\mathcal{E}) = \mu_J(s_J^{-1}(\mathcal{E}))$ for every measurable subset $\mathcal{E}$ of $Sel(F)$. In particular, if J contains all the points $x_1, \ldots, x_n$, then $s_J[\mu_J](\bigcap_{i=1}^{n}[x_i \to E_i]) = \mu_J\{\vec{y} \in \prod_{x \in J} F(x) \mid \forall i, 1 \leq i \leq n \cdot y_{x_i} \in E_i\}$ (since $s_J(\vec{y})(x_i) = f_{\vec{y}}(x_i) = y_{x_i}$, where we agree to write the tuple $\vec{y}$ with indices in J, namely, as $(y_x)_{x \in J}$) $= \mu_J(\bigcap_{i=1}^{n} \pi_{x_i}^{-1}(E_i))$. By definition of μ_J, this is equal to $\prod_{i=1}^{n} g(x_i)(E_i)$, hence to $\mu(\bigcap_{i=1}^{n}[x_i \to E_i])$.

In other words, $s_J[\mu_J]$ coincides with μ on all subsets of $\mathcal{A}$ of the form $\bigcap_{i=1}^{n}[x_i \to E_i]$ where every x_i is in J. It is certainly not the case in general that $s_J[\mu_J]$ and μ coincide! The condition that every x_i is in J is crucial. This condition is satisfied by A and every A_k, $k \in \mathbb{N}$, provided we take $J = (x_{ki})_{k \in \mathbb{N}, i \in I_k}$. Since $s_J[\mu_J]$ is σ-additive, the equation $\mu(A) = \sum_{k=0}^{+\infty} \mu(A_k)$ holds.

This construction also shows that μ indeed exists, something we had deferred the verification of. The problem was to show that if $\bigcap_{i=1}^{n}[x_i \to E_i] = \bigcap_{j=1}^{m}[x'_j \to E'_j]$ then $\prod_{i=1}^{n} g(x_i)(E_i) = \prod_{j=1}^{m}[x'_j \to E'_j]$. Take $J = \{x_1, \cdots, x_n\} \cup \{x'_1, \cdots, x'_m\}$. The inverse image by s_J of the set $\bigcap_{i=1}^{n}[x_i \to E_i] = \bigcap_{j=1}^{m}[x'_j \to E'_j]$ is equal to $\bigcap_{i=1}^{n} \pi_{x_i}^{-1}(E_i) = \bigcap_{j=1}^{m} \pi_{x'_j}^{-1}(E'_j)$, and its μ_J-measure is $\prod_{i=1}^{n} g(x_i)(E_i) = \prod_{j=1}^{m}[x'_j \to E'_j]$, by definition of the product probability measure.

The existence of ϖ_g follows directly from Carathéodory's theorem. It is clear that it is a probability measure. Uniqueness follows from the fact that probability measures are uniquely determined by their values on any π-system that generates the σ-algebra [2, Theorem 3.3]. A π-*system* is a collection of sets that is closed under binary intersections, and certainly $\mathcal{A}$ qualifies, as a semiring. $\qquad\square$

4   IT Automata

Consider the following simple form of non-deterministic automaton, which we call an *IT automaton* (for Ionescu-Tulcea automaton): a tuple (Z, Λ, F) where Z, Λ are measurable spaces, and $F\colon Z\Lambda^* \to \mathbb{P}^*(\Lambda)$ is a measurable multifunction with a measurable selection. Z can be thought as a space of input values, Λ as a space of states, and F as a generalized transition relation, which given a finite history $z\lambda_1\cdots\lambda_n$ produces a non-empty set $F(z\lambda_1\cdots\lambda_n)$ of possible next states. The idea is that the system starts with some input value z, goes to the first state $\lambda_1 \in F(z)$, then to a second state $\lambda_2 \in F(z\lambda_1)$, ..., to an nth state $\lambda_n \in F(z\lambda_1\cdots\lambda_{n-1})$, and so on. In other words, IT automata are just like non-deterministic automata, except on possibly infinite state spaces and with a history-dependent transition relation.

We use the notation Λ^* for the space of all finite words on the alphabet Λ, which we equate with the countable coproduct $\sum_{n\in\mathbb{N}} \Lambda^n$. We also write $Z\Lambda^*$ instead of $Z \times \Lambda^*$, and will more generally drop the $\times$ symbol in cases where this is not ambiguous. Accordingly, we write $z\lambda_1\cdots\lambda_n$ in word notation, instead of as the tuple $(z, \lambda_1, \cdots, \lambda_n)$. We have already done so before.

Since an IT automaton starting from input value z will produce infinitely many states λ_1, ..., λ_n, it is natural to study the space $Z\Lambda^{\mathbb{N}}$ of *infinite paths* of the automaton, where $\Lambda^{\mathbb{N}}$ is the product of countably infinitely many copies of Λ. (This is written Λ^ω in language theory.) The σ-algebra on $Z\Lambda^{\mathbb{N}}$ is generated by so-called *cylinders*, which are exactly the products $X_0\Lambda_1\Lambda_2\ldots\Lambda_n\Lambda^{\mathbb{N}}$ with $X_0 \in \Sigma_Z$ and $\Lambda_i \in \Sigma_\Lambda$, $1 \le i \le n$, and $n \in \mathbb{N}$.

A *randomized scheduler* for such an IT automaton is a measurable map $g\colon Z\Lambda^* \to \mathcal{P}(\Lambda)$ such that $g(z\lambda_1\cdots\lambda_n)$ is concentrated on $F(z\lambda_1\cdots\lambda_n)$ for every $z\lambda_1\cdots\lambda_n \in Z\Lambda^*$.

Given a measurable subset $\mathcal{E}$ of $Z\Lambda^{\mathbb{N}}$, and an input value $z \in |Z|$, the probability that the induced infinite path lies in $\mathcal{E}$, where λ_n is chosen at random with probability $g(z\lambda_1\cdots\lambda_{n-1})$ at each step, is given by Theorem 2, with $Q_0 = Z\Lambda^*$, $Q_i = \Lambda$ for every $i \ge 1$, and letting g_n be the restriction of g to $Z\Lambda^{n-1}$. Explicitly:

Proposition 2. *Let Z, Λ be two measurable spaces, and $g\colon Z\Lambda^* \to \mathcal{P}(\Lambda)$ be a measurable map. There is a unique map $P_g\colon Z\Lambda^* \to \mathcal{P}(Z\Lambda^{\mathbb{N}})$ such that $P_g(w)(X_0 \Lambda_1\Lambda_2\cdots\Lambda_n\cdots\Lambda_{n+m}\Lambda^{\mathbb{N}})$ is equal to:*

$$\chi_{X_0}(z)\chi_{\Lambda_1}(\lambda_1)\cdots\chi_{\Lambda_n}(\lambda_n)\int_{\lambda_{n+1}\in\Lambda_{n+1}} dg(w)\int_{\lambda_{n+2}\in\Lambda_{n+2}} dg(w\lambda_{n+1})\cdots\int_{\lambda_{n+m}\in\Lambda_{n+m}} dg(w\lambda_{n+1}\cdots\lambda_{n+m-1})$$

for all measurable subsets X_0 of Z, Λ_i of Λ ($1 \le i \le m$, $n \le m$), and elements $w = z\lambda_1\cdots\lambda_n \in Z\Lambda^*$. Moreover, P_g is measurable.

A *pure scheduler* for the IT automaton (Z, Λ, F) is just a measurable selection of F: given the history $z\lambda_1 \cdots \lambda_{n-1}$, pick a next state λ_n from $F(z\lambda_1 \cdots \lambda_{n-1})$.

Given a pure scheduler f, define $\tilde{f} \colon Z\Lambda^* \to Z\Lambda^{\mathbb{N}}$ so that $\tilde{f}(z\lambda_1 \cdots \lambda_n)$ is the unique infinite path that we obtain by starting with the history $z\lambda_1 \cdots \lambda_n$ and repeatedly computing next states, using f: $\tilde{f}(z\lambda_1 \cdots \lambda_n) = z\lambda_1 \cdots \lambda_n \lambda_{n+1} \cdots \lambda_{n+m} \cdots$ where $\lambda_{n+k+1} = f(z\lambda_1 \cdots \lambda_n \lambda_{n+1} \cdots \lambda_{n+k})$, for every $k \in \mathbb{N}$.

One might think of doing the following. Fix an IT automaton (Z, Λ, F), and a randomized scheduler g for this automaton. Pick a pure scheduler f at random, with respect to the probability measure ϖ_g given by Theorem 3, and show that the probability that $\tilde{f}(z)$ falls into any given measurable set $\mathcal{E}$ of infinite paths is equal to the probability $P_g(z)(\mathcal{E})$ given in Proposition 2. We did the computation, and checked that this indeed holds... except this is all wrong! This does not make sense unless the map $f \mapsto \tilde{f}(z)$ (for $z \in |Z|$ fixed) is measurable. We cannot dismiss the problem: this is *the* central question here.

To state it another way, the weak σ-algebra on the space of pure schedulers has to be replaced by a larger one: the $\widetilde{\text{weak } \sigma\text{-algebra}}$ on the set $|Sel(F)|$ of pure schedulers is the smallest that makes every map $f \in |Sel(F)| \mapsto \tilde{f}(z\lambda_1 \cdots \lambda_n)$ measurable, for every $z\lambda_1 \cdots \lambda_n \in Z\Lambda^*$. We write $\widetilde{Sel}(F)$ for $|Sel(F)|$ with the $\widetilde{\text{weak } \sigma\text{-algebra}}$.

For $w = z\lambda_1 \cdots \lambda_n \in Z\Lambda^*$, and a measurable subset $\mathcal{E}$ of $\Lambda^{\mathbb{N}}$, let us write $w\mathcal{E}$ for $\{z\} \times \{\lambda_1\} \times \cdots \times \{\lambda_n\} \times \mathcal{E}$. The $\widetilde{\text{weak } \sigma\text{-algebra}}$ is generated by the sets $[w \tilde{\to} w\mathcal{E}]$, defined as the set of pure schedulers f such that $\tilde{f}(w) \in w\mathcal{E}$.

We wish to define our probability measure $\tilde{\omega}_g$ on pure schedulers f by saying that the probability that $\tilde{f}(w) \in w\mathcal{E}$ (for any fixed w, $\mathcal{E}$) is exactly $P_g(w)(\mathcal{E})$, where P_g is given in Proposition 2: namely, $\tilde{\omega}_g([w \tilde{\to} w\mathcal{E}]) = P_g(w)(\mathcal{E})$. That cannot be enough to define $\tilde{\omega}_g$, and we need to at least define $\tilde{\omega}_g(\bigcap_{i=1}^n [w_i \tilde{\to} w_i\mathcal{E}_i])$ for all finite intersections of sets $[w_i \tilde{\to} w_i\mathcal{E}_i]$. Now there is a big difference with the case of random measurable selections (Section 3): the choices we make for $\tilde{f}(w_i)$ for different indices i *cannot in general be independent*. Indeed, imagine we have chosen $\tilde{f}(w_i)$, for some i, to be $w_i\lambda_{n+1} \cdots \lambda_{n+m} \cdots$: then we have no choice for $\tilde{f}(w_i\lambda_{n+1})$, and also for $\tilde{f}(w_i\lambda_{n+1}\lambda_{n+2})$, ..., which must all be equal to $w_i\lambda_{n+1} \cdots \lambda_{n+m} \cdots$. This is the *consistency* condition mentioned in the proof of Proposition 3 below: in general, if we have chosen $\tilde{f}(w_i)$ as $w_i\omega_i$ for some $\omega_i \in \Lambda^{\mathbb{N}}$, and later we need to choose $\tilde{f}(w_j)$ where w_i is a prefix of w_j, and w_j is a prefix of $w_i\omega_i$, then we must choose it as $\tilde{f}(w_j) = w_i\omega_i$.

We still proceed in a manner similar to Section 3. We now need an extra assumption: say that measurable space X *has measurable diagonal* if and only if the diagonal $\Delta = \text{Gr}(=) = \{(x, x) \mid x \in |X|\}$ is measurable in $X \times X$. Dravecký [9, Theorem 1] shows a number of equivalent conditions. One of these is that X has measurable diagonal if and only if there is a countable family $(E_n)_{n \in \mathbb{N}}$ of measurable subsets of X that separates points, that is, such that for any two distinct points x, y, there is an E_n that contains one and not the other. This is true for all Polish spaces, notably. Another one is that X has measurable diagonal if and only if for every measurable function f from an arbitrary measurable space

Y to X, the obvious multifunction $y \in Y \mapsto \{f(y)\}$ is measurable. This apparent tautology is wrong when X does not have measurable diagonal! The canonical counter-example is $Y = X$, $f = \mathrm{id}_X$: the corresponding multifunction is precisely the one whose graph is Δ. Finally, every one-element subset $\{x\}$ of a space X with measurable diagonal is measurable in X; indeed, $\{x\}$ is the vertical cut $\Delta_{|x}$.

Proposition 3. *Let (Z, Λ, F) be an IT automaton, with a randomized scheduler $g \colon Z\Lambda^* \to \mathcal{P}(\Lambda)$. Assume that Z and Λ both have measurable diagonals. Let $P_g \colon Z\Lambda^* \to \mathcal{P}(Z\Lambda^{\mathbb{N}})$ be the probability-on-paths map given in Proposition 2. There is a probability measure $\widetilde{\omega}_g$ on $\widetilde{Sel}(F)$ such that $\widetilde{\omega}_g([z \overset{\to}{\to} \mathcal{E}]) = P_g(z)(\mathcal{E})$ for all $z \in Z$ and $\mathcal{E} \in \Sigma_{Z\Lambda^{\mathbb{N}}}$.*

Proof. (Outline. The technical details are relegated to Appendix A.) As a notational help, we write w, possibly subscripted or primed, for finite words in $Z\Lambda^*$, and ω, possibly subscripted or primed, for infinite words in $\Lambda^{\mathbb{N}}$.

Let us write $\preceq$ for the prefix relation on finite and infinite words. Say that a set of words A of $Z\Lambda^*$ is *prefix-closed* if and only if for every $w \in Z\Lambda^*$ that is a prefix of some element of A, w is in A as well.

Let $\mathcal{A}$ be the semiring of all finite intersections of basic measurable subsets $\bigcap_{i=1}^{n}[w_i \overset{\to}{\to} w_i\mathcal{E}_i]$. By adding extra words if needed, we may assume that $\mathcal{I} = \{w_1, \ldots, w_n\}$ is prefix-closed.

It is hard to even attempt to describe explicitly the values of $\widetilde{\omega}_g$ on elements of $\mathcal{A}$. In the proof of Theorem 3, we had eventually shown that the value of the desired measure ϖ_g coincided with the image measure of some other measure defined on a product space for sufficiently small parts of the semiring. We define $\widetilde{\omega}_g$ on $\mathcal{A}(\mathcal{I})$ by a similar Łomnick-Ulam-like trick: through image measures of some measures $\mu_{\mathcal{W}}$ under maps $\alpha_{\mathcal{W}}$, for countably infinite subsets $\mathcal{W}$ of $Z\Lambda^*$.

Given a countably infinite set $\mathcal{W} = (w_j)_{j \in J}$ of words in $Z\Lambda^*$ (with $J = \mathbb{N} \setminus \{0\}$, say), let $Cst_{\mathcal{W}}$ be the subspace of $\prod_{w \in \mathcal{W}} Path_F(w)$ consisting of those tuples $(w_j\omega_j)_{j \in J}$ that are *consistent*: for all $i, j \in J$ such that $w_i \preceq w_j \preceq w_i\omega_i$, then $w_i\omega_i = w_j\omega_j$. In pictures, if the leftmost two zones are equal then the rightmost zones are equal, too:

Using Ionescu-Tulcea's Theorem 2, we build a probability measure $\mu_{\mathcal{W}}$ on $Cst_{\mathcal{W}}$ (Proposition 5 in Appendix A). Intuitively, $\mu_{\mathcal{W}}$ picks $w_1\omega_1$ at random using probability measure $P_g(w_1)$, then picks $w_2\omega_2, \ldots, w_j\omega_j, \ldots$, as follows. At step j, if $w_i \preceq w_j \preceq w_i\omega_i$ for some previous i, $1 \le i < j$, then we pick $w_j\omega_j$, deterministically, as equal to $w_i\omega_i$, enforcing consistency; otherwise, we pick $w_j\omega_j$ at random using probability measure $P_g(w_j)$. All this makes sense provided we sort $\mathcal{W}$ topologically, i.e., we choose the indexing scheme so that any w_i that is a prefix of w_j occurs before w_j, viz., $i \le j$ (Lemma 6 in Appendix A).

Given $w \in \mathcal{W}$, say $w = w_j$, let us write $[w : \mathcal{E}]$ for the set of consistent tuples $\overrightarrow{w\omega}$ in $Cst_{\mathcal{W}}$ such that $\omega_j \in \mathcal{E}$. One can show that, given any finite prefix-closed set $\{w_1, \ldots, w_n\}$ of words in $Z\Lambda^*$, the value $\mu_W(\bigcap_{i=1}^{n}[w_i : \mathcal{E}_i])$ is the same for *all* countably infinite subsets W that contain $w_1, \ldots, w_n$ (Lemma 9 in Appendix A). The formula we obtain simplifies when $n = 1$ and w_1 is a single letter $z \in Z$: $\mu_W([z : \mathcal{E}]) = P_g(z)(z\mathcal{E})$ (Lemma 10 in Appendix A).

We take the image measure of μ_W under a measurable map $\alpha_W \colon \overrightarrow{w\omega} \in Cst_W \mapsto f_{\overrightarrow{w\omega}} \in \widetilde{Sel}(F)$ that retrieves a canonical pure scheduler from a consistent set of tuples (Proposition 6 in Appendix A), defined in such a way that $\widetilde{f}_{\overrightarrow{w\omega}}(w_j) = w_j\omega_j$ for every $j \in J$. This is done by patching, similarly to Lemma 5.

Note that $\alpha_W[\mu_W](\bigcap_{i=1}^{n}[w_i \overset{\sim}{\rightarrow} w_i\mathcal{E}_i])$ is independent of W, provided that W contains the prefix-closed subset $\{w_1, \ldots, w_n\}$. Indeed, $\alpha_W[\mu_W](\bigcap_{i=1}^{n}[w_i \overset{\sim}{\rightarrow} w_i\mathcal{E}_i]) = \mu_W(\bigcap_{i=1}^{n}[w_i : \mathcal{E}_i])$, which we have shown independent of W. We can therefore define $\widetilde{\omega}_g$ as coinciding with $\alpha_W[\mu_W]$ on those elements $\bigcap_{i=1}^{n}[w_i \overset{\sim}{\rightarrow} w_i\mathcal{E}_i]$ of $\mathcal{A}$ with $\{w_1, \ldots, w_n\} \subseteq W$. As such, it is σ-additive on $\mathcal{A}$: as in the proof of Theorem 3, let $A_k = \bigcap_{w \in \mathcal{I}_k}[w \overset{\sim}{\rightarrow} \mathcal{E}_{wi}]$, $k \in \mathbb{N}$, be a countable family of disjoint elements of $\mathcal{A}$, where $\mathcal{I}_k$ is finite and prefix-closed, and $A = \bigcap_{w \in \mathcal{I}}[w \overset{\sim}{\rightarrow} \mathcal{E}_w]$ be their union, assumed in $\mathcal{A}$, with $\mathcal{I}$ prefix-closed again. Then $\widetilde{\omega}_g(A) = \sum_{k \in \mathbb{N}} \widetilde{\omega}_g(A_k)$, since $\widetilde{\omega}_g$ coincides with the measure $\alpha_W[\mu_W]$ on A and every A_k, for $W = \bigcup_{k \in \mathbb{N}} \mathcal{I}_k \cup \mathcal{I}$ (which is countable).

Finally, $\widetilde{\omega}_g([z \overset{\sim}{\rightarrow} z\mathcal{E}]) = \mu_W([z : \mathcal{E}]) = P_g(z)(z\mathcal{E})$, for every measurable subset $\mathcal{E}$ of $\Lambda^{\mathbb{N}}$. Since $P_g(z)$ is concentrated on $Path_F(z)$, hence on $z\Lambda^{\mathbb{N}}$, for every measurable subset $\mathcal{E}$ of $Z\Lambda^{\mathbb{N}}$, $\widetilde{\omega}_g([z \overset{\sim}{\rightarrow} \mathcal{E}]) = \widetilde{\omega}_g([z \overset{\sim}{\rightarrow} z\mathcal{E}_{|z}]) = P_g(z)(z\mathcal{E}_{|z}) = P_g(z)(\mathcal{E} \cap z\Lambda^{\mathbb{N}}) = P_g(z)(\mathcal{E})$. □

We can now integrate on infinite paths ω with respect to $P_g(z)$, or on pure schedulers, and this will give the same average value:

Proposition 4. *Under the assumptions of Proposition 3, let $h \colon Z\Lambda^{\mathbb{N}} \to \mathbb{R}^+$ be a measurable map, and $z \in Z$, then:*

$$\int_{f \in \widetilde{Sel}(F)} h(\widetilde{f}(z)) d\widetilde{\omega}_g = \int_{\omega \in Z\Lambda^{\mathbb{N}}} h(\omega) dP_g(z).$$

Proof. Let $\widetilde{\alpha}_z \colon f \mapsto \widetilde{f}(z)$. This is a measurable map, since $\widetilde{\alpha}_z^{-1}(\mathcal{E}) = [z \overset{\sim}{\rightarrow} \mathcal{E}]$. The left-hand side is $\int_{f \in \widetilde{Sel}(F)} h(\widetilde{\alpha}_z(f)) d\widetilde{\omega}_g = \int_{\omega \in Z\Lambda^{\mathbb{N}}} h(\omega) d\widetilde{\alpha}_z[\widetilde{\omega}_g]$ by the change of variables formula (1). Proposition 3 states precisely that $\widetilde{\alpha}_z[\widetilde{\omega}_g] = P_g(z)$. □

We may think of h as a *payoff* function on infinite paths. The above shows that the average payoff with respect to $P_g(z)$ is also the average of the individual payoffs $h(\widetilde{f}(z))$ one would get by drawing a pure scheduler f at random instead.

Fact 1 then implies that the value of the average payoff is bounded by the payoff evaluated on two pure schedulers f^- and f^+:

Corollary 1 (Randomness for Free). *Under the assumptions of Proposition 3, let $h \colon Z\Lambda^{\mathbb{N}} \to \mathbb{R}^+$ be a measurable map, and $z \in Z$. There are two pure schedulers f^- and f^+ in $Sel(F)$ such that:*

$$h(\widetilde{f^-}(z)) \leq \int_{\omega \in Z\Lambda^{\mathbb{N}}} h(\omega) dP_g(z) \leq h(\widetilde{f^+}(z)).$$

5 Conclusion

We have established a few "randomness for free" type theorems for measurable multifunctions first, for IT automata second. The results are pleasing, and our assumptions are fairly minimal. Our proofs use fairly simple ideas, too: there is the Chatterjee-Doyen-Gimbert-Henzinger idea of drawing measurable selections/pure schedulers at random first, and this makes sense because of a combination of patching, of Carathéodory's measure extension theorem, and of a Łomnik-Ulam type trick.

Where should we go next? One may push the results on IT automata to *partially observable IT automata*. Instead of a measurable transition multifunction $F\colon Z\Lambda^* \to \mathbb{P}^*(\Lambda)$, such automata have a measurable transition multifunction $F\colon Z\Lambda^* \to \mathbb{P}^*(\varXi)$, where $\varXi$ is an (additional) measurable space $\varXi$ of *actual states*. Such states q are mapped to *observable* states $\lambda \in \Lambda$ by a measurable map $\tau\colon Z\Lambda^* \times \varXi \to \Lambda$; this may depend on the past history $w \in Z\Lambda^*$, viz., $\lambda = \tau(w, q)$. We pick the next observable state λ after history w by picking q from $F(w)$, then computing $\tau(w, q)$. Modifying the notion of consistent paths as required, it seems feasible to prove an analogue of Proposition 3 for partially observable IT automata. The σ-algebra on $Sel(F)$ needs to be changed again! so that we cannot reuse Proposition 3 as is. Once this is done, we can proceed to stochastic automata [4]. Given a stochastic automaton with state space Q and action space L, take $Z = Q$, $\Lambda = (L \times Q)_\perp$ (writing $X_\perp$ for $X + \{\perp\}$), and $\varXi = \mathcal{P}(\Lambda)$. Taking g to be the second projection map π_2, Theorem 3 allows us to draw the observation maps $\tau\colon Q(L \times Q)^*_\perp \times \mathcal{P}((L \times Q)_\perp) \to (L \times Q)_\perp$ at random, and this will simulate the probabilistic choice of q' with respect to μ described in the introduction. Combining this with the alluded "randomness for free" result for partially observable IT automata, we hope that it would settle the "randomness for free" question for general, stochastic automata.

References

1. Adjé, A., Bouissou, O., Goubault-Larrecq, J., Goubault, É., Putot, S.: Analyzing numerical programs with partially known distributions. In: Cohen, E., Rybalchenko, A. (eds.) Proc. 5th IFIP TC2 WG 2.3 Conf. Verified Software—Theories, Tools, and Experiments (VSTTE 2013). LNCS, Springer (2013),
   http://www.lsv.ens-cachan.fr/Publis/PAPERS/PDF/ABGGP-vstte13.pdf
2. Billingsley, P.: Probability and Measure, 3rd edn. Wiley series in probability and mathematical statistics. John Wiley and Sons (1995)
3. Blackwell, D.H.: A Borel set not containing a graph. Annals of Mathematical Statistics 39, 1345–1347 (1968)
4. Cattani, S., Segala, R., Kwiatkowska, M., Norman, G.: Stochastic transition systems for continuous state spaces and non-determinism. In: Sassone, V. (ed.) FOSSACS 2005. LNCS, vol. 3441, pp. 125–139. Springer, Heidelberg (2005)
5. Chatterjee, K., Doyen, L., Gimbert, H., Henzinger, T.A.: Randomness for Free. In: Hliněný, P., Kučera, A. (eds.) MFCS 2010. LNCS, vol. 6281, pp. 246–257. Springer, Heidelberg (2010)

6. Desharnais, J., Edalat, A., Panangaden, P.: Bisimulation for labelled Markov processes. Information and Computation 179(2), 163–193 (2002)
7. Desharnais, J., Gupta, V., Jagadeesan, R., Panangaden, P.: Metrics for labelled Markov processes. Theoretical Computer Science 318, 323–354 (2004)
8. Doberkat, E.-E.: Stochastic Relations: Foundations for Markov Transition Systems. Studies in Informatics Series. Chapman and Hall/CRC (2007)
9. Dravecký, J.: Spaces with measurable diagonal. Matematický Časopis 25(1), 3–9 (1975)
10. Ferns, N., Panangaden, P., Precup, D.: Bisimulation metrics for continuous Markov decision processes. SIAM Journal on Computing 40(6), 1662–1714 (2011)
11. Giry, M.: A categorical approach to probability theory. In: Banaschewski, B. (ed.) Categorical Aspects of Topology and Analysis. LNM, vol. 915, pp. 68–85. Springer (1981)
12. Himmelberg, C.J.: Measurable relations. Fundamenta Mathematicae 87, 53–71 (1975)
13. Kallenberg, O.: Foundations of Modern Probability, 2nd edn. Probability and its Applications. Springer (2002)
14. Panangaden, P.: Probabilistic relations. In: Baier, C., Huth, M., Kwiatkowska, M., Ryan, M. (eds.) Proceedings of PROBMIV 1998, pp. 59–74 (1998), http://www.cs.bham.ac.uk/~mzk/probmiv/prelproc98/
15. Panangaden, P.: Probability and measure for concurrency theorists. Theoretical Computer Science 253(2), 287–309 (2001)
16. Panangaden, P.: Labelled Markov Processes. Imperial College Press (2009)
17. Srivastava, S.M.: A Course on Borel Sets. Graduate Texts in Mathematics, vol. 180. Springer (1998)
18. Wagner, D.H.: Survey of measurable selection theorems: An update. In: Measure Theory, Oberwolfach, 1979. LNM, vol. 794, pp. 176–219. Springer (1980)

A   Auxiliary Results Needed for Proposition 3

In the following, we fix an IT automaton (Z, Λ, F) with a randomized scheduler $g \colon Z\Lambda^* \to \mathcal{P}(\Lambda)$. We also assume that Z and Λ both have measurable diagonals. Let $P_g \colon Z\Lambda^* \to \mathcal{P}(Z\Lambda^{\mathbb{N}})$ be the probability-on-paths map given in Proposition 2.

In the rest of this section, we fix a countably infinite set $\mathcal{W}$ of words in $Z\Lambda^*$. We let $J = \{1, 2, \cdots\} = \mathbb{N} \smallsetminus \{0\}$. This will serve as an index set.

Lemma 6. *One can write $\mathcal{W}$ as a family $(w_j)_{j \in J}$ in such a way that for all $i, j \in J$, if $w_i \preceq w_j$ then $i \leq j$.*

In other words, one can sort the words topologically. Proving this involves showing that the order type of $\mathcal{W}$ under the prefix ordering is at most ω.

Proof. There are only countably many elements of Z, resp. Λ, that can occur in any word from $\mathcal{W}$, so one can attribute each of them a unique natural number. Equate each element with the corresponding natural number. Each word $w = z\lambda_1 \cdots \lambda_n$ can now be encoded as $p_1^z p_2^{\lambda_1} \cdots p_{n+1}^{\lambda_n}$, where $p_1, p_2, \ldots, p_n, \ldots$

enumerate the prime numbers. This way, if w_i is a proper prefix of w_j, then w_i will be encoded as a lower number than w_j. Now enumerate the words w_i in the order of their encodings. $\qquad\square$

In the rest of Section A, we assume such a topologically sorted indexing scheme $(w_j)_{j \in J}$ for $\mathcal{W}$.

Lemma 7. *For every $n \in J$, $P_g(w_n)$ is concentrated on $Path_F(w_n)$.*

Proof. Since Z and Λ have measurable diagonals, so does $Z\Lambda^*$. For each $n \in J$, $\{w_n\}$ is therefore measurable in $Z\Lambda^*$, so we can apply Lemma 2 and conclude that $P_g(w_n)$ is concentrated on $Path_F(w_n)$. $\qquad\square$

In particular, the restriction $P_g(w_n)_{|Path_F(w_n)}$ makes sense. To reduce visual clutter, we simply write $P_g(w_n)$ for the latter probability measure on $Path_F(w_n)$.

We shall write $w_n \Lambda^{\mathbb{N}}$ for the set of words in $Z\Lambda^{\mathbb{N}}$ that have w_n as a prefix. As the product $\{w_n\} \times \Lambda^{\mathbb{N}}$, this is a measurable set. (Recall that one-element sets are measurable, since Z and Λ have measurable diagonals.)

We wish to draw the values $w_j \omega_j$ of $\widetilde{f}(w_j)$, $j \in J$, in a consistent way, namely, if we have already mapped w_i to the value $w_i \omega_i$, and $w_i \preceq w_j \preceq w_i \omega_i$, then we have no choice and must choose to map w_j to $w_i \omega_i$ as well. We achieve this by using Ionescu-Tulcea's Theorem 2 on *another* probabilistic transition system, defined as follows, and which we only use as a mathematical helper. This is certainly no real-life, practical transition system, and is not meant to be.

We let $Q_0 = \{*\}$, $q_0 = *$, $Q_n = Path_F(w_n)$ for $n \in J$, and $g_n \colon \prod_{i=0}^{n-1} Q_i \to \mathcal{P}(Q_n)$ map $(*, w_1\omega_1, \cdots, w_{n-1}\omega_{n-1})$ to:

- the Dirac mass $\delta_{w_i \omega_i}$, where i is the least index in $1, \cdots, n-1$ such that $w_i \preceq w_n \preceq w_i \omega_i$, if one such index exists; this implements consistency;
- otherwise, the probability measure $P_g(w_n)$.

Both are probability measures on $Q_n = Path_F(w_n)$: the second one by Lemma 7, the first one because $w_i \omega_i$ is a path in $Path_F$ starting with w_n.

Lemma 8. *For every $n \in J$, g_n is measurable.*

Proof. For every i, $1 \leq i < n$, such that $w_i \preceq w_n$, the set A_i of infinite words $w_i \omega_i$ in $Path_F(w_i)$ that have w_n as a prefix is $Path_F(w_i) \cap w_n \Lambda^{\mathbb{N}}$, hence is measurable. By extension, when $w_i \not\preceq w_n$, write A_i for the empty set. Rephrasing the definition, $g_n(*, w_1\omega_1, \cdots, w_{n-1}\omega_{n-1})$ is defined as $\mathfrak{e}_{Path_F(w_n)}(w_i\omega_i)$ on $E_i = \{*\} \times \overline{A_1} \times \cdots \times \overline{A_{i-1}} \times A_i \times |Q_{i+1}| \times \cdots \times |Q_{n-1}|$, $1 \leq i < n$ (recall that $\mathfrak{e}$ is the monad unit, which is a measurable map), and as $P_g(w_n)$ on $\bigcup_{i=1}^{n-1} E_i$ (a constant map). So g_n is a patch of measurable maps, and is therefore measurable by Lemma 4. $\qquad\square$

We can now apply Ionescu-Tulcea's Theorem, as promised, and obtain a probability measure on $\{*\} \times \prod_{j \in J} Path_F(w_j) \cong \prod_{j=1}^{+\infty} Path_F(w_j)$ that we decide to call $\mu_{\mathcal{W}}$. (This is the measure $P_{g_*}(*)$ of Theorem 2, but we wish to avoid any visual confusion with P_g.)

Let $Cst_{\mathcal{W}}$ be the set of tuples of paths $(w_j\omega_j)_{j\in J}$ in $\prod_{j\in J} Path_F(w_j)$ that are consistent, i.e., such that for all $i, j \in J$ such that $w_i \preceq w_j \preceq w_i\omega_i$, $w_j\omega_j = w_i\omega_i$. These are the only ones we can ever hope to produce from a pure scheduler f, namely, that are of the form $(\tilde{f}(w_j))_{j\in J}$.

Proposition 5. *The probability measure $\mu_{\mathcal{W}}$ is concentrated on the set $Cst_{\mathcal{W}}$ of consistent tuples.*

Proof. Let $\mathcal{F}_n$ be the multifunction from $\prod_{j=1}^{n-1} Path_F(w_j)$ to $Path_F(w_n)$ defined, similarly to g_n, by letting $\mathcal{F}_n(w_1\omega_1, \cdots, w_{n-1}\omega_{n-1})$ be:

- $\{w_i\omega_i\}$, where i is the least index in $1, \cdots, n-1$ such that $w_i \preceq w_n \preceq w_i\omega_i$, if one such index exists;
- otherwise, $Path_F(w_n)$.

One checks easily that $\mathcal{F}_n$ is a measurable multifunction, and that $g_n(w_1\omega_1, \cdots, w_{n-1}\omega_{n-1})$ is concentrated on $\mathcal{F}_n(w_1\omega_1, \cdots, w_{n-1}\omega_{n-1})$. We apply Lemma 2 to $\mathcal{E}_n = \mathrm{Gr}\,\mathcal{F}_n$ and obtain that $\mu_{\mathcal{W}}$ is supported on $Path_{\mathcal{E}}$.

It remains to show that $Path_{\mathcal{E}} = Cst_{\mathcal{W}}$. Given any inconsistent tuple $(w_j\omega_j)_{j\in J}$, there must be two indices $i, j \in J$ such that $w_i \preceq w_j \preceq w_i\omega_i$ but $w_j\omega_j \neq w_i\omega_i$. Since $\mathcal{W}$ is topologically sorted, $i \leq j$. Take i minimal. Then $w_j\omega_j$ would be outside $\{w_i\omega_i\} = \mathcal{F}_j(w_1\omega_1, \cdots, w_{j-1}\omega_{j-1}) = (\mathcal{E}_j)_{|(w_1\omega_1,\cdots,w_{j-1}\omega_{j-1})}$, showing that the tuple is not in $Path_{\mathcal{E}}$. This establishes that $Path_{\mathcal{E}} \subseteq Cst_{\mathcal{W}}$. The converse inclusion is obvious. $\square$

The restriction $\mu_{\mathcal{W}|Cst_{\mathcal{W}}}$ therefore makes sense. Again, we simply write $\mu_{\mathcal{W}}$ for this restriction, and consider it as a probability measure on $Cst_{\mathcal{W}}$.

Given $w \in \mathcal{W}$, say $w = w_j$, let us write $[w : \mathcal{E}]$ for the set of consistent tuples $\overrightarrow{w\omega}$ in $Cst_{\mathcal{W}}$ such that $\omega_j \in \mathcal{E}$.

A set A of words in $Z\Lambda^*$ is *prefix-closed* if and only if, for every $w \in Z\Lambda^*$ and $\lambda \in \Lambda$, $w\lambda \in \mathcal{I}$ implies $w \in \mathcal{I}$.

Lemma 9. *Let $\{w'_1, \ldots, w'_n\}$ be a finite prefix-closed set of words in $Z\Lambda^*$. For all measurable subsets $\mathcal{E}_1, \ldots, \mathcal{E}_n$ of $\Lambda^{\mathbb{N}}$, the value of $\mu_{\mathcal{W}}(\bigcap_{i=1}^n [w'_i : \mathcal{E}_i])$ is independent of the countably infinite set $\mathcal{W}$, provided it is a superset of $\{w'_1, \ldots, w'_n\}$.*

Proof. Consider any countably infinite superset $\mathcal{W}$ of $\{w'_1, \ldots, w'_n\}$. Write $\mathcal{W}$ as $(w_j)_{j\in J}$, as usual. Let $w'_1 = w_{j_1}, \ldots, w'_n = w_{j_n}$. Up to permutation, we may assume that $j_1 < j_2 < \cdots < j_n$. Then $\bigcap_{i=1}^n [w'_i : \mathcal{E}_i]$ is equal to the intersection of $Cst_{\mathcal{W}}$ with $\prod_{j=1}^{j_1-1} Path_F(w_j) \times w_{j_1}\mathcal{E}_1 \times \prod_{j=j_1+1}^{j_2-1} Path_F(w_j) \times w_{j_2}\mathcal{E}_2 \times \cdots \times \prod_{j=j_{n-1}+1}^{j_n-1} Path_F(w_j) \times w_{j_n}\mathcal{E}_n \times \prod_{j=j_n+1}^{+\infty} Path_F(w_j)$. We now use formula (2). This requires an abbreviation for all the integrals $\int_{w_j\omega_j\in Z\Lambda^{\mathbb{N}}} dg_j(*, w_1\omega_1, \cdots, w_{j-1}\omega_{j-1})$ with $j \notin \{j_1, j_2, \cdots, j_n\}$—which will turn to be useless: write $\iint_{k\cdots\ell}\overrightarrow{\mathbf{dg}}$ for the list of symbols

$$\int_{w_k\omega_k\in Path_F(w_k)} dg_k(*, w_1\omega_1, \cdots, w_{k-1}\omega_{k-1}) \int_{w_{k+1}\omega_{k+1}\in Path_F(w_{k+1})} dg_{k+1}(*, w_1\omega_1, \cdots, w_k\omega_k)$$

$$\cdots \int_{w_\ell\omega_\ell\in Path_F(w_\ell)} dg_\ell(*, w_1\omega_1, \cdots, w_{\ell-1}\omega_{\ell-1}).$$

We can now write:

$$\mu_{\mathcal{W}}(\bigcap_{i=1}^{n}[w_i' : \mathcal{E}_i]) = \iint_{1\cdots j_1-1} \overrightarrow{\mathbf{dg}} \int_{w_{j_1}\omega_{j_1}\in w_{j_1}\mathcal{E}_1} dg_{j_1}(*, w_1\omega_1, \cdots, w_{j_1-1}\omega_{j_1-1}) \qquad (4)$$

$$\iint_{j_1+1\cdots j_2-1} \overrightarrow{\mathbf{dg}} \int_{w_{j_2}\omega_{j_2}\in w_{j_2}\mathcal{E}_2} dg_{j_2}(*, w_1\omega_1, \cdots, w_{j_2-1}\omega_{j_2-1})\cdots$$

$$\iint_{j_{n-1}+1\cdots j_n-1} \overrightarrow{\mathbf{dg}} \int_{w_{j_n}\omega_{j_n}\in w_{j_n}\mathcal{E}_n} dg_{j_n}(*, w_1\omega_1, \cdots, w_{j_n-1}\omega_{j_n-1}).$$

Since $\{w_1,\ldots,w_n\}$ is prefix-closed, for each i, $1 \leq i \leq n$, $g_{j_i}(*, w_1\omega_1, \cdots, w_{j_i-1}\omega_{j_i-1})$ can be written as a function $g_i'(w_{j_1}\omega_{j_1}, \cdots, w_{j_{i-1}}\omega_{j_{i-1}})$ of just those words $w_k\omega_k$ with $k \in \{j_1,\ldots,j_n\}, k < j_i$. Explicitly, $g_i'(w_{j_1}\omega_{j_1}, \cdots, w_{j_{i-1}}\omega_{j_{i-1}})$ is $\delta_{w_{j_\ell}\omega_{j_\ell}}$ where ℓ is the least index, $1 \leq \ell < i$, such that $w_{j_\ell} \preceq w_{j_i} \preceq w_{j_\ell}\omega_{j_\ell}$ if one exists, and $P_g(w_{j_i})$ otherwise. In (4), the final integral $\int_{w_{j_n}\omega_{j_n}\in w_{j_n}\mathcal{E}_n} dg_{j_n}(*, w_1\omega_1, \cdots, w_{j_n-1}\omega_{j_n-1}) = \int_{w_{j_n}\omega_{j_n}\in w_{j_n}\mathcal{E}_n} dg_n'(w_{j_1}\omega_{j_1}, \cdots, w_{j_{n-1}}\omega_{j_{n-1}})$ is independent of all the formal variables $w_{j_{n-1}+1}\omega_{j_{n-1}+1}, \ldots, w_{j_n-1}\omega_{j_n-1}$ that the integrals $\int_{w_j\omega\in Path_F(w_j)} dg_j(*, w_1\omega_1, \cdots, w_{j-1}\omega_{j-1})$ hidden in $\iint_{j_{n-1}+1\cdots j_n-1} \overrightarrow{\mathbf{dg}}$ introduce. Since all these integrals are with respect to probability measures, they merely contribute a factor of 1. We repeat the process, from right to left in (4), erasing all the notations $\iint_{j_{n-1}+1\cdots j_n-1} \overrightarrow{\mathbf{dg}}$, and obtain:

$$\mu_{\mathcal{W}}(\bigcap_{i=1}^{n}[w_i' : \mathcal{E}_i]) = \int_{w_{j_1}\omega_{j_1}\in w_{j_1}\mathcal{E}_1} dg_{j_1}'() \int_{w_{j_2}\omega_{j_2}\in w_{j_2}\mathcal{E}_2} dg_2'(w_{j_1}\omega_{j_1}) \cdots \int_{w_{j_n}\omega_{j_n}\in w_{j_n}\mathcal{E}_n} dg_n'(w_{j_1}\omega_{j_1}, \cdots, w_{j_{n-1}}\omega_{j_{n-1}}).$$

$$(5)$$

It is now evident that $\mu_{\mathcal{W}}(\bigcap_{i=1}^{n}[w_i' : \mathcal{E}_i])$ is independent of $\mathcal{W}$.   □

Applying (5) to the case $n = 1$, $w_{j_1} = z \in Z$ (which is automatically prefix-closed), and noticing that $g_{j_1}'() = P_g(w_{j_1}) = P_g(z)$, we obtain:

Lemma 10. *For every $z \in Z$, for every countably infinite set $\mathcal{W}$ of words of $Z\Lambda^*$ containing z, for every measurable subset $\mathcal{E}$ of $\Lambda^{\mathbb{N}}$, $\mu_{\mathcal{W}}([z : \mathcal{E}]) = P_g(z)(z\mathcal{E})$.*

Proposition 6. *For every consistent tuple $\overrightarrow{w\omega} = (w_j\omega_j)_{j\in J}$ in $Cst_{\mathcal{W}}$, there is a pure scheduler $f_{\overrightarrow{w\omega}}$ such that $\tilde{f}_{\overrightarrow{w\omega}}(w_j) = w_j\omega_j$ for every $j \in J$. Moreover, the map $\alpha_{\mathcal{W}} : \overrightarrow{w\omega} \in Cst_{\mathcal{W}} \mapsto f_{\overrightarrow{w\omega}} \in \widetilde{Sel}(F)$ is measurable.*

Proof. If $w \in Z\Lambda^*$ is a prefix of an infinite word in $\Lambda^{\mathbb{N}}$, then this infinite word can be written in a unique way as $w\lambda\omega$ for some $\lambda \in \Lambda$ and $\omega \in \Lambda^{\mathbb{N}}$: let us call λ the *letter after* w in the infinite word.

Since (Z, Λ, F) is an IT automaton, F has a measurable selection σ. For each consistent tuple $\overrightarrow{w\omega} = (w_j\omega_j)_{j\in J}$ in $Cst_{\mathcal{W}}$, we define a pure scheduler $f_{\overrightarrow{w\omega}}$ as follows. For every $w \in Z\Lambda^*$,

– either $w_j \preceq w \preceq w_j\omega_j$ for some $j \in J$, and $f_{\overrightarrow{\mathbf{w}\omega}}(w)$ is the letter after w in $w_j\omega_j$;

– or $w_j \preceq w \preceq w_j\omega_j$ for no $j \in J$, and $f_{\overrightarrow{\mathbf{w}\omega}}(w) = \sigma(w)$.

In the first case, it does not matter which $j \in J$ is picked, because of consistency. Imagine indeed that $w_i \preceq w \preceq w_i\omega_i$ and $w_j \preceq w \preceq w_j\omega_j$ for two indices $i, j \in J$. Since w_i and w_j are two prefixes of the same word w, one of them must be a prefix of the other, say $w_i \preceq w_j$. Then $w_j \preceq w \preceq w_i\omega_i$, and consistency entails that $w_i\omega_i = w_j\omega_j$, so the letter after w is the same in both infinite words.

For short, let us say that w is *stored in* $\overrightarrow{\mathbf{w}\omega}$ if and only if $w_j \preceq w \preceq w_j\omega_j$ for some $j \in J$. In this case, it is easy to see that $\tilde{f}_{\overrightarrow{\mathbf{w}\omega}}(w) = w_j\omega_j$. This implies our first claim, namely that $\tilde{f}_{\overrightarrow{\mathbf{w}\omega}}(w_j) = w_j\omega_j$ for every $j \in J$.

When w is not stored in $\overrightarrow{\mathbf{w}\omega}$, the situation is a bit more complicated. Let $\tilde{\sigma}_0(w) = w$, $\tilde{\sigma}_{k+1}(w) = \sigma_k(w)\sigma(\sigma_k(w))$, be the ever longer sequence of finite prefixes of $\tilde{\sigma}(w)$. If no $\tilde{\sigma}_k(w)$ is stored in $\overrightarrow{\mathbf{w}\omega}$, then $\tilde{f}_{\overrightarrow{\mathbf{w}\omega}}(w) = \tilde{\sigma}(w)$. But there may be a $k \in \mathbb{N}$ such that $\tilde{\sigma}_k(w)$ is stored in $\overrightarrow{\mathbf{w}\omega}$. Taking the least such k, it must be the case that $\tilde{f}_{\overrightarrow{\mathbf{w}\omega}}(w) = w_j\omega_j$ where j is any index of J such that $w_j \preceq \tilde{\sigma}_k(w) \preceq w_j\omega_j$. These remarks being made, let us proceed.

We wish to show that the map $\alpha_{\mathcal{W}} \colon \overrightarrow{\mathbf{w}\omega} \in Cst_{\mathcal{W}} \mapsto f_{\overrightarrow{\mathbf{w}\omega}} \in \widetilde{Sel}(F)$ is measurable. For now, fix $w \in Z\Lambda^*$.

The set S_k of tuples $\overrightarrow{\mathbf{w}\omega} \in Cst$ such that $\tilde{\sigma}_k(w)$ is stored in $\overrightarrow{\mathbf{w}\omega}$, is measurable. Indeed, $\tilde{\sigma}_k(w)$ is stored in $\overrightarrow{\mathbf{w}\omega}$ if and only if there is a $j \in J$ such that $w_j \preceq \tilde{\sigma}_k(w) \preceq w_j\omega_j$, so $S_k = \bigcup_{\substack{j \in J \\ w_j \preceq \tilde{\sigma}_k(w)}} \pi_j^{-1}(\tilde{\sigma}_k(w)\Lambda^{\mathbb{N}})$, where $\pi_j \colon Cst \to Path_F(w_i)$ is projection onto the jth component. It follows that the set $S_{=k} = S_k \setminus \bigcup_{\ell=0}^{k-1} S_\ell$ of tuples $\overrightarrow{\mathbf{w}\omega} \in Cst$ such that k is the least natural number such that $\tilde{\sigma}_k(w)$ is stored in $\overrightarrow{\mathbf{w}\omega}$ is also measurable. Let S_∞ be the (measurable) complement of $\bigcup_{k=0}^{+\infty} S_k$.

Let φ_k map each $\overrightarrow{\mathbf{w}\omega} \in S_k$ to $\tilde{f}_{\overrightarrow{\mathbf{w}\omega}}(w)$, i.e., to $w_j\omega_j$ where $j \in J$ is such that $w_j \preceq \tilde{\sigma}_k(w) \preceq w_j\omega_j$. For every measurable subset $\mathcal{E}$ of $Z\Lambda^{\mathbb{N}}$, $\varphi_k^{-1}(\mathcal{E}) = \bigcup_{\substack{j \in J \\ w_j \preceq \tilde{\sigma}_k(w)}} \tilde{\sigma}_k(w)\Lambda^{\mathbb{N}}$ is measurable, so φ_k is measurable. Also, the function φ_∞ that maps each $\overrightarrow{\mathbf{w}\omega} \in S_\infty$ to $\tilde{f}_{\overrightarrow{\mathbf{w}\omega}}(w) = \tilde{\sigma}(w)$ is measurable since constant. By patching φ_k, $k \in \mathbb{N}$, and φ_∞ (Lemma 4), we obtain that the map $\alpha_w \colon \overrightarrow{\mathbf{w}\omega} \in Cst \mapsto f_{\overrightarrow{\mathbf{w}\omega}}(w)$ is measurable.

We now observe that, for all $w \in Z\Lambda^*$ and $\mathcal{E} \in \Sigma_{\Lambda^{\mathbb{N}}}$, $\alpha_{\mathcal{W}}^{-1}([w \tilde{\to} w\mathcal{E}]) = \{\overrightarrow{\mathbf{w}\omega} \in Cst_{\mathcal{W}} \mid \tilde{f}_{\overrightarrow{\mathbf{w}\omega}}(w) \in w\mathcal{E}\} = \alpha_w^{-1}(w\mathcal{E})$, which is measurable: so $\alpha_{\mathcal{W}}$ is measurable.   □

A Final Coalgebra for k-regular Sequences

Helle Hvid Hansen[1,3], Clemens Kupke[2],
Jan Rutten[1,3], and Joost Winter[3]

[1] Radboud University Nijmegen
[2] University of Strathclyde
[3] CWI Amsterdam

Abstract. We study k-regular sequences from a coalgebraic perspective. Building on the observation that the set of streams over a semiring S can be turned into a final coalgebra, we obtain characterizations of k-regular sequences in terms of finite weighted automata, finite systems of behavioral differential equations, and recognizable power series. The latter characterization is obtained via an isomorphism of final coalgebras based on the k-adic numeration system.

Dedication

It is our greatest pleasure to dedicate this article to Prakash Panangaden on the occasion of his 60th birthday. There are not many subjects in our own research that have not been influenced by his work and ideas. Before the notion of finality in semantics became prominent in the early nineties of the previous century, Prakash was already writing [14] about infinite objects requiring " ... a limit construction and a final object ... ". Another early reference that is of direct relevance for the present paper is [16], published as a report in 1985, in which streams and stream functions play a key role. For these and many other similar such inspiring examples, we are immensely grateful.

1 Introduction

Infinite sequences, or streams, are much studied in the fields of number theory, analysis, combinatorics, formal languages and many more. Streams are also one of the best known examples of a final coalgebra [18]. Of particular interest is the classification of streams in terms of certain finite automata, or alternatively, stream differential equations of a certain form. The simplest such class consists of all eventually periodic streams (over a set S). They are generated by finite automata in which each state is assigned an output in S and a unique next state. Let us call these deterministic 1-automata as they are deterministic automata on a one-letter alphabet (with output in S). In terms of stream differential equations (cf. [18]), eventually periodic streams are defined by simple systems of stream differential equations such as $x(0) = 0$, $x' = y$, $y(0) = 1$, $y' = y$.

F. van Breugel et al. (Eds.): Panangaden Festschrift, LNCS 8464, pp. 363–383, 2014.

We consider two ways of generalizing deterministic 1-automata. One is by going from deterministic to weighted transitions. In this setting we must assume that the output set S has the structure of a semiring. The class of sequences generated by finite weighted 1-automata is known to be the class of rational power series on a 1-letter alphabet, or in the case that S is a field, the class of rational streams (cf. [18]). In terms of stream differential equations, rational streams are defined by equations such as $x(0) = 0$, $x' = x + y$, $y(0) = 1$, $y' = 2x$.

The other generalization is by going from a one-letter alphabet to a k-letter alphabet, for $k \in \mathbb{N}$. Here, finite deterministic k-automata generate exactly the k-automatic sequences [2]. It was shown in [9,12] that k-automatic sequences are defined by systems of equations involving the stream operation $\mathbf{zip}_k$, such as (for $k = 2$), $x(0) = 1$, $x = \mathbf{zip}_2(x,y)$, $y(0) = 2$, $y = \mathbf{zip}_2(y,y)$. (Note that the left-hand sides are x and y, and not x' and y'). These equations can be expressed using the **even** and **odd** stream operations, such as $x(0) = 1$, $\mathbf{even}(x) = y, \mathbf{odd}(x) = y$, $y(0) = 2$, $\mathbf{even}(y) = y, \mathbf{odd}(y) = x$. This approach generalizes easily to arbitrary $k \geq 2$.

In this paper we will show that (generalizing in both directions) finite weighted k-automata generate exactly the k-regular sequences. On the side of equational specification, we show that k-regular streams are defined by systems of linear $\mathbf{zip}_k$-behavioral differential equations, which are equations such as, e.g., $x(0) = 0$, $x' = \mathbf{zip}(x+y, 2y)$, $y(0) = 1$, $y' = \mathbf{zip}(2x, x+y)$. One way of summarising our insights in a slogan would be: k-regular sequences are to k-automatic sequences what rational streams are to eventually periodic ones. Our main characterization results are stated in Theorem 14.

Our approach is coalgebraic, although we use the more familiar terminology of automata. A seemingly small, technical difference with existing work is our use of the bijective k-adic numeration system as opposed to the non-bijective standard base k numeration. The advantage of using the bijective numeration system is that the automaton structure on streams obtained via the k-adic numeration yields immediately a final k-automaton, rather than a relatively final one for zero-consistent automata, as in [12]. Consequently, we obtain an isomorphism between the final k-automaton of sequences and the (classic) final k-automaton of formal power series, and this isomorphism restricts to one between k-regular sequences and rational formal power series. We also obtain a characterization of k-automatic sequences as those k-regular sequences that have finite output range. Another generalization with respect to [1] is the assumption that S is just a semiring, not a ring.

Finally, we provide a connection between our coalgebraic presentation of the k-regular sequences, and sequences attainable by so-called *divide-and-conquer recurrences* (see e.g. [10], [20]). We also note that linear $\mathbf{zip}_k$-behavioral differential equations give an easy way of specifying these sequences coinductively in the functional programming language Haskell.

Related work. The k-regular sequences were introduced in [1] as generalizations of k-automatic sequences, and are further treated in [3], Chapter 16 of [2], and Chapter 5 of [6]. Some open questions posed in the original paper [1] were solved in [5,15]. The work in this article builds on existing coalgebraic approaches to automatic sequences, which can be found in [12] and [9]. In particular, our systems of linear zip-behavioral differential equations can be seen as a linear generalization of the $\mathbf{zip}_k$-specifications in [9]. The isomorphism of final coalgebras presented in this paper is probably folklore, but we think its usefulness warrants an explicit inclusion in our paper. Finally, we remark that the definitions and results of section 3.2 on solutions to linear $\mathbf{zip}_k$-behavioral differential equations can be seen as instances of more general concepts in the theory of bialgebra (cf. [4]). Such an abstract presentation is, however, not necessary and would not improve the results of the paper.

Acknowledgements. We would like to thank Alexandra Silva for helpful discussions on weighted automata.

2   Preliminaries

2.1   Semirings and Semimodules

Throughout this paper, S denotes a semiring. A semiring $S = (S, +, \cdot, 0, 1)$ consists of a commutative monoid $(S, +, 0)$ and a monoid $(S, \cdot, 1)$ such that the following identities hold for all s, t, u in S: $0 \cdot s = s \cdot 0 = 0$, $s \cdot (t + u) = s \cdot t + s \cdot u$, $(s + t) \cdot u = s \cdot u + t \cdot u$. A *left-semimodule over S* is a commutative monoid $(M, +, 0)$ together with a left-action by S, i.e., a map $S \times M \to M$, denoted as scalar multiplication $(s, m) \mapsto sm$ for all $s \in S, m \in M$ which satisfies: $(st)m = s(tm)$, $s(m + n) = sm + sn$, $(s + t)m = sm + tm$, $0m = 0$, $1m = m$. A *left-linear map* between left-semimodules is a map $f \colon M \to N$ which respects scalar multiplication and sum: $f(sm) = sf(m)$ and $f(m_1 + m_2) = f(m_1) + f(m_2)$. Right-semimodules over S are defined similarly via a right action. If S is commutative, i.e., the multiplicative monoid $(S, \cdot, 1)$ is commutative, then left and right semimodules are the same.

We will work in the setting of left-semimodules over S and left-linear maps, which for simplicity we refer to as *S-semimodules* and *linear maps*. Note that S is itself an S-semimodule with the left action given by multiplication in S.

2.2   Stream Operations, Zip and Unzip

We will use some notation and terminology from coinductive stream calculus (see e.g. [18]). A *stream* (over the semiring S) is an (infinite) sequence $(\sigma(0), \sigma(1), \sigma(2), \dots)$ of elements from S, or more formally, a map $\sigma \colon \mathbb{N} \to S$, also written $\sigma \in S^{\mathbb{N}}$. We will use the terminology of *streams* and *sequences* interchangeably, and the notions can be regarded as synonymous.

The *initial value* and *derivative* of a stream $\sigma \in S^{\mathbb{N}}$ are $\sigma(0)$ and σ', respectively, where $\sigma'(n) = \sigma(n + 1)$ for all $n \in \mathbb{N}$. The initial value and derivative of σ are also known as $\mathbf{head}(\sigma)$ and $\mathbf{tail}(\sigma)$.

The streams $S^{\mathbb{N}}$ form an S-semimodule under pointwise addition and scalar multiplication which are the unique stream operations that satisfy the following *stream differential equations* (cf. [18]):

$$
\begin{array}{ll}
(\sigma + \tau)(0) = \sigma(0) + \tau(0), & (\sigma + \tau)' = \sigma' + \tau', \\
(a\sigma)(0) = a\sigma(0), & (a\sigma)' = a\sigma'.
\end{array} \tag{1}
$$

for all $\sigma, \tau \in S^{\mathbb{N}}$ and $a \in S$. Note that from the above equations it follows immediately that **head**: $S^{\mathbb{N}} \to S$ and **tail**: $S^{\mathbb{N}} \to S^{\mathbb{N}}$ are linear.

The *shift* operation $\mathfrak{X}$ is defined as $\mathfrak{X}\sigma = (0, \sigma(0), \sigma(1), \ldots)$, or equivalently, by the stream differential equation:

$$
(\mathfrak{X}\sigma)(0) = 0, \qquad (\mathfrak{X}\sigma)' = \sigma.
$$

We will use the so-called *fundamental theorem of stream calculus*(cf. [18])[1]:

$$
\text{for all } \sigma \in S^{\mathbb{N}}: \quad \sigma = \sigma(0) + \mathfrak{X}\sigma' \tag{2}
$$

Of central importance to us are the k-ary operations $\mathbf{zip}_k$. For $k \in \mathbb{N}$, $\mathbf{zip}_k$ zips together k streams $\sigma_0, \ldots, \sigma_{k-1}$ into one by taking elements in turn from its arguments. Formally, for $k \in \mathbb{N}$ and streams $\sigma_0, \ldots, \sigma_{k-1}$ the stream operation $\mathbf{zip}_k$ is defined by

$$
\mathbf{zip}_k(\sigma_0, \ldots, \sigma_{k-1})(i + nk) = \sigma_i(n) \qquad \forall n, i \in \mathbb{N}, 0 \le i < k \tag{3}
$$

or equivalently, by the stream differential equation:

$$
\begin{array}{l}
\mathbf{zip}_k(\sigma_0, \ldots, \sigma_{k-1})(0) = \sigma_0(0) \\
\mathbf{zip}_k(\sigma_0, \ldots, \sigma_{k-1})' = \mathbf{zip}_k(\sigma_1, \ldots, \sigma_{k-1}, \sigma_0').
\end{array} \tag{4}
$$

For example, for $k = 2$, we have $\mathbf{zip}_2(\sigma, \tau) = (\sigma(0), \tau(0), \sigma(1), \tau(1), \ldots)$.

Conversely, the unzipping operations are defined as follows for $k, j \in \mathbb{N}$ with $j < k$:

$$
\mathbf{unzip}_{j,k}(\sigma)(n) = \sigma(j + nk) \qquad \forall n \in \mathbb{N} \tag{5}
$$

For $k = 2$, $\mathbf{unzip}_{0,2}$ and $\mathbf{unzip}_{1,2}$ are also known as **even** and **odd**:

$$
\begin{array}{l}
\mathbf{unzip}_{0,2}(\sigma) = \mathbf{even}(\sigma) = (\sigma(0), \sigma(2), \sigma(4), \ldots) \\
\mathbf{unzip}_{1,2}(\sigma) = \mathbf{odd}(\sigma) = (\sigma(1), \sigma(3), \sigma(5), \ldots)
\end{array}
$$

It can easily be verified that

$$
\mathbf{zip}_k(\mathbf{unzip}_{0,k}(\sigma), \ldots, \mathbf{unzip}_{k-1,k}(\sigma)) = \sigma \tag{6}
$$

and conversely that (for i with $0 \le i < k$)

$$
\mathbf{unzip}_{i,k}(\mathbf{zip}_k(\sigma_0, \ldots, \sigma_{k-1})) = \sigma_i. \tag{7}
$$

In other words, $\mathbf{zip}_k \colon (S^{\mathbb{N}})^k \to S^{\mathbb{N}}$ is a bijection with inverse

$$
\mathbf{unzip}_k = (\mathbf{unzip}_{0,k}, \ldots, \mathbf{unzip}_{k-1,k}) \colon S^{\mathbb{N}} \to (S^{\mathbb{N}})^k.
$$

The **unzip**-operations relate to the more familiar notion of a k-kernel. The k-*kernel* of a stream σ can be defined as the closure of the set $\{\sigma\}$ under the operations $\mathbf{unzip}_{j,k}$ for $0 \le j < k$.

---

[1] Here $\sigma(0)$ is overloaded as the stream $(\sigma(0), 0, 0, \ldots)$

2.3 Automata as Coalgebras

We briefly recall some basic definitions on (weighted) automata (with weights in a semiring), and how these are modelled as coalgebras [17].

Automata and formal power series. Given a finite alphabet A and a semiring S, a *(linear) A-automaton (with output in S)* is a triple (Q, o, δ), where Q is an S-semimodule, $o \colon Q \to S$ is a linear map assigning an output value $o(q)$ to each $q \in Q$, and $\delta \colon Q \to Q^A$ is a linear map assigning to each $q \in Q$ and $a \in A$ a next state $\delta(q)(a)$ which we will also denote by q_a and refer to as the *a-derivative of q*. Note the absence of initial states or state vectors in this presentation. The transition function δ can be extended to a map $\delta^* \colon Q \to Q^{A^*}$ in the usual inductive manner: $\delta^*(q)(\epsilon) = q$ and $\delta^*(q)(wa) = \delta(\delta^*(q)(w))(a)$.

The set $S\langle\!\langle A \rangle\!\rangle$ of *formal power series over noncommuting variables from A with outputs in S* is the function space $S\langle\!\langle A \rangle\!\rangle = \{\rho \colon A^* \to S\}$ equipped with pointwise S-semimodule structure. We note that a formal power series $\rho \colon A^* \to S$ can also be seen as an S-weighted language.[2] The formal power series *generated by a state $q \in Q$* in an A-automaton (Q, o, δ) is the map $[\![q]\!]_L \colon A^* \to S$ defined by $[\![q]\!]_L(w) = o(\delta^*(q)(w))$.

An A-automaton is a coalgebra for the functor $S \times (-)^A$ on the category of S-semimodules and linear maps. The theory of universal coalgebra [17] now directly yields an associated notion of homomorphism. Diagrammatically, given A-automata (Q, o_Q, δ_Q) and (R, o_R, δ_R), a linear map $h \colon Q \to R$ is a *homomorphism* iff it makes the diagram

$$
\begin{array}{ccc}
Q & \xrightarrow{\;\;h\;\;} & R \\
{\scriptstyle (o_Q, \delta_Q)}\downarrow & & \downarrow{\scriptstyle (o_R, \delta_R)} \\
S \times Q^A & \xrightarrow{\;1_S \times h^A\;} & S \times R^A
\end{array}
$$

commute, or equivalently, iff for all $q \in Q$, $o_Q(q) = o_R(h(q))$ and $h(q_a) = h(q)_a$ for all $a \in A$. An *isomorphism of A-automata* is a bijective homomorphism.

The set $S\langle\!\langle A \rangle\!\rangle$ of formal power series is itself an A-automaton

$$\mathcal{L} = (S\langle\!\langle A \rangle\!\rangle, O, \Delta)$$

with O and Δ defined by

$$O(\rho) = \rho(\epsilon) \qquad \text{and} \qquad \Delta(\rho)(a)(w) = \rho(aw).$$

In fact, $(S\langle\!\langle A \rangle\!\rangle, O, \Delta)$ is known to be *final* in the category of A-automata (this follows from [18, Theorem 9.1] combined with the fact that all final mappings are linear). This means that given any A-automaton (Q, o, δ), there is a unique homomorphism $(Q, o, \delta) \to (S\langle\!\langle A \rangle\!\rangle, O, \Delta)$. This unique homomorphism is, in fact, the function $[\![-]\!]_L \colon Q \to S\langle\!\langle A \rangle\!\rangle$ which maps $q \in Q$ to the formal power series generated by q. We note that final coalgebras are unique only up to isomorphism.

---

[2] This explains our later use of L as subscript to indicate formal power series.

Recognizable formal power series are characterized in terms of weighted automata. We first introduce some notation. For a set X, we denote by S_ω^X the set of all functions $\phi: X \to S$ with finite support, i.e., $\phi(x) \neq 0$ for only finitely many $x \in X$. Equivalently, such a ϕ can be seen as a linear combination $a_1 x_1 + \cdots + a_n x_n$ where $a_i = \phi(x_i)$, $i = 1, \ldots, n$ and $\phi(x) = 0$ for all $x \notin \{x_1, \ldots, x_n\}$. The set X is included into S_ω^X via the map $\eta: X \hookrightarrow S_\omega^X$ defined as $\eta(x) = 1x$. Importantly, by taking pointwise S-semimodule structure, S_ω^X is the *free S-semimodule* over the set X, which means that for each function $f: X \to M$ into some S-semimodule M, there is a unique linear map $\hat{f}: S_\omega^X \to M$ extending f, i.e., $\hat{f} \circ \eta = f$.

Weighted automata. An *S-weighted A-automaton* is a triple (X, o, δ) where X is a set (of states), $o: X \to S$ is an output function, and $\delta: X \to (S_\omega^X)^A$ is a transition function. In terms of weighted transitions, $\delta(x)(a)(y) \in S$ is the weight of the a-transition from x to y. We say that (X, o, δ) is *finite* if X is finite. An S-weighted A-automaton is a coalgebra for the functor $S \times (S_\omega^-)^A$ on the category of sets and functions. We note that a nondeterministic automaton is a **2**-weighted automaton where $\mathbf{2} = (\{\bot, \top\}, \vee, \wedge, \bot, \top)$ is the Boolean semiring.

Determinization. Any S-weighted A-automaton (X, o, δ) can be *determinized* to an A-automaton, by constructing an A automaton $(S_\omega^X, \hat{o}, \hat{\delta})$, where $\hat{o}$ and $\hat{\delta}$ are the unique linear extensions of o and δ to the free semimodule S_ω^X, i.e. the unique linear mappings satisfying $\hat{o}(\eta(x)) = o(x)$ and $\hat{\delta}(\eta(x)) = \delta(x)$.

This construction can be summarized in the following diagram:

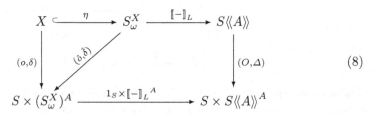

$$(8)$$

We say that a state x in an S-weighted A-automaton *generates* $\rho \in S\langle\!\langle A \rangle\!\rangle$ if $[\![\eta(x)]\!]_L = \rho$. When X is finite, the determinization has a finitely generated S-semimodule S_ω^X as its state space, but as a set S_ω^X is generally infinite. For further details on determinization and its categorical/coalgebraic setting we refer to [19].

The above now yields the following definition of the recognizable power series.

Definition 1. *A formal power series $\rho \in S\langle\!\langle A \rangle\!\rangle$ is recognizable if and only if there is a finite S-weighted A-automaton (X, o, δ) such that $\rho = [\![x]\!]_L$ for some $x \in X$.*

In other words, ρ is recognizable if and only if it is generated by some state in a finite S-weighted A-automaton. This definition is easily seen to correspond to the classic definition in e.g. [6].

2.4   Numeration Systems

For any natural number $k \in \mathbb{N}$ with $k \geq 1$, let A_k denote the alphabet of digits

$$A_k = \{\texttt{1}, \ldots, \texttt{k}\}$$

We emphasize the use of the digits as alphabet symbols by writing them in a fixed-width font. The map $\nu_k \colon A_k^* \to \mathbb{N}$, which assigns to each string of digits the natural number it represents, is defined inductively by:

$$\nu_k(\epsilon) = 0 \qquad \text{and} \qquad \nu_k(\texttt{i} \cdot w) = i + k \cdot \nu_k(w).$$

It is well-known and easy to see that ν_k is a bijection between natural numbers and their representation in the k-adic numeration system[3], with the least significant digit on the left. For example, the 2-adic numeration of the natural numbers starts as follows: $\epsilon, \texttt{1}, \texttt{2}, \texttt{11}, \texttt{21}, \texttt{12}, \texttt{22}, \texttt{111}, \ldots$

We contrast the bijective k-adic numeration system with the familiar *(standard) base k numeration system* which is defined as follows. The alphabet of digits is

$$B_k = \{\texttt{0}, \ldots, \texttt{k}-\texttt{1}\}$$

and, whenever $k \geq 2$, we can define a mapping $\xi_k \colon B_k^* \to \mathbb{N}$ inductively by

$$\xi_k(\epsilon) = 0 \qquad \text{and} \qquad \xi_k(\texttt{i} \cdot w) = i + k \cdot \xi_k(w)$$

This again gives us a presentation with the least significant digit on the left.[4] For example, standard base 2 is the reverse binary notation with zero represented by ϵ, i.e., starting as $\epsilon, \texttt{1}, \texttt{01}, \texttt{10}, \texttt{11}, \texttt{001}, \texttt{101}, \texttt{011}, \texttt{111}, \ldots$ The map ξ_k has the property that for all $w \in B_k^*$, $\xi(w) = \xi(w \cdot \texttt{0})$, and hence ξ_k is not bijective.

Finally, observe that, from the inductive definitions of the k-adic and standard base k numeration, we obtain that

$$\nu_k(\texttt{a}_0 \ldots \texttt{a}_\texttt{n}) = \sum_{i=0}^{n} a_i k^i \qquad \text{and} \qquad \xi_k(\texttt{b}_0 \ldots \texttt{b}_\texttt{n}) = \sum_{i=0}^{n} b_i k^i$$

for all $\texttt{a}_\texttt{i} \in A_k$ and $\texttt{b}_\texttt{i} \in B_k$, which can be taken as alternative definitions of the two numeration systems.

In most literature on k-automatic and k-regular sequences, the standard base k numeration system is employed. However, we prefer the bijective k-adic numeration system since it yields a bijective correspondence at the level of final coalgebras. Moreover, unlike in the standard base k numeration, there is a straightforward and intuitive bijective numeration for the case $k = 1$ given by $\epsilon, \texttt{1}, \texttt{11}, \texttt{111}, \ldots$

---

[3] Unrelated to and not to be confused with the p-adic numbers.

[4] For a more standard presentation with the most significant digit on the left, switch the inductive definition to $\xi_k(w \cdot \texttt{i}) = i + k \cdot \xi_k(w)$.

3 Characterizations of k-regular Sequences

The notion of a k-regular sequence with values in a ring was introduced in [1]. The following definition is (roughly) a direct generalization to sequences with values in a semiring. We discuss the more precise relationship in the remark below.

Definition 2. *A sequence $\sigma \in S^{\mathbb{N}}$ is k-regular iff the k-kernel of σ is contained in a finitely generated S-subsemimodule of $S^{\mathbb{N}}$, or equivalently, iff there is a finite set of generators $\Sigma = \{\sigma_0, \ldots \sigma_{n-1}\}$ with $\sigma \in \Sigma$, and an indexed family $a_{h,i,j}$ for all $h, i, j \in \mathbb{N}$ with $h < n$, $i < n$, $j < k$, such that for all $h < n$ and $j < k$*

$$\mathbf{unzip}_{j,k}(\sigma_h) = \sum_{i<n} a_{h,i,j}\sigma_i$$

or equivalently, for all $h < n$:

$$\sigma_h = \mathbf{zip}_k\left(\sum_{i<n} a_{h,i,0}\sigma_i, \ldots, \sum_{i<n} a_{h,i,k}\sigma_i\right). \qquad (9)$$

Remark 3. In [1], the definition of a k-regular sequence is as follows: *Let R be a ring and R' a (commutative) Noetherian ring contained in R. A sequence $\sigma \in R^{\mathbb{N}}$ is (R', k)-regular if each sequence in the k-kernel of σ is an R'-linear combination of some finite set of sequences $\sigma_1, \ldots, \sigma_n \in R^{\mathbb{N}}$.* In terms of modules, this is equivalent with saying that the k-kernel of σ is contained in a finitely generated R'-submodule of $R^{\mathbb{N}}$ (viewed as an R'-module). Since R' is assumed to be Noetherian, this in turn is equivalent with the k-kernel itself being a finitely generated R'-submodule of $R^{\mathbb{N}}$. For simplicity, we do not distinguish between the semiring S of values and a subsemiring $S' \subseteq S$ from which linear coefficients may be taken. Hence in the terminology of [1], our definition of k-regular could be phrased as (S, k)-regular. If we assume that S is a Noetherian semiring (cf. [8]), then our definition is equivalent to requiring that the k-kernel of σ is a finitely generated S-subsemimodule of $S^{\mathbb{N}}$.

In this section we will give characterizations of k-regular sequences in terms of finite weighted automata, finite systems of (certain) behavioral differential equations, and recognizable formal power series.

3.1 An Isomorphism between Final A_k-automata

We start by defining an A_k-automaton $\mathcal{S}$ with state space $S^{\mathbb{N}}$ as the composition of bijections:

$$S^{\mathbb{N}} \xrightarrow[\cong]{(\mathbf{head},\mathbf{tail})} S \times S^{\mathbb{N}} \xrightarrow[\cong]{1_S \times \mathbf{unzip}_k} S \times (S^{\mathbb{N}})^{A_k}$$

That is,

$$\mathcal{S} := (S^{\mathbb{N}}, \mathbf{head}, \mathbf{unzip}_k \circ \mathbf{tail})$$

In [9, Proposition 26] it was observed that S is a final A_k-automaton. This result will also follow from our Proposition 5 below, and the finality of $\mathcal{L}$.

Given an A_k-automaton (Q, o, δ), we let $[\![-]\!]_S \colon Q \to S^{\mathbb{N}}$ denote the unique mapping into the final A_k-automaton on $S^{\mathbb{N}}$:

$$
\begin{array}{ccc}
Q & \xrightarrow{\;[\![-]\!]_S\;} & S^{\mathbb{N}} \\[2pt]
{\scriptstyle (o,\delta)}\big\downarrow & & \big\downarrow{\scriptstyle (\textbf{head},\textbf{unzip}_k \circ \textbf{tail})} \\[4pt]
S \times Q^{A_k} & \xrightarrow{\;1_S \times [\![-]\!]_S^{A_k}\;} & S \times (S^{\mathbb{N}})^{A_k}
\end{array}
\tag{10}
$$

The commutativity of the above diagram means that for all $q \in Q$:

$$
\begin{aligned}
o(q) &= \textbf{head}([\![q]\!]_S) \\
[\![\delta(q)(\texttt{i})]\!]_S &= [\![q_{\texttt{i}}]\!]_S = \textbf{unzip}_{i-1,k}([\![q]\!]_S') \qquad \text{for all } \texttt{i} \in A_k
\end{aligned}
\tag{11}
$$

or, equivalently, using the $\textbf{zip}_k$-$\textbf{unzip}_k$ isomorphism (6),

$$
[\![q]\!]_S(0) = o(q), \qquad [\![q]\!]_S' = \textbf{zip}_k([\![\delta(q)(1)]\!]_S, \ldots, [\![\delta(q)(\texttt{k})]\!]_S).
\tag{12}
$$

We refer to $[\![q]\!]_S$ as the *stream semantics of* q. Conversely, we will say that q *generates the stream* $[\![q]\!]_S$.

As final coalgebras are unique up to isomorphism, it follows that S and $\mathcal{L} = (S\langle\!\langle A_k\rangle\!\rangle, O, \Delta)$ are isomorphic. We will show that the unique isomorphism between S and $\mathcal{L}$ is concretely given by k-adic numeration. First, we define a map from sequences to formal power series.

Definition 4. *We define the map* $h_L \colon S^{\mathbb{N}} \to S\langle\!\langle A_k\rangle\!\rangle$ *by*

$$
h_L(\sigma)(w) = \sigma(\nu_k(w)) \qquad \text{for all } w \in A_k^*
\tag{13}
$$

where $\nu_k \colon A_k^* \to \mathbb{N}$ *is the bijective k-adic numeration given in Section 2.4. For* $\sigma \in S^{\mathbb{N}}$ *we refer to* $h_L(\sigma)$ *as the formal power series corresponding to* σ *via k-adic numeration.*

Proposition 5. *The map* $h_L \colon S^{\mathbb{N}} \to S\langle\!\langle A_k\rangle\!\rangle$ *is an isomorphism of A_k-automata from S to $\mathcal{L}$, i.e., the following diagram commutes (where $h_S = h_L^{-1}$):*

$$
\begin{array}{ccc}
S^{\mathbb{N}} & \underset{h_S}{\overset{h_L}{\rightleftarrows}} & S\langle\!\langle A_k\rangle\!\rangle \\[2pt]
{\scriptstyle (\textbf{head},\textbf{unzip}_k \circ \textbf{tail})}\big\downarrow & & \big\downarrow{\scriptstyle (O,\Delta)} \\[4pt]
S \times (S^{\mathbb{N}})^{A_k} & \underset{1_S \times h_S^k}{\overset{1_S \times h_L^k}{\rightleftarrows}} & S \times S\langle\!\langle A_k\rangle\!\rangle^{A_k}
\end{array}
$$

Proof. We must show that h_L is a bijective homomorphism. From the fact that ν_k is a bijection it directly follows that h_L is a bijection. It remains to show that h_L is a homomorphism of A_k-automata.

For this, we first have to prove that $\mathbf{head}(\sigma) = O(h_L(\sigma))$, which holds because

$$\mathbf{head}(\sigma) = \sigma(0) = \sigma(\nu_k(\epsilon)) = h_L(\sigma)(\epsilon) = O(h_L(\sigma)).$$

Now, we have to show that $(h_L(\sigma))_{\mathtt{i}} = h_L(\sigma_{\mathtt{i}})$. This holds, because given any $w \in A_k^*$ and $\mathtt{i} \in A_k$, we have:

$$
\begin{aligned}
(h_L(\sigma))_{\mathtt{i}}(w) &= (h_L(\sigma))(\mathtt{i} \cdot w) \\
&= \sigma(\nu_k(\mathtt{i} \cdot w)) \\
&= \sigma(i + k \cdot \nu_k(w)) \\
&= \sigma'((i-1) + k \cdot \nu_k(w)) \\
&= \mathbf{unzip}_{i-1,k}(\sigma')(\nu_k(w)) \\
&= h_L((\mathbf{unzip}_{i-1,k} \circ \mathbf{tail})(\sigma))(w) \\
&= h_L(\sigma_{\mathtt{i}})(w)
\end{aligned}
$$

Finally, it can easily be verified that h_L is linear using (1). $\square$

In combination with the fact that homomorphisms to final automata are unique, this now directly leads to the following corollary:

Corollary 6. *For any A_k-automaton (Q, o, δ), we have*

$$h_L \circ [\![-]\!]_S = [\![-]\!]_L \quad and \quad h_S \circ [\![-]\!]_L = [\![-]\!]_S.$$

3.2 Systems of Linear zip-Behavioral Differential Equations

The finality of $\mathcal{S}$ gives rise to a coinduction principle for weighted automata. Namely, by defining an S-weighted A_k-automaton (X, o, δ) we are defining the streams $[\![\eta(x)]\!]_S \in S^{\mathbb{N}}$ for each $x \in X$, via determinization and finality as described in the following diagram, which is the analogue of (8) only with $\mathcal{S}$ instead of $\mathcal{L}$.

$$(14)$$

The existence of $[\![-]\!]_S$ and the commutativity of the above diagram immediately tells us the following fact:

Lemma 7. *A sequence σ is generated by a state in a finite weighted automaton if and only if there is a finite set of sequences $\Sigma = \{\sigma_0, \ldots, \sigma_{n-1}\}$ with $\sigma \in \Sigma$ such that for all $j < k$ and $i < n$, $\mathbf{unzip}_{j,k}(\sigma_i')$ is a linear combination of elements from Σ.*

We can formulate the uniqueness of $[\![-]\!]_S$ and the commutativity of (14) in terms of the $\mathbf{zip}_k$-operations by using the homomorphism condition (12).

Lemma 8. *Given an S-weighted A_k-automaton (X, o, δ), $[\![-]\!]_S$ is the unique linear mapping $S^X_\omega \to S^\mathbb{N}$ satisfying, for each $x \in X$, the equations*

$$[\![\eta(x)]\!]_S(0) = o(x) \qquad [\![\eta(x)]\!]'_S = \mathbf{zip}_k([\![\delta(x)(1)]\!]_S, \ldots, [\![\delta(x)(\mathtt{k})]\!]_S).$$

(Recall that $\eta \colon X \hookrightarrow S^X_\omega$ is the inclusion of states into the determinization.)

The above lemma justifies an alternative method of specifying streams via equations involving the $\mathbf{zip}_k$-operation. A *system of linear $\mathbf{zip}_k$-behavioral differential equations over a set (of variables) X* is a system of equations, one for each $x \in X$, of the form,

$$x(0) = a_x, \qquad x' = \mathbf{zip}_k(\alpha_{x,1}, \ldots, \alpha_{x,k}) \tag{15}$$

where $a_x \in S$ and $\alpha_{x,1}, \ldots, \alpha_{x,k}$ are S-linear combinations over X.

Linear $\mathbf{zip}_k$-behavioral differential equations and "plain" behavioral differential equations (using formal power series derivatives) both describe weighted automata, but the use of linear $\mathbf{zip}$-behavioral differential equations makes it explicit that we intend to apply the finality of S (to obtain streams), rather than the finality of $\mathcal{L}$ (to obtain formal power series). Explicitly, given a system of linear $\mathbf{zip}_k$-behavioral differential equations as in (15), the corresponding S-weighted A_k-automaton (X, o, δ) is given by $o(x) = a_x$, and $\delta(x)(\mathtt{i}) = \alpha_{x,i}$ for all $x \in X$ and $\mathtt{i} \in A_k$.

We illustrate with a small example. The streams specified in derivative form by the behavioral differential equations:

$$\begin{aligned} o(x) &= 1, & x_1 &= x + y, & x_2 &= 3x + y \\ o(y) &= 2, & y_1 &= y, & y_2 &= x + 2y \end{aligned}$$

are equivalently specified in terms of $\mathbf{unzip}_{0,2} \circ \mathbf{tail}$ and $\mathbf{unzip}_{1,2} \circ \mathbf{tail}$, i.e., by $\mathbf{even} \circ \mathbf{tail}$ and $\mathbf{odd} \circ \mathbf{tail}$:

$$\begin{aligned} o(x) &= 1, & \mathbf{even}(x') &= x + y, & \mathbf{odd}(x') &= 3x + y \\ o(y) &= 2, & \mathbf{even}(y') &= y, & \mathbf{odd}(y') &= x + 2y \end{aligned}$$

and equivalently (via the zip-unzip isomorphism (6)), by the system of linear $\mathbf{zip}_2$-behavioral differential equations:

$$\begin{aligned} o(x) &= 1, & x' &= \mathbf{zip}_2(x + y, 3x + y) \\ o(y) &= 2, & y' &= \mathbf{zip}_2(y, x + 2y) \end{aligned}$$

A *solution* to a system of linear $\mathbf{zip}_k$-behavioral differential equations over X with components given as in (15), is a map $f \colon X \to S^\mathbb{N}$ such that for all $x \in X$,

$$f(x)(0) = a_x, \qquad f(x)' = \mathbf{zip}_k(\hat{f}(\alpha_{x,0}), \ldots, \hat{f}(\alpha_{x,k-1})) \tag{16}$$

where $\hat{f} \colon S^X_\omega \to S^\mathbb{N}$ is the linear extension of f with respect to the semimodule structure on $S^\mathbb{N}$.

The basic fact that justifies viewing systems of linear $\mathbf{zip}_k$-behavioral differential equations as defining streams, is stated in the following lemma.

Lemma 9. *Every system of linear zip_k-behavioral differential equations has a unique solution given by the final stream semantics $[\![-]\!]_S \circ \eta$ of the corresponding weighted automaton (X, o, δ).*

Proof. By Lemma 8, the map $[\![-]\!]_S \circ \eta \colon X \to S^\mathbb{N}$ is a solution, and it is unique by uniqueness of $[\![-]\!]_S$. $\qquad\square$

We will say that a stream σ is *defined by system of linear zip_k-behavioral differential equations over X* if $\sigma = [\![\eta(x)]\!]_S$ for some $x \in X$ in such a system. In what follows, we will suppress η and simply write $[\![x]\!]_S$ instead of $[\![\eta(x)]\!]_S$.

3.3   Characterizations of k-regular Sequences

We will now show that k-regular sequences are obtained precisely by the stream semantics finite S-weighted A_k-automata. It will follow that k-regular sequences are in bijective correspondence with recognizable formal power series via k-adic numeration. This is an analogue of the result in [1] which shows that k-regular sequences over the integers $\mathbb{Z}$ correspond to *some* $\mathbb{Z}$-rational power series in noncommuting variables $\{0, \ldots, \mathrm{k} - 1\}$ via the standard base k numeration.

Proposition 10. *Given any $k \geq 2$, if $\sigma \in S^\mathbb{N}$ is a k-regular sequence, then there is a finite S-weighted A_k-automaton (X, o, δ) and an $x \in X$, such that $[\![x]\!]_S = \sigma$.*

Proof. If σ is k-regular, there is a finite set of sequences $\Sigma = \{\sigma_0, \ldots, \sigma_{n-1}\}$ with $\sigma \in \Sigma$, and values a_h and $b_{h,i,j}$ in S indexed over $h < n$, $i < n$, $j < k$, such that for all $h < n$:

$$\sigma_h = \mathrm{zip}_k \left(\sum_{i<n} a_{h,i,0}\sigma_i, \ldots, \sum_{i<n} a_{h,i,k-1}\sigma_i \right).$$

Taking the derivative and second derivative of each σ_h using (4), we obtain:

$$\sigma_h' = \mathrm{zip}_k \left(\sum_{i<n} a_{h,i,1}\sigma_i, \ldots, \sum_{i<n} a_{h,i,0}\sigma_i' \right)$$

$$\sigma_h'' = \mathrm{zip}_k \left(\sum_{i<n} a_{h,i,2}\sigma_i, \ldots, \sum_{i<n} a_{h,i,0}\sigma_i', \sum_{i\leq n} a_{h,i,1}\sigma_i' \right)$$

Hence, for each $\sigma \in \Sigma^+ := \Sigma \cup \{\sigma' \mid \sigma \in \Sigma\}$ and $j < k$, $\mathrm{unzip}_{j,k}(\sigma')$ is a linear combination of elements from Σ^+, and hence there is a finite S-weighted A_k-automaton (X, o, δ) and an $x \in X$, such that $[\![x]\!]_S = \sigma$ by Lemma 7. $\qquad\square$

Example 11. We illustrate Proposition 10 with a well-known 2-regular sequence, which the composer Per Nørgård used in a variety of his compositions, and which he called the *infinity sequence*[5] (A004718 on the Online Encyclopedia of Integer Sequences[6]):

$$\sigma = (0, 1, -1, 2, 1, 0, -2, 3, -1, 2, 0, 1, 2, \ldots) \in \mathbb{Z}^{\mathbb{N}}$$

This sequence can be characterized uniquely by the following equations:

$$o(x) = 0 \qquad\qquad x = \mathbf{zip}_2(-x, x + y)$$
$$o(y) = 1 \qquad\qquad y = \mathbf{zip}_2(y, y)$$

(with x denoting σ). The **zip**-equations on the right-hand side are a system in the format of (9) and hence the sequence is 2-regular. Taking derivatives and second derivatives of the **zip**-equations, we now get using (4):

$$x' = \mathbf{zip}_2(x + y, -x') \qquad\qquad x'' = \mathbf{zip}_2(-x', x' + y')$$
$$y' = \mathbf{zip}_2(y, y') \qquad\qquad y'' = \mathbf{zip}_2(y', y')$$

We can now compute the initial values of x' and y' as

$$o(x') = o(\mathbf{zip}_2(x + y, -x')) = o(x + y) = o(x) + o(y) = 1$$
$$o(y') = o(\mathbf{zip}_2(y, y)) = o(y) = 1.$$

Introducing new variables z and w representing x' and y' respectively, we now can specify a weighted automaton as the unique solution to the following system of **zip**-equations:

$$o(x) = 0 \qquad\qquad x' = \mathbf{zip}_2(x + y, -z)$$
$$o(y) = 1 \qquad\qquad y' = \mathbf{zip}_2(y, w)$$
$$o(z) = 1 \qquad\qquad z' = \mathbf{zip}_2(-z, z + w)$$
$$o(w) = 1 \qquad\qquad w' = \mathbf{zip}_2(w, w)$$

The final homomorphism $[\![-]\!]_S$ maps x to Nørgård's infinity sequence:

$$[\![x]\!]_S = (0, 1, -1, 2, 1, 0, -2, 3, -1, 2, 0, 1, 2, \ldots)$$

We remark, however, that this weighted automaton is not minimal, as y and w both are mapped onto the constant sequence of ones.

Example 12. Another example, which can be constructed in the same manner as the previous example, is given by the following $\mathbb{N}$-weighted A_2-automaton:

$$o(x) = 1 \qquad\qquad x' = \mathbf{zip}_2(x, x)$$
$$o(y) = 1 \qquad\qquad y' = \mathbf{zip}_2(2y, 2y + x)$$
$$o(z) = 1 \qquad\qquad z' = \mathbf{zip}_2(z, x + y)$$

---

[5] http://pernoergaard.dk/eng/strukturer/uendelig/uindhold.html
[6] http://oeis.org

Here, the final homomorphism $[\![-]\!]_S$ maps x onto the constant stream of ones, y onto the stream of natural numbers, and z onto Kimberling's sequence (A003602 on OEIS):

$$[\![z]\!]_S = (1,1,2,1,3,2,4,1,5,3,6,2,7,4,8,\ldots)$$

We now prove the converse of Proposition 10.

Proposition 13. *Given any $k \geq 2$, a finite S-weighted A_k-automaton (X, o, δ), and a state $x \in X$, $[\![x]\!]_S$ is k-regular.*

Proof. By Lemma 8, we have

$$[\![x]\!]_S' = \mathbf{zip}_k([\![\delta(x)(1)]\!]_S, \ldots, [\![\delta(x)(k)]\!]_S)$$

and by (4) and (2) that

$$[\![x]\!]_S = \mathbf{zip}_k(o(x) + \mathcal{X}[\![\delta(x)(k)]\!]_S, [\![\delta(x)(1)]\!]_S, \ldots, [\![\delta(x)(k\text{-}1)]\!]_S). \qquad (17)$$

Using the fact that $(\mathcal{X}[\![x]\!]_S)' = [\![x]\!]_S$, and applying again (4) and (2), we obtain:

$$\mathcal{X}[\![x]\!]_S = \mathbf{zip}_k(\mathcal{X}[\![\delta(x)(k\text{-}1)]\!]_S, o(x) + \mathcal{X}[\![\delta(x)(k)]\!]_S, [\![\delta(x)(1)]\!]_S, \ldots, [\![\delta(x)(k\text{-}2)]\!]_S) \qquad (18)$$

By defining the set of generators

$$\Sigma = \{[\![x]\!]_S \mid x \in X\} \cup \{\mathcal{X}[\![x]\!]_S \mid x \in X\} \cup \{(1,0,0,\ldots)\}$$

the equations (17) and (18) show (via the zip-unzip isomorphism (6)) that for each generator $\sigma \in \Sigma$ and $j < k$, $\mathbf{unzip}_{j,k}(\sigma)$ is a linear combination of the generators. It follows from the definition that $[\![x]\!]_S$ is k-regular for all $x \in X$. $\qquad \square$

We now can gather, from our previous results, the following equivalent characterizations of k-regular sequences, arriving at our main theorem:

Theorem 14. *The following are equivalent for any stream $\sigma \in S^{\mathbb{N}}$:*

1. *σ is k-regular.*
2. *σ is generated by a state in a finite weighted A_k-automaton.*
3. *σ is defined by a linear system of $\mathbf{zip}_k$-behavioral differential equations over a finite set of variables.*
4. *$h_L(\sigma) \in S\langle\!\langle A_k \rangle\!\rangle$ is a recognizable power series.*

Proof. $1 \Rightarrow 2$ is Proposition 10. $2 \Rightarrow 1$ is Proposition 13. $2 \Leftrightarrow 3$ follows from Lemma 8 and Lemma 9. $2 \Leftrightarrow 4$ follows from Proposition 5. $\qquad \square$

The equivalence $1 \Leftrightarrow 4$ in the above theorem combined with the fact that h_L is a bijection with inverse h_S directly yields the following corollary, establishing a bijective correspondence between k-regular power series and recognizable power series on the alphabet A_k:

Corollary 15. *For all formal power series $\rho \in S\langle\!\langle A_k \rangle\!\rangle$ over variables A_k, we have: ρ is recognizable if and only if the sequence $h_S(\rho) \in S^\mathbb{N}$ is k-regular.*

The equivalence $1 \Leftrightarrow 4$ of Theorem 14 is analogous to [1, Theorem 4.3], which says that $\sigma \in \mathbb{Z}^\mathbb{N}$ is k-regular if and only if the formal power series $\sum_{n<\omega} \sigma(n)\bar{\xi}(n)$ is rational (or equivalently, recognizable), where $\bar{\xi} \colon \mathbb{N} \to B_k^* \backslash B_k^* 0$ is the inverse of the bijection obtained by restricting the standard base k numeration $\xi \colon B_k^* \to \mathbb{N}$ to words not ending in 0. In contrast with our results, [1, Theorem 4.3] cannot be extended to a bijective correspondence between the classes of k-regular sequences and rational power series over variables in B_k as there are rational power series $\rho \in S\langle\!\langle B_k \rangle\!\rangle$ that do not correspond to any k-regular sequence via standard base k numeration. In other words, there is no analogue of our Corollary 15 in the presentation using standard base k numeration from [1].

Zero-consistent automata. It is also possible to characterize k-regular sequences by a class of weighted automata that read the numeration used in [1, Theorem 4.3] (standard base k backwards). This class of automata is provided by the so-called zero-consistent S-weighted B_k-automata which are a mild generalisation of the zero-consistent automata that have been described in [12].

The defining feature for zero-consistent automata is that the (immediate) output of the automaton does not change when reading letter 0. Intuitively, reading letter 0 corresponds to moving from a state generating stream σ to a state that generates the stream $\mathbf{unzip}_{0,k}(\sigma)$. Zero-consistency reflects the fact that the head of any stream σ is equal to the head of the stream $\mathbf{unzip}_{0,k}(\sigma)$. More generally, reading a letter j with $0 \leq j < k$ in this setting corresponds to moving from a state that represents a stream σ to a state that represents the stream $\mathbf{unzip}_{j,k}(\sigma)$. As Definition 2 of k-regular sequences is based on the $\mathbf{unzip}_k$-operations (and not on the $\mathbf{unzip}_k \circ \mathbf{tail}$-operations as used in the definition of S) it is rather straightforward to prove that the k-regular sequences are precisely the ones that can be generated using zero-consistent automata.

3.4   Connections to Automatic Sequences

All of the results that were presented earlier in this section can be seen as generalizations of corresponding results about the k-automatic sequences. In this subsection, we will state (without proofs, but with some explanations about the relationships) the corresponding theorems. We remark that, unlike in the case of k-regular sequences, the results for automatic sequences are, in this form, well-known in the literature. The following definition is analogous to Definition 2, but uses the more restrictive condition that the k-kernel is finite, rather than finitely generated:

Definition 16. *A sequence $\sigma \in S^\mathbb{N}$ is k-automatic iff the k-kernel of σ is finite, or equivalently, iff there is a finite set of sequences $\Sigma = \{\sigma_0, \ldots \sigma_{n-1}\}$ with $\sigma \in \Sigma$, such that for all $h, j \in \mathbb{N}$ with $h < n$ and $j < k$, $\mathbf{unzip}_{j,k}(\sigma_h) \in \Sigma$.*

We can now obtain the following results, using essentially the same techniques have been used in this paper for k-regular sequences:

- A sequence $\sigma \in S^{\mathbb{N}}$ is k-automatic if and only if there is a finite A_k-automaton (Q, o, δ) such that $\sigma = \llbracket q \rrbracket_S$ for some $q \in Q$. (Equivalent to [2, Theorem 5.2.7])
- A sequence σ is k-automatic iff it takes finitely many values $v_1, \ldots, v_n$, and for each v_i, the language $\{w \in A_k^* \mid h_L(\sigma)(w) = v_i\}$ is regular. (Analogous to [2, Lemma 5.2.6], with change from standard to bijective numeration)

4   Relation to Divide and Conquer Recurrences

Divide and conquer recurrences, which have been considered for example in [10] and [20], can somewhat informally be seen as—in the case of $k = 2$, to which we will restrict ourselves in this section—sequences σ where $\sigma(0)$ is given, and for each n, $\sigma(n)$ is defined in terms of $\sigma(\mathbf{floor}((n-1)/2))$, $\sigma(\mathbf{ceil}((n-1)/2))$, and polynomials in n.

In this section, we will establish a close link between divide and conquer recurrences satisfying a number of 'natural' conditions, and the k-regular sequences, by showing that their sequences occur as 2-regular sequences.

We will restrict ourselves to special (more precisely defined) restricted versions of divide and conquer recurrences. To be precise, we will consider recurrences of the form

$$\sigma(2n) = a_2\sigma(n-1) + a_3\sigma(n) + \tau_1(n) \qquad \sigma(2n+1) = a_1\sigma(n) + \tau_0(n)$$

where a_1, a_2, a_3 are scalars from the semiring S, and τ_1 and τ_0 are themselves 2-regular sequences. We furthermore hold the assumption that $\sigma(0) = 0$ (we will later see that this assumption can be relaxed).

Now observe that

$$\sigma(2n+2) = \sigma(2(n+1)) = a_2\sigma(n) + a_3\sigma(n+1) + \tau_1'(n).$$

As a result of the equalities $\sigma(2n+1) = (\mathbf{even}(\sigma'))(n)$ and $\sigma(2n+2) = (\mathbf{odd}(\sigma'))(n)$ we now derive

$$(\mathbf{even}(\sigma'))(n) = (a_1\sigma + \tau_0)(n)$$

$$\text{and} \qquad (\mathbf{odd}(\sigma'))(n) = (a_2\sigma + a_3\sigma' + \tau_1')(n)$$

and hence also

$$\mathbf{even}(\sigma') = a_1\sigma + \tau_0 \qquad \mathbf{odd}(\sigma') = a_2\sigma + a_3\sigma' + \tau_1'.$$

A large number of combinatorial problems can be expressed by means of divide and conquer recurrences of this type, and can be transformed using this construction, including problems such as the Josephus problem, the sequence of all numbers whose ternary representation does not contain the digit 1, or does not contain the digit 2, counting of quicksort insertions, and a variety of other combinatorial and graph-theoretic problems. An overview of many of these examples can be found at `http://oeis.org/somedcgf.html`. One of the questions asked here, is whether all the examples presented there are indeed 2-regular. We will soon see that this question can be answered positively.

Example 17. As an example illustrating the construction, the recurrence given by

$$\sigma(0) = 0 \qquad \sigma(2n) = \sigma(n) + \sigma(n-1) + 2n - 2 \qquad \sigma(2n+1) = 2\sigma(n) + 2n - 1$$

specifying the sorting numbers (OEIS A001855) can first be transformed into:

$$\sigma(0) = 0 \qquad \sigma(2n+1) = 2\sigma(n) + 2n - 1 \qquad \sigma(2n+2) = \sigma(n+1) + \sigma(n) + 2n$$

We can coinductively specify the streams **ones** and **nats** by

$$\begin{aligned} o(\textbf{ones}) &= 1, & \textbf{even}(\textbf{ones}') &= \textbf{ones}, & \textbf{odd}(\textbf{ones}') &= \textbf{ones} \\ o(\textbf{nats}) &= 1, & \textbf{even}(\textbf{nats}') &= 2 \cdot \textbf{nats}, & \textbf{odd}(\textbf{nats}') &= 2 \cdot \textbf{nats} + \textbf{ones} \end{aligned}$$

and now we can transform the earlier recurrence into the behavioral differential equation:

$$o(\sigma) = 0 \qquad \textbf{even}(\sigma') = 2\sigma + 2 \cdot \textbf{nats} - \textbf{ones} \qquad \textbf{odd}(\sigma') = \sigma' + \sigma + 2 \cdot \textbf{nats}$$

We can now establish that σ is again 2-regular:

Proposition 18. *Let τ_0 and τ_1 be 2-regular sequences over a semiring S, and let $a_0, a_1, a_2,$ and a_3 be elements of S. Then there is a unique sequence σ satisfying*

$$\sigma(0) = a_0 \qquad \textbf{even}(\sigma') = a_1\sigma + \tau_0 \qquad \textbf{odd}(\sigma') = a_2\sigma + a_3\sigma' + \tau_1$$

which is again 2-regular.

Proof. If τ_0 and τ_1 are 2-regular, there are finite weighted automata (X_0, o_0, δ_0) and (X_1, o_1, δ_1) with elements $x_0 \in X_0$, $x_1 \in X_1$ such that $[\![x_0]\!]_S = \tau_0$ and $[\![x_1]\!]_S = \tau_1$.

Observe that, if σ satisfies the above equation, we can directly derive:

$$\sigma'(0) = a_1 a_0 + o(\tau_0) \qquad \textbf{even}(\sigma'') = a_2\sigma + a_3\sigma' + \tau_1 \quad \textbf{odd}(\sigma'') = a_1\sigma' + \tau_0'$$

We thus specify a system $(X_0 \cup X_1 \cup \{y, z\}, o, \delta)$ satisfying the behavioral differential equations for X_0 and X_1 as before, and additionally:

$$\begin{aligned} o(y) &= a_0 & \textbf{even}(y') &= a_1 y + o(x_0) & \textbf{odd}(y') &= a_2 y + a_3 z + x_1 \\ o(z) &= a_1 a_0 + o(x_0) & \textbf{even}(z') &= a_2 y + a_3 z + x_1 & \textbf{odd}(z') &= a_1 z + x_0' \end{aligned}$$

By Lemma 9, this system has a unique solution, in which $[\![y]\!]_S$ satisfies the equations for σ and $[\![z]\!]_S = [\![y]\!]_S'$. Because, given systems for τ_0 and τ_1, any solution to the original equation for σ has to satisfy all equations in the composite system, the solution for σ also is unique. $\qquad\square$

This construction now leads to a large collection of examples, directly derivable from specifications of divide and conquer-recurrences. We have used the overview on `http://oeis.org/somedcgf.html` as a basis for the examples that follow. We will not explicitly specify the constructions used: however, all examples have been obtained by a combination of the constructions presented in Propositions 10 and 18. All the examples that follow are 2-regular over the ring $\mathbb{Z}$.

Example 19. We can now specify a large amount of sequences as 2-regular sequences. In most cases, we just need the three variables x (over the ring $\mathbb{Z}$, and in cases where no negative coefficients occur also over the semiring $\mathbb{N}$), **nats** and **ones**, where x represents the sequence itself; in some cases we need a fourth variable x', but these cases still fit in the format of Proposition 18.

$o(x)$	$\mathbf{even}(x')$	$\mathbf{odd}(x')$	OEIS
1	$4x$	$4x + \mathbf{ones}$	A000695
0	$2x + 2 \cdot \mathbf{nats} - \mathbf{ones}$	$x + x' + 2 \cdot \mathbf{nats}$	A001855
0	$2x + 2 \cdot \mathbf{nats}$	$x + x' + 2 \cdot \mathbf{nats} + \mathbf{ones}$	A003314
1	$-x$	$x + \mathbf{ones}$	A004718
1	$3x$	$3x + \mathbf{ones}$	A005836
1	$2x - 1$	$x + x'$	A006165
0	$x + \mathbf{ones}$	0	A007814
1	$-x$	$\mathbf{ones} - x$	A065359
0	$2(x + \mathbf{nats} + \mathbf{ones})$	$x + x' + 2(\mathbf{nats} + \mathbf{ones})$	A067699
1	$2x + \mathbf{ones}$	$2 \cdot \mathbf{nats} + \mathbf{ones}$	A086799

Here, for example, A000695 is the Moser-de Bruijn sequence which is the ordered sequence of all numbers that can be written as a sum of distinct powers of 4; the nth element of A001855 is the maximal number of comparisons needed to sort n elements by binary insertion; and A005836 is the ordered sequence of all numbers whose base 3 representation contains no 2. For specifications of the other examples, as well as a large amount of background information about these sequences, we refer to the entries at the OEIS.

Remark 20. All the examples in this section can be easily implemented in the functional programming language Haskell. Building on the existing work on stream calculus in Haskell (e.g. [13], [11], [21]), the **zip** operation of arbitrary arity can be specified in Haskell using the (behavioral differential) equation

```
xzip (s:t) = head s : xzip (t ++ [tail s])
```

allowing us to specify all of the above examples with a single line of Haskell code. For example, the (tail of the) Nørgård sequence can now be specified by:

```
n = 1 : xzip [-n, n + ones]
```

5   Conclusions and Future Work

We have given a coalgebraic (or automata-theoretic) as well as an algebraic characterization of k-regular sequences: the k-regular sequences are exactly the sequences that are generated by finite S-weighted automata over the k-letter alphabet A_k. They are also exactly the sequences that can be defined by a finite system of linear **zip**-behavioral differential equations. We also showed that there is an isomorphism between the final A_k-automaton of formal power series and the A_k-automaton of sequences (which is then also final). This isomorphism is given by bijective k-adic numeration, and we derived from it directly that k-regular sequences are in bijective correspondence with recognizable formal power series over A_k.

The following table gives an overview of the classes of sequences and formal power series that are generated by finite deterministic, respectively weighted, automata with respect to three different final automata:

S semiring	deterministic automata	weighted automata
1-letter $(S^{\mathbb{N}}, \textbf{head}, \textbf{tail})$	eventually periodic $(= 1\text{-automatic})$	recognizable $(= 1\text{-regular})$
k-letter $(S\langle\!\langle A_k\rangle\!\rangle, O, \Delta)$	S-simple power series	recognizable power series
k-letter $(S^{\mathbb{N}}, \textbf{head}, \textbf{unzip}_k \circ \textbf{tail})$	k-automatic sequences	k-regular sequences

If S is finite, then the right-hand "weighted" column collapses and becomes equal to the left-hand "deterministic" column. Hence every finite weighted automaton with output in a finite S is equivalent to a finite deterministic automaton with output in S.

Generalization to other numeration systems. The k-adic numeration system appears to be a 'nice' choice because it is bijective and hence gives a bijective correspondence between k-regular sequences and recognizable series (which is not the case with standard base k numeration). It may be interesting to investigate whether corresponding results can be obtained with respect to other (bijective or not) numeration systems.

Relation to S-algebraic sequences. In [7], a coalgebraic characterization of algebraic power series, which generalizes context-free languages, was provided. Here, (constructively) algebraic power series can be described using systems of behavioral differential equations over a finite set X, where each derivative is given as a polynomial over X with coefficients in S. As with weighted automata, such a system can be determinized into an automaton whose states are polynomials over X with coefficients in S.

It would be interesting to see if we can connect this notion of (constructive) algebraicity to the existing notion of k-context-free sequences (see e.g. [15]), using techniques analogous to the ones used in this paper to connect k-regular sequences, recognizable sequences and power series. As a final remark, we note that it is easily possible to use the isomorphism from Section 3.1 as the basis for a definition of k-context-freeness, however, we note that the product inherited on $S^{\mathbb{N}}$ from the convolution product on $S\langle\langle A_k \rangle\rangle$ differs from the standard convolution product on $S^{\mathbb{N}}$ except for the case where $k = 1$.

References

1. Allouche, J.-P., Shallit, J.O.: The ring of k-regular sequences. Theoretical Computer Science 98, 163–197 (1992)
2. Allouche, J.-P., Shallit, J.O.: Automatic Sequences – Theory, Applications, Generalizations. Cambridge University Press (2003)
3. Allouche, J.-P., Shallit, J.O.: The ring of k-regular sequences, II. Theoretical Computer Science 307, 3–29 (2003)
4. Bartels, F.: On Generalized Coinduction and Probabilistic Specification Formats. PhD thesis, Vrije Universiteit Amsterdam (2004)
5. Bell, J.P.: On the values attained by a k-regular sequence. Advances in Applied Mathematics 34, 634–643 (2005)
6. Berstel, J., Reutenauer, C.: Noncommutative Rational Series with Applications. Cambridge University Press (2011)
7. Bonsangue, M.M., Rutten, J., Winter, J.: Defining context-free power series coalgebraically. In: Pattinson, D., Schröder, L. (eds.) CMCS 2012. LNCS, vol. 7399, pp. 20–39. Springer, Heidelberg (2012)
8. Ésik, Z., Maletti, A.: The category of simulations for weighted tree automata. International Journal of Foundations of Computer Science (IJFCS) 22(8), 1845–1859 (2011)
9. Grabmayer, C., Endrullis, J., Hendriks, D., Klop, J.W., Moss, L.S.: Automatic sequences and zip-specifications. In: Proceedings of Logic in Computer Science (LICS 2012), pp. 335–344. IEEE Computer Society Press (2012)
10. Graham, R.L., Knuth, D.E., Patashnik, O.: Concrete Mathematics: A Foundation for Computer Science, 2nd edn. Addison-Wesley Longman Publishing Co., Inc. (1994)
11. Hinze, R.: Concrete stream calculus—an extended study. Journal of Functional Programming 20(5-6), 463–535 (2011)
12. Kupke, C., Rutten, J.J.M.M.: On the final coalgebra of automatic sequences. In: Constable, R.L., Silva, A. (eds.) Kozen Festschrift. LNCS, vol. 7230, pp. 149–164. Springer, Heidelberg (2012)
13. Douglas McIlroy, M.: The music of streams. Information Processing Letters 77(2-4), 189–195 (2001)
14. Mendler, N.P., Panangaden, P., Constable, R.L.: Infinite objects in type theory. In: Proceedings of Logic in Computer Science (LICS 1986), pp. 249–255. IEEE Computer Society Press (1986)
15. Moshe, Y.: On some questions regarding k-regular and k-context-free sequences. Theoretical Computer Science 400, 62–69 (2008)
16. Rajopadhye, S.V., Panangaden, P.: Verification of systolic arrays: A stream function approach. In: International Conference on Parallel Processing (ICPP 1986), pp. 773–775. IEEE Computer Society Press (1986)

17. Rutten, J.: Universal coalgebra: a theory of systems. Theoretical Computer Science 249(1), 3–80 (2000)
18. Rutten, J.: Behavioural differential equations: a coinductive calculus of streams, automata, and power series. Theoretical Computer Science 308(1-3), 1–53 (2003)
19. Silva, A., Bonchi, F., Bonsangue, M., Rutten, J.: Generalizing determinization from automata to coalgebras. Logical Methods in Computer Science 9(1) (2013)
20. Stephan, R.: Divide-and-conquer generating functions. Part I. Elementary sequences. ArXiv Mathematics e-prints (July 2003)
21. Winter, J.: QStream: a suite of streams. In: Heckel, R., Milius, S. (eds.) CALCO 2013. LNCS, vol. 8089, pp. 353–358. Springer, Heidelberg (2013)

Automata Learning: A Categorical Perspective

Bart Jacobs and Alexandra Silva

Institute for Computing and Information Sciences, Radboud University Nijmegen
{bart,alexandra}@cs.ru.nl
April 2, 2014

Dedicated to Prakash Panangaden on the occasion of his 60th birthday

Abstract. Automata learning is a known technique to infer a finite state machine from a set of observations. In this paper, we revisit Angluin's original algorithm from a categorical perspective. This abstract view on the main ingredients of the algorithm lays a uniform framework to derive algorithms for other types of automata. We show a straightforward generalization to Moore and Mealy machines, which yields an algorithm already know in the literature, and we discuss generalizations to other types of automata, including weighted automata.

1 Introduction

One of the topics Prakash Panangaden has always been interested in is learning. He is not only a great scholar himself, but he also has a great drive to spread his knowledge, through his many lectures and discussions with colleagues around the world. We, as authors, have enjoyed and been inspired by him and his work.

On a more technical level, learning is an active area in computer science, especially in artificial intelligence. It involves deducing a (minimal) machine from observations. In this paper we explore, redescribe, and generalize part of this research using a categorical perspective. In this way we apply Prakash's favourite language to an area that is close to him — since he is a member of McGill's Reasoning and Learning Lab.

Finite automata or state machines have a wide range of applications in Computer Science. One of their applications is in verification of software systems and security protocols. Typically, the behavior of the system is modeled by a finite state machine and then desired properties, encoded in an appropriate logic, are checked against the model. Models are unfortunately not always available and the rapid changes in the system require frequent adaptations. This has motivated a lot of research into inferring or learning a model from a given system just by observing its behavior or response to certain queries.

Automata learning, or regular inference [4], is a widely used technique for creating an automaton model from observations. The regular inference algorithms provide sequences of input, so called membership queries, to a system, and observe the responses to infer an automaton. In addition, equivalence queries check whether the inferred automaton is equivalent to the system being learned. The original algorithm [4], by Dana Angluin, works for deterministic finite automata, but since then it has been extended and generalized to other types of automata [5,21,1], including Mealy machines and I/O automata, and even a special class of context-free grammars.

F. van Breugel et al. (Eds.): Panangaden Festschrift, LNCS 8464, pp. 384–406, 2014.

Category theory provides an abstract framework to study structures in mathematics and computer science. Automata are prime examples of such structures and have been studied using both algebras and coalgebras (see *e.g.* [13] and [19]) in two somewhat independent research streams. In the last few years, strengths of both perspectives on automata are being combined fruitfully, leading to the derivation of new algorithms and results. In this paper, we again explore the power of abstraction and recast the main ingredients of Angluin's algorithm using basic categorical concepts, from algebra and coalgebra, which open the door to instantiations to other types of automata, in other categories, without having to reprove correctness of the algorithm.

In this paper we sketch the straightforward generalization from deterministic automata to Moore and Mealy machines, which yields an algorithm known in the literature [21,1,2], that had been developed inspired by Angluin's algorithm but without an explicit connection of the similarities and differences in both. Our abstract view provides this connection and opens the door to even further generalizations to other types of automata.

In the proof of minimality of the inferred automaton we have used a technique that goes back to Kalman and has since then been explored by a multitude of authors, including Prakash himself. Among his many research interests, Prakash's recent activities includes using Stone-type dualities to minimize automata. He has observed that there might be connections with automata learning (personal communication). This paper provides a first step towards exploring this connection.

Organization of the paper. The rest of the paper is organized as follows. In Section 2, we recall the basic ingredients of Angluin's algorithm for deterministic automata and show how we can recast them in a categorical language. In Section 3, we present the first generalization of the algorithm based on the categorical reformulation by varying the functor under consideration and provide a learning algorithm for Mealy/Moore automata. In Section 4, we show a different type of generalization by changing the base category in which the automata are considered from **Sets** to **Vect**, obtaining in this manner an algorithm for linear weighted automata.

2 Automata Learning: The Basic Algorithm

In this section, we explain the ingredients of Angluin's original algorithm for learning deterministic finite automata and rephrase them using basic categorical constructs as we proceed.

Let us first introduce some notation and basic definitions. Let A be a finite set of labels, often called an alphabet, and A^* the set of finite words or sequences of elements of A. We will use λ to denote the empty word and, given two words $u, v \in A^*$, uv denotes their concatenation.

A language over A is a subset of words in A^*, that is $L \in 2^{A^*}$. We will often switch between the equivalent representation of a language as a set of words and as its characteristic function. Given a language L and a word $u \in A^*$, we write $L(u)$ to denote 1 if $u \in L$ and 0 otherwise.

Given two languages U and V, we will denote by $U \cdot V$ (or simply UV) the concatenation of the two languages $U \cdot V = \{uv \mid u \in U, v \in V\}$. Given a language L and $a \in A$ we can define its right and left derivative by setting

$$a^{-1}L = \{u \mid au \in L\} \qquad \text{and} \qquad La^{-1} = \{u \mid ua \in L\}.$$

A language L is *prefix-closed* if $La^{-1} \subseteq L$, for all $a \in A$, and *suffix-closed* if $a^{-1}L \subseteq L$, for all $a \in A$. Note that every non-empty suffix or prefix-closed language must contain the empty word λ. We will use $\downarrow u$ (resp. $\uparrow u$) to denote the set of prefixes (resp. suffixes) of a word $u \in A^*$.

$$\downarrow u = \{w \in A^* \mid w \text{ is a prefix of } u\} \qquad \uparrow u = \{w \in A^* \mid w \text{ is a suffix of } u\}$$

For the rest of this paper we fix a language $\mathcal{L} \in 2^{A^*}$ to be learned: the master language. This learning means that we seek a finite deterministic automaton that accepts $\mathcal{L}$. Many definitions and results are parametric in $\mathcal{L}$ but we do not always make this explicit.

2.1 Observation Tables

The algorithm of Angluin incrementally constructs an observation table with Boolean entries. The rows are labelled by words in $S \cup S \cdot A$, where S is a non-empty finite prefix-closed language, and the columns by a non-empty finite *suffix-closed* language E. For arbitrary $U, V \subseteq A^*$, define $row: U \rightarrow 2^V$ by $row(u)(v) = \mathcal{L}(uv)$. Formally, an observation table is a triple (S, E, row), where $row: (S \cup S \cdot A) \rightarrow 2^E$. Note that $\cup$ here is used for language union and not coproduct. Since row is fully determined by the language $\mathcal{L}$ we will from now on refer to an observation table as a pair (S, E), leaving the language $\mathcal{L}$ implicit.

There are two crucial properties of the observation table that play a key role in the algorithm of [4] allowing for the construction of a deterministic automaton from an observation table: closedness and consistency.

Definition 1 (Closed and Consistent Table [4]). *An observation table (S, E) is* closed *if for all $t \in S \cdot A$ there exists an $s \in S$ such that $row(t) = row(s)$. An observation table (S, E) is* consistent *if whenever s_1 and s_2 are elements of S such that $row(s_1) = row(s_2)$, for all $a \in A$, $row(s_1 a) = row(s_2 a)$.*

In many categories each map $f : A \rightarrow B$ can be factored as $f = (A \twoheadrightarrow \bullet \rightarrowtail B)$, describing f as an epimorphism followed by a monomorphism. In the category **Sets** of sets and functions epimorphisms (resp. monomorphisms) are surjections (resp. injections). Using these factorizations we come to the following categorical reformulations.

Lemma 2. *An observation table (S, E) is closed (resp. consistent) if and only if there exists a necessarilly unique map i (resp. j) such that the diagram on the left (resp. right) commutes.*

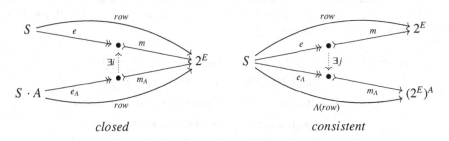

closed   *consistent*

Here, $\Lambda(row)$ is obtained by abstraction (Currying), so that $\Lambda(row)(s)(a) = row(sa)$.

Proof. Suppose the table is closed according to Definition 1. Then, for every $t \in S \cdot A$ there exists an $s \in S$ such that $row(s) = row(t)$. We define i by $i(e_A(t)) = e(s)$, using that e_A is epi/surjective. It remains to show that $m \circ i = m_A$.

$$
\begin{aligned}
(m \circ i)(e_A(t)) &= (m \circ e)(s) && \text{definition of } i \\
&= row(s) && \text{definition of } row \\
&= row(t) && \text{closedness assumption} \\
&= m_A(e_A(t)) && \text{definition of } row.
\end{aligned}
$$

The uniqueness of i is immediate using that m is monic.

Conversely, suppose that there exists i such that $m \circ i = m_A$ and let $t = ua \in S \cdot A$. Take s such that $e(s) = i(e_A(t))$ (which exists since e is epi). We need to show $row(s) = row(t)$.

$$
\begin{aligned}
row(s) &= m(e(s)) && \text{factorization of } row \\
&= m(i(e_A(t))) && \text{assumption } e(s) = i(e_A(t)) \\
&= m_A(e_A(t)) && \text{assumption } m \circ i = m_A \\
&= row(t) && \text{definition of } row.
\end{aligned}
$$

Suppose the table is consistent according to Definition 1. That is, if $s_1, s_2 \in S$ are such that $row(s_1) = row(s_2)$ then, for all $a \in A$, it holds that $row(s_1a) = row(s_2a)$. We define j by $j(e(s)) = e_A(s)$, using that e is epi. By definition, $j \circ e = e_A$. It remains to show that j is well-defined. Let s_1, s_2 be such that $e(s_1) = e(s_2)$. We need to show $e_A(s_1) = e_A(s_2)$.

$$
\begin{aligned}
e(s_1) = e(s_2) &\Rightarrow row(s_1) = row(s_2) && \text{definition of } row \\
&\Rightarrow \forall_{a \in A} \cdot row(s_1a) = row(s_2a) && \text{consistency assumption} \\
&\Rightarrow \Lambda(row)(s_1) = \Lambda(row)(s_2) && \text{definition of } \Lambda \\
&\Rightarrow (m_\Lambda \circ e_\Lambda)(s_1) = (m_\Lambda \circ e_\Lambda)(s_2) && \text{factorization of } \Lambda(row) \\
&\Rightarrow e_\Lambda(s_1) = e_\Lambda(s_2) && m_\Lambda \text{ is monic.}
\end{aligned}
$$

The uniqueness of j follows directly from using the fact that e is epi.

Conversely suppose that there exists j such that $j \circ e = e_\Lambda$ and let $s_1, s_2 \in S$ be such that $row(s_1) = row(s_2)$. We need to show $row(s_1 a) = row(s_2 a)$, for all $a \in A$ or, equivalently, $\Lambda(row)(s_1) = \Lambda(row)(s_2)$.

$$
\begin{aligned}
\Lambda(row)(s_1) &= m_\Lambda(e_\Lambda(s_1)) && \text{factorization of } \Lambda(row) \\
&= m_\Lambda(j(e(s_1))) && \text{assumption } e_\Lambda = j \circ e \\
&= m_\Lambda(j(e(s_2))) && \text{assumption } row(s_1) = row(s_2) \\
&= m_\Lambda(e_\Lambda(s_2)) && \text{assumption } e_\Lambda = j \circ e \\
&= \Lambda(row)(s_2) && \text{factorization of } \Lambda(row). \qquad \square
\end{aligned}
$$

Closed and consistent observation tables are important in the algorithm of [4] because they can be translated into a deterministic automaton. We first describe the construction concretely and subsequently more abstractly using our categorical reformulation.

Definition 3 (Automaton associated to a closed and consistent observation table [4]). *Given a closed and consistent table (S, E) one can construct a deterministic automaton $M(S, E) = (Q, q_0, \delta, F)$ where*

- *Q is a finite set of states, $F \subseteq Q$ is a set of final states and $q_0 \in Q$ is the initial state;*
- *$\delta: Q \times A \to Q$ is the transition function.*

These Q, F and δ are given by:

$$
\begin{aligned}
Q &= \{row(s) \mid s \in S\} & q_0 &= row(\lambda) \\
F &= \{row(s) \mid s \in S, row(s)(\lambda) = 1\} & \delta(row(s), a) &= row(sa).
\end{aligned}
$$

To see that this is a well-defined automaton we need to check three facts: that the initial state is indeed an element of Q; that F is a well-defined subset, or equivalently, a well-defined function of type $Q \to 2$; and that δ is a well-defined function of type $Q \times A \to Q$.

For the first, note that since S is a non-empty prefix-closed language, it must contain λ, so q_0 is an element of Q.

For the second and third points, suppose s_1 and s_2 are elements of S such that $(\star)\ row(s_1) = row(s_2)$. We must show that

$$
\lambda \in E \quad \text{and} \quad row(s_1) \in F \iff row(s_2) \in F \tag{1}
$$
$$
\delta(row(s_1), a) = \delta(row(s_2), a) \in Q, \text{ for all } a \in A. \tag{2}
$$

Since E is non-empty and suffix-closed, it must also contain λ. We also have :

$$
row(s_1) \in F \iff row(s_1)(\lambda) = 1 \overset{(\star)}{\iff} row(s_2)(\lambda) = 1 \iff row(s_2) \in F.
$$

This concludes the proof of (1) above. Since the observation table is consistent, we have for each $a \in A$, that $(\star)$ implies $row(s_1 a) = row(s_2 a)$ and hence we can calculate

$$
\delta(row(s_1), a) = row(s_1 a) = row(s_2 a) = \delta(row(s_2), a).
$$

It remains to show that $row(s_1a) \in Q$. Since the table is closed, there exists an $s \in S$ such that $row(s) = row(s_1a)$. Hence, $row(s_1a) \in Q$ and (2) above holds.

In our categorical reformulation of the construction of the automaton Q we use that epis/surjections and monos/injections in the category **Sets** form a factorization system (see *e.g.* [7]). This allows us to use the diagonal-fill-in property in the next result.

Lemma 4. *The transition function δ of the automaton associated with a closed and consistent observation table can be obtained as the unique diagonal in the following diagram,*

$$
\begin{array}{ccc}
S & \xrightarrow{\;e\;} & Q \\
{\scriptstyle\varphi}\downarrow & {\scriptstyle\delta}\nearrow & \downarrow{\scriptstyle\psi} \\
Q^A & \xrightarrow[m^A]{} & (2^E)^A
\end{array}
\qquad where \qquad
\begin{cases}
\varphi = \Lambda(i \circ e_A) \\
\psi = m_\Lambda \circ j.
\end{cases}
$$

Proof. The function δ obtained by diagonalization above satisfies:

$$\delta(e(s))(a) = \varphi(s)(a) = i(e_A(sa)).$$

This is the same as the above definition of δ, since $e(s)$ and $i(e_A(sa))$ represent, respectively, $row(s)$ and $row(sa)$. □

Definition 5 (Automaton associated with an observation table). *Let (S, E) be a closed and consistent observation table. The automaton $(Q, init, final, \delta)$ associated with the table is given in the following diagram.*

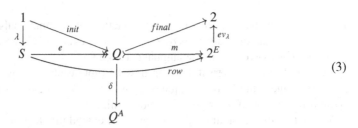

$$(3)$$

The initial state $init = row(\lambda)$ and the set of final states is given by evaluating the table in the λ column. These two functions exist because S and E are prefix- and suffix-closed, respectively. The transition function δ was defined in Lemma 4.

Next we give a categorical proof of the minimality result of [4].

Theorem 6. *The automaton associated with a closed and consistent observation table is minimal.*

Proof. An automaton is minimal if all states are reachable from the initial state and if no two different states recognize the same language (this property is referred to as observability).

Following a characterization that goes back to Kalman and then followed by other authors [16,6,9,8] these two properties can be nicely captured in the following diagram, where in the middle we have the automaton of Definition 5.

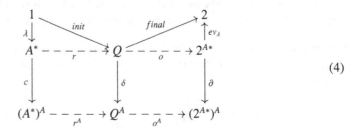

$$(4)$$

Let us define the unknown ingredients in the above diagram. On the left we have A^*, with a transition structure given by appending a letter to the end of the word:

$$c(u)(a) = ua.$$

The set A^*, together with the above transition structure, is the initial algebra of the functor $1 + A \times -$ on **Sets**. The map r exists and is unique by initiality; it sends $a_1 \cdots a_n$ to $\delta(\delta(\cdots \delta(init)(a_1) \cdots)(a_{n-1}))(a_n)$.

On the right we have 2^{A^*}, the set of languages over A, with a transition structure given by the Brzozowski/left derivative of a language:

$$\partial(L)(a) = a^{-1}L = \{u \mid au \in L\}.$$

The set 2^{A^*}, together with this transition function, is the final coalgebra of the functor $2 \times (-)^A$. The map of coalgebras $o: Q \to 2^{A^*}$ thus exists and is unique by finality. The map o assigns to every state the language it accepts.

Reachability and observability can now be rephrased in terms of properties of the functions r and o in (4): the automaton Q is *reachable* if $r: A^* \to Q$ is epic/surjective and it is *observable* if $o: Q \to 2^{A^*}$ is monic/injective.

To see that the automaton Q is minimal we extend the diagram above by including the auxiliary arrows of diagram (3). On the right we insert the mono $m: Q \rightarrowtail 2^E$ from Lemma 2 and complete the diagram:

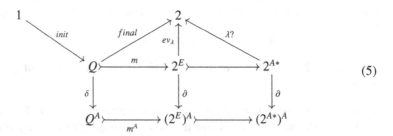

$$(5)$$

Note that the transition function $\partial: 2^E \to (2^E)^A$ given by $\partial(L)(a) = a^{-1}L$ is well-defined because the subset $E \subseteq A^*$ is suffix-closed. Moreover, note that the inclusion $E \hookrightarrow A^*$

gives rise to an injection $2^E \rightarrowtail 2^{A^*}$, by taking images (here 2^- is the covariant powerset functor). This is a map of coalgebras $2^E \rightarrowtail 2^{A^*}$, making $(2^E, \partial)$ a subcoalgebra of the final coalgebra. This transition map ∂ in (5) satisfies:

$$\partial \circ row = \Lambda(row) : S \longrightarrow (2^E)^A. \tag{6}$$

where $\Lambda(row)$ is as in Lemma 2. The proof is easy:

$$
\begin{aligned}
(\partial \circ row)(s)(a)(v) = \partial(row(s))(a)(v) = a^{-1}(row(s))(v) &= row(s)(av) \\
&= \mathcal{L}(sav) \\
&= row(sa)(v) \\
&= \Lambda(row)(s)(a)(v).
\end{aligned}
$$

One can now see that the rectangle on the left in (5) commutes by precomposing its two maps $Q \rightrightarrows (2^E)^A$ with the epi $e : S \twoheadrightarrow Q$ from Lemma 2:

$$(\partial \circ m) \circ e = \partial \circ row \overset{(6)}{=} \Lambda(row) = m_\Lambda \circ e_\Lambda = m_\Lambda \circ j \circ e = (m^A \circ \delta) \circ e,$$

where the last equation uses the definition of δ from Lemma 4. Therefore, the unique map $Q \rightarrowtail 2^{A^*}$ to the final coalgebra is a mono, being a composite of two monos in (5), and we can conclude that the automaton Q is observable.

It remains to show that the automaton Q is reachable. This means that we must show that the map $r : A^* \to Q$ in diagram (4) is surjective/epic. We are done if we can show that $r \circ n = e : S \twoheadrightarrow Q$, where we write n for the inclusion map $S \hookrightarrow A^*$.

We prove this equation $r \circ n = e$ via the mono $m : Q \rightarrowtail 2^E$, and show that $m \circ r \circ n = m \circ e = row : S \to 2^E$. We do so by induction on the length of strings $u \in S$. Thus:

$$
\begin{aligned}
(m \circ r \circ n)(\lambda) &= m(r(\lambda)) \\
&= m(init) \\
&= m(e(\lambda)) \\
&= row(\lambda) \\
(m \circ r \circ n)(ua) &= (m \circ r)(ua) \\
&= (m \circ r)(c(u)(a)) \\
&= ((m \circ r)^A)(c(u))(a) \\
&= ((m \circ r)^A \circ c)(u)(a) \\
&\overset{(4)}{=} (m^A \circ \delta \circ r)(u)(a) \\
&\overset{(5)}{=} (\partial \circ m \circ r)(u)(a) \\
&= (\partial \circ m \circ r \circ n)(u)(a) \qquad \text{since } S \text{ is prefix-closed} \\
&\overset{(IH)}{=} (\partial \circ row)(u)(a) \\
&\overset{(6)}{=} \Lambda(row)(u)(a) \\
&= row(ua).
\end{aligned}
$$

$\square$

2.2   The Learning Algorithm

We present the algorithm of [4] in Figure 1. In the algorithm, there is a teacher which has the capacity of answering two types of questions: yes/no to the query on whether a word belongs to the master language and yes/no to the question whether a certain guess of the automaton accepting the master language is correct. In the case of a negative answer of the latter question, the teacher also provides a counter-example. The learner builds an observation table by asking the teacher queries of membership of words of increasing length. Once the table is closed and consistent, the learner tries to guess the master language. We explain every step by means of an example, over the alphabet $A = \{a, b\}$.

Input: Minimally Adequate Teacher of the master language $\mathcal{L}$.
Output: Minimal automaton accepting $\mathcal{L}$.
1: **function** LEARNER
2:     $S \leftarrow \{\lambda\} ; E \leftarrow \{\lambda\}$.
3:     **repeat**
4:         **while** (S, E) is not closed or not consistent **do**
5:             **if** (S, E) is not consistent **then**
6:                 find $s_1, s_2 \in S$, $a \in A$, and $e \in E$ such that
7:                     $row(s_1) = row(s_2)$ and $\mathcal{L}(s_1ae) \neq \mathcal{L}(s_2ae)$
8:                     $E \leftarrow E \cup \{ae\}$.
9:             **end if**
10:             **if** (S, E) is not closed **then**
11:                 find $s_1 \in S$, $a \in A$ such that
12:                     $row(s_1a) \neq row(s)$, for all $s \in S$
13:                     $S \leftarrow S \cup \{s_1a\}$.
14:             **end if**
15:         **end while**
16:         Make the conjecture $M(S, E)$.
17:         **if** the Teacher replies **no** to the conjecture, with a counter-example t **then**
18:             $S \leftarrow S \cup \downarrow t$.
19:         **end if**
20:     **until** the Teacher replies **yes** to the conjecture $M(S, E)$.
21:     **return** $M(S, E)$.
22: **end function**

Fig. 1. Angluin's algorithm for deterministic finite automata [4]

Imagine the Learner receives as input a Teacher for the master language

$$\mathcal{L} = \{u \in \{a, b\}^* \mid \text{the number of } a\text{'s in } u \text{ is divisible by 3}\}.$$

In the first step of the while loop it builds a table for $S = \{\lambda\}$ and $E = \{\lambda\}$.

Step 1

	λ
$S \{$ λ	1
$S \cdot A \{$ a	0
b	1

(S, E) consistent? ✓
(S, E) closed? No, $row(a) = (\lambda \mapsto 0) \neq (\lambda \mapsto 1) = row(\lambda)$.
Then, $S \leftarrow S \cup \{a\}$ and we go to **Step 2**.

We extend the row index set S so we get a new observation table and we again check for closedness and consistency.

Step 2

	λ
λ	1
a	0
b	1
aa	0
ab	0

(S, E) consistent? ✓

(S, E) closed? ✓

Then, we guess the automaton:

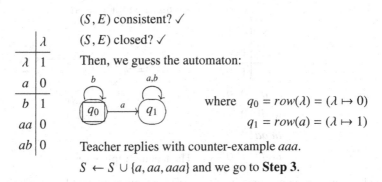

where $q_0 = row(\lambda) = (\lambda \mapsto 0)$

$q_1 = row(a) = (\lambda \mapsto 1)$

Teacher replies with counter-example aaa.

$S \leftarrow S \cup \{a, aa, aaa\}$ and we go to **Step 3**.

In the second step we managed to build a closed and consistent table which enabled us to make a first guess on the automaton. The guess was wrong so the teacher provided a counter-example, which we use to extend the row index set, generating a larger table.

Step 3

	λ
λ	1
a	0
aa	0
aaa	1
b	1
ab	0
aab	0
$aaaa$	0
$aaab$	1

(S, E) consistent?

No, $row(a) = row(aa)$ but $row(aa) \neq row(aaa)$.

Then $E \leftarrow E \cup \{a\}$ and we go to (**Step 3.1**).

In the third step the test of consistency failed for the first time and hence we extend the column index set E from $\{\lambda\}$ to $\{\lambda, a\}$. This extension allows to distinguish states (that is, rows of the table) that were indistinguishable in the previous step though they could be differentiated after an a step.

Step 3.1

	λ	a
λ	1	0
a	0	0
aa	0	1
aaa	1	0
b	1	0
ab	0	0
aab	0	1
aaaa	0	0
aaab	1	0

(S, E) consistent? ✓

(S, E) closed? ✓

We make another guess:

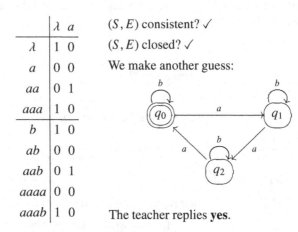

The teacher replies **yes**.

In the last step, we again constructed a closed and consistent table, which allowed us to make another guess of the automaton accepting the master language. This second guess yielded the expected automaton.

3   Angluin's Algorithm for Moore and Mealy Machines

Moore automata generalize deterministic automata by replacing the subset of final states by a function $o\colon Q \to B$, where B is a set of outputs. Mealy automata are another variation where instead of having the outputs associated to the states, each transition has an associated input and output letter. In a nutshell, here are the three types of automata we have encountered so far.

$o\colon Q \to 2$	$o\colon Q \to B$	
$\delta\colon Q \to Q^A$	$\delta\colon Q \to Q^A$	$\delta\colon Q \to (Q \times B)^A$
Deterministic automata	Moore automata	Mealy automata

Note that using the isomorphism $(Q \times B)^A \cong Q^A \times B^A$ we can also reduce Mealy to Moore automata (with a higher order output set).

Definition 1 of closedness and consistency does not depend at all on the fact that the output set is 2. More interestingly also the categorical proof of minimality (Theorem 6) is not specific for deterministic automata but can be carried over to an arbitrary Moore automaton. Hence, we can straightforwardly use all the categorical definitions and results above for an arbitrary output set B. This allow us to define what a closed and consistent table for a Moore/Mealy automaton is and derive the algorithm that infers the automaton recognizing the behavior of a Moore/Mealy automaton.

The master language $\mathcal{L}\colon A^* \to 2$ is now replaced by a weighted language (or formal power series) $\mathcal{L}\colon A^* \to B$. Here, we use our abstract view on automata. Every automaton has a canonical semantics and universe of behavior associated with it: the final coalgebra of the functor associated with the transition structure. In the case of deterministic automata the final coalgebra is precisely the set of formal languages 2^{A^*} and in the case of Moore automata it is the set of weighted languages B^{A^*}. Note that for Mealy machines where the output set is B^A we then get as semantics B^{A^+}, where A^+ denotes the set of non-empty words over A.

Apart from the change in the type of the master language, the rest of the algorithm in Figure 1 is precisely the same and also the proof of minimality carries over since we have phrased it in the general setting using the final coalgebra.

Let us now illustrate the algorithm for Mealy machines using as example the language $L: A^+ \to B$, with $A = \{a, b\}$ and $B = \mathbb{N}$, given by, for $w \in A^+$,

$$L(w) = \begin{cases} |w|_a \mod 3 & \text{if } w = ua \\ 3 & \text{if } w = aab(bb)^n, n \in \mathbb{N} \\ 0 & \text{otherwise} \end{cases}$$

where $u \in A^*$ and $|u|_a$ denotes the number of a's in the word u.

For notational convenience we will use the following equivalent representation of row

$$\frac{row: S \to (B^A)^E}{row: S \to B^{E \cdot A}}$$

This last representation, $row: S \to B^{E \cdot A}$, starting with $E \cdot A = \{\lambda\} \cdot A = A$, is precisely what can be found in the existing algorithms for Mealy machines [21,1,2].

In the first step of the algorithm we build a table for $S = \{\lambda\}$ and $E \cdot A = A$.

Step 1

	a	b
S { λ	1	0
$S \cdot A$ { a	2	0
b	1	0

(S, E) consistent? ✓

(S, E) closed? No, $row(a) \neq row(\lambda)$.

Then, $S \leftarrow S \cup \{a\}$ and we go to **Step 2**.

We extend the row index set S so we get a new observation table and we again check for closedness and consistency.

Step 2

	a	b
λ	1	0
a	2	0
b	1	0
aa	0	0
ab	2	0

(S, E) consistent? ✓

(S, E) closed? No, $row(aa) \neq row(\lambda)$ and $row(aa) \neq row(a)$.

Then, $S \leftarrow S \cup \{aa\}$ and we go to **Step 3**.

We again extend the row index set S we check for closedness and consistency of the new table.

Step 3

	a	b
λ	1	0
a	2	0
aa	0	0
b	1	0
ab	2	0
aaa	1	0
aab	0	0

(S, E) consistent? ✓

(S, E) closed? ✓

Then, we guess the Mealy automaton $\mathcal{A}$:

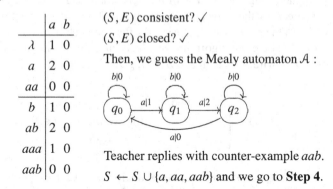

Teacher replies with counter-example aab.

$S \leftarrow S \cup \{a, aa, aab\}$ and we go to **Step 4**.

We process the teacher's counter-example and analyze the resulting table. Note that the above example is of minimal length, but in fact this is not guaranteed: the teacher can reply with an arbitrary counter-example. Shorter counter-examples do not necessarily imply less steps in the algorithm. For instance, the teacher could have replied with $aabb$ and, in fact, this would cause the algorithm to terminate in one less step than it will now.

Step 4

	a	b
λ	1	0
a	2	0
aa	0	0
aab	0	0
b	1	0
ab	2	0
aaa	1	0
$aaba$	1	0
$aabb$	0	3

(S, E) consistent?

No, $row(aa) = row(aab)$ and $row(aab) \neq row(aabb)$.

Then $E \leftarrow E \cup \{b\}$ and we go to (**Step 4.1**).

In the fourth step the consistency check failed and therefore $E \cdot A$, the column index, gets extended from $\{\lambda\} \cdot A$ to $\{\lambda, b\} \cdot A$.

Step 4.1

	a	b	ba	bb
λ	1	0	1	0
a	2	0	2	0
aa	0	0	0	0
aab	0	0	0	3
b	1	0	1	0
ab	2	0	2	0
aaa	1	0	1	0
$aaba$	1	0	1	0
$aabb$	0	3	0	0

(S, E) consistent? ✓

(S, E) closed? No, $row(aabb) \neq row(s)$, for all $s \in S$.

Then, $S \leftarrow S \cup \{aabb\}$ and we go to **Step 5**.

Step 5

	a	b	ba	bb
λ	1	0	1	0
a	2	0	2	0
aa	0	0	0	0
aab	0	0	0	3
$aabb$	0	3	0	0
b	1	0	1	0
ab	2	0	2	0
aaa	1	0	1	0
$aaba$	1	0	1	0
$aabba$	1	0	1	0
$aabbb$	0	0	0	0

(S, E) consistent? ✓

(S, E) closed? ✓

Then, we guess the Mealy automaton $\mathcal{A}$:

Teacher replies **yes**.

4   Further Generalizations: Linear Weighted Automata

In Section 3, we showed that the categorical perspective on Angluin's algorithm delivers without any extra effort an algorithm for Moore and Mealy automata. That generalization was obtained by observing that changes in the output set did not have any influence on the construction of the observation table. We will now consider yet another example in which we consider as output set $\mathbb{F}$, a field, and we construct an automaton where the state space is now a finite dimensional vector space over the field $\mathbb{F}$. These automata are known as linear weighted automata over the field $\mathbb{F}$ [12]. These examples illustrate two possible avenues of generalization: in the Mealy/Moore case we changed the functor type of the automaton whereas here we change also the underlying category (in other words the state space of the automaton) and consider automata in the category **Vect** of vector spaces and linear maps.

Formally, a linear weighted automaton over a finite alphabet A and with outputs in a field $\mathbb{F}$ is a quadruple (V, v_0, δ, ϕ) where

- V is a finite dimensional vector space over $\mathbb{F}$ and $v_0 \in V$ is the initial vector.
- $\delta \colon V \to V^A$ is a linear map determining the transition structure.
- $\phi \colon V \to \mathbb{F}$ is a linear map assigning outputs in $\mathbb{F}$ to states.

That is, a linear weighted automaton is just a Moore automaton in the category of vector spaces and linear maps.

Linear weighted automata recognize weighted languages $\mathbb{F}^{A^*}$ in the following way. A word $w = a_1 \cdots a_n \in A^*$ is assigned the weight $r \in \mathbb{F}$ if and only if

$$\phi(\delta(\delta(\cdots \delta(v_0)(a_1) \cdots)(a_{n-1}))(a_n)) = r.$$

More formally, $\mathbb{F}^{A^*}$ together with a transition structure given by

$$\lambda?(L) = L(\lambda) \qquad \partial(L)(a)(u) = L(au)$$

is the final coalgebra, in the category of vector spaces and linear maps, of the functor $\mathbb{F} \times (-)^A$. Note that $\mathbb{F}^{A^*}$ has a vector space structure given by pointwise sum and scalar multiplication. More precisely, given $r_1, \ldots, r_k \in \mathbb{F}$ and $L_1, \cdots, L_k \in \mathbb{F}^{A^*}$, we define:

$$(r_1 L_1 + \cdots + r_k L_k)(w) = r_1(L_1(w)) + \cdots + r_k(L_k(w))$$

where on the right the sum and scalar multiplication are the ones in $\mathbb{F}$. The functions $\lambda?$ and ∂ are also linear w.r.t. the vector space operations defined above.

Lemma 7. *Given a linear weighted automaton (V, v_0, δ, ϕ) there exists a unique linear homomorphism such that the following diagram commutes.*

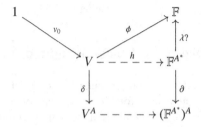

Proof. For any $v \in V$, we define $h(v)$ by induction on $w \in A^*$.

$$h(v)(\lambda) = \phi(v) \qquad\qquad h(v)(aw) = h(\delta(v)(a))(w)$$

It is easy to see that h makes the above diagram commute. It remains to show uniqueness and linearity of h. Linearity follows by induction and using the linearity of ϕ and δ.

$$
\begin{aligned}
h(r_1v_1 + \cdots + r_kv_k)(\lambda) \quad &= \phi(r_1v_1 + \cdots + r_kv_k) \\
&= r_1\phi(v_1) + \cdots + r_k\phi(v_k) &&\text{linearity of } \phi \\
&= r_1(h(v_1)(\lambda)) + \cdots + r_k(h(v_k)(\lambda)) \\
h(r_1v_1 + \cdots + r_kv_k)(aw) \quad &= h(\delta(r_1v_1 + \cdots + r_kv_k)(a))(w) \\
&= h(r_1\delta(v_1)(a) + \cdots + r_k\delta(v_k)(a))(w) &&\text{linearity of } \delta \\
&= r_1(h(\delta(v_1)(a))(w)) + \cdots + r_k(h(\delta(v_k)(a))(w)) &&\text{induction hyp.} \\
&= r_1(h(v_1)(aw)) + \cdots + r_k(h(v_k)(aw))
\end{aligned}
$$

For uniqueness, suppose there is a map g such that $\lambda? \circ g = \phi$ and $\partial \circ g = g^A \circ \delta$. Then, for any $v \in V$,

$$
g(v)(\lambda) = \lambda?(g(v)) = \phi(v) = h(v)(\lambda)
$$

and, for any $a \in A$,

$$
g(v)(aw) = (\partial(g(v))(a))(w) = g(\delta(v)(a))(w) \overset{\text{(IH)}}{=} h(\delta(v)(a))(w) = h(w)(aw). \quad \square
$$

The master language is now an element of the final coalgebra $\mathbb{F}^{A^*}$. Our categorical definitions and results – closedness, consistency and minimality – are valid also for linear weighted automata. In the definitions of closedness and consistency we use the usual epi-mono factorization of linear maps. Using the matrix representation of linear maps, epimorphisms (resp. monomorphisms) will correspond to matrices of full column (resp. row) rank.

For convenience, in the algorithm, we still want to use the sets S and E and we will represent the table as $S \to \mathbb{F}^E$. This is possible since all vector spaces are freely generated and a linear map is determined by its value on the basis vectors. First, we just recall the definitions of closedness and consistency and we instantiate them concretely. We need some notation. Let $V(S)$ denote the free vector space generated by S (if S is finite $V(S) = \mathbb{F}^S$), whose elements we will frequently denote by finite formal sums $\sum_I r_i s_i$.

An observation table is now a linear map $\overline{row}: V(S) \to \mathbb{F}^E$. An observation table (S,E) is closed (resp. consistent) if and only if there exist necessarilly unique linear maps i and j such that the diagram on the left (resp. right) commutes.

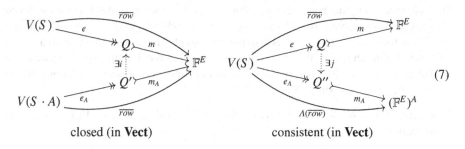

closed (in **Vect**)   consistent (in **Vect**)   (7)

The existence of the maps i and j induce a (linear) transition structure $\delta\colon Q \to Q^A$ as defined in Lemma 4.

More concretely, in the linear setting, a table is closed if for all $t \in S \cdot A$, there exist $s_i \in S$ such that

$$\overline{row}(t) = \sum_I r_i \times \overline{row}(s_i).$$

A table is consistent if whenever

$$\sum_I r_i \times \overline{row}(s_i) = \sum_J r_j \times \overline{row}(t_j),$$

for $s_i, t_j \in S$, then, for all $a \in A$,

$$\sum_I r_i \times \overline{row}(s_i a) = \sum_J r_j \times \overline{row}(t_j a).$$

The map $\overline{row}$ has some special properties, related with the fact that $V(-)$ is actually a monad. We defined V above on sets but, V can also be defined on maps $f\colon U \to T$ as $V(f)\colon V(U) \to V(T)$, where

$$V(f)\left(\sum_I r_i u_i \right) = \sum_I r_i f(u_i).$$

This $V(-)$ is a functor and, of interest to us, a monad where the unit $\eta\colon S \to V(S)$ is given by the trivial unit linear combination. Given a map $f\colon S \to V$, where V is a vector space, let $\overline{f}\colon V(S) \to V$ denote its linearization given by

$$\overline{f}\left(\sum_I r_i s_i \right) = \sum_I r_i f(s_i).$$

The linearization of f is the unique linear map satisfying $\overline{f} \circ \eta = f$. We will use this uniqueness property below in the proofs.

The map $\overline{row}$ in the diagrams above is the linearization of the map $row\colon S \to \mathbb{F}^E$ that only determines the values of the elements of S, which act as a basis for the vector space $V(S)$.

It might be interesting to observe that closedness in sets implies closedness in the linear setting, but consistency has a dual property: consistency in the linear setting implies consistency in sets. This is interesting for the algorithm: if closedness is already true for the table indexed by the basis vectors – $row\colon S \to \mathbb{F}^E$ - then it is also true for the linear view on the same table – $\overline{row}\colon V(S) \to \mathbb{F}^E$. On the other hand if $row\colon S \to \mathbb{F}^E$ is not consistent then $\overline{row}\colon V(S) \to \mathbb{F}^E$ is also not consistent. These correspondences are both interesting from an algorithmic point of view: note that the definition of consistency in the linear setting involves comparison of arbitrary linear combinations of rows, which is a rather expensive operation.

Lemma 8. *Let $row\colon S \to \mathbb{F}^E$ be an observation table and $\overline{row}\colon V(S) \to \mathbb{F}^E$ its linearization.*

① *If* row *is closed then so is* $\overline{row}$.

② *If* $\overline{row}$ *is consistent then so is* row.

Proof. Recall the definitions of $row: S \to \mathbb{F}^E$ being closed and consistent.

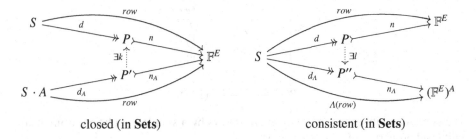

closed (in **Sets**)   consistent (in **Sets**)

First, observe that we can always define $h: V(P) \to Q$, with Q as given in (7), as follows:

$$
\begin{array}{ccc}
V(S) & \xrightarrow{\ V(d)\ } & V(P) \\
{\scriptstyle e}\downarrow & {\scriptstyle h}\nearrow & \downarrow{\scriptstyle \bar{n}} \\
Q & \xrightarrow[\ m\ ]{} & \mathbb{F}^E
\end{array}
$$

The commutativity of the above diagram follows by observing that

$$m \circ e \circ \eta = row \quad \text{and} \quad \bar{n} \circ V(d) \circ \eta = \bar{n} \circ \eta \circ d = n \circ d = row.$$

For ①, suppose row is closed and let us prove that $\overline{row}$ is also closed, according to (7). We define the map i as follows.

$$
\begin{array}{ccc}
V(S \cdot A) & \xrightarrow{\ e_A\ } & Q' \\
{\scriptstyle V(d_A)}\downarrow & & \\
V(P') & & \\
{\scriptstyle V(k)}\downarrow & {\scriptstyle i} & \downarrow{\scriptstyle m_A}\\
V(P) & & \\
{\scriptstyle h}\downarrow & & \\
Q & \xrightarrow[\ m\ ]{} & \mathbb{F}^E
\end{array}
$$

Remains to show the commutativity of the above diagram, which follows by observing that

$$m_A \circ e_A \circ \eta = \overline{row} \circ \eta = row \quad \text{and}$$
$$m \circ h \circ V(k) \circ V(d_A) \circ \eta = \bar{n} \circ \eta \circ k \circ d_A = n \circ k \circ d_A = n_A \circ d_A = row.$$

For ②, suppose $\overline{row}$ is consistent and let us now prove that row is also consistent. We define l as follows.

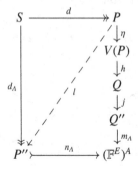

The commutativity of the above diagram follows by a simple calculation, using naturality and properties of j as given above.

$$
\begin{aligned}
m_\Lambda \circ j \circ h \circ \eta \circ d &= m_\Lambda \circ j \circ h \circ V(d) \circ \eta & &\text{naturality of } \eta \\
&= m_\Lambda \circ j \circ e \circ \eta & &h \circ V(d) = e \\
&= m_\Lambda \circ e_\Lambda \circ \eta & &j \circ e = e_\Lambda \\
&= \Lambda(\overline{row}) \circ \eta & &m_\Lambda \circ e_\Lambda = \Lambda(\overline{row}) \\
&= \Lambda(row) & & \\
&= n_\Lambda \circ d_\Lambda & &\text{factorization of } \Lambda(row) \qquad \square
\end{aligned}
$$

The converses of ① and ② in the above lemma fail, as we point out in the example below.

We will illustrate the algorithm in the linear setting with the following example over $\mathbb{F} = \mathbb{R}$.

$$
\mathcal{L}(u) = \begin{cases} 1 & \text{if } u = \lambda \\ 2|u|_b & \text{if } |u|_a \text{ is even} \\ 0 & \text{otherwise} \end{cases}
$$

In the first step we build a table for $S = \{\lambda\}$ and $E = \{\lambda\}$.

Step 1

(S, E) consistent? ✓

(S, E) closed? ✓

Then, we guess the automaton:

	λ
λ	1
a	0
b	2

$b|2$

$q_0/1$ where $q_0 = row(\lambda)$.

Teacher replies with counter-example aa.

$S \leftarrow S \cup \{a, aa\}$ and we go to **Step 2**.

Note that the above table would not be closed in **Sets** because $row(a) \neq row(\lambda)$. However, $row(a) = 0 \times row(\lambda)$ and hence the table is closed in the linear setting. Also note that we use the following conventions in the representation of the automaton:

q/r denotes $\phi(q) = row(q)(\lambda) = r$ (the output of the state) and $q \xrightarrow{\ a|r\ } q'$ denotes $\delta(q)(a) = r \times q'$.

Step 2

	λ
λ	1
a	0
aa	1
b	2
ab	0
aaa	0
aab	2

(S, E) consistent? ✓
(S, E) closed? ✓
Then, we guess the automaton:

where $q_0 = row(\lambda)$
$q_1 = row(a)$

Teacher replies with counter-example ab.
$S \leftarrow S \cup \{a, ab\}$ and we go to **Step 3**.

Step 3

	λ
λ	1
a	0
aa	1
ab	0
b	2
ab	0
aaa	0
aab	2
aba	2
abb	0

(S, E) consistent?
No, $row(a) = row(ab)$ and $row(aa) \neq row(aba)$.
Then $E \leftarrow E \cup \{a\}$ and we go to (**Step 3.1**).

Step 3.1

	λ	a
λ	1	0
a	0	1
aa	1	0
ab	0	2
b	2	0
ab	0	2
aaa	0	1
aab	2	0
aba	2	0
abb	0	4

(S, E) consistent? ✓

(S, E) closed? ✓

Then, we guess the automaton:

where $q_0 = row(\lambda)$

$q_1 = row(a)$

Teacher replies **yes**.

5 Discussion

We have presented the first steps towards a categorical understanding and generalization of Angluin's learning algorithm, originally defined for deterministic finite automata. The categorical reformulation enables us to explore two avenues of generalization: varying the functor (giving for instance different input/output for the automaton) and varying the category under study (changing for instance the type of computations involved). The variations we concretely considered in this paper were rather mild but interestingly enough yielded algorithms for Mealy/Moore automata and linear weighted automata.

The possibilities of further generalizations are vast. We would like to provide a categorical proof of termination of the algorithm. This requires a categorical understanding on how the algorithm determines which rows/columns need to be added in order for the observation table to be closed/consistent. Once this is understood, we expect that further variations on the functor can also be considered. We conjecture that there is a deep connection with the construction of the initial and final sequence of two functors. On the one hand, the functor whose initial algebra determines the *experiments/queries* needed to build the observation table. In the case we studied here the functor in question is $1 + A \times -$. On the other hand, the functor whose final coalgebra determines the notion of behavior. In all our examples, this was the functor $B \times (-)^A$. We expect that the connection between the two functors can be explained by duality, as Prakash also hinted to us (personal communication).

The production of counter-examples can also be a good place for exploring enhancements of the algorithm. In this subject, recent work on bisimulations will be of use, as we explain next. In order to return a counter-example, the teacher essentially tries to build a bisimulation relation between the guessed automaton $M(S, E)$ and the actual minimal automaton recognizing the master language. The latter is $\langle \mathcal{L} \rangle$ the subcoalgebra of the final coalgebra $(2^{A^*}, \langle \lambda?, \partial \rangle)$ generated by $\mathcal{L}$. Let us illustrate this with the first counter-example generated on page 393 in the example for deterministic automata. The procedure of constructing a bisimulation containing the initial state of the guessed automaton and the master language is depicted below. The pairs added to the bisimulation are connected by a dashed line.

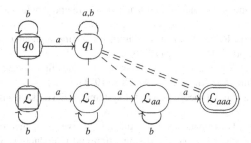

The double dashed line, in red, shows the first contradiction in the bisimulation construction: q_1 has to be related to $\mathcal{L}_{aaa}$ but they differ in their output. Hence the path leading to these implies that aaa is the counter-example returned.

Finding counter-examples can be optimized by using enhancements of the bisimulation method. In the case of deterministic automata, this was first observed in the 70's by Hopcroft, Karp, and Tarjan [14,3,15] and has since then been improved (see e.g [22,11,18]) and explored in other contexts, notably in concurrency theory [20,17,10].

References

1. Aarts, F.: Inference and abstraction of communication protocols. Master's thesis, Radboud University Nijmegen and Uppsala University (2009)
2. Aarts, F., de Ruiter, J., Poll, E.: Formal models of bank cards for free. In: ICST Workshops, pp. 461–468. IEEE (2013)
3. Aho, A.V., Hopcroft, J.E., Ullman, J.D.: The Design and Analysis of Computer Algorithms. Addison-Wesley (1974)
4. Angluin, D.: Learning regular sets from queries and counterexamples. Inf. Comput. 75(2), 87–106 (1987)
5. Angluin, D., Csürös, M.: Learning Markov chains with variable memory length from noisy output. In: Freund, Y., Schapire, R.E. (eds.) COLT, pp. 298–308. ACM (1997)
6. Arbib, M.A., Manes, E.G.: Adjoint machines, state-behavior machines, and duality. Journal of Pure and Applied Algebra 6(3), 313–344 (1975)
7. Barr, M., Wells, C.: Toposes, Triples and Theories. Springer, Berlin (1985) Revized and corrected version available from URL,
   www.cwru.edu/artsci/math/wells/pub/ttt.html
8. Bonchi, F., Bonsangue, M., Hansen, H., Panangaden, P., Rutten, J., Silva, A.: Algebra-coalgebra duality in Brzozowski's minimization algorithm. In: ACM Transactions on Computational Logic (TOCL) (2014)
9. Bonchi, F., Bonsangue, M.M., Rutten, J.J.M.M., Silva, A.: Brzozowski's Algorithm (Co)Algebraically. In: Constable, R.L., Silva, A. (eds.) Kozen Festschrift. LNCS, vol. 7230, pp. 12–23. Springer, Heidelberg (2012)
10. Bonchi, F., Caltais, G., Pous, D., Silva, A.: Brzozowski's and up-to algorithms for must testing. In: Shan, C.-C. (ed.) APLAS 2013. LNCS, vol. 8301, pp. 1–16. Springer, Heidelberg (2013)

11. Bonchi, F., Pous, D.: Checking nfa equivalence with bisimulations up to congruence. In: Giacobazzi, R., Cousot, R. (eds.) POPL, pp. 457–468. ACM (2013)
12. Boreale, M.: Weighted bisimulation in linear algebraic form. In: Bravetti, M., Zavattaro, G. (eds.) CONCUR 2009. LNCS, vol. 5710, pp. 163–177. Springer, Heidelberg (2009)
13. Eilenberg, S.: Automata, languages, and machines. Pure and Applied Mathematics. Elsevier Science (1974)
14. Hopcroft, J.E., Karp, R.M.: A linear algorithm for testing equivalence of finite automata. Technical report, Cornell University (1979)
15. Hopcroft, J.E.: An n log n algorithm for minimizing states in a finite automaton. Technical report, Stanford University, Stanford, CA, USA (1971)
16. Kalman, R.: On the general theory of control systems. IRE Transactions on Automatic Control 4(3), 110–110 (1959)
17. Pous, D., Sangiorgi, D.: Enhancements of the coinductive proof method. In: Sangiorgi, D., Rutten, J. (eds.) Advanced Topics in Bisimulation and Coinduction. Cambridge Tracts in Theoretical Computer Science, vol. 52, Cambridge University Press (November 2011)
18. Rot, J., Bonsangue, M., Rutten, J.M.M.: Coalgebraic bisimulation-up-to. In: van Emde Boas, P., Groen, F.C.A., Italiano, G.F., Nawrocki, J., Sack, H. (eds.) SOFSEM 2013. LNCS, vol. 7741, pp. 369–381. Springer, Heidelberg (2013)
19. Rutten, J.J.M.M.: Automata and coinduction (an exercise in coalgebra). In: Sangiorgi, D., de Simone, R. (eds.) CONCUR 1998. LNCS, vol. 1466, pp. 194–218. Springer, Heidelberg (1998)
20. Sangiorgi, D.: Beyond Bisimulation: The "up-to" Techniques. In: de Boer, F.S., Bonsangue, M.M., Graf, S., de Roever, W.-P. (eds.) FMCO 2005. LNCS, vol. 4111, pp. 161–171. Springer, Heidelberg (2006)
21. Shahbaz, M., Groz, R.: Inferring Mealy Machines. In: Cavalcanti, A., Dams, D.R. (eds.) FM 2009. LNCS, vol. 5850, pp. 207–222. Springer, Heidelberg (2009)
22. De Wulf, M., Doyen, L., Henzinger, T.A., Raskin, J.-F.: Antichains: A new algorithm for checking universality of finite automata. In: Ball, T., Jones, R.B. (eds.) CAV 2006. LNCS, vol. 4144, pp. 17–30. Springer, Heidelberg (2006)

Optimal Coin Flipping

Dexter Kozen

Department of Computer Science
Cornell University
Ithaca, New York 14853-7501, USA
kozen@cs.cornell.edu
http://www.cs.cornell.edu/~kozen

In honor of Prakash Panangaden on the occasion of his sixtieth birthday

Abstract. This paper studies the problem of simulating a coin of arbitrary real bias q with a coin of arbitrary real bias p with minimum loss of entropy. We establish a lower bound that is strictly greater than the information-theoretic bound. We show that as a function of q, it is an everywhere-discontinuous self-similar fractal. We provide efficient protocols that achieve the lower bound to within any desired accuracy for $(3 - \sqrt{5})/2 < p < 1/2$ and achieve it exactly for $p = 1/2$.

Keywords: probabilistic protocols, randomness, entropy.

1 Introduction

A *discrete simulation protocol* is any procedure that maps a stream of digits from one alphabet to a stream of digits from another alphabet. If the input sequence comes from a random process, then the statistical properties of the input stream impart statistical properties to the output stream, and we can think of the protocol as a reduction from one random source to another.

The *efficiency* of the simulation is the rate of entropy produced per unit of entropy consumed [4,7]. The efficiency measures the amount of randomness lost in the conversion. By general information-theoretic considerations, this value cannot exceed unity [1,6]. In general, the efficiency may not exist, or it may exist but vary with time.

A paradigmatic example is the simulation of a coin of arbitrary real bias q with a coin of arbitrary real bias p. Here, both the input and output alphabets are binary, the input is a sequence of i.i.d. bias-p coin flips, $0 < p < 1$, and the output is a sequence of i.i.d. bias-q coin flips, $0 \le q \le 1$. We call this a *p, q-simulation protocol*. For such protocols, the efficiency is

$$\frac{H(q) \cdot E_{\text{prod}}}{H(p) \cdot E_{\text{cons}}},$$

where H is the Shannon entropy

$$H(p) = -p \log p - (1 - p) \log(1 - p)$$

F. van Breugel et al. (Eds.): Panangaden Festschrift, LNCS 8464, pp. 407–426, 2014.
© Springer International Publishing Switzerland 2014

and E_{prod} and E_{cons} are, respectively, the expected number of output digits produced and the expected number of input digits consumed in one round of the protocol. If $E_{\text{prod}} = 1$, this gives an information-theoretic lower bound

$$E_{\text{cons}} \geq \frac{H(q)}{H(p)}$$

on the number of bias-p coin flips required by the protocol to produce one output digit. To maximize the efficiency of the simulation, we should minimize this quantity.

A classical example of a $p, \frac{1}{2}$-simulation protocol is the *von Neumann trick* [9]. The bias-p coin is flipped twice. If the outcome is HT, the protocol halts and declares H for the fair coin. If the outcome is TH, the protocol halts and declares T. On any other outcome, the process is repeated. This protocol has the advantage that it is oblivious to the bias of the input coin, but its efficiency is quite poor even for p close to $1/2$. For example, for $p = 1/3$, the von Neumann trick consumes 4.5 input digits per output digit, whereas the Shannon bound is only $1/(\log 3 - 2/3) \approx 1.083 \cdots$.

More efficient simulations and enhancements have been studied in [4,7,13]. It is known that any discrete i.i.d. process can simulate any other discrete i.i.d. process with efficiency asymptotically approaching 1, provided the protocol is allowed unbounded *latency*; that is, it may wait and produce arbitrarily long strings of output digits at once. Unbounded latency is exploited in [7] to simulate a fair coin with an arbitrary coin with asymptotically optimal efficiency. The technique is a generalization of the von Neumann trick. In the other direction, [6, Theorem 5.12.3] shows that a fair coin can in principle generate one output digit of an arbitrary coin with expected consumption at most two more than the entropy. In conjunction with [6, Theorem 5.4.2], this yields a method for generating a sequence of i.i.d. bias-q coins from a fair coin with efficiency asymptotically approaching 1, again allowing unbounded latency.

In this paper we consider *non-oblivious, one-bit output* protocols: those that output exactly one output digit in each round but take advantage of the knowledge of p. For fixed $0 < p < 1$, let $E_{\text{opt}}(q)$ be the infimum of E_{cons} over all one-bit output p, q-simulation protocols. We show:

- The function $E_{\text{opt}}(q)$ is an everywhere-discontinuous self-similar fractal. For all but finitely many points, it is strictly larger than the Shannon bound $H(q)/H(p)$. A graph of E_{opt} compared to the Shannon bound for $p = 1/3$ is shown in Fig. 1.
- For all $0 \leq q \leq 1$, there exists a p, q-simulation protocol that achieves $E_{\text{opt}}(q)$. Previously, this was known only for $p = 1/2$ [6].
- There exists a single *residual probability protocol* that is optimal for all q. A *residual probability protocol* is a protocol whose state set is the closed unit interval $[0, 1]$ and the probability of halting and reporting heads (respectively, tails) starting from state q is q (respectively, $1 - q$). It is optimal for all q in the sense that $E_{\text{cons}}(q) = E_{\text{opt}}(q)$. The protocol is nondeterministic, and it is not known whether it can be made deterministic in general, even for rational p and q.

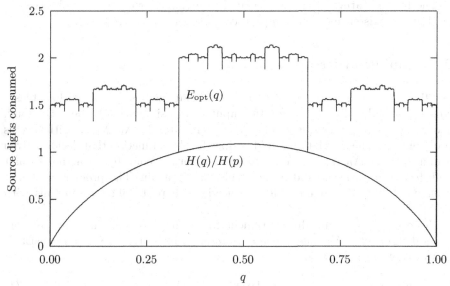

Fig. 1. Comparison of $E_{opt}(q)$ and the Shannon bound $H(q)/H(p)$ for $p = 1/3$

- For $(3 - \sqrt{5})/2 < p \leq 1/2$, we exhibit a family of deterministic, efficiently computable[1] residual probability protocols that achieve $E_{opt}(q) + \varepsilon$ for any desired degree of accuracy $\varepsilon > 0$ and all q.
- For a fair input coin ($p = 1/2$), we show that the optimal residual probability protocol is computable, and determine the values of $E_{opt}(q)$ exactly. A similar protocol for $p = 1/2$ was proposed in [6] but without proof, and the values of E_{opt} were not established.

Some of the proof techniques we use are somewhat nonstandard. One particular innovation is the coalgebraic formulation of stochastic simulation protocols introduced in Section 2. In contrast to the usual formulation of stochastic processes as sequences of random variables, this approach gives a powerful technique for reasoning about various functions defined as fixpoints of recursive equations.

1.1   Other Related Work

There is a large body of interesting work on extracting randomness from weak random sources (e.g. [10,11,14,15]). These models typically work with imperfect knowledge of the input source and provide only approximate guarantees on the quality of the output. In this paper, however, we assume that the statistical properties of the input and output are known completely, and simulations must be exact.

---

[1] As we are computing with real numbers, we assume unit-time real arithmetic and comparison of real numbers. These assumptions are not necessary if computation is restricted to the rationals.

The fractal nature of certain residual probability protocols was observed in [8], but the existence of optimal protocols was left unresolved.

2   Simulation Protocols

Let $0 < p \leq 1/2$ and $0 \leq q \leq 1$. To simulate a bias-q coin with a bias-p coin, we would ordinarily define the input to the simulation to be a Bernoulli process consisting of a sequence of i.i.d. random variables $X_0, X_1, \ldots$ with success probability p. The simulation would be specified by a function that decides, given a finite history $X_0, X_1, \ldots, X_{n-1}$ of previous bias-p coin flips, whether to halt and declare heads, halt and declare tails, or flip again. The process must halt with probability 1 and must declare heads with probability q and tails with probability $1 - q$.

However, it is technically convenient to specify protocols in terms of more general state sets. We thus define a *protocol* to be a triple (S, β, s_0) consisting of a coalgebra (S, β), where

$$\beta : S \; \rightarrow \; \{\text{H}, \text{T}\} + (\{0, 1\} \rightarrow S), \tag{1}$$

and a distinguished *start state* $s_0 \in S$.[2] Intuitively, depending on the current state, the protocol decides either

- to halt immediately and return H or T, thereby declaring the result of the bias-q coin flip to be heads or tails, respectively; or
- to consume a random bias-p coin flip (0 or 1), and based on that information, enter a new state.

A protocol is a p, q-*simulation protocol* if, when it is started in its start state s_0 with the input stream generated by a Bernoulli process with success probability p, it halts with probability 1, declaring H with probability q and T with probability $1 - q$.

The protocol is *computable* if the function β is.

Example 1. A traditional choice for the state set would be $\{0, 1\}^*$, the history of outcomes of previous bias-p coin flips. The transition function would be

$$\beta : \{0, 1\}^* \rightarrow \{\text{H}, \text{T}\} + (\{0, 1\} \rightarrow \{0, 1\}^*),$$

and the start state would be the empty history $\varepsilon \in \{0, 1\}^*$. The next step of the protocol is determined by the previous history. If this history is $X_0, \ldots, X_{n-1}$ and the protocol decides to halt and declare heads or tails, then $\beta(X_0, \ldots, X_{n-1})$ would be H or T, respectively. If on the other hand the protocol decides not to halt, and the result of the next bias-p coin flip is X_n, then $\beta(X_0, \ldots, X_{n-1})(X_n) = X_0, \ldots, X_n$.

---

[2] For clarity, we are using different symbols to distinguish the input coin (heads = 0, tails = 1) from the output coin (heads = H, tails = T).

Example 2. The following example is a slight modification of one from [8]. The state set is the closed real interval $[0,1]$. If $q \in \{0,1\}$, then $\beta(q) \in \{\text{H}, \text{T}\}$, otherwise $\beta(q) \in \{0,1\} \to [0,1]$. The values are

$$\beta(q) = \begin{cases} \text{H} & \text{if } q = 1 \\ \text{T} & \text{if } q = 0 \end{cases}$$

$$\beta(q)_X = \begin{cases} 0 & \text{if } 0 < q \le p \text{ and } X = 1 \\ \dfrac{q}{p} & \text{if } 0 < q \le p \text{ and } X = 0 \\ \dfrac{q-p}{1-p} & \text{if } p < q < 1 \text{ and } X = 1 \\ 1 & \text{if } p < q < 1 \text{ and } X = 0. \end{cases}$$

Intuitively, if $p < q < 1$ and the bias-p coin flip returns heads (0), which occurs with probability p, then we halt and output heads; this gives a fraction p/q of the desired probability q of heads of the simulated bias-q coin. If the bias-p coin returns tails (1), which occurs with probability $1 - p$, we rescale the problem to condition on that outcome, setting the state to $(q - p)/(1 - p)$ because that is the residual probability of heads, and repeat. Similarly, if $0 < q \le p$ and the bias-p coin returns tails, then we halt and output tails; and if not, we rescale appropriately and repeat.

Example 3. The final coalgebra $(\mathcal{C}, \delta)$ of the type (1) is the set of binary prefix codes for the two-element alphabet $\{\text{H}, \text{T}\}$. Each such code consists of a pair of disjoint sets $H, T \subseteq \{0,1\}^*$ such that the elements of $H \cup T$ are pairwise prefix-incomparable. The operation δ is defined by

$$\delta(H,T) = \begin{cases} \text{H} & \text{if } \varepsilon \in H \\ \text{T} & \text{if } \varepsilon \in T \\ \lambda a \in \{0,1\}.(D_a(H), D_a(T)) & \text{otherwise,} \end{cases}$$

where D_a is the Brzozowski derivative $D_a(A) = \{x \mid ax \in A\}$.

The coalgebra $(\mathcal{C}, \delta)$ is *final* in the sense that from any other coalgebra (S, β), there is a unique coalgebra homomorphism code : $(S, \beta) \to (\mathcal{C}, \delta)$, defined by: $\text{code}(s) = (H_s, T_s)$, where H_s (respectively, T_s) is the set of strings $x \in \{0,1\}^*$ such that running the protocol starting from s results in output H (respectively, T) after consuming input digits x. The function code is a coalgebra homomorphism in that $\delta(\text{code}(s)) = \beta(s)$ if $\beta(s) \in \{\text{H}, \text{T}\}$, otherwise $\beta(s) : \{0,1\} \to S$ and $\delta(\text{code}(s)) = \text{code} \circ (\beta(s)) : \{0,1\} \to \mathcal{C}$.

In the definition of $\mathcal{C}$, the sets H, T must be disjoint to ensure that δ is well-defined. They need not be nonempty; in fact, if $\beta(s) = \text{H}$, then $h(s) = (\{\varepsilon\}, \varnothing)$. There is no other possible choice for $h(s)$ due to the requirement that H, T be disjoint and elements of $H \cup T$ be pairwise prefix-incomparable. The single-state subcoalgebra $(\varnothing, \varnothing)$ represents protocols that never halt.

2.1   Residual Probability Protocols

Intuitively, the *residual probability* of a state s of a p, q-simulation protocol is the probability r that the protocol halts and declare heads when started in state s. In order to halt with probability 1 from that state, it should also halt and declare tails with probability $1 - r$. It is conceivable that a protocol might want to take different actions in two different states, even if the residual probabilities are the same.

Formally, a *residual probability protocol* is a protocol whose state space is the closed unit interval $[0, 1]$ and whose probability of halting and declaring heads (resp., tails) when started in state q is q (resp., $1 - q$). Thus the next action of the protocol depends only on the residual probability. Example 2 is an example of a residual probability protocol. Theorem 4 below says that when searching for an optimal protocol, we can restrict our attention to residual probability protocols without loss of generality.

2.2   Impatient Protocols

A protocol (S, β) is *impatient* if in every state s, the probability of halting in at most one step is nonzero; that is, either $\beta(s)$, $\beta(\beta(s)_0)$, or $\beta(\beta(s)_1 \in \{\text{H}, \text{T}\})$. Assuming computable real arithmetic and comparison of real numbers[3], every p, q has a computable impatient protocol; for example, the protocol of Example 2, as well as others described in [8], are computable and impatient. Every impatient protocol has at most one infinite computation path starting from any state, which occurs with probability 0.

Impatient strategies are not necessarily optimal. Example 2 is not: in that example, $\beta(1 - p)_0 = 1$ and $\beta(1 - p)_1 = (1 - 2p)/(1 - p)$, whereas a better choice would be $\beta(1 - p)_0 = 0$ and $\beta(1 - p)_1 = 1$.

2.3   Greedy Protocols

Greedy protocols are a special class of impatient residual probability protocols. Intuitively, a protocol is locally greedy at a state if it attempts to optimize in the next step by halting as early as possible with the maximum allowable probability. To define this formally, we start with the special case

$$(1 - p)^2 \le p \le 1 - p; \tag{2}$$

equivalently, $(3 - \sqrt{5})/2 \le p \le 1/2$. In this case, let us define the *ambiguous region* as the open interval $(p, 1 - p)$. A greedy protocol must halt immediately when $q \in \{0, 1\}$, declaring heads for the q-coin if $q = 1$ and tails if $q = 0$. Otherwise, if q is not in the ambiguous region, it must flip the p-coin and halt if the outcome is tails, which occurs with probability $1 - p$, declaring either tails or heads for the q-coin, depending on whether $q \le p$ or $q \ge 1 - p$, respectively. If q is in the ambiguous region, it must flip the p-coin and halt if the outcome is heads,

---

[3] If p and q are rational, this assumption is not needed.

which occurs with probability p, but there is a choice whether to declare heads or tails for the q-coin, leading to two possible greedy strategies. If it declares heads when the p-coin returns heads, then it must rescale to $(q - p)/(1 - p)$ when the p-coin returns tails. If it declares tails when the p-coin returns heads, then it must rescale to $q/(1 - p)$ when the p-coin returns tails. It is not immediately clear which action will ultimately be better.

The significance of the restriction (2) is that the protocol exits the ambiguous region after only one step, and that is the case that we will focus on in this paper. More generally, let $k = \lfloor -1/\log_2(1 - p) \rfloor$, the least positive integer such that $(1-p)^{k+1} < 1/2$. The *ambiguous region* for p is the open interval $(b, 1-b)$, where b is either $1-(1-p)^k$ or $(1-p)^{k+1}$, depending on which interval is smaller. Under the restriction (2), $k = 1$. In this more general situation, a protocol is *greedy* if it moves so as to enter one of the regions $q \leq p$ or $q \geq 1 - p$ as quickly as possible; this is determined except when q is in the ambiguous region.

Greedy strategies are not necessarily optimal. For example, let p be a transcendental number satisfying (2). There is an uncountable nowhere-dense set of points on which the greedy strategy achieves its best running time $1/(1 - p)$; that is, the protocol never enters the ambiguous region. It can be shown that these are exactly finite and infinite alternating sums of increasing integer powers of p:

$$J = \{p^{k_0} - p^{k_1} + p^{k_2} - p^{k_3} + \cdots \mid k_i \in \mathbb{Z},\ 0 \leq k_0 < k_1 < \cdots\}.$$

Consider $q = 2p(1-p)$. Then $p < q < 1-p$, so q is in the ambiguous region. After one greedy step in either direction, it is easily checked that the resulting image of q is not in J. Moreover, there must subsequently be an infinite computation path, because otherwise p would be algebraic. Thus the expectation of any greedy protocol is strictly larger than $p + (1-p)(1 + 1/(1-p)) = 2$. A better strategy is to flip the p-coin twice, declaring heads if the outcome is 10 or 01, tails otherwise. The expectation is 2, and this is optimal.

3   Coalgebras and Fixpoint Induction

Technically, coalgebras of type (1) are F-coalgebras, where $F : \mathsf{Set} \to \mathsf{Set}$ is the polynomial functor $FX = \mathbb{1} + \mathbb{1} + X^2$. Given an F-coalgebra (S, β), many interesting functions can be specified by providing an F-algebra (A, α) with some extra order structure allowing for the existence of least fixpoints. The function defined is the least fixpoint of the map

$$f \mapsto \alpha \circ Ff \circ \beta, \qquad (3)$$

that is, the least f such that the following diagram commutes:

$$
\begin{array}{ccc}
S & \xrightarrow{\ f\ } & A \\
{\scriptstyle \beta}\downarrow & & \uparrow{\scriptstyle \alpha} \\
FS & \xrightarrow[\ Ff\ ]{} & FA
\end{array}
$$

Intuitively, the destructor $\beta : S \to FS$ computes the arguments to a recursive call, the map $Ff : FS \to FA$ is the recursive call, and the constructor $\alpha : FA \to A$ is the construction applied to the returned element. This general scheme for recursively defined functions has been previously studied in [2,3,5].

If A is a chain-complete partially ordered set and α order-continuous, then the map (3) is monotone and order-continuous on functions $S \to A$ under the pointwise order, therefore by the Knaster–Tarski theorem has a unique least fixpoint.

Example 4. The *outcome* $O(s)$ of the simulation starting from state s is a random variable defined on the probability space $\{0,1\}^\omega$ taking values in $\{H, T, \bot\}$. The value $\bot$ signifies nonhalting. Formally,

$$O : S \to \{0,1\}^\omega \to \{H, T, \bot\}$$

is the least fixpoint of the equation

$$O(s)(X \cdot \sigma) = \begin{cases} \beta(s) & \text{if } \beta(s) \in \{H, T\} \\ O(\beta(s)_X)(\sigma) & \text{if } \beta(s) \in \{0,1\} \to S. \end{cases}$$

This would be specified by the F-algebra (A, α), where

$$A = \{0,1\}^\omega \to \{H, T, \bot\}$$

$$\alpha(f) = \begin{cases} \lambda\sigma \in \{0,1\}^\omega.f & \text{if } f \in \{H, T\} \\ \lambda\sigma \in \{0,1\}^\omega.f(\text{head } \sigma)(\text{tail } \sigma) & \text{if } f \in \{0,1\} \to A \end{cases}$$

under the pointwise ordering on A induced by $\bot < H$ and $\bot < T$.

Example 5. Define $P(s) = \Pr(O(s) = H)$, the probability that the outcome is heads starting in state s. This is specified by the F-algebra on $[0,1]$ with constructor

$$X \mapsto \begin{cases} 1 & \text{if } X = H \\ 0 & \text{if } X = T \\ p \cdot X(0) + (1-p) \cdot X(1) & \text{if } X \in \{0,1\} \to [0,1] \end{cases}$$

and the natural order on $[0,1]$.

Example 6. The expected consumption of input digits starting from state s satisfies the equation

$$E(s) = \begin{cases} 0 & \text{if } \beta(s) \in \{H, T\} \\ 1 + p \cdot E(\beta(s)_0) + (1-p) \cdot E(\beta(s)_1) & \text{if } \beta(s) \in \{0,1\} \to S. \end{cases}$$

The function E is specified by the F-algebra on $\mathbb{R}^+ = \{x \in \mathbb{R} \mid x \geq 0\} \cup \{\infty\}$ with constructor

$$X \mapsto \begin{cases} 0 & \text{if } X \in \{H, T\} \\ 1 + p \cdot X(0) + (1-p) \cdot X(1) & \text{if } X \in \{0,1\} \to \mathbb{R}^+ \end{cases}$$

and the natural order on $\mathbb{R}^+$.

An important property for our purposes, due to Adámek, Milius, and Velebil [3], is that the least fixpoint construction is *natural* in S the sense that if f and f' are the least solutions of (3) in the F-coalgebras S and S', respectively, and if $h : S \to S'$ is an F-coalgebra morphism, then $f = f' \circ h$ (Theorem 1 below). The significance of this property is that a function defined by (3), such as the probability of heads or the expected comsumption of input digits, is the same whether measured in S or any quotient of S by a bisimulation. In particular, if $s \in S$ is a start state of a protocol and $\mathsf{code}(s) \in C$ is its image in the final coalgebra, then the expected consumption of input digits starting in state s is just the expected codeword length $\sum_{x \in \mathsf{code}(s)} \Pr(x) \cdot |x|$ if $P(s) = 1$, or ∞ if $P(s) < 1$.[4]

Theorem 1 ([3], Proposition 3.5). *Let (A, α) be an ordered F-algebra such that A is a chain-complete and α order-continuous. The construction of the least fixpoint of (3) is natural in S; that is, if $h : S \to S'$ is an F-coalgebra morphism, then $f_S = f_{S'} \circ h$.*

Proof. Let τ_S be the map (3) on functions $S \to A$. The assumptions on A and α imply that τ_S is monotone and order-continuous under the pointwise order on $S \to A$. Let $\perp$ be the bottom element of A. The map $\lambda s \in S.\perp$ is the bottom element of $S \to A$. If $h : S \to S'$ is an F-coalgebra morphism, then clearly $\lambda s \in S.\perp = (\lambda s \in S'.\perp) \circ h$, therefore the selection of $\lambda s \in S.\perp$ is natural in S. Moreover, it is easily argued that τ_S is also natural in S. By induction, $\tau_S^n(\lambda s \in S.\perp)$ is natural in S for all n. By continuity, the least fixpoint is $\sup_n \tau_S^n(\lambda s \in S.\perp)$, and the result follows from the observation that suprema are preserved by composition with h on the right.   □

3.1   Fixpoint Induction

The construction of the least fixpoint of the monotone map τ_S admits the use of the following *fixpoint induction rule* [12]: If $f : S \to A$ is the least fixpoint of τ_S, and if $\tau_S(g) \leq g$, then $f \leq g$.

3.2   Two Metrics

A popular metric on streams defines the distance between two streams to be 2^{-n} if n is the length of their maximal common prefix, or 0 if the streams are equal. There is an analogous metric on codes. We say that binary codes $s = (H, T)$ and $t = (H', T')$ *agree to length n* if for all words $x \in \{0, 1\}^*$ of length n or less, $x \in H$ iff $x \in H'$ and $x \in T$ iff $x \in T'$. We define $d'(s, t) = p^n$ if n is the maximum number such that s and t agree to length n, or 0 if they are equal. We use p^n

---

[4] Here $\Pr(x) = p^{\#0(x)}(1 - p)^{\#1(x)}$, where $\#a(x)$ is the number of occurrences of a in x for $a \in \{0, 1\}$ and $x \in \{0, 1\}^*$. We write $x \in \mathsf{code}(s)$ for $x \in H \cup T$, where $\mathsf{code}(s) = (H, T)$ is the image of state s under the unique F-coalgebra morphism to the final F-coalgebra C.

instead of 2^{-n} for technical reasons, but the difference is of no consequence, as the same topology is generated. The metric d' satisfies the recurrence

$$d'(s,t) = \begin{cases} 1 & \text{if either } \delta(s), \delta(t) \in \{\text{H}, \text{T}\} \text{ and } \delta(s) \neq \delta(t) \\ 0 & \text{if both } \delta(s), \delta(t) \in \{\text{H}, \text{T}\} \text{ and } \delta(s) = \delta(t) \\ p \cdot \max(d'(\delta(s)_0, \delta(t)_0), d'(\delta(s)_1, \delta(t)_1)) \\ \quad \text{if } \delta(s), \delta(t) \in \{0, 1\} \to \mathcal{C}, \end{cases}$$

and in fact this can be taken as a formal definition according to (3). A similar map d' is induced on the states of any protocol by $d'(s,t) = d'(\text{code}(s), \text{code}(t))$, where code is the unique F-coalgebra morphism to $\mathcal{C}$. On arbitrary protocols, the map d' is not a metric in general, but only a pseudometric.

Alternatively, we might consider two protocols similar if, when run simultaneously, they halt at the same time and produce the same output with high probability. Thus we define $d : S \times T \to [0,1]$ to be the least solution of the equation

$$d(s,t) = \begin{cases} 1 & \text{if either } \beta(s), \beta(t) \in \{\text{H}, \text{T}\} \text{ and } \beta(s) \neq \beta(t) \\ 0 & \text{if both } \beta(s), \beta(t) \in \{\text{H}, \text{T}\} \text{ and } \beta(s) = \beta(t) \\ p \cdot d(\beta(s)_0, \beta(t)_0) + (1-p) \cdot d(\beta(s)_1, \beta(t)_1) \\ \quad \text{if } \beta(s), \beta(t) \in \{0, 1\} \to S. \end{cases}$$

Formally, d can be specified in curried form $d(s,t) = d(s)(t)$ by an F-algebra on $T \to [0,1]$ as above. We could also define an F-coalgebra on $S \times T$ with

$$(s,t) \mapsto \begin{cases} \text{H} & \text{if either } \beta(s), \beta(t) \in \{\text{H}, \text{T}\} \text{ and } \beta(s) \neq \beta(t) \\ \text{T} & \text{if both } \beta(s), \beta(t) \in \{\text{H}, \text{T}\} \text{ and } \beta(s) = \beta(t) \\ \lambda a \in \{0,1\}.(\beta(s)_a, \beta(t)_a) & \text{if } \beta(s), \beta(t) \in \{0, 1\} \to S \end{cases}$$

and take $d(s,t) = \Pr(O(s,t) = \text{H})$.

Symmetry and the triangle inequality are easy to verify, thus any protocol S is a pseudometric space under the distance functions d and d'.

Lemma 1. *Let S and T be F-coalgebras, $s \in S$, and $t \in T$. The following are equivalent:*

1. $d(s,t) = 0$
2. $d'(s,t) = 0$
3. s and t are bisimilar.

Proof. The states s and t are bisimilar iff they have the same image in the final coalgebra, and d and d' are also preserved. Thus if s and t are bisimilar, then $d(s,t) = d'(s,t) = 0$. Conversely, any two distinct prefix codes must differ on some codeword $x \in \{0,1\}^*$, in which case both $d(s,t), d'(s,t) \geq p^{|x|}$. □

Lemma 2. *Every d'-open set is d-open. If $E(s) < \infty$, then every d-open neighborhood of s is d'-open.*

Proof. If s and t disagree on x, then the probability of disagreement is at least $p^{|x|}$, thus $d(s,t) \geq d'(s,t)$, so every basic d'-open set $\{t \mid d'(s,t) < \varepsilon\}$ contains the basic d-open set $\{t \mid d(s,t) < \varepsilon\}$, thus is also d-open.

Conversely, suppose $E(s) < \infty$. If $d'(s,t) \leq p^n$, then the codes s and t agree to length n, thus s and t differ with probability at most $\Pr(|x| > n) \leq E(s)/n$ by the Markov inequality. Thus $d(s,t) \leq E(s)/n$. We conclude that $d(s,t) \leq E(s)/\log_p d'(s,t)$. $\square$

Lemma 2 says that d generates a finer topology than d' on $\mathcal{C}$. They are not the same: an example of a d-open set that is not d'-open is the ε-neighborhood of $s = (\varnothing, \varnothing)$ in the d-metric for any $0 < \varepsilon < 1$. For $s_n = (\{0,1\}^n, \varnothing)$, $d(s,s_n) = 1$ but $d'(s,s_n) = p^n$.

In the final F-coalgebra $\mathcal{C}$, $d(s,t) = 0$ implies $s = t$, since bisimilar states of $\mathcal{C}$ are equal. Thus $\mathcal{C}$ is a metric space under d. However, it is not complete, even restricted to points with finite expectation. For example, the sequence $(\{0,1\}^n, \varnothing)$ has no limit point. However, the subspace of points with expected running time bounded by any constant b is compact, thus complete, as we will now show.

Theorem 2. *Let $\mathcal{C}_b$ be the subspace of points $s \in \mathcal{C}$ such that $E(s) \leq b$. Then $\mathcal{C}_b$ is a compact, hence complete, metric space under d.*

Proof. We have argued that $\mathcal{C}_b$ is a metric space, thus it remains to show compactness. Certainly $\mathcal{C}_b$ is compact under d'. By Lemma 2, d and d' generate the same topology on $\mathcal{C}_b$, therefore $\mathcal{C}_b$ is also compact under d. $\square$

Recall that $P(s) = \Pr(O(s) = \text{H})$.

Lemma 3. *The map P is continuous with respect to d on $\mathcal{C}$.*

Proof.

$$\begin{aligned}
|P(s) - P(t)| &= |\Pr(O(s) = \text{H} \wedge O(t) \neq \text{H}) - \Pr(O(t) = \text{H} \wedge O(s) \neq \text{H})| \\
&\leq \Pr(O(s) = \text{H} \wedge O(t) \neq \text{H}) + \Pr(O(t) = \text{H} \wedge O(s) \neq \text{H}) \\
&\leq \Pr(O(s) \neq O(t)) \\
&= d(s,t).
\end{aligned}$$

$\square$

The map E is not continuous at any point in either metric, not even restricted to $\mathcal{C}_b$. However, we have the following.

Lemma 4. *Let $A \subseteq \mathcal{C}$ and let $\text{cl}'(A)$ denote the closure of A under the d' metric. Then $\sup\{E(s) \mid s \in \text{cl}'(A)\} \leq \sup\{E(t) \mid t \in A\}$.*

Proof. Recall that for points s in the final coalgebra, $E(s) = \sum_{x \in s} \Pr(x) \cdot |x|$ if $\sum_{x \in s} \Pr(x) = 1$, and ∞ if $\sum_{x \in s} \Pr(x) < 1$. Let $s \in \text{cl}'(A)$. If $\sum_{x \in s} \Pr(x) < 1$, then that is also true for some $t \in A$, so in that case both suprema are ∞; so assume that $\sum_{x \in s} \Pr(x) < 1$.

For $\varepsilon > 0$, let n be large enough that

$$\sum_{\substack{x \in s \\ |x| \leq n}} \Pr(x) \cdot |x| \geq \begin{cases} E(s) - \varepsilon & \text{if } E(s) < \infty, \\ 1/\varepsilon & \text{if } E(s) = \infty \end{cases}$$

and choose a point $t \in A$ such that s and t agree to length n. Then

$$E(t) \geq \sum_{\substack{x \in t \\ |x| \leq n}} \Pr(x) \cdot |x| = \sum_{\substack{x \in s \\ |x| \leq n}} \Pr(x) \cdot |x| \geq \begin{cases} E(s) - \varepsilon & \text{if } E(s) < \infty, \\ 1/\varepsilon & \text{if } E(s) = \infty, \end{cases}$$

thus $\sup\{E(t) \mid t \in A\} \geq E(s)$. As s was arbitrary, the conclusion follows. □

4   Residual Probability Protocols Are Optimal

Let $E_{\text{opt}}(q)$ be the infimum of expectations of all p, q-simulation protocols. There exist protocols with expectation at most $1/p$ (e.g., Example 2), so $E_{\text{opt}}(q) \leq 1/p$. A p, q-simulation protocol with start state s is *optimal* if $E_{\text{cons}}(s) = E_{\text{opt}}(q)$.

Theorem 3. *For every p, q such that $0 < p \leq 1/2$ and $0 \leq q \leq 1$, there exists an optimal p, q-simulation protocol.*

Proof. We show that $E_{\text{opt}}(q)$ is attained at a state in the final F-coalgebra $\mathcal{C}$. Let $s_0, s_1, \ldots$ be a sequence of start states of p, q-protocols such that $E(s_n)$ is decreasing and $\lim_n E(s_n) = E_{\text{opt}}(q)$. Since $E(s)$ and $P(s)$ are preserved under morphisms, the images of these states in $\mathcal{C}$ are also start states of p, q-protocols in $\mathcal{C}$ and their expectations are the same, thus we can assume without loss of generality that the s_n are states of $\mathcal{C}_{1/p}$. Since $\mathcal{C}_{1/p}$ is compact, there exists a convergent subsequence with limit $u_q \in \mathcal{C}_{1/p} \in \mathcal{C}_{1/p}$. Since P is continuous (Lemma 3), $P(u_q) = q$, thus u_q is the start state of a p, q-protocol. By Lemma 4, $E(u_q) = E_{\text{opt}}(q)$. □

Theorem 4. *For every p, there is a residual probability protocol U_p that is optimal for every q.*

Proof. Let u_q be the optimal p, q-protocol constructed in Theorem 3. Consider the coalgebra $U_p = ([0, 1], v)$, where

$$v(q) = \begin{cases} \delta(u_q) & \text{if } \delta(u_q) \in \{\text{H}, \text{T}\} \\ \lambda X \in \{0, 1\}. \Pr(O(\delta(u_q)_X) = \text{H}) & \text{if } \delta(u_q) \in \{0, 1\} \rightarrow \mathcal{C}. \end{cases}$$

We claim that for all q,

$$E_U(q) = E_{\text{opt}}(q) \qquad \Pr(O(q) = \text{H}) = q \qquad \Pr(O(q) = \text{T}) = 1 - q, \qquad (4)$$

thus U_p with start state q is an optimal p, q-simulation protocol. We first show that

$$E_U(q) \leq E_{\text{opt}}(q) \qquad \Pr(O(q) = \text{H}) \leq q \qquad \Pr(O(q) = \text{T}) \leq 1 - q \quad (5)$$

by fixpoint induction.

For the first inequality of (5), define a property φ on S to be *hereditary* if $\varphi(\beta(s)_0)$ and $\varphi(\beta(s)_1)$ whenever $\beta(s) \in \{0, 1\} \to S$ and $\varphi(s)$. The property

$$E(s) = E_{\text{opt}}(P(s)) \tag{6}$$

is hereditary, because it says that s is an optimal protocol for its residual probability. But if s is, then so must be its successors; if not, then we could replace them by a better protocol and thereby improve $E(s)$ as well.

Now we proceed by fixpoint induction to show that $E_U(q) \leq E_{\text{opt}}(q)$. It suffices to show that E_{opt} is a fixpoint of the defining equation τ for E_U.

$$\tau(E_{\text{opt}})(q)$$

$$= \begin{cases} 0 & \text{if } v(q) \in \{\text{H}, \text{T}\} \\ 1 + p \cdot E_{\text{opt}}(v(q)_0) + (1 - p) \cdot E_{\text{opt}}(v(q)_1) & \text{if } v(q) \in [0, 1]^2 \end{cases} \tag{7}$$

$$= \begin{cases} 0 & \text{if } \delta(u_q) \in \{\text{H}, \text{T}\} \\ 1 + p \cdot E_{\text{opt}}(P(\delta(u_q)_0)) + (1 - p) \cdot E_{\text{opt}}(P(\delta(u_q)_1)) & \text{if } \delta(u_q) \in \mathcal{C}^2 \end{cases} \tag{8}$$

$$= \begin{cases} 0 & \text{if } \delta(u_q) \in \{\text{H}, \text{T}\} \\ 1 + p \cdot E_{\mathcal{C}}(u_{P(\delta(u_q)_0)}) + (1 - p) \cdot E_{\mathcal{C}}(u_{P(\delta(u_q)_1)}) & \text{if } \delta(u_q) \in \mathcal{C}^2 \end{cases} \tag{9}$$

$$= \begin{cases} 0 & \text{if } \delta(u_q) \in \{\text{H}, \text{T}\} \\ 1 + p \cdot E_{\mathcal{C}}(\delta(u_q)_0) + (1 - p) \cdot E_{\mathcal{C}}(\delta(u_q)_1) & \text{if } \delta(u_q) \in \mathcal{C}^2 \end{cases} \tag{10}$$

$$= E_{\mathcal{C}}(u_q) \tag{11}$$

$$= E_{\text{opt}}(q). \tag{12}$$

Inference (7) is by the definition of τ. Inference (8) is by the definition of $v(q)$. Inference (9) is from the construction of Theorem 3. Inference (10) is by the fact that $\delta(u_q)_1$ and $\delta(u_q)_0$ satisfy property (6), since u_q does and the property is hereditary, therefore

$$E_{\mathcal{C}}(\delta(u_q)_X) = E_{\text{opt}}(P(\delta(u_q)_X)) = E_{\mathcal{C}}(u_{P(\delta(u_q)_X)})$$

for $X \in \{0, 1\}$. Inference (11) is by the definition of $E_{\mathcal{C}}$. Inference (12) is by Theorem 3.

For the second inequality of (5), writing $P(q)$ for $\Pr(O(q) = \text{H})$, it suffices to show that the identity function on $[0, 1]$ is a fixpoint of the defining equation τ

for P.

$$\tau(\lambda q.q)(q) = \begin{cases} 1 & \text{if } \upsilon(q) = \text{H} \\ 0 & \text{if } \upsilon(q) = \text{T} \\ p \cdot (\lambda q.q)(\upsilon(q)_0) + (1-p) \cdot (\lambda q.q)(\upsilon(q)_1) & \text{if } \upsilon(q) \in [0,1]^2 \end{cases} \tag{13}$$

$$= \begin{cases} 1 & \text{if } \upsilon(q) = \text{H} \\ 0 & \text{if } \upsilon(q) = \text{T} \\ p \cdot \upsilon(q)_0 + (1-p) \cdot \upsilon(q)_1 & \text{if } \upsilon(q) \in [0,1]^2 \end{cases} \tag{14}$$

$$= \begin{cases} 1 & \text{if } \delta(u_q) = \text{H} \\ 0 & \text{if } \delta(u_q) = \text{T} \\ p \cdot P(\delta(u_q)_0) + (1-p) \cdot P(\delta(u_q)_1) & \text{if } \delta(u_q) \in \mathcal{C}^2 \end{cases} \tag{15}$$

$$= P(u_q) \tag{16}$$

$$= q. \tag{17}$$

Inference (13) is by definition of τ. Inference (14) is by the application of the identity function. Inference (15) is by definition of υ. Inference (16) is by definition of $P(u_q)$. Inference (17) is by the fact that u_q is the start state of a p, q-protocol.

The proof of the third inequality of (5) is symmetric.

Now we argue that all the inequalities (5) are actually equalities (4). By the first inequality, the probability of halting is 1, since E_U is finite. Since the last two inequalities hold and the left-hand sides sum to 1, the last two inequalities must be equalities. Then U with start state q is a p, q-simulation protocol, thus $E_{\text{opt}}(q) \leq E_U(q)$, therefore the first inequality of (5) is an equality as well. $\qquad \square$

5   Properties of E_{opt}

We assume throughout this section and the next that $(3 - \sqrt{5})/2 \leq p \leq 1/2$; equivalently, $(1 - p)^2 \leq p \leq 1 - p$.

For fixed p, a real number $q \in [0, 1]$ is *exceptional of degree d* if it has a finite binary prefix code with probabilities $p, 1 - p$ whose longest codeword is of length d. The number q is *exceptional* if it is exceptional of some finite degree.

If q is exceptional of degree d, then so is $1 - q$, and the pair of codes form a finite loop-free p, q-protocol with maximum running time d. In this case q and $1 - q$ are polynomial functions of p of degree d. The twelve exceptional values of degree at most 2 are shown in Table 1.

Some rows of Table 1 collapse for certain degenerate values of p. For $p = 1/2$, rows (iii), (iv), and (v) collapse and rows (ii) and (vi) collapse. For $p = (3-\sqrt{5})/2$, rows (ii) and (v) collapse. These are the only two degenerate values that cause collapse. Rows (v) and (vii) would collapse for $p = 1/3$, but this case is ruled out by the assumption $p \geq (3 - \sqrt{5})/2 \approx .382$.

Table 1. Exceptional values of degree at most 2

	q	$1-q$	degree	E_{opt}
(i)	0	1	0	$=0$
(ii)	p	$1-p$	1	$=1$
(iii)	$p(1-p)$	$1-p+p^2$	2	$=1+p$
(iv)	p^2	$1-p^2$	2	$=1+p$
(v)	$(1-p)^2$	$p+p(1-p)$	2	$=2-p$
(vi)	$2p(1-p)$	$p^2+(1-p)^2$	2	≤ 2

The exceptional points form a countable dense subset of the unit interval. The set is countable because there are countably many polynomials in p with integer coefficients. It is dense because for any $0 \leq a < b \leq 1$, for sufficiently large n (viz., $n > \log_{1-p} b-a$), $\Pr(x) \leq (1-p)^n < b-a$ for all binary strings x of length n, therefore $a \leq \sum_{x \in A} \Pr(x) \leq b$ for some $A \subseteq \{0,1\}^n$.

Lemma 5. *Let $([0,1], \beta)$ be a greedy residual probability protocol with expectation E. If $(3 - \sqrt{5})/2 \leq p < 1/2$, then*

1. *For $q \leq p$ or $1 - p \leq q$, $E(q) < 2$.*
2. *For $p < q < 1 - p$, $E(q) < (2 - p)/(1 - p + p^2)$.*

If $p = 1/2$, then $E(q) \leq 2$.

Proof. For $q \in [0,p] \cup [1-p, 1]$, either $\beta(q) \in \{\text{H}, \text{T}\}$ or $\beta(q)_1 \in \{\text{H}, \text{T}\}$, thus the protocol takes at most one step with probability at least $1-p$. For $q \in (p, 1-p)$, either $\beta(q)_0 = \text{H}$ and $\beta(q)_1 = (q-p)/(1-p)$ or $\beta(q)_0 = \text{T}$ and $\beta(q)_1 = q/(1-p)$. In the former case, $q < 1 - p \leq 1 - (1-p)^2$ so $\beta(q)_1 = (q-p)/(1-p) < p$. In the latter case, $(1-p)^2 \leq p < q$ so $\beta(q)_1 = q/(1-p) > 1 - p$. In either case, the protocol reenters the region $[0,p] \cup [1-p, 1]$ in the next step. Thus $E(q)$ is bounded by M for $q \in [0,p] \cup [1-p, 1]$ and by N for $(p, 1-p)$, where M and N satisfy the system of recurrences

$$M = (1-p) + p(1+N) = 1 + pN$$
$$N = p + (1-p)(1+M) = 1 + (1-p)M. \tag{18}$$

The unique bounded solution is

$$M = \frac{1+p}{1-p+p^2} \qquad N = \frac{2-p}{1-p+p^2},$$

thus

$$E(q) \leq \begin{cases} \dfrac{1+p}{1-p+p^2} & \text{if } q \leq p \text{ or } 1-p \leq q \\[2mm] \dfrac{2-p}{1-p+p^2} & \text{if } p < q < 1-p. \end{cases}$$

In the case $q \leq p$ or $1 - p \leq q$, the value is 2 for $p = 1/2$ and strictly less than 2 if $p < 1/2$. The inequality is also strict in the case $p < q < 1 - p$ if $p < 1/2$, since it is governed by the system (18). $\square$

We show that for $p < 1/2$, the function E_{opt} has a dense set of discontinuities on the unit interval, and the function is self-similar but for a discrete set of exceptions.

Lemma 6. *For all non-exceptional q, $E_{\text{opt}}(q) \geq 1/(1-p)$, and for $p < q < 1-p$, $E_{\text{opt}} \geq 2$.*

Proof. We will show in Lemma 8 that greedy is optimal on non-exceptional q, and non-exceptionality is preserved by greedy steps. Thus the optimal protocol is purely greedy on non-exceptional q. The remainder of the proof is similar to the proof of the corresponding inequalities (3.14) and (3.15) of [8].

The first inequality follows from the observation that a greedy protocol can do no better than to halt with probability $1 - p$ in every step, giving the same expectation as a Bernoulli process with success probability $1 - p$.

For the second, if $p < q < 1 - p$, then after one greedy step, the residual probability is either $q' = q/(1-p) > 1 - p$ or $q' = (q-p)/(1-p) < p$. In either case, by the previous argument, $E_{\text{opt}}(q') \geq 1/(1-p)$. Thus

$$E_{\text{opt}}(q) = 1 + (1-p)E_{\text{opt}}(q') \geq 1 + (1-p)\frac{1}{1-p} = 2.$$

$\square$

Theorem 5. *For $p < 1/2$, the function E_{opt} is everywhere discontinuous; that is, every open subinterval of the closed unit interval contains a discontinuity.*

Proof. The argument is very similar to one given in [8], with minor modifications to account for exceptional points.

It follows from Lemmas 5 and 6 that E_{opt} has discontinuities at p and $1 - p$. By Lemma 6, all non-exceptional q approaching p from above have $E_{\text{opt}}(q) \geq 2$; by Lemma 5, all non-exceptional q approaching p from below have $E_{\text{opt}}(q) \leq (1+p)/(1-p+p^2) < 2$; and $E_{\text{opt}}(p) = 1$.

Now we show that every nonempty open interval (a, b) contains a discontinuity. If the interval (a, b) is entirely contained in one of the three regions $(0, p)$, $(p, 1-p)$, or $(1-p, 1)$, then a greedy step maps the non-exceptional elements of (a, b) conformally to a larger subinterval. For example, if $(a, b) \subseteq (0, p)$, then

$$E_{\text{opt}}(q) = 1 + pE_{\text{opt}}(q/p)$$

for non-exceptional $a < q < b$, thus

$$E_{\text{opt}}(q/p) = (E_{\text{opt}}(q) - 1)/p$$

for $a/p < q/p < b/p$, so the non-exceptional elements of (a, b) are mapped conformally onto the interval $(a/p, b/p)$. But the length of this interval is $(b-a)/p$, thus we have produced a longer interval.

A similar argument holds if (a, b) is contained in one of the intervals $(p, 1-p)$ or $(1-p, 1)$. In each of these three cases, we can produce an interval of continuity that is longer than (a, b) by a factor of at least $1/(1 - p)$. This process can be repeated at most $\log_{1-p}(b - a)$ steps before the interval must contain one of the discontinuities p or $1 - p$. As the mappings were conformal on non-exceptional points, the original interval (a, b) must have contained a discontinuity. $\square$

6   Algorithms

Throughout this section, as in the last, we assume that $(3 - \sqrt{5})/2 \leq p \leq 1/2$.

Lemma 7. *For residual probability protocols, a greedy step is optimal at all but finitely many exceptional q.*

Proof. Suppose first that $p < 1/2$. By Lemma 5, $E_{\mathrm{opt}} \leq 2 - \varepsilon$ for some $\varepsilon > 0$. Suppose we have a residual probability protocol that is not greedy at q for some $0 < q \leq p$ or $1 - p \leq q < 1$. If the protocol generates an infinite computation path from q, then

$$E(q) \geq p + (1 - p)(1 + \frac{1}{1 - p}) = 2.$$

This is the minimum possible expectation with at least one an infinite path if the protocol does not halt with probability at least $1 - p$ in the first step. Truncating at depth k, the running time would be

$$2 - p^{k-1}(1 - p)(k + \frac{1}{1 - p}) = 2 - p^{k-1}(k(1 - p) + 1),$$

and this is greater than any $2 - \varepsilon$ for sufficiently large k. By Lemma 5, any protocol that is not greedy in the first step and generates a computation path of length at least k cannot be optimal. But the only q that can generate computation trees of depth k or less are the exceptional q of degree at most k, and there are only finitely many of these.

If $p = 1/2$, the situation is even simpler. By Lemma 5, $E_{\mathrm{opt}} \leq 2$. In this case, however, any impatient protocol is greedy. If the protocol is not impatient at q, then all computation paths are of length at least 2. The only way this can be optimal is if q is exceptional of degree 2, and all computation paths are of length exactly 2. But according to Table 1, this is impossible: row (vi) collapses to row (ii) for $p = 1/2$, so there is no such optimal computation.

Now let us consider the case $p < q < 1 - p$. Any strategy that is not greedy in the first step must take at least 2 steps in all instances; it cannot halt in one step with probability $1 - p$, because that probability is too big to assign either H or T. If the protocol generates an infinite computation path from q, then it takes time at least

$$2 + p^2(2 + \frac{1}{1 - p}).$$

But N is less than this for $p \geq (3 - \sqrt{5})/2$:

$$\frac{2 - p}{1 - p + p^2} \leq 2 + p^2(2 + \frac{1}{1 - p}).$$

This can be shown by comparing derivatives. The derivative of the left-hand side is negative for all points greater than $2 - \sqrt{3}$, and $2 - \sqrt{3} < (3 - \sqrt{5})/2 \leq p$.

The derivative of the right-hand side is positive for all p. The inequality holds at $3/8$, where the values are $104/49$ and $401/160$, respectively, and $2 - \sqrt{3} < 3/8 < (3 - \sqrt{5})/2$.

As above, by Lemma 5, any protocol that is not greedy but generates a computation path of sufficient length k cannot be optimal. So if the optimal protocol is not greedy at q, then q must be exceptional of degree at most k.   □

Lemma 8. *Assume $(3 - \sqrt{5})/2 \leq p \leq 1/2$. At all non-exceptional points, greedy is globally optimal.*

Proof. By Lemma 7, the optimal local strategy at all but finitely many exceptional points is greedy. But it is not difficult to show that a greedy step preserves non-exceptionality, therefore for non-exceptional points, greedy is globally optimal as well.   □

Theorem 6. *For $p = 1/2$, $E_{\mathrm{opt}}(q) = 2$ but for the following exceptional values: $E_{\mathrm{opt}}(k/2^n) = (2^n - 1)/2^{n-1}$, $k \leq 2^n$ odd. Greedy is optimal for all q.*

Proof. Lemmas 5 and 7 establish that $E_{\mathrm{opt}}(q) \leq 2$ for all q and that $E_{\mathrm{opt}}(q) = 2$ for all nonexceptional q. Any non-greedy strategy takes at least two steps on all computation paths, thus greedy is optimal for all q. For the exceptional points mentioned in the statement of the theorem, it is easily checked inductively that the greedy strategy behaves as stated. Moreover, all exceptional points are of this form.   □

6.1   An Approximation Algorithm

Were it not for the ambiguous region $(p, 1 - p)$, we would be done. We could check in each step whether q is one of finitely many exceptional values; if so, obtain the optimal strategy by table lookup, and if not, take a greedy step. Note that this gives an optimal protocol for $p = 1/2$, as the ambiguous region is empty.

Unfortunately, for q in the ambiguous region $(p, 1 - p)$, there are always two choices, and we do not know which will ultimately be the better choice. To approximate the globally optimal expectation E_{opt} to within any desired $\varepsilon > 0$, we will simulate all possible greedy choices down to a fixed depth k depending on ε.

Let d be a bound on the degree of those exceptional points for which a local greedy step is not optimal, as guaranteed by Lemma 7. Let G be the set of exceptional points of degree at most $d + k$. As G is a finite set, whenever $q \in G$ during the execution of the protocol, we can obtain the optimal local action by table lookup and take that action.

Otherwise, on input $q \notin G$, if q is not in the ambiguous region $(p, 1 - p)$, we take the unique possible greedy step. This is optimal, by Lemma 7. If $q \in (p, 1 - p)$, we have two greedy choices. We know that one of them is optimal, but we do not know which. In this case we simulate all possible greedy paths down to depth k. This involves branching when q is in the ambiguous region $(p, 1 - p)$ to simulate the two possible greedy steps. No greedy path ever encounters a $q \in G$ by choice of G, so we know that some greedy path is optimal down to depth k.

At depth k, we have several paths x that are currently being simulated. One of them is optimal. For each such x, let E_x be the expected time to halt before reaching the end of x, given that the path x is taken; that is, E_x is the expected length of a shortest path prefix-incomparable to x. Let $f_x(q) \in [0,1]$ be the residual probability after following path x if the computation has not halted by then. Then

$$E_{\text{opt}}(q) = \min_x (E_x + \Pr(x) \cdot (k + E_{\text{opt}}(f_x(q))))$$
$$\geq \min_x (E_x + \Pr(x) \cdot k).$$

But for any such x, continuing from x with a purely greedy strategy yields an expectation no worse than

$$E_x + \Pr(x) \cdot (k + 2) \tag{19}$$

by Lemma 5, and

$$\min_x (E_x + \Pr(x) \cdot (k + 2)) \leq \min_x (E_x + \Pr(x) \cdot k) + 2(1 - p)^k$$
$$\leq E_{\text{opt}}(q) + \varepsilon,$$

provided k is large enough that $(1 - p)^k \leq \varepsilon/2$, that is, $k \geq \log_{1-p}(\varepsilon/2)$. Thus the greedy strategy x that minimizes (19) will be within ε of optimal.

6.2 Analysis

The algorithm constructs a tree with $2^{k/2}$ nodes in the worst case, where $k = \log_{1-p}(\varepsilon/2)$. It is $2^{k/2}$ and not 2^k because branching occurs at most once every two steps. The algorithm thus runs in time bounded by $2^{k/2} \leq (\varepsilon/2)^{1/\log(1-p)^2}$. The exponent $1/\log(1 - p)^2$ ranges between $-.72$ and $-.5$ for p in the range $(3 - \sqrt{5})/2 \leq p \leq 1/2$, thus the algorithm is better than linear in $1/\varepsilon$.

7 Conclusion

Several questions present themselves for further investigation.

Our analysis gives a worst-case time bound less than linear in $1/\varepsilon$, but empirical evidence suggests that the true time bound is exponentially better and that we actually achieve the optimal on all but a very sparse set. In the many experiments we have tried, the size of the set of candidate greedy paths x does not grow beyond two if demonstrably suboptimal paths are pruned along the way, and the algorithm invariably exits the loop with one candidate, which must be optimal.

The restriction $p \geq (3 - \sqrt{5})/2$ was made to simplify many of the proofs, but it should be possible to eliminate it.

Most importantly, it would be nice to know whether the optimal protocol is computable for all rational p and q.

Acknowledgments. Much of this work was conducted while the author was visiting the University of Warsaw, the University of Copenhagen, and the University of Aarhus during April and May 2009. Special thanks to Mikołaj Bojańczyk, Fritz Henglein, Bobby Kleinberg, Eryk Kopczyński, Peter Bro Miltersen, Damian Niwiński, Nicholas Ruozzi, Michael Schwartzbach, Anna Talarczyk, Paweł Urzyczyn, Aaron Wagner, Anna Zdunik, and the anonymous referees. This work was supported by NSF grant CCF-0635028.

References

1. Adamek, J.: Foundations of Coding. Wiley (1991)
2. Adámek, J., Lücke, D., Milius, S.: Recursive coalgebras of finitary functors. Theoretical Informatics and Applications 41, 447–462 (2007)
3. Adámek, J., Milius, S., Velebil, J.: Elgot algebras. Log. Methods Comput. Sci. 2(5:4), 1–31 (2006)
4. Blum, M.: Independent unbiased coin flips from a correlated biased source: a finite state Markov chain. Combinatorica 6(2), 97–108 (1986)
5. Capretta, V., Uustalu, T., Vene, V.: Corecursive algebras: A study of general structured corecursion. In: Oliveira, M.V.M., Woodcock, J. (eds.) SBMF 2009. LNCS, vol. 5902, pp. 84–100. Springer, Heidelberg (2009)
6. Cover, T.M., Thomas, J.A.: Elements of Information Theory. Wiley-Interscience (August 1991)
7. Elias, P.: The efficient construction of an unbiased random sequence. Ann. Math. Stat. 43(3), 865–870 (1992)
8. Kozen, D.: Coinductive proof principles for stochastic processes. Logical Methods in Computer Science 3(4:8) (2007)
9. von Neumann, J.: Various techniques used in connection with random digits. In: Forsythe, G.E. (ed.) National Bureau of Standards. Applied Math Series, vol. 12, pp. 36–38 (1951); reprinted in: von Neumann's Collected Works, vol. 5, pp. 768–770. Pergamon Press (1963)
10. Nisan, N., Ta-shma, A.: Extracting randomness: A survey and new constructions. Journal of Computer and System Sciences 58, 148–173 (1999)
11. Nisan, N., Zuckerman, D.: Randomness is linear in space. Journal of Computer and System Sciences 52, 43–52 (1996)
12. Park, D.M.R.: Fixpoint induction and proofs of program properties. In: Meltzer, B., Michie, D. (eds.) Machine Intelligence, vol. 5, pp. 59–78. Edinburgh University Press (1969)
13. Peres, Y.: Iterating von Neumann's procedure for extracting random bits. Ann. Stat. 20(1), 590–597 (1992)
14. Srinivasan, A., Zuckerman, D.: Computing with very weak random sources. SIAM J. Computing 28, 264–275 (1999)
15. Ta-shma, A.: On extracting randomness from weak random sources. In: Proc. 28th ACM Symp. Theory of Computing, pp. 276–285 (1996)

Entanglement, Flow and Classical Simulatability in Measurement Based Quantum Computation

Damian Markham[1] and Elham Kashefi[2]

[1] CNRS LTCI, Département Informatique et Réseaux, Telecom ParisTech,
23 avenue d'Italie, CS 51327, 75214 Paris CEDEX 13, France
[2] School of Informatics, University of Edinburgh, 10 Crichton Street,
Edinburgh EH8 9AB, UK

Abstract. The question of which and how a particular class of entangled resource states (known as graph states) can be used for measurement based quantum computation (MBQC) recently gave rise to the notion of *Flow* and its generalisation *gFlow*. That is a causal structure for measurements guaranteeing deterministic computation. Furthermore, gFlow has proven itself to be a powerful tool in studying the difference between the measurement-based and circuit models for quantum computing, as well as analysing cryptographic protocols. On the other hand, entanglement is known to play a crucial role in MBQC. In this paper we first show how gFlow can be used to directly give a bound on the classical simulation of an MBQC. Our method offers an interpretation of the gFlow as showing how information flows through a computation, giving rise to an information light cone.We then establish a link between entanglement and the existence of gFlow for a graph state. We show that the gFlow can be used to upper bound the entanglement width and what we call the *structural entanglement* of a graph state. In turn this gives another method relating the gFlow to upper bound on how efficiently a computation can be simulated classically. These two methods of getting bounds on the difficulty of classical simulation are different and complementary and several known results follow. In particular known relations between the MBQC and the circuit model allow these results to be translated across models.

Measurement Based Quantum Computing (MBQC) [1] has attracted attention recently as one of the main competitors for a realisation of a quantum computer, its role in understanding the power and significance of entanglement for computation [2, 3], and that it plays a key role in the development of cryptographic protocols [4, 5]. In MBQC one starts off with a large multiparty entangled resource state and the computation is driven by a series of local measurements, the choice of which can depend on the result of previous measurements in the series. The formal language for MBQC was jointly developed by Prakash Panangaden in [6]. In this work we are interested in the question of how to recognise or characterise a 'good' resource for measurement based quantum computing. Given the fact that after the generation of the state, all operations are local, it is natural to expect entanglement to play a key role. Indeed it has been shown

F. van Breugel et al. (Eds.): Panangaden Festschrift, LNCS 8464, pp. 427–453, 2014.
© Springer International Publishing Switzerland 2014

that the entanglement of a resource state must be sufficiently high for it to be universal and not classically simulatable (we note that these two properties are currently not known to be equivalent, though it is broadly expected that they are) [7–11].

A related question to universality is that of the ability, or not, for a resource to allow any unitary MBQC computation on it at all. This question is addressed by what is called Flow or its generalisation referred hereinafter gFlow [12, 13], for a particular class of resource states, called graph states [14] and its extension open graph state [12] (see also below). There exist efficient algorithms [15, 16] to find gFlow if it exists, and once gFlow is found, it gives an explicit measurement pattern which gives a unitary computation across the resource graph state in hand. Subsequently gFlow has been a useful tool for exploring many aspects of MBQC such as efficient translation between MBQC and the circuit model [12, 17], analysing cryptographic protocols [18], direct pattern design in MBQC [19], proving bounds on depth complexity [13, 20] and from a more fundamental perspective, the arrival of causal order in MBQC [20–23].

In this paper we show that gFlow also gives a bound on the difficulty in classically simulating MBQC, and how it can be interpreted as a flow of information. This leads to the observation that the *causal forward cone* (the 'forward cone' given by the qubits who's corrections directly or indirectly depend on that qubit's measurement results) is equal to the information cone (the cone of qubits where the information spreads to through the computation). We then establish an intuitive link between gFlow on the one hand, and entanglement of a resource state on the other. We further make this connection explicit by showing how gFlow can be used to give bounds on the entanglement of a graph state. In this way we will see that properties of simulateability of MBQC on a resource in terms of entanglement can be translated to conditions in terms of gFlow. Via a known relationship between the circuit model and MBQC these results can also lead to conditions on simulateability of circuits. One such example is a rederivation of the result by Jozsa [24].

The organisation of this papers is as follows. In Section 1 we mention basic observations about entanglement conditions for any good resource state for MBQC which will be then linked to gFlow. In Section 2 we introduce graph states and review the notion gFlow and several preliminary notions necessary for the rest of the paper. In Section 3 we prove that gFlow can be used to give bounds on direct simulation of a MBQC. In Section 4 we discuss how gFlow can be used to see how information flows through a resource in an MBQC, giving in particular an information light cone which coincides with the causal cone as defined in [22, 23]. In Section 5 we show how gFlow can be used to upper bound the entanglement of a resource state which gives a new route to bounding simulatability of a MBQC, which is different and complementary to the direct simulation in Section 3. We finish with discussions.

1   Entanglement and Determinism in MBQC

In measurement-based quantum computing one starts with a large entangled resource state $|\Psi_{RES}\rangle$ on n qubits. We identify a two sets of qubits, I which will represent the inputs, and O which will represent the outputs of the computation, with $n \geq |O| \geq |I|$. Generally one can consider three types of computation using this resource, one with a classical input and a classical output (let's call this CC), one with a quantum input and a classical output QC and one with a quantum input and a quantum output QQ. Clearly QQ is the most general, since one can always encode classical information onto quantum states. In this work we focus on QQ. When considering a quantum input $|\psi\rangle_S$ (on a system S of $|I|$ qubits) the first step is to teleport the input system qubits S onto I on the resource state, by some global map on I and S. This can be done for example by entangling I with S (using, say, a control-Z gate) and performing Pauli X measurements on S then appropriate corrections (see e.g. [25] for graph state resources). The computation then proceeds by a series of measurements on individual qubits, followed by corrections, then further measurements and corrections and so on until the computation is complete. We call the sequence of measurements and corrections the *measurement pattern* (see [6] for formal definitions). The outputs qubits, labelled O are those qubits which at the end of the computation are not measured. In this way the computation uses the resource state to transfer the input from I to O, in a kind of involved teleportation, at the same time performing some unitary over the input.

We begin with the following definitions.

Definition 1. *A resource state $|\Psi_{RES}\rangle$ on n qubits, with defined input qubits I and output qubits O is D-Happy if for all bi-partitions A, B such that $I \in A$ and $O \in B$ we have*

$$E_{A,B}(|\Psi_{RES}\rangle) \geq |I|, \tag{1}$$

where $E_{A,B}(|\Psi_{RES}\rangle) = S(\rho_A) = S(\rho_B)$ is the entropy of entanglement across partition A, B, where $S(\rho_A) = -\ln Tr(\rho_A)$ is the von Neumann entropy of the reduced state $\rho_A = Tr_B(\rho_{AB})$.

Definition 2. *A MBQC pattern is called* unitary *if for all inputs the returned state of the output is an encoding of a unitary acting on the input.*

A similar notion was defined in [26] as *information preserving pattern*. We present a simple but important observation about the link between the above two definitions.

Theorem 1. *There exists a unitary MBQC pattern on a resource state $|\Psi_{Res}\rangle$ on n qubits, with input qubits I and output qubits O only if it is D-Happy.*

Proof. We start by noting that in order to teleport a state $|\psi\rangle \in \mathbb{C}_2^{\otimes |I|}$ perfectly across $|\Psi_{Res}\rangle$ to its output space, it is necessary to have $E_{A,B} \geq |I|$, for all bi-partitions A, B such that $I \in A$ and $O \in B$. To see this is true one can consider the state to be teleported as half a maximally entangled state. After the teleportation one would end up with a state with entanglement $E_{A,B} = |I|$. Since all operations are local, and it is not possible to increase entanglement in the process, this implies that we started with $E_{A,B} \geq |I|$. We then note that any measurement based computation can be considered as a teleportation across any cut which divides the inputs from the outputs - since all operations are local to each qubit, they are certainly local to any cut.   □

Recall that a MBQC computation evolves through various branches, depending on the measurement outcomes. In a unitary MBQC pattern as defined above, it is possible that different branch implements different unitary operators. A weaker notion of unitary computation is given below.

Definition 3. *A measurement based quantum computation is called* deterministic *if for all inputs the returned state of the output is an encoding of a fixed unitary acting on the input independent of the branch of the computation.*

Other types of determinism and their connections can be found in [13, 26]. In this paper we only consider the above central notions as they can be directly linked to the concept of structural entanglement as we present later. Moreover it is known that for graph states with $|I| = |O|$ the two definitions of unitary and determinism, defined above, are equivalent [26]. This will allow us to link the concept of gFlow to D-Happy as we discussed next.

2   Preliminaries: Graph States, Flow and gFlow

In the previous section we presented a necessary condition for computation across a resource state based on entanglement. The simple idea there was that if information can be transferred across a resource state, that state must be maximally entangled across each cut. We did not say anything about how this can be done however. This is where the ideas of Flow [12] and its generalisation gFlow[13] play a role, where a constructive definition together with efficient algorithm could be obtained for particular class of resource states of many qubits - graph states (defined below), with chosen input I and output O. If a graph state has gFlow, it implies that a unitary computation can be carried out across it [12, 13]. Not only that, gFlow also gives instructions how to do it, and tells you what class of computations will be carried out (which unitaries). We will show that we can further use gFlow to give a simple bound on classical simulation of the computation based on the size of the forward cones implied by the measurement patters. This gives rise to an interpretation of the gFlow as showing us how information is 'spread' across the resource state throughout the computation, in an information light cone (which coincides with the causal forward cone in MBQC [20, 22, 23]). In this section we review the definitions of graph states [14], open graph states, Flow [12] and gFlow [13] and related concepts.

We start by defining the resource states considered, graph states [14]. A graph state is a multipartite state $|G\rangle$ of n qubits, in one to one correspondence to a simple undirected graph G, with vertices V and edges E. Every vertex is associated to a qubit, and every edge can be understood as an entangling operation between qubits which have been initialised in the state

$$|+\rangle := (|0\rangle + |1\rangle)/\sqrt{2}.$$

We then have

$$|G\rangle_V := \prod_{i,j \in E} CZ_{i,j} |+ + \cdots +\rangle_V,\qquad(2)$$

where $CZ_{i,j}$ is the control-Z operation between qubits i and j. It is clear from this definition that the entanglement across a cut A, B is bounded by the number of edges cutting it, denoted $C_{A,B}$, i.e. $E_{A,B} \leq C_{A,B}$.

Graph states can equivalently be defined by their stabiliser operators [14], a set of n operators, each associated to one vertex defined as

$$K_i := X_i \otimes_{j \in N(i)} Z_j,\qquad(3)$$

where X and Z are the Pauli operators (and $Y = iZX$). The graph state $|G\rangle$ is the unique state satisfying all the eigenvalue equations (also called stabiliser relations or equations)

$$K_i |G\rangle_V = |G\rangle_V.$$

The above relation is the key to how gFlow works - gFlow tells us how to apply the stabilisers to correct for measurements.

When used as a resource state for MBQC we assign some vertices as inputs $I \in V$ and some as outputs $O \in V$. In order to preserve the space we have that the size of the input set $|I| \leq |O|$. We call the graph, with these assignments an *open graph* denoted as $G(I, O, V)$. The associated state is slightly different, the input vertices are no longer prepared in the $|+\rangle$ state, but can be arbitrary input qubits $|\psi\rangle_I$. The rest of the vertices are prepared as normal, and again, every edge corresponds to a control-Z operation. We denote such a state as $|G(\psi)\rangle$

$$|G(\psi)\rangle_V := \prod_{i,j \in E} CZ_{i,j} |\psi\rangle_I |+ \cdots +\rangle_{V/I}\qquad(4)$$

where the state only depends on the inputs so different open graphs may have the same open graph state if they share the same set I, and graph G even if they have different assigned outputs O. The stabilisers are now reduced to those only on the non-inputs (we denote this set I^c)

$$K_i |G(\psi)\rangle_V = |G(\psi)\rangle \quad \forall i \in I^c.\qquad(5)$$

Here the stabilisers define a space (of dimension $2^{|I|}$) of states such that this equation holds. The open graph state defined in Equation 4 is equivalent to starting in the standard graph state Equation 2 and teleporting an input $|\psi\rangle_S$ over system S (of $|I|$ qubits) onto the input vertices I by performing control-Zs between S and I, followed by Pauli X measurements on the S qubits and corrections (see e.g. [25]).

In the standard model of MBQC [6, 27] measurements are performed in one of the equatorial planes defined by the $X - Y$, $X - Z$ or $Y - Z$ planes, and correction operations are local Pauli operators. By the end of the computation all vertices will be measured except the outputs (we denote this O^C). The gFlow assigns a set of correction operators for each of these measurements.

Before giving the definition of gFlow, we give the intuition to how it works for measurements in the $X - Y$ plane. This corresponds to measuring in the basis $|\pm^\theta\rangle := (|0\rangle \pm e^{i\theta}|1\rangle)/\sqrt{2}$. We denote the projections associated to results ± 1 as $P^{\pm,\theta} := |\pm^\theta\rangle\langle\pm^\theta|$. For later use we denote the results in binary form as $r_i = 0$ for $+1$ and $r_i = 1$ for -1 outcomes. When measuring a state $|\psi\rangle$, in quantum mechanics the result is random (in fact normally in MBQC the probabilities are $1/2$ and $1/2$), which takes the resulting state to one of two branches, either the positive branch $P^{+,\theta}|\psi\rangle/p_+$ with probability $p_+ = \langle\psi|P^{+,\theta}|\psi\rangle$, or the negative branch $P^{-,\theta}|\psi\rangle/p_-$ with probability $p_- = \langle\psi|P^{-,\theta}|\psi\rangle$.

Clearly to perform a deterministic computation U, we need to recover a deterministic evolution, hence corrections need to be applied. By convention we take the positive branch to be the ideal branch (note that of course $P^{+,\theta}|\psi\rangle/p_+$ is not in general a unitary embedding, this is an additional requirement which is also satisfied for our case). The task is then to find a correction operator to take the state when projected onto the -1 result to that of the $+1$ result (possibly ignoring the state of the measured qubit, since it is no longer used). The starting point is to notice that for all measurements in the $X - Y$ plane, the projections are related to each other by a Pauli Z operator (for the other planes it is similarly the orthogonal Pauli operator) $P^{+,\theta} = ZP^{-,\theta}Z$. Imagine if it were possible to know the outcome of the measurement before it was performed (for example by traveling back in time after the measurement was performed and telling yourself), instead of correcting after the event, if we knew that we were about to get -1, we could cheat and apply a Pauli Z operator - then the 'measurement' (projection) would take us onto the projection we wanted, the positive branch. Obviously this is not possible without time travel since in quantum mechanics the results of measurements are random and cannot be known beforehand (we can only predict probabilities). However, we can use the stabilisers to *simulate* this strategy.

Imagine we applied the measurement on qubit i, then our time-travelling correction strategy for the -1 result would be to perform a Pauli Z operator on qubit i. Now, if we take a neighbour $j \notin I$, the stabiliser condition (Equation 5) tells us that

$$Z_i|G(\psi)\rangle = Z_i K_{j=N(i)}|G(\psi)\rangle \tag{6}$$
$$= \mathbf{1}_i \otimes X_j \otimes_{k\in N(j)\neq i} Z_k|G(\psi)\rangle. \tag{7}$$

Since $X_j \otimes_{k\in N(j)\neq i} Z_k$ are on different systems from the measured qubit i, it does not matter when they are performed (they commute with the measurement). In this way, applying $X_j \otimes_{k\in N(j)\neq i} Z_k$ correction operator after the measurement, is the same as applying a Z correction before the measurement - so that it has exactly the same effect. The latter is sometimes called an 'anachronical correction', since it is as if we could go back in time and correct the measurement before it happened. The same works if a product of stabilisers is used in Equation 6 as long as their product results in one Pauli Z operator on qubit i, and we call the vertices associated to these stabilisers as the *correcting set*. Graphically this condition is ensured if the total number of edges between the correcting set and the vertex being measured is 1 modulo 2. This motivates the definition of the odd neighbourhood of a set of vertices K, denoted $Odd(K) := \{\mu \mid |N(\mu) \cap K| = 1 \bmod 2\}$, which will be used in the definition of gFlow below.

Using this idea, gFlow plays the role of making sure it is possible to make a good choice of which neighbour (or set of neighbours) to choose in a consistent way - so that corrections do not somehow contradict or interfere with one another. Indeed, gFlow is composed of a time order $\prec$ (partial order over vertices) and a choice of neighbouring sets (correcting sets) for each measured vertex i, denoted $g(i)$ with this in mind. Firstly the time order should be consistent, so that corrections happen after the assigned measurements - this appears as *(g1)* in the Definition 4 below. Secondly, the correction should not invalidate or affect earlier corrections. This is true if no Pauli Z operators appear in the past when applying the stabiliser corrections, i.e. the correcting set is not oddly connected to the past - this appears as *(g2)* in Definition 4. Finally the correcting set should correct for the measurement it is assigned to. For measurements in the $X - Y$ plane this corresponds to the application of a Pauli Z operator when the correcting stabilisers are applied, which means the correcting set should be oddly connected to the measured vertex - which appears as *(g3)* in the definition below (the analogous corrections for the other planes appear after).

Definition 4. *An open graph state $G(I, O, V)$ has gFlow if there exists a map $g : O^c \to \mathcal{P}^{I^c}$ (from measured qubits to a subset of prepared qubits) and a partial order $\prec$ over V such that for all $i \in O^c$*
(g1) if $j \in g(i)$ and $i \neq j$ then $i \prec j$
(g2) if $j \prec i$ and $i \neq j$ then $j \notin Odd(g(i))$
(g3) for measurements in the $X - Y$ plane, $i \notin g(i)$ and $i \in Odd(g(i))$
(g4) for measurements in the $X - Z$ plane, $i \in g(i)$ and $i \in Odd(g(i))$
(g5) for measurements in the $Y - Z$ plane, $i \in g(i)$ and $i \notin Odd(g(i))$
Flow is a special case of gFlow, when all measurements are performed on the $X - Y$ plane, and the correction sets $g(i)$ have only one element.

In this way the product of $\prod_{j \in g(i)} K_j$ applies the appropriate 'anachronical' correction on vertex i, whilst not affecting other previous corrections. The associated computation can be carried out as follows. First generate the open graph state, then go through round by round (in the order given by $\prec$), measureing each qubit i, denoting the binary form of the outcome r_i, followed by the correction given by

$$\left(\sigma_i \prod_{j \in g(i)} K_j \right)^{r_i} \tag{8}$$

where σ_i is the Pauli Z_i, Y_i or X_i for measurement on qubit i done on the $X - Y$, $X - Z$ or $Y - Z$ planes respectively, so that Equation 8 is trivial over i and non-trivial only on future qubits of i i.e. on j such that $i \prec j$.

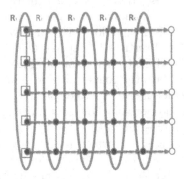

Fig. 1. An example of gFlow for the two dimensional clusters state as an open graph state. Following the convention in [12, 13] inputs are identified by vertices with squares around them, and outputs are identified as hollow vertices (hence all non-hollow vertices will be measured in the computation). The choice of gFlow for a given vertex is indicated by red dotted arrows from the vertex to its gFlow (these are called *gFlow paths*, see Definition 5. Note that gFlow paths need not follow graph edges, as in Figure 3b). The induced measurement rounds are highlighted in grey, (see Definition 8).

In [13] it is shown that gFlow is a necessary and sufficient condition for an open graph state to allow a *uniform* unitary, deterministic computation to be performed across it, where uniform means that each qubit can be measured at an arbitrary angle on one of the planes. Hence the existence of gFlow implies the resource is also D-Happy. Intuitively on can think that the existence of gFlow guarantees that the entanglement of the graph state is such that the random effects of local measurements can be absorbed and countered by yet unmeasured qubits. The following definitions will be used to discuss how information travels throughout the computation [20].

Definition 5. *A gFlow path starting from a vertex μ, denoted as $gPath(\mu)$, is an ordered set of vertices such that for each pair (i,j) we have $j \in g(i)$ and the first element of the set is μ.*

Definition 6. *An influencing path starting from a vertex μ, denoted as $gInf(\mu)$, is an ordered set of vertices such that each pair (i,j) is on a gFlow path or is preceded immediately by a pair on a gFlow path.*

Definition 7. *The forward cone $F_C(\mu)$ of a vertex μ is the set of all vertices touched by all influencing paths from μ.*

The concept of the forward cone appears in [20, 22, 23] and can be understood as a causal light cone, as described in [22, 23]. The partial order $\prec$ in a gFlow defines time order for the rounds of measurements. We say a vertex μ is in a round R_x if it is measured in round x.

Definition 8. *The set R_x, denotes the set of vertices which are measured in the xth round of measurements according to the gFlow.*

The best way to understand these definitions is through some examples. The gFlow (which is also a Flow in this case) is illustrated for the 2D cluster state in Figure 1. In Figure 2 we show examples of influencing paths and their union, which make up the forward cone for the 2D cluster state.

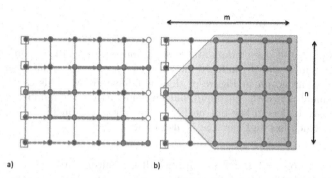

Fig. 2. a) The bold red lines are examples of two possible influencing paths from the central input vertex (see Definition 6), for the gFlow paths given by the red dotted arrows. An influencing path is path which follows gFlow paths and no more than one edge between gFlow paths. b) The collection of all influencing paths identifies the set of vertices (in red) in the *forward cone* (see Definition 7). The maximum size of forward cone for the 2D cluster state is indicated by the red shaded region (for the same gFlow). For an $n \times m$ 2D cluster state the maximum forward cone is of size $|F_{C_{max}}| = nm - n^2/4$. This gives a bound on classical simulation for a computation, in Theorem 2. The same region has an interpretation as an information light cone (see Section 4).

Before moving on to the interpretations of gFlow with respect to simulation and information flow, we review some examples which illustrate its power as a tool for analysing entanglement (as potential resources for MBQC), and in accessing the tradeoff between classical processing and number of measurement rounds (depth [20]). We start with an example of an open graph for which there is no gFlow in Figure 3 a). It can easily be seen that there is no possible assignment of correction sets $g(i)$ and time order satisfying the conditions in gFlow for any measurement axes. Indeed its inability to act as a resource for computation across it follows directly from the fact that the entanglement across it is less then the number of inputs (hence it is not D-happy). We note however that there are examples of graph states which are D-happy, but do not allow a gFlow [13, 26]. All such known examples still do allow computation across them. The second example is one where there exists a gFlow, but it necessarily has some correction sets which have more than one member - i.e. there is no Flow, as shown in Figure 3 b). The associated gFlow is give by assignments $g(1) = 4$, $g(2) = 5$ and $g(3) = 4, 5, 6$, with partial order given by the ordered measurement

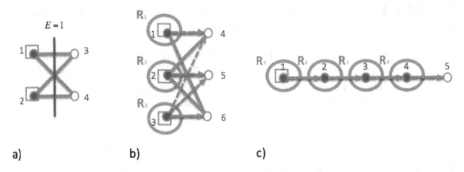

Fig. 3. Examples of open graph states with and without gFlow. The gFlow paths are red dotted lines, and the induced measurement rounds are highlighted in grey (see Definition 8). The graph in Figure a) does not have gFlow. This can be seen since the entanglement across the cut input/output is lower than the number of inputs. The graph in Figure b) has gFlow but no Flow [13]. The graph in Figure c) is the linear cluster state which has a gFlow that is also Flow.

rounds $R_1 = 1$, $R_2 = 2$ and $R_3 = 3$, and all measurements in the $X - Y$ plane. Note here that a gFlow path need not lie on a graph edge as for the gFlow path $(3, 4)$. The third example is the simple linear cluster state in Figure 3 c), where gFlow follows along the line.

A final example illustrates how gFlow can be used to find advantages in the number of rounds needed in a computation (taken from [13]). In Figure 4 the same open graph can have different gFlows. In the first case, Figure 4 a), the gFlow has correcting sets of size one, hence it is a Flow ($g(i) = i + 4$), and the number of rounds is the number of inputs (in the example this is four, but it easily extends to arbitrary size). More complicated gFlows can be found by increasing the size of some correcting sets, with the benefit of reducing the number of rounds. Figure 4 b), we set the correcting sets as $g(1) = 5, 6, 7, 8$, $g(2) = 6, 7, 8$, $g(3) = 7, 8$ and $g(4) = 8$. It can easily be checked that this assigment allows all measurements to be done in the same round since for every vertex i, the correcting set $g(i)$ is oddly connected only to i.

The above example illustrates a general scheme that could be understood as a tradeoff between rounds of computation and the amount of classical processing needed, but we have not yet talked about classical processing. To see how it works, we should think back again at what the gFlow does. Recall that gFlow tells us on which sets of vertices we should apply corrections (Equation 8). In particular, for a vertex i, the correction associated to its measurement result (r_i, where $r_i = 0$ corresponds to the ideal branch and $r_i = 1$ to that which needs to be corrected) is the application of the product of the stabilisers of all the vertices in $g(i)$ (minus the Z_i) - i.e. the correction is $(Z_i \prod_{j \in g(i)} K_j)^{r_i}$. Thus, if a vertex l is in the gFlow (or is a neighbour to a gFlow vertex) of another vertex i, then it will receive an $X_l^{r_i}$ (or $Z_l^{r_i}$) correction. The total number of corrections for a vertex depends on how many gFlow set (or neighbourhoods of gFlow set) to which that vertex belongs to. In the example Figure 4 b), vertex 8 has corrections

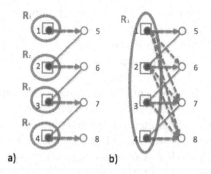

Fig. 4. This open graph state has several possible gFlows, and illustrates how gFlow can be used to find advantages in terms of the number of rounds needed (depth) in a computation. a) is a gFlow with one correcting vertex per qubit, hence it is also a flow. This requires a number of rounds scaling with the number of inputs. b) is a gFlow which has largest size scaling with the number of inputs, but all measurements can be done in one round. Indeed all intermediary tradeoffs are also possible. This exemplifies the tradeoff between classical computation required and the number of measurement rounds needed.

from all inputs - hence it must receive the correction $X_8^{r_1 \oplus r_2 \oplus r_3 \oplus r_4}$ (where $\oplus$ is the sum mod 2). In general, to calculate the Pauli X correction that should be applied on qubit j requires calculating the parity of all the r_is where $j \in g(i)$ and for the Pauli Z correction the parity of all the r_is where j is a neighbour of $g(i)$. We assume this is done classically (since it is a simple calculation), however, by increasing the size of the gFlows (in order to reduce the depth), we necessarily increase the size of this classical computation. This tradeoff has recently been translated to a tradeoff between the degree of the initial Hamiltonian and time of computation in the adiabatic model [28].

This tradeoff, a particular feature of the measurement based model, gave rise to a distinction in the power of measurement based quantum computation compared to the circuit model with respect to the number of time steps required [29]. The first example of a depth separation between quantum circuit and MBQC was proven for the calculation of parity function (where depth is defined to be the minimum number of rounds for a computation) [20]. Indeed this is a general feature that the depth of MBQC can be logarithmically better than the circuit model, where the difference is absorbed into the classical processing. More concretely it was shown that the depth complexity of MBQC is equal to the depth complexity for the circuit model with the addition of unbounded fan out gates [29].

3   Direct Simulation from gFlow

We will now see how we can derive a simple classical simulation, by tracking the stabilisers and logical operators. This idea is exactly how one can understand the Gottesman Knill theorem for the efficient classical simulatability of computations including only Clifford operations [30]. The proof follows from tracking stabilisers operator since they are an efficient way to describe a stabiliser state (such as a graph state), and Clifford operations, by definition transform stabiliser states to other stabiliser states, so computations can be simply tracked and described [30].

In what follows we will represent the computation in terms of the evolution of a set of logical operators. In physics there are two main, equivalent, ways that one views quantum evolution. One method (more common in quantum information) is where we look at how a state develops, and keep track of it as it evolves. This is known as the Schrödinger representation. Equivalently, one can view the state as having not altered, but the operators defining measurements having changed. This picture is known as the Heisenberg representation of evolution. In between these two pictures lies another way of representing evolution, which has been developed for quantum information - the so called 'logical Heisenberg' representation [30, 31]. In this method we track the evolution of a complete set of logical operators - in this case the Pauli operators. To recover the Shrödinger representation, we remember that any state density matrix can be decomposed into Pauli operators (see Equations (13 and 14 in the next section). The logical Heisenberg representation has proved a very instructive way to view the evolution of MBQC [27, 32], and as we will see leads to a simple bound on the cost of classical simulation.

Our simulation will follow the main treatment of [27, 32], with the addition that we will consider rotated operators and their decomposition into Pauli operators, and we will use gFlow to instruct our procedure for updating the operators, which eventually leads to our main theorem. Our main tool will be the stabiliser formalism [30]. As mentioned in Section 2, for an open graph state the stabilisers define a space. Generally we talk in terms of a stabiliser group S, which is a subgroup of the Pauli operators. In the case of the open graph states, the generators of the stabiliser group are given by the operators K_i (Equation 3), so that $S = \langle \{K_i\}_{i=1}^n \rangle$. These are not the unique generators, indeed multiplying each of these by any one generator gives a new set of generators. The stabiliser group defines a space (the stabiliser space, or 'code' in error correction terminology) by a set of eigen equations - it is the space of states which are unchanged by the group. For the open graph states this is given by Equation 5 that is for all $i \in I^c : K_i|G(\psi)\rangle = |G(\psi)\rangle$, which implies all products of K_i (i.e. all elements of S) leave the states unchanged. We say the states $|G(\psi)\rangle$ are *stabilised* by the group S. In general if the stabiliser group for n qubits is generated by k elements, then the stabiliser space is of dimension 2^{n-k}. Essentially the stabilisers act like the identity over the this space, defining the space itself. In addition to tracking the logical operators, we will also track the stabiliser operators - indeed this will be a key tool for the former.

One can picture the whole of the computation in a high level as follows. The attaching of the input to the graph state (forming the open graph state), encodes the input space onto the many qubit state. The information is in some sense 'spread' over the large entangled state (we will talk more about computation as spreading of information in the next section). We call this encoded space the logical space. During the computation the information is pushed forward through the measurements towards the outputs, so that after the final measurements the logical space sits on only the output qubits. During this push the logical space is also rotated around, resulting in unitary computation. One can think of the stabilisers as keeping track of where the logical space is sitting, and the logical operators as telling you how the space has been rotated (in a sense the logical operators track both).

If state $|\psi\rangle$ in the stabiliser space, with stabiliser group $S = \{S_i\}$ evolves under unitary U, the new state $U|\psi\rangle$ is clearly stabilised by $\{US_iU^\dagger\}$, giving the updated stabiliser group. Under measurement things are slightly more complicated. In this work we use only single qubit projective measurements, which we write as two outcome measurements of the form $A_i = P_i^+ - P_i^-$ where $P_i^\pm$ are the projectors onto the ± 1 outcomes where i indicates the qubit measured. As usual we denote r_i as the binary representation of the measurement outcome with $r_i = 0$ when the outcome is $+1$ and $r_i = 1$ when the outcome is -1. If it is possible to find a set of generators such that only one anticommutes with the measurement, call it S_i, and the rest commute, the update simply replaces S_i with $-1^{r_i}A_i$. It is not hard to see that this group will stabilise the state after measurement [30]. The projection from the measurement will not change the eigenvalue relation of commuting operators, and the projected state is clearly a $+1$ eigen state of the operator $-1^{r_i}A_i$. The trick is to find a suitable set of generators allowing for such an update (i.e. such that one and only one anticommutes with the measurement) - which is where the gFlow comes in.

So how should we describe the evolution of our logical operators? We want them to describe the information as it evolves. Talking in terms of pure states (which suffices for our discussion) if $|\psi\rangle \to |\tilde{\psi}\rangle$, we want that our logical operators evolve $L \to \tilde{L}$ so that their expectation is preserved, that is we demand $\langle\psi|L|\psi\rangle = \langle\tilde{\psi}|\tilde{L}|\tilde{\psi}\rangle$. In this way, the new operators $\tilde{L}$ genuinely reflect the information of the evolved space (see [27, 31, 32] for more details). Under a unitary evolution $|\psi\rangle \to U|\psi\rangle$, we then have $L \to ULU^\dagger$, clearly satisfying our requirement. For measurements, the trick will be to ensure that the logical operators commute with the measurement operators, in which case, they remain unchanged (measuring commuting observables cannot affect their expectation). The way of doing this will be to multiply by stabilisers - which act as identity on the logical space, so can be introduced without affecting the validity of the logical operators.

We will now see how we can track the evolution of the stabilisers and logical operators through the computation. This will be done in three steps. Note that our procedure does not exactly reflect the step by step process of the computation, as we do not consider corrections, rather it reflects the update as if all measurements had the outcome $+1$ - which is indeed the role of the corrections in the first place. In our discussion below we focus on measurements on the $X - Y$ plane, similar arguments simply apply to the other planes.

Step 1: The first step is to prepare the stabilisers in a form that will allow us to simulate the measurements through the computation more easily. Physically it corresponds to the unitary process of applying the control-Z operators generating the open graph states (Equation 4), followed by simplifying the measurement operator by applying first the appropriate local rotation. The stabilisers of the open graph state are already given in Equation 3. For each input i an informationally complete set of operators is given by the Pauli operators X_i, Z_i and $Y_i = iZ_iX_i$. If we know X_i and Z_i we can calculate Y_i, hence we concentrate only on these two, and denote them as L_{X_i} and L_{Z_i} as we trace them through the computation. The control-Z operators generating the open graph state is unitary, thus after being attached to the graph the logical operators become $L_{X_i} = X_i \otimes_{j \in N(i)} Z_j$ and $L_{Z_i} = Z_i$ (using the relation $L \to ULU^\dagger$ where U is the control-Z operator, see also [6]).

Now we want to put these in a form ready to simulate measurements. The idea is based on the fact that a measurement in the $X - Y$ plane is equivalent to first rotating around the Z axis, followed by measurement in the X basis (similar relations are true for the other two planes used). We initialise all the stabilisers and logical operators by doing this rotation, and consider Pauli X measurements afterwards. The resulting state is sometimes called a rotated graph state. At the same time we replace the individual stabilisers by products given by the gFlow. We thus start with stabilisers

$$S = \left\langle \left\{ S_i := \prod_{j \in g(i)} K_j^{\theta_j} \right\}_{i \in O_C}, \{G_i\}_i \in O \right\rangle, \qquad (9)$$

where $K_i^{\theta_i} := e^{i\theta_i/2Z_i} K_i e^{-i\theta_i/2Z_i} = \cos\theta_i X_i \otimes_{j \in N(i)} Z_j + i\sin\theta_i Z_i X_i \otimes_{j \in N(i)} Z_j$ are the rotated graph state stabilisers and θ_i is the angle of the measurement for qubit i. The set $\{G_i\}_i \in O$ are there simply to complete the set of generators in the case that $|I| < |O|$, chosen such that $[G_i, X_j] = 0$, $\forall j \notin O$. Such a set can always be found as follows, take an arbitrary set of operators which complete a generating set (note that the operators S_i above are by definition all independent and so can form part of a generating set, then there is always some set of operators in S which complete this set of generators). To ensure commutation relation, we go round by round, starting from R_1, we go through each vertex ν in the round, and check if it commutes with X_ν- and if not we multiply it by S_i. These operators are still valid generators and they commute with all the X_ν measurements. At the same time, by applying the local unitary Phase rotations, the logical operators are initialised to

$$L_{X_i} = e^{i\theta_i/2Z_i} X_i e^{-i\theta_i/2Z_i} \otimes_{j \in N(i)} Z_j,$$
$$= \cos\theta_i X_i \otimes_{j \in N(i)} Z_j + i\sin\theta_i Z_i X_i \otimes_{j \in N(i)} Z_j,$$
$$L_{Z_i} = Z_i. \qquad (10)$$

Step 2: The second step is to take the logical operators to a form which is convenient for measuring X_ν on all the non-outputs - by making sure that they commute with X_ν $\forall \nu \in O_C$. This update does not actually reflect any physical operation, rather it is just rewriting by multiplication of logical identities, i.e. stabilisers. However it is this step where the cost of the simulation arises, both in time and space of simulation. Although this is not a physical update we will trace through what would happen in the computation to see how our update can be carried out to ensure consistency in maintaining commutation.

We first expand logical operators in terms of products of Pauli operators

$$L_\alpha = \sum_i a_i M_i^\alpha, \tag{11}$$

where M_i^α is some product of Pauli operators, this is always possible since the Pauli operators forms a complete operator basis. Then, starting in R_1 with the stabilisers (Equation 9) and logical operators (Equation 10), we proceed with each round as follows, going from the first to the final round in sequence. In round R_x we update each Pauli term M_i^α in each L_α as follows:

$$\forall \mu \in R_x : \quad \text{If } [M_i^\alpha, X_\mu] = 0, \quad M_i^\alpha \to M_i^\alpha$$
$$\text{If } \{M_i^\alpha, X_\mu\} = 0, \quad M_i^\alpha \to S_\mu M_i^\alpha$$

After this step is complete, by the properties of gFlow it is easy to see that all the L_α will commute with all X_ν, i.e. $[L_\alpha, X_\nu] = 0 \ \forall \alpha, \forall \nu \in O_c$.

Step 3: The third step reflects the measurement of the computation, however with the unphysical condition that all outcomes are plus one. Although this does not really reflect measurement, it reflects the computation, since corrections are made so that this is always the final state. We first update the stabilisers and then use these to update the logical operators so that they are trivial (identity) everywhere except the outputs. The stabilisers are replaced with

$$S = \langle \{X_i\}_{i \in O_C}, \{G_i\}_i \in O \rangle. \tag{12}$$

One can picture this as measurements with fixed $+1$ outcome as follows. We first notice that $\{S_i, X_i\} = 0$ and $[S_{i>j}, X_j] = 0$, as can easily be seen from the definition of gFlow. To update the stabiliser operators to arrive at Equation 12 from Equation 9, again we start in R_1 and proceed with each round, going from the first to the final round, and in each round R_x, we replace all the stabilisers $S_{i \in R_x}$ with X_i (corresponding to measuring X_i and getting result $+1$). Because of the condition $\{S_i, X_i\} = 0$ and $[S_{i>j}, X_j] = 0$, this reflects exactly measurement with the $+1$ outcomes, and finally we end up with the stabilisers (Equation 12).

The next part is to use these new stabilisers to update the logical operators one final time. Again we do so term by term in the decomposition into Pauli operators. If a term M_i^α has an X_μ for $\mu \notin O$, it is multiplied by X_μ (which is now a stabiliser, hence a logical identity). The remaining logical operators are trivial (i.e. identity) on everything except the outputs, and they encode the unitary evolution of the computation $L_\alpha \to U^\dagger L_\alpha U$. This completes the classical simulation.

The efficiency of this procedure is dominated by the size of the logical operators (the number of terms occurring in the expansion). The stabilisers are updated efficiently (nothing in the initialisation or the update scales larger than

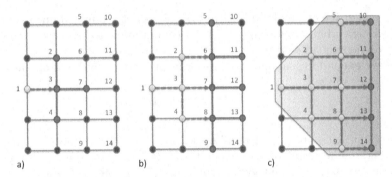

Fig. 5. Illustration of Step 2 the update procedure for the cluster state. The red vertices represent the qubits where the logical operators L_{X_1} and L_{Z_1} are non-trivial. a) is the point directly after X_1 has been considered. b) is the point directly after qubits in the first and second round have been measured. c) is the point after qubits in the third round have been considered. See text for details. The number of qubits touched in the update procedure is equal to the size of the forward cone F_C, which gives an upper bound to the size of the final logical operators, and hence the cost of directly simulating the computation (see Theorem 2). The F_C also acts as a light cone for the information spread throughout the computation.

$O(n)$ where n is the size of the pattern). Similarly the initialisation of the logical operators is efficient, however, during each update step, each term in the expansion into Pauli operators must be checked and possibly updated. When an S_μ is added to the term in the second step, the size increases by $2^{|g(\mu)|}$, where $|g(\mu)|$ is the number of vertices in the correcting set. This is necessary for every Pauli Z operator introduced by previous updates. Starting from R_1 these Pauli Z operators are introduced on all the neighbours of the correcting sets - that is along the gFlow path and one graph edge further. Thus they follow along all possible influencing paths. Some terms may cancel out, so the total number of terms is less than equal to $2^{|F_C(\nu)|}$. From this we get the following theorem.

Theorem 2. *An MBQC over an open graph state with gFlow can be simulated classically in $O(n \exp(|F_{C_{max}}|))$ where $F_{C_{max}}$ is the maximum forward cone over all the inputs. More explicitly the logical operators L_α associated to vertex μ can be updated with $O(\exp(|F_C(\nu)|))$.*

As mentioned, the above simulation does not take into account correction (since it is unnecessary in terms of simulating the computation). One may wonder given the simulation above where would the corrections come in at all. The answer is in the last step - when measureing X_i, and getting result r_i, instead of replacing by X_i, we should replace by $-1^{r_i} X_i$. This would add signs throughout the logical operators which in general could not be undone by simply products of Pauli operators. With the exception being the case when each logical operator only has one M^α in expansion (Equation 11), i.e. is just a product of Pauli operators, which occurs when the angles $\theta_i = 0, \pi$, i.e. measurements onto Pauli operators only. Then the minus signs can be all flipped coherently by multiplying by stabilisers. This is another way of seeing that if only Pauli measurements are made, all corrections can be made at the end. In such a case one can also see that the size of the logical operators becomes small - only one term each - so that this simulation itself is efficient. This simple observation will allow us to derive the equivalent of Gottesman-Knill Theorem directly in MBQC. As all the Clifford operates can be implemented in MBQC using only Pauli operators. Having removed any dependency as described above will lead to an efficient classical simulation of any MBQC pattern implementing Clifford operators and Pauli measurements. This interplay between efficiency and the angles of measurement is something not taken into account in the above theorem, and offers more potential for better bounds. We leave this to future work for now.

To see the updating which truly corresponds to a computation, i.e. including corrections, one can combine steps 2 and 3 to get rid of the X_is round by round by applying the post measurement stabilisers $-1^{r_i} X_i$ and in addition perform the correction operation (given by the gFlow) to remove the -1^{r_i}. The effect is that one can simply remove the measured X_is whilst tracing through the computation.

For clarity we go through the example for the first few rounds on the 2D cluster state. For input of qubit 1 before being attached to the graph it is described entirely by two logical operators $L_{X_1} = X_1$ and $L_{Z_1} = Z_1$. After Step 1 initialisation (joining to the open graph state and 'rotating' each qubit according to the measurement basis), these become

$$L_{X_1} = e^{i\theta_1 Z_1} X_1 Z_3$$
$$L_{Z_1} = Z_1.$$

Here we have abbreviated the terms coming from the rotated basis into the exponent $e^{\theta_1 Z_1} = \cos\theta_1 \mathbb{1} + i\sin\theta_1 Z_1$, and for ease of notation we remove the tensor product symbol.

We next consider Step 2, starting with round R_1 and operator X_1 that anticommutes with Z_1s, hence for those terms in the L_α where this occurs we are required to multiply by $S_1 = K_3^{\theta_3} = Z_1 \otimes Z_2 \otimes e^{i\theta_3 Z_3} X_3 \otimes Z_4 \otimes Z_7$. This is equivalent to putting it up into the exponent, so that the logical operators become

$$L_{X_1} = X_1 e^{i\theta_1 Z_2} e^{i\theta_3 Z_3} X_3 Z_4 Z_7 Z_3$$
$$L_{Z_1} = Z_2 e^{i\theta_3 Z_3} X_3 Z_4 Z_7.$$

In Step 3 the X_1s are removed (since after measurement and correction the X_1 are a logical identity), and the logical operators are thus non-trivial on qubits $2, 3, 4, 7$ after R_1, as illustrated in Figure 5a). In the second round R_2, X_ν on qubits 2, 3 and 4 are considered. We update the logical operators by considering these one by one, starting with X_2 (any order in the same round gives the same final result). This anti commutes with Z_2 - which comes from the application of $S_1 = K_3^{\theta_3}$ in the previous round. Indeed this is how the updates are affected along all influencing paths. When the Z_2 occurs we are forced to multiply the term by $S_2 = K_6^{\theta_6} = Z_2 \otimes Z_5 \otimes e^{\theta_6 Z_6} X_6 \otimes Z_7 \otimes Z_{11}$. This takes the logical operators to

$$L_{X_1} = e^{i\theta_1} e^{i\theta_3 Z_3} X_3 Z_4 Z_5 e^{\theta_6 Z_6} X_6 Z_{11} Z_3$$

$$L_{Z_1} = e^{i\theta_3 Z_3} X_3 Z_4 Z_5 e^{\theta_6 Z_6} X_6 Z_{11}.$$

Note that here, two Z_7 operators have cancelled out - they came from two occasions where qubit 7 was a neighbour of one of the correcting sets. In this way, it is possible that some qubits in the set of influencing paths cancel out - this happens if the number of influencing paths they sit in as gFlow paths is even, and the number arriving from non-gFlow paths is also even (at this point in our calculation the number of times it is on a gFlow path is zero, and it is in 2 influencing paths not as a gFlow).

After qubits X_3 and X_4 are also considered, we have to do the same trick to get rid of the Z_3 and Z_4s, by multiplying the terms where they occur by $S_3 = K_7^{\theta_7}$ and $S_4 = K_8^{\theta_8}$ respectively. Finally we end up with logical operators

$$L_{X_1} = e^{i\theta_1} e^{i\theta_3 Z_6} e^{i\theta_7 Z_7} X_7 Z_8 Z_{12} X_3 Z_5 e^{\theta_6 Z_6} X_6 Z_7 e^{i\theta_8 Z_8} X_8 Z_9 Z_{11} Z_{13} Z_6 e^{i\theta_7 Z_7} X_7 Z_8 Z_{12}$$

$$L_{Z_1} = X_3 e^{i\theta_3 Z_6} e^{i\theta_7 Z_7} X_7 Z_8 Z_{12} Z_5 e^{\theta_6 Z_6} X_6 Z_7 e^{i\theta_8 Z_8} X_8 Z_{11} Z_{12}.$$

Again, in Step 3 we get rid of the X_2, X_3, X_4s, hence after the measurements in round R_2 the logical operators are non-trivial only on qubits $6, 7, 8, 9, 11, 12, 13$, as indicated in Figure 5b). It is then clear how after the third round of measurements we will be left with logical operators that sit on the highlighted qubits 10, 11, 12, 13 and 14, as indicated in Figure 5c).

For any graph and any measurement pattern with gFlow, each time a Pauli Z operator is added, unless it is in the output set, we will have to multiply that term by a stabiliser - which will add a splitting of two. During the update procedure, Pauli Z operators are added along every influencing path. Sometimes these will cancel out, depending on the graph, but sometimes not, so that this gives an upper bound to the complexity for the direct update procedure which is the content of Theorem 2.

We thus see an initial way to go from a gFlow to a classical simulation. However, for certain examples this bound can be very bad. We have already mentioned that this is the case where all of the measurement angles are zero or $\pi/2$ - i.e. measuring the Pauli operators - there is no splitting of the logical operators, and only one term is needed for each logical operator, hence this simulation becomes efficient, which is not captured by Theorem 2 (where we effectively assume the worst case for the angles). Another example is a computation across along a 1D graph state, with one input, say on the left, and an output on the right (see Fig. 3c). There the gFlow simply follows the line, thus the influencing volume is big, however, this is always a simple one qubit computation, and indeed all computations on a 1D cluster state are easy to simulate classically [33]. In Section 5 we will see how connections to entanglement allow us to make tighter bounds on classical simulatability which will work well for this example and many others. Before we do that however, in the next section we will discuss how the update above can be interpreted as information flow, in tern giving the interpretation of the forward cone F_C as a light cone for the information.

4   Flow of Information and F_C as Information Light Cones

The gFlow gives a causal structure on top of a graph state induced by the correction procedure, called the forward cone F_C (Definition 7). In this section we will also look at how the same cone can be understood as a forward cone of information, and moreover a light cone (so that information cannot travel beyond this cone).

The forward cone can be viewed as an information forward cone directly from the simulation procedure described in the previous section, and the interpretation of the logical Heisenberg representation as showing us where information sits (see for example [34]). Consider a density matrix of some input i

$$\rho_i = \frac{1}{2} \left(\mathbf{1} + \eta_x X_i + \eta_y Y_i + \eta_z Z_i \right) \tag{13}$$

The state is totally described by the coefficients η_i. The logical Heisenberg representation ensures that at any time the evolved state, denoted as $\tilde{\rho}$, which now can be sitting over many systems, is described as

$$\tilde{\rho} = \frac{1}{2} \left(\mathbf{1} + \eta_x \tilde{L_{X_i}} + \eta_y \tilde{L_{Y_i}} + \eta_z \tilde{L_{Z_i}} \right), \tag{14}$$

where the $\tilde{L_\alpha}$ are the updated logical operators of α corresponding to the evolution.

The information is then preserved, but 'spread' over to different operators in the following sense. To recover the information encoded on the original system i (i.e. recover the η_i), we should measure the logical operators $\tilde{L_{X_i}}$, $\tilde{L_{Y_i}}$ and $\tilde{L_{Z_i}}$. Thus the information can be said to have spread over the range of the logical operators. From the simulation in the previous section, it is clear that the logical operators, and hence the information of input qubit i spreads out over the causal forward cone F_C defined by the gFlow (see Figure 5).

One may then ask if this is all that is allowed, or could we understand the information as having spread further than the influencing cone (after all, this is not the only way one may simulate a computation)? The answer (at least for patterns where we wish uniform determinism, i.e. that all measurements on a Pauli equator are allowed) is no, in that the spread must be balanced by consistency amongst all measurements, which is the function of the gFlow, which defines the cone F_C.

Let us first return back to Step 2 in the simulation above, which is where this spread of information occurs in the simulation. The trick is simply multiplying the logical operators by a logical identity (i.e. the stabilisers). This part however is clearly not restricted to the cone. One could easily expand a logical operator to cover practically all qubits in this way. The reason we do not allow this is because we want to do measurements, and we want to do them over all qubits not outputs so that all logical operators are preserved (this is what we mean by consistency). Say one did this for operator L_{X_i}, so that its extent was over many qubits. Taking its expansion into products of Pauli operators (as in Equation 11), one would have a sum of many terms, including Pauli Z and X operators on any given qubit in its range. When measuring qubits, to ensure the survival of the information, we asked that the logical operators be taken to a form which commutes with the measurement - this was the role of Step 2 in the above. If one did not have this, information may be lost. This can only be the case if in each term $M_j^{X_i}$ of the expansion of L_{X_i}, the part of $M_j^{X_i}$ on vertex μ is the same (say σ_μ) or identity for all terms.

One could have, for example, that this is indeed the case, i.e. for a particular L_{X_i}, extended so that it touched many qubits, that over each such qubit μ all the terms in the expansion of L_{X_i} were either the same Pauli σ_μ or identity. In such a case, one could happily measure those qubits in the Pauli basis σ_μ, the information would be preserved, and the logical operators could be calculated (if we wanted to consider the information over the outputs we would then follow Step 3 to leave them as identity everywhere else, though one would have potentially different evolutions for different branches). In this way its final spread may indeed be beyond the light cone given by gFlow. The problem with this would be that we want to transfer the logical operators not just of one input i, but of *all* the inputs. It is shown in [27] that to achieve this, in such a way that every measured vertex one can choose amongst a set of measurements across one of the planes, the only way to do it is via a gFlow. Hence, for an input η, not only is the forward cone $F_C(\eta)$ also an information cone, but to transfer all the information at the same time, it is a light cone for the information contained η - that is, the information can not spread beyond it, and the computation be consistent for all inputs.

From the perspective of information flow, theorem 2 says, unsurprisingly, that the more information is spread through a computation, the more costly it is to simulate. However, again we should be careful to note that the true cost of simulation depends on the angles of the measurements, which is not captured by the size of F_C, hence not by theorem 2. As we saw, for angles $0, \pi$, the simulation is simple, however the spread of information is still large.

5   Bounds on Entanglement from gFlow

In this section we will show a connection between gFlow on the one hand, and entanglement conditions for both the universality of a resource state and the classical simulatability of a computation on the given resource, on the other. More precisely we will show that the Flow and gFlow can be used to upper bound the entanglement of the graph state, in terms of the entanglement width [9] and what we will call the *structural entanglement* (though not explicitly defined, it can be understood from [8], see also [7]). Conditions of universal family of resource states, and for classical simulatability are known for both these measures, which can be translated to conditions about the gFlow through our bounds [9–11]. Several known results can then be derived for both the measurement based model and circuit based model (through the known maps between the two models [12, 17]). For example we reproduce the result by Josza stating that a circuit which any wire are touched by at most logarithmic (in the size of the input) many number of two qubit gates can be classically simulated efficiently [24].

Let us first define the entanglement measures we are interested in. The entanglement width [9] of a pure state $|\psi\rangle$ is defined as

$$\chi_{wd}(|\psi\rangle) := \min_{T} \max_{e} \chi_{T,e}^{bi}(|\psi\rangle), \tag{15}$$

where $\chi_{T,e}^{bi}(|\psi\rangle)$ is the the log-Schmidt rank across the bipartite cut defined by T and e where T is a sub-cubic graph with n leaves and e is an edge of T. Each leaf corresponds to a qubit of the state $|\psi\rangle$. The bipartite cut is defined by removing edge e to give two separate trees. The leaves of one tree correspond to one side of the cut, and the other tree the other side of the cut. It was shown that if the entanglement of a family of resource state does not scale polynomially with the size of its input space then, that family cannot be a universal [9, 11] (in the case of QQ computations, note that this is not the same as asking for universality in the CC case). It was also shown that any MBQC can be simulated in $O(npoly(2^{\chi_{wd}}))$ [10].

Motivated by the proofs in [8], we define the *structural entanglement* as

$$E_{struc}(|\psi\rangle) := \min_{\substack{Order \\ 1,\dots,n}} \max_{\substack{cut\ k \\ A=1,\dots,k \\ B=k+1,\dots,n}} \chi_{AB}(|\psi\rangle), \tag{16}$$

where the minimum is taken over all orderings (labelings) of the qubits $1,\dots,n$, and the max is taken over a cut defined for a given ordering by taking all qubits $1,\dots,k$ on partition A and the rest on partition B and $\chi_{AB}(|\psi\rangle)$ being the log-Schmidt rank over cut AB. Although not explicitly stated in terms of this measure, in [8] it is shown that any MBQC pattern can be simulated classically in $O(n^2 poly(2^{E_{struc}}))$.

It is easy to see that the tree in Figure 6 defines a set of cuts such that any cut either splits the graph in two with all leaves below or equal to a value k on the left, and above k on the right (as per the optimisation for E_{struc}), or else it just identifies one leaf. This clearly implies that

$$E_{struc} \geq \chi_{wd}. \tag{17}$$

We will now see how the Flow and gFlow can be used to upper bound E_{struc}, and in turn χ_{wd}. We start by considering Flow, which is simpler to picture, but the ideas easily extend to gFlow. The idea is that they both can be used to define a natural order, which gives a simple bound to E_{struc} which comes from induced disjoint input-output paths. Indeed, if an open graph state has Flow, following the image of the Flow function, f, (Definition 4) from each of the inputs leads to disjoint lines to the outputs, which cover all the non outputs [15, 17] (called *Flow wires*).

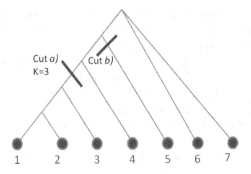

Fig. 6. This tree defines a set of cuts showing that $E_{struc} \geq \chi_{wd}$. Any cut either sits in the same set of cuts as that optimised for E_{struc} - effectively choosing a k such that all qubits of number less than or equal to k are in partition A and all higher qubits are in partition B (cut a)), or else it singles out one qubit (cut b)), which can never be the unique maximum.

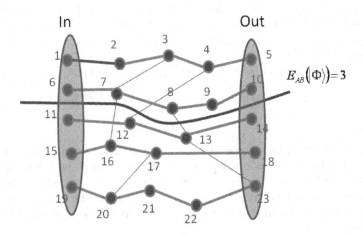

Fig. 7. Flow defines a natural ordering from top left to right across each Flow wire from top to bottom as shown. This is used to define cuts by a number k where partition A consists of all qubits below qubit k in the ordering and partition B consists of all qubits above k in the ordering. The entanglement across any cut is upper bounded by the number of edges cut (in this case $k = 10$ and the entanglement is exact).

We consider first the case where $|I| = |O|$. The numbering goes as follows. We start with an arbitrary input going along the image of the Flow function of that vertex till we reach an output qubit. Next we choose another not selected input qubit and carry on in this fashion, till we cover all the inputs, see Figure 7 and Figure 8. Note that based on the definition of Flow, no input qubit could belong to the image of the Flow function of another input qubits hence on each such Flow wire there will sit only one input qubit and hence we have $|I|$ such wires. To calculate the entanglement we note the fact that the entanglement across a

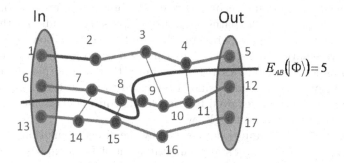

Fig. 8. In the worst case the number of edges cut by a line equals $1 + 2C_F$. This implies a bound on the structural entanglement (see text).

cut C for a graph state can be bounded by the number of edges crossing the cut $E_{A,B} \leq C$. This is clear since in preparation of the state each edge corresponds to a control-Z operator, and C such operators can create at most C e-bits.

Definition 9. *For an open graph with flow, we denote C_F the maximum number of edges crossing between Flow wires.*

We easily see that a cut between two Flow wires gives entanglement at most C_F (see Figure 7). This can be at most doubled (plus one) by choosing a lower number to cut at (thus potentially increasing the number of edges cut) (see Figure 8). We thus have that $E_{struc} \leq 1 + 2C_F$.

To extend this to the case where $|O| > |I|$ we must consider the worst case, for which each extra qubit adds one unit of entanglement. In this general setting we now call C_F the maximum number of edges crossing between Flow wires, when the output qubits not in a Flow wire are ignored along with their edges. We also call $\Delta := |O| - |I|$. We then have the following observation.

Theorem 3. *A graph state with Flow has structural entanglement*

$$E_{struc} \leq 1 + 2C_F + \Delta. \tag{18}$$

Thus any computation can be simulated in at least $O(n^2 poly(2^{2C_F + \Delta}))$.

We note that any computation which can be done with a number of outputs greater than $|I|$ can be done with $|I| = |O|$ without changing the Flow or C_F by simply removing the extra Δ output qubits from the graph resource. Thus $\Delta = 0$ for most interesting cases. This is clear since the existence of Flow is robust against losing the extra outputs, and this guarantees the computation.

This result can be extended to open graphs with gFlow by using gFlow to find disjoint input-output lines as follows. As we saw earlier, it is clear that if an open graph state has gFlow, then it is necessarily D-happy (from Theorem 1 and the fact that gFlow implies unitary computation). This in turn means that any cut which separates the input and the output goes through at least $C \geq |I|$ edges.

Taking a result from graph theory, Menger's theorem [35] says that this implies there are at least n parallel paths going from inputs to outputs. Furthermore it can be shown that there are parallel paths which sit along gFlow paths and can be found systematically also [36]. This can be used to give a natural order to the graph as for Flow, but with the possibility that non-output qubits do not sit in the disjoint paths, and so should be added to the Δ term. Again the size of Δ may be reduced or removed by considering equivalent smaller graphs, but this is less well understood for gFlow.

This result covers examples not covered by the direct simulation from Section 3, for example the $1D$ cluster state. The statement of Theorem 3 is a very similar sounding statement to Jozsa's [24] which states that a quantum computation on a circuit can be simulated in $O(npoly(2^D))$ where D is the maximum number of gates that touch or cross a circuit wire.

6   Conclusions

We have seen that gFlow can be used for two complementary approaches for giving bounds on the efficiency of classical simulation for MBQC. In Theorem 2 we saw that classical simulation is possible with resources scaling as exponential in the size of the largest causal future cone defined by the gFlow. On the other hand in Theorem 3 gFlow can be used to upper bound the entanglement, and hence give bounds on resources for classical simulation in terms of the number of edges crossing gFlow wires (parallel wires from input to output induced by gFlow). Simple and straightforward, but illuminating conditions for entanglement of general resource states are described in Theorem 1. Furthermore the causal future cone induced by gFlow is seen to be at the same time a light cone for information spreading.

The results on classical simulation combine two of the main approaches for bounding the cost of classical simulation for quantum computation - explicit tracking of the computation using an efficient form (used for example in the Gottesman Knill [30] theorem and related results (e.g. [37])) and bounds coming from entanglement (used for simulating computation [7–11] as well as many body physics (e.g. [38])). This offers the perspective of bridging these two approaches via gFlow. A natural question is the interplay between the angle of measurement and efficiency of simulation via the gFlow update procedure presented here. Setting all angles to zero or π makes the simulation efficient (as per the Gottesman Knill theorem), however for general angles it is not efficient(indeed Theorem 2 represents this worst case situation). The in between ground, combined with bounds by entanglement may present new classes of computation admitting efficient simulation for example. Furthermore, we may gain more insight into how efficiency of classical simulation is related to other features of computation illuminated by the study of gFlow.

It is also interesting in itself that from Flow and gFlow one can derive bounds on the entanglement of a graph state. Since there exist efficient algorithms to calculate the Flow and gFlow of graphs [15, 16], and given Flow and gFlow one can easily bound the entanglement, one may use this to upper bound the entanglement of general graph states. This is both important for recognising good resources (since the existence of Flow does not talk about universality, whereas the entanglement gives bounds on this also [9, 11]), and more generally as entanglement represents important resource in other areas of quantum information.

The two approaches to classical simulateability can also be understood as arising from two notions of 'spreading' of information. We have seen in Section 4 that the forward cone given by gFlow bounding the cost of classical simulation can also be interpreted as a spread of information - so that the more spread the information is, the more costly the simulation. The bounds arising from entanglement ([7], [8] e.t.c) which lead to Theorem 3, can also be understood as assigning a cost to the spread of information as follows. The entanglement measure key to these results is a bipartite measure, the Schmidt measure of entanglement, which counts the minimum number of product states (with respect to a particular cut) needed to describe the state. This may be interpreted as saying how 'spread' across product bases the state is. Indeed it is exactly the rank of the reduced density matrix of one cut, so in a sense says the size of the space in which it must be understood to sit (in this sense the 'spread' is over the state space rather than precisely the parties). The trick of [7] and subsequent work is to find an efficient form to describe the state and its updating through a computation based on this minimum decomposition. Again, the smaller the 'spread' in this sense, the smaller the cost of this simulation. As we have also seen in Sections 3 and 4, a big 'spread' of information is however not enough to imply that a computation is difficult to simulate - MBQC with only Pauli measurements is efficient to simulate, but the spread of information is large (in both senses - the future cone is large, and the entanglement is large). To capture the difficulty in simulation, one must also include something about how this 'spread' of information is used. In the case of MBQC studied here, universality (and presumably the difficulty in simulation) is given by the use of arbitrary angles for the measurements, using the spread of information in the most universal way. This balance between spread of information through entanglement and how it can be used also plays a key role in analogies between MBQC and thermodynamics and in particular phase transitions [39, 40]. It is an exciting prospect that these pictures may be unified from the perspective of gFlow or similar notions.

Acknowledgements. The authors would like to thank Bobby Antonio, Simon Perdrix and Einar Pius for useful discussions and feedback. We are particularly grateful to Vincent Danos and Prakash Panangaden for many discussions on the topics of this paper which gave rise to many of the ideas mentioned here directly and indirectly. DM is funded by the FREQUENCY (ANR-09-BLAN-0410), HIPER-COM (2011-CHRI-006) projects, and by the Ville de Paris Emergences program, project CiQWii. EK is funded by UK Engineering and Physical Sciences Research Council grant EP/E059600/1.

References

1. Raussendorf, R., Briegel, H.J.: A one-way quantum computer. Physical Review Letters 86, 5188 (2001)
2. Anders, J., Browne, D.E.: Computational power of correlations. Physical Review Letters 102, 050502 (2009)
3. Hobanand, M.J., Wallman, J.J, Browne, D.E.: Generalized bell-inequality experiments and computation. Physical Review A 84(6), 062107 (2011)
4. Broadbent, A., Fitzsimons, J., Kashefi, E.: Universal blind quantum computing. In: Proceedings of the 50th Annual IEEE Symposium on Foundations of Computer Science (FOCS 2009), p. 517 (2009)
5. Markham, D., Sanders, B.C.: Graph states for quantum secret sharing. Physical Review A 78, 042309 (2008)
6. Danos, V., Kashefi, E., Panangaden, P.: The measurement calculus. Journal of ACM 54, 8 (2007)
7. Vidal, G.: Efficient classical simulation of slightly entangled quantum computations. Physical Review Letters 91(14), 147902 (2003)
8. Yoran, N., Short, A.J.: Efficient classical simulation of the approximate quantum fourier transform. Physical Review A 76(4), 042321 (2007)
9. Van den Nest, M., Miyake, A., Dür, W., Briegel, H.J.: Universal resources for measurement-based quantum computation. Physical Review Letters 97(15), 150504 (2006)
10. Van den Nest, M., Dür, W., Vidal, G., Briegel, H.J.: Classical simulation versus universality in measurement-based quantum computation. Physical Review A 75(1), 012337 (2007)
11. Van den Nest, M., Dür, W., Miyake, A., Briegel, H.J.: Fundamentals of universality in one-way quantum computation. New Journal of Physics 9(6), 204 (2007)
12. Danos, V., Kashefi, E.: Determinism in the one-way model. Physical Review A 74, 052310 (2006)
13. Browne, D., Kashefi, E., Mhalla, M., Perdrix, S.: Generalized flow and determinism in measurement-based quantum computation. New Journal of Physics 9, 250 (2007)
14. Hein, M., Eisert, J., Briegel, H.J.: Multi-party entanglement in graph states. Physical Review A 69(6), 062311 (2004)
15. de Beaudrap, N.: Finding flows in the one-way measurement model. Phys. Rev. A 77, 022328, 8 (2006, 2008)
16. Mhalla, M., Perdrix, S.: Finding optimal flows efficiently. In: Aceto, L., Damgård, I., Goldberg, L.A., Halldórsson, M.M., Ingólfsdóttir, A., Walukiewicz, I. (eds.) ICALP 2008, Part I. LNCS, vol. 5125, pp. 857–868. Springer, Heidelberg (2008)
17. de Beaudrap, N.: Unitary-circuit semantics for measurement-based computations. International Journal of Quantum Information (IJQI) 8, 1 (2009)
18. Kashefi, E., Markham, D., Mhalla, M., Perdrix, S.: Information flow in secret sharing protocols. Electron. Notes Theor. Comput. Sci. 9 (2009)
19. de Beaudrap, N., Danos, V., Kashefi, E., Roetteler, M.: Quadratic form expansions for unitaries. In: Kawano, Y., Mosca, M. (eds.) TQC 2008. LNCS, vol. 5106, pp. 29–46. Springer, Heidelberg (2008)
20. Broadbent, A., Kashefi, E.: Parallelizing quantum circuits. Theoretical Computer Science 410(26), 2489 (2009)
21. Kashefi., E., da Silva, R.D., Galvao, E.F.: Closed timelike curves in measurement-based quantum computation. Phys. Rev. A 83, 012316 (2011)

22. Raussendorf, R., Sarvepalli, P., Wei, T.-C., Haghnegahdar, P.: Measurement-based quantum computation–a quantum-meachanical toy model for spacetime? arXiv:1108.5774 (2011)
23. Raussendorf, R., Sarvepalli, P., Wei, T.-C., Haghnegahdar, P.: Symmetry constraints on temporal order in measurement-based quantum computation. Electronic Proceedings in Theoretical Computer Science (EPTCS) 95, 219 (2012)
24. Jozsa, R.: On the simulation of quantum circuits. quant-ph/0603163 (2006)
25. Perdrix, S., Marin, A., Markham, D.: Access structure in graphs in high dimension and application to secret sharing. In: Proceedings of the Eighth Conference on the Theory of Quantum Computation, Communication and Cryptography (2013)
26. Mhalla, M., Murao, M., Perdrix, S., Someya, M., Turner, P.S.: Which graph states are useful for quantum information processing? In: Proceedings of the Sixth Conference on the Theory of Quantum Computation, Communication and Cryptography (2011)
27. Browne, D., Briegel, H.J.: One-way quantum computation. In: Lectures on Quantum Information, p. 359. Wiley Online Library (2006)
28. Antonio, B., Markham, D., Anders, J.: Trade-off between computation time and hamiltonian degree in adiabatic graph-state quantum computation. arXiv:1309.1443 (2013)
29. Browne, D.E., Kashefi, E., Perdrix, S.: Computational depth complexity of measurement-based quantum computation. In: Proceedings of the Fifth Conference on the Theory of Quantum Computation, Communication and Cryptography, p. 35 (2010)
30. Gottesman, D.: Stabilizer Codes and Quantum Error Correction. PhD thesis, California Institute of Technology (1997)
31. Gottesman, D.: The heisenberg representation of quantum computers, talk at. In: International Conference on Group Theoretic Methods in Physics (1998)
32. Raussendorf, R., Browne, D.E., Briegel, H.J.: Measurement-based quantum computation with cluster states. Physical Review A 68, 022312 (2003)
33. Nielsen, M.A.: Cluster-state quantum computation. Reports on Mathematical Physics 57, 147 (2005)
34. Deutsch, D., Hayden, P.: Information flow in entangled quantum systems. In: Proceedings of the Royal Society of London. Series A: Mathematical, Physical and Engineering Sciences, vol. 456, p. 1759 (2000)
35. Diestel, R.: Graph Theory, 4th edn. Springer (2010)
36. Murao, M., Miyazaki, J., Hajdušek, M.: Translating measurement-based quantum computation with gflow into quantum circuit. arXiv:1306.2724 (2013)
37. Linden, N., Clark, S., Jozsa, R.: Generalised clifford groups and simulation of associated quantum circuits. arXiv:0701103 (2007)
38. Vidal, G.: A class of quantum many-body states that can be efficiently simulated. Phys. Rev. Lett. 101(110501) (2008)
39. Anders, J., Hajdušek, M., Markham, D., Vedral, V.: How much of one-way computation is just thermodynamics? Found. Phys. 38 (2008)
40. Markham, D., Anders, J., Hajdušek, M., Vedral, V.: Measurement based quantum computation on fractal lattices. In: EPTCS 26 (2010)

Euclidean Representations of Quantum States

Keye Martin, Johnny Feng, and Tanner Crowder

Naval Research Laboratory
Washington, DC 20375
{keye.martin,johnny.feng,tanner.crowder}@nrl.navy.mil

Abstract. The Bloch representation of a qubit is a beautiful way to picture quantum states. We prove that such a representation does not exist in higher dimensions by uncovering some of the structure required of a general Euclidean representation of quantum states.

1 Introduction

The Bloch representation [9] of a qubit offers a beautiful way to imagine quantum states and the quantum channels that act on them. Its simplicity enables the calculation of things like capacity and scope [7], as well as the development of physically realizable protocols for achieving them [5], things that appear difficult if not impossible when formulated in the standard Kraus representation. This raises the question of whether or not Bloch representations exist in higher dimensions. In this paper, we show that they do not, at least not the kind one would ideally hope for, such as a ball in Euclidean space.

On the other hand, analogues of the idea in higher dimensions [4] do exist, but obtaining a clear and insightful characterization of which points in the unit ball actually represent quantum states has thus far proven difficult. Our results show that to some extent this difficulty is intrinsic to the problem of representing quantum states. It is hoped that the more abstract perspective taken in this paper, one that emphasizes the dependence of convex structure on the set of extreme points, will help serve as a guide in one day uncovering a higher dimensional Euclidean representation of quantum states.

2 The Degree of an Extreme Point

Throughout, $\mathbb{E}$ and $\mathbb{F}$ will denote nonempty compact convex subsets of $\mathbb{R}^n$, the unit ball will be written $\mathbb{B}^n := \{x \in \mathbb{R}^n : |x| \leq 1\}$ and Ω^n is the set of $n \times n$ quantum states on an n dimensional complex Hilbert space. Note that Ω^n can be regarded as a compact convex subset of $\mathbb{R}^{2n^2}$.

A point $e \in \mathbb{E}$ is *extreme* when $e = px + (1-p)y$ implies that $e = x = y$ for all $x, y \in \mathbb{E}$ and $p \in (0, 1)$. The extreme points of a line segment/triangle/rectangle are its vertices, while the extreme points of Ω^n are the pure states (projections). By the Minkowski-Caratheodory result, if $\mathbb{E}$ has dimension k, then each point of $\mathbb{E}$ is a convex sum of at most $k + 1$ extreme points. We now introduce an idea that to the best of our knowledge is new:

F. van Breugel et al. (Eds.): Panangaden Festschrift, LNCS 8464, pp. 454–457, 2014.

Definition 1. The *degree* $d(x)$ of an element $x \in \mathbb{E}$ is the smallest number of extreme points which can appear in a convex sum equal to x.

Notice that a convex linear bijection $[\![\cdot]\!] : \mathbb{E} \to \mathbb{F}$ must preserve degree: $d([\![x]\!]) = d(x)$ for all $x \in \mathbb{E}$. Here in turn is how to calculate the degree of a quantum state:

Proposition 1. *The degree of a quantum state is equal to the number of nonzero eigenvalues that it possesses.*

Proof. Given a quantum state $\rho \in \Omega^n$, we first write it in terms of its spectral decomposition

$$\rho = \sum_{i=1}^{d} \lambda_i |i\rangle \langle i|$$

where the λ_i are its nonzero eigenvalues arranged in decreasing order and $d \leq n$. If it were possible to write ρ as a convex sum of less than d pure states, then we would have

$$\rho = \sum_{i=1}^{k} p_i |\psi_i\rangle \langle \psi_i|$$

where the p_i are again arranged in decreasing order and $k < d$. By [8], (p_i) is majorized by (λ_i) i.e.

$$(\forall j \in \{1, \ldots, k\}) \sum_{i=1}^{j} p_i \leq \sum_{i=1}^{j} \lambda_i$$

But $\sum_{i=1}^{k} p_i = 1$, which implies that $\lambda_{k+1} = 0$, contradicting $\lambda_{k+1} > 0$. $\square$

To illustrate, at least n pure states are required to write the completely mixed state I/n as a convex sum. This turns out to be an essential reason for the difficulty in obtaining 'simple' representations of quantum states in higher dimensions.

3   Euclidean Representations of Quantum States

To help motivate what follows, let us first recall the Bloch representation of a qubit, which uses the spin operators

$$I = \begin{pmatrix} 1 & 0 \\ 0 & 1 \end{pmatrix} \quad \sigma_x = \begin{pmatrix} 0 & 1 \\ 1 & 0 \end{pmatrix} \quad \sigma_y = \begin{pmatrix} 0 & -i \\ i & 0 \end{pmatrix} \quad \sigma_z = \begin{pmatrix} 1 & 0 \\ 0 & -1 \end{pmatrix}$$

to uniquely rewrite a quantum state $\rho \in \Omega^2$ as

$$\rho = \frac{1}{2}(I + r_x \sigma_x + r_y \sigma_y + r_z \sigma_z)$$

where $r = (r_x, r_y, r_z) \in \mathbb{B}^3$. In addition, if one begins with any such $r \in \mathbb{B}^3$, the expression above yields a quantum state, so that a bijection $[\![\cdot]\!] : \Omega^2 \to \mathbb{B}^3$ results. This bijection has many important properties. Here are three of particular interest:

- $[\![I/2]\!] = 0$
- $[\![I - \rho]\!] + [\![\rho]\!] = 0$ for all $\rho \in \Omega^2$,
- $[\![(1 - x)\rho + x\sigma]\!] = (1 - x)[\![\rho]\!] + x[\![\sigma]\!]$ for all $\rho, \sigma \in \Omega^n$ and $x \in [0, 1]$.

The second property implies that Ω^2 has a convex linear involution whose only fixed point is $I/2$, while the third simply says that the Bloch representation is convex linear.

In general, when n is a power of two, a convex linear injection $[\![\cdot]\!] : \Omega^n \to \mathbb{B}^m$ exists [4] with $[\![I/n]\!] = 0$. In addition, Ω^n can admit many convex linear involutions that fix I/n. However, for $n > 2$, such an injection $[\![\cdot]\!]$ can *never* be surjective and such an involution must have fixed points other than I/n:

Theorem 1. *For the set of quantum states Ω^n, the following are equivalent:*

(i) *There is a convex linear bijection $[\![\cdot]\!] : \Omega^n \to \mathbb{B}^m$ for some m.*
(ii) *The dimension of the underlying state space is $n = 2$.*
(iii) *There is a convex linear involution $* : \Omega^n \to \Omega^n$ whose only fixed point is I/n.*

Proof. (i) $\equiv$ (ii). For (i) $\Rightarrow$ (ii), the degree of any point in $\mathbb{B}^m$ never exceeds two, which one sees by noting that each point in the ball is on a line that travels from one unit vector to another, and that the unit vectors in $\mathbb{B}^m$ are the extreme points. But a convex linear bijection must preserve degree, so we have $n = d(I/n) \leq 2$, which gives $n = 2$. For (i) $\Leftarrow$ (ii), use the Bloch representation.

(ii) $\Rightarrow$ (iii): Take $m = 3$ and use the map defined at the start of this section. The desired involution takes ρ to $I - \rho$. In the Bloch representation, this is the antipodal map $a(r) = -r$ on $\mathbb{B}^3$.

(iii) $\Rightarrow$ (ii): Let ρ be a pure state. The involution is a convex linear bijection, so it maps extreme points to extreme points, which implies ρ^* is also extreme. By the convex linearity and involutive property of $*$, $(\rho + \rho^*)/2$ is a fixed point of $*$. But then $I/n = (\rho + \rho^*)/2$, which implies that $d(I/n) \leq 2$, proving $n = 2$. $\qquad\square$

Corollary 1. *Let $[\![\cdot]\!] : \Omega^n \to \mathbb{B}^m$ be a convex linear injection with $[\![I/n]\!] = 0$. If ρ is a pure state and $x = [\![\rho]\!]$, then $-x \in \mathrm{Im}[\![\cdot]\!]$ iff $n = 2$. Put another way, the antipode of a pure Bloch vector is not a Bloch vector unless $n = 2$.*

Proof. For ($\Rightarrow$), we again derive $d(I/n) = 2$; for ($\Leftarrow$), use $((I-\rho)+\rho)/2 = I/2$. $\quad\square$

Notice that any Euclidean representation of n dimensional states must contain points of degree i, for each $1 \leq i \leq n$.

References

1. Bourdon, P.S., Williams, H.T.: Unital quantum operations on the Bloch ball and Bloch region. Phys. Rev. A. 69, 022314 (2004)
2. Brünig, E., et al.: Parametrizations of density matrices. J. Mod. Opt. 59(1), 1–20 (2012)
3. Byrd, S., Khaneja, N.: Characterization of the positivity of the density matrix in terms of the coherence vector representation. Phys. Rev. A 68, 062322 (2003)
4. Crowder, T.: Representations of quantum channels. Ph.D. Thesis (2013), http://gradworks.umi.com/35/91/3591941.html
5. Feng, J.: A domain of unital channels. Found. Phys. 42(7), 959–975 (2012)
6. Kimura, G.: The bloch vector for N-level systems. Phys. Lett. A 314, 339–349 (2003)
7. Martin, K.: The scope of a quantum channel. In: Proceedings of Symposia in Applied Mathematics, American Math Society, vol. 71, pp. 213–233 (2012)
8. Nielsen, M.A.: Probability distributions consistent with a mixed state. Physical Review A 61(6), 064301 (2000)
9. Nielsen, M., Chuang, I.: Quantum computation and quantum information. Cambridge University Press (2000)

TCC, with History

Vijay Saraswat[1], Vineet Gupta[2], and Radha Jagadeesan[3,*]

[1] IBM TJ Watson Research Center
[2] Google, Inc.
[3] DePaul University

Abstract. Modern computer systems are awash in a sea of asynchronous events. There is an increasing need for a declarative language that can permit business users to specify complex event-processing rules. Such rules should be able to correlate different event streams, detect absence of events (negative information), permit aggregations over sliding windows, specify dependent sliding windows etc. For instance it should be possible to precisely state a rule such as "Every seventh trading session that DowJones has risen consecutively, and IBM's stock is off 3% over its average in this period, evaluate IBM position", "Declare the sensor as faulty if no reading has been received for 500 ms", etc. Further, the language should be implementable efficiently in an event-driven fashion.

We propose the Timed (Default) Concurrent Constraint, TCC, programming framework as a foundation for such complex event processing. The framework (developed in the mid 90s) interprets computation as deduction in a fragment of linear temporal logic. It permits the programmer to write rules that can re-act instantaneously to incoming events and determine the "resumption" that will respond to subsequent events. The framework is very powerful in that it permits instantaneous pre-emption, and allows user-definable temporal operators ("multi-form time").

However, the TCC framework "forgets" information from one instant to the next. We make two extensions. First, we extend the TCC model to carry the store from previous time instants as "past" information in the current time instant. This permits rules to be written with rich queries over the past. Second, we show that many of the powerful properties of the agent language can be folded into the query language by permitting agents and queries to be defined mutually recursively, building on the testing interpretation of intuitionistic logic described in RCC [21]. We show that this permits queries to move "back and forth" in the past, e.g. "Order a review if the last time that IBM stock price dropped by 10% in a day, there was more than 20% increase in trading volume for Oracle the following day."

We provide a formal semantics for TCC + Histories and establish some basic properties.

Keywords: synchronous programming, concurrent constraint programming, RCC, TCC, HCC, complex event processing.

* Radha Jagadeesan was supported by NSF 0916741.

F. van Breugel et al. (Eds.): Panangaden Festschrift, LNCS 8464, pp. 458–475, 2014.

1   Introduction

1.1   Timed Concurrent Constraint Programming

From about 1985 to about 1995, the programming languages/embedded systems community worked out a very robust programming model for time-based systems, under the framework of "synchronous languages", such as Esterel, Signal and Lustre ([4,19,3,7]). In particular, the authors developed the *Timed (Default) Concurrent Constraint Programming Framework*, [30], based on the simple idea of extending "across time" the ideas of concurrent constraint programming, using the Synchrony Hypothesis of Berry [4].[1] One thinks of a reactive system as lying inert, waiting for a stimulus from the outside world. On each stimulation, the system computes an instantaneous response, and prepares itself for further interaction (by computing a resumption). The system is *amnesiac* in that its past state is flushed, only the resumption is kept. Thus the system has an internal notion of time that corresponds to its periodic interaction with the outside world.

This notion of time can be made explicit through certain temporal combinators within the language used to program these agents. TCC is built on just six orthogonal basic combinators:[2]

(Agents) A,B ::= c | if G then A | if c else A | A and A | some V in A | hence A
  | Z | mu Z in A
(Goals) G ::= c | G and G

Above, c ranges over constraints; X,V over first-order variables used in constraints; Z over Agent variables; A,B ranges over Agent formulas, and G over Goal formulas.

The TCC framework is parametric on an underlying notion of *constraint system C* [30]: essentially such a system specifies pieces of partial information, called *tokens* or *constraints*, and an *entailment* relation which specifies which tokens follow from which other sets of tokens. The (tell) c agent adds the constraint c to a shared store of constraints. The (positive ask) agent if c then A reduces to A if the store is strong enough to entail c. The (negative ask) agent if c else A reduces to A only if the final store (at this time instant) will not be strong enough to entail c (this circularity – the final store is defined in terms of the final store – is characteristic of defaults [28]). The (parallel composition) agent A and B behaves as both A and B. The agent some X in A introduces a new local variable X in A. The agent hence A is the only agent with temporal behavior – it reduces to A at every time instant after the current instant. The agent mu z A (taken from the modal mu calculus) behaves like A with occurrences of z replaced by mu Z A.

This language is powerful enough to be the basis for a rich algebra of temporal control constructs. For instance, one can define:

---

[1] In the rest of this paper we will use the acronym TCC to stand for Timed Default Concurrent Constraint Programming.

[2] We introduce recursion explicitly through mu; in fact recursion is definable in TCC.

1. `always A` (run A at every time step);
2. `do A watching c` (run A until such time instant as the condition c is true, at which point abort the remainder of A);
3. `next A` (run A only at the next time step);
4. `time A on c` (run A but on a clock derived from the basic clock by only passing through those ticks at which the condition c is true).

The last combinator in particular is very powerful – it realizes the idea of "multi-form" time, the notion that the basic clock on which an agent is defined may itself be defined by another agent [30].

TCC (and its continuous time extension, HCC, [17]) have been used in modeling complex electro-mechanical systems (photo copiers [18], robots [1]) and biological systems [6]. They have a very well-developed theory – semantic foundations, reasoning framework, implementation techniques, compilation into finite state automata, abstract interpretation, etc. (see Related Works section below).

Unlike the other systems mentioned above (Esterel, and other reactive languages), TCC, and its parent framework, Concurrent Constraint Programming (CCP) are declarative and rule-based. Computation can be interpreted as deduction corresponding to certain "agent" formulas in linear time temporal logic, defined over a certain notion of *defaults* [30]. Defaults play a crucial role in permitting agents to detect the *absence* of information. This is critical for faithfully modeling such computational phenomena as time-outs and strong pre-emption. This logical reading extends the understanding of CCP [35] as computation in intuitionistic logic [32].

1.2   Event Processing

Over the last decade a new and interesting application area has emerged, *event processing*, [23]. The basic computational problem in event processing is to implement a powerful "sense, analyze, respond" system. The system should be capable of receiving multiple (usually discrete) time-varying signals, correlating them in potentially complex ways involving detecting the absence of events, maintaining sliding windows, computing statistics over sliding windows (averages, max values, etc), and comparing these values. If the desired temporal pattern is detected, then appropriate programmer specified action (e.g. issuing an alert) needs to be taken.

For an event processing language to be useful, it should be capable of expressing complex patterns of temporal interactions. For example, it should be possible to support rules of the form:

1. Every tenth time the price drops within an hour emit volatility warning.
2. Every seventh trading session that DowJones rises consecutively, and IBM stock has fallen over this interval, evaluate IBM position.
3. Declare the sensor is faulty if no reading has been received in the past 10s.
4. Declare the room is too cold if the average temperature over the last 100s is below a threshold.
5. Ignore an over global limit notification on an account if an over global limit notification was sent on this account in the past two days.

6. If the merchant has been tenured less than 90 days, and the sum of the transactions in the last 7 days is much higher than the 7 day average for the last 90 days, then investigate a 7 day hit and run possibility.

We also desire a language in which programs can be understood declaratively as "rules". Ultimately the language needs to mesh well with an OPS-like rule language, such as ILOG JRules and ILOG Business Rules. We desire that the programmer should be able to reason rigorously (if informally) about such rules. We require that the rules should be compilable efficiently. For example queries involving sliding window averages should be implemented in an incremental forward-driven fashion (with a rolling average being maintained).

1.3 TCC for Event Processing

Given the many valuable properties of TCC, it is interesting to consider it as a basis for complex event processing. Incoming events can be represented as atoms to be added to the constraint store. As events arrive, they are buffered while the system is active (executing events it received at the previous tick). Once the system quiesces, and the buffer is not empty, the system is advanced to the next time unit, and all buffered events added.

A fundamental limitation of TCC for complex event processing, however, is that TCC computations do not maintain history. All rules must be written in a "forward looking" fashion, responding to the current events received, and whatever state has been explicitly stored from past interactions. For instance, to express the rule "Trigger an alert whenever it is the case that the stock price of company A falls over 10%, while that of company B has risen over the past 7 days", the programmer must write code that maintains in the current store the value of the proposition "the stock of company B has risen over the past 7 days". Now on receipt of a notification that the stock price of A has fallen, a check can be made for the value of the proposition and an alert emitted if necessary.

But this way of writing rules is awkward. In essence, the programmer is being made to work like a compiler – figure out how to write the rule in such a way that it is always event-driven and forward looking. In many cases it is very natural instead to simply write a query over the past that "looks back" and checks if the desired condition is true, on demand.

Our basic move, therefore, is to augment TCC with *history*. When moving from time step t to $t + 1$, we propose to retain the constraints computed at t, and time-stamp them with t. Thus the store will contain not just the current constraint, but also, separately and equally, past constraints, each tagged with the time at which they were computed.

A simple way to accomodate this view in TCC is simply to work over the constraint system $H(C)$ built from C as follows. The tokens in $H(C)$ are of the form $\texttt{time}^i c$ for some i, where c is a token of C. A multiset Γ of such tokens entails $\texttt{time}^k d$ only if Γ_k entails d (in C), where Γ_k is the set of all constraints c such that $\texttt{time}^k c \in \Gamma$. Given k, by abuse of notation we will say that Γ entails $\texttt{prev}\ c$ at k if Γ_{k-1} entails c (in C). Using $H(C)$, the user can write ask agents that query the past. Tell agents must still be prevented from modifying the past by ensuring that they can only assert constraints about the instantaneous state. The operational semantics is now modified to carry past constraints automatically in the constraint store, and to tag tell constraints with the current time step.

Example 1 (Querying the past in TCC($H(C)$)). The rule:

```
always if ((prev price(IBM)) > price(IBM)) then signalIBMDrop
```

will trigger if there is a drop in price of IBM stock over successive time instants.

Unfortunately, this simple technique is not powerful enough. What if we wanted to trigger a rule if the current price is less than half the price at any point in the past when MSFT stock was above a certain threshold? In other words, it is natural to require *recursive* computations in our queries, capable of examining the past at arbitrary depth.

[21] in fact develops such a rich framework for CCP, called RCC. RCC is based on the idea that a judgement $A_0, \ldots, A_{n-1} \vdash G$ can be regarded as asking whether the system of concurrently interacting agents A_i ($i < n$) satisfy the query or goal G. Appendix A provides more details.) Queries are internalized in the agent language through the production A::= if G then A.

Queries are not restricted to primitive constraints c. Recursive queries are permitted. *Universal* queries, G::= all V in G, are permitted, where V is a first-order variable. Such a query can be thought of as succeeding only when the query G succeeds, where V is a brand-new variable that does not occur in the agents. *Hypothetical* queries are also permitted: G::= if A then G can be thought of as temporarily augmenting the system currently being tested with A and asking if the augmented system satisfies G. If so, the guard is satisfied and execution continues, with the temporary augmentation discarded. Hypothetical queries permit "what if" reasoning and allow for a compact representation of very powerful idioms. To implement this, the underlying infrastructure must support the notion of copying the entire concurrent assembly of agents.

[21] shows that the computational interpretation is sound and complete with respect to the obvious logical interpretation of the queries.

TCC with deep guards. We now consider how to apply these ideas to TCC. Clearly, we need to augment the power of guards, G. To add recursion across time, we introduce G ::= hitherto G (analogously to A ::= hence A). The query hitherto G is intended to be true if G is true at every point in the past (excluding the current one). We also introduce recursive queries, G ::= mu X G, and require that X be guarded in G (occur inside a hitherto). Similarly, we introduce universal queries G ::= all V in G.

We could introduce hypothetical queries, G::= if A then G. However, we can do something richer. Note that A is not permitted to operate in the "past", i.e. it is not possible for an agent to spawn an agent to "change the past". (Concretely, A::= hitherto A is not allowed. This is fundamental to the basic idea that computation always moves ahead in time.) However, within the scope of hypothetical execution, it does make sense to add agents to the past – these agents are free to participate in "what if" reasoning, exploring what might have been. Therefore we introduce a new category of *nested agents*, B, which is the same as A except that it permits B::= hitherto B, and add G::= if B then G.

With these constructs it is possible for a query to "move back in time" arbitrarily deeply, spawn agents in the past, and ask queries of the modified system. Still, the nested agents and queries are asymmetric: nested agents can move back in time (hitherto B) as well as forward (hence B), but queries can only move backwards. Logically, it makes sense, then, to permit queries to also move forward in time; we add G::= hence G. This permits us to express a query that checks whether the day after the last time IBM stock fell 10% it was the case that MSFT stock rose 10%. The natural formulation of this query would involve moving back in time and then forward.

Table 1 summarizes the language being considered, which we name "TCC, with history". The basic picture of computation supported by this language is as follows: The system interacts with incoming events in a synchronous fashion. The rate at which events arrive is controlled by the environment and not by the system. Each interaction marks the progression of the system down a time-line. At each instant, the state of the system carries the entire state of past interactions. This is accessible to be queried in a very rich way through a query language which permits computations to move backwards and forwards in the past, and also spawn hypothetical queries. However, querying cannot change the actual past, only read it.

This paper takes the first step in studying this language. Section 2 discusses how some interesting idioms can be expressed in this language. For reasons of space we omit standard extensions of the query language with "bag of" operators that permit the collection of some statistic over all answers to a query (these are very important in practice). Section 3 formalizes the informal reasoning presented here. We conclude with an outline of the work that lies ahead.

Table 1. TCC, with history

(Agents) A ::= c | if G then A | if c else A | A and A | some V in A | hence A
 | X | mu X in A

(Goals) G ::= c | if B then G | G and G | G or G | all V in G | hence G
 | hitherto G | X | mu X in G

(Nested Agents) B ::= c | if G then B | if c else B | B and B | some V in B
 | hence B | hitherto B | X | mu X in B

Agents A are those B's which do not have any occurrence of the hitherto combinator.

Contributions. The contributions of this paper may be summarized as follows:

- We motivate the use of TCC for complex temporal event processing. TCC is capable of handling the absence of information.
- We extend TCC with a way to capture the past history of the system. This permits a natural declarative style of querying the past.
- We motivate the introduction of *recursive queries* in TCC. This permits recursive queries that can reach arbitrarily deeply into the past.

- Motivated by [21] and the testing interpretation of intuitionistic logic, we further introduce "hypothetical" queries if B then G that ask if the current system augmented with the agent B can answer the query G. Unlike TCC, we also permit such nested agents to move backwards in time, allowing speculative augmentation of the past. Together, these two capabilities permit B to move backwards and forwards in time, while confined to the past.
- We provide a formal operational semantics for the language, based on an interpretation of programs as formulas in linear time temporal logic, and computation as deduction.
- We establish that the semantics of this language is conservative over TCC. That is, the behaviors of a program in this language that is also expressible in TCC are exactly the same as in TCC.

1.4 Related Work

Several authors have explored the properties of TCC in the last two decades, extending it in various directions. [37] shows that the synchronous languages Lustre and Argos can be embedded in TCC. Expressiveness is further discussed in [24]: different variants that express recursiveness in different ways are discussed and related. It is shown that equivalence of programs with replication (or parameterless recursive procedures) is decidable. [26,25] propose an extension to TCC (ntcc) that can handle asynchronous communication, and nondeterministic behavior, by providing a guarded-choice operator and an unbounded but finite delay operator. A denotational semantics, and a proof system for temporal properties are presented. Another approach to reasoning about TCC programs is provided in [11,10]. More decidability results for TCC and ntcc are presented in [39]: strongest post-condition equivalence for "locally independent" ntcc programs is shown to be decidable. This language is capable of specifying certain kinds of infinite-state reactive systems. [5] discusses a variant capable of dealing with "soft" constraints and preferences; the intended application area is a collection of agents negotiating over quality of service. Abstract diagnosis for a variant of TCC is considered in [9].

In terms of implementation, [31] describes an initial implementation in Java, for reactive computation. This is currently being extended to an implementation of the language discussed in this paper, on top of X10 [29,8].

1.5 Our Research with Prakash

This paper is offered as a contribution to Prakash Panangaden's Festschrift. During 1990, Prakash reached out to Vijay, having heard of Concurrent Constraint Programming, and apparently being intrigued by the idea of computing (in parallel) with partial information. This led to a visit to Vijay at Xerox PARC, leading to an invitation to join the work that eventually was published in POPL91 [33]. Prakash was particularly intrigued by the development of constraint systems through cylindric algebras and later pursued this work [27]. During the 90s the authors further developed CCP with discrete time (TCC) and continuous time (HCC), and started exploring the integration of probabilistic information into CCP [16]. This led to another collaboration between Prakash and Radha and Vineet [14]. Thus it is fair to say that parallel computing with partial information – in its many forms – has been a shared interest between the authors and Prakash.

Radha Jagadeesan was a Ph.D student of Prakash at Cornell from 1987-91. He claims that he continues to remain a student of Prakash to this day, and holds Prakash responsible for all inadequacies in Radha's own view towards logic and semantics! In addition to constraint programming, Radha has also worked with Prakash and Vineet on probabilistic processes, focussing primarily on metric based approximate reasoning for discrete [13,12] and real time [15] probablistic processes.

2 Programming in TCC with Histories

We now consider how several idioms of practical interest can be expressed in this language.

2.1 A Concrete TCC Language, V

To fix intuitions, we work on top of a constraint system which permits (sorted) function and predicate symbols, with equality ("="). Amongst the sorts available are Boolean and Int. Sorts are closed under products and function space, i.e. if S1, S2 are sorts, then so are S1 × S2 and S1 => S2.

(Terms) s,t ::= X | f(t1,...,tn)
(Constraints) A,B ::= s=t | p(t1,..., tn) | c,c

The equality predicate is interpreted as a congruence relation (it is symmetric, reflexive and transitive, and equal terms can be substituted for each other in all contexts). A set of constraints c1,..., cn entails p(t1,..., tk) if and only if it entails s1=t1, ..., sk=tk (for some terms s1,..., sk) and for some i, ci is p(s1,..., sk).

For convenience, we will also permit linear arithmetic constraints, and arithmetic inequality, <, <=.

We will also find it convenient to permit prev(t) as a term, when t is a term. A constraint store can establish prev(u)=v at time t if it can establish u=v at time $t-1$, and v is *rigid*, i.e. does not change value with time. The only rigid terms are the constants – we assume they denote the same value at every time instant.

We shall adopt the convention of specifying named agents through agent clauses of the form a -: A, and named goals through goal clauses of the form g :- G, where a and g are atomic formulas. The predicate names for agents, goals and primitive constraints are understood as being drawn from disjoint spaces.

2.2 Programming in V

Example 2 (past G, next G). We define the query past(X=Y), intended to be true at query time i precisely if X=Y is true at query time $i-1$.

```
past(X=Y) :- all U in if (hitherto hitherto U=true) then hitherto (U=true or X=Y)
```

Here is how we understand it. To establish the goal past(X=Y) in a configuration Γ at query time i, we are permitted to assume U=true (at all times) in $[0, i-2]$, for a brand-new variable U. In turn, we must establish either U=true or X=Y in $[0, i-1]$. Clearly, the assumption establishes the desired goal in $[0, i-2]$. Hence we are left with time $i-1$. No agent in Γ knows about U. Therefore the only way past(X=Y) can be established is if at the previous time instant X=Y can be established.

The past predicate can be defined in a similar way for other constraints of interest, e.g.:

```
past(X>Y) :- all U in if (hitherto hitherto U=true) then hitherto (U=true or X>Y)
```

The code next(X=Y) is the dual:

```
next(X=Y) :- all U in if (hence hence U=true) then hence (U=true or X=Y)
```

Example 3 (once G). We express the query once G that succeeds only if G can be established at some point in the past. We illustrate for G of the form X=Y.

```
once(X=Y) :- X=Y or past(once(X=Y)).
```

This goal can be established in a configuration at i only if G can be established at i, or, recursively, the goal can be established at $i-1$.

If arithmetic is available, and recursion with parameters, one can program within t do G:

Example 4 (within t do G). We require G to be established within t time units in the past:

```
within T do X=Y :- X=Y or (T > 0 and past(within T-1 do X=Y)).
```

We show that the query language has enough power to internalize else.

Example 5 (not(X=Y)). We express the query not(X=Y). This query succeeds only if X=Y cannot be established:

```
not(X=Y) :- all U in if (if X=Y else U=false) then U=false
```

This goal can be established in a configuration Γ only if Γ, if X=Y else U=false can establish U=false. But this can happen only if Γ can evolve in such a way that X=Y cannot be established (per the semantics of the TCC if/else).

Example 6 (last X then G). We would like to express that G is true at the last time instant at which X was true (assuming there is a time instant at which X is true):

```
last X=Y then U=V :- prev last1 X=Y then U=V.
last1 X=Y then U=V :- (X=Y and U=V) or (not(X=Y) and prev(last1 X=Y then U=V)).
```

Intuitively, at the last time instant, a check is made for X=Y. If it is true, then U=V must be true, else the goal will fail. If it is not known to be true, then the goal succeeds provided that the same goal can be established at the previous time instant.

We turn now to using these general constructions to show how a complex event query can be formulated.

Example 7. An example of the use of this goal is the query that returns the previous price of a stock. We shall imagine that if in a time instant an event arrives that specifies the price P of a stock S, then the constraint price(P)=S is added to the store. Note that many stock price events may arrive at the same time instant – we assume that all are for different stocks. It is not necessary that each time instant contains a constraint about the price of a given stock S. In this case, we may wish to determine the previous prices of the stock S, which is the price of the stock at the first instant before the current one at which a price event was received.

```
prevPriceOfStock(S)=P :-
    (prev(price(S))>0 and P=prev(price(S))) or
    (not(prev(price(S))>0) and prev(prevPriceOfStock(S)=P)).
```

Now one can use this query to determine whether the price has dropped. The query checks that there is a price event at the current time instant, and the price it specifies for the stock S is less than the previously known price for S.

```
priceDropped(S) :- prevPriceOfStock(S) > price(S).
```

Such a query can now be used to time an agent. The agent

```
time next^10 emitVolatilityWarning on priceDropped(S)
```

will emit a volatility warning at the tenth time instant at which the price has dropped. Using standard TCC idioms, this agent can be packaged up thus:

```
every hour
  do time next^10 emitVolatilityWarning
     on priceDropped(s)
  watching hour.
```

to precisely capture the rule "Every tenth time the price drops within an hour, emit a volatility warning".

The above provides a flavor of the richness of this system.

3 Semantic Model

3.1 Transition Relations

The central problem we address is the temporal evolution of (mutually dependent) agents and guards.

We add a new formula $B ::= \text{time}^i B$, to keep track of formulas that are intended to hold at a point in time in the past. We abuse notation slightly by permitting $\text{time}^0 B$ and treating it indistinguishably from B. Below, Γ, Δ, Π range over (possibly empty) multisets of B formulas. For a multiset of formulas $\Delta = B_0, \ldots, B_{n-1}$ we let $\text{time}^i \Delta$ stand for $\text{time}^i B_0, \ldots, \text{time}^i B_{n-1}$. Similary for hence Δ and hitherto Δ.

We define three transition relations. All of them are indexed with the *current* time instant j and the *query* time instant i (with $i \leq j$). The main relation of interest is $\Gamma \Vdash_b^{i,j} G$ (read: "Γ *proves* G at (past) query time i (with quiescent store b) when the current time is j ($i \leq j$)"). We need to carry j in the relation because in order to prove a goal of the form hence G we only need to consider time steps upto j. Note that Γ will, in general, contain formulas active at different time instants $k \leq j$ (i.e. Γ will contain formulas of the form $\mathtt{time}^k B$). G however, is never explicitly timed, since at query time i we care only about queries holding at time i.

To define this relation, we need two auxiliary relations that define evolution *within* time instants in the past ($\Gamma \xrightarrow{i,j}_b \Gamma'$), and *across* time instants in the past ($\Gamma \rightsquigarrow^{i,j} \Gamma'$). Note that these auxiliary relations may work with hypothetical pasts, since they may reflect the presence of assumptions B made by the goal G being solved at j.

We let $\sigma^i(\Gamma)$ stand for the set of all formulas c s.t. $\mathtt{time}^i c \in \Gamma$, i.e. the subset of constraints known to be in effect at time i.

The Provability Relation for Goals. The logical rules are straightforward, and correspond to RHS rules for the appropriate logical connective, in a sequent-style presentation. Rule 1 uses σ^i to pick out the constraints in effect at query time i from the current configuration. Rule 2 ensures that in order to prove a goal of the form if B then G at query time i, the assumption B is added at time i to the current configuration.

$$\frac{\sigma^i(\Gamma) \vdash c}{\Gamma \Vdash^{i,j}_b c} \qquad \frac{\Gamma \Vdash^{i,j}_b G[\mu X\, G/X]}{\Gamma \Vdash^{i,j}_b \mu X\, G} \tag{1}$$

$$\frac{\Gamma \Vdash^{i,j}_b G_0 \quad \Gamma \Vdash^{i,j}_b G_1}{\Gamma \Vdash^{i,j}_b G_0 \text{ and } G_1} \qquad \frac{\Gamma, \mathtt{time}^i B \Vdash^{i,j}_b G}{\Gamma \Vdash^{i,j}_b \text{ if } B \text{ then } G} \tag{2}$$

$$\frac{\Gamma \Vdash^{i,j}_b G_0}{\Gamma \Vdash^{i,j}_b G_0 \text{ or } G_1} \qquad \frac{\Gamma \Vdash^{i,j}_b G_1}{\Gamma \Vdash^{i,j}_b G_0 \text{ or } G_1} \tag{3}$$

$$\frac{\Gamma \Vdash^{i,j}_b G[t/V]}{\Gamma \Vdash^{i,j}_b \text{ some } V \text{ in } G} \qquad \frac{\Gamma \Vdash^{i,j}_b G \quad (V \text{ not free in } \Gamma)}{\Gamma \Vdash^{i,j}_b \text{ all } V \text{ in } G} \tag{4}$$

Note that b is not used in these rules; we will see later that it is used when specifying how the LHS evolves within a time instant. In Rule 4 t is some term.

We consider now the temporal rules. These rules have a (finite) set of assumptions, indicated by the for all quantifier. A goal $\mathtt{hitherto}\ G$ can be proved at query time i if it can be proven at every time in $[0,i)$. A goal $\mathtt{hence}\ G$ can be proved at query time i if it can be proven at every time in $(i,j]$.

$$\frac{\Gamma \Vdash^{0,j}_{b_0} G}{\Gamma \Vdash^{1,j}_b \mathtt{hitherto}\ G} \qquad \frac{\Gamma \Vdash^{i-2,j}_{b_0} \mathtt{hitherto}\ G \quad \Gamma \Vdash^{i-1,j}_{b_1} G}{\Gamma \Vdash^{i,j}_b \mathtt{hitherto}\ G} \tag{5}$$

$$\frac{\Gamma \Vdash^{i+1,j}_b G \quad \Gamma \Vdash^{i+1,j}_b \mathtt{hence}\ G}{\Gamma \Vdash^{i,j}_b \mathtt{hence}\ G} \qquad \frac{\Gamma \Vdash^{j,j}_{b_0} G}{\Gamma \Vdash^{j-1,j}_b \mathtt{hence}\ G} \tag{6}$$

Now, since we permit B's to occur on the LHS, and these could evolve, we must also have the following rules. In Rule 7, the configuration is partitioned into three groups of formulas – Γ_{i-} which are all the formulas $\mathtt{time}^k B$ with $k < i$, Γ_{i+} which are all the formulas $\mathtt{time}^k B$ with $k \geq i$, and $\mathtt{hitherto}\ \Delta$. The rule captures the notion that to prove a formula G at time i one must "go back" to time 0 in order to account for the effects of any $\mathtt{hitherto}\ B$ formulas in Γ. In this traversal into the past, $\mathtt{hitherto}\ B$ agents are carried "backwards" (exactly as $\mathtt{hence}\ B$ agents are carried forward in TCC, see Rule 15), together with "past" state. The recursion is stopped by Rule 8.

$$\frac{\Gamma_{i-},\mathtt{time}^{i-1}\Delta,\mathtt{hitherto}\ \Delta \leadsto^{i-1,j} \Gamma' \quad \Gamma',\Gamma_{i+} \longrightarrow_b^{i,j} \Gamma'' \quad \Gamma'' \left\|\frac{i,j}{b}\right. G \quad i > 0}{\Gamma_{i-},\Gamma_{i+},\mathtt{hitherto}\ \Delta \left\|\frac{i,j}{b}\right. G} \tag{7}$$

$$\frac{\Gamma \longrightarrow_b^{0,j} \Gamma' \quad \Gamma' \left\|\frac{0,j}{b}\right. G}{\Gamma \left\|\frac{0,j}{b}\right. G} \tag{8}$$

The Evolution Relation. The rules for $\longrightarrow$ (evolution within a time instant) are as in [30], changed in an appropriate way to consider the more general notion of execution at possibly past time points. (The special case of TCC execution is obtained by considering the relation $\Gamma \longrightarrow_b^{j,j} \Gamma'$ and restricting ask agents to check primitive constraints.) At query time i, an agent $\mathtt{time}^i$ if G then B can be reduced to $\mathtt{time}^i B$ provided that the goal G can be proved from the current configuration. To reduce an $\mathtt{time}^i$ if a else B agent, we use the quiescent information b (associated with query time i), as usual for Default CC.

$$\frac{\Gamma \left\|\frac{i,j}{b}\right. G}{\Gamma,\mathtt{time}^i \text{ if } G \text{ then } B \longrightarrow_b^{i,j} \Gamma,\mathtt{time}^i B} \tag{9}$$

$$\frac{b \nvdash a}{\Gamma,\mathtt{time}^i \text{ if } a \text{ else } B \longrightarrow_b^{i,j} \Gamma,\mathtt{time}^i B} \tag{10}$$

$$\frac{}{\Gamma,\mathtt{time}^i B_0 \text{ and } B_1 \longrightarrow_b^{i,j} \Gamma,\mathtt{time}^i B_0,\mathtt{time}^i B_1} \tag{11}$$

$$\frac{(Y \text{ not free in } B,\Gamma,\Pi)}{\Gamma,\mathtt{time}^i \text{ some } V \text{ in } B \longrightarrow_b^{i,j} \Gamma,\mathtt{time}^i B[Y/V]} \tag{12}$$

$$\frac{}{\Gamma,\mathtt{time}^i \mu X\, B \longrightarrow_b^{i,j} \Gamma,\mathtt{time}^i B[\mu X\, B/X]} \tag{13}$$

Note that there is no rule for $\mathtt{hitherto}\ B$ – it does not contribute to instantaneous evolution, or to the step relation. It is of use in the proves relation when moving backwards in time.

The rules for evolution across time instances are as follows. They differ from the rules for TCC only in that the final constraint at the previous time step is explicitly carried forward into the configuration at the next time step ($\mathtt{time}^i b$) with the appropriate time index (i) to distinguish it from the constraints that will be generated at other time steps. The first rule is used to advance time in a query computation, the second to advance time for the overall (top-level) computation. Below, let Π consist of formulas of the form $\texttt{time}^i\, c$ for some i, and Γ' does not contain such formulas.

$$\frac{\Gamma \xrightarrow{\;\;*\;\;}_b^{i,j} \Gamma', \Pi, \text{hence } \Delta \quad \Gamma', \Pi, \text{hence } \Delta \not\longrightarrow_b^{i,j} \quad \sigma^i(\Pi) = b \quad i < j}{\Gamma \rightsquigarrow^{i+1,j} \Pi, \texttt{time}^{i+1}\Delta, \text{hence } \Delta} \tag{14}$$

$$\frac{\Gamma \xrightarrow{\;\;*\;\;}_b^{i,j} \Gamma', \Pi, \text{hence } \Delta \quad \Gamma', \Pi, \text{hence } \Delta \not\longrightarrow_b^{i,j} \quad \sigma^i(\Pi) = b \quad i = j}{\Gamma \rightsquigarrow^{j+1,j+1} \Pi, \texttt{time}^{j+1}\Delta, \text{hence } \Delta} \tag{15}$$

Definition 1 (Execution). *We say that a sequence of agents $\Gamma_0, \Gamma_1 \ldots, \Gamma_n, \ldots$ is an execution if for all $i > 0$, $\Gamma_i \rightsquigarrow^{i+1,i+1} \Gamma_{i+1}$.*

Let Γ be a multiset of agents. Then $\Gamma \upharpoonright i$ is the set of all formulas B such that $\texttt{time}^i\, B \in \Gamma$.

Proposition 1 (TCC+history does not change the past.). *Let $\Gamma_0, \Gamma_1 \ldots, \Gamma_n, \ldots$ be an execution. Then for every $j > 0$ and $m, n > j$ it is the case that $\Gamma_m \upharpoonright j$ and $\Gamma_n \upharpoonright j$ are multisets of constraints that are equivalent.*

The following proposition relies on the fact that a multiset of nested TCC+history agents cannot have sub-formulas of the form if G then B (unless G is a constraint), if B then G, hence G, hitherto G, hitherto B. Therefore in any proof of the judgement $\Gamma \Vdash_b^{j,j} c$ only judgements of the form $\Gamma' \Vdash_b^{j,j} c'$ are generated. No "travel" in time is possible.

Proposition 2. *Suppose Γ is a multiset of nested TCC+history agents such that $\Gamma \upharpoonright j$ is a multiset of TCC agents. Then $\Gamma \Vdash_b^{j,j} c$ iff $\Gamma \upharpoonright j \vdash_b c$, where $\vdash_b$ represents the TCC entailment relation with b the final resting point.*

Theorem 1. *TCC + History is conservative over TCC.*

4   Conclusion

This paper represents the first step in the study of TCC, augmented with history and a rich notion of queries. A number of areas of work open up.

Expressiveness. Does this language realize the intuition that queries can be multi-form in time, just as agents can be multi-form in time? Is it semantically meaningful to consider deep negative guards?

Denotational semantics. The basic semantic intuition is that this language permits rich querying of the past, with a deep interplay between agents and guards. Since the past is not modified it should be possible to adapt the denotational semantics of [30] (based on prefix-closed sets of traces) to this setting.

Finitary implementations. For many uses of the language, it would be valuable to bound the amount of past information that needs to be carried in the state. Does this language admit of finite state compilability (a la TCC)? If not, what restrictions need to be placed to achieve finite state compilability?

References

1. Alenius, L., Gupta, V.: Modeling an AERCam: A case study in modeling with concurrent constraint languages. In: CP 1998 Workshop on Modeling and Computation in the Concurrent Constraint Languages (October 1998)
2. Andreoli, J.-M.: Logic programming with focusing proofs in linear logic. Journal of Logic and Computation 2, 297–347 (1992)
3. Benveniste, A., Guernic, P.L., Jacquemot, C.: Synchronous programming with events and relations: the signal language and its semantics. Science of Computer Programming 16(2), 103–149 (1991)
4. Berry, G., Gonthier, G.: The esterel synchronous programming language: design, semantics, implementation. Sci. Comput. Program. 19(2), 87–152 (1992)
5. Bistarelli, S., Gabbrielli, M., Meo, M.C., Santini, F.: Timed soft concurrent constraint programs. In: Lea, D., Zavattaro, G. (eds.) COORDINATION 2008. LNCS, vol. 5052, pp. 50–66. Springer, Heidelberg (2008)
6. Bockmayr, A., Courtois, A.: Using hybrid concurrent constraint programming to model dynamic biological systems. In: Stuckey, P.J. (ed.) ICLP 2002. LNCS, vol. 2401, pp. 85–99. Springer, Heidelberg (2002)
7. Caspi, P., Pilaud, D., Halbwachs, N., Plaice, J.A.: Lustre: a declarative language for real-time programming. In: Proceedings of the 14th ACM SIGACT-SIGPLAN Symposium on Principles of Programming Languages, POPL 1987, pp. 178–188. ACM, New York (1987)
8. Charles, P., Grothoff, C., Saraswat, V., Donawa, C., Kielstra, A., Ebcioglu, K., von Praun, C., Sarkar, V.: X10: an object-oriented approach to non-uniform cluster computing. In: Proceedings of the 20th Annual ACM SIGPLAN Conference on Object-Oriented Programming, Systems, Languages, and Applications, OOPSLA 2005, pp. 519–538. ACM, New York (2005)
9. Comini, M., Titolo, L., Villanueva, A.: Abstract diagnosis for timed concurrent constraint programs. CoRR, abs/1109.1587 (2011)
10. de Boer, F.S., Gabbrielli, M., Meo, M.C.: Proving correctness of timed concurrent constraint programs. In: Nielsen, M., Engberg, U. (eds.) FOSSACS 2002. LNCS, vol. 2303, pp. 37–51. Springer, Heidelberg (2002)
11. de Boer, F.S., Gabbrielli, M., Meo, M.C.: A temporal logic for reasoning about timed concurrent constraint programs. In: Temporal Representation and Reasoning, TIME 2001, pp. 227–233. IEEE (2001)
12. Desharnais, J., Gupta, V., Jagadeesan, R., Panangaden, P.: Approximating labelled markov processes. Inf. Comput. 184(1), 160–200 (2003)
13. Desharnais, J., Gupta, V., Jagadeesan, R., Panangaden, P.: Metrics for labelled markov processes. Theor. Comput. Sci. 318(3), 323–354 (2004)
14. Gupta, V., Jagadeesan, R., Panangaden, P.: Stochastic processes as concurrent constraint programs. In: Proceedings of the 26th ACM SIGPLAN-SIGACT on Principles of Programming Languages, POPL 1999, San Antonio, TX, January 20-22, pp. 189–202. ACM Press, New York (1999)
15. Gupta, V., Jagadeesan, R., Panangaden, P.: Approximate reasoning for real-time probabilistic processes. Logical Methods in Computer Science 2(1) (2006)

16. Gupta, V., Jagadeesan, R., Saraswat, V.: Probabilistic concurrent constraint programming. In: Mazurkiewicz, A., Winkowski, J. (eds.) CONCUR 1997. LNCS, vol. 1243, pp. 243–257. Springer, Heidelberg (1997)

17. Gupta, V., Jagadeesan, R., Saraswat, V.A.: Computing with continuous change. Sci. Comput. Program. 30(1-2), 3–49 (1998)

18. Gupta, V., Struss, P.: Modeling a copier paper path: A case study in modeling transportation processes. In: Proceedings of the 9th International Workshop on Qualitative Reasoning, pp. 74–83 (1995)

19. Harel, D.: Statecharts: A visual formalism for complex systems. Sci. Comput. Program. 8(3), 231–274 (1987)

20. Jaffar, J., Lassez, J.-L.: Constraint Logic Programming. In: Proceedings of the 14th Annual ACM Symposium on Principles of Programming Languages, POPL 1987, Munich, Germany, pp. 111–119. ACM Press, New York (1987)

21. Jagadeesan, R., Nadathur, G., Saraswat, V.: Testing concurrent systems: an interpretation of intuitionistic logic. In: Sarukkai, S., Sen, S. (eds.) FSTTCS 2005. LNCS, vol. 3821, pp. 517–528. Springer, Heidelberg (2005)

22. Liang, C., Miller, D.: Focusing and polarization in intuitionistic logic. In: Duparc, J., Henzinger, T.A. (eds.) CSL 2007. LNCS, vol. 4646, pp. 451–465. Springer, Heidelberg (2007)

23. Luckham, D.C. (ed.): The power of events: an introduction to complex event processing in distributed enterprise systems. Addison Wesley (2002)

24. Nielsen, M., Palamidessi, C., Valencia, F.D.: On the expressive power of temporal concurrent constraint programming languages. In: Proceedings of the 4th ACM SIGPLAN International Conference on Principles and Practice of Declarative Programming, PPDP 2002, pp. 156–167. ACM, New York (2002)

25. Nielsen, M., Palamidessi, C., Valencia, F.D.: Temporal concurrent constraint programming: Denotation, logic and applications. Nord. J. Comput. 9(1), 145–188 (2002)

26. Palamidessi, C., Valencia, F.: A temporal concurrent constraint programming calculus. In: Walsh, T. (ed.) CP 2001. LNCS, vol. 2239, pp. 302–316. Springer, Heidelberg (2001)

27. Panangaden, P., Saraswat, V.A., Scott, P.J., Seely, R.A.G.: A hyperdoctrinal view of concurrent constraint programming. In: de Bakker, J.W., de Roever, W.-P., Rozenberg, G. (eds.) REX 1992. LNCS, vol. 666, pp. 457–476. Springer, Heidelberg (1993)

28. Reiter, R.: A logic for default reasoning. Artificial Intelligence 13, 81–137 (1980)

29. Saraswat, V., Jagadeesan, R.: Concurrent clustered programming, pp. 353–367. Springer, London (2005)

30. Saraswat, V., Jagadeesan, R., Gupta, V.: Timed default concurrent constraint programming. Journal of Symbolic Computation 22(5-6), 475–520 (1996)

31. Saraswat, V.A., Jagadeesan, R., Gupta, V.: jcc: Integrating timed default concurrent constraint programming into java. In: Pires, F.M., Abreu, S.P. (eds.) EPIA 2003. LNCS (LNAI), vol. 2902, pp. 156–170. Springer, Heidelberg (2003)

32. Saraswat, V., Lincoln, P.: Higher-order Linear Concurrent Constraint Programming. Technical report, Xerox PARC (1992)

33. Saraswat, V., Panangaden, P., Rinard, M.: Semantics of concurrent constraint programming. In: Proceedings of the 18th ACM SIGPLAN-SIGACT on Principles of programming languages, POPL 1991. ACM Press, New York (1991)

34. Saraswat, V.A.: The concurrent logic programming language cp: Definition and operational semantics. In: Proceedings of the 14th ACM SIGACT-SIGPLAN Symposium on Principles of Programming Languages, POPL 1987, pp. 49–62. ACM, New York (1987)

35. Saraswat, V.A.: Concurrent constraint programming. MIT Press, Cambridge (1993)
36. Shapiro, E., Fuchi, K. (eds.): Concurrent Prolog. MIT Press, Cambridge (1988)
37. Tini, S.: On the expressiveness of timed concurrent constraint programming. Electronic Notes in Theoretical Computer Science 27, 3–17 (1999)
38. Ueda, K.: Guarded Horn Clauses. In: Logic Programming - Japanese Conference, pp. 168–179 (1985)
39. Valencia, F.D.: Decidability of infinite-state timed ccp processes and first-order ltl. Theoretical Computer Science 330(3), 577–607 (2005)

A   Background

The basic idea of TCC may be summarized as follows:

TCC = CCP + Synchrony hypothesis

CCP, concurrent constraint programming, is a simple view of parallel computation that arises from multiple interacting agents sharing a common store of constraints. Constraints are expressions (such as X >= Y + Z) over a finite set of free variables. Each constraint is associated with a *solution set*, a set of mappings from variables to values (called *valuations*) that makes the constraint "true". e.g. the set of valuations that makes X >= Y + Z true is the set of valuations T s.t. T(X), T(Y) and T(Z) are numbers satisfying T(X) >= T(Y) + T(Z).

Two fundamental operations on constraints are used in CCP – tell c (add c to the current store), and ask c (check if c is entailed by the current store). Note that addition is conjunctive – the solution set of c,d is the intersection of the solution sets of c and d. Say that c *entails* d if the solution set of c is contained in that of d (that is, if v is a solution for c, then it is a solution for d) and *disentails* d if the solution sets of c and d are disjoint. The operation ask c succeeds if the store entails c, fails if the store disentails c, and suspends otherwise.

In CCP, the programmer specifies a set of agents over shared variables that interact with each other by telling and asking constraints on the shared variables. The fundamental property of CCP is that computations are determinate – the result is the same, regardless of the order in which agents are executed. Furthermore, programs have a declarative interpretation, they can be read as formulas in logic and have the property that if a program P logically entails a constraint c, then execution of P will result in a store that entails c.

CCP is a rich and powerful framework for (asynchronous) concurrent computation. TCC arises from CCP by "extending" CCP across time. We add the new control construct next: if A is an agent, then so is next A. The intuitive idea is that computation progresses in a series of steps. In each step, some input is received from the environment (an "event"), and added to the store. The program is then run to quiescence. This will yield a store of constraints, this provides the "instantaneous response". In addition it will yield a set of next A1, ..., next An agents. (Note some of these agents can be simple constraints.) These are precisely the agents that are used to respond to the next event, at the next time instant.

Notice that this view is concerned with a logical notion of time – time is just a sequence of ticks arriving from the environment (with additional input). There is no intrinsic association of this sequence of ticks with "real" time, e.g. msecs. This is the powerful insight that underlies the notion of multiform time. This notion says that the temporal constructs in the language can all be used for any user-defined notion of time, not just the "built-in" notion of time. In TCC, this is captured by the time A on B combinator. For the agent A, the agent B defines the notion of time – only those time ticks that "pass" the test B are passed on to A. Thus A is executed with a "programmer supplied" clock. Of course, these constructs can be nested, thus time time time A on B1 on B2 will supply to A only those time ticks that pass B1 and B2.

This flexibility of the basic formalism permits a large number of combinators to be definable by the user. Combinators such as the following are definable in A and B:

do A watching c: Execute A, across time instants but abort it as soon as there is a time instant which satisfies the constraint c.

suspend c activate d A: Execute A, across time instants, suspending it as soon as a time instant is reached in which c is true. Then activate it as soon as a time instant is reached in which d is true.

A.1   RCC– Combining Agent Execution and Testing

The key intuition was the recognition that CCP corresponds to "computation on the left", or forward chaining, and (definite clause) logic programming corresponds to backward chaining. This is illustrated by the following characterization of CCP agents as formulas in intuitionistic logic:

(Agents) $A ::= c \mid G \Rightarrow A \mid E \mid$ some V in A
(Goals) $G ::= c \mid$ all G and G
(Clauses) $P ::= E \Rightarrow D \mid$ all P and P

Computation is initiated on the presentation of an initial agent, A, and progresses in the "forward" direction. One thinks of a sequent $A_1, A_n \rightarrow$ as a multiset of interacting agents operating on a store of constraints (the subset of the A_i that are constraints). If the store is powerful enough to entail the condition G of an agent $G \Rightarrow A$, then $G \Rightarrow A$ can be replaced by A. This corresponds to the application of the left hand rule for implication. Recursive calls E are replaced by the body D of their defining clauses $E \Rightarrow D$. Computation terminates when no more implication can be discharged.

This is logically sound. Clearly if we start computing with an agent A and terminate in a state with the subset of constraints σ then we have $A \vdash \sigma$, where $\vdash$ represents provability in Intuitionistic Logic (IL), augmented with axioms from the underlying constraint system, C. Is this logically complete? Indeed – [32] shows that if there is a constraint d that is entailed by A, then in fact it is entailed by σ the constraint store of the final configuration obtained by executing A as a CCP agent. Hence CCP operational semantics is sound and complete with respect to entailment of constraints.

Note that this language corresponds to "flat" guards. In the early development of concurrent logic programming languages [36,34,38] a lot of attention was paid to "deep" guards. How can deep guards be integrated into CCP?

One idea is to look at definite clause logic programming. The logical picture here is well known.

(Goals) G ::= c | all G and G | all G or G | H | some V in G
(Clauses) P ::= H ⇒ G | all P and P

Computation corresponds to posing a query, or a goal, against a database of clauses of the form $H \Rightarrow G$ [20]. A configuration consists of a collection of G formulas. In each step, a conjunction is replaced by its components, an existential some V in G by G, with V a "new" variable, and an atom H by the body G of a clause $H \Rightarrow G$ from the program. A disjunct is non-deterministically replaced by one of its disjuncts. Computation terminates when the configuration contains only constraints. Of particular interest are terminal configurations in which the constraints are jointly satisfiable, these correspond to answers for the original query.

Is there a reasonable way to combine the two? In fact, it is possible to do this, and a lot more. It is possible to give an intuitive operational semantics for the following system of agents and goals.

(Agents) D ::= c | G ⇒ D | E | E ⇒ D | all D and D | some V in D | all V in D
(Goals)   G ::= c | A ⇒ G | H | G ⇒ H | all G and G | all G or G | some V in G
          | all V in G
(Clauses) P ::= H ⇒ G | E ⇒ D | all P and P

Note that richness of interplay between agents and goals – agents can be defined in terms of goals, and goals can be defined in terms of agents.

What is the underlying programming intuition? We think of D as representing a concurrent, interacting system of agents (interacting through a shared constraint store). We think of G as a *test* of such a system. We think of a sequent D ⊢ G as establishing that the system D *passes* the test G. With this interpretation, we can think of an agent $G \Rightarrow D$ as saying: if the current system of agents can pass the test G, then reduce to D. Conversely, one thinks of the goal $D \Rightarrow G$ as a "what if" test: Suppose the existing system is augmented with the agent D. Does it now pass the test G? Similarly, all V in G is a *generic* goal: it asks the question "Does the system pass the test G" for some completely unknown variable V (hence for all possible values of V).

We showed further that this semantics is sound and complete with respect to the interpretation of agents and goals as formulas in Intuitionistic Logic (IL). The key insight is to "segregate" the atomic formulas that occur in agents (E) and in goals (H) – these must come from disjoint vocabularies. Therefore the "left hand side" (LHS) and the "right hand side" (RHS) of a sequent can no longer communicate through the application of identity rules ($\Gamma, A \vdash A$). Rather the replacement constraint inference rule must be used. Computation can be performed in potentially arbitrary combinations of LHS steps and RHS steps, corresponding to evolution of the concurrent agent and simplifcation of the test, respectively.

Subsequent work by Liang and Miller [22] established a connection between [21] and the notion of *focussing* proofs developed by Andreoli [2]. Indeed, the notion of combining forward and backward chaining in the very flexible way described above has seen significant recent work.

Probabilistic and Quantum Event Structures

Glynn Winskel

Computer Laboratory, University of Cambridge, UK

Abstract. A mathematical theory of probabilistic and quantum event structures is developed. It has some claim to providing fundamental models of distributed probabilistic and quantum systems, and has formed the basis for distributed probabilistic and quantum games.

1 Introduction

Prakash Panangaden has been drawn to conceptual problems in computer science, logic and computation, how to structure and understand probabilistic computation, and the boundaries of computer science with physics. I hope here to be dealing with subjects close to Prakash's heart.

Event structures have emerged as a fundamental model of distributed computation, a model in which the traditional view of a history as a sequence of events is replaced by a view of a history as a partial order of events. This article studies the mathematics needed to take event structures into the realm of distributed probabilistic and distributed quantum computation. The lack of a sufficiently general definition of probabilistic event structure became apparent in work on concurrent games and strategies, in extending concurrent strategies to probabilistic strategies—see the companion work [1]. The description of a probabilistic event structure here meets that need and extends previous definitions, summarised in [2].

A probabilistic event structure essentially comprises an event structure together with a continuous valuation on the Scott open sets of its domain of configurations. The continuous valuation assigns a probability to each open set. However open sets are several levels removed from the events of an event structure, so a more workable definition is obtained by considering the probabilities of basic open sets, generated by single finite configurations; for each finite configuration this specifies the probability of a result which extends the finite configuration. Such valuations on configuration determine the continuous valuations from which they arise, and can be characterised through the device of "drop functions." The characterisation yields a workable definition of probabilistic event structure.

In a quantum event structure events are interpreted as unitary or projection operators in a Hilbert space. Unitary operators are associated with events of preparation, such as a change of coordinates with which to make a measurement or a time period over which the system is allowed to evolve undisturbed. Projection operators are associated with events of elementary tests. Causally

F. van Breugel et al. (Eds.): Panangaden Festschrift, LNCS 8464, pp. 476–497, 2014.

independent (*i.e.* concurrent) events are interpreted by commuting operators. A configuration of the event structure is thought of as a distributed quantum experiment; it describes which events of preparation and tests to perform and their (partial) order of dependency. Once given an initial state as a density operator, a quantum event structure assigns an intrinsic weight to each finite configuration. This does not make the whole event structure into a probabilistic event structure, but it does do so locally: under each configuration there is a probabilistic event structure giving the probabilities over the outcomes of the experiment the configuration describes. Quantum theory is often described as a contextual theory, in that it is only sensible to consider outcomes w.r.t. a specified measurement context [3]. In a quantum event structure configurations assume the role of measurement contexts; w.r.t. a measurement context expressed as a configuration, the sub-configurations constitute the possible outcomes.

2 Event Structures

An *event structure* comprises $(E, \leq, \mathrm{Con})$, consisting of a set E, of *events* which are partially ordered by $\leq$, the *causal dependency relation*, and a nonempty *consistency relation* Con consisting of finite subsets of E, which satisfy

$$\{e' \mid e' \leq e\} \text{ is finite for all } e \in E,$$
$$\{e\} \in \mathrm{Con} \text{ for all } e \in E,$$
$$Y \subseteq X \in \mathrm{Con} \implies Y \in \mathrm{Con}, \text{ and}$$
$$X \in \mathrm{Con} \ \& \ e \leq e' \in X \implies X \cup \{e\} \in \mathrm{Con}.$$

The *configurations* $\mathcal{C}^\infty(E)$ of an event structure E consist of those subsets $x \subseteq E$ which are

(Consistent) $\forall X \subseteq x. \ X$ is finite $\Rightarrow X \in \mathrm{Con} \ x \in \mathrm{Con}$, and
(Down-closed) $\forall e, e'. \ e' \leq e \in x \implies e' \in x.$

Often we shall be concerned with just the finite configurations, $\mathcal{C}(E)$.

We say an event structure is *elementary* when the consistency relation consists of all finite subsets of events. Two events e, e' which are both consistent and incomparable w.r.t. causal dependency in an event structure are regarded as *concurrent*, written $e \, co \, e'$. We shall occasionally say events are in *conflict* when they are they are not consistent. For $X \subseteq E$ we write $[X]$ for $\{e \in E \mid \exists e' \in X. \ e \leq e'\}$, the down-closure of X; note if $X \in \mathrm{Con}$, then $[X] \in \mathrm{Con}$ so is a configuration.

Notation 1. Let E be an event structure. We use $x{-\!\!\!\subset} y$ to mean y covers x in $\mathcal{C}^\infty(E)$, *i.e.* $x \subsetneq y$ in $\mathcal{C}^\infty(E)$ with nothing in between, and $x \overset{e}{-\!\!\!\subset} y$ to mean $x \cup \{e\} = y$ for $x, y \in \mathcal{C}^\infty(E)$ and event $e \notin x$. We use $x \overset{e}{-\!\!\!\subset}$, expressing that event e is enabled at configuration x, when $x \overset{e}{-\!\!\!\subset} y$ for some y. We write $\{x_i \mid i \in I\}{\uparrow}$ to indicate that a subset of configurations is compatible, *i.e.* bounded above by a configuration.

3   Probabilistic Event Structures

A probabilistic event structure comprises an event structure $(E, \leq, \mathrm{Con})$ with a continuous valuation on its Scott open sets of configurations. Recall a *continuous valuation* is a function w from the Scott-open subsets of $\mathcal{C}^{\infty}(E)$ to $[0, 1]$ which is

(normalized) $w(\mathcal{C}^{\infty}(E)) = 1;$ (strict) $w(\varnothing) = 0;$
(monotone) $U \subseteq V \implies w(U) \leq w(V);$
(modular) $w(U \cup V) + w(U \cap V) = w(U) + w(V);$ and
(continuous) $w(\bigcup_{i \in I} U_i) = \sup_{i \in I} w(U_i)$ for *directed* unions $\bigcup_{i \in I} U_i$.

The value $w(U)$ of a continuous valuation w specifies the probability of a result in open set U. Continuous valuations traditionally play the role of elements in probabilistic powerdomains [4]. Continuous valuations are determined by their restrictions to basic open sets

$$\widehat{x} =_{\mathrm{def}} \{ y \in \mathcal{C}^{\infty}(E) \mid x \subseteq y \},$$

for x a finite configuration. A characterisation of such restrictions yields an equivalent, more workable definition of probabilistic event structure, that we present in Section 3.2. As preparation we first develop some machinery for assigning values to "general intervals."

3.1   General Intervals and Drop Functions

Throughout this section assume E is an event structure and $v : \mathcal{C}(E) \to \mathbb{R}$. Extend $\mathcal{C}(E)$ to a lattice $\mathcal{C}(E)^{\top}$ by adjoining an extra top element $\top$. Write its order as $x \sqsubseteq y$ and its finite join operations as $x \vee y$ and $\bigvee_{i \in I} x_i$. Extend v to $v^{\top} : \mathcal{C}(E)^{\top} \to \mathbb{R}$ by taking $v^{\top}(\top) = 0$.

We are concerned with drops in value across general intervals $[y; x_1, \cdots, x_n]$, where $y, x_1, \cdots, x_n \in \mathcal{C}(E)^{\top}$ with $y \sqsubseteq x_1, \cdots, x_n$ in $\mathcal{C}(E)^{\top}$. The interval is thought of as specifying the set of configurations $\widehat{y} \smallsetminus (\widehat{x}_1 \cup \cdots \cup \widehat{x}_n)$, *viz.* those configurations above or equal to y and not above or equal to any $x_1, \cdots, x_n$. As such the intervals form a basis of the Lawson topology on $\mathcal{C}^{\infty}(E)^{\top}$.

Define the *drop functions* $d_v^{(n)}[y; x_1, \cdots, x_n] \in \mathbb{R}$ for $y, x_1, \cdots, x_n \in \mathcal{C}(E)^{\top}$ with $y \sqsubseteq x_1, \cdots, x_n$ in $\mathcal{C}(E)^{\top}$, by induction, taking

$$d_v^{(0)}[y;] =_{\mathrm{def}} v^{\top}(y) \text{ and}$$

$$d_v^{(n)}[y; x_1, \cdots, x_n] =_{\mathrm{def}} d_v^{(n-1)}[y; x_1, \cdots, x_{n-1}] - d_v^{(n-1)}[x_n; x_1 \vee x_n, \cdots, x_{n-1} \vee x_n],$$

for $n > 0$.

The following proposition shows how drop functions assign to general intervals $[y; x_1, \cdots, x_n]$ the value of being in $\widehat{y}$ minus the value of being in $\widehat{x}_1 \cup \cdots \cup \widehat{x}_n$, and that the latter is calculated using the inclusion-exclusion principle for sets; notice that an overlap $\bigcap_{i \in I} \widehat{x}_i$ equals $\widehat{\bigvee_{i \in I} x_i}$, where $\varnothing \neq I \subseteq \{1, \cdots, n\}$.

Proposition 1. *Let $n \in \omega$. For $y, x_1, \cdots, x_n \in C(E)^\top$ with $y \sqsubseteq x_1, \cdots, x_n$,*

$$d_v^{(n)}[y; x_1, \cdots, x_n] = v(y) - \sum_{\emptyset \neq I \subseteq \{1, \cdots, n\}} (-1)^{|I|+1} v(\bigvee_{i \in I} x_i).$$

For $y, x_1, \cdots, x_n \in C(E)$ with $y \subseteq x_1, \cdots, x_n$,

$$d_v^{(n)}[y; x_1, \cdots, x_n] = v(y) - \sum_I (-1)^{|I|+1} v(\bigcup_{i \in I} x_i),$$

where the index I ranges over sets satisfying $\emptyset \neq I \subseteq \{1, \cdots, n\}$ s.t. $\{x_i \mid i \in I\}\uparrow$.

Proof. We prove the first statement by induction on n. For the basis, when $n = 0$, $d_v^{(n)}[y;] = v(y)$, as required. For the induction step, with $n > 0$, we reason

$$\begin{aligned} d_v^{(n)}[y; x_1, \cdots, x_n] =_{\text{def}} & \; d_v^{(n-1)}[y; x_1, \cdots, x_{n-1}] - d_v^{(n-1)}[x_n; x_1 \vee x_n, \cdots, x_{n-1} \vee x_n] \\ = & \; v(y) - \sum_{\emptyset \neq I \subseteq \{1, \cdots, n-1\}} (-1)^{|I|+1} v(\bigvee_{i \in I} x_i) \\ & - v(x_n) + \sum_{\emptyset \neq J \subseteq \{1, \cdots, n-1\}} (-1)^{|I|+1} v(\bigvee_{j \in J} x_i \vee x_n), \end{aligned}$$

making use of the induction hypothesis. Consider subsets K for which $\emptyset \neq K \subseteq \{1, \cdots, n\}$. Either $n \notin K$, in which case $\emptyset \neq K \subseteq \{1, \cdots, n-1\}$, or $n \in K$, in which case $K = \{n\}$ or $J =_{\text{def}} K \setminus \{n\}$ satisfies $\emptyset \neq J \subseteq \{1, \cdots, n-1\}$. From this observation, the sum above amounts to

$$v(y) - \sum_{\emptyset \neq K \subseteq \{1, \cdots, n\}} (-1)^{|K|+1} v(\bigvee_{k \in K} x_k),$$

as required to maintain the induction hypothesis.

The second expression of the proposition is got by discarding all terms $v(\bigvee_{i \in I} x_i)$ for which $\bigvee_{i \in I} x_i = \top$ which leaves the sum unaffected as they contribute 0. □

Corollary 1. *Let $n \in \omega$ and $y, x_1, \cdots, x_n \in C(E)^\top$ with $y \sqsubseteq x_1, \cdots, x_n$. For ρ an n-permutation,*

$$d_v^{(n)}[y; x_{\rho(1)}, \cdots, x_{\rho(n)}] = d_v^{(n)}[y; x_1, \cdots, x_n].$$

Proof. As by Proposition 1, the value of $d_v^{(n)}[y; x_1, \cdots, x_n]$ is insensitive to permutations of its arguments. □

In the following results we lay out the fundamental properties of drop functions for later use.

Proposition 2. *Assume $n \geq 1$ and $y, x_1, \cdots, x_n \in C(E)^\top$ with $y \sqsubseteq x_1, \cdots, x_n$. If $y = x_i$ for some i with $1 \leq i \leq n$ then $d_v^{(n)}[y; x_1, \cdots, x_n] = 0$.*

Proof. By Corollary 1, it suffices to show $d_v^{(n)}[y; x_1, \cdots, x_n] = 0$ when $y = x_n$. In this case,

$$\begin{aligned} d_v^{(n)}[y; x_1, \cdots, x_n] &= d_v^{(n-1)}[y; x_1, \cdots, x_{n-1}] - d_v^{(n-1)}[x_n; x_1 \vee x_n, \cdots, x_{n-1} \vee x_n] \\ &= d_v^{(n-1)}[y; x_1, \cdots, x_{n-1}] - d_v^{(n-1)}[y; x_1, \cdots, x_{n-1}] \\ &= 0. \end{aligned}$$

$\square$

Corollary 2. *Assume* $n \geq 1$ *and* $y, x_1, \cdots, x_n \in C(E)^\top$ *with* $y \sqsubseteq x_1, \cdots, x_n$. *If* $x_i \sqsubseteq x_j$ *for distinct* i, j *with* $1 \leq i, j \leq n$ *then*

$$d_v^{(n)}[y; x_1, \cdots, x_n] = d_v^{(n-1)}[y; x_1, \cdots, x_{j-1}, x_{j+1}, \cdots, x_n].$$

Proof. By Corollary 1, it suffices to show

$$d_v^{(n)}[y; x_1, \cdots, x_{n-1}, x_n] = d_v^{(n-1)}[y; x_1, \cdots, x_{n-1}]$$

when $x_{n-1} \sqsubseteq x_n$. Then,

$$\begin{aligned} d_v^{(n)}[y; x_1, \cdots, x_n] &= d_v^{(n-1)}[y; x_1, \cdots, x_{n-1}] - d_v^{(n-1)}[x_n; x_1 \vee x_n, \cdots, x_{n-1} \vee x_n] \\ &= d_v^{(n-1)}[y; x_1, \cdots, x_{n-1}] - d_v^{(n-1)}[x_n; x_1 \vee x_n, \cdots, x_{n-2}, x_n] \\ &= d_v^{(n-1)}[y; x_1, \cdots, x_{n-1}] - 0, \end{aligned}$$

by Proposition 2.

$\square$

Proposition 3. *Assume* $n \in \omega$ *and* $y, x_1, \cdots, x_n \in C(E)^\top$ *with* $y \sqsubseteq x_1, \cdots, x_n$. *Then,* $d_v^{(n)}[y; x_1, \cdots, x_n] = 0$ *if* $y = \top$ *and* $d_v^{(n)}[y; x_1, \cdots, x_n] = d_v^{(n-1)}[y; x_1, \cdots, x_{i-1}, x_{i+1}, \cdots, x_n]$ *if* $x_i = \top$ *with* $1 \leq i \leq n$.

Proof. When $n = 0$, $d_v^{(0)}[\top;] = v^\top(\top) = 0$. When $n \geq 1$, $d_v^{(n)}[\top; x_1, \cdots, x_n] = 0$ by Proposition 2 as *e.g.* $x_n = \top$. For the remaining statement, w.l.o.g. we may assume $i = n$ and that $x_n = \top$, yielding

$$\begin{aligned} &d_v^{(n)}[y; x_1, \cdots, \top] = \\ &d_v^{(n-1)}[y; x_1, \cdots, x_{n-1}] - d_v^{(n-1)}[\top; x_1 \vee \top, \cdots, x_{n-1} \vee \top] = d_v^{(n-1)}[y; x_1, \cdots, x_{n-1}]. \end{aligned}$$

$\square$

It will be important that drops across general intervals can be reduced to sums of drops across intervals based on coverings, as explained in the next two results.

Lemma 1. *Let* $n \geq 1$. *Let* $y, x_1, \cdots, x_n, x_n' \in C(E)^\top$ *with* $y \sqsubseteq x_1, \cdots, x_n$. *Assume* $x_n \sqsubseteq x_n'$. *Then,*

$$d_v^{(n)}[y; x_1, \cdots, x_n'] = d_v^{(n)}[y; x_1, \cdots, x_n] + d_v^{(n)}[x_n; x_1 \vee x_n, \cdots, x_{n-1} \vee x_n, x_n'].$$

Proof. By definition,

the r.h.s. $= d_v^{(n-1)}[y; x_1, \cdots, x_{n-1}] - d_v^{(n-1)}[x_n; x_1 \vee x_n, \cdots, x_{n-1} \vee x_n]$

$\qquad + d_v^{(n-1)}[x_n; x_1 \vee x_n, \cdots, x_{n-1} \vee x_n] - d_v^{(n-1)}[x_n'; x_1 \vee x_n', \cdots, x_{n-1} \vee x_n']$

$\quad = d_v^{(n-1)}[y; x_1, \cdots, x_{n-1}] - d_v^{(n-1)}[x_n'; x_1 \vee x_n', \cdots, x_{n-1} \vee x_n']$

$\quad = d_v^{(n)}[y; x_1, \cdots, x_{n-1}, x_n']$

$\quad = $ the l.h.s..

$\square$

Lemma 2. *Let* $y \subseteq x_1, \cdots, x_n$ *in* $\mathcal{C}(E)$. *Then,* $d_v^{(n)}[y; x_1, \cdots, x_n]$ *is expressible as a sum of terms* $d_v^{(k)}[u; w_1, \cdots, w_k]$ *where* $y \subseteq u \multimap w_i$ *in* $\mathcal{C}(E)$ *and* $w_i \subseteq x_1 \cup \cdots \cup x_n$, *for all* i *with* $1 \leq i \leq k$. *(The set* $x_1 \cup \cdots \cup x_n$ *need not be in* $\mathcal{C}(E)$.)

Proof. Define the *weight* of a term $d_v^{(n)}[y; x_1, \cdots, x_n]$, where $y \subseteq x_1, \cdots, x_n$ in $\mathcal{C}(E)$, to be the product $|x_1 \smallsetminus y| \times \cdots \times |x_n \smallsetminus y|$.

Assume $y \subseteq x_1, \cdots, x_n'$ in $\mathcal{C}(E)$. By Proposition 2, if y equals x_n' or some x_i, then $d_v^{(n)}[y; x_1, \cdots, x_n'] = 0$, so may be deleted as a contribution to a sum. Otherwise, if $y \not\subseteq x_n \not\subseteq x_n'$, by Lemma 1 we can rewrite $d_v^{(n)}[y; x_1, \cdots, x_n']$ to the sum

$$d_v^{(n)}[y; x_1, \cdots, x_n] + d_v^{(n)}[x_n; x_1 \vee x_n, \cdots, x_{n-1} \vee x_n, x_n'],$$

where we further observe

$$|x_n \smallsetminus y| < |x_n' \smallsetminus y|, \qquad |x_n' \smallsetminus x_n| < |x_n' \smallsetminus y|$$

and

$$|(x_i \cup x_n) \smallsetminus x_n| \leq |x_i \smallsetminus y|,$$

whenever $x_i \vee x_n \neq \top$. Using Proposition 3 we may tidy away any mentions of $\top$. This reduces $d_v^{(n)}[y; x_1, \cdots, x_n']$ to the sum of at most two terms, each of lesser weight. For notational simplicity we have concentrated on the nth argument in $d_v^{(n)}[y; x_1, \cdots, x_n']$, but by Corollary 1 an analogous reduction is possible w.r.t. any argument.

Repeated use of the reduction, rewrites $d_v^{(n)}[y; x_1, \cdots, x_n]$ to a sum of terms of the form

$$d_v^{(k)}[u; w_1, \cdots, w_k]$$

where $k \leq n$ and $u \multimap w_1, \cdots, w_k \subseteq x_1 \cup \cdots \cup x_n$. This justifies the claims of the lemma. $\square$

3.2 Probabilistic Event Structures

A probabilistic event structure is an event structure associated with a $[0, 1]$-valuation on configurations, normalised to 1 at the emptyset, such that no general interval has a negative drop.

Definition 1. Let E be an event structure. A *configuration-valuation* on E is function $v : \mathcal{C}(E) \to [0,1]$ such that $v(\varnothing) = 1$ and which satisfies the *drop condition*

$$d_v^{(n)}[y; x_1, \cdots, x_n] \geq 0$$

for all $n \geq 1$ and $y, x_1, \cdots, x_n \in \mathcal{C}(E)$ with $y \subseteq x_1, \cdots, x_n$. A *probabilistic event structure* comprises an event structure E together with a configuration-valuation $v : \mathcal{C}(E) \to [0,1]$.[1]

Proposition 4. *Let E be an event structure. Let $v : \mathcal{C}(E) \to [0,1]$. Then, v is a configuration-valuation iff $d_v^{(n)}[y; x_1, \cdots, x_n] \geq 0$ for all $n \in \omega$ and $y, x_1, \cdots, x_n \in \mathcal{C}(E)^\top$ with $y \subseteq x_1, \cdots, x_n$. If v is a configuration-valuation, then*

$$y \subseteq x \implies v^\top(y) \geq v^\top(x),$$

for all $x, y \in \mathcal{C}(E)^\top$.

Proof. By Proposition 3 and as $d_v^{(1)}[y; x] = v^\top(y) - v^\top(x)$. □

By Lemma 2, in showing we have a probabilistic event structure it suffices to verify the "drop condition" only for special general intervals $[y; x_1, \cdots, x_n]$ in which the configurations $x_1, \cdots, x_n$ cover y.

Proposition 5. *Let E be an event structure. Let $v : \mathcal{C}(E) \to [0,1]$. v is a configuration-valuation iff $v(\varnothing) = 1$ and*

$$d_v^{(n)}[y; x_1, \cdots, x_n] \geq 0$$

for all $n \geq 1$ and $y {-\!\subset} x_1, \cdots, x_n$ in $\mathcal{C}(E)$.

4   The Characterisation

Our goal is to prove that probabilistic event structures correspond to event structures with a continuous valuation. It is clear that a continuous valuation w on the Scott-open subsets of an event structure E gives rise to a configuration-valuation v on E: take $v(x) =_{\text{def}} w(\widehat{x})$, for $x \in \mathcal{C}(E)$. We will show that this construction has an inverse, that a configuration-valuation determines a continuous valuation. For this we need a combinatorial lemma:[2]

---

[1] Samy Abbes has pointed out that the "drop condition" appears in early work of the Russian mathematician V.A.Rohlin [5](as relation (6) of Section 3, p.7), and Klaus Keimel that functions satisfying the "drop condition" are called "totally convex" or "completely monotone" in the literature [6]. The rediscovery of the "drop condition" and its reuse in the context of event structures was motivated by Lemma 2, tying it to occurrences of events.

[2] The proof of the combinatorial lemma, due to the author, appears with acknowledgement as Lemma 6.App.1 in [7], the PhD thesis of my former student Daniele Varacca, whom I thank, both for the collaboration and the latex.

Lemma 3. *For all finite sets* I, J,

$$\sum_{\substack{\emptyset \neq K \subseteq I \times J \\ \pi_1(K)=I, \pi_2(K)=J}} (-1)^{|K|} = (-1)^{|I|+|J|-1}.$$

Proof. W.l.o.g. we can take $I = \{1, \ldots, n\}$ and $J = \{1, \ldots, m\}$. Also observe that a subset $K \subseteq I \times J$ such that $\pi_1(K) = I, \pi_2(K) = J$ is in fact a surjective and total relation between the two sets, pictured below.

Let

$$t_{n,m} =_{\text{def}} \sum_{\substack{\emptyset \neq K \subseteq I \times J \\ \pi_1(K)=I, \pi_2(K)=J}} (-1)^{|K|};$$

$$t^o_{n,m} =_{\text{def}} |\{\emptyset \neq K \subseteq I \times J \mid |K| \text{ odd}, \pi_1(K) = I, \pi_2(K) = J\}|;$$

$$t^e_{n,m} := |\{\emptyset \neq K \subseteq I \times J \mid |K| \text{ even}, \pi_1(K) = I, \pi_2(K) = J\}|.$$

Clearly $t_{n,m} = t^e_{n,m} - t^o_{n,m}$. We want to prove that $t_{n,m} = (-1)^{n+m+1}$. We do this by induction on n. It is easy to check that this is true for $n = 1$. In this case, if m is even then $t^e_{1,m} = 1$ and $t^o_{1,m} = 0$, so that $t^e_{1,m} - t^o_{1,m} = (-1)^{1+m+1}$. Similarly if m is odd.

Now assume that $t_{n,p} = (-1)^{n+p+1}$, for every p, and compute $t_{n+1,m}$. To evaluate $t_{n+1,m}$ we count all surjective and total relations K between I and J together with their "sign." Consider the pairs in K of the form $(n+1, h)$ for $h \in J$. The result of removing them is a a total surjective relation between $\{1, \ldots, n\}$ and a subset J_K of $\{1, \ldots, m\}$.

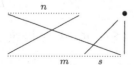

Consider first the case where $J_K = \{1, \ldots, m\}$. Consider the contribution of such K's to $t_{n+1,m}$. There are $\binom{m}{s}$ ways of choosing s pairs of the form $(n+1, h)$. For every such choice there are $t_{n,m}$ (signed) relations. Adding the pairs $(n+1, h)$ possibly modifies the sign of such relations. In all the contribution amounts to

$$\sum_{1 \leq s \leq m} \binom{m}{s}(-1)^s t_{n,m}.$$

Suppose now that J_K is a proper subset of $\{1, \ldots, m\}$ leaving out r elements.

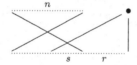

Since K is surjective, all such elements h must be in a pair of the form $(n+1, h)$. Moreover there can be s pairs of the form $(n+1, h')$ with $h' \in J_K$. What is the contribution of such K's to $t_{n,m}$? There are $\binom{m}{r}$ ways of choosing the elements that are left out. For every such choice and for every s such that $0 \le s \le m - r$ there are $\binom{m-r}{s}$ ways of choosing the $h' \in J_K$. And for every such choice there are $t_{n,m-r}$ (signed) relations. Adding the pairs $(n+1, h)$ and $(n+1, h')$ possibly modifies the sign of such relations. In all, for every r such that $1 \le r \le m-1$, the contribution amounts to

$$\binom{m}{r} \sum_{1 \le s \le m-r} \binom{m}{s} (-1)^{s+r} t_{n,m-n} .$$

The (signed) sum of all these contribution will give us $t_{n+1,m}$. Now we use the induction hypothesis and we write $(-1)^{n+p+1}$ for $t_{n,p}$.

Thus,

$$t_{n+1,m} = \sum_{1 \le s \le m} \binom{m}{s} (-1)^s t_{n,m}$$
$$+ \sum_{1 \le r \le m-1} \binom{m}{r} \sum_{0 \le s \le m-r} \binom{m-r}{s} (-1)^{s+r} t_{n,m-r}$$
$$= \sum_{1 \le s \le m} \binom{m}{s} (-1)^{s+n+m+1}$$
$$+ \sum_{1 \le r \le m-1} \binom{m}{r} \sum_{0 \le s \le m-r} \binom{m-r}{s} (-1)^{s+n+m+1}$$
$$= (-1)^{n+m+1} \left(\sum_{1 \le s \le m} \binom{m}{s} (-1)^s \right.$$
$$+ \left. \sum_{1 \le r \le m-1} \binom{m}{r} \sum_{0 \le s \le m-r} \binom{m-r}{s} (-1)^s \right) .$$

By the binomial formula, for $1 \le r \le m-1$ we have

$$0 = (1-1)^{m-r} = \sum_{0 \le s \le m-r} \binom{m-r}{s} (-1)^s .$$

So we are left with

$$t_{n+1,m} = (-1)^{n+m+1} \left(\sum_{1 \le s \le m} \binom{m}{s} (-1)^s \right)$$
$$= (-1)^{n+m+1} \left(\sum_{0 \le s \le m} \binom{m}{s} (-1)^s - \binom{m}{0} (-1)^0 \right)$$
$$= (-1)^{n+m+1} (0 - 1)$$
$$= (-1)^{n+1+m+1} ,$$

as required.   □

Theorem 1. *A configuration-valuation v on an event structure E extends to a unique continuous valuation w_v on the open sets of $C^\infty(E)$, so that $w_v(\widehat{x}) = v(x)$, for all $x \in C(E)$.*

Conversely, a continuous valuation w on the open sets of $C^\infty(E)$ restricts to a configuration-valuation v_w on E, assigning $v_w(x) = w(\widehat{x})$, for all $x \in C(E)$.

Proof. The proof is inspired by the proofs in the appendix of [2] and the thesis [7].

First, a continuous valuation w on the open sets of $C^\infty(E)$ restricts to a configuration-valuation v defined as $v(x) =_{\mathrm{def}} w(\widehat{x})$ for $x \in C(E)$. Note that any extension of a configuration-valuation to a continuous valuation is bound to be unique by continuity.

To show the converse we first define a function w from the basic open sets $Bs =_{\mathrm{def}} \{\widehat{x_1} \cup \cdots \cup \widehat{x_n} \mid x_1, \cdots, x_n \in C(E)\}$ to $[0,1]$ and show that it is normalised, strict, monotone and modular. Define

$$w(\widehat{x_1} \cup \cdots \cup \widehat{x_n}) =_{\mathrm{def}} 1 - d_v^{(n)}[\varnothing; x_1, \cdots, x_n]$$
$$= \sum_{\varnothing \neq I \subseteq \{1,\cdots,n\}} (-1)^{|I|+1} v(\bigvee_{i \in I} x_i)$$

—this can be shown to be well-defined using Corollaries 1 and 2.

Clearly, w is normalised in the sense that $w(C^\infty(E)) = w(\widehat{\varnothing}) = 1$ and strict in that $w(\varnothing) = 1 - v(\varnothing) = 0$.

To see that it is monotone, first observe that

$$w(\widehat{x_1} \cup \cdots \cup \widehat{x_n}) \leq w(\widehat{x_1} \cup \cdots \cup \widehat{x_{n+1}})$$

as

$$w(\widehat{x_1} \cup \cdots \cup \widehat{x_{n+1}}) - w(\widehat{x_1} \cup \cdots \cup \widehat{x_n}) = d_v^{(n)}[\varnothing; x_1, \cdots, x_n] - d_v^{(n+1)}[\varnothing; x_1, \cdots, x_{n+1}]$$
$$= d_v^{(n)}[x_{n+1}; x_1 \vee x_{n+1}, \cdots, x_n \vee x_{n+1}] \geq 0.$$

By a simple induction (on m),

$$w(\widehat{x_1} \cup \cdots \cup \widehat{x_n}) \leq w(\widehat{x_1} \cup \cdots \cup \widehat{x_n} \cup \widehat{y_1} \cup \cdots \cup \widehat{y_m}).$$

Suppose that $\widehat{x_1} \cup \cdots \cup \widehat{x_n} \subseteq \widehat{y_1} \cup \cdots \cup \widehat{y_m}$. Then $\widehat{y_1} \cup \cdots \cup \widehat{y_m} = \widehat{x_1} \cup \cdots \cup \widehat{x_n} \cup \widehat{y_1} \cup \cdots \cup \widehat{y_m}$. By the above,

$$w(\widehat{x_1} \cup \cdots \cup \widehat{x_n}) \leq w(\widehat{x_1} \cup \cdots \cup \widehat{x_n} \cup \widehat{y_1} \cup \cdots \cup \widehat{y_m})$$
$$= w(\widehat{y_1} \cup \cdots \cup \widehat{y_m}),$$

as required to show w is monotone.

To show modularity we require

$$w(\widehat{x_1} \cup \cdots \cup \widehat{x_n}) + w(\widehat{y_1} \cup \cdots \cup \widehat{y_m})$$
$$= w(\widehat{x_1} \cup \cdots \cup \widehat{x_n} \cup \widehat{y_1} \cup \cdots \cup \widehat{y_m}) + w((\widehat{x_1} \cup \cdots \cup \widehat{x_n}) \cap (\widehat{y_1} \cup \cdots \cup \widehat{y_m})).$$

Note

$$(\widehat{x_1} \cup \cdots \cup \widehat{x_n}) \cap (\widehat{y_1} \cup \cdots \cup \widehat{y_m}) = (\widehat{x_1} \cap \widehat{y_1}) \cup \cdots \cup (\widehat{x_i} \cap \widehat{y_j}) \cdots \cup (\widehat{x_n} \cap \widehat{y_m})$$
$$= \widehat{x_1 \vee y_1} \cup \cdots \cup \widehat{x_i \vee y_j} \cdots \cup \widehat{x_n \vee y_m}.$$

From the definition of w we require

$$w(\widehat{x_1} \cup \cdots \cup \widehat{x_n} \cup \widehat{y_1} \cup \cdots \cup \widehat{y_m})$$
$$= \sum_{\varnothing \neq I \subseteq \{1,\cdots,n\}} (-1)^{|I|+1} v(\bigvee_{i \in I} x_i) + \sum_{\varnothing \neq J \subseteq \{1,\cdots,m\}} (-1)^{|J|+1} v(\bigvee_{j \in J} y_j)$$
$$- \sum_{\varnothing \neq R \subseteq \{1,\cdots,n\} \times \{1,\cdots,m\}} (-1)^{|R|+1} v(\bigvee_{(i,j) \in R} x_i \vee y_j). \qquad (1)$$

Consider the definition of $w(\widehat{x_1} \cup \cdots \cup \widehat{x_n} \cup \widehat{y_1} \cup \cdots \cup \widehat{y_m})$ as a sum. Its components are associated with indices which either lie entirely within $\{1,\cdots,n\}$, entirely within $\{1,\cdots,m\}$, or overlap both. Hence

$$w(\widehat{x_1} \cup \cdots \cup \widehat{x_n} \cup \widehat{y_1} \cup \cdots \cup \widehat{y_m})$$
$$= \sum_{\varnothing \neq I \subseteq \{1,\cdots,n\}} (-1)^{|I|+1} v(\bigvee_{i \in I} x_i) + \sum_{\varnothing \neq J \subseteq \{1,\cdots,m\}} (-1)^{|J|+1} v(\bigvee_{j \in J} y_j)$$
$$+ \sum_{\varnothing \neq I \subseteq \{1,\cdots,n\},\varnothing \neq J \subseteq \{1,\cdots,m\}} (-1)^{|I|+|J|+1} v(\bigvee_{i \in I} x_i \vee \bigvee_{j \in J} y_j). \qquad (2)$$

Comparing (1) and (2), we require

$$- \sum_{\varnothing \neq R \subseteq \{1,\cdots,n\} \times \{1,\cdots,m\}} (-1)^{|R|+1} v(\bigvee_{(i,j) \in R} x_i \vee y_j)$$
$$= \sum_{\varnothing \neq I \subseteq \{1,\cdots,n\},\varnothing \neq J \subseteq \{1,\cdots,m\}} (-1)^{|I|+|J|+1} v(\bigvee_{i \in I} x_i \vee \bigvee_{j \in J} y_j). \qquad (3)$$

Observe that

$$\bigvee_{(i,j) \in R} x_i \vee y_j = \bigvee_{i \in I} x_i \vee \bigvee_{j \in J} y_j$$

when $I = R_1 =_{\text{def}} \{i \in I \mid \exists j \in J.\ (i,j) \in R\}$ and $J = R_2 =_{\text{def}} \{j \in J \mid \exists i \in I.\ (i,j) \in R\}$ for a relation $R \subseteq \{1,\cdots,n\} \times \{1,\cdots,m\}$. With this observation we see that equality (3) follows from the combinatorial lemma, Lemma 3 above. This shows modularity.

Finally, we can extend w to all open sets by taking an open set U to $\sup_{b \in Bs\ \&\ b \subseteq U} w(b)$. The verification that w is indeed a continuous valuation extending v is now straightforward. $\square$

The above theorem also holds (with the same proof) for Scott domains. Now, by [8], Corollary 4.3:

Theorem 2. *For a configuration-valuation v on E there is a unique probability measure μ_v on the Borel subsets of $C^\infty(E)$ extending w_v.*

When x a finite configuration has $v(x) > 0$ and $\mu_v(\{x\}) = 0$ we can understand x as being a transient configuration on the way to a final with probability $v(x)$. In general, there is a simple expression for the probability of terminating at a finite configuration, helpful in the examples that follow.

Proposition 6. *Let E, v be a probabilistic event structure. For any finite configuration $y \in \mathcal{C}(E)$, the singleton set $\{y\}$ is a Borel subset with probability measure*

$$\mu_v(\{y\}) = \inf\{d_v^{(n)}[y; x_1, \cdots, x_n] \mid n \in \omega \ \& \ y \nsubseteq x_1, \cdots, x_n \in \mathcal{C}(E)\}.$$

Proof. Let $y \in \mathcal{C}(E)$. Then $\{y\} = \hat{y} \setminus U_y$ is clearly Borel as $U_y =_{\text{def}} \{x \in \mathcal{C}^\infty(E) \mid y \nsubseteq x\}$ is open. Let w be the continuous valuation extending v. Then

$$w(U_y) = \sup\{w(\hat{x}_1 \cup \cdots \cup \hat{x}_n) \mid y \nsubseteq x_1, \cdots, x_n \in \mathcal{C}(E)\}$$

as U_y is the directed union $\bigcup\{\hat{x}_1 \cup \cdots \cup \hat{x}_n \mid y \nsubseteq x_1, \cdots, x_n \in \mathcal{C}(E)\}$. Hence

$$\mu_v(\{y\}) = v(y) - w(U_y) = v(y) - \sup\{w(\hat{x}_1 \cup \cdots \cup \hat{x}_n) \mid y \nsubseteq x_1, \cdots, x_n \in \mathcal{C}(E)\}$$

$$= \inf\{v(y) - \sum_{\varnothing \neq I \subseteq \{1, \cdots, n\}} (-1)^{|I|+1} v(\bigvee_{i \in I} x_i) \mid y \nsubseteq x_1, \cdots, x_n \in \mathcal{C}(E)\}$$

$$= \inf\{d_v^{(n)}[y; x_1, \cdots, x_n] \mid n \in \omega \ \& \ y \nsubseteq x_1, \cdots, x_n \in \mathcal{C}(E)\}.$$

$\square$

Example 1. Consider the event structure comprising two concurrent events e_1, e_2 with configuration-valuation v for which $v(\varnothing) = 1, v(\{e_1\}) = 1/3, v(\{e_2\}) = 1/2$ and $v(\{e_1, e_2\}) = 1/12$. This means in particular that there is a probability of $1/3$ of a result within the Scott open set consisting of both the configuration $\{e_1\}$ and the configuration $\{e_1, e_2\}$. In other words, there is a probability of $1/3$ of observing e_1 (possibly with or possibly without e_2). The induced probability measure p assigns a probability to any Borel set, in this simple case any subset of configurations, and is determined by its value on single configurations: $p(\varnothing) = 1 - 4/12 - 6/12 + 1/12 = 3/12, p(\{e_1\}) = 4/12 - 1/12 = 3/12, p(\{e_2\}) = 6/12 - 1/12 = 5/12$ and $p(\{e_1, e_2\}) = 1/12$. Thus there is a probability of $3/12$ of observing neither e_1 nor e_2, and a probability of $5/12$ of observing just the event e_2 (and not e_1). There is a drop $d_v^{(0)}[\varnothing; \{e_1\}, \{e_2\}] = 1 - 4/12 - 6/12 + 1/12 = 3/12$ corresponding to the probability of remaining at the empty configuration and not observing any event. Sometimes it's said that probability "leaks" at the empty configuration, but it's more accurate to think of this leak in probability as associated with a non-zero chance that the initial observation of no events will not improve. $\square$

Example 2. Consider the event structure with events $\mathbb{N}^+$ with causal dependency $n \leq n + 1$, with all finite subsets consistent. It is not hard to check that all subsets of $\mathcal{C}^\infty(\mathbb{N}^+)$ are Borel sets. Consider the ensuing probability distributions w.r.t. the following configuration-valuations:
(i) $v_0(x) = 1$ for all $x \in \mathcal{C}(\mathbb{N}^+)$. The resulting probability distribution assigns probability 1 to the singleton set $\{\mathbb{N}^+\}$, comprising the single infinite configuration $\mathbb{N}^+$, and 0 to $\varnothing$ and all other singleton sets of configurations.
(ii) $v_1(\varnothing) = v_1(\{1\}) = 1$ and $v_1(x) = 0$ for all other $x \in \mathcal{C}(\mathbb{N}^+)$. The resulting probability distribution assigns probability 0 to $\varnothing$ and probability 1 to the singleton set $\{1\}$, and 0 to all other singleton sets of configurations.
(iii) $v_2(\varnothing) = 1$ and $v_2(\{1, \cdots, n\}) = (1/2)^n$ for all $n \in \mathbb{N}^+$. The resulting probability distribution assigns probability $1/2$ to $\varnothing$ and $(1/2)^{n+1}$ to each singleton

$\{\{1,\cdots,n\}\}$ and 0 to the singleton set $\{\mathbb{N}^+\}$, comprising the single infinite configuration $\mathbb{N}^+$.   □

Remark. There is a seeming redundancy in the definition of purely probabilistic event structures, in that there are two different ways to say, for example, that events e_1 and e_2 do not occur together at a finite configuration y where $y\overset{e_1}{-\!\!\subset}x_1$ and $y\overset{e_2}{-\!\!\subset}x_2$: either through $y\cup\{e_1,e_2\}\notin\text{Con}$; or via the configuration-valuation v through $v(x_1\cup x_2)=0$. However, when we mix probability with nondeterminism [1], we make use of both consistency and the valuation. In the next section, for a quantum event structure, consistency will be important in determining when there is a sensible intrinsic probability distribution on a family of configurations, even though the probability of the union of the configurations ends up being zero.

5   Quantum Event Structures

Event structures are a model of distributed computation in which the causal dependence and independence of events is made explicit. By associating events with the most basic operators on a Hilbert space, *viz.* projection and unitary operators, so that independent (*i.e.* concurrent) events are associated with independent (*i.e.* commuting) operators, we obtain quantum event structures.

An event associated with a projection is thought of as an elementary positive test; its occurrence leaves the system in the eigenspace associated with eigenvalue 1 (rather than 0) of the projection. An event associated with a unitary operator is an event of preparation; the preparation might be a change of the direction in which to make a measurement, or the undisturbed evolution of the system over a time interval. A configuration is thought of as specifying a distributed quantum experiment. As we shall see, w.r.t. an initial state given as a density operator, each configuration w of a quantum event structure determines a probabilistic event structure, giving a probability distribution on its sub-configurations—the possible results of the experiment w.

Throughout let $\mathcal{H}$ be a separable Hilbert space over the complex numbers. For operators A, B on $\mathcal{H}$ we write $[A, B] =_{\text{def}} AB - BA$.

5.1   Events as Operators

Formally, we obtain a quantum event structure from an event structure by interpreting its events as unitary or projection operators which must commute when events are concurrent.

Definition 2. A *quantum event structure* (over $\mathcal{H}$) comprises an event structure $(E, \leq, \text{Con})$ together with an assignment Q_e of projection or unitary operators on $\mathcal{H}$ to events $e \in E$ such that for all $e_1, e_2 \in E$,

$$e_1 \: co \: e_2 \implies [Q_{e_1}, Q_{e_2}] = 0\,.$$

Given a finite configuration, $x \in \mathcal{C}(E)$, define the operator A_x to be the composition $Q_{e_n} Q_{e_{n-1}} \cdots Q_{e_2} Q_{e_1}$ for some covering chain

$$\varnothing \xrightarrow{e_1} x_1 \xrightarrow{e_2} x_2 \cdots \xrightarrow{e_n} x_n = x$$

in $\mathcal{C}(E)$. This is well-defined as for any two covering chains up to x the sequences of events are Mazurkiewicz trace equivalent, *i.e.* obtainable, one from the other, by successively interchanging concurrent events. In particular $A_\varnothing$ is the identity operator on $\mathcal{H}$. An *initial state* is given by a density operator ρ on $\mathcal{H}$.

Interpretation. Consider first the simpler situation where in a quantum event structure E, Q the event structure E is elementary (*i.e.* all finite subsets are consistent). We regard E, Q as specifying a, possibly distributed, quantum experiment. The experiment says which unitary operators (events of preparation) and projection operators (elementary positive tests) to apply and in which order. The order being partial permits commuting operators to be applied concurrently, independently of each other, perhaps in a distributed fashion.

For a quantum event structure, E, Q, in general, an individual configuration $w \in \mathcal{C}^\infty(E)$ inherits the order of the ambient event structure E to become an elementary event structure, and can itself be regarded as a quantum experiment. The quantum event structure E, Q represents a collection of quantum experiments which may extend or overlap each other: when $w \subseteq w'$ in $\mathcal{C}^\infty(E)$ the experiment w' extends the experiment w, or equivalently w is a restriction of the experiment w'. In this sense a quantum event structure in general represents a nondeterministic quantum experiment. The extra generality will be crucial later in interpreting probabilistic quantum experiments.

5.2   From Quantum to Probabilistic

Consider a quantum event structure with initial state. A configuration w stands for an experiment and specifies which tests and preparations to try and in which order. In general, not all the tests in w need succeed, yielding as final result a possibly proper sub-configuration x of w. Theorem 3 below explains how there is an inherent probability distribution q_w over such final results. So an experiment provides a context for measurement w.r.t. which there is an intrinsic probability distribution over the possible outcomes. In particular, when the event structure is elementary it itself becomes a probabilistic event structure. (Below, by an unnormalised density operator we mean a positive, self-adjoint operator with trace less than or equal to one.)

Theorem 3. *Let E, Q be a quantum event structure with initial state ρ. Each configuration $x \in \mathcal{C}(E)$ is associated with an unnormalised density operator $\rho_x =_{\mathrm{def}} A_x \rho A_x^\dagger$ and a value in $[0, 1]$ given by $v(x) =_{\mathrm{def}} \mathrm{Tr}(\rho_x) = \mathrm{Tr}(A_x^\dagger A_x \rho)$. For any $w \in \mathcal{C}^\infty(E)$, the function v restricts to a configuration-valuation v_w on the elementary event structure w (viz. the event structure with events w, and causal dependency and (trivial) consistency inherited from E); hence v_w extends to a probability measure q_w on $\mathcal{F}_w =_{\mathrm{def}} \{x \in \mathcal{C}^\infty(E) \mid x \subseteq w\}$.*

Proof. We show v restricts to a configuration-valuation on $\mathcal{F}_w$. As $A_\varnothing = \mathrm{id}_{\mathcal{H}}$, $v(\varnothing) = \mathrm{Tr}(\rho) = 1$. By Lemma 2, we need only to show $d_v^{(n)}[y; x_1, \cdots, x_n] \geq 0$ when $y \xrightarrow{e_1} x_1, \cdots, y \xrightarrow{e_n} x_n$ in $\mathcal{F}_w$.

First, observe that if for some event e_i the operator Q_{e_i} is unitary, then $d_v^{(n)}[y; x_1, \cdots, x_n] = 0$. W.l.o.g. suppose e_n is assigned the unitary operator U. Then, $A_{x_n} = U A_y$ so

$$v(x_n) = \mathrm{Tr}(A_{x_n}^\dagger A_{x_n} \rho) = \mathrm{Tr}(A_y^\dagger U^\dagger U A_y \rho) = \mathrm{Tr}(A_y^\dagger A_y \rho) = v(y).$$

Let $\varnothing \neq I \subseteq \{1, \cdots, n\}$. Then, either $\bigcup_{i \in I} x_i = \bigcup_{i \in I} x_i \cup x_n$ or $\bigcup_{i \in I} x_i \xrightarrow{e_n} \bigcup_{i \in I} x_i \cup x_n$. In the either case—in the latter case by an argument similar to that above,

$$v\left(\bigcup_{i \in I} x_i\right) = v\left(\bigcup_{i \in I} x_i \cup x_n\right).$$

Consequently,

$$
\begin{aligned}
d_v^{(n)}[y; x_1, \cdots, x_n] &= d_v^{(n-1)}[y; x_1, \cdots, x_{n-1}] - d_v^{(n-1)}[x_n; x_1 \cup x_n, \cdots, x_{n-1} \cup x_n] \\
&= v(y) - \sum_I (-1)^{|I|+1} v\left(\bigcup_{i \in I} x_i\right) - v(x_n) + \sum_I (-1)^{|I|+1} v\left(\bigcup_{i \in I} x_i \cup x_n\right) \\
&= 0
\end{aligned}
$$

—above index I is understood to range over sets for which $\varnothing \neq I \subseteq \{1, \cdots, n\}$.

It remains to consider the case where all events e_i are assigned projection operators P_{e_i}. As $x_1, \cdots, x_n \subseteq w$ we must have that all the projection operators $P_{e_1}, \cdots, P_{e_n}$ commute.

As $[P_{e_i}, P_{e_j}] = 0$, for $1 \leq i, j \leq n$, we can assume an orthonormal basis which extends the sub-basis of eigenvectors of all the projection operators P_{e_i}, for $1 \leq i \leq n$. Let $y \subseteq x \subseteq \bigcup_{1 \leq i \leq n} x_i$. Define P_x to be the projection operator got as the composition of all the projection operators P_e for $e \in x \setminus y$—this is a projection operator, well-defined irrespective of the order of composition as the relevant projection operators commute. Define B_x to be the set of those basis vectors fixed by the projection operator P_x. In particular, P_y is the identity operator and B_y the set of all basis vectors. When $x, x' \in \mathcal{C}(E)$ with $y \subseteq x \subseteq \bigcup_{1 \leq i \leq n} x_i$ and $y \subseteq x' \subseteq \bigcup_{1 \leq i \leq n} x_i$,

$$B_{x \cup x'} = B_x \cap B_{x'}.$$

Also,

$$P_x |\psi\rangle = \sum_{i \in B_x} \langle i | \psi \rangle |i\rangle,$$

so

$$\langle \psi | P_x | \psi \rangle = \sum_{i \in B_x} \langle i | \psi \rangle \langle \psi | i \rangle = \sum_{i \in B_x} |\langle i | \psi \rangle|^2,$$

for all $|\psi\rangle \in \mathcal{H}$.

Assume $\rho = \sum_k p_k |\psi_k\rangle\langle\psi_k|$, where the ψ_k are normalised and all the p_k are positive with sum $\sum_k p_k = 1$. For x with $y \subseteq x \subseteq \bigcup_{1 \leq i \leq n} x_i$,

$$
\begin{aligned}
v(x) &= \mathrm{Tr}(A_x^\dagger A_x \rho) \\
&= \mathrm{Tr}(A_y^\dagger P_x^\dagger P_x A_y \rho) \\
&= \mathrm{Tr}(A_y^\dagger P_x A_y \sum_k p_k |\psi_k\rangle\langle\psi_k|) \\
&= \sum_k p_k \,\mathrm{Tr}(A_y^\dagger P_x A_y |\psi_k\rangle\langle\psi_k|) \\
&= \sum_k p_k \langle A_y \psi_k | P_x | A_y \psi_k \rangle \\
&= \sum_{i \in B_x} \sum_k p_k |\langle i | A_y \psi_k\rangle|^2 = \sum_{i \in B_x} r_i \,,
\end{aligned}
$$

where we define $r_i =_{\mathrm{def}} \sum_k p_k |\langle i | A_y \psi_k\rangle|^2$, necessarily a non-negative real for $i \in B_x$.

We now establish that

$$
d_v^{(n)}[y; x_1, \cdots, x_n] = \sum_{i \in B_y \setminus B_{x_1} \cup \cdots \cup B_{x_n}} r_i \,,
$$

for all $n \in \omega$, by mathematical induction—it then follows directly that its value is non-negative.

The base case of the induction, when $n = 0$, follows as

$$
d_v^{(0)}[y;] = v(y) = \sum_{i \in B_y} r_i \,,
$$

a special case of the result we have just established.

For the induction step, assume $n > 0$. Observe that

$$
B_y \setminus B_{x_1} \cup \cdots \cup B_{x_{n-1}} = (B_y \setminus B_{x_1} \cup \cdots \cup B_{x_n}) \uplus (B_{x_n} \setminus B_{x_1 \cup x_n} \cup \cdots \cup B_{x_{n-1} \cup x_n}) \,,
$$

where as signified the outer union is disjoint. Hence,

$$
\sum_{i \in B_y \setminus B_{x_1} \cup \cdots \cup B_{x_{n-1}}} r_i = \sum_{i \in B_y \setminus B_{x_1} \cup \cdots \cup B_{x_n}} r_i + \sum_{i \in B_{x_n} \setminus B_{x_1 \cup x_n} \cup \cdots \cup B_{x_{n-1} \cup x_n}} r_i \,,
$$

By definition,

$$
d_v^{(n)}[y; x_1, \cdots, x_n] =_{\mathrm{def}} d_v^{(n-1)}[y; x_1, \cdots, x_{n-1}] - d_v^{(n-1)}[x_n; x_1 \cup x_n, \cdots, x_{n-1} \cup x_n]
$$

—making use of the fact that we are only forming unions of compatible configurations. From the induction hypothesis,

$$
d_v^{(n-1)}[y; x_1, \cdots, x_{n-1}] = \sum_{i \in B_y \setminus B_{x_1} \cup \cdots \cup B_{x_{n-1}}} r_i
$$

and $d_v^{(n-1)}[x_n; x_1 \cup x_n, \cdots, x_{n-1} \cup x_n] = \sum_{i \in B_{x_n} \setminus B_{x_1 \cup x_n} \cup \cdots \cup B_{x_{n-1} \cup x_n}} r_i$.

Hence

$$d_v^{(n)}[y; x_1, \cdots, x_n] = \sum_{i \in B_y \setminus B_{x_1} \cup \cdots \cup B_{x_n}} r_i \,,$$

ensuring $d_v^{(n)}[y; x_1, \cdots, x_n] \geq 0$, as required.

By Theorem 2, the configuration-valuation v_w extends to a unique probability measure on $\mathcal{F}_w$. $\qquad\square$

Corollary 3. *Let E, Q be a quantum event structure in which E is elementary. Assume an initial state ρ. Then, $x \mapsto \mathrm{Tr}(A_x^\dagger A_x \rho)$, for $x \in \mathcal{C}(E)$, is a configuration-valuation on E. It extends to a probability measure on the Borel sets of $\mathcal{C}^\infty(E)$.*

Theorem 3 is reminiscent of the consistent-histories approach to quantum theory [9] once we understand configurations as partial-order histories. The traditional decoherence/consistency conditions on histories, saying when a family of histories supports a probability distribution, have been replaced by $\subseteq$-compatibility.

Example 3. Let E comprise the quantum event structure with two concurrent events e_0 and e_1 associated with projectors P_0 and P_1, where necessarily $[P_0, P_1] = 0$. Assume an initial state $|\psi\rangle\langle\psi|$, corresponding to the pure state $|\psi\rangle$. The configuration $\{e_0, e_1\}$ is associated with the following probability distribution. The probability that e_0 succeeds is $\|P_0|\psi\rangle\|^2$, that e_1 succeeds $\|P_1|\psi\rangle\|^2$, and that both succeed is $\|P_1 P_0|\psi\rangle\|^2$.

In the case where P_0 and P_1 commute because $P_0 P_1 = P_1 P_0 = 0$, the events e_0 and e_1 are mutually exclusive in the sense that there is probability zero of both events e_0 and e_1 succeeding, probability $\|P_0|\psi\rangle\|^2$ of e_0 succeeding, $\|P_1|\psi\rangle\|^2$ of e_1 succeeding, and probability $1 - \|P_0|\psi\rangle\|^2 - \|P_1|\psi\rangle\|^2$ of getting stuck at the empty configuration where neither event succeeds.

A special case of this is the measurement of a qubit in state ψ, the measurement of 0 where $P_0 = |0\rangle\langle0|$, and the measurement of 1 where $P_1 = |1\rangle\langle1|$, though here $\|P_0|\psi\rangle\|^2 + \|P_1|\psi\rangle\|^2 = 1$, as a measurement of the qubit will determine a result of either 0 or 1. $\qquad\square$

Example 4. Let E comprise the event structure with three events e_1, e_2, e_3 with trivial causal dependency and consistency relation generated by taking $\{e_1, e_2\} \in$ Con and $\{e_2, e_3\} \in$ Con—so $\{e_1, e_3\} \notin$ Con. To be a quantum event structure we must have $[Q_{e_1}, Q_{e_2}] = 0$, $[Q_{e_2}, Q_{e_3}] = 0$. The maximal configurations are $\{e_1, e_2\}$ and $\{e_2, e_3\}$. Assume an initial state $|\psi\rangle\langle\psi|$. The first maximal configuration is associated with a probability distribution where e_1 occurs with probability $\|Q_{e_1}|\psi\rangle\|^2$ and e_2 occurs with probability $\|Q_{e_2}|\psi\rangle\|^2$. The second maximal configuration is associated with a probability distribution where e_2 occurs with probability $\|Q_{e_2}|\psi\rangle\|^2$ and e_3 occurs with probability $\|Q_{e_3}|\psi\rangle\|^2$. $\qquad\square$

5.3   Measurement

To support measurements yielding values we associate values with configurations of a quantum event structure E, Q, in the form of a measurable function, $V :$

$\mathcal{C}^\infty(E) \to \mathbb{R}$. If the experiment results in $x \in \mathcal{C}^\infty(E)$ we obtain $V(x)$ as the measurement value resulting from the experiment. By Theorem 3, assuming an initial state given by a density operator ρ, we obtain a probability measure q_w on the sub-configurations of $w \in \mathcal{C}^\infty(E)$. This is interpreted as giving a probability distribution on the final results of an experiment w. Accordingly, w.r.t. an experiment $w \in \mathcal{C}^\infty(E)$, the expected value is

$$\mathbf{E}_w(V) =_{\text{def}} \int_{x \in \mathcal{F}_w} V(x) \, dq_w(x).$$

Traditionally quantum measurement is associated with an Hermitian operator A on $\mathcal{H}$ where the possible values of a measurement are eigenvalues of A. How is this realized by a quantum event structure? Suppose the Hermitian operator has spectral decomposition

$$A = \sum_{i \in I} \lambda_i P_i$$

where orthogonal projection operators P_i are associated with eigenvalue λ_i. The projection operators satisfy $\sum_{i \in I} P_i = \text{id}_\mathcal{H}$ and $P_i P_j = 0$ if $i \neq j$.

Form the quantum event structure with concurrent events e_i, for $i \in I$, and $Q(e_i) = P_i$. Because the projection operators are orthogonal, $[P_i, P_j] = 0$ when $i \neq j$, so we do indeed obtain a quantum event structure. Let $V(\{e_i\}) = \lambda_i$, and take arbitrary values on all other configurations. The event structure has a single, maximum configuration $w =_{\text{def}} \{e_i \mid i \in I\}$. It is the experiment w which will correspond to traditional measurement via A. Assume an initial state $|\psi\rangle\langle\psi|$. It can be checked that the probability ascribed to each of the singleton configurations $\{e_i\}$ is $\langle\psi|P_i|\psi\rangle$, and is zero elsewhere. Consequently,

$$\mathbf{E}_w(V) = \sum_{i \in I} \lambda_i \langle\psi|P_i|\psi\rangle = \langle\psi|A|\psi\rangle$$

—the well-known expression for the expected value of the measurement A on pure state $|\psi\rangle$.

Example 5. The spin state of a spin-1/2 particle is an element of two-dimensional Hilbert space, $\mathcal{H}_2$. Traditionally the Hermitian operator for measuring spin in a particular fixed direction is

$$|\!\uparrow\rangle\langle\uparrow| - |\!\downarrow\rangle\langle\downarrow|.$$

It has eigenvectors $|\!\uparrow\rangle$ (spin up) with eigenvalue $+1$ and $|\!\downarrow\rangle$ (spin down) with eigenvalue -1. Accordingly, its quantum event structure comprises the two concurrent events u associated with projector $|\!\uparrow\rangle\langle\uparrow|$ and d with projector $|\!\downarrow\rangle\langle\downarrow|$. Its configurations are: $\varnothing$, $\{u\}$, $\{d\}$ and $\{u, d\}$. The value associated with the configuration $\{u\}$ is $+1$, and that with $\{d\}$ is -1. Given an initial pure state $a|\!\uparrow\rangle + b|\!\downarrow\rangle$, the probability of the experiment $\{u, d\}$ yielding value $+1$ is $|a|^2$ and that of yielding -1 is $|b|^2$. The probability that the experiment ends in configurations $\varnothing$ or $\{u, d\}$ is zero. Its expected value is $|a|^2 - |b|^2$. This would be the average value resulting from measuring the spin of a large number of particles initially in the pure state. $\qquad\square$

An Event Logic. One way to assign values to configurations is via logic of which the assertions will be true (taken as 1) or false (0) at a configuration. Given a countable event structure E, we can build terms for events and assertions in a straightforward way. Event terms are given by $\epsilon ::= e \in E \mid v \in \text{Var}$, where Var is a set of variables over events, and assertions by

$$L ::= \epsilon \mid T \mid F \mid L_1 \wedge L_2 \mid L_1 \vee L_2 \mid \neg L \mid \forall v.L \mid \exists v.L.$$

W.r.t. an environment $\zeta : \text{Var} \to E$, an assertion L denotes $[\![L]\!]\zeta$, a Borel subset of $C^{\infty}(E)$, for example:

$$[\![e]\!]\zeta = \{x \in C^{\infty}(E) \mid e \in x\} \qquad [\![v]\!]\zeta = \{x \in C^{\infty}(E) \mid \zeta(v) \in x\}$$
$$[\![\forall v.L]\!]\zeta = \{x \in C^{\infty}(E) \mid \forall e \in x.\ x \in [\![L]\!]\zeta[e/v]\}$$
$$[\![\exists v.L]\!]\zeta = \{x \in C^{\infty}(E) \mid \exists e \in x.\ x \in [\![L]\!]\zeta[e/v]\}$$

with T, F, $\wedge$, $\vee$ and $\neg$ interpreted standardly by the set of all configurations, the emptyset, intersection, union and complement. In this logic, for example, $\neg(a{\downarrow} \wedge b{\downarrow}) \wedge \neg(a{\uparrow} \wedge b{\uparrow})$ could express the anti-correlation of the spin of particles a and b.

W.r.t. a quantum event structure with initial state, for an experiment the configuration w, the probability of the result of the quantum experiment satisfying L, a closed assertion of the logic, is

$$q_w(L \cap \mathcal{F}_w),$$

which coincides with the expected value of the characteristic function for L.

5.4   Probabilistic Quantum Experiments

It can be useful, or even necessary, to allow the choice of which quantum measurements to perform to be made probabilistically. For example, experiments to invalidate the Bell inequalities, to demonstrate the non-locality of quantum physics, may make use of probabilistic quantum experiments.

Formally, a probability distribution over quantum experiments can be realized by a total map of event structures $f : P \to E$ where P, v is a probabilistic event structure and E, Q is a quantum event structure; the configurations of E correspond to quantum experiments assigned probabilities through P. Through the map f we can integrate the probabilistic and quantum features. Via the map f, the event structure E inherits a configuration valuation, making it itself a probabilistic event structure; we can see this indirectly by noting that if v_o is a continuous valuation on the open sets of P then $v_o f^{-1}$ is a continuous valuation on the open sets of E. On the other hand, via f the event structure P becomes a quantum event structure; an event $p \in P$ is interpreted as operation $Q(f(p))$. Of course, f can be the identity map, as is so in Example 6 below.

Suppose E, Q is a quantum event structure with initial state ρ and a measurable value function $V : \mathcal{C}^\infty(E) \to \mathbb{R}$. Recall, from Section 5.3, that the expected value of a quantum experiment $w \in \mathcal{C}^\infty(E)$ is

$$\mathbf{E}_w(V) =_{\text{def}} \int_{x \in \mathcal{F}_w} V(x) \, dq_w(x),$$

where q_w is the probability measure induced on $\mathcal{F}_w$ by Q and ρ. The expected value of a probabilistic quantum experiment $f : P \to E$, where P, v is a probabilistic event structure is

$$\int_{w \in \mathcal{C}^\infty(E)} \mathbf{E}_w(V) \, d\mu f^{-1}(w),$$

where μ is the probability measure induced on $\mathcal{C}^\infty(P)$ by the configuration-valuation v. Specialising the value function to the characteristic function of a Borel subset $L \subseteq \mathcal{C}^\infty(E)$, perhaps given by an assertion of the event logic of Section 5.3, the probability of the result of the probabilistic experiment satisfying L is

$$\int_{w \in \mathcal{C}^\infty(E)} q_w(L \cap \mathcal{F}_w) \, d\mu f^{-1}(w).$$

The following example illustrates how a very simple form of probabilistic quantum experiment (in which the event structure has a discrete partial order of causal dependency) provides a basis for the analysis of Bell and EPR experiments.

Example 6. Imagine an observer who randomly chooses between measuring spin in a first fixed direction $\mathbf{a_1}$ or in a second fixed direction $\mathbf{a_2}$. Assume that the probability of measuring in the $\mathbf{a_1}$-direction is p_1 and in the $\mathbf{a_2}$-direction is p_2, where $p_1 + p_2 = 1$. The two directions $\mathbf{a_1}$ and $\mathbf{a_2}$ correspond to choices of bases $|{\uparrow}a_1\rangle, |{\downarrow}a_1\rangle$ and $|{\uparrow}a_2\rangle, |{\downarrow}a_2\rangle$ in $\mathcal{H}_2$. We describe this scenario as a probabilistic quantum experiment. The quantum event structure has four events, ${\uparrow}a_1, {\downarrow}a_1, {\uparrow}a_2, {\downarrow}a_2$, in which ${\uparrow}a_1, {\downarrow}a_1$ are concurrent, as are ${\uparrow}a_2, {\downarrow}a_2$; all other pairs of events are in conflict. The event ${\uparrow}a_1$ is associated with measuring spin up in direction $\mathbf{a_1}$ and the event ${\downarrow}a_1$ with measuring spin down in direction $\mathbf{a_1}$. Similarly, events ${\uparrow}a_2$ and ${\downarrow}a_2$ correspond to measuring spin up and down, respectively, in direction $\mathbf{a_2}$. Correspondingly, we associate events with the following projection operators:

$$Q({\uparrow}a_1) = |{\uparrow}a_1\rangle\langle{\uparrow}a_1|, \qquad Q({\downarrow}a_1) = |{\downarrow}a_1\rangle\langle{\downarrow}a_1|,$$
$$Q(u_2) = |{\uparrow}a_2\rangle\langle{\uparrow}a_2|, \qquad Q(d_2) = |{\downarrow}a_2\rangle\langle{\downarrow}a_2|.$$

The configurations of the event structure take the form

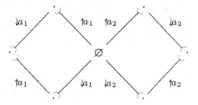

where we have taken the liberty of inscribing the events just on the covering intervals. Measurement in the $\mathbf{a_1}$-direction corresponds to the configuration $\{\uparrow a_1, \downarrow a_1\}$—the configuration to the far left in the diagram—and in the $\mathbf{a_2}$-direction to the configuration $\{\uparrow a_2, \downarrow a_2\}$—that to the far right. To describe that the probability of the measurement in the $\mathbf{a_1}$-direction is p_1 and that in the $\mathbf{a_2}$-direction is p_2, we assign a configuration valuation v for which

$$v(\{\uparrow a_1, \downarrow a_1\}) = v(\{\uparrow a_1\}) = v(\{\downarrow a_1\}) = p_1,$$
$$v(\{\uparrow a_2, \downarrow a_2\}) = v(\{\uparrow a_2\}) = v(\{\downarrow a_2\}) = p_2 \text{ and } v(\varnothing) = 1.$$

Such a probabilistic quantum experiment is not very interesting on its own. But imagine that there are two similar observers A and B measuring the spins in directions $\mathbf{a_1}$, $\mathbf{a_2}$ and $\mathbf{b_1}$, $\mathbf{b_2}$, respectively, of two particles created so that together they have zero angular momentum, ensuring they have a total spin of zero in any direction. Then quantum mechanics predicts some remarkable correlations between the observations of A and B, even at distances where their individual choices of what directions to perform their measurements could not possibly be communicated from one observer to another. For example, were both observers to choose the same direction to measure spin, then if one measured spin up then other would have to measure spin down even though the observers were light years apart.

We can describe such scenarios by a probabilistic quantum experiment which is essentially a simple parallel composition of two versions of the (single-observer) experiment above. In more detail, make two copies of the single-observer event structure: that for A, the event structure E_A, has events $\uparrow a_1, \downarrow a_1, \uparrow a_2, \downarrow a_2$, while that for B, the event structure E_B, has events $\uparrow b_1, \downarrow b_1, \uparrow b_2, \downarrow b_2$. Assume they possess configuration valuations v_A and v_B, respectively, determined by the probabilistic choices of directions made by A and B. Write Q_A and Q_B for the respective assignments of projection operators to events of E_A and E_B. The probabilistic event structure for the two observers together is got as $E_A \| E_B$, their simple parallel composition got by juxtaposition, with configuration valuation $v(x) = v_A(x_A) \times v_B(x_B)$, for $x \in \mathcal{C}(E_A \| E_B)$, where x_A and x_B are projections of x to configurations of A and B. In this compound system an event such as e.g. $\uparrow a_1$ is interpreted as the projection operator $Q_A(\uparrow a_1) \otimes \mathrm{id}_{\mathcal{H}_2}$ on the Hilbert space $\mathcal{H}_2 \otimes \mathcal{H}_2$, where the combined state of the two particles belongs. We can capture the correlation or anti-correlation of the observers' measurements of spin through a value function on configurations, given by

$$V(\{\uparrow a_i, \uparrow b_j\}) = V(\{\downarrow a_i, \downarrow b_j\}) = 1, \quad V(\{\uparrow a_i, \downarrow b_j\}) = V(\{\downarrow a_i, \uparrow b_j\}) = -1, \text{ and}$$
$$V(x) = 0 \text{ otherwise},$$

and study their expectations under various initial states and choices of measurement. In this way probabilistic quantum experiments, as formalised through probabilistic and quantum event structures, provide a basis for the analysis of Bell or EPR experiments.   □

The ideas of probabilistic and quantum event structures carry over to probabilistic and quantum games and their strategies [1]; the result of the play of quantum strategy against a counterstrategy is a probabilistic event structure. This is yielding operations and languages which should be helpful in a structured development and analysis of experiments on quantum systems.

Acknowledgments. Congratulations Prakash on your 60th birthday—stay young at heart!
Discussion with Samy Abbes, Samson Abramsky, Nathan Bowler, Peter Hines, Ohad Kammar, Klaus Keimel, Mike Mislove and Prakash has been helpful. Daniele Varacca deserves special thanks for our earlier work on probabilistic event structures. I gratefully acknowledge the ERC Advanced Grant ECSYM.

References

1. Winskel, G.: Distributed probabilistic and quantum strategies. In: MFPS 2013. Electr. Notes Theor. Comput. Sci. (2013)
2. Varacca, D., Völzer, H., Winskel, G.: Probabilistic event structures and domains. Theor. Comput. Sci. 358(2-3), 173–199 (2006)
3. Abramsky, S., Brandenburger, A.: A unified sheaf-theoretic account of non-locality and contextuality. CoRR abs/1102.0264 (2011)
4. Jones, C., Plotkin, G.: A probabilistic powerdomain of valuations. In: LICS 1989. IEEE Computer Society (1989)
5. Rohlin, V.A.: On the fundamental ideas of measure theory. Amer. Math. Soc. Translation 1952(71), 55 (1952)
6. Goubault-Larrecq, J., Keimel, K.: Choquet-Kendall-Matheron theorems for non-hausdorff spaces. Mathematical Structures in Computer Science 21(3), 511–561 (2011)
7. Varacca, D.: Probability, nondeterminism and concurrency. PhD Thesis, Aarhus University (2003)
8. Alvarez-Manilla, M., Edalat, A., Saheb-Djahromi, N.: An extension result for continuous valuations. Journal of the London Mathematical Society 61(2), 629–640 (2000)
9. Griffiths, R.B.: Consistent quantum theory. CUP (2002)

Author Index